Computational Statics and Dynamics

Andreas Öchsner

Computational Statics and Dynamics

An Introduction Based on the Finite Element Method

Second Edition

Springer

Andreas Öchsner
Esslingen University of Applied Sciences
Esslingen am Neckar, Baden-Württemberg
Germany

ISBN 978-981-15-1280-3 ISBN 978-981-15-1278-0 (eBook)
https://doi.org/10.1007/978-981-15-1278-0

This Springer imprint is published by the registered company Springer Nature Singapore Pte Ltd.
The registered company address is: 152 Beach Road, #21-01/04 Gateway East, Singapore 189721, Singapore

Der Vorteil der Klugheit besteht darin,
dass man sich dumm stellen kann.
Das Gegenteil ist schon schwieriger.
—Kurt Tucholsky (1890–1935)

Preface to the Second Edition

The pedagogical methodology proposed in the first edition has been well accepted by students and the scientific community. The second edition has been extended by approximately 150 pages. Many additional examples to practice the finite element approach to engineering problems and a new chapter on shear deformable plate elements have been included. Furthermore, the weighted residual method has been introduced and consistently applied to allow illustrating the same method for all element types. In the appendix, a new section on the triangular element formulation has been included. Last but not least, the entire content and the graphical illustrations have been thoroughly revised and updated.

Esslingen, Germany
October 2019

Andreas Öchsner

Preface to the First Edition

This book results from the core course *Computational Statics and Dynamics* (6522ENG) held at Griffith University, Australia, in the scope of the 'Bachelor of Engineering with Honours in Mechanical Engineering' degree program. This 13-week course comprises three hours of lectures and two hours of tutorials per week. An additional lab component based on a commercial finite element program is not covered in this textbook. This course is based on the finite element method and located in the third year of the study program. It relies heavily on the fundamental knowledge of the first years of engineering education, i.e. higher mathematics, materials science, applied mechanics, design, and programming skills. This is the reason why many students find this topic difficult to master and this textbook should provide some guidance for success in this course.

All derivations in the following chapters follow a common approach: First, the three fundamental equations of continuum mechanics, i.e. the kinematics equation, the constitutive equation, and the equilibrium equation, are combined to construct the partial differential equation. Subsequently, the weighted residual method, a universal approach to derive any of the classical approximation methods, is applied to derive the principal finite element equation for each element type.

Chapter 1 illustrates the finite element method in the context of engineering practice and academic education. Chapter 2 covers the simplest one-dimensional element type, i.e. the rod/bar element. First, this element type is considered in pure one-dimensional structures and then, the case of plane truss structures is covered. Chapter 3 covers the simplest one-dimensional beam formulation according to Euler–Bernoulli. This element is also generalized by superposition with a rod element (so-called generalized beam element) and its arrangement in plane frame structures is treated. Chapter 4 introduces a higher beam bending theory according to Timoshenko. This theory considers the contribution of the shear force on the deformation. Chapter 5 introduces two-dimensional plane elasticity elements for the plane stress and plane strain state. Chapter 6 covers classical plate bending elements which can be seen as the two-dimensional generalization of Euler–Bernoulli beams. Chapter 7 treats three-dimensional solids on the example of hexahedral elements. Chapters 8 and 9 briefly introduce the topic of transient analysis.

In order to deepen the understanding of the derived equations and theories, each technical chapter collects at its end supplementary problems. These supplementary problems start with fundamental knowledge questions on the theory of the chapter and are followed by calculation problems. In total over 80 of such additional calculation problems are provided and a short solution for each problem is included in this book. It should be noted that these short solutions contain major steps for the solution of the problem and not only, for example, a numerical value for the final result. This should ensure that students are able to successfully master these problems. I hope that students find this book a useful complement to many classical textbooks. I look forward to receive their comments and suggestions.

Southport, Australia
January 2016

Andreas Öchsner

Acknowledgements

It is important to highlight the contribution of many undergraduate and postgraduate students which helped to finalize the content of this book. Their questions and comments during different lectures and their work in the scope of final year projects helped to compile this book. The help and support of my tutors and Ph.D. students Leonhard Hitzler and Zia Javanbakht is also gratefully acknowledged. Furthermore, I would like to express my sincere appreciation to the Springer-Verlag, especially to Dr. Christoph Baumann, for giving me the opportunity to realize this book. A professional publishing company with the right understanding was the prerequisite to complete this comprehensive project. Finally, I would like to thank my family for the understanding and patience during the preparation of this book. I would like to especially thank Marco for his corrections and suggestions.

Contents

Symbols and Abbreviations

Latin Symbols (Capital Letters)

A	Area, cross-sectional area
$\boldsymbol{A}$	Matrix, cf. derivation of interpolation functions
$\boldsymbol{B}$	Matrix which contains derivatives of interpolation functions
C_{ijkl}	Fourth-order elasticity tensor
$\boldsymbol{C}$	Elasticity matrix
D_{b}	Bending rigidity (plate)
D_{s}	Shear rigidity (plate)
$\boldsymbol{D}$	Compliance matrix, plate elasticity matrix
$\boldsymbol{D}_{\mathrm{b}}$	Bending rigidity matrix (plate)
$\boldsymbol{D}_{\mathrm{s}}$	Shear rigidity matrix (plate)
E	YOUNG's modulus
EA	Tensile stiffness
EI	Bending stiffness
F	Force
G	Shear modulus
GA	Shear stiffness
I	Second moment of area, abbreviation for an integral statement
I_{p}	Polar second moment of area
$\mathbf{I}$	Identity matrix (diagonal matrix), $\mathbf{I} = \lceil 1\,1\,1 \ldots \rfloor$
J	Jacobian determinant, cf. coordinate transformation
$\boldsymbol{J}$	Jacobian matrix
$\boldsymbol{K}$	Global stiffness matrix
$\boldsymbol{K}^{\mathrm{e}}$	Elemental stiffness matrix
$\boldsymbol{K}_{\mathrm{T}}$	Tangent stiffness matrix
L	Element length
M	Moment
$\boldsymbol{M}$	Mass matrix
N	Normal force (internal), interpolation function

$\boldsymbol{N}$	Column matrix of interpolation functions
$\boldsymbol{N}_i$	3 × 3 Matrix of interpolation functions for node i
$\overline{N}$	Shape function
P	LEGENDRE polynomial, point
Q	Shear force (internal)
$\dot{Q}$	Heat transfer rate
R	Equivalent nodal force, radius of curvature of a curve, stress ratio
$\boldsymbol{T}$	Transformation matrix
U	Perimeter
V	Volume
W	Weight function
W^*	Fundamental solution
$\boldsymbol{W}$	Column matrix of weight functions
X	Global Cartesian coordinate
Y	Global Cartesian coordinate
Z	Global Cartesian coordinate

Latin Symbols (Small Letters)

a	Acceleration, basis coefficient, geometric dimension
$\boldsymbol{a}$	Column matrix of basis coefficients
b	Coefficient, function, geometric dimension
$\boldsymbol{b}$	Column matrix of body forces acting per unit volume
c	Constant of integration, coefficient, geometric dimension
c_{D}	Drag coefficient
d	Coefficient, geometric dimension
$\boldsymbol{e}$	Column matrix of generalized strains
f	Body force, scalar function
$\boldsymbol{f}$	Column matrix of loads
g	Scalar function, standard gravity
h	Geometric dimension
i	Iteration index, node number
j	Iteration index
k	Auxiliary function, elastic embedding modulus, elastic foundation modulus, stiffness, spring constant, thermal conductivity
k_{s}	Shear correction factor
m	Distributed moment, element number, mass
$\boldsymbol{m}$	Matrix function
n	Node number, increment number
n_j	Components of the normal vector
$\boldsymbol{n}$	Normal vector
p	Distributed load in x-direction

q Distributed load in y-direction, internal variable (hardening)
$\dot{q}$ Heat flux
r Residual
$\boldsymbol{r}$ Residual column matrix
$\boldsymbol{s}$ Column matrix of stress deviator components, column matrix of generalized stresses
t Time, traction force
t_i Components of the traction force vector
t_{end} Convergence value
$\boldsymbol{t}$ Column matrix of traction forces
u Displacement
u^0 Exact solution
$\dot{u}$ Velocity, $\dot{u} = v$
$\ddot{u}$ Acceleration, $\ddot{u} = \dot{v} = a$
$\boldsymbol{u}$ Column matrix of displacements, column matrix of nodal unknowns
v Auxiliary function, velocity
$\boldsymbol{v}$ Variable matrix
w Weight for numerical integration
x Cartesian coordinate
$\boldsymbol{x}$ Column matrix of Cartesian coordinates
y Cartesian coordinate
z Cartesian coordinate

Greek Symbols (Capital Letters)

Γ Boundary
Λ Factor
Ω Domain

Greek Symbols (Small Letters)

α Parameter, rotation angle
β Parameter
γ Shear strain (engineering definition), parameter, specific weight per unit volume, $\gamma = \varrho g$
δ Geometric dimension
ε Strain
ε^{el} Elastic strain
ε_{ij} Second-order strain tensor
ε_{c} Elastic limit strain in compression

$\boldsymbol{\varepsilon}$ Column matrix of strain components
ζ Natural coordinate
η Natural coordinate
$\dot{\eta}$ Rate of energy generation per unit volume
κ Curvature, isotropic hardening parameter
λ LAMÉ's constant
μ LAMÉ's constant
ν POISSON's ratio
ξ Natural coordinate
$\dot{\o}$ Rate of energy generation per unit length
ϱ Mass density
σ Stress, normal stress
σ_{ij} Second-order stress tensor
$\boldsymbol{\sigma}$ Column matrix of stress components
τ Shear stress
ϕ Rotation (TIMOSHENKO beam)
φ Basis function, rotation (BERNOULLI beam)
$\boldsymbol{\chi}$ Column matrix of basis functions
ψ Basis function

Mathematical Symbols

$\times$	Multiplication sign (used where essential)
$[\ldots]$	Matrix
$\lceil\ldots\rfloor$	Diagonal matrix
$[\ldots]^{\mathrm{T}}$	Transpose
$\langle\ldots\rangle$	MACAULAY's bracket
$\langle\ldots,\ldots\rangle$	Inner product
$\mathcal{L}\{\ldots\}$	Differential operator
$\boldsymbol{\mathcal{L}}$	Matrix of differential operators
$\deg(\ldots)$	Degree of a polynomial
$\mathrm{sgn}(\ldots)$	Signum (sign) function
∂	Partial derivative symbol (rounded *d*)
$\mathbb{R}$	Set of real numbers
δ	DIRAC delta function
$\mathbf{1}$	Identity column matrix, $\mathbf{1} = [1\,1\,1\,0\,0\,0]^{\mathrm{T}}$
$\mathbf{L}$	Diagonal scaling matrix, $\mathbf{L} = \lceil 1\,1\,1\,0\,0\,0 \rfloor$
$O(\ldots)$	Order of

Indices, Superscripted

$\ldots^{e}$	Element
$\ldots^{el}$	Elastic
$\ldots^{pl}$	Plastic
$\ldots^{R}$	Reaction

Indices, Subscripted

$\ldots_{b}$	Bending
$\ldots_{c}$	Center, compression
$\ldots_{lim}$	Limit
$\ldots_{p}$	Nodal value ('point')
$\ldots_{s}$	Shear, spring
$\ldots_{t}$	Tensile

Abbreviations

1D	One-dimensional
2D	Two-dimensional
3D	Three-dimensional
a.u.	Arbitrary unit
BC	Boundary condition
BD	Backward difference
BEM	Boundary element method
CAD	Computer-aided design
CD	Centered difference
const.	Constant
dim.	Dimension
DE	Differential equation
DOF	Degree(s) of freedom
EBT	EULER–BERNOULLI beam theory (elementary beam theory)
FD	Forward difference
FDM	Finite difference method
FEM	Finite element method
FGM	Functionally graded material
FVM	Finite volume method
inc	Increment
max	Maximum
PDE	Partial differential equation

RVE	Representative volume element
SI	International system of units
sym.	Symmetric
TBT	TIMOSHENKO beam theory
WRM	Weighted residual method

Some Standard Abbreviations

ca.	About, approximately (from Latin 'circa')
cf.	Compare (from Latin 'confer')
ead.	The same (woman) (from Latin 'eadem')
e.g.	For example (from Latin 'exempli gratia')
et al.	And others (from Latin 'et alii')
et seq.	And what follows (from Latin 'et sequens')
etc.	And others (from Latin 'et cetera')
i.a.	Among other things (from Latin 'inter alia'), in the absence of (from Latin 'in absentia')
ibid.	In the same place (the same), used in citations (from Latin 'ibidem')
id.	The same (man) (from Latin 'idem')
i.e.	That is (from Latin 'id est')
loc. cit.	In the place cited (from Latin 'loco citato')
N.N.	Unknown name, used as a placeholder for unknown names (from Latin 'nomen nescio')
op. cit.	In the work cited (from Latin 'opere citato')
pp.	Pages
q.e.d.	Which had to be demonstrated (from Latin 'quod erat demonstrandum')
viz.	Namely, precisely (from Latin 'videlicet')
vs.	Against (from Latin 'versus')

Chapter 1
Introduction to the Finite Element Method

Abstract The first chapter classifies the content as well as the focus of this textbook. The importance of computational methods in the modern design process is highlighted. In engineering practice, the description of processes is centered around partial differential equations, and the finite element method is introduced as an approximation method to solve these equations.

Complex engineering structures have been successfully built for a long time, even without the use of any computer-based design and simulation tools, see the example shown in Fig. 1.1. However, computer-based design and analysis is becoming more and more important in all high-technology areas. This computational focus enables engineering companies to realize significant cost reductions in the design and development process due to the reduced need for physical models and real experiments. A significant contribution to the success of this approach is based on the development of powerful computer hardware and software in the last few decades. As a recent example, the design of commercial aircrafts reflects this trend.

Moreover, the landscape of the engineering profession is dynamically changing and the new requirements of the digital revolution in engineering, i.e., to work in the new area of integrated design and simulation, requires a stronger focus on computer-based analyses tools. In traditional engineering approaches, the two areas of design and simulation would be represented by different departments in a company. However, the development of advanced design and simulation software packages and powerful computer hardware merges these areas into a new virtual, computer-based environment. Employees with these skills are necessary in the modern day engineering context all over the world (the 'global village'), where technologies such as 'cloud computing' are a part of the daily routine.

The interaction between design and simulation is mainly represented software-wise by computer-aided design (CAD) programs which allow the modeling of the geometry of an engineering structure, and simulation packages, for example based on the finite element method (FEM). This might be done by different programs or an incorporation of both packages under a common interface. Figures 1.2 and 1.3 illustrate this process where first a geometrical representation of an engineering structure is shown. In a second step, this geometry is approximated based on smaller geometrical entities, so-called finite elements.

A. Öchsner, *Computational Statics and Dynamics*,
https://doi.org/10.1007/978-981-15-1278-0_1

Fig. 1.1 Sydney Harbour Bridge—the world's largest steel arch bridge, Australia (construction period 1924–1932)

In the context of engineering education, it must be stated that courses on the finite element method require a certain foundation (see Fig. 1.4) which are normally provided during the first years of study. This may imply that students face some difficulties compared to the early foundation courses because the comprehensive treatment of this method assembles a considerable amount of engineering knowledge.

To study the finite element method in tertiary education is a challenging task and different approaches are available. This ranges from classical lectures and the corresponding textbooks [1, 2, 18, 19] to the classical tutorials with 'hand calculations' [8]. Some books focus only on one-dimensional elements to reduce the requirements on the mathematical framework [9, 14]. In a more modern academic context, so-called problem or project based approaches are also common in some countries [11]. It is also common that the more theoretical sessions are accompanied by computer based laboratories where a commercial finite element package is introduced [15]. As an alternative, a real programming language can be used to teach the computer implementation and to develop own routines [7, 17] or in a simpler approach the application of a computer algebra system [12, 13]. The scope of this book is to provide a solid foundation in the theory of the method, whereas a focus is the foundation of continuum mechanics and the weighted residual method to derive the principal finite element equation. All the provided concepts and finite element formulations are highlighted based on a significant number of exercises.

Engineers describe physical phenomena and processes typically by equations, particularly by partial differential equations [3, 5, 16]. In this context, the derivation and the solution of these differential equations (see Fig. 1.5) is the task of engineers,

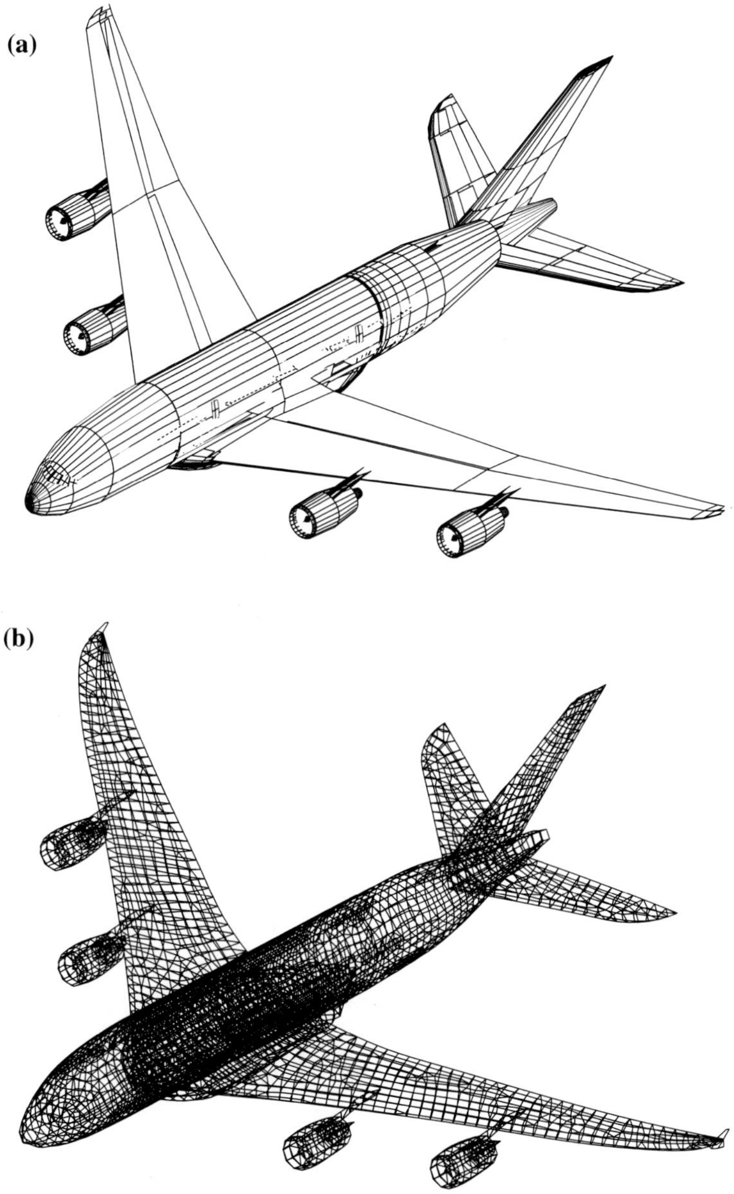

Fig. 1.2 Airbus A380 **a** geometry and **b** finite element mesh

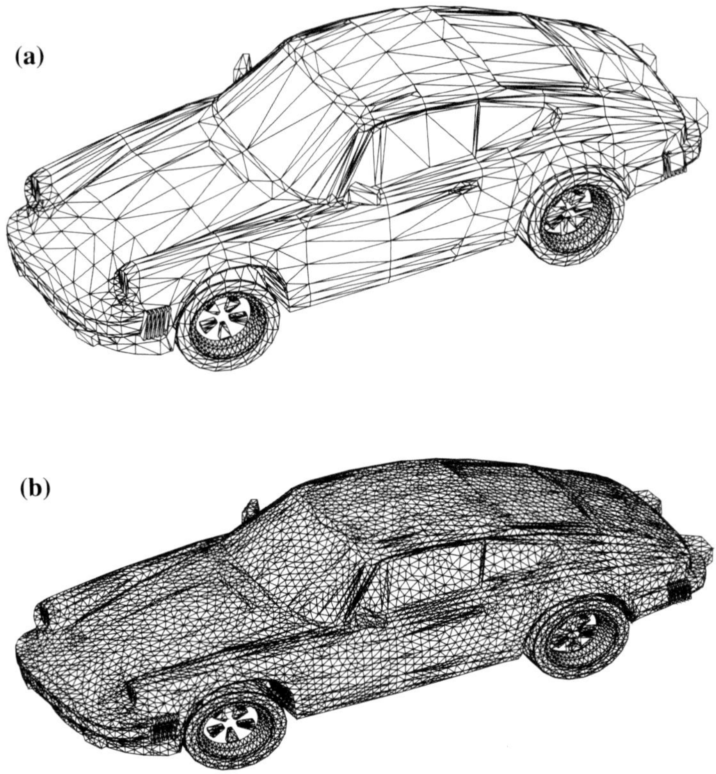

Fig. 1.3 Porsche 911 **a** geometry and **b** finite element mesh

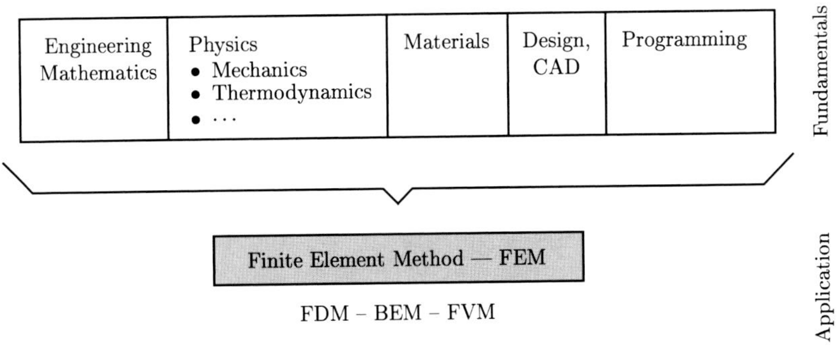

Fig. 1.4 Finite element method in the context of engineering education

obviously requiring fundamental knowledge from physics and engineering mathematics.

The importance of partial differential equations is clearly represented in the following quote: 'For more than 250 years partial differential equations have been clearly the most important tool available to mankind in order to understand a large

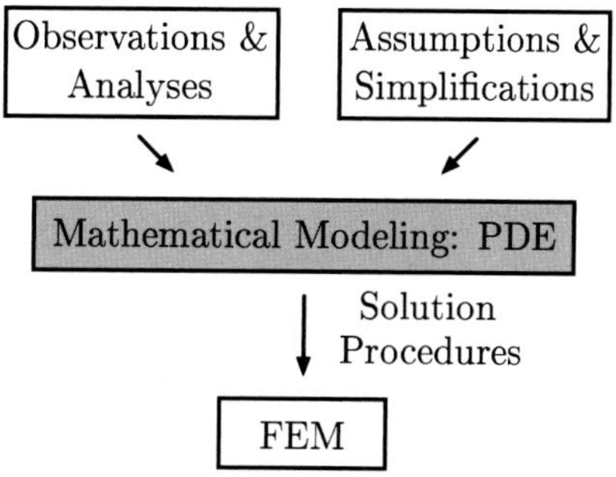

Fig. 1.5 Modeling based on partial differential equations

variety of phenomena, natural at first and then those originating from human activity and technological development. Mechanics, physics and their engineering applications were the first to benefit from the impact of partial differential equations on modeling and design, ...' [6].

In the one-dimensional case, a physical problem can be generally described in a spatial domain Ω by the differential equation

$$\mathcal{L}\{y(x)\} = b \quad (x \in \Omega) \tag{1.1}$$

and by the conditions which are prescribed on the boundary Γ. The differential equation is also called the *strong form* or the *original statement* of the problem. The expression 'strong form' comes from the fact that the differential equation describes exactly each point x in the domain of the problem. The operator $\mathcal{L}\{\dots\}$ in Eq. (1.1) is an arbitrary differential operator which can take, for example, the following forms:

$$\mathcal{L}\{\dots\} = \frac{\mathrm{d}^2}{\mathrm{d}x^2}\{\dots\}, \tag{1.2}$$

$$\mathcal{L}\{\dots\} = \frac{\mathrm{d}^4}{\mathrm{d}x^4}\{\dots\}, \tag{1.3}$$

$$\mathcal{L}\{\dots\} = \frac{\mathrm{d}^4}{\mathrm{d}x^4}\{\dots\} + \frac{\mathrm{d}}{\mathrm{d}x}\{\dots\} + \{\dots\}. \tag{1.4}$$

Furthermore, variable b in Eq. (1.1) is a given function, and in the case of $b = 0$, the equation reduces to the *homogeneous differential equation*: $\mathcal{L}\{y(x)\} = 0$. More specific expressions of Eqs. (1.3)–(1.4) can take the following form [10]:

$$a\frac{\mathrm{d}^2 y(x)}{\mathrm{d}x^2} = b\,, \tag{1.5}$$

$$a\frac{\mathrm{d}^4 y(x)}{\mathrm{d}x^4} = b\,, \tag{1.6}$$

and will be used to describe the behavior of rods and beams in the following sections.

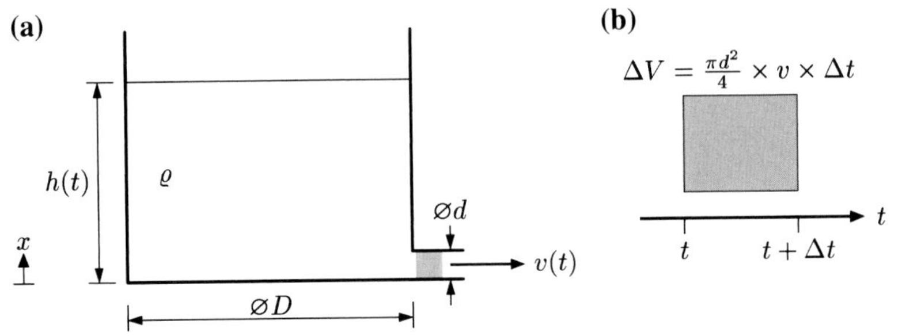

Fig. 1.6 Draining of a water tank: **a** general configuration ; **b** small fluid volume at the outlet

1.1 Example: Draining of a water tank

Given is a cylindrical water tank[1] (diameter D) which has at $x = 0$ a small hole of diameter d, see Fig. 1.6a. The initial water-level is $h(t = t_0) = h_0$ and the drain velocity is denoted by $v(t)$. Derive the differential equation which describes the draining of the water tank as a function of time. Assume that the water density ϱ is constant in the entire tank.

1.1 Solution

The water volume in the tank is $V(t) = h(t)A = h(t)\dfrac{\pi D^2}{4}$. The rate of change of this volume can be expressed as

$$\frac{\mathrm{d}V}{\mathrm{d}t} = -\frac{\pi d^2}{4} \times v(t)\,, \tag{1.7}$$

where the minus sign indicates a decrease in the tank volume. The velocity $v(t)$ can be determined from the conservation of energy, i.e.

$$\frac{1}{2}m(v(t))^2 = mgh(t)\,, \tag{1.8}$$

or

$$v(t) = \sqrt{2gh(t)}\,. \tag{1.9}$$

Introducing this expression for the velocity in Eq. (1.7) gives:

$$\frac{\mathrm{d}V}{\mathrm{d}t} = -\frac{\pi d^2}{4} \times \sqrt{2gh(t)}\,, \tag{1.10}$$

$$\frac{\mathrm{d}\left(h(t)\frac{\pi D^2}{4}\right)}{\mathrm{d}t} = -\frac{\pi d^2}{4} \times \sqrt{2gh(t)}\,, \tag{1.11}$$

[1]This example is adopted from [4].

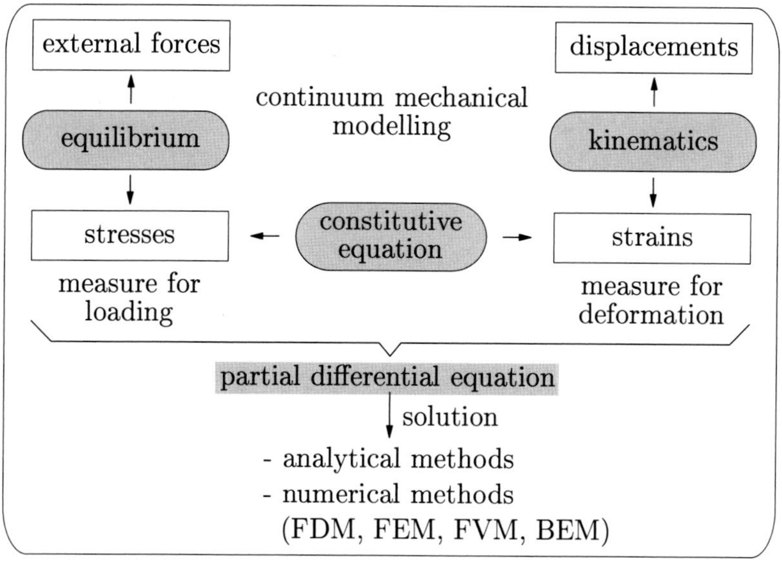

Fig. 1.7 Continuum mechanical modelling

or finally

$$\frac{\mathrm{d}h(t)}{\mathrm{d}t} = -\left(\frac{d}{D}\right)^2 \times \sqrt{2g} \times \sqrt{h(t)}\,. \tag{1.12}$$

In a more formal way, this differential equation can be written as:

$$\mathcal{L}\{h(t)\} = -\left(\frac{d}{D}\right)^2 \times \sqrt{2g} \times \sqrt{h(t)}\,. \tag{1.13}$$

It should be noted here that the solution of this differential equation can be obtained by separation of variables as [4]:

$$h(t) = \left(\sqrt{h_0} - \left(\frac{d}{D}\right)^2 \times \sqrt{2g} \times \frac{t}{2}\right)^2\,. \tag{1.14}$$

Let us highlight at the end of this section that the derivations in the following chapters follow a common approach, see Fig. 1.7.

A combination of the kinematics equation (i.e., the relation between the strains and displacements) with the constitutive equation (i.e., the relation between the stresses and strains) and the equilibrium equation (i.e., the equilibrium between the internal

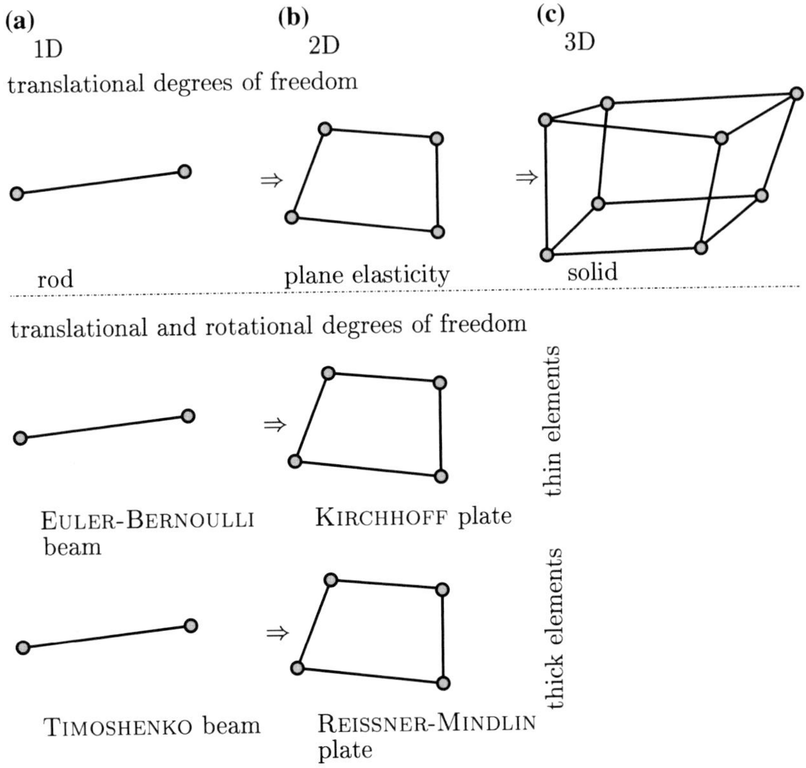

Fig. 1.8 Classification of considered finite elements: **a** one-dimensional, **b** two-dimensional, and **c** three-dimensional elements

reactions and the external loads) results in a partial differential equation. Limited to simple cases, analytical solutions are only covered in the scope of supplementary problems. The focus of this book remains on approximate solutions based on the finite element method (FEM).

The considered finite element types are shown in Fig. 1.8 and can be classified either according to their dimensionality (i.e., 1D, 2D, or 3D) or according to their degrees of freedom (i.e., only translational degrees of freedom or together with rotational degrees). The applied approach to derive the principal finite element equations is the same for all elements and presented in a consistent manner. Despite the fact that there are many possible approaches, this book solely relies on the weighted residual method, a universal approach to derive any of the classical approximation methods.

Another element classification is based on the number of nodes. The one-dimensional elements shown in Fig. 1.8 are composed of two nodes and commonly referred to as 'line 2' elements. The two-dimensional elements in Fig. 1.8 are composed of four nodes and referred to as 'quad 4' elements while the eight-node three-dimensional elements are called 'hex 8'. The manifoldness of common finite elements is much beyond this small selection of representative elements and the interested reader is referred to, for example, the textbooks [18, 19] to discover more one-dimensional (e.g. line 3), two-dimensional (e.g. quad 8, quad 9, triangular shaped elements tria 3 or tria 6), and three-dimensional (e.g. hex 20, tetrahedrons tet 4 or tet 10, pyramids pyr 5 or pyr 15, or wedges/prisms wedge 6 or wedge 15) elements.

It should be noted here that the understanding of one-dimensional elements is essential and allows a simple transformation to elements of higher dimensionality: The approach for the rod element can easily be generalized to plane and solid elements, while the beam formulations have their analogon as plate elements.

References

1. Bathe K-J (1996) Finite element procedures. Prentice-Hall, Upper Saddle River
2. Cook RD, Malkus DS, Plesha ME, Witt RJ (2002) Concepts and applications of finite element analysis. Wiley, New York
3. Debnath L (2012) Nonlinear partial differential equations for scientists and engineers. Springer, New York
4. Edelstein-Keshet L (2010) Integral calculus: mathematics 103. http://ugrad.math.ubc.ca/coursedoc/math103/site2010/keshet.notes/Chapter9.pdf. Accessed 1 Dec 2014
5. Formaggia L, Saleri F, Veneziani A (2012) Solving numerical PDEs: problems, applications, exercises. Springer, Milan
6. Glowinski R, Neittaanmi P (eds) (2008) Partial differential equations: modelling and numerical simulation. Springer, Dordrecht
7. Javanbakht Z, Öchsner A (2017) Advanced finite element simulation with MSC Marc: application of user subroutines. Springer, Cham
8. Javanbakht Z, Öchsner A (2018) Computational statics revision course. Springer, Cham
9. Merkel M, Öchsner A (2014) Eindimensionale Finite Elemente: Ein Einstieg in die Methode. Springer Vieweg, Wiesbaden
10. Öchsner A (2014) Elasto-plasticity of frame structure elements: modeling and simulation of rods and beams. Springer, Berlin
11. Öchsner A (2018) A project-based introduction to computational statics. Springer, Cham
12. Öchsner A, Makvandi R (2019) Finite elements for truss and frame structures: an introduction based on the computer algebra system maxima. Springer, Cham
13. Öchsner A, Makvandi R (2020) Finite elements using Maxima: theory and routines for rods and beams. Springer, Cham
14. Öchsner A, Merkel M (2018) One-dimensional finite elements: an introduction to the FE method. Springer, Cham
15. Öchsner A, Öchsner M (2018) A first introduction to the finite element analysis program MSC Marc/Mentat. Springer, Cham
16. Salsa S (2008) Partial differential equations in action: from modelling to theory. Springer, Milano
17. Trapp M, Öchsner A (2018) Computational plasticity for finite elements: a Fortran-based introduction. Springer, Cham
18. Zienkiewicz OC, Taylor RL (2000) The finite element method. Vol. 1: the basis. Butterworth-Heinemann, Oxford
19. Zienkiewicz OC, Taylor RL (2000) The finite element method. Vol. 2: solid mechanics. Butterworth-Heinemann, Oxford

Chapter 2
Rods and Trusses

Abstract This chapter starts with the analytical description of rod/bar members. Based on the three basic equations of continuum mechanics, i.e., the kinematics relationship, the constitutive law and the equilibrium equation, the partial differential equation, which describes the physical problem, is derived. The weighted residual method is then used to derive the principal finite element equation for rod elements. Assembly of elements and the consideration of boundary conditions is treated in detail. The chapter concludes with the spatial arrangements of rod elements in a plane to form truss structures.

2.1 Introduction

A rod is defined as a prismatic body whose axial dimension is much larger than its transverse dimensions. This structural member is only loaded in the direction of the main body axes, see Fig. 2.1a. As a result of this loading, the deformation occurs only along its main axis.

The following derivations are restricted to some simplifications:

- only applying to straight rods,
- displacements are (infinitesimally) small,
- strains are (infinitesimally) small,
- material is linear-elastic (homogeneous and isotropic).

The ultimate goal of the finite element approach is to replace the continuum description of the structural member (partial differential equation) by a discretized description based on finite elements (denoted by Roman numerals) where the nodes (denoted by Arabic numbers) now play a major role for the evaluation of the primary quantities, see Fig. 2.1b. It should be noted here that the alternatively nomenclature 'bar' is also found in scientific literature to describe a rod member. Details on the continuum mechanical description of rods can be found in [2, 3] and the basic equations are derived in detail in the following sections.

A. Öchsner, *Computational Statics and Dynamics*,
https://doi.org/10.1007/978-981-15-1278-0_2

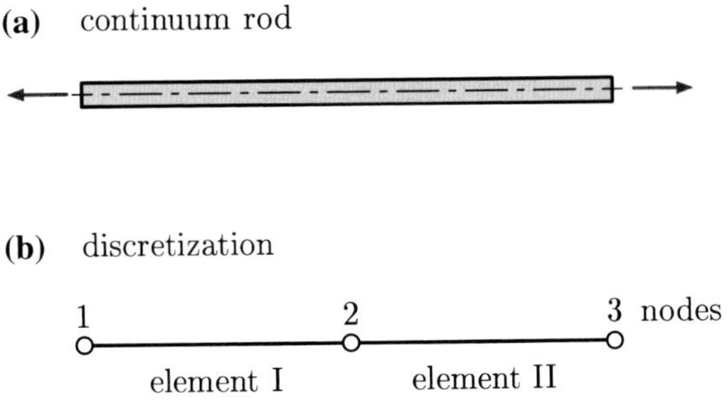

Fig. 2.1 **a** Continuum rod and **b** discretization with two finite elements

2.2 Derivation of the Governing Differential Equation

2.2.1 *Kinematics*

To derive the strain-displacement relation (kinematics relation), an axially loaded rod is considered as shown in Fig. 2.2. The length of the member is equal to L and the constant axial tensile stiffness is equal to EA. The load is either given as a single force F_x and/or as a distributed load $p_x(x)$.

This distributed load has the unit of force per unit length. In the case of a body force f_x (unit: force per unit volume), the distributed load takes the form $p_x(x) = f_x(x)A(x)$ where A is the cross-sectional area of the rod. A typical example for a body force would be the dead weight, i.e. the mass under the influence of gravity. In the case of a traction force t_x (unit: force per unit area), the distributed load can be written as $p_x(x) = t_x(x)U(x)$ where $U(x)$ is the perimeter of the cross section. Typical examples are frictional resistance, viscous drag and surface shear.

Let us now consider a differential element dx of such a rod as shown in Fig. 2.3. Under an acting load, this element deforms as indicated in Fig. 2.3b where the initial point at the position x is displaced by u_x and the end point at the position $x + \mathrm{d}x$

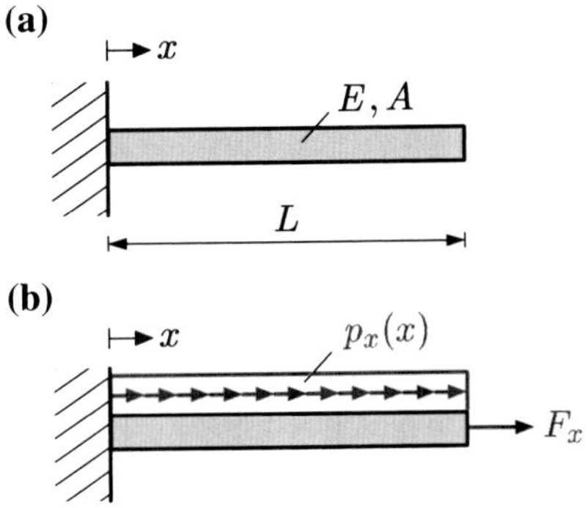

Fig. 2.2 General configuration of an axially loaded rod: **a** geometry and material property; **b** prescribed loads

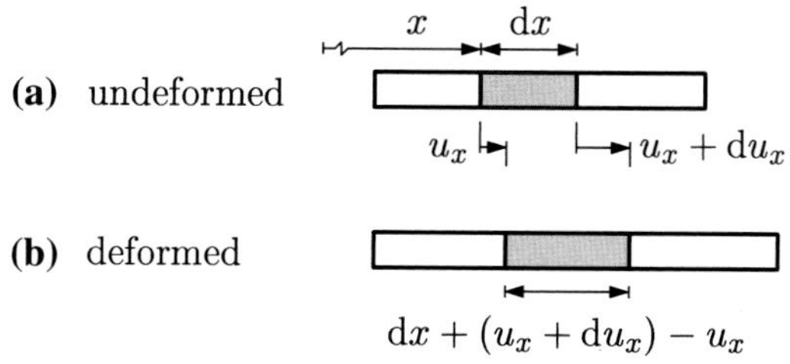

Fig. 2.3 Elongation of a differential element of length dx: **a** undeformed configuration; **b** deformed configuration

is displaced by $u_x + \mathrm{d}u_x$. Thus, the differential element which has a length of dx in the unloaded state elongates to a length of $\mathrm{d}x + (u_x + \mathrm{d}u_x) - u_x$.

The engineering strain, i.e., the increase in length related to the original length, can be expressed as

$$\varepsilon_x = \frac{(\mathrm{d}x + (u_x + \mathrm{d}u_x) - u_x) - (\mathrm{d}x)}{\mathrm{d}x}, \tag{2.1}$$

or finally as:

$$\varepsilon_x(x) = \frac{\mathrm{d}u_x(x)}{\mathrm{d}x}. \tag{2.2}$$

The last equation is often expressed in a less mathematical way (non-differential) as $\varepsilon_x = \frac{\Delta L}{L}$ where ΔL is the change in length of the entire rod element.

2.2.2 *Constitutive Equation*

The constitutive equation, i.e., the relation between the stress σ_x and the strain ε_x, is given in its simplest form as HOOKE's law[1]

$$\sigma_x(x) = E\varepsilon_x(x), \tag{2.3}$$

where the YOUNG's modulus[2] E is in the case of linear elasticity a material constant. For the considered rod element, the normal stress and strain is constant over the cross section as shown in Fig. 2.4.

2.2.3 *Equilibrium*

The equilibrium equation between the external forces and internal reactions can be derived for a differential element of length dx as shown in Fig. 2.5. It is assumed for

[1] Robert HOOKE (1635–1703), English natural philosopher, architect and polymath.

[2] Thomas YOUNG (1773–1829), English polymath.

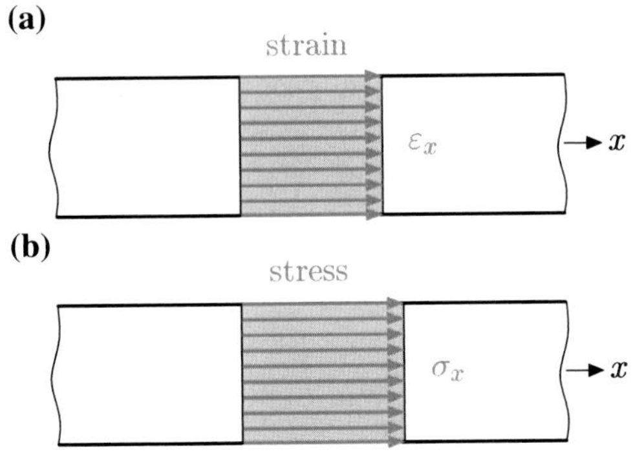

Fig. 2.4 Axially loaded rod: **a** strain and **b** stress distribution

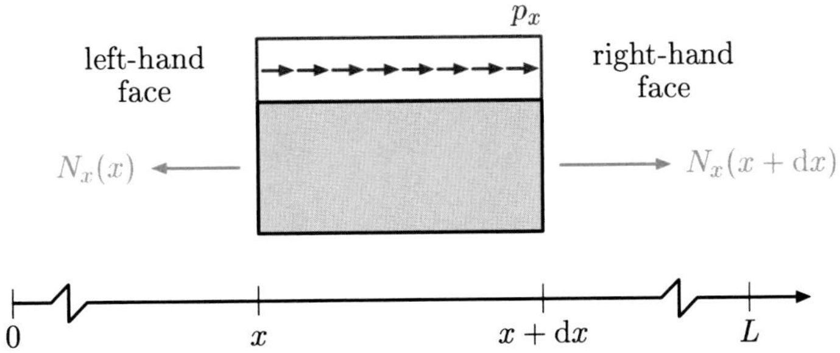

Fig. 2.5 Differential element of a rod with internal reactions and constant external distributed load

simplicity that the distributed load p_x and the cross-sectional area A are constant in this figure. The internal reactions N_x are drawn in their positive directions, i.e., at the left-hand face in the negative and at the right-hand face in the positive x-direction.

The force equilibrium in the x-direction for a static configuration requires that

$$-N_x(x) + p_x \mathrm{d}x + N_x(x + \mathrm{d}x) = 0 \tag{2.4}$$

holds. A first-order TAYLOR's[3] series expansion (cf. Appendix A.10) of the normal force $N_x(x + \mathrm{d}x)$ around point x, i.e.

$$N_x(x + \mathrm{d}x) \approx N_x(x) + \left.\frac{\mathrm{d}N_x}{\mathrm{d}x}\right|_x \mathrm{d}x\,, \tag{2.5}$$

[3]Brook TAYLOR (1685–1731), English mathematician.

Table 2.1 Fundamental governing equations of a rod for deformation along the x-axis

Expression	Equation
Kinematics	$\varepsilon_x(x) = \dfrac{\mathrm{d}u_x(x)}{\mathrm{d}x}$
Equilibrium	$\dfrac{\mathrm{d}N_x(x)}{\mathrm{d}x} = -p_x(x)$
Constitution	$\sigma_x(x) = E\varepsilon_x(x)$

allows to finally express Eq. (2.4) as:

$$\frac{\mathrm{d}N_x(x)}{\mathrm{d}x} = -p_x(x)\,. \tag{2.6}$$

The three fundamental equations to describe the behavior of a rod element are summarized in Table 2.1.

A slightly different derivation of the equilibrium equation is obtained as follows: Equation (2.4) can be expressed based on the normal stresses as:

$$-\sigma_x(x)A + p_x\mathrm{d}x + \sigma_x(x+\mathrm{d}x)A = 0\,. \tag{2.7}$$

A first-order TAYLOR's series expansion of the stress $\sigma_x(x+\mathrm{d}x)$ around point x, i.e.

$$\sigma_x(x+\mathrm{d}x) \approx \sigma_x(x) + \left.\frac{\mathrm{d}\sigma_x}{\mathrm{d}x}\right|_x \mathrm{d}x\,, \tag{2.8}$$

allows to finally express Eq. (2.7) as:

$$\frac{\mathrm{d}\sigma_x(x)}{\mathrm{d}x} + \frac{p_x(x)}{A} = 0\,. \tag{2.9}$$

The last equation with $\sigma_x = \frac{N_x}{A}$ immediately gives Eq. (2.6).

2.2.4 *Differential Equation*

To derive the governing partial differential equation, the three fundamental equations given in Table 2.1 must be combined. Introducing the kinematics relation (2.2) into HOOKE's law (2.3) gives:

$$\sigma_x(x) = E\frac{\mathrm{d}u_x}{\mathrm{d}x}\,. \tag{2.10}$$

Considering in the last equation that a normal stress is defined as an acting force N_x over a cross-sectional area A:

$$\frac{N_x}{A} = E\frac{\mathrm{d}u_x}{\mathrm{d}x}. \tag{2.11}$$

The last equation can be differentiated with respect to the x-coordinate to give:

$$\frac{\mathrm{d}N_x}{\mathrm{d}x} = \frac{\mathrm{d}}{\mathrm{d}x}\left(EA\frac{\mathrm{d}u_x}{\mathrm{d}x}\right), \tag{2.12}$$

where the derivative of the normal force can be replaced by the equilibrium equation (2.6) to obtain in the general case:

$$\frac{\mathrm{d}}{\mathrm{d}x}\left(E(x)A(x)\frac{\mathrm{d}u_x(x)}{\mathrm{d}x}\right) = -p_x(x). \tag{2.13}$$

The general case in the formulation with the internal normal force distribution reads:

$$E(x)A(x)\frac{\mathrm{d}u_x(x)}{\mathrm{d}x} = N_x(x). \tag{2.14}$$

Thus, to obtain the displacement field $u_x(x)$, one may start from Eq. (10.9) or from Eq. (2.14). The first approach requires to state the distribution of the distributed load $p_x(x)$ while for the second approach one requires the internal normal force distribution $N_x(x)$.

If the axial tensile stiffness EA is constant, the formulation (10.9) can be simplified to:

$$EA\frac{\mathrm{d}^2u_x(x)}{\mathrm{d}x^2} = -p_x(x). \tag{2.15}$$

Some common formulations of the governing partial differential equation are collected in Table 2.2. It should be noted here that some of the different cases given in Table 2.2 can be combined. The last case in Table 2.2 refers to the case of elastic embedding of a rod where the embedding modulus k has the unit of force per unit area. Analytical solutions for different loading and support conditions can be found, for example, in [6].

If we replace the common formulation of the first order derivative, i.e. $\frac{\mathrm{d}(\ldots)}{\mathrm{d}x}$, by a formal operator symbol, i.e. $\mathcal{L}_1(\ldots)$, the basic equations can be stated in a more formal way as given in Table 2.3. Such a formulation is advantageous in the two- and three-dimensional cases. It should be noted here that the transposed ('T'), i.e., $\mathcal{L}_1^{\mathrm{T}} = \left(\frac{\mathrm{d}(\ldots)}{\mathrm{d}x}\right)^{\mathrm{T}}$, is only used to show later similar structures of the equations in the two- and three-dimensional case.

Table 2.2 Different formulations of the partial differential equation for a rod (x-axis: right facing)

Configuration	Partial differential equation
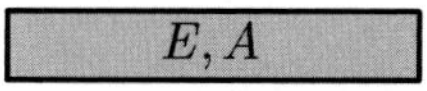	$EA\dfrac{\mathrm{d}^2u_x}{\mathrm{d}x^2} = 0$
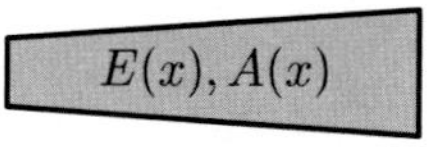	$\dfrac{\mathrm{d}}{\mathrm{d}x}\left(E(x)A(x)\dfrac{\mathrm{d}u_x}{\mathrm{d}x}\right) = 0$
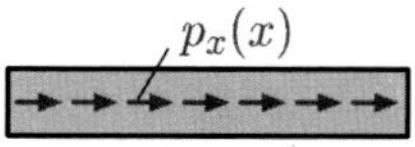	$EA\dfrac{\mathrm{d}^2u_x}{\mathrm{d}x^2} = -p_x(x)$
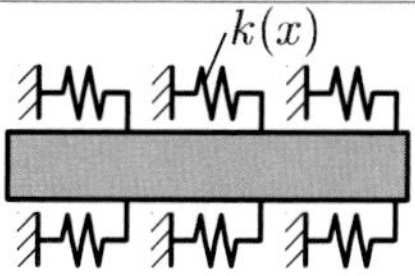	$EA\dfrac{\mathrm{d}^2u_x}{\mathrm{d}x^2} = k(x)u_x$

Table 2.3 Different formulations of the basic equations for a rod (x-axis along the principal rod axis). E: YOUNG's modulus; A: cross-sectional area; p_x: length-specific distributed normal load; $\mathcal{L}_1 = \frac{\mathrm{d}(\ldots)}{\mathrm{d}x}$: first-order derivative; b: volume-specific distributed normal load

Specific formulation	General formulation
Kinematics	
$\varepsilon_x(x) = \dfrac{\mathrm{d}u_x(x)}{\mathrm{d}x}$	$\varepsilon_x(x) = \mathcal{L}_1\left(u_x(x)\right)$
Constitution	
$\sigma_x(x) = E\varepsilon_x(x)$	$\sigma_x(x) = C\varepsilon_x(x)$
Equilibrium	
$\dfrac{\mathrm{d}\sigma_x(x)}{\mathrm{d}x} + \dfrac{p_x(x)}{A} = 0$	$\mathcal{L}_1^{\mathrm{T}}\left(\sigma_x(x)\right) + b = 0$
PDE (A = const.)	
$\dfrac{\mathrm{d}}{\mathrm{d}x}\left(E(x)\dfrac{\mathrm{d}u_x}{\mathrm{d}x}\right) + \dfrac{p_x(x)}{A} = 0$	$\mathcal{L}_1^{\mathrm{T}}\left(C\mathcal{L}_1\left(u_x(x)\right)\right) + b = 0$
	or $\mathcal{L}_1^{\mathrm{T}}\left(EA\mathcal{L}_1\left(u_x(x)\right)\right) + p_x = 0$

2.3 Finite Element Solution

2.3.1 Derivation of the Principal Finite Element Equation

Let us consider in the following the governing differential equation according to Eq. (2.15). This formulation assumes that the axial tensile stiffness EA is constant and we obtain

$$EA\frac{\mathrm{d}^2u^0(x)}{\mathrm{d}x^2}+p(x)=0\,, \tag{2.16}$$

where $u^0(x)$ represents the *exact* solution of the problem. The last equation, which contains the exact solution of the problem, is fulfilled at each location x of the rod and is called the *strong formulation* of the problem. Replacing the exact solution in Eq. (2.16) by an approximate solution $u(x)$, a residual r is obtained:

$$r(x)=EA\frac{\mathrm{d}^2u(x)}{\mathrm{d}x^2}+p(x)\neq 0\,. \tag{2.17}$$

As a consequence of the introduction of the approximate solution $u(x)$, it is in general no longer possible to satisfy the differential equation at each location x of the rod. It is alternatively requested in the following that the differential equation is fulfilled over a certain length (and no longer at each location x) and the following integral statement[4] is obtained

$$\int_0^L W^{\mathrm{T}}(x)\left(EA\frac{\mathrm{d}^2u(x)}{\mathrm{d}x^2}+p(x)\right)\mathrm{d}x\overset{!}{=}0\,, \tag{2.18}$$

which is called the *inner product*.[5] The function $W(x)$ in Eq. (2.18) is called the weight function which distributes the error or the residual in the considered domain. Alternatively, the weight function is sometimes called the test function.

Integrating by parts[6] of the first expression in the brackets of Eq. (2.18) gives

$$\int_0^L \underbrace{W^{\mathrm{T}}}_{f}\,EA\underbrace{\frac{\mathrm{d}^2u(x)}{\mathrm{d}x^2}}_{g'}\mathrm{d}x=EA\left[W^{\mathrm{T}}\frac{\mathrm{d}u(x)}{\mathrm{d}x}\right]_0^L-EA\int_0^L\frac{\mathrm{d}W^{\mathrm{T}}(x)}{\mathrm{d}x}\frac{\mathrm{d}u(x)}{\mathrm{d}x}\mathrm{d}x\,. \tag{2.19}$$

Under consideration of Eq. (2.18), the so-called *weak formulation* of the problem is obtained as:

$$EA\int_0^L\frac{\mathrm{d}W^{\mathrm{T}}(x)}{\mathrm{d}x}\frac{\mathrm{d}u(x)}{\mathrm{d}x}\mathrm{d}x=EA\left[W^{\mathrm{T}}(x)\frac{\mathrm{d}u(x)}{\mathrm{d}x}\right]_0^L+\int_0^L W^{\mathrm{T}}(x)p(x)\,\mathrm{d}x\,. \tag{2.20}$$

---

[4]The use of the transposed 'T' for the scalar weight function W is not obvious at the first glance. However, the following matrix operations will clarify this approach.

[5]The general formulation of the inner product states the integration over the volume V, see Eq. (8.20). For this integration, the strong form (2.16) must be written as $E\frac{\mathrm{d}^2u^0(x)}{\mathrm{d}x^2}+\frac{p(x)}{A}$ at which the distributed load is now given as force per unit volume.

[6]A common representation of integration by parts of two functions $f(x)$ and $g(x)$ is: $\int fg'\mathrm{d}x=fg-\int f'g\,\mathrm{d}x$.

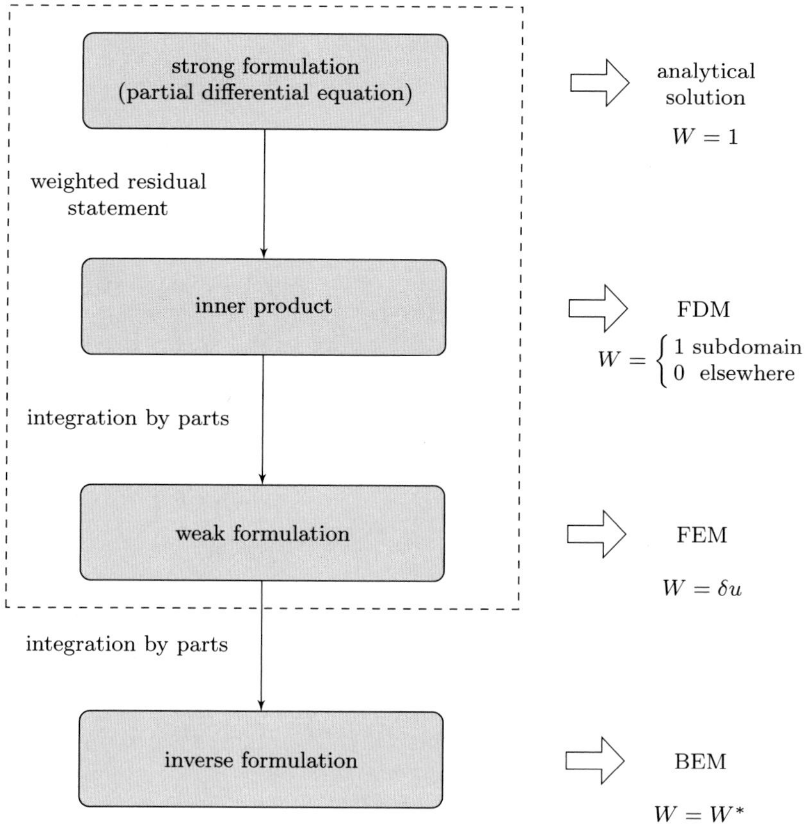

Fig. 2.6 Some classical approximation methods in the context of the weighted residual method

Looking at the weak formulation, it can be seen that the integration by parts shifted one derivative from the approximate solution to the weight function and a symmetrical formulation with respect to the derivatives is obtained. This symmetry with respect to the derivatives of the approximate solution and the weight function will guarantee in the following that a symmetric stiffness matrix is derived for the rod element. Figure 2.6 illustrates some common approximation methods in the context of the weighted residual method.

In order to continue the derivation of the principal finite element equation, the displacement $u(x)$ and the weight function $W(x)$ must be expressed by some functions. The common way to express the unknown function $u(x)$ in the scope of the finite element method is the so-called nodal approach. This approach states that the unknown function within an element (superscript 'e') is given by

$$u^{\mathrm{e}}(x) = \boldsymbol{N}^{\mathrm{T}}(x)\,\boldsymbol{u}_{\mathrm{p}}^{\mathrm{e}} = \begin{bmatrix} N_1 & N_2 & \cdots & N_n \end{bmatrix} \times \begin{bmatrix} u_1 \\ u_2 \\ \vdots \\ u_n \end{bmatrix}, \tag{2.21}$$

where $\boldsymbol{u}_{\mathrm{p}}^{\mathrm{e}}$ is the column matrix of n nodal unknowns and $\boldsymbol{N}(x)$ is the column matrix of the *interpolation functions*. Thus, the displacement at any point inside an element is approximated based on nodal values and interpolation functions which distribute these displacements between the nodes in a certain way. Equation (2.21) illustrates a basic idea of the finite element method where the unknown function is not approximated over the entire domain of the problem (in general Ω) but in a sub-domain (Ω^{e}), the so-called finite element. In a similar way as the unknown function, the weight function is approximated as

$$W_x(x) = \boldsymbol{N}(x)^{\mathrm{T}}\delta\boldsymbol{u}_{\mathrm{p}} = \begin{bmatrix} N_1 & N_2 & \cdots & N_n \end{bmatrix} \times \begin{bmatrix} \delta u_1 \\ \delta u_2 \\ \vdots \\ \delta u_n \end{bmatrix}, \tag{2.22}$$

where δu_i represents the so-called arbitrary or virtual displacements. It will be shown in the following that the virtual displacements occur on both sides of Eq. (2.20) and can be eliminated. Thus, these virtual displacements do not need a deeper consideration at this point of the derivation. Equation (2.20) requires the derivatives of $u(x)$ and $W(x)$ which can be written on the element level as:

$$\frac{\mathrm{d}u^{\mathrm{e}}(x)}{\mathrm{d}x} = \frac{\mathrm{d}}{\mathrm{d}x}\left(\boldsymbol{N}^{\mathrm{T}}(x)\,\boldsymbol{u}_{\mathrm{p}}\right) = \frac{\mathrm{d}\boldsymbol{N}^{\mathrm{T}}(x)}{\mathrm{d}x}\boldsymbol{u}_{\mathrm{p}}, \tag{2.23}$$

$$\frac{\mathrm{d}W(x)}{\mathrm{d}x} = \frac{\mathrm{d}}{\mathrm{d}x}\left(\boldsymbol{N}^{\mathrm{T}}(x)\,\delta\boldsymbol{u}_{\mathrm{p}}\right) = \frac{\mathrm{d}\boldsymbol{N}^{\mathrm{T}}(x)}{\mathrm{d}x}\delta\boldsymbol{u}_{\mathrm{p}}. \tag{2.24}$$

It should be noted here that the nodal unknowns and their virtual counterparts are constant values, i.e. not a function of x, and are therefore not affected by the differential operator. It is common in some references (e.g. [1, 10]) to introduce the matrix which contains the derivatives of the interpolation functions as a matrix denoted by $\boldsymbol{B} = \frac{\mathrm{d}\boldsymbol{N}(x)}{\mathrm{d}x}$. Thus, the derivatives can be be written as:

$$\frac{\mathrm{d}u^{\mathrm{e}}(x)}{\mathrm{d}x} = \boldsymbol{B}^{\mathrm{T}}\,\boldsymbol{u}_{\mathrm{p}}, \tag{2.25}$$

$$\frac{\mathrm{d}W(x)}{\mathrm{d}x} = \boldsymbol{B}^{\mathrm{T}}\delta\boldsymbol{u}_{\mathrm{p}}. \tag{2.26}$$

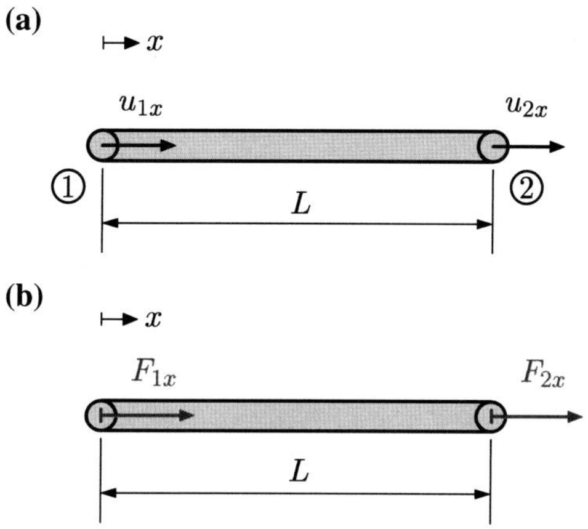

Fig. 2.7 Definition of the one-dimensional linear rod element: **a** deformations; **b** external loads. The nodes are symbolized by the two circles at the ends (◯)

2.3.1.1 Linear Element Formulation

Let us consider in the following a rod element which is composed of two nodes as schematically shown in Fig. 2.7. Each node has only one degree of freedom, i.e., a displacement in the direction of the principal axis (cf. Fig. 2.7a) and each node can be only loaded by a single force acting in x-direction (cf. Fig. 2.7b).

Since there are only two nodes with two unknowns, the equation for the unknown displacement in the element and its virtual counterpart (cf. Eqs. (2.21) and (2.22)) are simplified to the following expressions:

$$u^{\mathrm{e}}(x) = \boldsymbol{N}^{\mathrm{T}}(x)\,\boldsymbol{u}_{\mathrm{p}} = \begin{bmatrix} N_1 & N_2 \end{bmatrix} \times \begin{bmatrix} u_1 \\ u_2 \end{bmatrix}, \tag{2.27}$$

and

$$W(x) = \boldsymbol{N}(x)^{\mathrm{T}}\delta\boldsymbol{u}_{\mathrm{p}} = \begin{bmatrix} N_1 & N_2 \end{bmatrix} \times \begin{bmatrix} \delta u_1 \\ \delta u_2 \end{bmatrix}, \tag{2.28}$$

or for the transposed of the weight function:

$$W^{\mathrm{T}}(x) = \left(\boldsymbol{N}(x)^{\mathrm{T}}\delta\boldsymbol{u}_{\mathrm{p}}\right)^{\mathrm{T}} = \delta\boldsymbol{u}_{\mathrm{p}}^{\mathrm{T}}\boldsymbol{N}(x), \tag{2.29}$$

$$\frac{\mathrm{d}W^{\mathrm{T}}(x)}{\mathrm{d}x} = \delta\boldsymbol{u}_{\mathrm{p}}^{\mathrm{T}}\frac{\mathrm{d}\boldsymbol{N}(x)}{\mathrm{d}x}. \tag{2.30}$$

Let us first consider in the following only the left-hand side of Eq. (2.20) in order to derive the expression for the elemental stiffness matrix $\boldsymbol{K}^{\mathrm{e}}$ of the linear rod element. Introduction of expressions (2.27) and (2.28) in the weak form gives

$$EA \int_0^L \left(\delta \boldsymbol{u}_\mathrm{p}^\mathrm{T} \frac{\mathrm{d}\boldsymbol{N}(x)}{\mathrm{d}x} \right) \left(\frac{\mathrm{d}\boldsymbol{N}^\mathrm{T}(x)}{\mathrm{d}x} \boldsymbol{u}_\mathrm{p} \right) \mathrm{d}x \,, \tag{2.31}$$

or under consideration that the column matrix of the nodal unknowns can be considered as constant as:

$$\delta \boldsymbol{u}_\mathrm{p}^\mathrm{T} \underbrace{EA \int_0^L \left(\frac{\mathrm{d}\boldsymbol{N}(x)}{\mathrm{d}x} \right) \left(\frac{\mathrm{d}\boldsymbol{N}^\mathrm{T}(x)}{\mathrm{d}x} \right) \mathrm{d}x}_{\boldsymbol{K}^\mathrm{e}} \, \boldsymbol{u}_\mathrm{p} \,. \tag{2.32}$$

It will be seen in the following that the expression $\delta \boldsymbol{u}_\mathrm{p}^\mathrm{T}$ can be 'canceled' with an identical expression on the right-hand side of Eq. (2.20) and $\boldsymbol{u}_\mathrm{p}$ represents the column matrix of the unknown nodal displacements. Under consideration of the $\boldsymbol{B}$-matrix, the stiffness matrix can be expressed in a more general way for constant tensile stiffness EA as:

$$\boldsymbol{K}^\mathrm{e} = EA \int_0^L \boldsymbol{B}\boldsymbol{B}^\mathrm{T} \mathrm{d}x \,. \tag{2.33}$$

In order to further evaluate Eq. (2.32), we can introduce the components of the derivatives to give:

$$EA \int_0^L \begin{bmatrix} \dfrac{\mathrm{d}N_1(x)}{\mathrm{d}x} \\ \dfrac{\mathrm{d}N_2(x)}{\mathrm{d}x} \end{bmatrix} \begin{bmatrix} \dfrac{\mathrm{d}N_1(x)}{\mathrm{d}x} & \dfrac{\mathrm{d}N_2(x)}{\mathrm{d}x} \end{bmatrix} \mathrm{d}x \,, \tag{2.34}$$

or after the matrix multiplication as:

$$EA \int_0^L \begin{bmatrix} \dfrac{\mathrm{d}N_1(x)}{\mathrm{d}x}\dfrac{\mathrm{d}N_1(x)}{\mathrm{d}x} & \dfrac{\mathrm{d}N_1(x)}{\mathrm{d}x}\dfrac{\mathrm{d}N_2(x)}{\mathrm{d}x} \\ \dfrac{\mathrm{d}N_2(x)}{\mathrm{d}x}\dfrac{\mathrm{d}N_1(x)}{\mathrm{d}x} & \dfrac{\mathrm{d}N_2(x)}{\mathrm{d}x}\dfrac{\mathrm{d}N_2(x)}{\mathrm{d}x} \end{bmatrix} \mathrm{d}x \,. \tag{2.35}$$

Any further evaluation of this equation requires now that the functional expressions $N_1(x)$ and $N_2(x)$ are known. The simplest assumption that can be done is that the nodal values are linearly distributed within the element, from its value at the node to zero at the opposite node. For such a linear superposition, the interpolation functions can be assumed as shown in Fig. 2.8a, b.

The graphical interaction of interpolation functions with the nodal displacement values is shown in Fig. 2.9.

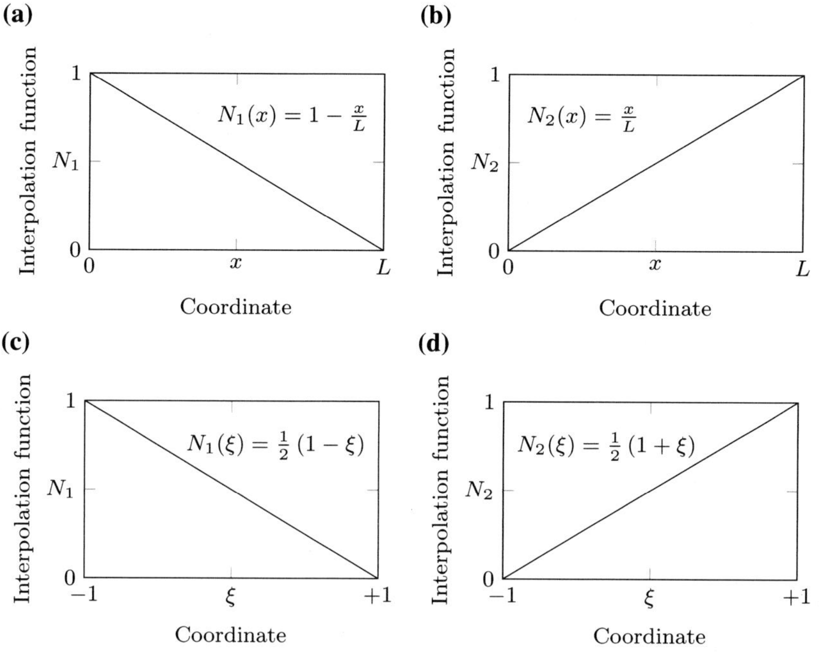

Fig. 2.8 Interpolation functions for the linear rod element: **a** and **b** physical coordinate (x); **c** and **d** natural coordinate (ξ)

The derivatives of the interpolation functions can easily be calculated as

$$\frac{dN_1(x)}{dx} = -\frac{1}{L}, \quad \frac{dN_2(x)}{dx} = \frac{1}{L}, \tag{2.36}$$

$$\frac{dN_1(\xi)}{d\xi} = -\frac{1}{2}, \quad \frac{dN_2(\xi)}{d\xi} = \frac{1}{2}. \tag{2.37}$$

Thus, the $\boldsymbol{B}$-matrix given in Eq. (2.25) takes the form:

$$\boldsymbol{B} = \frac{1}{L}\begin{bmatrix} -1 \\ 1 \end{bmatrix}. \tag{2.38}$$

The derivatives introduced into Eq. (2.39) give

$$EA\int_0^L \begin{bmatrix} \frac{1}{L^2} & -\frac{1}{L^2} \\ -\frac{1}{L^2} & \frac{1}{L^2} \end{bmatrix} dx = \frac{EA}{L^2}\int_0^L \begin{bmatrix} 1 & -1 \\ -1 & 1 \end{bmatrix} dx. \tag{2.39}$$

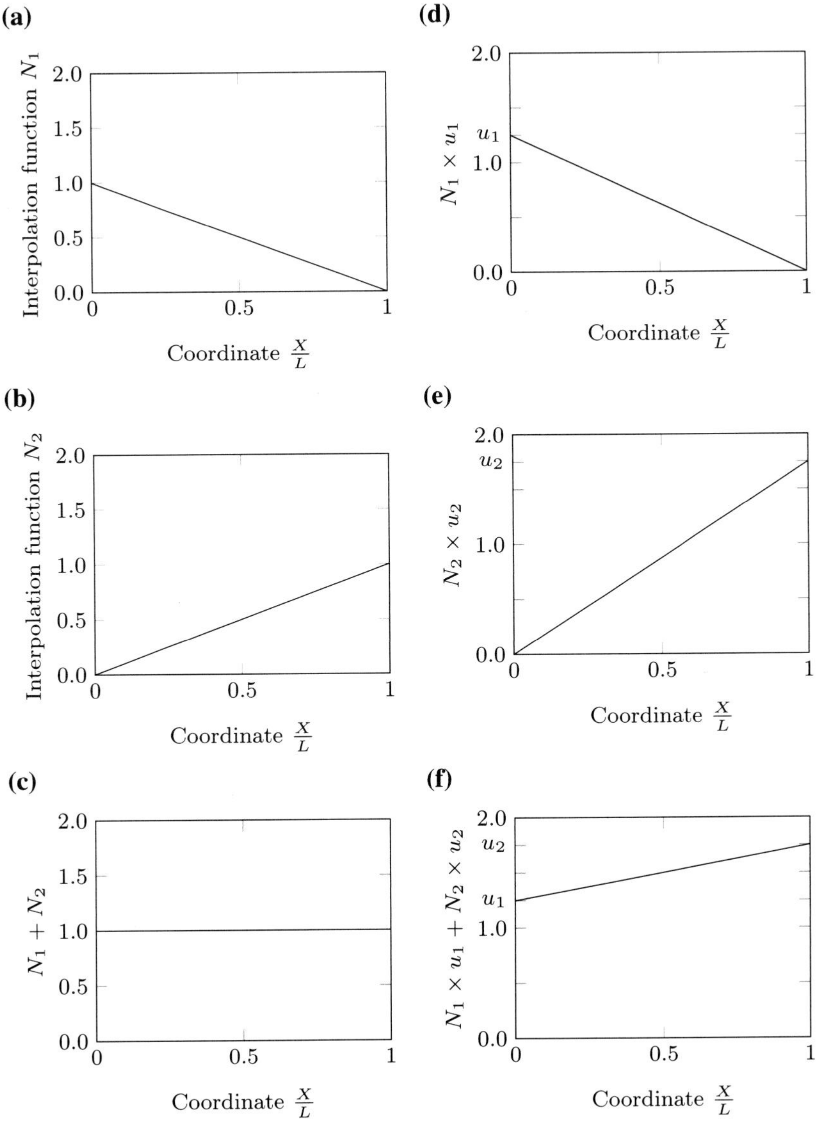

Fig. 2.9 Graphical interpretation of interpolation functions: **a–c** 'pure' interpolation functions; **d–e** weighted with nodal values

The integral in the last equation can be analytically integrated to obtain

$$\frac{EA}{L^2}\begin{bmatrix} x & -x \\ -x & x \end{bmatrix}\Bigg|_0^L = \frac{EA}{L^2}\begin{bmatrix} L & -L \\ -L & L \end{bmatrix} \tag{2.40}$$

and the stiffness matrix for a linear rod element is given by:

$$\boldsymbol{K}^{\mathrm{e}} = \frac{EA}{L}\begin{bmatrix} 1 & -1 \\ -1 & 1 \end{bmatrix}. \tag{2.41}$$

It must be noted here that an analytical integration as performed to obtain Eq. (2.40) cannot be performed in commercial finite element codes since they are written in traditional programming languages such as FORTRAN. Instead of the analytical integration, a numerical integration is performed (cf. Appendix A.9) where the integral is approximated by the evaluation and weighting of functional values at so-called integration or GAUSS[7] points. To this end, the Cartesian coordinate x is transformed to the natural coordinate ξ ranging from -1 to 1. Depending on the origin of the Cartesian coordinate system, the transformation can be performed based on the relations given in Table 2.4.

The integral in Eq. (2.39) can be written in terms of the natural coordinate ξ and approximated in terms of a GAUSS-LEGENDRE[8] quadrature as:

$$\boldsymbol{K}^{\mathrm{e}} = \frac{EA}{L^2}\int\limits_{-1}^{1}\begin{bmatrix} 1 & -1 \\ -1 & 1 \end{bmatrix}\frac{L}{2}\,\mathrm{d}\xi \approx \frac{EA}{L^2}\sum_{i=1}^{n}\begin{bmatrix} \cdots\cdots \\ \cdots\cdots \end{bmatrix}(\xi_i)w(\xi_i)\,, \tag{2.42}$$

where the matrix is to be evaluated at the n integration points and multiplied by certain weights w, cf. Appendix A.9. Since the matrix is in this simple case only composed of constant values, it is sufficient to consider a one-point integration rule ($\xi = 0$, $w = 2$) to achieve the analytical result[9] as:

$$\boldsymbol{K}^{\mathrm{e}} = \frac{EA}{2L}\begin{bmatrix} 1 & -1 \\ -1 & 1 \end{bmatrix}\Bigg|_{\xi=0} \times \underbrace{2}_{w} = \frac{EA}{L}\begin{bmatrix} 1 & -1 \\ -1 & 1 \end{bmatrix}. \tag{2.43}$$

The transformation between Cartesian (x) and natural coordinates (ξ) as indicated in Table 2.4 can be further generalized. Let us assume for this purpose that the Cartesian coordinate can be interpolated in the following manner:

---

[7] Johann Carl Friedrich GAUSS (1777–1855), German mathematician and physical scientist.

[8] Adrien-Marie LEGENDRE (1752–1833), French mathematician.

[9] It must be noted here that in the general case only an *approximation* of the integral can be obtained and that the exact, i.e. analytical solution, is reserved for simple cases.

Table 2.4 Transformation between Cartesian (x) and natural coordinates (ξ)

Configuration	Transformation
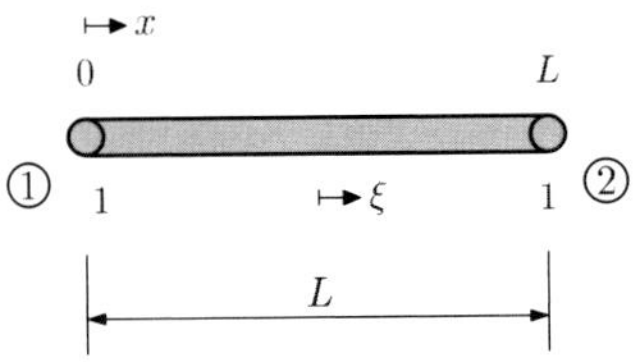	$\xi = \frac{2x}{L} - 1\,,$ $\frac{\mathrm{d}\xi}{\mathrm{d}x} = \frac{2}{L}\,.$
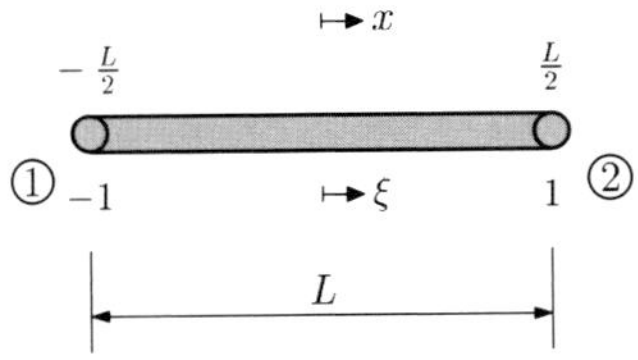	$\xi = \frac{2x}{L}\,,$ $\frac{\mathrm{d}\xi}{\mathrm{d}x} = \frac{2}{L}\,.$
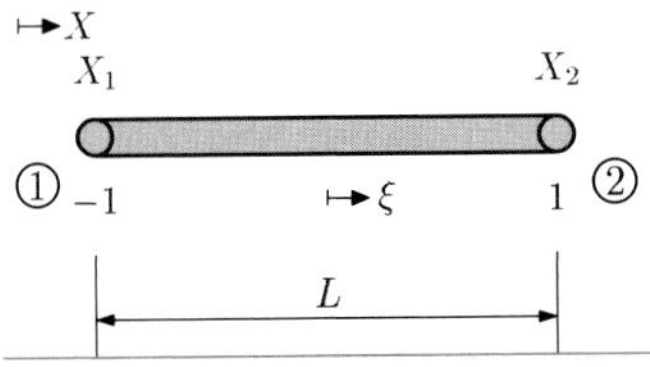	$\xi = \frac{2}{X_2 - X_1}(X - X_1) - 1\,,$ $\frac{\mathrm{d}\xi}{\mathrm{d}X} = \frac{2}{L}\,.$

$$x(\xi) = \overline{N}_1(\xi)x_1 + \overline{N}_2(\xi)x_2\,, \tag{2.44}$$

where x_1 and x_2 are the coordinates of the start and end node in the elemental Cartesian coordinate system. The interpolation functions $\overline{N}_i(\xi)$ are—in the case of the coordinate approximation—called shape functions because they describe the geometry or shape of the element. Considering the shape functions in natural coordinates as given in Fig. 2.8 for the displacement interpolation (a so-called isoparametric formulation), the following expression for the derivative of the Cartesian coordinate with respect to the natural coordinate is obtained:

$$\frac{\mathrm{d}x(\xi)}{\mathrm{d}\xi} = \frac{\mathrm{d}\overline{N}_1(\xi)}{\mathrm{d}\xi}x_1 + \frac{\mathrm{d}\overline{N}_2(\xi)}{\mathrm{d}\xi}x_2 = -\frac{1}{2}x_1 + \frac{1}{2}x_2\,. \tag{2.45}$$

The last equation allows to reproduce the geometrical derivatives given in Table 2.4 or for any other location of the elemental Cartesian coordinate system. Equation (2.45) is also known as the general form of the Jacobian determinant and allows to perform the numerical integration of the stiffness matrix in natural coordinates as outlined in Eq. (A.41). The choice of the shape functions in Eq. (2.44) allows to distinguish different element formulations. If the degree of the shape functions is equal to the degree of the interpolation functions, i.e. $\deg(\overline{N}) = \deg(N)$, a so-called isoparametric element formulation is obtained. If the degree of the shape functions is smaller than the degree of the interpolation functions, i.e. $\deg(\overline{N}) < \deg(N)$, a so-called sub-parametric element formulation is obtained. A larger degree of the shape functions compared to the interpolation functions, i.e. $\deg(\overline{N}) > \deg(N)$, gives a so-called superparametric element formulation.

Let us summarize here in a systematic manner the major steps which are required to calculate the elemental stiffness matrix of a linear rod element.

❶ Introduce an elemental coordinate system (x).

❷ Express the coordinates (x_i) of the corner nodes i ($i = 1, 2$) in this elemental coordinate system.

❸ Calculate the partial derivative of the Cartesian (x) coordinate with respect to the natural (ξ) coordinate, see Eq. (2.45):

$$\frac{\mathrm{d}x(\xi)}{\mathrm{d}\xi} = J = -\frac{1}{2}x_1 + \frac{1}{2}x_2 .$$

❹ Calculate the partial derivative of the natural (ξ) coordinate with respect to the Cartesian (x) coordinate, see Eq. (A.50):

$$\frac{\mathrm{d}\xi}{\mathrm{d}x} = \frac{1}{J} .$$

❺ Calculate the $\boldsymbol{B}$-matrix and its transposed, see Eqs. (2.33)–(2.34):

$$\boldsymbol{B}^{\mathrm{T}} = \left[\frac{\mathrm{d}N_1(x)}{\mathrm{d}x}\ \frac{\mathrm{d}N_2(x)}{\mathrm{d}x}\right] ,$$

where the partial derivatives are $\frac{\mathrm{d}N_1(x)}{\mathrm{d}x} = \frac{\mathrm{d}N_1(\xi)}{\mathrm{d}\xi}\frac{\mathrm{d}\xi}{\mathrm{d}x}, \ldots$ and the derivatives of the interpolation functions are given in Eq. (2.37), i.e., $\frac{\partial N_1(\xi)}{\partial \xi} = -\frac{1}{2}, \ldots$

❻ Calculate the triple matrix product $\boldsymbol{B}\boldsymbol{C}^{\mathrm{T}}\boldsymbol{B}$, where the elasticity matrix $\boldsymbol{C}$ is given in this special case as the scalar YOUNG's modulus E.

❼ Perform the numerical integration based on a 1-point integration rule:

$$\int_V (\boldsymbol{B}\boldsymbol{C}\boldsymbol{B}^{\mathrm{T}})\mathrm{d}V = \boldsymbol{B}E\boldsymbol{B}^{\mathrm{T}}J \times 2 \times A\Big|_{(0)} .$$

❽ $\boldsymbol{K}$ obtained.

Let us now consider the right-hand side of Eq. (2.20) in order to derive the expression for the elemental load column matrix $\boldsymbol{f}^{\mathrm{e}}$ of the linear rod element. The first part of the right-hand side, i.e.

$$EA\left[W^{\mathrm{T}}(x)\frac{\mathrm{d}u(x)}{\mathrm{d}x}\right]_0^L \tag{2.46}$$

results with the definition of the weight function according to Eq. (2.28) in

$$EA\left[\delta\boldsymbol{u}_{\mathrm{p}}^{\mathrm{T}}\boldsymbol{N}(x)\frac{\mathrm{d}u(x)}{\mathrm{d}x}\right]_0^L, \tag{2.47}$$

or in components

$$\delta\boldsymbol{u}_{\mathrm{p}}^{\mathrm{T}}EA\left[\begin{bmatrix}N_1\\N_2\end{bmatrix}\frac{\mathrm{d}u(x)}{\mathrm{d}x}\right]_0^L. \tag{2.48}$$

The virtual displacements $\delta\boldsymbol{u}_{\mathrm{p}}^{\mathrm{T}}$ in the last equation can be 'canceled' with a corresponding expression in Eq. (2.32). Furthermore, the last equation constitutes a system of two equations which must be evaluated at the integration boundaries, i.e. at $x = 0$ and $x = L$. The first equation reads:

$$\left(N_1 EA\frac{\mathrm{d}u}{\mathrm{d}x}\right)_{x=L} - \left(N_1 EA\frac{\mathrm{d}u}{\mathrm{d}x}\right)_{x=0}. \tag{2.49}$$

This gives under consideration of the boundary values of the interpolation functions, i.e. $N_1(L) = 0$ and $N_1(0) = 1$, the following statement:

$$-EA\left.\frac{\mathrm{d}u}{\mathrm{d}x}\right|_{x=0} \overset{(2.11)}{=} -N_x(x=0). \tag{2.50}$$

A corresponding expression can be derived for the second equation as:

$$EA\left.\frac{\mathrm{d}u}{\mathrm{d}x}\right|_{x=L} \overset{(2.11)}{=} N_x(x=L). \tag{2.51}$$

It must be noted here that the forces N_x are the internal reactions according to Fig. 2.5. The external loads with their positive directions according to Fig. 2.7b can be obtained from the internal loads by inverting the sign at the left-hand boundary and by maintaining the positive direction of the internal reaction at the right-hand boundary. This can easily be shown by balancing the internal and external forces at each boundary node. Thus, the contribution to the load matrix due to single *external* forces F_i at the nodes is expressed by:

$$\boldsymbol{f}_F^{\mathrm{e}} = \begin{bmatrix}F_{1x}\\F_{2x}\end{bmatrix}. \tag{2.52}$$

The second part of Eq. (2.20), i.e. after 'canceling' of the virtual displacements $\delta\boldsymbol{u}^{\mathrm{T}}$

Table 2.5 Equivalent nodal loads for a linear rod element (x-axis: right facing)

Loading	Axial force
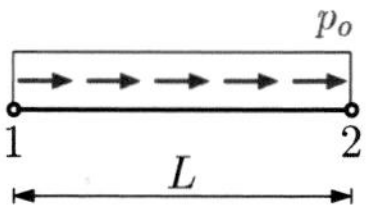	$F_{1x} = \frac{p_0 L}{2}$
	$F_{2x} = \frac{p_0 L}{2}$
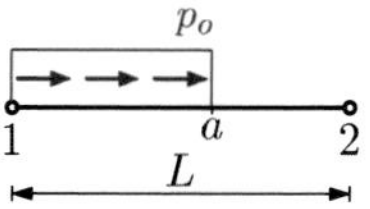	$F_{1x} = -\frac{p_0 a^2}{2L} + p_0 a$
	$F_{2x} = \frac{p_0 a^2}{2L}$
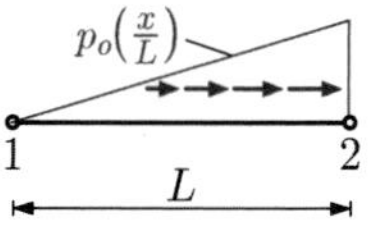	$F_{1x} = \frac{p_0 L}{6}$
	$F_{2x} = \frac{p_0 L}{3}$
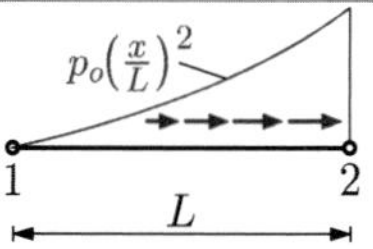	$F_{1x} = \frac{p_0 L}{12}$
	$F_{2x} = \frac{p_0 L}{4}$
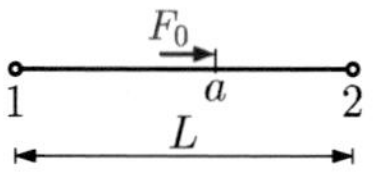	$F_{1x} = \frac{F_0 (L-a)}{L}$
	$F_{2x} = \frac{F_0 a}{L}$

$$\int_0^L N(x)\, p(x)\, \mathrm{d}x \tag{2.53}$$

represents the general rule to determine equivalent nodal loads in the case of arbitrarily distributed loads $p(x)$. As an example, the evaluation of Eq. (2.53) for a constant load p results in the following load matrix:

$$\boldsymbol{f}_p^{\mathrm{e}} = p \int_0^L \begin{bmatrix} N_1 \\ N_2 \end{bmatrix} \mathrm{d}x = \frac{pL}{2} \begin{bmatrix} 1 \\ 1 \end{bmatrix}. \tag{2.54}$$

Further expressions for equivalent nodal loads can be taken from Table 2.5. Let us remind ourselves at this step that in the scope of the finite element method any type of load can be only introduced at nodes into the discretized structure.

Table 2.6 Comments on the accuracy of the finite element solution for a single cantilever linear rod element

Configuration	
Axial tensile stiffness and loading	Accuracy of $u(x)$
$EA =$ const.; loaded by single force F at node 2	FE gives analytical solution at nodes and between nodes
$EA =$ const.; displacement BC u at node 2	FE gives exact nodal values and analytical solution between nodes
$EA =$ const.; distributed load p	FE gives analytical solution at nodes but only approximate solution between nodes
$EA \neq$ const.; loaded by single force F at node 2	FE gives approximate solution at nodes and approximate solution between nodes
$EA \neq$ const.; displacement BC u at node 2	FE gives exact nodal values but only approximate solution between nodes

Based on the derived results, the principal finite element equation for a single linear rod element with constant axial tensile stiffness EA can be expressed in a general form as

$$\boldsymbol{K}^{\mathrm{e}}\boldsymbol{u}_{\mathrm{p}}^{\mathrm{e}} = \boldsymbol{f}^{\mathrm{e}}\,, \tag{2.55}$$

or in components as:

$$\frac{EA}{L}\begin{bmatrix} 1 & -1 \\ -1 & 1 \end{bmatrix}\begin{bmatrix} u_{1x} \\ u_{2x} \end{bmatrix} = \begin{bmatrix} F_{1x} \\ F_{2x} \end{bmatrix} + \int_0^L \begin{bmatrix} N_1 \\ N_2 \end{bmatrix} p_x(x)\,\mathrm{d}x\,. \tag{2.56}$$

At the end of this derivation, a few comments on the accuracy of a linear rod element should be given, cf. Table 2.6. As can be seen, the linear rod element gives under certain conditions the exact, i.e. the analytical solution. This is illustrated by several examples in the section 'Solved Rod Problems' and 'Supplementary Problems'.

The following description is related to a more formalized derivation of the principal finite element equation (this approach will be consistently used to derive the principal finite element equation for two- and three-dimensional elements). Based on the general formulation of the partial differential equation given in Table 2.3, the strong formulation can be written as[10]:

$$\mathcal{L}_1^{\mathrm{T}} EA \mathcal{L}_1 u_x^0 + p_x = 0\,. \tag{2.57}$$

[10]The use of the transposed 'T' for the scalar operator $\mathcal{L}_1$ is not obvious at the first glance. However, the following matrix operations will clarify this approach.

Replacing the exact solution u_x^0 by an approximate solution u_x, a residual r is obtained:

$$r = \mathcal{L}_1^{\mathrm{T}} EA\mathcal{L}_1 u_x + p_x \neq 0\,. \tag{2.58}$$

The inner product is obtained by weighting the residual and integration as

$$\int_L W_x^{\mathrm{T}} \left(\mathcal{L}_1^{\mathrm{T}} EA\mathcal{L}_1 u_x + p_x\right) \mathrm{d}L \overset{!}{=} 0\,, \tag{2.59}$$

where $W_x(x)$ is the scalar weight function. Application of the GREEN–GAUSS theorem (cf. Sect. A.7) gives the weak formulation as:

$$\int_L (\mathcal{L}_1 W_x)^{\mathrm{T}} EA(\mathcal{L}_1 u_x)\mathrm{d}L = \int_s W_x^{\mathrm{T}} \underbrace{(EA\mathcal{L}_1 u_x)}_{N_x=\sigma_x A} n_x \mathrm{d}s + \int_L W_x^{\mathrm{T}} p_x \mathrm{d}L\,. \tag{2.60}$$

Any further development of Eq. (2.60) requires that the general expressions for the displacement and weight functions, i.e. u_x and W_x, are now approximated by some functional representations. With the nodal approaches for the displacements u_x (2.21) and the weight function W_x (2.22), the weak formulation reads:

$$\int_L (\mathcal{L}_1 \boldsymbol{N}^{\mathrm{T}} \delta \boldsymbol{u}_{\mathrm{p}})^{\mathrm{T}} EA(\mathcal{L}_1 \boldsymbol{N}^{\mathrm{T}} \boldsymbol{u}_{\mathrm{p}})\mathrm{d}L = \int_s \delta \boldsymbol{u}_{\mathrm{p}}^{\mathrm{T}} \boldsymbol{N} N_x n_x \mathrm{d}s + \int_L \delta \boldsymbol{u}_{\mathrm{p}}^{\mathrm{T}} \boldsymbol{N} p_x \mathrm{d}L\,. \tag{2.61}$$

Application of Eq. (A.131), i.e. $(\mathcal{L}_1 \boldsymbol{N}^{\mathrm{T}} \delta \boldsymbol{u}_{\mathrm{p}})^{\mathrm{T}} = ((\mathcal{L}_1 \boldsymbol{N}^{\mathrm{T}})\delta \boldsymbol{u}_{\mathrm{p}})^{\mathrm{T}} = \delta \boldsymbol{u}_{\mathrm{p}}^{\mathrm{T}} (\mathcal{L}_1 \boldsymbol{N}^{\mathrm{T}})^{\mathrm{T}}$, allows to express the weak formulation as:

$$\delta \boldsymbol{u}_{\mathrm{p}}^{\mathrm{T}} \int_L (\mathcal{L}_1 \boldsymbol{N}^{\mathrm{T}})^{\mathrm{T}} EA(\mathcal{L}_1 \boldsymbol{N}^{\mathrm{T}})\mathrm{d}L \boldsymbol{u}_{\mathrm{p}} = \delta \boldsymbol{u}_{\mathrm{p}}^{\mathrm{T}} \int_s \boldsymbol{N} N_x n_x \mathrm{d}s + \delta \boldsymbol{u}_{\mathrm{p}}^{\mathrm{T}} \int_L \boldsymbol{N} p_x \mathrm{d}L\,. \tag{2.62}$$

The virtual deformations can be eliminated from both sides of the last equations and the general form of the principal finite element equations is obtained:

$$\int_L (\mathcal{L}_1 \boldsymbol{N}^{\mathrm{T}})^{\mathrm{T}} EA(\mathcal{L}_1 \boldsymbol{N}^{\mathrm{T}})\mathrm{d}L \boldsymbol{u}_{\mathrm{p}}^{\mathrm{e}} = \int_s \boldsymbol{N} N_x n_x \mathrm{d}s + \int_L \boldsymbol{N} p_x \mathrm{d}L\,. \tag{2.63}$$

Based on the general formulation of the partial differential equation given in Table 2.3 and the general derivations presented in second part of Sect. 2.3.1.1, the major steps to transform the partial differential equation into the principal finite element equation are summarized in Table 2.7.

Alternatively, we may base our derivation on the other general formulation of the partial differential equation given in Table 2.3. Thus, the strong formulation can be

Table 2.7 Summary: derivation of the principal finite element equation for linear rod elements; general approach (version 1)

Strong formulation
$\mathcal{L}_1^{\mathrm{T}} EA \mathcal{L}_1 u_x^0 + p_x = 0$
Inner product
$\int_L W^{\mathrm{T}} \left(\mathcal{L}_1^{\mathrm{T}} EA \mathcal{L}_1 u_x + p_x\right) \mathrm{d}L = 0$
Weak formulation
$\int_L (\mathcal{L}_1 W)^{\mathrm{T}} EA (\mathcal{L}_1 u_x) \mathrm{d}L = \int_s W^{\mathrm{T}} N_x n_x \mathrm{d}s + \int_L W^{\mathrm{T}} p_x \mathrm{d}L$
Principal finite element equation (line 2 with 2 DOF)
$\underbrace{\int_L \underbrace{(\mathcal{L}_1 \boldsymbol{N}^{\mathrm{T}})^{\mathrm{T}}}_{\boldsymbol{B}} EA \underbrace{(\mathcal{L}_1 \boldsymbol{N}^{\mathrm{T}})}_{\boldsymbol{B}^{\mathrm{T}}} \mathrm{d}L}_{\boldsymbol{K}^{\mathrm{e}}} \begin{bmatrix} u_{1x} \\ u_{2x} \end{bmatrix} = \begin{bmatrix} F_{1x} \\ F_{2x} \end{bmatrix} + \int_L \boldsymbol{N} p_x \mathrm{d}L$

written as[11]:

$$\mathcal{L}_1^{\mathrm{T}} C \mathcal{L}_1 u_x^0 + \underbrace{b}_{p_x/A} = 0 \,. \tag{2.64}$$

Replacing the exact solution u_x^0 by an approximate solution u_x, a residual r is obtained:

$$r = \mathcal{L}_1^{\mathrm{T}} C \mathcal{L}_1 u_x + b \neq 0 \,. \tag{2.65}$$

The inner product is now obtained by weighting the residual and integration *over the volume* as

$$\int_V W_x^{\mathrm{T}} \left(\mathcal{L}_1^{\mathrm{T}} C \mathcal{L}_1 u_x + b\right) \mathrm{d}V \stackrel{!}{=} 0 \,, \tag{2.66}$$

where $W_x(x)$ is the scalar weight function. Application of the GREEN–GAUSS theorem (cf. Sect. A.7) gives the weak formulation as:

$$\int_V (\mathcal{L}_1 W_x)^{\mathrm{T}} C (\mathcal{L}_1 u_x) \mathrm{d}V = \int_A W_x^{\mathrm{T}} t_x \mathrm{d}A + \int_V W_x^{\mathrm{T}} b \mathrm{d}V \,, \tag{2.67}$$

where the traction force t_x can be understood as the expression $(C\mathcal{L}_1 u_x)^{\mathrm{T}} n_x = \sigma_x^{\mathrm{T}} n_x$. With the nodal approaches for the displacements u_x (2.21) and the weight function W_x (2.22), the weak formulation finally reads:

[11] The use of the transposed 'T' for the scalar operator $\mathcal{L}_1$ is not obvious at the first glance. However, the following matrix operations will clarify this approach.

Table 2.8 Summary: derivation of the principal finite element equation for linear rod elements; general approach (version 2)

Strong formulation
$\mathcal{L}_1^{\mathrm{T}} C \mathcal{L}_1 u_x^0 + b = 0$
Inner product
$\int_V W^{\mathrm{T}} \left(\mathcal{L}_1^{\mathrm{T}} C \mathcal{L}_1 u_x + b\right) \mathrm{d}V = 0$
Weak formulation
$\int_V (\mathcal{L}_1 W)^{\mathrm{T}} C (\mathcal{L}_1 u_x) \mathrm{d}V = \int_A W^{\mathrm{T}} t_x \mathrm{d}A + \int_V W^{\mathrm{T}} b \mathrm{d}V$
Principal finite element equation (line 2 with 2 DOF)
$\underbrace{\int_V \underbrace{(\mathcal{L}_1 \boldsymbol{N}^{\mathrm{T}})^{\mathrm{T}}}_{\boldsymbol{B}} C \underbrace{(\mathcal{L}_1 \boldsymbol{N}^{\mathrm{T}})}_{\boldsymbol{B}^{\mathrm{T}}} \mathrm{d}V}_{\boldsymbol{K}^{\mathrm{e}}} \begin{bmatrix} u_{1x} \\ u_{2x} \end{bmatrix} = \begin{bmatrix} F_{1x} \\ F_{2x} \end{bmatrix} + \int_V \boldsymbol{N} b \, \mathrm{d}V$

$$\int_V (\mathcal{L}_1 \boldsymbol{N}^{\mathrm{T}})^{\mathrm{T}} C (\mathcal{L}_1 \boldsymbol{N}^{\mathrm{T}}) \mathrm{d}V \boldsymbol{u}_{\mathrm{p}}^{\mathrm{e}} = \int_A \boldsymbol{N} t_x \mathrm{d}A + \int_V \boldsymbol{N} b \mathrm{d}V \,. \tag{2.68}$$

Considering $\mathrm{dV} = A\mathrm{d}L$ for constant cross sections, the last representation can be transformed to formulation (2.63). The major steps to transform the partial differential equation into the principal finite element equation based on this most general approach are summarized in Table 2.8.

2.3.1.2 Quadratic Element Formulation

Let us consider now a rod element which is composed of three nodes as schematically shown in Fig. 2.10. Each node has again only one degree of freedom, i.e. a displacement in x-direction and each node can be only loaded by a single force acting along the x-axis. It is assumed in the following that the second node is exactly located in the middle, i.e. at $x = \frac{L}{2}$, of the element.

Since there are now three nodes with three unknowns, the equation for the unknown displacement in the element and its virtual counterpart (cf. Eqs. (2.21) and (2.22)) are now given by the expressions:

$$u^{\mathrm{e}}(x) = \boldsymbol{N}^{\mathrm{T}}(x)\, \boldsymbol{u}_{\mathrm{p}} = \begin{bmatrix} N_1 & N_2 & N_3 \end{bmatrix} \times \begin{bmatrix} u_1 \\ u_2 \\ u_3 \end{bmatrix} , \tag{2.69}$$

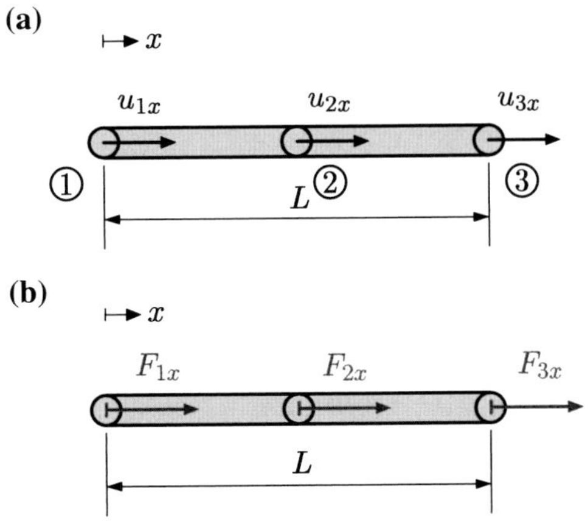

Fig. 2.10 Definition of the one-dimensional quadratic rod element: **a** deformations; **b** external loads. The nodes are symbolized by circles at the ends and in the middle (○)

and

$$W(x) = \boldsymbol{N}(x)^{\mathrm{T}}\delta\boldsymbol{u}_{\mathrm{p}} = \begin{bmatrix} N_1 & N_2 & N_3 \end{bmatrix} \times \begin{bmatrix} \delta u_1 \\ \delta u_2 \\ \delta u_3 \end{bmatrix}. \tag{2.70}$$

Similar as in Eq. (2.39), the elemental stiffness matrix can be expressed before evaluating the integral as:

$$\boldsymbol{K}^{\mathrm{e}} = EA \int_0^L \begin{bmatrix} \frac{\mathrm{d}N_1(x)}{\mathrm{d}x}\frac{\mathrm{d}N_1(x)}{\mathrm{d}x} & \frac{\mathrm{d}N_1(x)}{\mathrm{d}x}\frac{\mathrm{d}N_2(x)}{\mathrm{d}x} & \frac{\mathrm{d}N_1(x)}{\mathrm{d}x}\frac{\mathrm{d}N_3(x)}{\mathrm{d}x} \\ \frac{\mathrm{d}N_2(x)}{\mathrm{d}x}\frac{\mathrm{d}N_1(x)}{\mathrm{d}x} & \frac{\mathrm{d}N_2(x)}{\mathrm{d}x}\frac{\mathrm{d}N_2(x)}{\mathrm{d}x} & \frac{\mathrm{d}N_2(x)}{\mathrm{d}x}\frac{\mathrm{d}N_3(x)}{\mathrm{d}x} \\ \frac{\mathrm{d}N_3(x)}{\mathrm{d}x}\frac{\mathrm{d}N_1(x)}{\mathrm{d}x} & \frac{\mathrm{d}N_3(x)}{\mathrm{d}x}\frac{\mathrm{d}N_2(x)}{\mathrm{d}x} & \frac{\mathrm{d}N_3(x)}{\mathrm{d}x}\frac{\mathrm{d}N_3(x)}{\mathrm{d}x} \end{bmatrix} \mathrm{d}x\,. \tag{2.71}$$

The interpolation functions N_i in this case[12] are given by quadratic equations as shown in Fig. 2.11 in physical and natural coordinates.

From the functional expressions given in Fig. 2.11, the derivatives are obtained as $\frac{\mathrm{d}N_1}{\mathrm{d}x} = -\frac{3}{L} + \frac{4x}{L^2}$, $\frac{\mathrm{d}N_2}{\mathrm{d}x} = \frac{4}{L} - \frac{8x}{L^2}$, and $\frac{\mathrm{d}N_3}{\mathrm{d}x} = -\frac{1}{L} + \frac{4x}{L^2}$ and Eq. (2.71) can be evaluated by analytical or numerical integration to give the elemental stiffness matrix of the quadratic rod element as:

$$\boldsymbol{K}^{\mathrm{e}} = \frac{EA}{3L}\begin{bmatrix} 7 & -8 & 1 \\ -8 & 16 & -8 \\ 1 & -8 & 7 \end{bmatrix}. \tag{2.72}$$

[12]A formal derivation of the functional expressions is presented in Sect. 2.3.2.

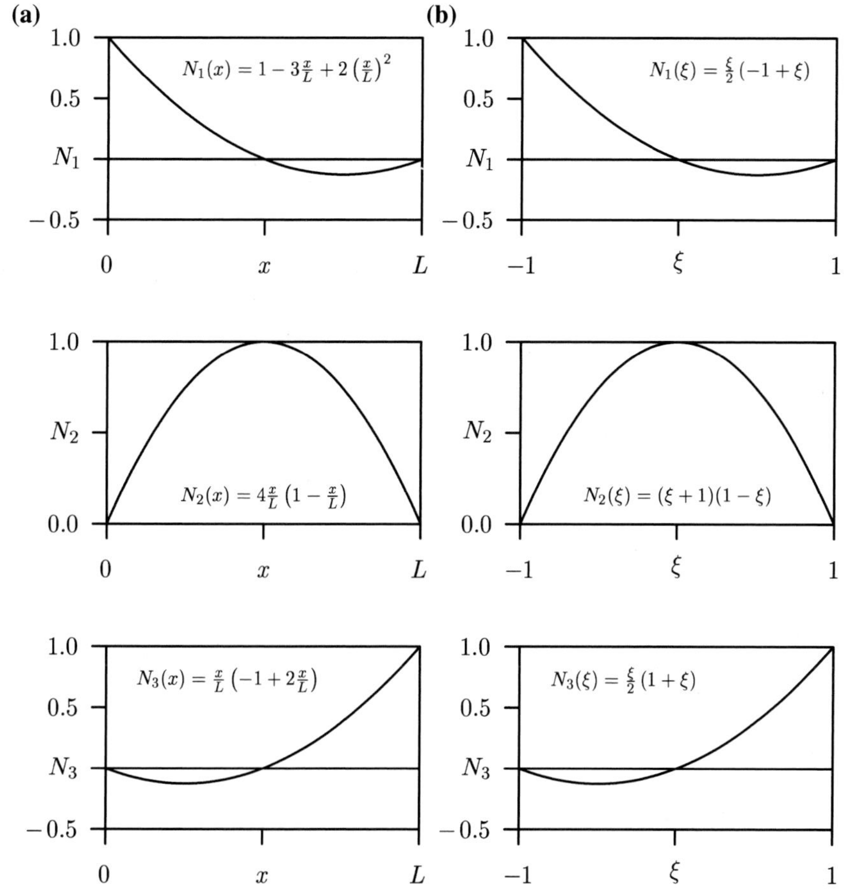

Fig. 2.11 Interpolation functions for the quadratic rod element with equidistant nodes: **a** physical coordinate (x); **b** natural coordinate (ξ)

Furthermore, it should be noted that the $\boldsymbol{B}$-matrix, cf. Eq. (2.25), takes the following form for the quadratic rod element:

$$\boldsymbol{B} = \frac{1}{L}\begin{bmatrix} -3 + \frac{4x}{L} \\ 4 - \frac{8x}{L} \\ -1 + \frac{4x}{L} \end{bmatrix} = \frac{1}{L}\begin{bmatrix} -1 + 2\xi \\ -4\xi \\ 1 + 2\xi \end{bmatrix}. \tag{2.73}$$

The right-hand side of Eq. (2.20) can be treated in a similar way as in Sect. 2.3.1.1 to obtain the elemental load vector in the form of:

$$\boldsymbol{f}^{\mathrm{e}} = \boldsymbol{f}^{\mathrm{e}}_{F} + \boldsymbol{f}^{\mathrm{e}}_{p} = \begin{bmatrix} F_{1x} \\ F_{2x} \\ F_{3x} \end{bmatrix} + \int_0^L \begin{bmatrix} N_1 \\ N_2 \\ N_3 \end{bmatrix} p(x)\,\mathrm{d}x\,. \tag{2.74}$$

Based on the derived results, the principal finite element equation for a single quadratic rod element with constant axial tensile stiffness EA can be expressed in components as:

$$\frac{EA}{3L}\begin{bmatrix} 7 & -8 & 1 \\ -8 & 16 & -8 \\ 1 & -8 & 7 \end{bmatrix}\begin{bmatrix} u_{1x} \\ u_{2x} \\ u_{3x} \end{bmatrix} = \begin{bmatrix} F_{1x} \\ F_{2x} \\ F_{3x} \end{bmatrix} + \int_0^L \begin{bmatrix} N_1 \\ N_2 \\ N_3 \end{bmatrix} p(x)\,\mathrm{d}x\,. \tag{2.75}$$

This formulation of the principal finite element equation can be alternatively expressed by eliminating the expression for the second node. The system given in Eq. (2.75) can be written in the form of single equations as:

$$\frac{EA}{3L}(7u_{1x} - 8u_{2x} + 1u_{3x}) = F_{1x} + I_1\,, \tag{2.76}$$

$$\frac{EA}{3L}(-8u_{1x} + 16u_{2x} - 8u_{3x}) = F_{2x} + I_2\,, \tag{2.77}$$

$$\frac{EA}{3L}(1u_{1x} - 8u_{2x} + 7u_{3x}) = F_{3x} + I_3\,, \tag{2.78}$$

where I_i is the abbreviation for the integral with the distributed load, e.g. $I_1 = \int N_1(x)p(x)\mathrm{d}x$. The second equation can be rearranged for u_{2x}, i.e.

$$u_{2x} = \frac{1}{2}u_{1x} + \frac{1}{2}u_{3x} + \frac{1}{16}\frac{3L}{EA}(F_{2x} + I_2)\,, \tag{2.79}$$

which can be introduced into Eqs. (2.76) and (2.78) to obtain:

$$\frac{EA}{3L}(3u_{1x} - 3u_{3x}) = F_{1x} + I_1 + \frac{1}{2}(F_{2x} + I_2)\,, \tag{2.80}$$

$$\frac{EA}{3L}(-3u_{1x} + 3u_{3x}) = F_{3x} + I_3 + \frac{1}{2}(F_{2x} + I_2)\,. \tag{2.81}$$

The last two equations can be written again in matrix form as:

$$\frac{EA}{L}\begin{bmatrix} 1 & -1 \\ -1 & 1 \end{bmatrix}\begin{bmatrix} u_{1x} \\ u_{3x} \end{bmatrix} = \begin{bmatrix} F_{1x} \\ F_{3x} \end{bmatrix} + \begin{bmatrix} I_1 \\ I_3 \end{bmatrix} + \begin{bmatrix} \frac{1}{2}F_{2x} \\ \frac{1}{2}F_{2x} \end{bmatrix} + \begin{bmatrix} \frac{1}{2}I_2 \\ \frac{1}{2}I_2 \end{bmatrix}\,. \tag{2.82}$$

Table 2.9 Equivalent nodal loads for a quadratic rod element (x-axis: right facing)

Loading	Axial force
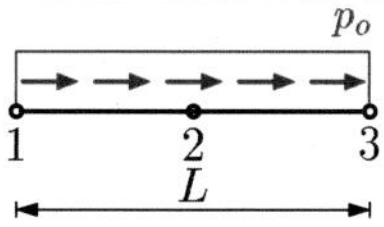	$F_{1x} = \dfrac{p_0 L}{6}$, $F_{2x} = \dfrac{2 p_0 L}{3}$, $F_{3x} = \dfrac{p_0 L}{6}$
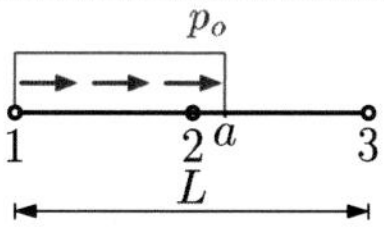	$F_{1x} = \dfrac{2p_0 a^3}{3L^2} - \dfrac{3p_0 a^3}{2L} + p_0 a$, $F_{2x} = -\dfrac{4p_0 a^3}{3L^2} + \dfrac{2p_0 a^2}{L}$, $F_{3x} = \dfrac{2p_0 a^3}{3L^2} - \dfrac{p_0 a^2}{2L}$
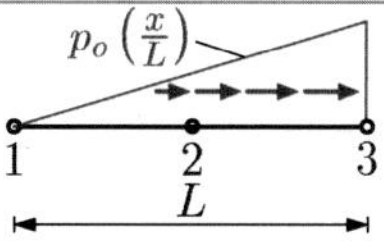	$F_{1x} = 0$, $F_{2x} = \dfrac{p_0 L}{3}$, $F_{3x} = \dfrac{p_0 L}{6}$
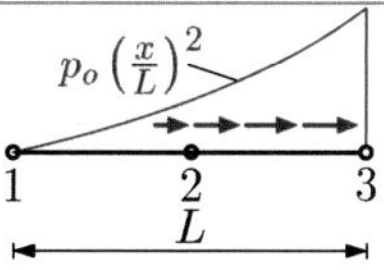	$F_{1x} = -\dfrac{p_0 L}{60}$, $F_{2x} = \dfrac{p_0 L}{5}$, $F_{3x} = \dfrac{3 p_0 L}{20}$
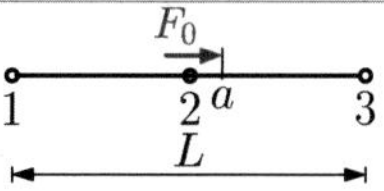	$F_{1x} = F_0 N_1(a)$, $F_{2x} = F_0 N_2(a)$, $F_{3x} = F_0 N_3(a)$

This formulation looks similar to the expression for the linear rod element in Eq. (2.56). However, the right-hand side contains here in addition the contribution of the load from the middle load and it should be not forgotten that the distribution of the displacement $u^{\mathrm{e}}(x)$ inside the element is of quadratic shape. The values of equivalent nodal loads, i.e. the evaluation of the integral in Eq. (2.75), is given for some standard cases in Table 2.9. The reader should here pay attention to the fact that these equivalent nodal loads are different to those in the case of the linear rod element, cf. Table 2.5.

At the end of this section again a few words on the accuracy of the quadratic rod element will be given. As can be seen in Table 2.10, the accuracy is for the investigated cases at least in the range of the linear element if we compare the general statements without investigating specific numerical values. For a constant distributed load, the quadratic element reproduces not only at the nodes but also between the nodes the analytical solution. However, it must be highlighted here that these results are element specific and that the finite element method calculates in the general case—even at nodes—only approximate solutions. Nevertheless, the comments presented in Table 2.10 can be helpful in special cases where a mesh refinement would not increase the accuracy but the computation time and the size of

Table 2.10 Comments on the accuracy of the finite element solution for a single cantilever quadratic rod element

Configuration	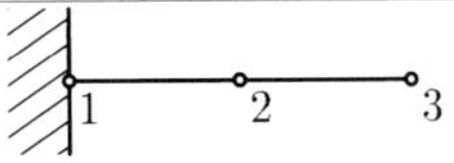
Axial tensile stiffness and loading	Accuracy of $u(x)$
EA = const.; loaded by single force F at node 3	FE gives analytical solution at nodes and between nodes
EA = const.; displacement BC u at node 3	FE gives exact nodal values and analytical solution between nodes
EA = const.; distributed load p = const.	FE gives analytical solution at nodes and between nodes
EA = const.; distributed load $p(x)$ = linear	FE gives analytical solution at nodes but only approximate solution between nodes
$EA \neq$ const.; loaded by single force F at node 3	FE gives approximate solution at nodes and approximate solution between nodes

the results file. If the problem is such that the exact solution is obtained at the nodes, a mesh refinement is in all likelihood not required in this case.

Let us summarize at the end of this section the major steps that were undertaken to transform the partial differential equation into the principal finite element equation, see Table 2.11.

2.3.2 *Derivation of Interpolation Functions*

A more general concept based on basis functions will be introduced in the following in order to derive the complete set of interpolation functions.[13] To this end, let us just assume that the shape of the displacement distribution $u^{\mathrm{e}}(\xi)$ within an element is without reference to the nodal values. It is obvious that this choice must be conform to the physical problem under consideration. We may assume that the distribution is given for an element with n nodes by a polynomial of the form

$$u^{\mathrm{e}}(\xi) = a_0 + a_1\xi + a_2\xi^2 + a_3\xi^3 + \cdots + a_{n-1}\xi^{n-1}\,, \tag{2.83}$$

which can be expressed in matrix notation as:

[13]This approach is presented in Ref. [5] in a general way.

Table 2.11 Summary: derivation of principal finite element equation for rod elements

Strong formulation
$$EA\frac{\mathrm{d}^2u^0(x)}{\mathrm{d}x^2}+p(x)=0$$
Inner product
$$\int_0^L W^{\mathrm{T}}(x)\left(EA\frac{\mathrm{d}^2u(x)}{\mathrm{d}x^2}+p(x)\right)\mathrm{d}x \stackrel{!}{=} 0$$
Weak formulation
$$EA\int_0^L \frac{\mathrm{d}W^{\mathrm{T}}(x)}{\mathrm{d}x}\frac{\mathrm{d}u(x)}{\mathrm{d}x}\mathrm{d}x = EA\left[W^{\mathrm{T}}(x)\frac{\mathrm{d}u(x)}{\mathrm{d}x}\right]_0^L+\int_0^L W^{\mathrm{T}}(x)p(x)\,\mathrm{d}x$$
Principal finite element equation
$$\frac{EA}{L}\begin{bmatrix}1 & -1\\ -1 & 1\end{bmatrix}\begin{bmatrix}u_{1x}\\ u_{2x}\end{bmatrix}=\begin{bmatrix}F_{1x}\\ F_{2x}\end{bmatrix}+\int_0^L\begin{bmatrix}N_1\\ N_2\end{bmatrix}p_x(x)\,\mathrm{d}x\ \text{(lin.)}$$
$$\frac{EA}{3L}\begin{bmatrix}7 & -8 & 1\\ -8 & 16 & -8\\ 1 & -8 & 7\end{bmatrix}\begin{bmatrix}u_{1x}\\ u_{2x}\\ u_{3x}\end{bmatrix}=\begin{bmatrix}F_{1x}\\ F_{2x}\\ F_{3x}\end{bmatrix}+\int_0^L\begin{bmatrix}N_1\\ N_2\\ N_3\end{bmatrix}p(x)\,\mathrm{d}x\ \text{(quad.)}$$

$$u^{\mathrm{e}}(\xi)=\boldsymbol{\chi}^{\mathrm{T}}\boldsymbol{a}=\begin{bmatrix}1 & \xi & \xi^2 & \xi^3 & \cdots & \xi^{n-1}\end{bmatrix}\begin{bmatrix}a_0\\ a_1\\ a_2\\ a_3\\ \vdots\\ a_{n-1}\end{bmatrix}. \tag{2.84}$$

The elements of $\boldsymbol{\chi}$ will be called *basis functions* and the elements of $\boldsymbol{a}$ will be called *basis coefficients*. If we assume that the number of basis functions equals the number of nodal variables associated with u, then the relationship between the basis coefficients $\boldsymbol{a}$ and the nodal values $\boldsymbol{u}_{\mathrm{p}}$ can be expressed as

$$\boldsymbol{a}=\boldsymbol{A}\boldsymbol{u}_{\mathrm{p}}, \tag{2.85}$$

where $\boldsymbol{A}$ is a square matrix of constants. Equalizing the nodal approach given in Eq. (2.21) with the new expression in (2.84) and considering (2.85) results in:

$$\boldsymbol{N}^{\mathrm{T}}\boldsymbol{u}_{\mathrm{p}}=\boldsymbol{\chi}^{\mathrm{T}}\boldsymbol{a}\quad\text{or}\quad\boldsymbol{N}^{\mathrm{T}}=\boldsymbol{\chi}^{\mathrm{T}}\boldsymbol{A}. \tag{2.86}$$

Thus, the row matrix of the interpolation functions $\boldsymbol{N}^{\mathrm{T}}$ can be factored into a row vector of basis functions $\boldsymbol{\chi}^{\mathrm{T}}$ and a square matrix $\boldsymbol{A}$ of constant coefficients.

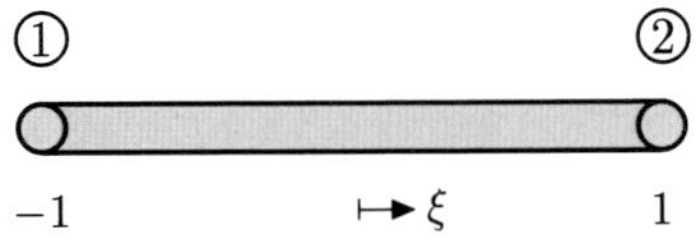

Fig. 2.12 Linear rod element described based on the natural coordinate (ξ)

To illustrate the procedure, let us have a look at a linear rod element as shown in Fig. 2.12 where the natural coordinate is used.

If the physical problem supports the assumption of a linear distribution of the displacement, the following linear description of the displacement field can be introduced:

$$u^{\mathrm{e}}(\xi) = a_0 + a_1\xi\,, \tag{2.87}$$

where the column matrix of the basis functions is given by $\chi = \begin{bmatrix}1 & \xi\end{bmatrix}^{\mathrm{T}}$ and the column matrix of the basis coefficients by $\boldsymbol{a} = \begin{bmatrix}a_0 \; a_1\end{bmatrix}^{\mathrm{T}}$. Evaluation of this function at both nodes gives:

$$\text{Node 1: } u_1 = u^{\mathrm{e}}(\xi = -1) = a_0 - a_1\,, \tag{2.88}$$

$$\text{Node 2: } u_2 = u^{\mathrm{e}}(\xi = +1) = a_0 + a_1\,. \tag{2.89}$$

The last two equations can be expressed in matrix notation according to Eq. (2.85) as:

$$\begin{bmatrix}u_1\\u_2\end{bmatrix} = \underbrace{\begin{bmatrix}1 & -1\\1 & 1\end{bmatrix}}_{\boldsymbol{A}^{-1}}\begin{bmatrix}a_0\\a_1\end{bmatrix}. \tag{2.90}$$

Solving this system of equations for the unknown basis functions a_i gives

$$\underbrace{\begin{bmatrix}a_0\\a_1\end{bmatrix}}_{\boldsymbol{a}} = \underbrace{\frac{1}{2}\begin{bmatrix}1 & 1\\-1 & 1\end{bmatrix}}_{\boldsymbol{A}}\underbrace{\begin{bmatrix}u_1\\u_2\end{bmatrix}}_{\boldsymbol{u}_{\mathrm{p}}}, \tag{2.91}$$

and the matrix of the interpolation functions results according to Eq. (2.86) as:

$$\boldsymbol{N}^{\mathrm{T}} = \chi^{\mathrm{T}}\boldsymbol{A} = \begin{bmatrix}1 & \xi\end{bmatrix}\frac{1}{2}\begin{bmatrix}1 & 1\\-1 & 1\end{bmatrix} = \begin{bmatrix}\frac{1}{2}(1-\xi) & \frac{1}{2}(1+\xi)\end{bmatrix} = \begin{bmatrix}N_1 \; N_2\end{bmatrix}. \tag{2.92}$$

Alternatively, one may use the Cartesian coordinate (x) to derive the interpolation functions based on the same approach. Assuming that the x-coordinate is in the range $0 \le x \le L$ and that the same ordinate values as given by Eq. (2.87) are maintained at the nodes, the following linear description of the displacement field can be introduced:

$$u^{\mathrm{e}}(x) = (a_0 - a_1) + \frac{2a_1}{L} \times x\,, \tag{2.93}$$

where the column matrix of the basis functions is given by $\chi = \begin{bmatrix}1 & x\end{bmatrix}^{\mathrm{T}}$ and the column matrix of the basis coefficients by $\boldsymbol{a} = \begin{bmatrix}(a_0 - a_1) & \frac{2a_1}{L}\end{bmatrix}^{\mathrm{T}}$. Evaluation of this function at both nodes gives:

$$\text{Node 1:}\ u_1 = u^{\mathrm{e}}(x=0) = a_0 - a_1\,, \tag{2.94}$$

$$\text{Node 2:}\ u_2 = u^{\mathrm{e}}(x=L) = a_0 + a_1 = (a_0 - a_1) + \tfrac{2a_1}{L}L\,. \tag{2.95}$$

The last two equations can be expressed in matrix notation according to Eq. (2.85) as:

$$\begin{bmatrix}u_1\\u_2\end{bmatrix} = \underbrace{\begin{bmatrix}1 & 0\\1 & L\end{bmatrix}}_{\boldsymbol{A}^{-1}}\begin{bmatrix}a_0 - a_1\\\frac{2a_1}{L}\end{bmatrix}. \tag{2.96}$$

Solving this system of equations for the unknown basis functions a_i gives

$$\underbrace{\begin{bmatrix}a_0 - a_1\\\frac{2a_1}{L}\end{bmatrix}}_{\boldsymbol{a}} = \underbrace{\frac{1}{L}\begin{bmatrix}L & 0\\-1 & 1\end{bmatrix}}_{\boldsymbol{A}}\underbrace{\begin{bmatrix}u_1\\u_2\end{bmatrix}}_{\boldsymbol{u}_{\mathrm{p}}}, \tag{2.97}$$

and the matrix of the interpolation functions results according to Eq. (2.86) as:

$$\boldsymbol{N}^{\mathrm{T}} = \chi^{\mathrm{T}}\boldsymbol{A} = \begin{bmatrix}1 & x\end{bmatrix}\frac{1}{L}\begin{bmatrix}L & 0\\-1 & 1\end{bmatrix} = \begin{bmatrix}\frac{1}{L}(L-x) & \frac{1}{L}(x)\end{bmatrix} = \begin{bmatrix}N_1 & N_2\end{bmatrix}. \tag{2.98}$$

If the Cartesian coordinate (x) is used based on a different set of ordinate values, the following linear description of the displacement field can be introduced:

$$u^{\mathrm{e}}(x) = a_0 + a_1 \times x\,, \tag{2.99}$$

where the column matrix of the basis functions is given by $\chi = \begin{bmatrix}1 & x\end{bmatrix}^{\mathrm{T}}$ and the column matrix of the basis coefficients by $\boldsymbol{a} = \begin{bmatrix}a_0 & a_1\end{bmatrix}^{\mathrm{T}}$. Evaluation of this function at both nodes gives:

$$\text{Node 1:}\ u_1 = u^{\mathrm{e}}(x=0) = a_0\,, \tag{2.100}$$

$$\text{Node 2:}\ u_2 = u^{\mathrm{e}}(x=L) = a_0 + a_1 L\,, \tag{2.101}$$

which can be expressed as in Eq. (2.96) and the same interpolation functions as presented in Eq. (2.98) are obtained.

In the case of an element with n equally spaced nodes, one can generalize Eqs. (2.88)–(2.89) by evaluation Eq. (2.83) at each node in the following manner:

$$\text{Node 1:}\quad u_1 = u^{\text{e}}(\xi = -1) = a_0 - a_1 \pm \dots, \tag{2.102}$$

$$\text{Node 2:}\quad u_2 = u^{\text{e}}\left(\xi = -1 + (2-1)\tfrac{2}{n-1}\right) = a_0 \pm \dots a_1, \tag{2.103}$$

$$\vdots \qquad\qquad \vdots$$

$$\text{Node } i:\quad u_i = u^{\text{e}}\left(\xi = -1 + (i-1)\tfrac{2}{n-1}\right) = a_0 \pm \dots, \tag{2.104}$$

$$\vdots \qquad\qquad \vdots$$

$$\text{Node } n:\quad u_n = u^{\text{e}}(\xi = +1) = a_0 + a_1 + \dots + a_{n-1}. \tag{2.105}$$

Rearranging this system of equations in matrix notation, i.e. $\boldsymbol{u}_{\text{p}} = \boldsymbol{A}^{-1}\boldsymbol{a}$, allows finally to solve for the n interpolation functions via $\boldsymbol{N}^{\text{T}} = \boldsymbol{\chi}^{\text{T}}\boldsymbol{A}$, where $\boldsymbol{\chi}^{\text{T}}$ is given in Eq. (2.84).

2.3.3 Assembly of Elements and Consideration of Boundary Conditions

Real structures of complex geometry (cf. Figs. 1.2b and 1.3b) require the application of many finite elements in order to discretize the geometry. Thus, it is necessary to assemble the single elemental equations $\boldsymbol{K}^{\text{e}}\boldsymbol{u}_{\text{p}}^{\text{e}} = \boldsymbol{f}^{\text{e}}$ to a global system of equations which can be symbolically written as $\boldsymbol{K}\boldsymbol{u}_{\text{p}} = \boldsymbol{f}$, where $\boldsymbol{K}$ is the global stiffness matrix, $\boldsymbol{u}_{\text{p}}$ the global column matrix of unknowns, and $\boldsymbol{f}$ the global column matrix of loads.

Let us illustrate the process to assemble the global system of equations for a three-element axial structure as shown in Fig. 2.13. As can be seen in Fig. 2.13a, each element has its own coordinate system x_i with $i = \text{I, II, III}$ and its own nodal displacements u^i_{1x} and u^i_{2x}. In order to assemble the single elements to a connected structure as shown in Fig. 2.13b, it is useful to introduce a global coordinate X and global nodal displacements denoted by u_{iX}. Comparing the elemental and global nodal displacements shown in Fig. 2.13, the following mapping between the local and global displacements can be derived:

$$u_{1X} = u^{\text{I}}_{1x}, \tag{2.106}$$

$$u_{2X} = u^{\text{I}}_{2x} = u^{\text{II}}_{1x}, \tag{2.107}$$

$$u_{3X} = u^{\text{II}}_{2x} = u^{\text{III}}_{1x}, \tag{2.108}$$

$$u_{4X} = u^{\text{III}}_{2x}. \tag{2.109}$$

One possible way to assemble the elemental stiffness matrices to the global system will be illustrated in the following. In a first step, each single element is considered separately and its elemental stiffness matrix is written as, for example, given in

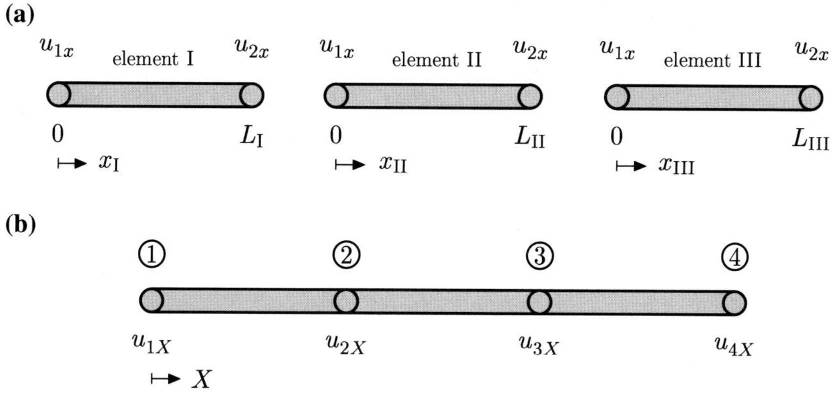

Fig. 2.13 Relationship between **a** elemental and **b** global nodes and displacements in a horizontal rod structure

Eq. (2.41). In addition, the corresponding *global* nodal displacements are written over the matrix and on the right-hand side which gives the following expressions:

$$\boldsymbol{K}^{\mathrm{e}}_{\mathrm{I}} = \frac{EA}{L} \begin{array}{c} \begin{matrix} u_{1X} & u_{2X} \end{matrix} \\ \begin{bmatrix} 1 & -1 \\ -1 & 1 \end{bmatrix} \end{array} \begin{matrix} u_{1X} \\ u_{2X} \end{matrix}, \tag{2.110}$$

$$\boldsymbol{K}^{\mathrm{e}}_{\mathrm{II}} = \frac{EA}{L} \begin{array}{c} \begin{matrix} u_{2X} & u_{3X} \end{matrix} \\ \begin{bmatrix} 1 & -1 \\ -1 & 1 \end{bmatrix} \end{array} \begin{matrix} u_{2X} \\ u_{3X} \end{matrix}, \tag{2.111}$$

$$\boldsymbol{K}^{\mathrm{e}}_{\mathrm{III}} = \frac{EA}{L} \begin{array}{c} \begin{matrix} u_{3X} & u_{4X} \end{matrix} \\ \begin{bmatrix} 1 & -1 \\ -1 & 1 \end{bmatrix} \end{array} \begin{matrix} u_{3X} \\ u_{4X} \end{matrix}. \tag{2.112}$$

By indicating the global unknowns in the described manner at each elemental stiffness matrix, it is easy to assign to each element in a matrix a unique index. For example, the upper right element of the stiffness matrix $\boldsymbol{K}^{\mathrm{e}}_{\mathrm{I}}$ has the index[14] (u_{1X}, u_{2X}) and the value $-\frac{EA}{L}$. The next step consists in indicating the structure of the global stiffness matrix with its correct dimension. To this end, the total number of global unknowns[15] must be determined. In general, the global number of unknowns is given by the number

[14]We follow here the convention where the first expression specifies the row and the second one the column: (row, column).

[15]The total number of unknowns is alternatively named the total number of degrees of freedom (DOF).

of nodes multiplied by the degrees of freedom per node. Thus, the number of global unknowns for a structure of rod elements is simply the total number of nodes in the assembled structure. It should be noted here that the determination of the unknowns at this step of the procedure is without any consideration of boundary conditions. For the problem shown in Fig. 2.13b, the number of nodes is four which equals the number of unknowns. Thus, the dimensions of the global stiffness matrix are given by (number global unknowns $\times$ number global unknowns) or for our example as (4×4) and the structure can be written as:

$$
\boldsymbol{K} = \begin{array}{c} \begin{array}{cccc} u_{1X} & u_{2X} & u_{3X} & u_{4X} \end{array} \\ \left[\begin{array}{c|c|c|c} \phantom{u_{1X}} & \phantom{u_{2X}} & \phantom{u_{3X}} & \phantom{u_{4X}} \\ \hline & & & \\ \hline & & & \\ \hline & & & \end{array}\right] \end{array} \begin{array}{c} u_{1X} \\ u_{2X} \\ u_{3X} \\ u_{4X} \end{array} . \tag{2.113}
$$

It is now required to indicate the global unknowns over the empty global stiffness matrix and on its right-hand side. Any order can be chosen but it is common for the problem under consideration to start with u_{1X} and simply move to the next node. The scheme for this consecutive use of the global unknowns from the lowest to the highest number is drawn on the matrix in Eq. (2.113). Each cell of the global stiffness matrix has now its unique index expressed by the global unknowns. Or in other words, each cell of each elemental stiffness matrix has a cell in the global stiffness matrix with the same index and each element of the elemental stiffness matrix must be placed in the global matrix based on this unique index scheme. As an example, the upper right element of the stiffness matrix $\boldsymbol{K}_{\mathrm{I}}^{\mathrm{e}}$ with the index (u_{1X}, u_{2X}) must be placed in the global stiffness matrix in the first row and the second column. If each entry of the elemental stiffness matrices is inserted into the global matrix based on the described index scheme, the assembly of the global stiffness matrix is completed. The process for the consecutive use of global unknowns is illustrated in Fig. 2.14a. As can be seen in this figure, there is an interaction at nodes where elements are connected and the corresponding entries of the elemental stiffness matrices are summed up. This interaction is illustrated in a different way in Fig. 2.14b where it can be seen that at each inner node two interpolation functions are acting, i.e. one from the left-hand element and one from the right-hand element.

A further important property of the global stiffness matrix can be seen in Fig. 2.14a. If an appropriate node numbering is chosen,[16] the global stiffness matrix reveals a strong band structure where all entries are grouped around the main diagonale and major parts grouped in the form of triangles contain only zeros. If there is such a clear boundary between the non-zero and the zero components, the border line is called the *skyline* of the matrix. As the elemental stiffness matrices, the global stiffness matrix

[16] Commercial finite element codes offer an option which is called 'bandwidth optimization' to achieve this structure. This is important if a direct solver is used in order to minimize the solution time and the amount of storage.

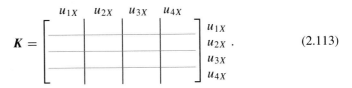

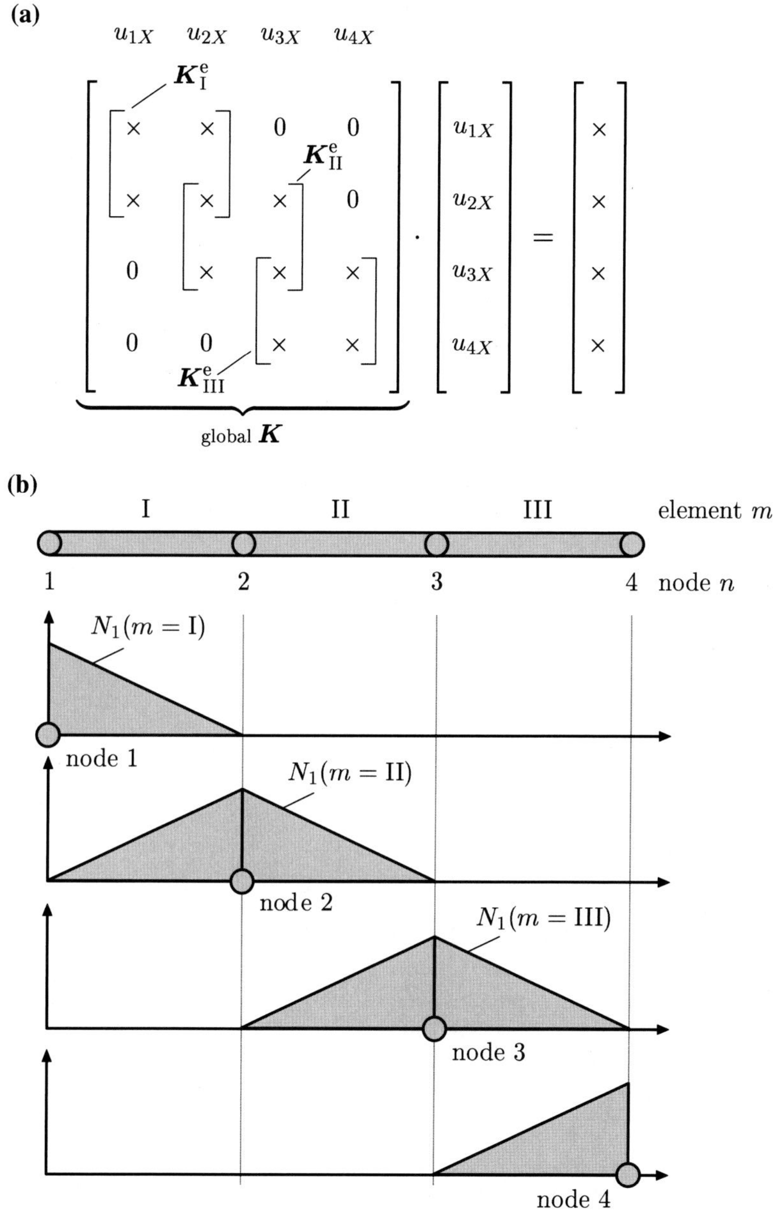

Fig. 2.14 Assembly process to the global stiffness matrix: **a** composition of the elemental stiffness matrices to the global system; **b** interaction of interpolation functions at common nodes

is symmetric and commercial finite element codes store only half of the entries in order to reduce the requirements for data storage.

An alternative way of assembling the global stiffness matrix can be based on the following strategy. The elemental stiffness matrices (2.110)–(2.112) can be written right from the beginning in a matrix with the dimension of the global stiffness matrix:

$$\boldsymbol{K}_{\text{I}}^{\text{e}} = \frac{EA}{L} \begin{array}{c} \begin{matrix} u_{1X} & u_{2X} & u_{3X} & u_{4X} \end{matrix} \\ \begin{bmatrix} 1 & -1 & 0 & 0 \\ -1 & 1 & 0 & 0 \\ 0 & 0 & 0 & 0 \\ 0 & 0 & 0 & 0 \end{bmatrix} \end{array} \begin{matrix} u_{1X} \\ u_{2X} \\ u_{3X} \\ u_{4X} \end{matrix}, \tag{2.114}$$

$$\boldsymbol{K}_{\text{II}}^{\text{e}} = \frac{EA}{L} \begin{array}{c} \begin{matrix} u_{1X} & u_{2X} & u_{3X} & u_{4X} \end{matrix} \\ \begin{bmatrix} 0 & 0 & 0 & 0 \\ 0 & 1 & -1 & 0 \\ 0 & -1 & 1 & 0 \\ 0 & 0 & 0 & 0 \end{bmatrix} \end{array} \begin{matrix} u_{1X} \\ u_{2X} \\ u_{3X} \\ u_{4X} \end{matrix}, \tag{2.115}$$

$$\boldsymbol{K}_{\text{III}}^{\text{e}} = \frac{EA}{L} \begin{array}{c} \begin{matrix} u_{1X} & u_{2X} & u_{3X} & u_{4X} \end{matrix} \\ \begin{bmatrix} 0 & 0 & 0 & 0 \\ 0 & 0 & 0 & 0 \\ 0 & 0 & 1 & -1 \\ 0 & 0 & -1 & 1 \end{bmatrix} \end{array} \begin{matrix} u_{1X} \\ u_{2X} \\ u_{3X} \\ u_{4X} \end{matrix}. \tag{2.116}$$

Then, the global stiffness matrix is obtained as the sum of the three elemental matrices.

In order to complete the assembly of the global finite element equation, the global load vector $\boldsymbol{f}$ must be composed. Here, it is more advantageous to look from the beginning at the assembled structure and fill the external single loads F_i, which are acting at nodes, in the proper order in the column matrix $\boldsymbol{f}$. A bit care must be taken if distributed loads were converted to equivalent nodal loads. For this case, components f_i from both elements must be summed up at inner nodes:

$$\boldsymbol{f} = \begin{bmatrix} F_1 \\ F_2 \\ F_3 \\ F_4 \end{bmatrix} + \begin{bmatrix} f_{1,\text{I}} \\ f_{2,\text{I}} + f_{2,\text{II}} \\ f_{3,\text{II}} + f_{3,\text{III}} \\ f_{4,\text{III}} \end{bmatrix}. \tag{2.117}$$

The global system of equations for the problem shown in Fig. 2.13b is finally obtained as:

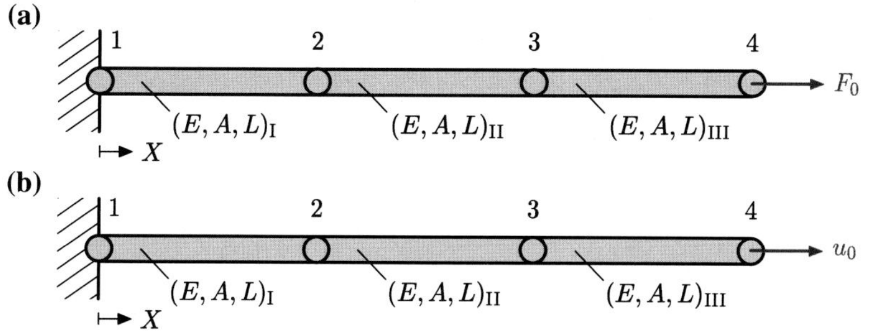

Fig. 2.15 Consideration of boundary conditions for a cantilever rod structure: **a** force boundary condition; **b** displacement boundary condition at the right-hand boundary node

$$\frac{EA}{L}\begin{bmatrix} 1 & -1 & 0 & 0 \\ -1 & 2 & -1 & 0 \\ 0 & -1 & 2 & -1 \\ 0 & 0 & -1 & 1 \end{bmatrix}\begin{bmatrix} u_{1X} \\ u_{2X} \\ u_{3X} \\ u_{4X} \end{bmatrix} = \begin{bmatrix} \cdots \\ \cdots \\ \cdots \\ \cdots \end{bmatrix}, \tag{2.118}$$

where the right-hand side is not specified since nothing on the loading is indicated in Fig. 2.13. This system of equations without consideration of any boundary conditions is called the non-reduced system. For this system, the global stiffness matrix $\boldsymbol{K}$ is still singular and cannot be inverted in order to solve the global system of equations. Boundary conditions must be introduced in order to make this matrix regular and thus invertible.

For the rod elements under consideration, two types of boundary conditions must be distinguished. The DIRICHLET boundary condition[17] specifies the displacement u at a node while the NEUMANN boundary condition[18] assigns a force F (i.e., $EA\frac{du}{dx}$) at a node. The different ways to handle these different types of boundary conditions will be explained in the following based on the problem shown in Fig. 2.15 where a cantilever rod structure has different boundary conditions at its right-hand end node.

The consideration of the homogeneous DIRICHLET boundary condition, i.e. $u_{1X} = u(X=0) = 0$, is the simplest case. To incorporate this boundary condition in the system (3.150), the first row and the first column can be canceled to obtain a reduced system as:

$$\frac{EA}{L}\begin{bmatrix} 2 & -1 & 0 \\ -1 & 2 & -1 \\ 0 & -1 & 1 \end{bmatrix}\begin{bmatrix} u_{2X} \\ u_{3X} \\ u_{4X} \end{bmatrix} = \begin{bmatrix} \cdots \\ \cdots \\ \cdots \end{bmatrix}. \tag{2.119}$$

In general we can state that a homogenous DIRICHLET boundary condition at node n ($u_{nX} = 0$) can be considered in the non-reduced system of equations by eliminating the nth row and nth column of the system. Let us consider next the case shown in

[17] Alternatively known as 1st kind, essential, geometric or kinematic boundary condition.

[18] Alternatively known as 2nd kind, natural or static boundary condition.

Fig. 2.15a where the right-hand end node is subjected to a force F_0. This external force can simply be specified on the right-hand side and since no other external forces are acting, the reduced system of equations is finally obtained as:

$$\frac{EA}{L}\begin{bmatrix} 2 & -1 & 0 \\ -1 & 2 & -1 \\ 0 & -1 & 1 \end{bmatrix}\begin{bmatrix} u_{2X} \\ u_{3X} \\ u_{4X} \end{bmatrix} = \begin{bmatrix} 0 \\ 0 \\ F_0 \end{bmatrix}. \tag{2.120}$$

This system of equations can be solved, e.g. by inverting the reduced stiffness matrix and solving for the unknown nodal displacements in the form $\boldsymbol{u}_{\text{p}} = \boldsymbol{K}^{-1}\boldsymbol{f}$:

$$\begin{bmatrix} u_{2X} \\ u_{3X} \\ u_{4X} \end{bmatrix} = \frac{LF_0}{EA}\begin{bmatrix} 1 \\ 2 \\ 3 \end{bmatrix}. \tag{2.121}$$

To incorporate a non-homogeneous DIRICHLET boundary condition ($u \neq 0$) as shown in Fig. 2.15b, three different strategies can be mentioned. The first one modifies the system shown in Eq. (2.120) in such a way that the boundary condition, i.e., $u_{4X} = u_0$, is directly introduced:

$$\frac{EA}{L}\begin{bmatrix} 2 & -1 & 0 \\ -1 & 2 & -1 \\ 0 & 0 & 1 \times \frac{L}{EA} \end{bmatrix}\begin{bmatrix} u_{2X} \\ u_{3X} \\ u_{4X} \end{bmatrix} = \begin{bmatrix} 0 \\ 0 \\ u_0 \end{bmatrix}, \tag{2.122}$$

where the last equation gives immediately the boundary condition as $u_{4X} = u_0$. The solution of the system of equations given in Eq. (2.135) can be obtained by inverting the coefficient matrix and multiplying it with the vector on the right-hand side as:

$$\begin{bmatrix} u_{2X} \\ u_{3X} \\ u_{4X} \end{bmatrix} = \begin{bmatrix} \frac{1}{3}u_0 \\ \frac{2}{3}u_0 \\ u_0 \end{bmatrix}. \tag{2.123}$$

In general we can state that a non-homogeneous DIRICHLET boundary condition at node n can be introduced in the system of equations by modifying the nth row in such a way that at the position of the nth column a '1' is obtained while all other entries of the nth row are set to zero. On the right-hand side, the given value is introduced at the nth position of the column matrix.

The second way of considering a non-homogenous DIRICHLET boundary condition consists in the following step: The column of the stiffness matrix, which corresponds to the node where the boundary condition is given, is multiplied by the given displacement. In other words, if the boundary condition is specified at node n, the nth column of the stiffness matrix is multiplied by the given value u_0:

$$\frac{EA}{L}\begin{bmatrix} 2 & -1 & 0\times u_0 \\ -1 & 2 & -1\times u_0 \\ 0 & -1 & 1\times u_0 \end{bmatrix}\begin{bmatrix} u_{2X} \\ u_{3X} \\ u_{4X} \end{bmatrix} = \begin{bmatrix} 0 \\ 0 \\ \cdots \end{bmatrix}. \tag{2.124}$$

Now we bring the nth column of the stiffness matrix to the right-hand side of the system

$$\frac{EA}{L}\begin{bmatrix} 2 & -1 \\ -1 & 2 \\ 0 & -1 \end{bmatrix}\begin{bmatrix} u_{2X} \\ u_{3X} \\ u_{4X} \end{bmatrix} = \begin{bmatrix} 0 \\ 0 \\ \cdots \end{bmatrix} - \frac{EA}{L}\begin{bmatrix} 0\times u_0 \\ -1\times u_0 \\ 1\times u_0 \end{bmatrix}, \tag{2.125}$$

and delete the nth row of the system:

$$\frac{EA}{L}\begin{bmatrix} 2 & -1 \\ -1 & 2 \end{bmatrix}\begin{bmatrix} u_{2X} \\ u_{3X} \end{bmatrix} = \begin{bmatrix} 0 \\ 0 \end{bmatrix} - \frac{EA}{L}\begin{bmatrix} 0\times u_0 \\ -1\times u_0 \end{bmatrix}. \tag{2.126}$$

As a result of this second approach, the dimension of the system of equations could be reduced compared to the first approach. However, this smaller matrix was not obtained for free since more steps have to be performed compared to the first possibility. The solution of Eq. (2.126) can be stated as:

$$\begin{bmatrix} u_{2X} \\ u_{3X} \end{bmatrix} = \begin{bmatrix} \frac{1}{3}u_0 \\ \frac{2}{3}u_0 \end{bmatrix}. \tag{2.127}$$

A third possible approach should be mentioned here since often the question arises by students why not simply replace in the column matrix of unknowns, i.e. on the left-hand side, the variable of the nodal value with the given value. This can be done but requires that the corresponding reaction force[19] is introduced on the right-hand side:

$$\frac{EA}{L}\begin{bmatrix} 2 & -1 & 0 \\ -1 & 2 & -1 \\ 0 & -1 & 1 \end{bmatrix}\underbrace{\begin{bmatrix} u_{2X} \\ u_{3X} \\ u_0 \end{bmatrix}}_{\boldsymbol{u}_{\text{p}}} = \begin{bmatrix} 0 \\ 0 \\ -F_4^{\text{R}} \end{bmatrix}. \tag{2.128}$$

However, the column matrix of the nodal displacements $\boldsymbol{u}_{\text{p}}$ contains now unknown quantities (u_{2X}, u_{3X}) and the given nodal boundary condition (u_0). On the other hand, the right-hand side contains the unknown reaction force F_4^{R}. Thus, the structure of the linear system of equations is unfavorable for the solution. To rearrange the system to the classical structure where all unknowns are collected on the left and given quantities on the right-hand side, it is advised to write out the three single equations as:

[19] Let us assume in the following that the reaction force F_4^{R} is oriented in the negative X-direction.

$$\frac{EA}{L}(2u_{2X} - u_{3X}) = 0\,, \tag{2.129}$$

$$\frac{EA}{L}(-u_{2X} + 2u_{3X} - u_0) = 0\,, \tag{2.130}$$

$$\frac{EA}{L}(-u_{3X} + u_0) = -F_4^{\mathrm{R}}\,. \tag{2.131}$$

After collecting unknown quantities on the left-hand side and known quantities on the right-hand side, one gets

$$\frac{EA}{L}(2u_{2X} - u_{3X}) = 0\,, \tag{2.132}$$

$$\frac{EA}{L}(-u_{2X} + 2u_{3X}) = \frac{EA}{L}u_0\,, \tag{2.133}$$

$$\frac{EA}{L}\left(-u_{3X} + \frac{L}{EA}F_4^{\mathrm{R}}\right) = -\frac{EA}{L}u_0\,, \tag{2.134}$$

or in matrix notation:

$$\frac{EA}{L}\begin{bmatrix} 2 & -1 & 0 \\ -1 & 2 & 0 \\ 0 & -1 & \frac{L}{EA} \end{bmatrix} \underbrace{\begin{bmatrix} u_{2X} \\ u_{3X} \\ F_4^{\mathrm{R}} \end{bmatrix}}_{\text{unknown}} = \underbrace{\frac{EA}{L}\begin{bmatrix} 0 \\ u_0 \\ -u_0 \end{bmatrix}}_{\text{given}}\,. \tag{2.135}$$

The solution of the last system of equations is obtained as:

$$\begin{bmatrix} u_{2X} \\ u_{3X} \\ F_4^{\mathrm{R}} \end{bmatrix} = \begin{bmatrix} \frac{1}{3}u_0 \\ \frac{2}{3}u_0 \\ -\frac{1}{3}\frac{EA}{L}u_0 \end{bmatrix}\,. \tag{2.136}$$

It should be noted here that this third approach is not the common way within the finite element method and is only shown for the sake of completeness. At this stage, let us summarize the considered boundary conditions, see Table 2.12.

A special type of 'boundary condition' can be realized by attaching a spring to a rod element as shown in Fig. 2.16. Let us have first a look at the configuration where the spring is attached to node 1 as shown in Fig. 2.16a. Assuming that node 2

Table 2.12 Different types of boundary conditions

DIRICHLET	NEUMANN
$u = 0$ (homogeneous)	F
$u \neq 0$ (non-homogeneous)	

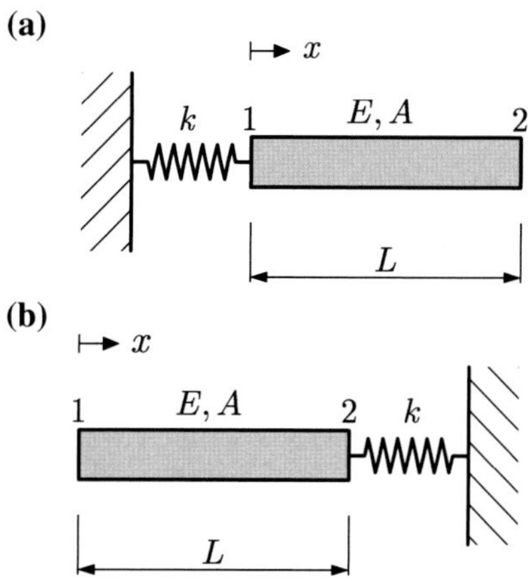

Fig. 2.16 Consideration of a spring in a rod structure: **a** spring attached to node 1 or **b** to node 2

is moved to the positive x-direction, the spring will cause a force on the rod element which can be expressed as $F_s = -ku_1$, where k is the spring constant and u_1 the displacement of node 1, i.e. where the spring is attached to the rod element.[20] It should be mentioned here that the required force to elongate the spring by u_1 in the positive x-direction is equal to ku_1 but the force acting on the rod is oriented in the negative x-direction.

Thus, the principal finite element equation for the rod element can be written as:

$$\frac{EA}{L}\begin{bmatrix} 1 & -1 \\ -1 & 1 \end{bmatrix}\begin{bmatrix} u_1 \\ u_2 \end{bmatrix} = \begin{bmatrix} F_s \\ F_2 \end{bmatrix} = \begin{bmatrix} -ku_1 \\ F_2 \end{bmatrix}. \tag{2.137}$$

Looking at Eq. (2.137), it can be concluded that the expression $-ku_1$ on the right-hand side should be shifted to the left-hand side where the expressions with the nodal unknowns are collected. Thus, one can obtain the following expression:

$$\frac{EA}{L}\left[\begin{array}{c|c} 1 + \frac{L}{EA}k & -1 \\ \hline -1 & 1 \end{array}\right]\begin{bmatrix} u_1 \\ u_2 \end{bmatrix} = \begin{bmatrix} 0 \\ F_2 \end{bmatrix}. \tag{2.138}$$

It can be seen from the last equation that a spring can simply be considered by adding the spring constant in the cell of the stiffness matrix with the index of the degree of freedom where the spring is attached, i.e. in our example the cell (u_1, u_1). If the spring would be attached at the second node, cf. Fig. 2.16b, the spring constant should be added in the cell (u_2, u_2) and the principal finite element equation for this case would finally read:

[20] It is assumed here that the spring is in its unstrained state in the sketched configuration, i.e. without the application of any force or displacement boundary conditions at the nodes of the rod.

$$\frac{EA}{L}\left[\begin{array}{c|c} 1 & -1 \\ \hline -1 & 1+\frac{L}{EA}k \end{array}\right]\begin{bmatrix} u_1 \\ u_2 \end{bmatrix} = \begin{bmatrix} F_1 \\ 0 \end{bmatrix}. \tag{2.139}$$

If we like to consider that the springs shown in Fig. 2.16 are pre-strained,[21] i.e. elongated or compressed by a displacement of magnitude u_s, the force which acts on the rod element is given[22] by $F_s = -k(u_1 - u_s)$ or $F_s = -k(u_2 - u_s)$ and the principal finite element equations given in (2.138) and (2.139) are modified to:

$$\frac{EA}{L}\left[\begin{array}{c|c} 1+\frac{L}{EA}k & -1 \\ \hline -1 & 1 \end{array}\right]\begin{bmatrix} u_1 \\ u_2 \end{bmatrix} = \begin{bmatrix} ku_s \\ F_2 \end{bmatrix}, \tag{2.140}$$

$$\frac{EA}{L}\left[\begin{array}{c|c} 1 & -1 \\ \hline -1 & 1+\frac{L}{EA}k \end{array}\right]\begin{bmatrix} u_1 \\ u_2 \end{bmatrix} = \begin{bmatrix} F_1 \\ ku_s \end{bmatrix}. \tag{2.141}$$

2.3.4 *Post-Computation: Determination of Strain, Stress and Further Quantities*

The previous section explained how to compose the global system of equations from which the primary unknowns, i.e. the nodal displacements, can be obtained. After the solution for the nodal unknowns, further quantities can be calculated in a post-computational step. Based on the kinematics relationship for the continuum rod according to Eq. (2.2) together with the nodal approach (2.21) and the definition of the $\boldsymbol{B}$-matrix (2.25), the following expression for the strain distribution inside a rod can be obtained:

$$\varepsilon_x^{\mathrm{e}}(x) = \frac{\mathrm{d}}{\mathrm{d}x}u^{\mathrm{e}}(x) = \frac{\mathrm{d}}{\mathrm{d}x}\boldsymbol{N}^{\mathrm{T}}(x)\,\boldsymbol{u}_{\mathrm{p}} = \boldsymbol{B}^{\mathrm{T}}\boldsymbol{u}_{\mathrm{p}}. \tag{2.142}$$

Considering the specific formulations of the $\boldsymbol{B}$-matrices for a linear and a quadratic rod element according to Eqs. (2.38) and (2.73), the strain distribution can be expressed as

---

[21] Such a pre-strained spring has its analogon in one-dimensional heat conduction in the form of a convective boundary condition: NEWTON's cooling law, i.e. $\dot{q} = h(T_\infty - T)$ where $\dot{q}$ is the heat flux in $\frac{\mathrm{W}}{\mathrm{m}^2}$, h is the heat transfer coefficient in $\frac{\mathrm{W}}{\mathrm{m}^2\,\mathrm{K}}$, T_∞ is the temperature of the environment and T is the temperature of the object's surface, is in a similar manner treated as this type of spring. See also Table 2.14.

[22] Setting $u_s = 0$ results in an unstrained spring.

$$\varepsilon_x^{\mathrm{e}}(x) = \frac{1}{L}(-u_1 + u_2) \quad (\text{lin.}), \tag{2.143}$$

$$\varepsilon_x^{\mathrm{e}}(x) = \frac{1}{L}\left(\left(-3 + \frac{4x}{L}\right)u_1 + \left(4 - \frac{8x}{L}\right)u_2 + \left(-1\frac{4x}{L}\right)u_3\right) \quad (\text{quad.}), \tag{2.144}$$

or expressed in the natural coordinate ξ:

$$\varepsilon_x^{\mathrm{e}}(\xi) = \frac{1}{L}(-u_1 + u_2) \quad (\text{lin.}), \tag{2.145}$$

$$\varepsilon_x^{\mathrm{e}}(\xi) = \frac{1}{L}((-1 + 2\xi)\,u_1 + (-4\xi)\,u_2 + (1 + 2\xi)\,u_3) \quad (\text{quad.}). \tag{2.146}$$

Based on the obtained strain distribution, HOOKE's law (2.3) permits the calculation of the stress distribution inside a rod element as

$$\sigma_x^{\mathrm{e}}(x) = E\frac{\mathrm{d}}{\mathrm{d}x}u^{\mathrm{e}}(x) = E\frac{\mathrm{d}}{\mathrm{d}x}\boldsymbol{N}^{\mathrm{T}}(x)\,\boldsymbol{u}_{\mathrm{p}} = E\boldsymbol{B}^{\mathrm{T}}\boldsymbol{u}_{\mathrm{p}}, \tag{2.147}$$

or based on the nodal displacements for a linear and quadratic rod element as a function of the natural coordinate ξ:

$$\sigma_x^{\mathrm{e}}(\xi) = \frac{E}{L}(-u_1 + u_2) \quad (\text{lin.}), \tag{2.148}$$

$$\sigma_x^{\mathrm{e}}(\xi) = \frac{E}{L}((-1 + 2\xi)\,u_1 + (-4\xi)\,u_2 + (1 + 2\xi)\,u_3) \quad (\text{quad.}). \tag{2.149}$$

The internal normal force N_x has been defined in Eq. (2.11) and can be calculated based on Eqs. (2.148) and (2.149):

$$N_x^{\mathrm{e}}(\xi) = \frac{EA}{L}(-u_1 + u_2) \quad (\text{lin.}), \tag{2.150}$$

$$N_x^{\mathrm{e}}(\xi) = \frac{EA}{L}((-1 + 2\xi)\,u_1 + (-4\xi)\,u_2 + (1 + 2\xi)\,u_3) \quad (\text{quad.}). \tag{2.151}$$

A final task is often to calculate the reaction forces at the supports or nodes of prescribed displacements. To explain the procedure, let us return to the example shown in Fig. 2.15. The free-body diagram of the problem can be sketched as shown in Fig. 2.17.

Based on the indicated reaction forces, the global (non-reduced) system of equations can be stated for the configuration in Fig. 2.17a as

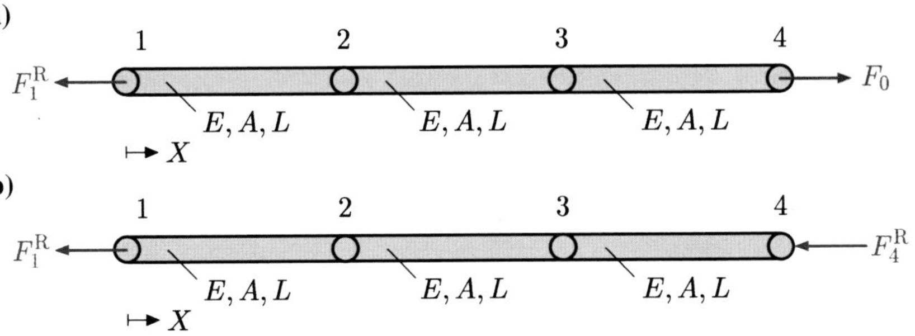

Fig. 2.17 Free-body diagram of the cantilever rod structure shown in Fig. 2.15

$$\frac{EA}{L}\begin{bmatrix} 1 & -1 & 0 & 0 \\ -1 & 2 & -1 & 0 \\ 0 & -1 & 2 & -1 \\ 0 & 0 & -1 & 1 \end{bmatrix}\begin{bmatrix} u_{1X} \\ u_{2X} \\ u_{3X} \\ u_{4X} \end{bmatrix} = \begin{bmatrix} -F_1^R \\ 0 \\ 0 \\ F_0 \end{bmatrix}, \tag{2.152}$$

or for Fig. 2.17b as:

$$\frac{EA}{L}\begin{bmatrix} 1 & -1 & 0 & 0 \\ -1 & 2 & -1 & 0 \\ 0 & -1 & 2 & -1 \\ 0 & 0 & -1 & 1 \end{bmatrix}\begin{bmatrix} u_{1X} \\ u_{2X} \\ u_{3X} \\ u_{4X} \end{bmatrix} = \begin{bmatrix} -F_1^R \\ 0 \\ 0 \\ -F_4^R \end{bmatrix}. \tag{2.153}$$

Knowing all nodal displacements, the support reaction F_1^R can be obtained for both cases by evaluating the first equation of the linear system as:

$$F_1^R = -\frac{EA}{L}(u_{1X} - u_{2X}) \,. \tag{2.154}$$

For the second case as shown in Fig. 2.17b, the reaction force F_4^R is obtained by evaluating the fourth equation of the the linear system (2.153) as:

$$F_4^R = -\frac{EA}{L}(-u_{3X} + u_{4X}) \,. \tag{2.155}$$

In general we can state that reactions forces are obtained from the non-reduced system of equations based on the prior to this calculated nodal displacements. Special attention must be given to the consideration of the reactions on the right-hand side of the system of equations since the pure calculation of the nodal displacements did not require an exact mentioning of these quantities. At the end of this section, let us highlight the different nature of the evaluated quantities as indicated in Table 2.13. It is important to realize that the elemental values are evaluated at integration points of the element.

Table 2.13 Evaluation of different quantities

Quantity	Nodal value	Elemental value
Displacement	X	
Strain		X
Stress		X
Reaction force	X	

Table 2.14 Comparison of analogous properties in one-dimensional heat conduction and solid mechanics. p_0: load per unit length in $\frac{\text{N}}{\text{m}}$; γ_0: load per unit volume in $\frac{\text{N}}{\text{m}^3}$; $\dot{\eta}_0$: rate of energy generation per unit volume in $\frac{\text{W}}{\text{m}^3}$; $\varnothing_0$: rate of energy generation per unit length in $\frac{\text{W}}{\text{m}}$; $\dot{q}_x$: heat flux in $\frac{\text{W}}{\text{m}^2}$; $\dot{Q}_x$: heat transfer rate in W; k: thermal conductivity in $\frac{\text{W}}{\text{m K}}$

Solid mechanics	Heat conduction
Partial differential equation	
$EA\dfrac{\mathrm{d}^2 u_x}{\mathrm{d}x^2} = -p_0$	$k\dfrac{\mathrm{d}^2 T}{\mathrm{d}x^2} = -\dot{\eta}_0$
$\left(E\dfrac{\mathrm{d}^2 u_x}{\mathrm{d}x^2} = -\gamma_0\right)$	$\left(kA\dfrac{\mathrm{d}^2 T}{\mathrm{d}x^2} = -\dot{\varnothing}_0\right)$
Primary variable	
Displacement u_x	Temperature T
Derivative of primary variable	
Strain $\varepsilon_x = \frac{\mathrm{d}u_x}{\mathrm{d}x}$	Temperature gradient $\frac{\mathrm{d}T}{\mathrm{d}x}$
Stress $\sigma_x = E\frac{\mathrm{d}u_x}{\mathrm{d}x}$	Heat flux $\dot{q}_x = -k\frac{\mathrm{d}T}{\mathrm{d}x}$
Force $F_x = EA\frac{\mathrm{d}u_x}{\mathrm{d}x}$	Heat transfer rate $\dot{Q}_x = -kA\frac{\mathrm{d}T}{\mathrm{d}x}$
Principal finite element equation	
$\dfrac{EA}{L}\begin{bmatrix} 1 & -1 \\ -1 & 1 \end{bmatrix}\begin{bmatrix} u_{1x} \\ u_{2x} \end{bmatrix} = \begin{bmatrix} F_{1x} \\ F_{2x} \end{bmatrix}$	$\dfrac{k}{L}\begin{bmatrix} 1 & -1 \\ -1 & 1 \end{bmatrix}\begin{bmatrix} T_1 \\ T_2 \end{bmatrix} = \begin{bmatrix} \dot{q}_{1x} \\ \dot{q}_{2x} \end{bmatrix}$
$\left(\dfrac{E}{L}\begin{bmatrix} 1 & -1 \\ -1 & 1 \end{bmatrix}\begin{bmatrix} u_{1x} \\ u_{2x} \end{bmatrix} = \begin{bmatrix} \sigma_{1x} \\ \sigma_{2x} \end{bmatrix}\right)$	$\left(\dfrac{kA}{L}\begin{bmatrix} 1 & -1 \\ -1 & 1 \end{bmatrix}\begin{bmatrix} T_1 \\ T_2 \end{bmatrix} = \begin{bmatrix} \dot{Q}_{1x} \\ \dot{Q}_{2x} \end{bmatrix}\right)$

2.3.5 Analogies to Other Field Problems

Further analogies to other field problems can be found, for example, in [8]. A comparison between solid mechanics and heat conduction is presented in Table 2.14.

2.3.6 Solved Rod Problems

2.1 Example: Rod structure fixed at both ends

Given is a rod structure as shown in Fig. 2.18. The structure is composed of two rods of different cross-sectional areas A_{I} and A_{II}. Length L and YOUNG's modulus E are the same for both rods. The structure is fixed at both ends and loaded by

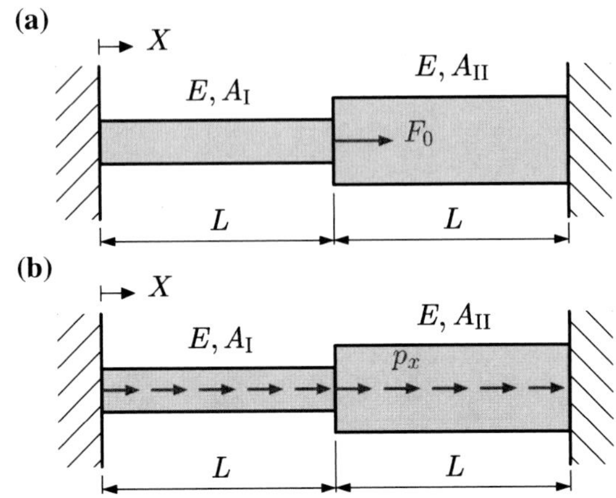

Fig. 2.18 Rod structure fixed at both ends: **a** axial point load; **b** load per length

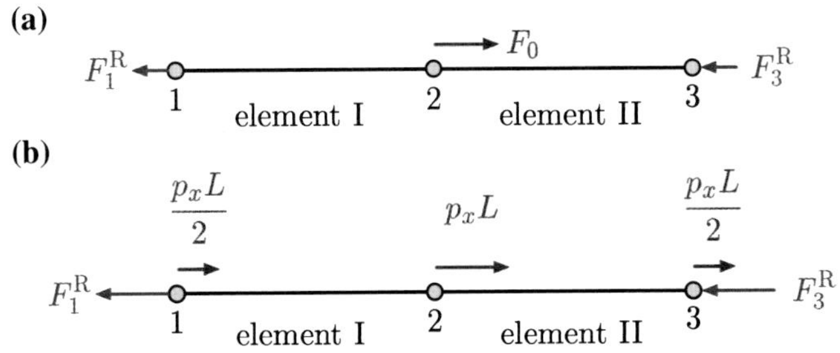

Fig. 2.19 Discretized rod structure: **a** point load and reaction forces; **b** equivalent nodal loads and reaction forces

(a) a point load F_0 in the middle and
(b) a uniform distributed load p_x, i.e. a force per unit length.

Model the rod structure with two linear finite elements and determine for both cases

- the displacement $u_2 = u(X = L)$ in the middle of the structure,
- the stresses and strains in both elements,
- the average stress and strain in the middle of the structure at $X = L$,
- the reaction forces at the supports and check the global force equilibrium.

Simplify all the results obtained for the special case of $A_{\text{I}} = A_{\text{II}} = A$.

2.1 Solution

The finite element discretization and all acting forces are shown in Fig. 2.19.
Case (a) point load:

• Displacement in the middle of the structure
The elemental stiffness matrix for each element is given by

$$\frac{EA_i}{L}\begin{bmatrix} 1 & -1 \\ -1 & 1 \end{bmatrix} \quad \text{with} \quad i = \text{I}, \text{II} \tag{2.156}$$

and can be assembled to obtain the global finite element equation:

$$\frac{E}{L}\begin{bmatrix} A_\text{I} & -A_\text{I} & 0 \\ -A_\text{I} & A_\text{I}+A_\text{II} & -A_\text{II} \\ 0 & -A_\text{II} & A_\text{II} \end{bmatrix}\begin{bmatrix} u_1 \\ u_2 \\ u_3 \end{bmatrix} = \begin{bmatrix} -F_1^\text{R} \\ F_0 \\ -F_3^\text{R} \end{bmatrix}. \tag{2.157}$$

Consideration of the boundary conditions, i.e. $u_1 = u_3 = 0$, in the last system of equations allows to solve for the unknown displacement in the middle of the structure:

$$u_2 = \frac{F_0 L}{E(A_\text{I}+A_\text{II})}. \tag{2.158}$$

• Stresses and strains in both elements
Based on the general definition of the strain in a rod element, i.e. $\varepsilon = \frac{1}{L}(u_\text{right} - u_\text{left})$, the constant strains in both elements can be derived under consideration of the boundary conditions as:

$$\varepsilon_\text{I} = \frac{1}{L}(u_2 - 0) = \frac{F_0}{E(A_\text{I}+A_\text{II})}, \tag{2.159}$$

$$\varepsilon_\text{II} = \frac{1}{L}(0 - u_2) = -\frac{F_0}{E(A_\text{I}+A_\text{II})}. \tag{2.160}$$

Application of HOOKE's law, i.e. $\sigma = E\varepsilon$, gives the constant stresses in each element:

$$\sigma_\text{I} = E\varepsilon_\text{I} = \frac{F_0}{(A_\text{I}+A_\text{II})}, \tag{2.161}$$

$$\sigma_\text{II} = E\varepsilon_\text{II} = -\frac{F_0}{(A_\text{I}+A_\text{II})}. \tag{2.162}$$

• Average stress and strain in the middle of the structure
As in the case of many finite element codes, the average stress and strain at the middle node can be calculated by the following averaging rule as:

$$\varepsilon_2 = \frac{\varepsilon_\text{I}+\varepsilon_\text{II}}{2} = 0, \tag{2.163}$$

$$\sigma_2 = \frac{\sigma_\text{I}+\sigma_\text{II}}{2} = 0. \tag{2.164}$$

As can be seen from this result, stress and strain values displayed at nodes should be taken with care.

• Reaction forces at the supports and check of the global force equilibrium
Evaluation of the first and third equation of the system (2.157) for known nodal displacements gives:

$$F_1^{\mathrm{R}} = \frac{EA_{\mathrm{I}}}{L} \times u_2 = \frac{A_{\mathrm{I}}}{A_{\mathrm{I}} + A_{\mathrm{II}}} \times F_0 \,, \tag{2.165}$$

$$F_3^{\mathrm{R}} = \frac{EA_{\mathrm{II}}}{L} \times u_2 = \frac{A_{\mathrm{II}}}{A_{\mathrm{I}} + A_{\mathrm{II}}} \times F_0 \,, \tag{2.166}$$

and the global force equilibrium

$$F_0 - \frac{A_{\mathrm{I}}}{A_{\mathrm{I}} + A_{\mathrm{II}}} \times F_0 - \frac{A_{\mathrm{II}}}{A_{\mathrm{I}} + A_{\mathrm{II}}} \times F_0 = 0 \tag{2.167}$$

is fulfilled.
Case (b) distributed load:

• Displacement in the middle of the structure
The global finite element equation results under consideration of the equivalent nodal loads, cf. Fig. 2.19, as

$$\frac{E}{L}\begin{bmatrix} A_{\mathrm{I}} & -A_{\mathrm{I}} & 0 \\ -A_{\mathrm{I}} & A_{\mathrm{I}} + A_{\mathrm{II}} & -A_{\mathrm{II}} \\ 0 & -A_{\mathrm{II}} & A_{\mathrm{II}} \end{bmatrix}\begin{bmatrix} u_1 \\ u_2 \\ u_3 \end{bmatrix} = \begin{bmatrix} -F_1^{\mathrm{R}} + \frac{p_x L}{2} \\ p_x L \\ -F_3^{\mathrm{R}} + \frac{p_x L}{2} \end{bmatrix}, \tag{2.168}$$

from which the displacement at node 2 follows under consideration of the boundary conditions:

$$u_2 = \frac{(p_x L)L}{E(A_{\mathrm{I}} + A_{\mathrm{II}})} \,. \tag{2.169}$$

• Stresses and strains in both elements and at the middle node
Based on the procedure given in (a), the constant strains and stresses are given by:

$$\varepsilon_{\mathrm{I}} = \frac{p_x L}{E(A_{\mathrm{I}} + A_{\mathrm{II}})} \,, \quad \sigma_{\mathrm{I}} = \frac{p_x L}{A_{\mathrm{I}} + A_{\mathrm{II}}} \,, \tag{2.170}$$

$$\varepsilon_{\mathrm{II}} = -\frac{p_x L}{E(A_{\mathrm{I}} + A_{\mathrm{II}})} \,, \quad \sigma_{\mathrm{II}} = -\frac{p_x L}{A_{\mathrm{I}} + A_{\mathrm{II}}} \,, \tag{2.171}$$

$$\varepsilon_2 = 0 \,, \quad \sigma_2 = 0 \,. \tag{2.172}$$

• Reaction forces at the supports and check of the global force equilibrium
Evaluation of the first and third equation of the system (2.168) for known nodal displacements gives:

Table 2.15 Results of the problem shown in Fig. 2.18 for the special case $A_{\mathrm{I}} = A_{\mathrm{II}} = A$

Quantity	Point load F	Distributed load p_x
u_2	$\frac{1}{2}\frac{F_0 L}{EA}$	$\frac{1}{2}\frac{(p_x L)L}{EA}$
ε_{I}	$\frac{F_0}{2EA}$	$\frac{p_x L}{2EA}$
$\varepsilon_{\mathrm{II}}$	$-\frac{F_0}{2EA}$	$-\frac{p_x L}{2EA}$
σ_{I}	$\frac{F_0}{2A}$	$\frac{p_x L}{2A}$
σ_{II}	$-\frac{F_0}{2A}$	$-\frac{p_x L}{2A}$
σ_2	0	0
ε_2	0	0
F_1^{R}	$\frac{F_0}{2}$	$p_x L$
F_3^{R}	$\frac{F_0}{2}$	$p_x L$

$$F_1^{\mathrm{R}} = \frac{p_x L}{2} + \frac{EA_{\mathrm{I}}}{L} \times u_2 = \left(\frac{1}{2} + \frac{A_{\mathrm{I}}}{A_{\mathrm{I}} + A_{\mathrm{II}}}\right) \times p_x L\,, \tag{2.173}$$

$$F_3^{\mathrm{R}} = \frac{p_x L}{2} + \frac{EA_{\mathrm{II}}}{L} \times u_2 = \left(\frac{1}{2} + \frac{A_{\mathrm{II}}}{A_{\mathrm{I}} + A_{\mathrm{II}}}\right) \times p_x L\,, \tag{2.174}$$

and the global force equilibrium is fulfilled. It can be concluded from this exercise that the equivalent loads applied at the supports do not influence the strains and stresses inside the rods but contribute to the reaction forces at the supports. Results for the special case $A_{\mathrm{I}} = A_{\mathrm{II}} = A$ are summarized in Table 2.15.

2.2 Example: Rod structure with gap Given is a rod structure as shown in Fig. 2.20. The structure is composed of a rod with cross-sectional area A, length L, and YOUNG's modulus E. The structure is fixed at the left-hand end and a gap of distance δ is between the right-hand end and a rigid wall. The structure is loaded by

(a) a point load F_0 in the middle and
(b) a point load F_0 at the right-hand end.

Model the rod structure with two linear finite elements and determine for both cases:

- the displacement $u_2 = u(X = L)$ in the middle of the structure for the case of no contact and contact,
- the reaction forces at the supports and check the global force equilibrium,

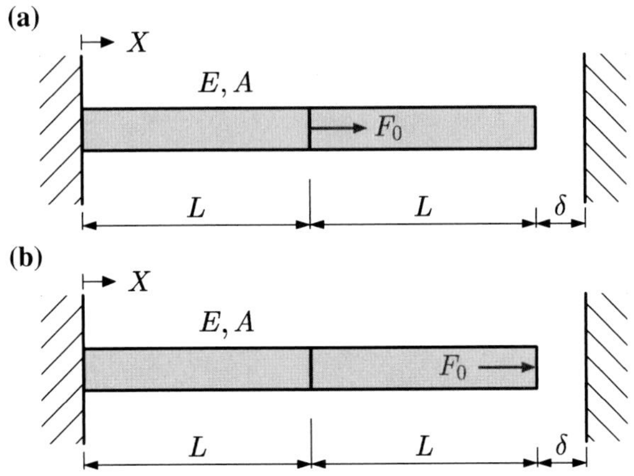

Fig. 2.20 Rod structure with a gap at the right end: **a** axial point load in the middle; **b** axial point load at the right end

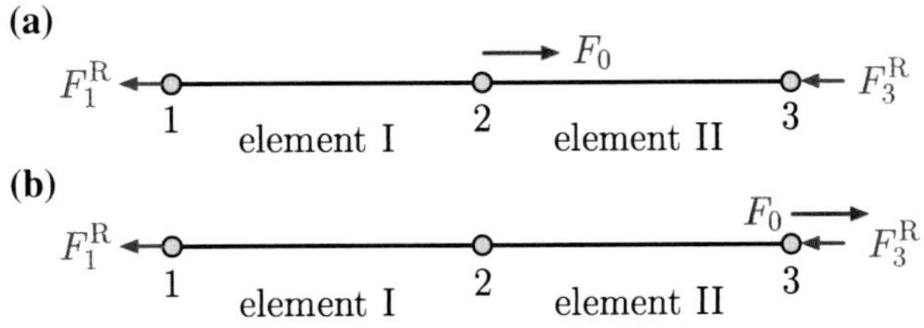

Fig. 2.21 Discretized rod structure: **a** axial point load in the middle; **b** axial point load at the right end. The reaction force R_3 is only acting in the case of contact

- the stress distribution in the rod structure for increasing force F_0.

2.2 Solution

The finite element discretization and all acting forces are shown in Fig. 2.21.
Case (a) point load in the middle:

- Displacement in the middle of the structure

In the case that there is no contact, element II is not acting, i.e. $u_2 = u_3$, or contributing to the global stiffness matrix and the problem can be described by

$$\frac{EA}{L}\begin{bmatrix} 1 & -1 \\ -1 & 1 \end{bmatrix}\begin{bmatrix} u_1 \\ u_2 \end{bmatrix} = \begin{bmatrix} -F_1^R \\ F \end{bmatrix}, \tag{2.175}$$

from which the displacement at node 2 can be obtained under consideration of the boundary condition ($u_1 = 0$) as:

$$u_2 = \frac{F_0 L}{EA}. \tag{2.176}$$

If the force F_0 is further increased to a value of $F_0 = \frac{EA\delta}{L}$, contact occurs, i.e. $u_2 = u_3 = \delta$, and the situation for the global system is different. Now, both elements contribute to the global system:

$$\frac{EA}{L}\begin{bmatrix} 1 & -1 & 0 \\ -1 & 2 & -1 \\ 0 & -1 & 1 \end{bmatrix}\begin{bmatrix} u_1 \\ u_2 \\ u_3 \end{bmatrix} = \begin{bmatrix} -F_1^{\text{R}} \\ F_0 \\ -F_3^{\text{R}} \end{bmatrix}. \tag{2.177}$$

Consideration of the boundary conditions at the left- and right-hand end, i.e. $u_1 = 0$ and $u_3 = \delta$, gives

$$\frac{EA}{L}\begin{bmatrix} 2 & -1 \\ 0 & 1 \end{bmatrix}\begin{bmatrix} u_2 \\ u_3 \end{bmatrix} = \begin{bmatrix} F_0 \\ \frac{EA\delta}{L} \end{bmatrix}, \tag{2.178}$$

and the displacement in the middle of the structure is obtained as:

$$u_2 = \frac{LF_0}{2EA} + \frac{\delta}{2}. \tag{2.179}$$

• Reaction forces
Based on the known values of the nodal displacements, Eq. (2.177) can be evaluated for the reaction forces:

$$F_1^{\text{R}} = \left(\frac{F_0}{2} + \frac{EA\delta}{2L}\right), \tag{2.180}$$

$$F_3^{\text{R}} = \left(\frac{F_0}{2} - \frac{EA\delta}{2L}\right). \tag{2.181}$$

It should be noted that both reaction forces are directed to the negative X-direction.

• Stress distribution in the rod structure for increasing force F_0
Since the nodal displacements are known, the strains can be obtained based on the general definition $\varepsilon = \frac{1}{L}(u_{\text{right}} - u_{\text{left}})$ and HOOKE's law gives the stresses. The results for the stress σ in both elements as a function of the applied external force F_0 are shown in Fig. 2.22. As can be seen from this figure, the global stiffness changes as soon as the gap is closed.
Case (b) point load at the right end:

• Displacement in the middle of the structure
In the case that there is no contact, the displacement of the middle node is simply half of the displacement obtained at the node of the right-hand end. Under the condition of contact, the global system of equations reads as:

$$\frac{EA}{L}\begin{bmatrix} 1 & -1 & 0 \\ -1 & 2 & -1 \\ 0 & -1 & 1 \end{bmatrix}\begin{bmatrix} u_1 \\ u_2 \\ u_3 \end{bmatrix} = \begin{bmatrix} -F_1^{\text{R}} \\ 0 \\ F_0 - F_3^{\text{R}} \end{bmatrix}. \tag{2.182}$$

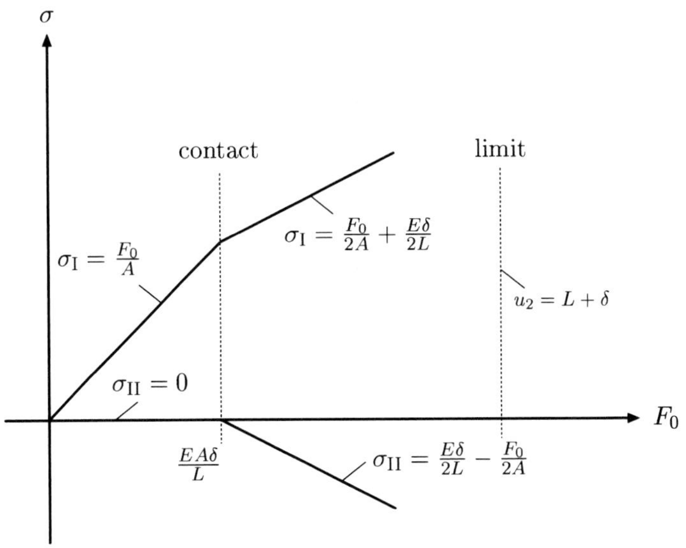

Fig. 2.22 Stress distribution in the rod structure as a function of the external point load F_0

Consideration of the boundary conditions at the left- and right-hand end, i.e. $u_1 = 0$ and $u_3 = \delta$, gives

$$\frac{EA}{L}\begin{bmatrix} 2 & -1 \\ 0 & 1 \end{bmatrix}\begin{bmatrix} u_2 \\ u_3 \end{bmatrix} = \begin{bmatrix} 0 \\ \frac{EA\delta}{L} \end{bmatrix}, \tag{2.183}$$

and the displacement in the middle of the structure is obtained as:

$$u_2 = \frac{\delta}{2}. \tag{2.184}$$

• Reaction forces
Based on the known values of the nodal displacements, Eq. (2.182) can be evaluated for the reaction forces:

$$F_1^{\mathrm{R}} = \frac{EA\delta}{2L}, \tag{2.185}$$

$$F_3^{\mathrm{R}} = F - \frac{EA\delta}{2L}. \tag{2.186}$$

• Stress distribution in the rod structure for increasing force F_0.
As mentioned in (a), strains and stresses can be calculated based on the known nodal displacements. The graphical representation of the stress in the rod is given in Fig. 2.23. It can be concluded that any additional force after closing the gap is absorbed by the support and does not affect the stress state in the rod.

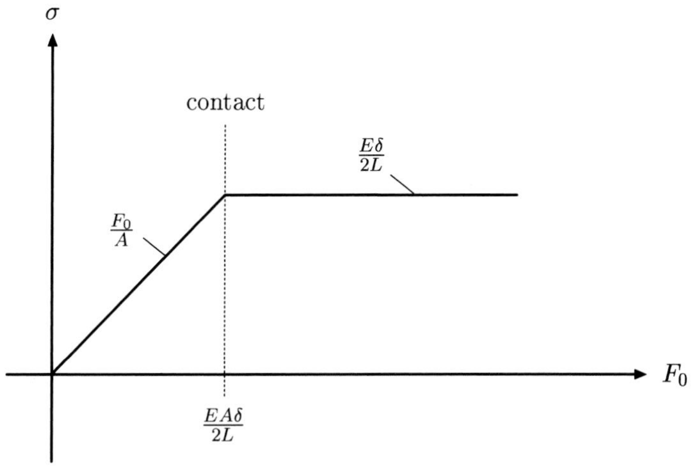

Fig. 2.23 Stress distribution in the rod structure as a function of the external point load F_0

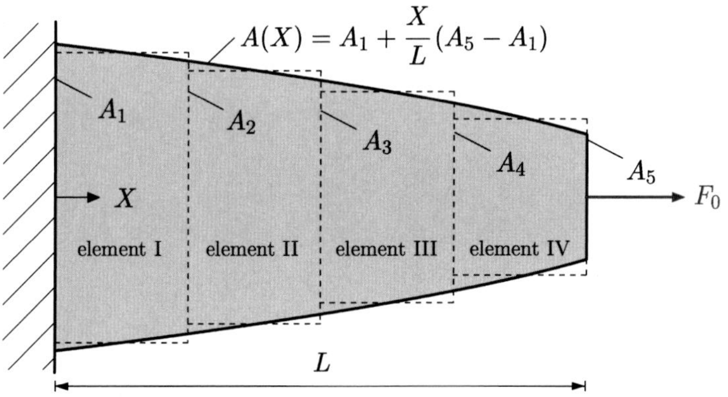

Fig. 2.24 Rod with changing cross-sectional area $A = A(x)$

2.3 Example: Rod with changing cross-sectional area

Given is a rod structure as shown in Fig. 2.24. The structure reveals a linear changing cross-sectional area $A(x)$ while the YOUNG's modulus E is assumed to be constant. The structure is fixed at the left-hand end and loaded by a single force F_0 at the right-hand end. The ratio between the area A_5 and A_1 is given by the factor a.

Model the rod structure with four linear finite elements of constant cross-sectional area and determine for the stepped rod the nodal displacements. Each element should have the same length $\frac{L}{4}$ and the cross section should be the average of the cross-sectional area at the left- and right-hand end of each step.

2.3 Solution

Given the ratio between the cross section at the right- and left-hand end as $\frac{A_5}{A_1} = a$ and the functional dependency of the cross-sectional area as

$$A(X) = A_1 + \frac{X}{L}(A_5 - A_1)\,, \tag{2.187}$$

the areas A_i $(i = 2, \ldots, 5)$ can be expressed as:

$$A_2 = \frac{3+a}{4}A_1\,,\; A_3 = \frac{1+a}{2}A_1\,,\; A_4 = \frac{1+3a}{4}A_1\,,\; A_5 = aA_1\,. \tag{2.188}$$

Based on these area relations, the averaged area for each element $A_{ij} = \frac{A_i+A_j}{2}$ is obtained as:

$$A_{12} = \frac{7+a}{8}A_1\,,\; A_{23} = \frac{5+3a}{8}A_1\,,\; A_{34} = \frac{3+5a}{8}A_1\,,\; A_{45} = \frac{1+7a}{8}A_1\,, \tag{2.189}$$

and the global stiffness matrix can be assembled to:

$$\frac{EA_1}{2L}\begin{bmatrix} 7+a & -(7+a) & 0 & 0 & 0 \\ -(7+a) & 12+4a & -(5+3a) & 0 & 0 \\ 0 & -(5+3a) & 8+8a & -(3+5a) & 0 \\ 0 & 0 & -(3+5a) & 4+12a & -(1+7a) \\ 0 & 0 & 0 & -(1+7a) & 1+7a \end{bmatrix}. \tag{2.190}$$

Considering the boundary condition at the left-hand end, i.e. $u_1 = 0$, the system of equations is obtained as

$$\frac{EA_1}{2L}\begin{bmatrix} 12+4a & -(5+3a) & 0 & 0 \\ -(5+3a) & 8+8a & -(3+5a) & 0 \\ 0 & -(3+5a) & 4+12a & -(1+7a) \\ 0 & 0 & -(1+7a) & 1+7a \end{bmatrix}\begin{bmatrix} u_2 \\ u_3 \\ u_4 \\ u_5 \end{bmatrix} = \begin{bmatrix} 0 \\ 0 \\ 0 \\ F_0 \end{bmatrix}, \tag{2.191}$$

from which the vector of nodal displacements can be calculated:

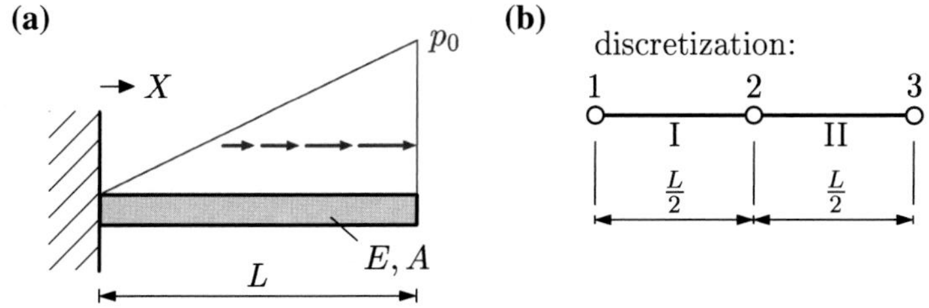

Fig. 2.25 Cantilever rod with triangular shaped distributed load: **a** geometry and boundary conditions and **b** discretization

$$\begin{bmatrix} u_1 \\ u_2 \\ u_3 \\ u_4 \\ u_5 \end{bmatrix} = \frac{LF_0}{EA_1} \begin{bmatrix} 0 \\ \dfrac{2}{7+a} \\ \dfrac{8(3+a)}{3a^2+26a+35} \\ \dfrac{2(71+98a+23a^2)}{105+253a+139a^2+15a^3} \\ \dfrac{32(11+53a+53a^2+11a^3)}{(1+7a)(105+253a+139a^2+15a^3)} \end{bmatrix} . \tag{2.192}$$

2.4 Example: Rod with linearly increasing distributed load

The following Fig. 2.25a shows a cantilever rod structure of length L which is loaded with a triangular shaped distributed load (maximum value of q_0 at $X = L$).

Use two linear rod elements of equal length $\frac{L}{2}$ (see Fig. 2.25b) and:

- Calculate for each element separately the vector of the equivalent nodal loads based on the general statement $\int \boldsymbol{N} p(x)\mathrm{d}x$. Use classical analytical integration for this task.
- Assemble the global system of equations without consideration of the boundary conditions at the fixed support.
- Obtain the reduced system of equations (the solution of the system of equations is not required).

2.4 Solution

The separated elements and the corresponding distributed loads are shown in Fig. 2.26.

Special consideration requires the elemental length $\frac{L}{2}$ since the interpolation functions and integrals are defined from $0 \ldots L$. Thus, let us calculate the equivalent nodal loads first for the length L and at the end we substitute $L := \frac{L}{2}$.

• Let us look in the following first separately at each element. The load vector for element I can be written as:

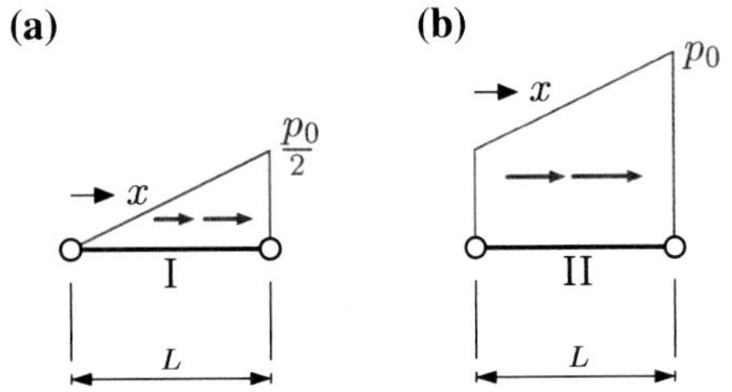

Fig. 2.26 Single elements and corresponding distributed loads

$$f_{\text{I}} = \int_0^L \boldsymbol{N}(x)p(x)\,\mathrm{d}x = \int_0^L \begin{bmatrix} N_{1u}(x) \\ N_{2u}(x) \end{bmatrix} \left(\frac{p_0}{2}\frac{x}{L} \right) \mathrm{d}x = \frac{p_0}{2L}\int_0^L \begin{bmatrix} 1-\frac{x}{L} \\ \frac{x}{L} \end{bmatrix} x\,\mathrm{d}x$$

$$= \frac{p_0}{2L}\begin{bmatrix} \frac{x^2}{2} - \frac{x^3}{3L} \\ \frac{x^3}{3L} \end{bmatrix}_0^L = \frac{p_0}{2}\begin{bmatrix} \frac{L}{6} \\ \frac{L}{3} \end{bmatrix}. \tag{2.193}$$

$$L := \frac{L}{2} \;\Rightarrow\; f_{\text{I}} = \frac{p_0}{4}\begin{bmatrix} \frac{L}{6} \\ \frac{L}{3} \end{bmatrix}. \tag{2.194}$$

In a similar way we obtain for element II:

$$f_{\text{II}} = \int_0^L \boldsymbol{N}(x)p(x)\,\mathrm{d}x = \int_0^L \begin{bmatrix} N_{1u}(x) \\ N_{2u}(x) \end{bmatrix} \left(p_0 \left[\frac{1}{2} + \frac{x}{2L} \right] \right) \mathrm{d}x$$

$$= \frac{p_0}{2}\int_0^L \begin{bmatrix} 1-\frac{x}{L} \\ \frac{x}{L} \end{bmatrix} \left(1+\frac{x}{L}\right) \mathrm{d}x = \frac{p_0}{2}\begin{bmatrix} x - \frac{x^3}{3L^2} \\ \frac{x^2}{2L} + \frac{x^3}{3L^2} \end{bmatrix}_0^L = \frac{p_0}{2}\begin{bmatrix} \frac{2L}{3} \\ \frac{5L}{6} \end{bmatrix}. \tag{2.195}$$

$$L := \frac{L}{2} \;\Rightarrow\; f_{\text{II}} = \frac{p_0}{4}\begin{bmatrix} \frac{2L}{3} \\ \frac{5L}{6} \end{bmatrix}. \tag{2.196}$$

Check: Simple superposition based on tabled values (see Table 2.5) for an element of length L, see Fig. 2.27.

Fig. 2.27 Superposition of simple load cases

$$\begin{bmatrix} \frac{p_0 L}{12} \\ \frac{p_0 L}{6} \end{bmatrix} + \begin{bmatrix} \frac{p_0 L}{4} \\ \frac{p_0 L}{4} \end{bmatrix} = \begin{bmatrix} \frac{p_0 L}{3} \\ \frac{5 p_0 L}{12} \end{bmatrix} \checkmark \tag{2.197}$$

- The principal finite element equation for element I reads:

$$\frac{EA}{\frac{L}{2}} \begin{bmatrix} 1 & -1 \\ -1 & 1 \end{bmatrix} \begin{bmatrix} u_{1X} \\ u_{2X} \end{bmatrix} = \frac{p_0}{4} \begin{bmatrix} \frac{L}{6} \\ \frac{L}{3} \end{bmatrix} + \begin{bmatrix} F_1^{\mathrm{R}} \\ 0 \end{bmatrix}. \tag{2.198}$$

and for element II:

$$\frac{EA}{\frac{L}{2}} \begin{bmatrix} 1 & -1 \\ -1 & 1 \end{bmatrix} \begin{bmatrix} u_{2X} \\ u_{3X} \end{bmatrix} = \frac{p_0}{4} \begin{bmatrix} \frac{2L}{3} \\ \frac{5L}{6} \end{bmatrix}. \tag{2.199}$$

Global system of equations:

$$\frac{EA}{\frac{L}{2}} \begin{bmatrix} 1 & -1 & 0 \\ -1 & 1+1 & -1 \\ 0 & -1 & 1 \end{bmatrix} \begin{bmatrix} u_{1X} \\ u_{2X} \\ u_{3X} \end{bmatrix} = \frac{p_0}{4} \begin{bmatrix} \frac{L}{6} \\ \frac{L}{3} + \frac{2L}{3} \\ \frac{5L}{6} \end{bmatrix} + \begin{bmatrix} F_1^{\mathrm{R}} \\ 0 \\ 0 \end{bmatrix}. \tag{2.200}$$

- Reduced system of equations: $u_{1X} = 0$

$$\frac{EA}{\frac{L}{2}} \begin{bmatrix} 2 & -1 \\ -1 & 1 \end{bmatrix} \begin{bmatrix} u_{2X} \\ u_{3X} \end{bmatrix} = \frac{p_0}{4} \begin{bmatrix} L \\ \frac{5L}{6} \end{bmatrix}. \tag{2.201}$$

2.4 Assembly of Elements to Plane Truss Structures

2.4.1 Rotational Transformation in a Plane

Let us consider in the following a rod element which can deform in the global X-Y plane. The local x-coordinate is rotated by an angle α against the global coordinate system (X, Y), cf. Fig. 2.28.

Each node has now in the global coordinate system two degrees of freedom, i.e. a displacement in the X- and a displacement in the Y-direction. These two global displacements at each node can be used to calculate the displacement in the direction of the rod axis, i.e. in the direction of the local x-axis. Based on the right-angled triangles shown in Fig. 2.28, the displacements in the local coordinate system are given based on the global displacements as:

$$u_{1x} = \cos\alpha\, u_{1X} + \sin\alpha\, u_{1Y}\,, \tag{2.202}$$

$$u_{2x} = \cos\alpha\, u_{2X} + \sin\alpha\, u_{2Y}\,. \tag{2.203}$$

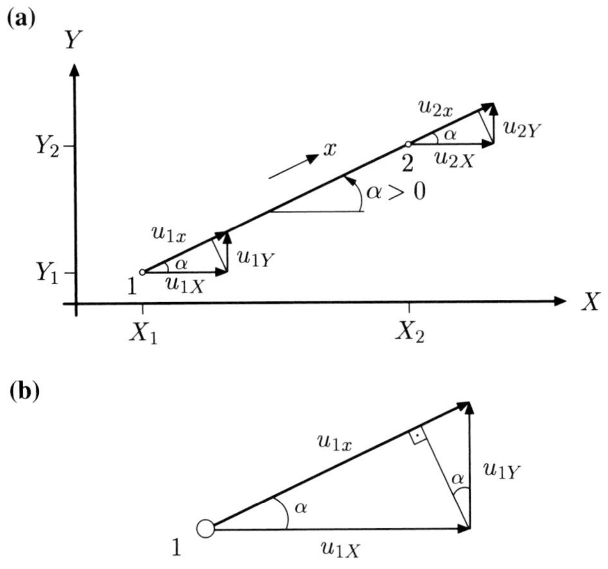

Fig. 2.28 Rotational transformation of a rod element in the X-Y plane: **a** total view and **b** detail for node 1

It is possible to derive in a similar way the global displacements based on the local displacements as:

$$u_{1X} = \cos\alpha \, u_{1x}\,, \qquad u_{2X} = \cos\alpha \, u_{2x}\,, \tag{2.204}$$

$$u_{1Y} = \sin\alpha \, u_{1x}\,, \qquad u_{2Y} = \sin\alpha \, u_{2x}\,. \tag{2.205}$$

The last relationships between the global and local displacements can be written in matrix notation as

$$\begin{bmatrix} u_{1X} \\ u_{1Y} \\ u_{2X} \\ u_{2Y} \end{bmatrix} = \begin{bmatrix} \cos\alpha & 0 \\ \sin\alpha & 0 \\ 0 & \cos\alpha \\ 0 & \sin\alpha \end{bmatrix} \begin{bmatrix} u_{1x} \\ u_{2x} \end{bmatrix}, \tag{2.206}$$

or in abbreviated matrix notation as:

$$\boldsymbol{u}_{XY} = \boldsymbol{T}^{\mathrm{T}} \boldsymbol{u}_{xy}\,, \tag{2.207}$$

where $\boldsymbol{u}_{XY}$ is the displacement column matrix in the global coordinate system and $\boldsymbol{u}_{xy}$ the local displacement column matrix. The last equation can be solved for the displacements in the local coordinate system and inverting[23] the transformation matrix gives

[23] Since the transformation matrix is orthogonal, it follows that $\boldsymbol{T}^{\mathrm{T}} = \boldsymbol{T}^{-1}$.

$$\boldsymbol{u}_{xy} = \boldsymbol{T}\boldsymbol{u}_{XY}\,, \tag{2.208}$$

or in components:

$$\begin{bmatrix} u_{1x} \\ u_{2x} \end{bmatrix} = \underbrace{\begin{bmatrix} \cos\alpha & \sin\alpha & 0 & 0 \\ 0 & 0 & \cos\alpha & \sin\alpha \end{bmatrix}}_{\boldsymbol{T}} \begin{bmatrix} u_{1X} \\ u_{1Y} \\ u_{2X} \\ u_{2Y} \end{bmatrix}. \tag{2.209}$$

It is possible to transform in a similar way the matrix of the external loads as:

$$\boldsymbol{f}_{XY} = \boldsymbol{T}^{\mathrm{T}}\boldsymbol{f}_{xy}\,, \tag{2.210}$$

$$\boldsymbol{f}_{xy} = \boldsymbol{T}\boldsymbol{f}_{XY}\,. \tag{2.211}$$

Considering the transformation of the local displacements and loads in the principal finite element equation according to Eq. (2.55), the transformation of the stiffness matrix into the global coordinate system is given as:

$$\underbrace{(\boldsymbol{T}^{\mathrm{T}}\boldsymbol{K}^{\mathrm{e}}_{xy}\boldsymbol{T})}_{\boldsymbol{K}^{\mathrm{e}}_{XY}}\,\boldsymbol{T}^{\mathrm{T}}\boldsymbol{u}_{xy} = \boldsymbol{T}^{\mathrm{T}}\boldsymbol{f}_{xy}\,, \tag{2.212}$$

or in components

$$\begin{bmatrix} \cos\alpha & 0 \\ \sin\alpha & 0 \\ 0 & \cos\alpha \\ 0 & \sin\alpha \end{bmatrix} \frac{EA}{L}\begin{bmatrix} 1 & -1 \\ -1 & 1 \end{bmatrix}\begin{bmatrix} \cos\alpha & \sin\alpha & 0 & 0 \\ 0 & 0 & \cos\alpha & \sin\alpha \end{bmatrix}. \tag{2.213}$$

The evaluation of this triple matrix product results finally in the stiffness matrix in the global X-Y coordinate system as:

$$\frac{AE}{L}\begin{bmatrix} \cos^2\alpha & \cos\alpha\,\sin\alpha & -\cos^2\alpha & -\cos\alpha\,\sin\alpha \\ \cos\alpha\,\sin\alpha & \sin^2\alpha & -\cos\alpha\,\sin\alpha & -\sin^2\alpha \\ -\cos^2\alpha & -\cos\alpha\,\sin\alpha & \cos^2\alpha & \cos\alpha\,\sin\alpha \\ -\cos\alpha\,\sin\alpha & -\sin^2\alpha & \cos\alpha\,\sin\alpha & \sin^2\alpha \end{bmatrix}\begin{bmatrix} u_{1X} \\ u_{1Y} \\ u_{2X} \\ u_{2Y} \end{bmatrix} = \begin{bmatrix} F_{1X} \\ F_{1Y} \\ F_{2X} \\ F_{2Y} \end{bmatrix}. \tag{2.214}$$

To simplify the solution of simple truss structures, Table 2.16 collects expressions for the global stiffness matrix for some common angles α.

It should be noted that the sine and cosine functions as well as the length L can be expressed based on the nodal coordinates (see Fig. 2.28a) as follows:

Table 2.16 Elemental stiffness matrices for truss elements in the X-Y plane given for different rotation angles α, cf. Eq. (2.214)

0°	180°
$\frac{EA}{L}\begin{bmatrix} 1 & 0 & -1 & 0 \\ 0 & 0 & 0 & 0 \\ -1 & 0 & 1 & 0 \\ 0 & 0 & 0 & 0 \end{bmatrix}$	$\frac{EA}{L}\begin{bmatrix} 1 & 0 & -1 & 0 \\ 0 & 0 & 0 & 0 \\ -1 & 0 & 1 & 0 \\ 0 & 0 & 0 & 0 \end{bmatrix}$
−30°	30°
$\frac{EA}{L}\begin{bmatrix} \frac{3}{4} & -\frac{1}{4}\sqrt{3} & -\frac{3}{4} & \frac{1}{4}\sqrt{3} \\ -\frac{1}{4}\sqrt{3} & \frac{1}{4} & \frac{1}{4}\sqrt{3} & -\frac{1}{4} \\ -\frac{3}{4} & \frac{1}{4}\sqrt{3} & \frac{3}{4} & -\frac{1}{4}\sqrt{3} \\ \frac{1}{4}\sqrt{3} & -\frac{1}{4} & -\frac{1}{4}\sqrt{3} & \frac{1}{4} \end{bmatrix}$	$\frac{EA}{L}\begin{bmatrix} \frac{3}{4} & \frac{1}{4}\sqrt{3} & -\frac{3}{4} & -\frac{1}{4}\sqrt{3} \\ \frac{1}{4}\sqrt{3} & \frac{1}{4} & -\frac{1}{4}\sqrt{3} & -\frac{1}{4} \\ -\frac{3}{4} & -\frac{1}{4}\sqrt{3} & \frac{3}{4} & \frac{1}{4}\sqrt{3} \\ -\frac{1}{4}\sqrt{3} & -\frac{1}{4} & \frac{1}{4}\sqrt{3} & \frac{1}{4} \end{bmatrix}$
−45°	45°
$\frac{EA}{L}\begin{bmatrix} \frac{1}{2} & -\frac{1}{2} & -\frac{1}{2} & \frac{1}{2} \\ -\frac{1}{2} & \frac{1}{2} & \frac{1}{2} & -\frac{1}{2} \\ -\frac{1}{2} & \frac{1}{2} & \frac{1}{2} & -\frac{1}{2} \\ \frac{1}{2} & -\frac{1}{2} & -\frac{1}{2} & \frac{1}{2} \end{bmatrix}$	$\frac{EA}{L}\begin{bmatrix} \frac{1}{2} & \frac{1}{2} & -\frac{1}{2} & -\frac{1}{2} \\ \frac{1}{2} & \frac{1}{2} & -\frac{1}{2} & -\frac{1}{2} \\ -\frac{1}{2} & -\frac{1}{2} & \frac{1}{2} & \frac{1}{2} \\ -\frac{1}{2} & -\frac{1}{2} & \frac{1}{2} & \frac{1}{2} \end{bmatrix}$
−90°	90°
$\frac{EA}{L}\begin{bmatrix} 0 & 0 & 0 & 0 \\ 0 & 1 & 0 & -1 \\ 0 & 0 & 0 & 0 \\ 0 & -1 & 0 & 1 \end{bmatrix}$	$\frac{EA}{L}\begin{bmatrix} 0 & 0 & 0 & 0 \\ 0 & 1 & 0 & -1 \\ 0 & 0 & 0 & 0 \\ 0 & -1 & 0 & 1 \end{bmatrix}$

$$\sin\alpha_{XY} = \frac{Y_2 - Y_1}{L}, \tag{2.215}$$

$$\cos\alpha_{XY} = \frac{X_2 - X_1}{L}, \tag{2.216}$$

$$L = \sqrt{(X_2 - X_1)^2 + (Y_2 - Y_1)^2}\,. \tag{2.217}$$

Let us consider in the following a slightly different configuration in which a rod element can now deformed in the global X-Z plane, Fig. 2.29. The local x-coordinate is rotated by an angle α against the global coordinate system (X, Z).

For this case, the global displacements can be expressed based on the local displacements as:

$$\underbrace{u_{1X}}_{>0} = \underbrace{\cos\alpha}_{>0}\ \underbrace{u_{1x}}_{>0}, \qquad \underbrace{u_{2X}}_{>0} = \underbrace{\cos\alpha}_{>0}\ \underbrace{u_{2x}}_{>0}, \tag{2.218}$$

$$\underbrace{u_{1Z}}_{>0} = -\underbrace{\sin\alpha}_{<0}\ \underbrace{u_{1x}}_{>0}, \qquad \underbrace{u_{2Z}}_{>0} = -\underbrace{\sin\alpha}_{<0}\ \underbrace{u_{2x}}_{>0}. \tag{2.219}$$

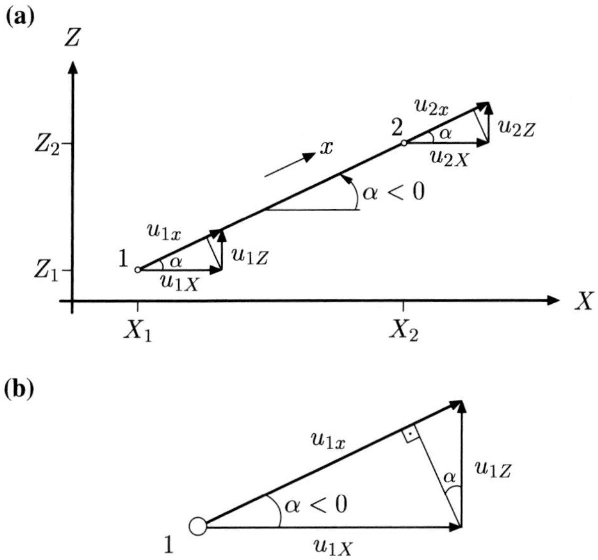

Fig. 2.29 Rotational transformation of a rod element in the X-Z plane: **a** total view and **b** detail for node 1

The last relationships between the global and local displacements can be written in matrix notation as

$$\begin{bmatrix} u_{1X} \\ u_{1Z} \\ u_{2X} \\ u_{2Z} \end{bmatrix} = \begin{bmatrix} \cos\alpha & 0 \\ -\sin\alpha & 0 \\ 0 & \cos\alpha \\ 0 & -\sin\alpha \end{bmatrix} \begin{bmatrix} u_{1x} \\ u_{2x} \end{bmatrix}, \tag{2.220}$$

or in abbreviated matrix notation as:

$$\boldsymbol{u}_{XZ} = \boldsymbol{T}^{\mathrm{T}} \boldsymbol{u}_{xz} \,. \tag{2.221}$$

The last equation can be solved for the displacements in the local coordinate system and inverting[24] the transformation matrix gives

$$\boldsymbol{u}_{xz} = \boldsymbol{T} \boldsymbol{u}_{XZ} \,, \tag{2.222}$$

[24]Since the transformation matrix is orthogonal, it follows that $\boldsymbol{T}^{\mathrm{T}} = \boldsymbol{T}^{-1}$.

or in components:

$$\begin{bmatrix} u_{1x} \\ u_{2x} \end{bmatrix} = \underbrace{\begin{bmatrix} \cos\alpha & -\sin\alpha & 0 & 0 \\ 0 & 0 & \cos\alpha & -\sin\alpha \end{bmatrix}}_{\boldsymbol{T}} \begin{bmatrix} u_{1X} \\ u_{1Z} \\ u_{2X} \\ u_{2Z} \end{bmatrix}. \tag{2.223}$$

It is possible to transform in a similar way the matrix of the external loads as:

$$\boldsymbol{f}_{XZ} = \boldsymbol{T}^{\mathrm{T}} \boldsymbol{f}_{xz}, \tag{2.224}$$

$$\boldsymbol{f}_{xz} = \boldsymbol{T} \boldsymbol{f}_{XZ}. \tag{2.225}$$

Considering the transformation of the local displacements and loads in the principal finite element equation according to Eq. (2.55), the transformation of the stiffness matrix into the global coordinate system is given as:

$$\underbrace{(\boldsymbol{T}^{\mathrm{T}} \boldsymbol{K}^{\mathrm{e}}_{xz} \boldsymbol{T})}_{\boldsymbol{K}^{\mathrm{e}}_{XZ}} \boldsymbol{T}^{\mathrm{T}} \boldsymbol{u}_{xz} = \boldsymbol{T}^{\mathrm{T}} \boldsymbol{f}_{xz}, \tag{2.226}$$

or in components

$$\begin{bmatrix} \cos\alpha & 0 \\ -\sin\alpha & 0 \\ 0 & \cos\alpha \\ 0 & -\sin\alpha \end{bmatrix} \frac{EA}{L} \begin{bmatrix} 1 & -1 \\ -1 & 1 \end{bmatrix} \begin{bmatrix} \cos\alpha & -\sin\alpha & 0 & 0 \\ 0 & 0 & \cos\alpha & -\sin\alpha \end{bmatrix}. \tag{2.227}$$

The evaluation of this triple matrix product results finally in the stiffness matrix in the global X-Z coordinate system as:

$$\frac{EA}{L} \begin{bmatrix} \cos^2\alpha & -\cos\alpha\sin\alpha & -\cos^2\alpha & \cos\alpha\sin\alpha \\ -\cos\alpha\sin\alpha & \sin^2\alpha & \cos\alpha\sin\alpha & -\sin^2\alpha \\ -\cos^2\alpha & \cos\alpha\sin\alpha & \cos^2\alpha & -\cos\alpha\sin\alpha \\ \cos\alpha\sin\alpha & -\sin^2\alpha & -\cos\alpha\sin\alpha & \sin^2\alpha \end{bmatrix} \begin{bmatrix} u_{1X} \\ u_{1Z} \\ u_{2X} \\ u_{2Z} \end{bmatrix} = \begin{bmatrix} F_{1X} \\ F_{1Z} \\ F_{2X} \\ F_{2Z} \end{bmatrix}. \tag{2.228}$$

To simplify the solution of simple truss structures in the X-Z plane, Table 2.17 collects expressions for the global stiffness matrix for some common angles α.

Let us note again that the sine and cosine functions as well as the length L can be expressed based on the nodal coordiantes (see Fig. 2.29a) as follows:

Table 2.17 Elemental stiffness matrices for truss elements in the X-Z plane given for different rotation angles α, cf. Eq. (2.228)

0°	180°
$\frac{EA}{L}\begin{bmatrix} 1 & 0 & -1 & 0 \\ 0 & 0 & 0 & 0 \\ -1 & 0 & 1 & 0 \\ 0 & 0 & 0 & 0 \end{bmatrix}$	$\frac{EA}{L}\begin{bmatrix} 1 & 0 & -1 & 0 \\ 0 & 0 & 0 & 0 \\ -1 & 0 & 1 & 0 \\ 0 & 0 & 0 & 0 \end{bmatrix}$
−30°	30°
$\frac{EA}{L}\begin{bmatrix} \frac{3}{4} & \frac{1}{4}\sqrt{3} & -\frac{3}{4} & -\frac{1}{4}\sqrt{3} \\ \frac{1}{4}\sqrt{3} & \frac{1}{4} & -\frac{1}{4}\sqrt{3} & -\frac{1}{4} \\ -\frac{3}{4} & -\frac{1}{4}\sqrt{3} & \frac{3}{4} & \frac{1}{4}\sqrt{3} \\ -\frac{1}{4}\sqrt{3} & -\frac{1}{4} & \frac{1}{4}\sqrt{3} & \frac{1}{4} \end{bmatrix}$	$\frac{EA}{L}\begin{bmatrix} \frac{3}{4} & -\frac{1}{4}\sqrt{3} & -\frac{3}{4} & \frac{1}{4}\sqrt{3} \\ -\frac{1}{4}\sqrt{3} & \frac{1}{4} & \frac{1}{4}\sqrt{3} & -\frac{1}{4} \\ -\frac{3}{4} & \frac{1}{4}\sqrt{3} & \frac{3}{4} & -\frac{1}{4}\sqrt{3} \\ \frac{1}{4}\sqrt{3} & -\frac{1}{4} & -\frac{1}{4}\sqrt{3} & \frac{1}{4} \end{bmatrix}$
−45°	45°
$\frac{EA}{L}\begin{bmatrix} \frac{1}{2} & \frac{1}{2} & -\frac{1}{2} & -\frac{1}{2} \\ \frac{1}{2} & \frac{1}{2} & -\frac{1}{2} & -\frac{1}{2} \\ -\frac{1}{2} & -\frac{1}{2} & \frac{1}{2} & \frac{1}{2} \\ -\frac{1}{2} & -\frac{1}{2} & \frac{1}{2} & \frac{1}{2} \end{bmatrix}$	$\frac{EA}{L}\begin{bmatrix} \frac{1}{2} & -\frac{1}{2} & -\frac{1}{2} & \frac{1}{2} \\ -\frac{1}{2} & \frac{1}{2} & \frac{1}{2} & -\frac{1}{2} \\ -\frac{1}{2} & \frac{1}{2} & \frac{1}{2} & -\frac{1}{2} \\ \frac{1}{2} & -\frac{1}{2} & -\frac{1}{2} & \frac{1}{2} \end{bmatrix}$
−90°	90°
$\frac{EA}{L}\begin{bmatrix} 0 & 0 & 0 & 0 \\ 0 & 1 & 0 & -1 \\ 0 & 0 & 0 & 0 \\ 0 & -1 & 0 & 1 \end{bmatrix}$	$\frac{EA}{L}\begin{bmatrix} 0 & 0 & 0 & 0 \\ 0 & 1 & 0 & -1 \\ 0 & 0 & 0 & 0 \\ 0 & -1 & 0 & 1 \end{bmatrix}$

$$\sin\alpha_{XZ} = -\frac{Z_2 - Z_1}{L}, \tag{2.229}$$

$$\cos\alpha_{XZ} = \frac{X_2 - X_1}{L}, \tag{2.230}$$

$$L = \sqrt{(X_2 - X_1)^2 + (Z_2 - Z_1)^2}\,. \tag{2.231}$$

2.4.2 Solved Truss Problems

2.5 Example: Truss structure arranged as an equilateral triangle

Given is the two-dimensional truss structure as shown in Fig. 2.30 where the trusses are arranged in the form of an equilateral triangle (all internal angles $\beta = 60°$). The three trusses have the same length L, the same YOUNG's modulus E, and the same cross-sectional area A. The structure is loaded by

(a) a horizontal force F_0 at node 2,
(b) a prescribed displacement u_0 at node 2.

Determine for both cases

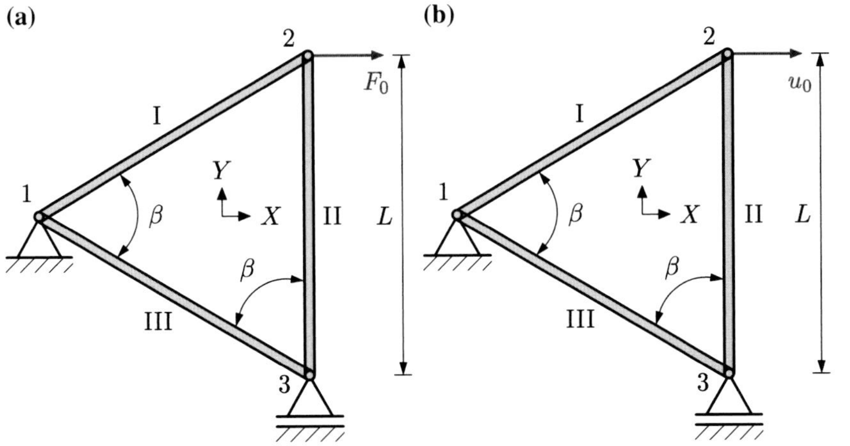

Fig. 2.30 Truss structure in the form of an equilateral triangle: **a** force boundary condition; **b** displacement boundary condition

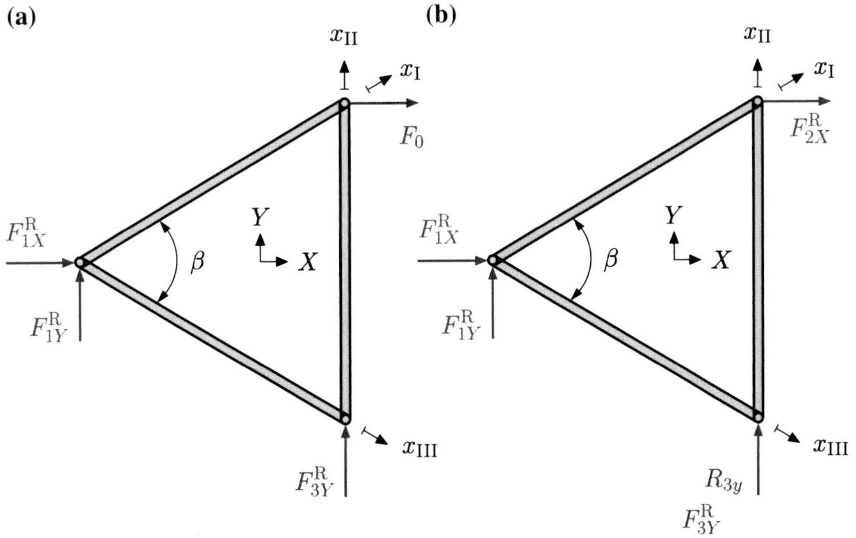

Fig. 2.31 Free-body diagram of the truss structure: **a** force boundary condition; **b** displacement boundary condition

- the global system of equations,
- the reduced system of equations,
- all nodal displacements,
- all reaction forces, and
- the force in each rod.

Table 2.18 Angles of rotation α_i and sine and cosine values for the problem shown in Fig. 2.31

Element	Angle of rotation	sine	cosine
I	30°	$\frac{1}{2}$	$\frac{\sqrt{3}}{2}$
II	90°	1	0
III	330°	$-\frac{1}{2}$	$\frac{\sqrt{3}}{2}$

2.5 Solution

The free-body diagram and the local coordinate axes of each element are shown in Fig. 2.31. From this figure, the rotational angles from the global to the local coordinate system can be determined and the sine and cosine values calculated as given in Table 2.18.

(a) Force boundary condition
Based on Eq. (2.214) and the values given in Table 2.18, the elemental stiffness matrices can be calculated as:

$$\boldsymbol{K}_{\mathrm{I}}^{\mathrm{e}} = \underbrace{\frac{EA}{L}}_{k_{\mathrm{I}}} \begin{bmatrix} \frac{3}{4} & \frac{\sqrt{3}}{4} & -\frac{3}{4} & -\frac{\sqrt{3}}{4} \\ \frac{\sqrt{3}}{4} & \frac{1}{4} & -\frac{\sqrt{3}}{4} & -\frac{1}{4} \\ -\frac{3}{4} & -\frac{\sqrt{3}}{4} & \frac{3}{4} & \frac{\sqrt{3}}{4} \\ -\frac{\sqrt{3}}{4} & -\frac{1}{4} & \frac{\sqrt{3}}{4} & \frac{1}{4} \end{bmatrix}, \tag{2.232}$$

$$\boldsymbol{K}_{\mathrm{II}}^{\mathrm{e}} = \underbrace{\frac{EA}{L}}_{k_{\mathrm{II}}} \begin{bmatrix} 0 & 0 & 0 & 0 \\ 0 & 1 & 0 & -1 \\ 0 & 0 & 0 & 0 \\ 0 & -1 & 0 & 1 \end{bmatrix}, \tag{2.233}$$

$$\boldsymbol{K}_{\mathrm{III}}^{\mathrm{e}} = \underbrace{\frac{EA}{L}}_{k_{\mathrm{III}}} \begin{bmatrix} \frac{3}{4} & -\frac{\sqrt{3}}{4} & -\frac{3}{4} & \frac{\sqrt{3}}{4} \\ -\frac{\sqrt{3}}{4} & \frac{1}{4} & \frac{\sqrt{3}}{4} & -\frac{1}{4} \\ -\frac{3}{4} & \frac{\sqrt{3}}{4} & \frac{3}{4} & -\frac{\sqrt{3}}{4} \\ \frac{\sqrt{3}}{4} & -\frac{1}{4} & -\frac{\sqrt{3}}{4} & \frac{1}{4} \end{bmatrix}, \tag{2.234}$$

which can be assembled to the global stiffness matrix as:

$$\frac{EA}{L}\begin{bmatrix} \frac{3}{4}+\frac{3}{4} & \frac{\sqrt{3}}{4}-\frac{\sqrt{3}}{4} & -\frac{3}{4} & -\frac{\sqrt{3}}{4} & -\frac{3}{4} & \frac{\sqrt{3}}{4} \\ \frac{\sqrt{3}}{4}-\frac{\sqrt{3}}{4} & \frac{1}{4}+\frac{1}{4} & -\frac{\sqrt{3}}{4} & -\frac{1}{4} & \frac{\sqrt{3}}{4} & -\frac{1}{4} \\ -\frac{3}{4} & -\frac{\sqrt{3}}{4} & \frac{3}{4} & \frac{\sqrt{3}}{4} & 0 & 0 \\ -\frac{\sqrt{3}}{4} & -\frac{1}{4} & \frac{\sqrt{3}}{4} & \frac{1}{4}+1 & 0 & -1 \\ -\frac{3}{4} & \frac{\sqrt{3}}{4} & 0 & 0 & \frac{3}{4} & -\frac{\sqrt{3}}{4} \\ \frac{\sqrt{3}}{4} & -\frac{1}{4} & 0 & -1 & -\frac{\sqrt{3}}{4} & 1+\frac{1}{4} \end{bmatrix} \begin{matrix} u_{1X} \\ u_{1Y} \\ u_{2X} \\ u_{2Y} \\ u_{3X} \\ u_{3Y} \end{matrix} . \tag{2.235}$$

Introducing the boundary conditions, i.e. $u_{1X} = u_{1Y} = u_{3Y} = 0$, gives the reduced system of equations as:

$$\frac{EA}{L}\begin{bmatrix} \frac{3}{4} & \frac{\sqrt{3}}{4} & 0 \\ \frac{\sqrt{3}}{4} & \frac{5}{4} & 0 \\ 0 & 0 & \frac{3}{4} \end{bmatrix}\begin{bmatrix} u_{2X} \\ u_{2Y} \\ u_{3X} \end{bmatrix} = \begin{bmatrix} F_0 \\ 0 \\ 0 \end{bmatrix} . \tag{2.236}$$

The solution of this system can be obtained, for example, by inverting the reduced stiffness matrix to give the reduced result matrix as:

$$\begin{bmatrix} u_{2X} \\ u_{2Y} \\ u_{3X} \end{bmatrix} = \frac{L}{EA}\begin{bmatrix} \frac{5}{3} & -\frac{\sqrt{3}}{3} & 0 \\ -\frac{\sqrt{3}}{3} & 1 & 0 \\ 0 & 0 & \frac{4}{3} \end{bmatrix}\begin{bmatrix} F_0 \\ 0 \\ 0 \end{bmatrix} = \frac{LF_0}{EA}\begin{bmatrix} \frac{5}{3} \\ -\frac{\sqrt{3}}{3} \\ 0 \end{bmatrix} . \tag{2.237}$$

The reaction forces can be obtained by multiplying the stiffness matrix according to Eq. (2.235) with the total displacement matrix, i.e.

$$\boldsymbol{u}^{\mathrm{T}} = \left[0\ 0\ u_{2X}\ u_{2Y}\ u_{3X}\ 0\right] , \tag{2.238}$$

to give:

$$F_{1X}^{\mathrm{R}} = -F_0\ ,\ \ F_{1Y}^{\mathrm{R}} = -\frac{\sqrt{3}\,F_0}{3}\ ,\ \ F_{3X}^{\mathrm{R}} = 0\ ,\ \ F_{3Y}^{\mathrm{R}} = \frac{\sqrt{3}\,F_0}{3} . \tag{2.239}$$

The rod forces can be obtained from the global coordinates as:

$$F_{\mathrm{I}} = k_{\mathrm{I}}\left(-\cos\alpha_{\mathrm{I}}u_{1X} - \sin\alpha_{\mathrm{I}}u_{1Y} + \cos\alpha_{\mathrm{I}}u_{2X} + \sin\alpha_{\mathrm{I}}u_{2Y}\right) = \frac{2\sqrt{3}\,F_0}{3} , \tag{2.240}$$

$$F_{\text{II}} = k_{\text{II}}\,(-\cos\alpha_{\text{II}} u_{3X} - \sin\alpha_{\text{II}} u_{3Y} + \cos\alpha_{\text{II}} u_{2X} + \sin\alpha_{\text{II}} u_{2Y}) = \frac{\sqrt{3}\,F_0}{3}, \quad (2.241)$$

$$F_{\text{III}} = k_{\text{III}}\,(-\cos\alpha_{\text{III}} u_{1X} - \sin\alpha_{\text{III}} u_{1Y} + \cos\alpha_{\text{III}} u_{3X} + \sin\alpha_{\text{III}} u_{3Y}) = 0\,. \quad (2.242)$$

(b) Displacement boundary condition
Considering the displacement boundary condition u at node 2, the reduced system of equations reads:

$$\frac{EA}{L}\begin{bmatrix} \frac{L}{EA} & 0 & 0 \\ \frac{\sqrt{3}}{4} & \frac{5}{4} & 0 \\ 0 & 0 & \frac{3}{4} \end{bmatrix}\begin{bmatrix} u_{2X} \\ u_{2Y} \\ u_{3X} \end{bmatrix} = \begin{bmatrix} u_0 \\ 0 \\ 0 \end{bmatrix}. \quad (2.243)$$

Inverting the reduced stiffness matrix can be used to calculate the unknown displacements as:

$$\begin{bmatrix} u_{2X} \\ u_{2Y} \\ u_{3X} \end{bmatrix} = \frac{L}{EA}\begin{bmatrix} \frac{EA}{L} & 0 & 0 \\ -\frac{\sqrt{3}EA}{5L} & \frac{4}{5} & 0 \\ 0 & 0 & \frac{4}{3} \end{bmatrix}\begin{bmatrix} u_0 \\ 0 \\ 0 \end{bmatrix} = u_0\begin{bmatrix} 1 \\ -\frac{\sqrt{3}}{5} \\ 0 \end{bmatrix}. \quad (2.244)$$

Reaction and rod forces can be obtained as described in part (a) as:

$$F^{\text{R}}_{1X} = -\frac{3}{5}\times\frac{EAu_0}{L}, \quad F^{\text{R}}_{1Y} = -\frac{\sqrt{3}}{5}\times\frac{EAu_0}{L}, \quad F^{\text{R}}_{2X} = \frac{3}{5}\times\frac{EAu_0}{L}, \quad (2.245)$$

$$F^{\text{R}}_{3X} = 0\,, \quad F^{\text{R}}_{3Y} = \frac{\sqrt{3}}{5}\times\frac{EAu_0}{L}. \quad (2.246)$$

$$F_{\text{I}} = \frac{2\sqrt{3}}{5}\times\frac{EAu_0}{L}, \quad F_{\text{II}} = \frac{\sqrt{3}}{5}\times\frac{EAu_0}{L}, \quad F_{\text{III}} = 0\,. \quad (2.247)$$

2.6 Example: Plane truss structure with two rod elements

The following Fig. 2.32 shows a two-dimensional truss structure. The two rod elements have the same cross-sectional area A and Young's modulus E. The length of each element can be calculated based on the given dimensions in the figure. The structure is loaded by prescribed displacements u_X and u_Y at node 2.
Determine:

- The global system of equations without consideration of the boundary conditions at node 1 and 3.
- The reduced system of equations.
- All nodal displacements.
- The elemental forces in each rod.

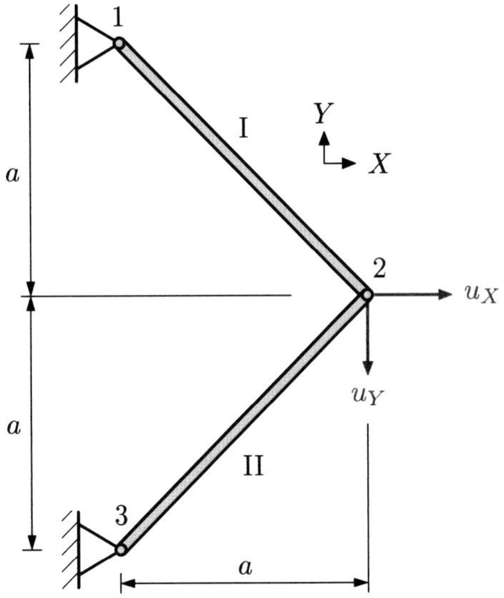

Fig. 2.32 Two-element truss structure with displacement boundary condition

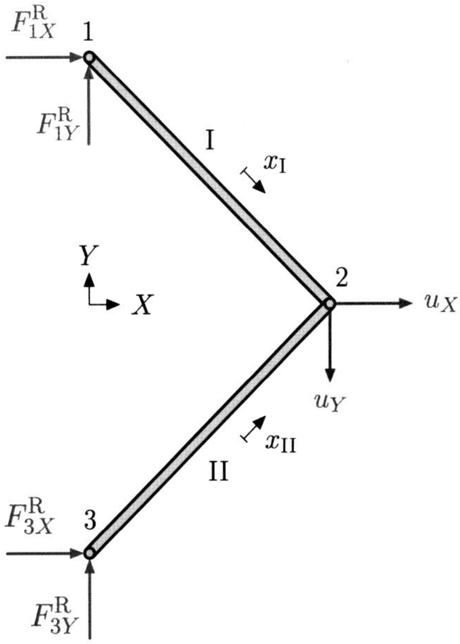

Fig. 2.33 Free-body diagram of the truss structure problem

2.6 Solution

The free-body diagram is shown in Fig. 2.33. Both elements have the same length of $L = \sqrt{2}a$

• Let us look in the following first separately at each element. The stiffness matrix for element I ($\alpha = -45°$) can be written as:

$$
K_I^e = \frac{EA}{\sqrt{2}a}
\begin{array}{c}
\begin{array}{cccc} u_{1X} & u_{1Y} & u_{2X} & u_{2Y} \end{array} \\
\begin{bmatrix}
\frac{1}{2} & -\frac{1}{2} & -\frac{1}{2} & \frac{1}{2} \\
-\frac{1}{2} & \frac{1}{2} & \frac{1}{2} & -\frac{1}{2} \\
-\frac{1}{2} & \frac{1}{2} & \frac{1}{2} & -\frac{1}{2} \\
\frac{1}{2} & -\frac{1}{2} & -\frac{1}{2} & \frac{1}{2}
\end{bmatrix}
\begin{array}{c} u_{1X} \\ u_{1Y} \\ u_{2X} \\ u_{2Y} \end{array}
\end{array}.
\tag{2.248}
$$

In the same way, the stiffness matrix for element II ($\alpha = +45°$) reads as:

$$
K_{II}^e = \frac{EA}{\sqrt{2}a}
\begin{array}{c}
\begin{array}{cccc} u_{3X} & u_{3Y} & u_{2X} & u_{2Y} \end{array} \\
\begin{bmatrix}
\frac{1}{2} & \frac{1}{2} & -\frac{1}{2} & -\frac{1}{2} \\
\frac{1}{2} & \frac{1}{2} & -\frac{1}{2} & -\frac{1}{2} \\
-\frac{1}{2} & -\frac{1}{2} & \frac{1}{2} & \frac{1}{2} \\
-\frac{1}{2} & -\frac{1}{2} & \frac{1}{2} & \frac{1}{2}
\end{bmatrix}
\begin{array}{c} u_{3X} \\ u_{3Y} \\ u_{2X} \\ u_{2Y} \end{array}
\end{array}.
\tag{2.249}
$$

The global system of equations without consideration of the boundary conditions is obtained as:

$$
\frac{EA}{\sqrt{2}a}
\begin{bmatrix}
\frac{1}{2} & -\frac{1}{2} & -\frac{1}{2} & \frac{1}{2} & 0 & 0 \\
-\frac{1}{2} & \frac{1}{2} & \frac{1}{2} & -\frac{1}{2} & 0 & 0 \\
-\frac{1}{2} & \frac{1}{2} & \frac{1}{2}+\frac{1}{2} & -\frac{1}{2}+\frac{1}{2} & -\frac{1}{2} & -\frac{1}{2} \\
\frac{1}{2} & -\frac{1}{2} & -\frac{1}{2}+\frac{1}{2} & \frac{1}{2}+\frac{1}{2} & -\frac{1}{2} & -\frac{1}{2} \\
0 & 0 & -\frac{1}{2} & -\frac{1}{2} & \frac{1}{2} & \frac{1}{2} \\
0 & 0 & -\frac{1}{2} & -\frac{1}{2} & \frac{1}{2} & \frac{1}{2}
\end{bmatrix}
\begin{bmatrix}
u_{1X} \\ u_{1Y} \\ u_{2X} \\ u_{2Y} \\ u_{3X} \\ u_{3Y}
\end{bmatrix}
=
\begin{bmatrix}
F_{1X}^R \\ F_{1Y}^R \\ 0 \\ 0 \\ F_{3X}^R \\ F_{3Y}^R
\end{bmatrix}.
\tag{2.250}
$$

- Introduction of the boundary conditions, i.e. $u_{1X} = u_{1Y} = u_{3X} = u_{3Y} = 0$, gives the following reduced system of equations:

$$
\frac{EA}{\sqrt{2}a}
\begin{bmatrix} 1 & 0 \\ 0 & 1 \end{bmatrix}
\begin{bmatrix} u_{2X} \\ u_{2Y} \end{bmatrix}
=
\begin{bmatrix} 0 \\ 0 \end{bmatrix}.
\tag{2.251}
$$

- All nodal displacements
  - $u_{1X} = u_{1Y} = u_{3X} = u_{3Y} = 0$,
  - $u_{2X} = u_X$, $u_{2Y} = u_Y$.

- Elemental forces in each rod

General: $\sigma = \dfrac{E}{L}(-u_1 + u_2) \Rightarrow F = \dfrac{EA}{L}(-u_1 + u_2)$.

Thus: $F = \dfrac{EA}{L}(-\cos(\alpha)u_{1X} - \sin(\alpha)u_{1Y} + \cos(\alpha)u_{2X} + \sin(\alpha)u_{2Y})$.

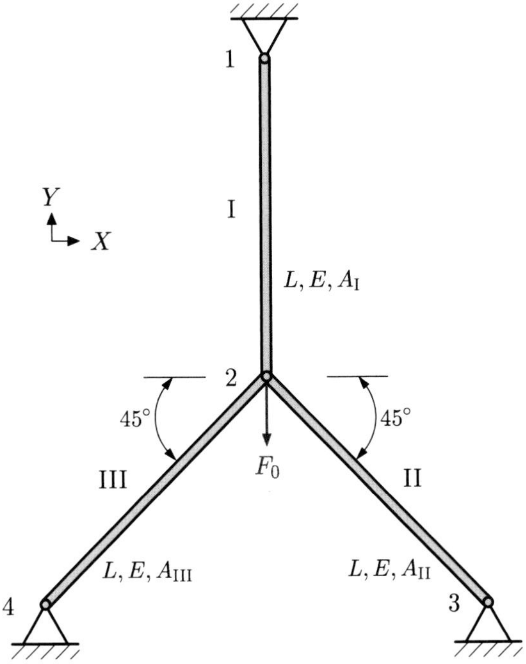

Fig. 2.34 Three-element truss structure with force boundary condition

Our case:

$$F_{\mathrm{I}} = \frac{EA}{\sqrt{2}a}\left(+\frac{1}{2}\sqrt{2}u_X - \frac{1}{2}\sqrt{2}u_Y\right) = \frac{EA}{2a}(u_X - u_Y)\,, \tag{2.252}$$

$$F_{\mathrm{II}} = \frac{EA}{\sqrt{2}a}\left(+\frac{1}{2}\sqrt{2}u_X + \frac{1}{2}\sqrt{2}u_Y\right) = \frac{EA}{2a}(u_X + u_Y)\,. \tag{2.253}$$

2.7 Example: Plane truss structure with three rod elements

The following Fig. 2.34 shows a two-dimensional truss structure. The three rod elements have the same YOUNG's modulus E and length L. However, the cross-sectional areas A_i $(i = \mathrm{I}, \mathrm{II}, \mathrm{III})$ are different from rod to rod. The structure is loaded by a point load F_0 at node 2.
Determine:

- the free body diagram,
- the global stiffness matrix,
- the reduced system of equations under consideration of the boundary conditions,
- the nodal displacements at node 2.
- Simplify the nodal displacements at node 2 for the special case $A_{\mathrm{I}} = A_{\mathrm{II}} = A_{\mathrm{III}} = A$.
- Calculate for the simplified case the vertical reaction forces at nodes 1, 3 and 4.

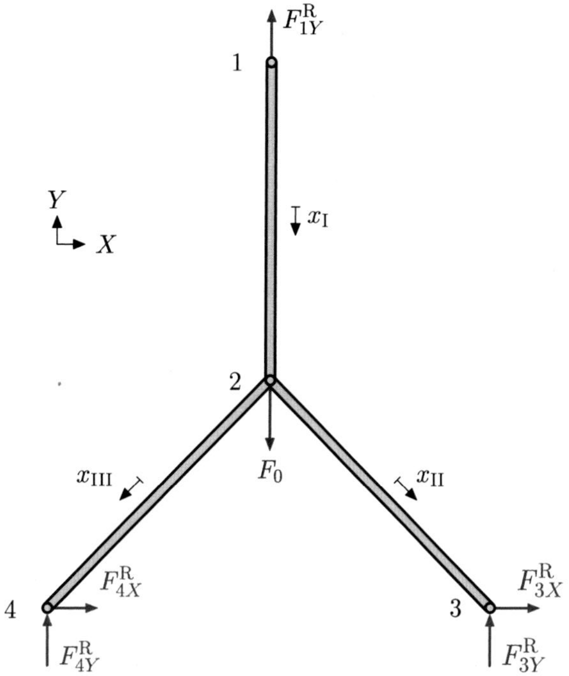

Fig. 2.35 Free body diagram of the truss structure problem

2.7 Solution

- The free body diagram is shown in Fig. 2.35.
- Let us look in the following first separately at each element. The stiffness matrix for element I ($\alpha = -90°$) can be written as:

$$\boldsymbol{K}^{\rm e}_{\rm I} = \frac{EA_{\rm I}}{L} \begin{matrix} u_{1X} & u_{1Y} & u_{2X} & u_{2Y} \\ \end{matrix} \begin{bmatrix} 0 & 0 & 0 & 0 \\ 0 & 1 & 0 & -1 \\ 0 & 0 & 0 & 0 \\ 0 & -1 & 0 & 1 \end{bmatrix} \begin{matrix} u_{1X} \\ u_{1Y} \\ u_{2X} \\ u_{2Y} \end{matrix} . \tag{2.254}$$

In the same way, the stiffness matrix for element II ($\alpha = -45°$) reads as:

$$\boldsymbol{K}^{\rm e}_{\rm II} = \frac{EA_{\rm II}}{L} \begin{matrix} u_{2X} & u_{2Y} & u_{3X} & u_{3Y} \\ \end{matrix} \begin{bmatrix} \frac{1}{2} & -\frac{1}{2} & -\frac{1}{2} & \frac{1}{2} \\ -\frac{1}{2} & \frac{1}{2} & \frac{1}{2} & -\frac{1}{2} \\ -\frac{1}{2} & \frac{1}{2} & \frac{1}{2} & -\frac{1}{2} \\ \frac{1}{2} & -\frac{1}{2} & -\frac{1}{2} & \frac{1}{2} \end{bmatrix} \begin{matrix} u_{2X} \\ u_{2Y} \\ u_{3X} \\ u_{3Y} \end{matrix} . \tag{2.255}$$

In the same way, the stiffness matrix for element III ($\alpha = 225°$) reads as:

$$\boldsymbol{K}^{\text{e}}_{\text{III}} = \frac{EA_{\text{III}}}{L} \begin{array}{c} \begin{matrix} u_{2X} & u_{2Y} & u_{4X} & u_{4Y} \end{matrix} \\ \begin{bmatrix} \frac{1}{2} & \frac{1}{2} & -\frac{1}{2} & -\frac{1}{2} \\ \frac{1}{2} & \frac{1}{2} & -\frac{1}{2} & -\frac{1}{2} \\ -\frac{1}{2} & -\frac{1}{2} & \frac{1}{2} & \frac{1}{2} \\ -\frac{1}{2} & -\frac{1}{2} & \frac{1}{2} & \frac{1}{2} \end{bmatrix} \end{array} \begin{matrix} u_{2X} \\ u_{2Y} \\ u_{4X} \\ u_{4Y} \end{matrix} . \tag{2.256}$$

The global stiffness matrix can be assembled as:

$$\boldsymbol{K} = \frac{E}{L} \begin{bmatrix} 0 & 0 & 0 & 0 & 0 & 0 & 0 & 0 \\ 0 & A_{\text{I}} & 0 & -A_{\text{I}} & 0 & 0 & 0 & 0 \\ 0 & 0 & \frac{A_{\text{II}}}{2} + \frac{A_{\text{III}}}{2} & -\frac{A_{\text{II}}}{2} + \frac{A_{\text{III}}}{2} & -\frac{A_{\text{II}}}{2} & \frac{A_{\text{II}}}{2} & -\frac{A_{\text{III}}}{2} & -\frac{A_{\text{III}}}{2} \\ 0 & -A_{\text{I}} & -\frac{A_{\text{II}}}{2} + \frac{A_{\text{III}}}{2} & A_{\text{I}} + \frac{A_{\text{II}}}{2} + \frac{A_{\text{III}}}{2} & \frac{A_{\text{II}}}{2} & -\frac{A_{\text{II}}}{2} & -\frac{A_{\text{III}}}{2} & -\frac{A_{\text{III}}}{2} \\ 0 & 0 & -\frac{A_{\text{II}}}{2} & \frac{A_{\text{II}}}{2} & \frac{A_{\text{II}}}{2} & -\frac{A_{\text{II}}}{2} & 0 & 0 \\ 0 & 0 & \frac{A_{\text{II}}}{2} & -\frac{A_{\text{II}}}{2} & -\frac{A_{\text{II}}}{2} & \frac{A_{\text{II}}}{2} & 0 & 0 \\ 0 & 0 & -\frac{A_{\text{III}}}{2} & -\frac{A_{\text{III}}}{2} & 0 & 0 & \frac{A_{\text{III}}}{2} & \frac{A_{\text{III}}}{2} \\ 0 & 0 & -\frac{A_{\text{III}}}{2} & -\frac{A_{\text{III}}}{2} & 0 & 0 & \frac{A_{\text{III}}}{2} & \frac{A_{\text{III}}}{2} \end{bmatrix} . \tag{2.257}$$

- Introduction of the boundary conditions, i.e. $u_{1X} = u_{1Y} = u_{3X} = u_{3Y} = u_{4X} = u_{4Y} = 0$, gives the following reduced system of equations:

$$\frac{E}{L} \begin{bmatrix} \frac{A_{\text{II}}}{2} + \frac{A_{\text{III}}}{2} & -\frac{A_{\text{II}}}{2} + \frac{A_{\text{III}}}{2} \\ -\frac{A_{\text{II}}}{2} + \frac{A_{\text{III}}}{2} & A_{\text{I}} + \frac{A_{\text{II}}}{2} + \frac{A_{\text{III}}}{2} \end{bmatrix} \begin{bmatrix} u_{2X} \\ u_{2Y} \end{bmatrix} = \begin{bmatrix} 0 \\ -F_0 \end{bmatrix} . \tag{2.258}$$

The solution of this system can be obtained, for example, by inverting the reduced stiffness matrix to give the reduced result matrix as:

$$\begin{bmatrix} u_{2X} \\ u_{2Y} \end{bmatrix} = \begin{bmatrix} -\dfrac{L\,(A_{\text{II}} - A_{\text{III}})\,F_0}{E\,(A_{\text{II}}\,A_{\text{I}} + A_{\text{III}}\,A_{\text{I}} + 2\,A_{\text{II}}\,A_{\text{III}})} \\ -\dfrac{L\,(A_{\text{II}} + A_{\text{III}})\,F_0}{E\,(A_{\text{II}}\,A_{\text{I}} + A_{\text{III}}\,A_{\text{I}} + 2\,A_{\text{II}}\,A_{\text{III}})} \end{bmatrix} . \tag{2.259}$$

- Special case $A_{\text{I}} = A_{\text{II}} = A_{\text{III}} = A$:

$$\begin{bmatrix} u_{2X} \\ u_{2Y} \end{bmatrix} = \begin{bmatrix} 0 \\ -\dfrac{L F_0}{2EA} \end{bmatrix} . \tag{2.260}$$

- Vertical reaction forces at nodes 1, 3 and 4 for the special case:

The evaluation of the second, sixth and eight equation of the non-reduced system of equations (see Eq. (2.257)) gives:

$$F_{1Y}^{\mathrm{R}} = \frac{F_0}{2}, \quad F_{3Y}^{\mathrm{R}} = F_{4Y}^{\mathrm{R}} = \frac{F_0}{4}. \tag{2.261}$$

Let us summarize at the end of this section the recommended steps for a linear finite element solution ('hand calculation'):

❶ Sketch the free-body diagram of the problem, including a global coordinate system.

❷ Subdivide the geometry into finite elements. Indicate the node and element numbers, local coordinate systems, and equivalent nodal loads.

❸ Write separately all elemental stiffness matrices expressed in the global coordinate system. Indicate the nodal unknowns on the right-hand sides and over the matrices.

❹ Determine the dimensions of the global stiffness matrix and sketch the structure of this matrix with global unknowns on the right-hand side and over the matrix.

❺ Insert step-by-step the values of the elemental stiffness matrices into the global stiffness matrix.

❻ Add the column matrix of unknowns and external loads to complete the global system of equations.

❼ Introduce the boundary conditions to obtain the reduced system of equations.

❽ Solve the reduced system of equations to obtain the unknown nodal deformations.

❾ Post-computation: determination of reaction forces, stresses and strains.

❿ Check the global equilibrium between the external loads and the support reactions.

It should be noted that some steps may be combined or omitted depending on the problem and the experience of the finite element user. The above steps can be seen as an initial structured guide to master a finite element problem. A comprehensive collection of exercises, which are solved based on these 10 steps, is given in [4, 7].

2.5 Supplementary Problems

2.8 Knowledge questions on rods and trusses

- How many material parameters are required for the one-dimensional HOOKE's law?
- State the one-dimensional HOOKE's law for a pure normal stress and strain state.
- HOOKE's law can be written as $\sigma(x) = E\varepsilon(x)$ for a special case. State two assumptions for this formulation.
- Explain the assumptions for (a) an 'isotropic' and (b) a 'homogeneous' material.
- State the major characteristic of an *elastic* material.
- State a common value for the YOUNG's modulus of steel, aluminum, and titanium.
- The following Fig. 2.36 shows two structural members of the same length L. State for this problem the condition between the spring constant k and properties E, A and L to obtain the same mechanical response.
- Consider again the two structural members shown in Fig. 2.36. Assume linear-elastic material behavior and sketch an ideal force-displacement diagram for the spring and a stress-strain diagram for the rod. Indicate the material constants in the diagrams. What is represented by the areas under the graphs?
- State the three (3) basic equations of continuum mechanics which are required to derive the partial differential equation of a static problem.
- Explain the meaning of the kinematics, constitutive and equilibrium equations.
- Name the primary unknown in the partial differential equation of a rod member.
- Explain in words the meaning of the strong formulation, the inner product, and the weak formulation in the scope of the weighted residual method.
- Given is a differential equation of the form $\mathrm{d}^2 f(x)/\mathrm{d}x^2 - a = 0$. State (a) the strong formulation and (b) the inner product of the problem.
- Given is a differential equation of the form $\mathrm{d}^2 y(x)/\mathrm{d}x^2 - c(x) = 0$. State (a) the strong formulation and (b) the inner product of the problem.
- State the difference between the (a) analytical and (b) the finite element solution of a problem described based on a partial differential equation.
- State in words the definition of a rod.
- Characterize in words the stress and strain distribution in an elastic rod.

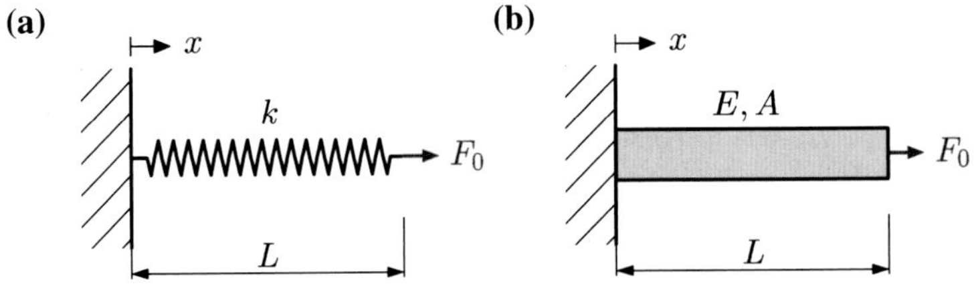

Fig. 2.36 Simple structural members: **a** spring and **b** rod

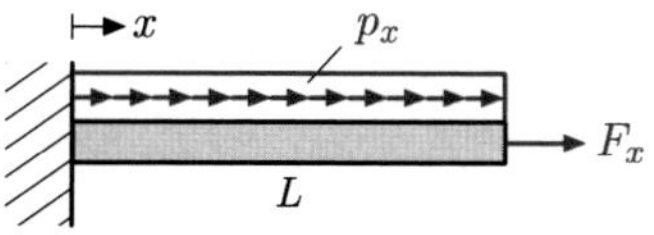

Fig. 2.37 Axially loaded continuum rod

- Sketch (a) the normal strain and (b) the normal stress distribution in the square cross section of a rod under tensile load. Explain in words how these quantities are connected.
- The following Fig. 2.37 shows a rod of length L and constant cross-sectional area A. The structure is loaded by a point load F_0 and a constant distributed load p_x. State for this problem three boundary conditions and the appropriate differential equation under the assumption that the YOUNG'S modulus E is a function of the spatial coordinate x.
- Which general types of 'load conditions' did we distinguish for rod problems?
- State the required (a) geometrical parameters and (b) material parameters to define a rod element.
- Sketch the interpolation functions $N_1(x)$ and $N_2(x)$ of a linear rod element.
- State four (4) characteristics of a finite element stiffness matrix.
- The stiffness matrix for a rod element can be stated as

$$\boldsymbol{K}^{\mathrm{e}} = \frac{EA}{L}\begin{bmatrix} 1 & -1 \\ -1 & 1 \end{bmatrix}.$$

  Which assumptions does this equation involve in regards to the (a) material and (b) the geometry?
- State the DOF per node for a truss element in a plane (2D) problem.
- State the DOF per node for a truss element in a 3D problem.
- The following Fig. 2.38a shows a *plane* truss structure which is composed of 15 rod (E, A) elements.
  State the size of the stiffness matrix of the *non-reduced* system of equations, i.e. without consideration of the boundary conditions. What is the size of the stiffness matrix of the *reduced* system of equations, i.e. under consideration of the boundary conditions? Consider now Fig. 2.38b where the rod element 1–2 (length L) has been replaced by a spring of stiffness $k = \frac{EA}{L}$. How does the overall stiffness of the truss structure change?
- The following Fig. 2.39 shows a *plane* truss structure which is composed of 2 rod elements.
  State the dimensions of the global stiffness matrix of the *non-reduced* system of equations, i.e. without consideration of the support conditions. What are the dimensions of the stiffness matrix of the *reduced* system of equations, i.e. under consideration of the support conditions? Which rod is higher loaded (consider only the magnitude)?

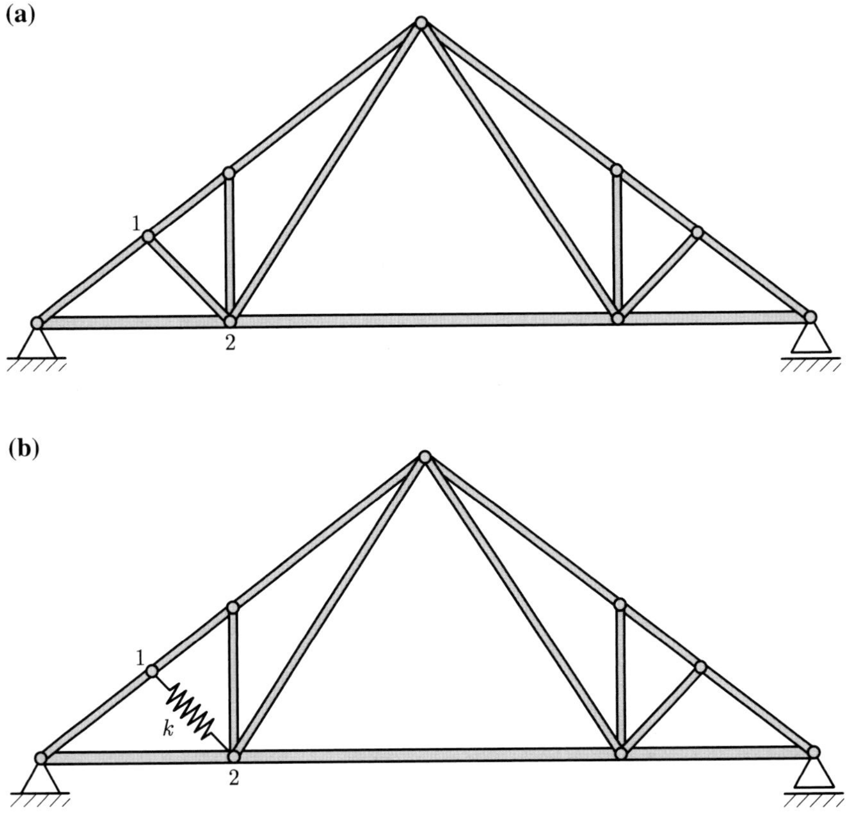

Fig. 2.38 Plane truss structure: **a** only rod elements, **b** one rod replaced by a spring. Nodes are symbolized by circles (◯)

Fig. 2.39 Plane truss structure. Nodes are symbolized by circles (∘)

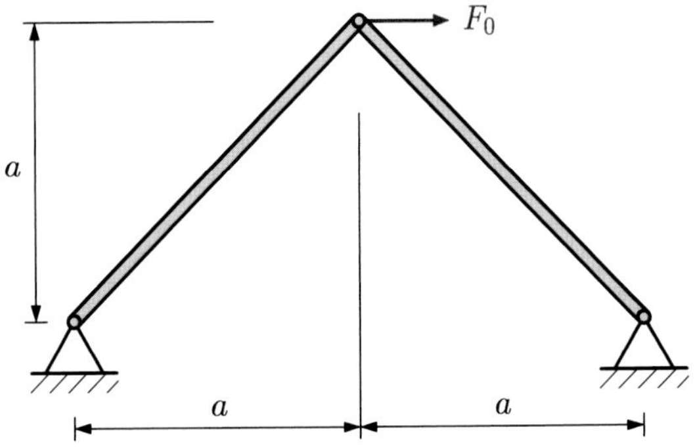

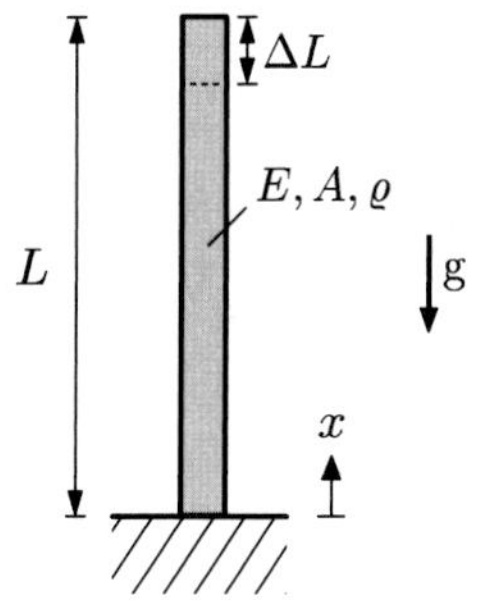

Fig. 2.40 Simplified model of a tower loaded under its dead weight

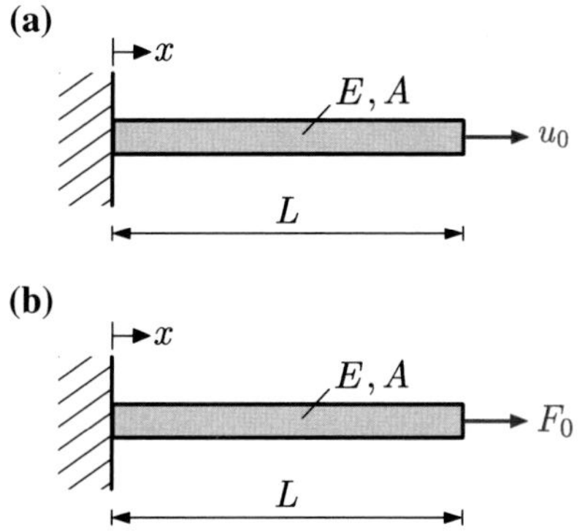

Fig. 2.41 Rod under different loading conditions: **a** displacement and **b** force

2.9 Simplified model of a tower under dead weight (analytical approach)

Given is a simplified model of a tower which is deforming under the influence of its dead weight, cf. Fig. 2.40. The tower is of original length L, cross-sectional area A, YOUNGs modulus E, and mass density ϱ. The standard gravity is given by g. Calculate

- The stress distribution $\sigma_x(x)$ in the member.
- The reduced length $L' = L - u_x(L)$ due to the acting dead weight.
- The maximum length $L_{\max}$ if a given stress limit $\sigma_{\max}$ at the foundation ($x = 0$) cannot be exceeded.

2.10 Analytical solution for a rod problem

Given is a rod of length L and constant axial tensile stiffness EA as shown in Fig. 2.41. At the left-hand side there is a fixed support and the right-hand side is either elongated by a displacement u_0 (case a) or loaded by a single force F_0 (case b). Determine the analytical solution for the elongation $u(x)$, the strain $\varepsilon(x)$, and stress $\sigma(x)$ along the rod axis.

2.11 Weighted residual method based on general formulation of partial differential equation

Derive the weak formulation for a rod based on the general formulation of the partial differential equation:

$$\mathcal{L}_1\left(C\mathcal{L}_1\left(u_x(x)\right)\right) + b = 0\,, \tag{2.262}$$

where $\mathcal{L}_1 = \frac{\mathrm{d}}{\mathrm{d}x}$, $C = E$ and $b = \frac{p_x(x)}{A}$. Simplify the GREEN–GAUSS theorem as given in Eq. (A.27) to derive the solution.

2.12 Weighted residual method with arbitrary distributed load for a rod
Derive the principal finite element equation for a rod element based on the weighted residual method. Starting point should be the partial differential equation with an arbitrary distributed load $p_x(x)$. In addition, it can be assumed that the axial tensile stiffness EA is constant.

2.13 Numerical integration and coordinate transformation
The derivation of the principal finite element equation involves numerical integration and coordinate transformation. The Cartesian coordinate range $x_1 \leq x \leq x_2$ is transformed to the natural coordinate range $-1 \leq \xi \leq 1$. The general transformation between these two coordinates is illustrated in Table 2.4 and given by:

$$\xi = \frac{2}{x_2 - x_1}(x - x_1) - 1\,. \tag{2.263}$$

Derive this relationship between the Cartesian and the natural coordinate.

2.14 Finite element solution for a rod problem
Given is a rod of length L and constant axial tensile stiffness EA as shown in Fig. 2.42. At the left-hand side there is a fixed support and the right-hand side is either loaded by a single force F_0 (case a) or elongated by a displacement u_0 (case b). Determine the finite element solution based on a single rod element for the elongation $u(x)$, the strain $\varepsilon(x)$, and stress $\sigma(x)$ along the rod axis.

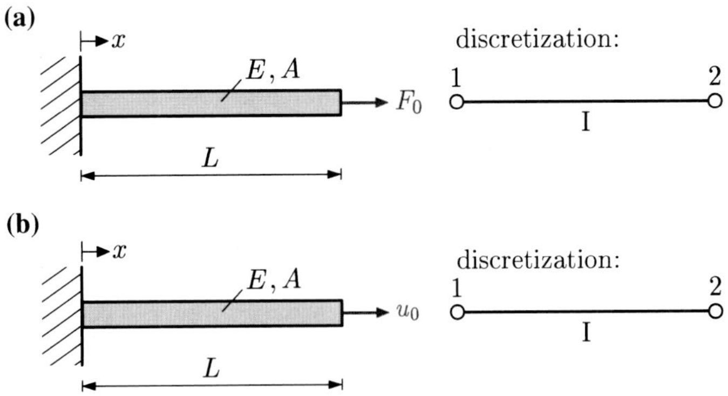

Fig. 2.42 Rod under different loading conditions: **a** force and **b** displacement boundary condition

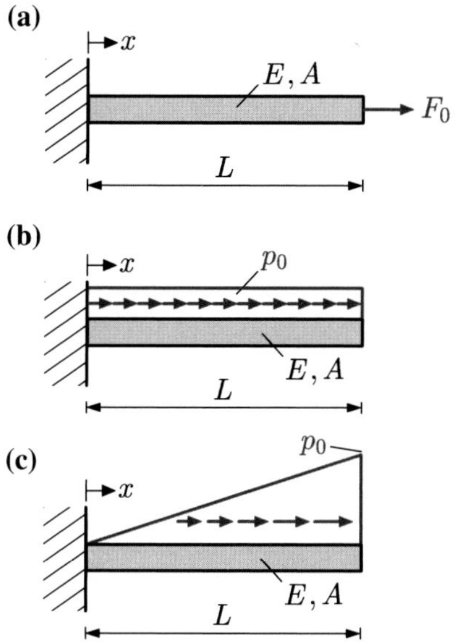

Fig. 2.43 Finite element approximation with a single element for different load cases: **a** single force, **b** constant distributed load, and **c** linear distributed load

2.15 Finite element approximation with a single linear rod element
Given is a rod with different load cases as shown in Fig. 2.43. The axial tensile stiffness EA is constant and the length is equal to L. Derive the finite element solution based on a single linear element and compare the elongation $u_x(x)$ and $u_x(L)$ with the analytical solution.

2.16 Different formulations for the displacement field of a linear rod element
The nodal approach allows to express the displacement field within an element based on the nodal displacement values as

$$u^{\mathrm{e}}(x) = N_1(x)u_1 + N_2(x)u_2 \,. \tag{2.264}$$

Alternatively, one may use the more general linear formulation:

$$u^{\mathrm{e}}(x) = a_0 + a_1 x \,. \tag{2.265}$$

Transform the nodal approach into the second formulation.

2.17 Finite element approximation with a single quadratic rod element
Solve problem 2.15 with a single quadratic rod element.

2.18 Equivalent nodal loads for a quadratic distribution (linear rod element)
Given are the following two formulations for a distributed quadratic load. Calculate the equivalent nodal loads for a linear rod element, cf. Fig. 2.44.

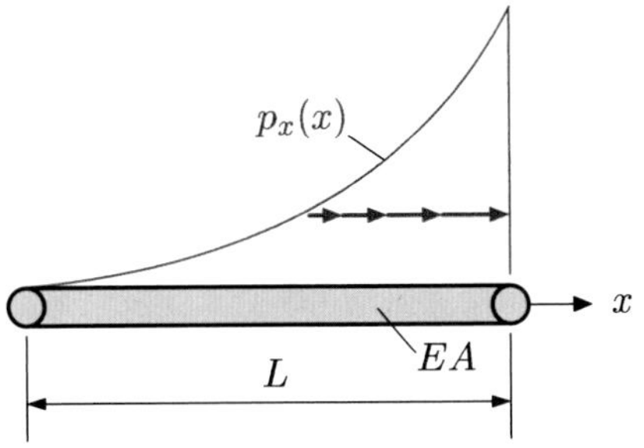

Fig. 2.44 Distributed load with quadratic function

(a) $p_x(x) = p_0^* x^2$,

(b) $p_x(x) = p_0 \left(\frac{x}{L}\right)^2$.

The dimension of the constant p_0^* is equal to force per unit length to power 3 while the dimension of p_0 is force per unit length.

2.19 Derivation of interpolation functions for a quadratic rod element
Derive the three interpolation functions for a quadratic rod element under the following assumption for the displacement field:

$$u^{\mathrm{e}}(\xi) = a_0 + a_1\xi + a_2\xi^2 . \tag{2.266}$$

Assume for the derivation that the third node is exactly in the middle of the element ($\xi = 0$). Plot the three interpolation functions in dependence of the natural coordinate ξ.

2.20 Derivation of the Jacobian determinant for a quadratic rod element
Consider a quadratic rod element with the second node located exactly in the middle of the element. Use the following nodal approach for the Cartesian coordinate to calculate the Jacobian determinant for the case that the elemental coordinate system is located in node 1:

$$x(\xi) = \overline{N}_1(\xi)x_1 + \overline{N}_2(\xi)x_2 + \overline{N}_3(\xi)x_3 . \tag{2.267}$$

2.21 Comparison of the stress distribution for a linear and quadratic rod element with linear increasing load

Given is a rod with linear increasing load as shown in Fig. 2.45. The axial tensile stiffness EA is constant and the length is equal to L. Calculate and compare the stress distribution based on:

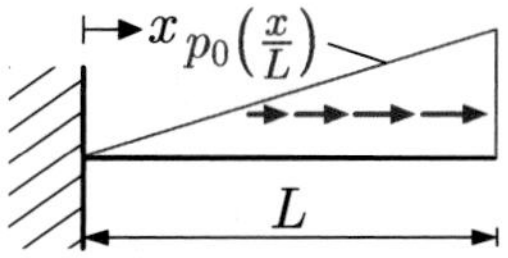

Fig. 2.45 Rod element with linear increasing load

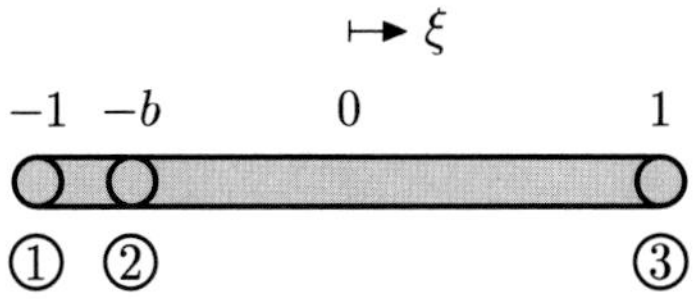

Fig. 2.46 Quadratic rod element with unevenly distributed nodes

(a) the analytical solution,
(b) a single *linear* rod element and
(c) a single *quadratic* rod element.

2.22 Derivation of interpolation functions and stiffness matrix for a quadratic rod element with unevenly distributed nodes
Derive the three interpolation functions for a quadratic rod element with unevenly distributed nodes (cf. Fig. 2.46) under the following assumption for the displacement field:

$$u^{\mathrm{e}}(\xi) = a_0 + a_1\xi + a_2\xi^2 \,. \tag{2.268}$$

Derive the stiffness matrix as a function of position b for an axial stiffness EA and length L of the element. Wich problems can occur if the second node is close to the boundary, e.g. $\xi = -0.9$?

2.23 Derivation of interpolation functions for a cubic rod element
Derive the four interpolation functions for a cubic rod element under the following assumption for the displacement field:

$$u^{\mathrm{e}}(\xi) = a_0 + a_1\xi + a_2\xi^2 + a_3\xi^3 \,. \tag{2.269}$$

Assume for the derivation that the nodes are equally spaced.

2.24 Structure composed of three linear rod elements
Calculate for the two structures shown in Fig. 2.47 the unknown displacement matrix and the reaction forces for $L_{\mathrm{I}} = L_{\mathrm{II}} = L_{\mathrm{III}} = L$ and

$$(EA)_{\mathrm{I}} = 3EA \,, \tag{2.270}$$

$$(EA)_{\mathrm{II}} = 2EA \,, \tag{2.271}$$

$$(EA)_{\mathrm{III}} = 1EA \,. \tag{2.272}$$

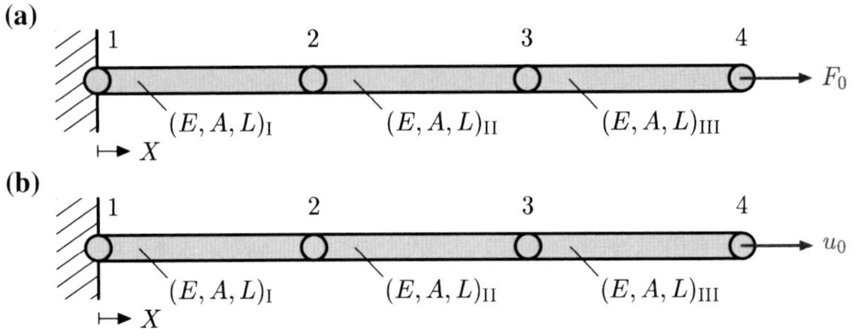

Fig. 2.47 Structure composed of three rod elements: **a** force boundary condition; **b** displacement boundary condition

Fig. 2.48 Rod structure discretized by four elements

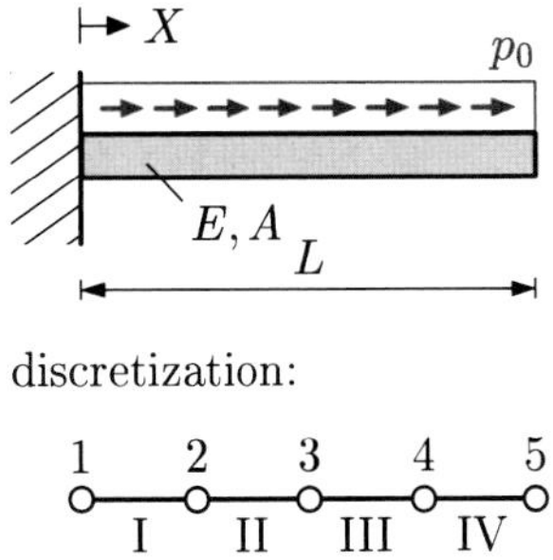

2.25 Finite element approximation of a rod with four elements: comparison of displacement, strain and stress distribution with analytical solution

Given is a rod of length L and tensile stiffness EA which is loaded by a constant distributed load p_0 as shown in Fig. 2.48. Use four linear rod elements of length $\frac{L}{4}$ to discretize the rod structure and calculate the nodal displacements, strains and stresses. Compare the results with the analytical solution and sketch the normalized finite element and analytical solutions $u(X)/\frac{p_0 L^2}{EA}$, $\varepsilon(X)/\frac{p_0 L}{EA}$, and $\sigma(X)/\frac{p_0 L}{A}$ over the normalized coordinate X/L.

2.26 Elongation of a bi-material rod: finite element solution and comparison with analytical solution

Given is a rod as shown in Fig. 2.49 which is made of two different sections with axial stiffness $k_{\mathrm{I}} = E_{\mathrm{I}} A_{\mathrm{I}}$ and $k_{\mathrm{II}} = E_{\mathrm{II}} A_{\mathrm{II}}$. Each section is of length L and in the left-hand section, i.e. $0 \leq x \leq L$, is a constant distributed load p_0 acting while the right-hand end is elongated by u_0. Use four linear rod elements of length $\frac{L}{4}$ to discretize the rod and calculate the nodal displacements, strains, and stresses. Compare the results with the analytical solution and sketch the distributions $u(x)$, $\varepsilon(x)$ and $\sigma(x)$ for the case $k_{\mathrm{I}} = 2k_{\mathrm{II}} = 1$, $L_{\mathrm{I}} = L_{\mathrm{II}} = 1$, $p_0 = 1$, $u_0 = 1$, and $E_{\mathrm{I}} = 2E_{\mathrm{II}} = 1$.

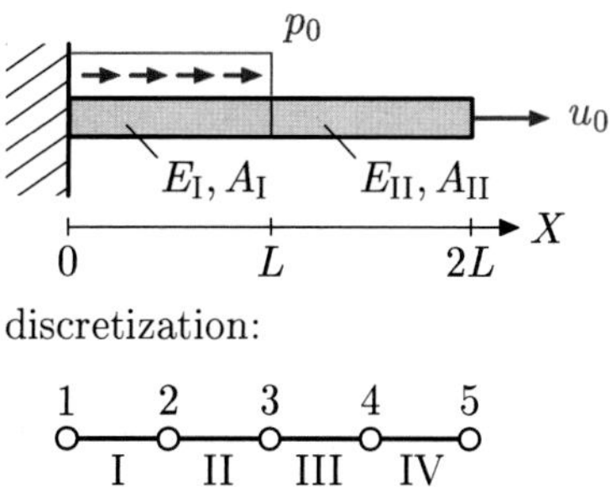

Fig. 2.49 Bi-material rod discretized by four elements

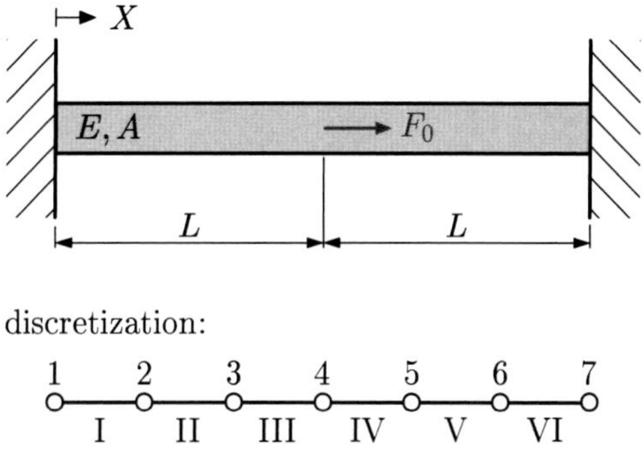

Fig. 2.50 Rod structure fixed at both ends

2.27 Stress distribution for a fixed-fixed rod structure

Given is a rod structure as shown in Fig. 2.50. The structure is of length $2L$, cross-sectional area A, and YOUNG's modulus E. The structure is fixed at both ends and loaded by a point load F_0 in the middle, i.e. $X = L$. Calculate the stress distribution based on six finite elements of length $\frac{L}{3}$. Show the difference between the elemental stress values and the averaged nodal values.

2.28 Linear rod element with variable cross section: derivation of stiffness matrix

Determine the elemental stiffness matrix for a linear rod element with changing cross-sectional area as shown in Fig. 2.51. Consider the following two relationships for a linear changing diameter and a linear changing area:

$$\text{(a)}\;\; d(x) = d_1 + \frac{x}{L}(d_2 - d_1)\,, \tag{2.273}$$

$$\text{(b)}\;\; A(x) = A_1 + \frac{x}{L}(A_2 - A_1)\,. \tag{2.274}$$

Use analytical integration to obtain the stiffness matrix and compare the results with a two-point GAUSS integration rule. A circular cross section can be assumed in case (a).

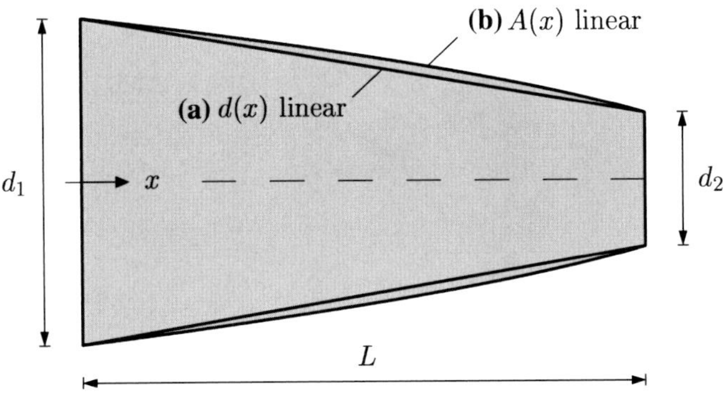

Fig. 2.51 Rod element with variable cross section: **a** linear changing diameter; **b** linear changing cross-sectional area

2.29 Quadratic rod element with variable cross section: derivation of stiffness matrix
Solve problem 2.28 with a single quadratic rod element.

2.30 Linear rod element with variable cross-section: comparison of displacements between FE and analytical solution for a single element
Determine for the rod element shown in Fig. 2.52 the end displacement based on a single finite element. The elemental stiffness matrix from problem 2.28 can be used. Distinguish two different cases of boundary conditions, i.e. a single force F_0 or a prescribed displacement u_0 at the right-hand end. Compare the results obtained with the analytical solution.

2.31 Quadratic rod element with variable cross section: comparison of end displacement between FE and analytical solution for single element
Recalculate problem 2.30 (a) for a single quadratic rod element.

2.32 Subdivided structure with variable cross section: comparison of displacements and stresses between FE and analytical solution for four elements
Calculate for the rod shown in Fig. 2.53 the distribution of the elongation $u(X)$ and the stress $\sigma(X)$. To this end, subdivide the structure in four elements of length $\frac{L}{4}$ and use the expression for the elemental stiffness matrix which was derived in problem 2.28. The ratio between the end and initial cross section area is equal to $\frac{A_5}{A_1} = 0.2$. Compare the results obtained with the analytical solution.

2.33 Submodel of a structure with variable cross section
To increase the accuracy of a stress approximation, the submodeling technique can be applied. Let us come back to Problem 2.32 and assume that the area of interest is near the left-hand support, i.e. $X \rightarrow 0$. In the first step of a submodeling analysis, the structure is simulated with a coarse mesh as indicated in Fig. 2.53. In the next

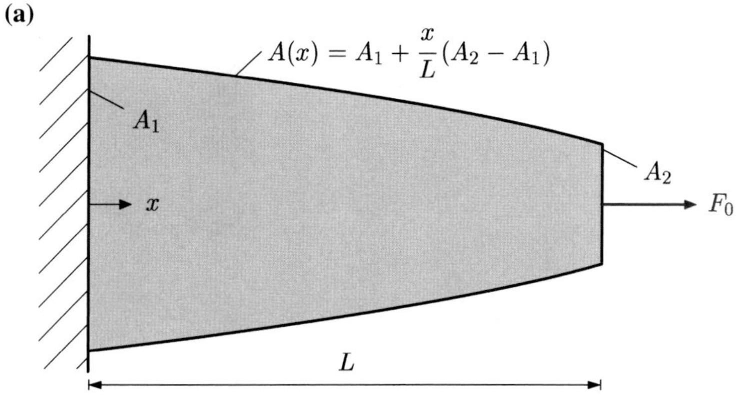

Fig. 2.52 Rod element with variable cross-section and different boundary conditions at the right end: **a** external force F_0; **b** displacement u_0

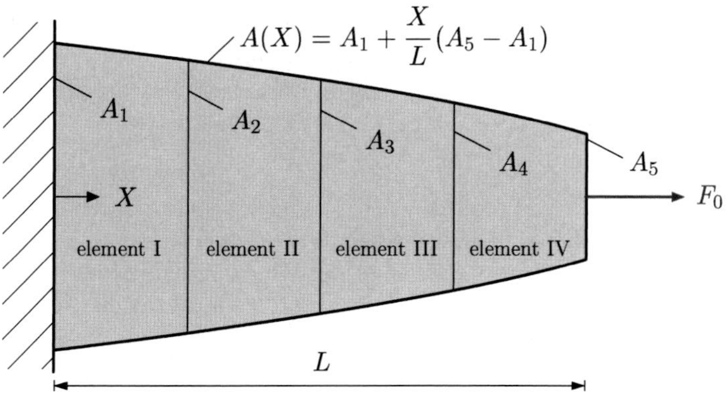

Fig. 2.53 Rod element with variable cross-section discretized by four finite elements

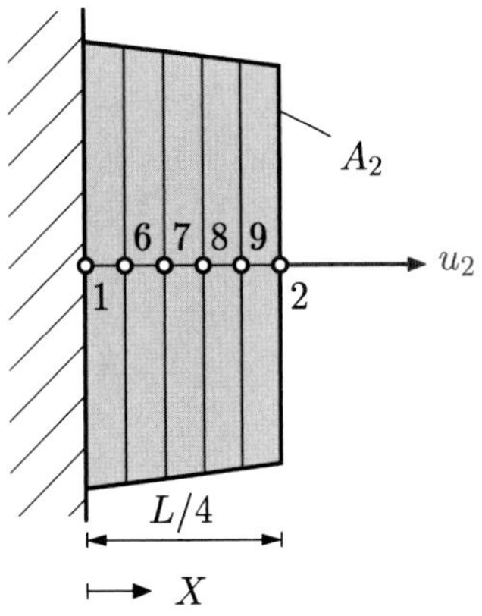

Fig. 2.54 Submodel of the structure with variable cross section, cf. Fig. 2.53

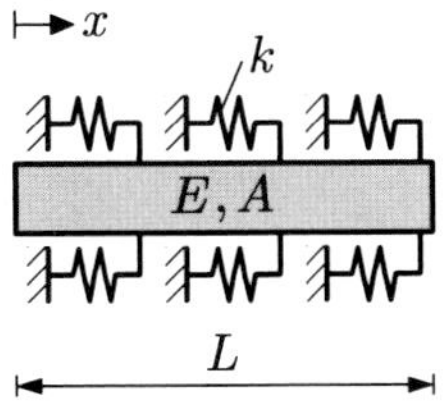

Fig. 2.55 Rod with elastic embedding

step, the area of element I is separately considered and discretized with a finer mesh. The nodal displacement u_2 from Problem 2.32 is applied as boundary condition at the right-hand end of the submodel. Calculate the stress distribution based on the submodel and compare with the analytical solution and the result from the coarse mesh (see Fig. 2.54). Further details on the submodeling technique can be found, for example, in [9].

2.34 Rod with elastic embedding: stiffness matrix

A rod with elastic embedding is schematically shown in Fig. 2.55. Derive the elemental stiffness matrix for a rod element with (a) linear and (b) quadratic interpolation functions under the assumption that the elastic modulus k is constant. The describing partial differential equation can be taken from Table 2.2.

2.35 Rod with elastic embedding: single force case

A cantilever rod with elastic embedding is loaded by a single force F_0 as shown in Fig. 2.56. Assume that the elastic modulus k and the axial tensile stiffness EA are constant. Use a single (a) linear and (b) quadratic rod element to determine

- the reduced system of equations,
- the elongation of the rod at $x = L$,
- simplify your result for the special case $k = 0$,

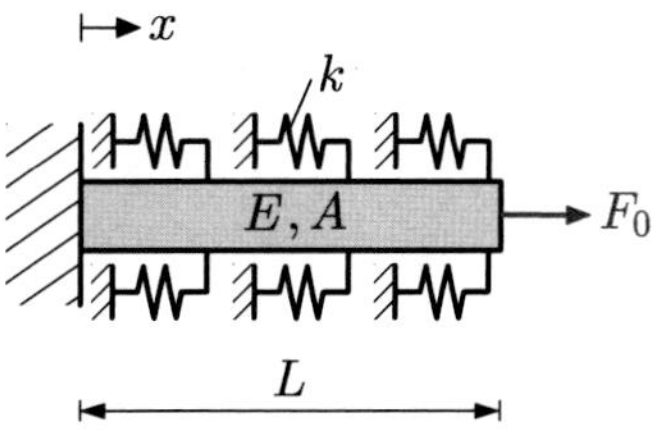

Fig. 2.56 Rod with elastic embedding loaded by a single force

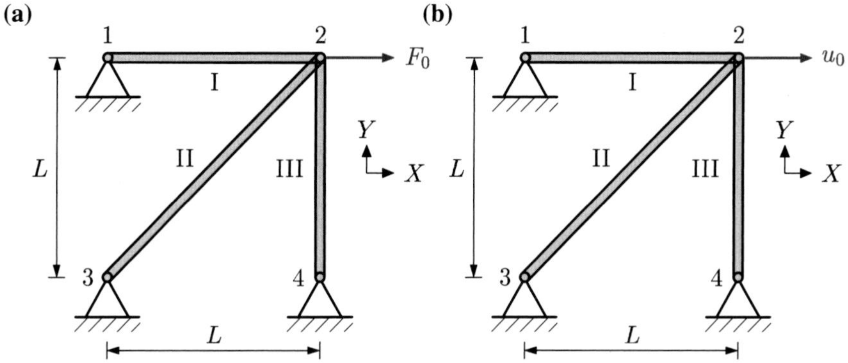

Fig. 2.57 Three-element truss structure with different external loading: **a** force boundary condition; **b** displacement boundary condition

- simplify your result for the special case $EA = 0$.
- Compare the finite element solution with the analytical solution for the case $k = 3$, $EA = 1$ and $L = 1$.

2.36 Plane truss structure arranged in a square

Given is the two-dimensional truss structure as shown in Fig. 2.57. The three truss elements have the same cross-sectional area A and YOUNG's modulus E. The length of each element can be taken from the dimensions given in the figure. The structure is loaded by

(a) a horizontal force F_0 at node 2,
(b) a prescribed horizontal displacement u_0 at node 2.

Determine for both cases:

- the global system of equations,
- the reduced system of equations,
- all nodal displacements,
- all reaction forces, and
- the force in each rod.

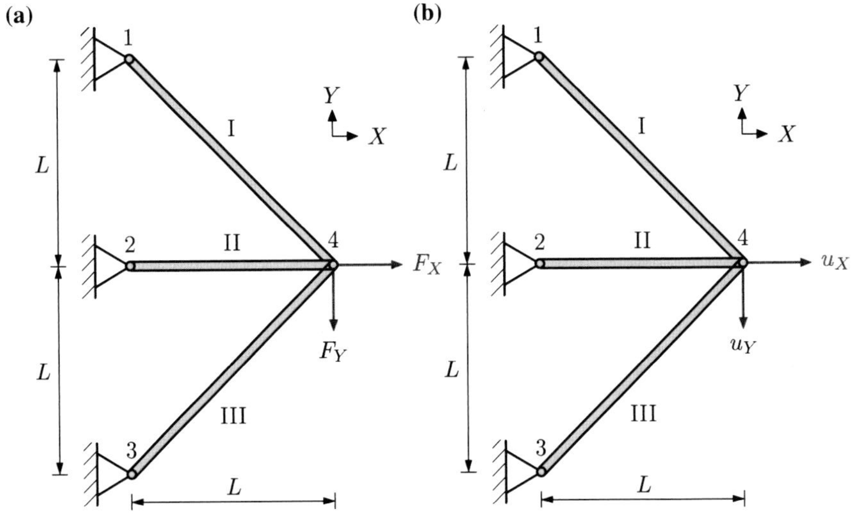

Fig. 2.58 Three-element truss structure with different external loading: **a** force boundary conditions; **b** displacement boundary conditions

2.37 Plane truss structure arranged in a triangle

Given is the two-dimensional truss structure as shown in Fig. 2.58. The three truss elements have the same cross-sectional area A and YOUNG's modulus E. The length of each element can be taken from the dimensions given in the figure. The structure is loaded by

(a) single forces F_X and F_Y at node 4,
(b) prescribed displacements u_X and u_Y at node 4.

Determine for both cases:

- the global system of equations,
- the reduced system of equations,
- all nodal displacements,
- all reaction forces, and
- the force in each rod.

2.38 Plane truss structure with two rod elements

Given is the two-dimensional truss structure as shown in Fig. 2.59. The two truss elements have the same cross-sectional area A and YOUNG's modulus E. The length of each element can be taken from the dimensions given in the figure. The structure is loaded by

a single forces F_0 at node 2 in X-direction and
a prescribed displacements u_0 at node 2 in Y-direction.

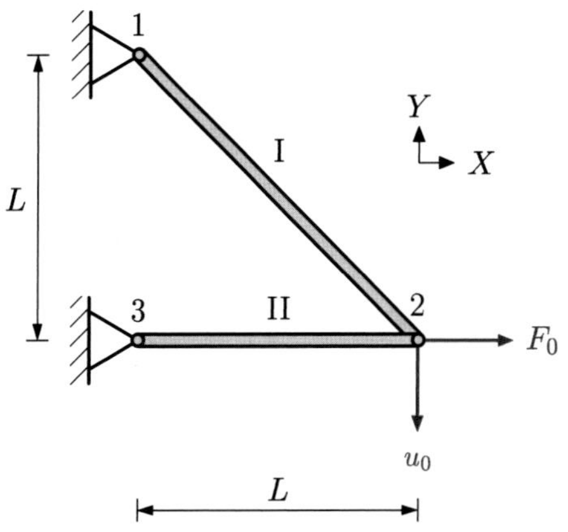

Fig. 2.59 Two-element truss structure with 'mixed' boundary conditions

Consider two linear truss (bar) finite elements and determine:

- the free body diagram,
- the global system of equations,
- the reduced system of equations under consideration of the boundary conditions,
- the nodal displacements at node 2,
- all reaction forces.
- Check if the global force equilibrium is fulfilled.

2.39 Truss structure in star formation

The following Fig. 2.60 shows a two-dimensional truss structure. The three rod elements have the same YOUNG's modulus E and length L. However, the cross-sectional areas A_i $(i = \text{I, II, III})$ are different from the vertical rod (A_I) to those of the 45° inclined rods $(A_\text{III} = A_\text{II})$. The structure is loaded by a point load F_0 at node 2. Develop a simplified (i.e., reduced number of rod elements) finite element truss structure under the consideration of the symmetry of the problem. Determine (do *not* consider three rod elements for the following questions):

- the equivalent statical system under the consideration of symmetry,
- the free-body diagram,
- the global stiffness matrix,
- the reduced system of equations under consideration of the boundary conditions,
- the nodal displacement at node 2, and
- simplify the nodal displacement at node 2 for the special case $A_\text{I} = A_\text{II} = A_\text{III} = A$.
- The vertical reaction forces at nodes 1 and 3 for the simplified model.

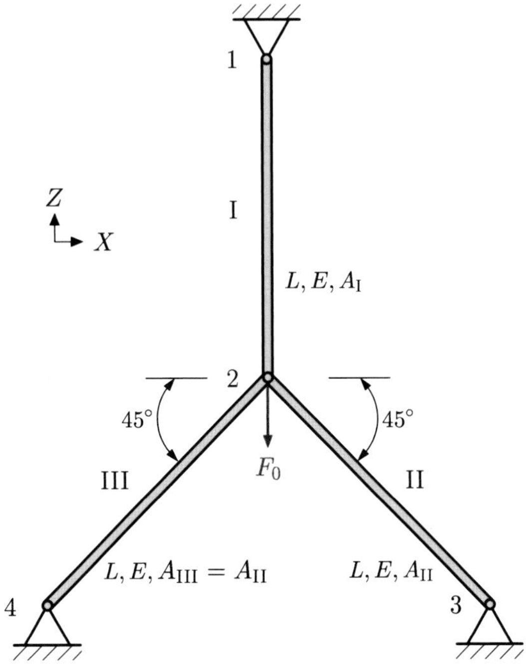

Fig. 2.60 Three-element truss structure with force boundary condition

References

1. Cook RD, Malkus DS, Plesha ME, Witt RJ (2002) Concepts and applications of finite element analysis. John Wiley & Sons, New York
2. Gross D, Hauger W, Schrder J, Wall WA, Bonet J (2011) Engineering mechanics 2: mechanics of materials. Springer, Berlin
3. Hartsuijker C, Welleman JW (2007) Engineering mechanics volume 2: stresses, strains, displacements. Springer, Dordrecht
4. Javanbakht Z, Öchsner A (2018) Computational statics revision course. Springer, Cham
5. MacNeal RH (1994) Finite elements: their design and performance. Marcel Dekker, New York
6. Öchsner A (2014) Elasto-plasticity of frame structure elements: modeling and simulation of rods and beams. Springer, Berlin
7. Öchsner A (2018) A project-based introduction to computational statics. Springer, Cham
8. Reddy JN (2006) An introduction to the finite element method. McGraw Hill, Singapore
9. da Silva LFM, Öchsner A, Adams RD (2018) Handbook of adhesion technology. Springer, Cham
10. Zienkiewicz OC, Taylor RL (2000) The finite element method. Vol. 1: the basis. Butterworth-Heinemann, Oxford

Chapter 3
Euler–Bernoulli Beams and Frames

Abstract This chapter starts with the analytical description of beam members. Based on the three basic equations of continuum mechanics, i.e., the kinematics relationship, the constitutive law and the equilibrium equation, the partial differential equation, which describes the physical problem, is derived. The weighted residual method is then used to derive the principal finite element equation for beam elements. Assembly of elements and the consideration of boundary conditions is treated in detail as well as the post-computation of some quantities. Furthermore, the classical beam element is generalized by the superposition of a beam and rod element. The chapter concludes with the spatial arrangements of generalized beam elements in a plane to form frame structures.

3.1 Introduction

A beam is defined as a long prismatic body as schematically shown in Fig. 3.1a. The following derivations are restricted to some simplifications:

- only applying to straight beams,
- no elongation along the x-axis,
- no torsion around the x-axis,
- deformations in a single plane, i.e. symmetrical bending,
- small deformations, and
- simple cross sections.

The external loads, which are considered within this chapter, are the single forces F_z, single moments M_y, distributed loads $q_z(x)$, and distributed moments $m_y(x)$. These loads have in common that their line of action (force) or the direction of the momentum vector are orthogonal to the center line of the beam and cause its bending. This is a different type of deformation compared to the rod element from Chap. 2, see Table 3.1. It should be noted here that these basic types of deformation can be superposed to account for more complex loading conditions [2].

A. Öchsner, *Computational Statics and Dynamics*,
https://doi.org/10.1007/978-981-15-1278-0_3

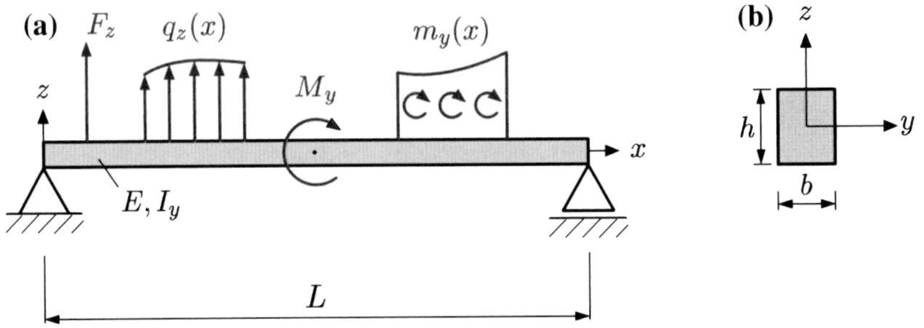

Fig. 3.1 General configuration for beam problems: **a** example of boundary conditions and external loads; **b** cross-sectional area

Table 3.1 Differentiation between rod and beam element; center line parallel to the x-axis

	Rod	Beam
Force	Along the rod axis	Perpendicular to the beam axis
Unknown	Displacement along rod axis u_x	Displacement perpendicular and rotation perpendicular to the beam axis u_z, φ_y

The classic theories of beam bending distinguish between shear-rigid and shear-flexible models. The shear rigid-beam, also called the BERNOULLI[1] beam,[2] neglects the shear deformation from the shear forces. This theory implies that a cross-sectional plane which was perpendicular to the beam axis before the deformation remains in the deformed state perpendicular to the beam axis, see Fig. 3.2a. Furthermore, it is assumed that a cross-sectional plane stays plane and unwarped in the deformed state. These two assumptions are also known as BERNOULLI's hypothesis. Altogether one imagines that cross-sectional planes are rigidly fixed to the center line of the beam[3] so that a change of the center line affects the entire deformation. Consequently, it is also assumed that the geometric dimensions[4] of the cross-sectional planes do not change.

In the case of a shear-flexible beam, also called the TIMOSHENKO[5] beam, the shear deformation is considered in addition to the bending deformation and cross-sectional planes are rotated by an angle γ compared to the perpendicular line, see Fig. 3.2b. For beams for which the length is 10–20 times larger than a characteristic dimension of the cross section, the shear fraction is usually disregarded in the first approximation. The different load types, meaning pure bending moment loading or shear due to shear force, lead to different stress fractions in a beam. In the case of a BERNOULLI beam,

[1] Jakob I. BERNOULLI (1655–1705), Swiss mathematician and physicist.

[2] More precisely, this beam is known as the EULER-BERNOULLI beam. A historical analysis of the development of the classical beam theory and the contribution of different scientists can be found in [8].

[3] More precisely, this is the neutral fibre or the bending line.

[4] Consequently, the width b and the height h of a, for example, rectangular cross section remain the same, see Fig. 3.1b.

[5] Stepan Prokopovych TYMOSHENKO (1878–1972), Ukrainian/US engineer.

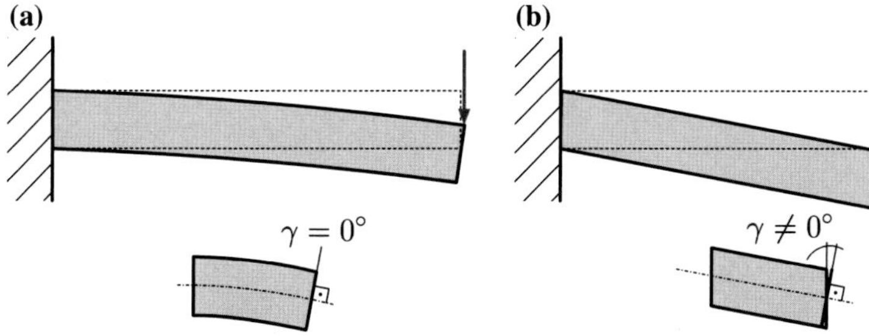

Fig. 3.2 Different deformation modes of a bending beam: **a** shear-rigid; **b** shear-flexible. Adapted from [7]

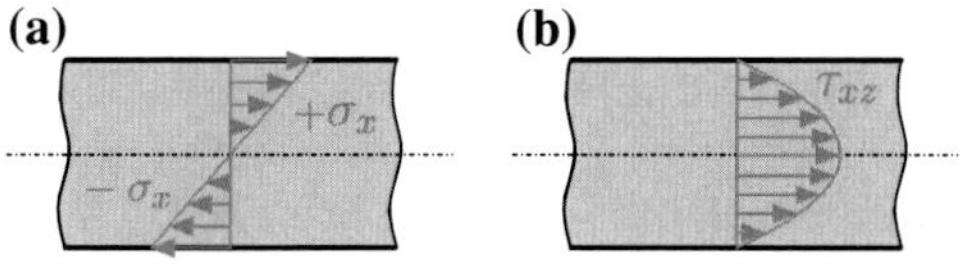

Fig. 3.3 Different stress distributions of a beam with rectangular cross section and linear-elastic material behavior: **a** normal stress and **b** shear stress

Table 3.2 Analogies between the beam and plate theories

	Beam theory	Plate theory
Dimensionality	1D	2D
Shear-rigid	BERNOULLI beam	KIRCHHOFF plate
Shear-flexible	TIMOSHENKO beam	REISSNER–MINDLIN plate

deformation occurs solely through normal forces, which are linearly distributed over the cross section. Consequently, a tension—alternatively a compression maximum on the bottom face—maximum on the top face occurs, see Fig. 3.3a. In the case of symmetric cross sections, the zero crossing[6] occurs in the middle of the cross section. The shear stress distribution for a rectangular cross section is parabolic at which the maximum occurs at the neutral axis and is zero at both the top and bottom surface, see Fig. 3.3b. This shear stress distribution can be calculated for the BERNOULLI beam but is not considered for the derivation of the deformation.

Finally, it needs to be noted that the one-dimensional beam theories have corresponding counterparts in two-dimensional space, see Table 3.2. In plate theories, the BERNOULLI beam corresponds to the shear-rigid KIRCHHOFF[7] plate and the TIMOSHENKO beam corresponds to the shear-flexible REISSNER[8]–MINDLIN[9] plate, [1, 5, 13].

[6]The sum of all points with $\sigma = 0$ along the beam axis is called the neutral fiber.

[7]Gustav Robert KIRCHHOFF (1824–1887), German physicist.

[8]Eric REISSNER (1913–1996), German/US engineer.

[9]Raymond David MINDLIN (1906–1987), US engineer.

Further details regarding the beam theory and the corresponding basic definitions and assumptions can be found in Refs. [4, 6, 9, 12]. In the following Sect. 3.2, only the BERNOULLI beam is considered. Consideration of the shear part takes place in Chap. 4.

3.2 Derivation of the Governing Differential Equation

3.2.1 Kinematics

For the derivation of the kinematics relation, a beam with length L is under constant moment loading $M_y(x) = \text{const.}$, meaning under *pure* bending, is considered, see Fig. 3.4. One can see that both external single moments at the left- and right-hand boundary lead to a positive bending moment distribution M_y within the beam. The vertical position of a point with respect to the center line of the beam *without action*

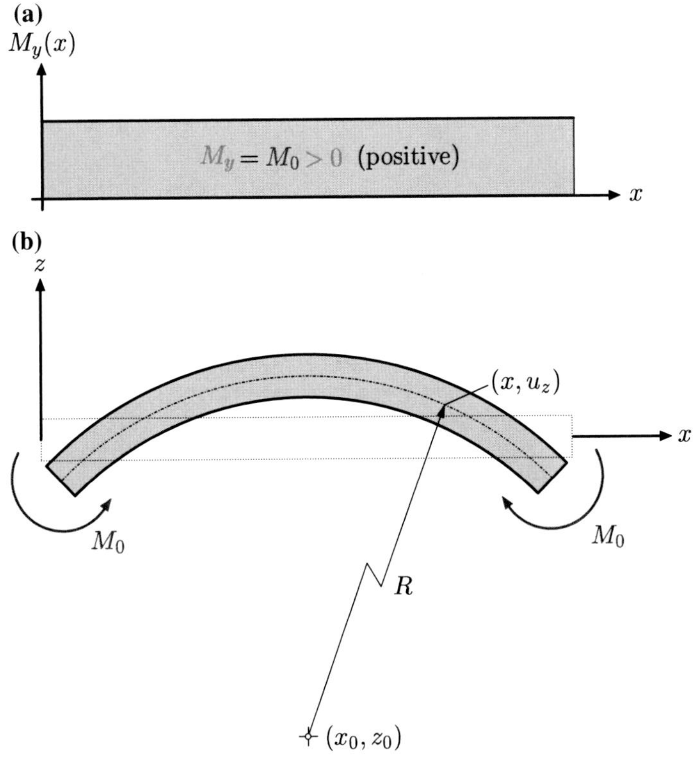

Fig. 3.4 Beam under pure bending in the x-z plane: **a** moment distribution; **b** deformed beam. Note that the deformation is exaggerated for better illustration. For the deformations considered in this chapter the following applies: $R \gg L$

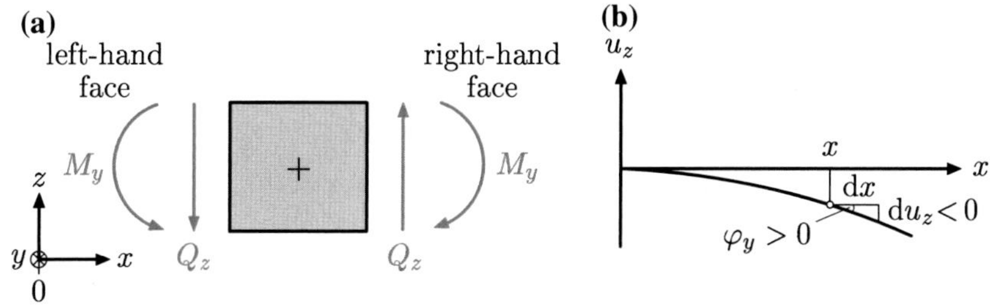

Fig. 3.5 Positive definition of **a** internal reactions and **b** rotation (but negative slope)

of an external load is described through the z-coordinate. The vertical *displacement* of a point on the center line of the beam, meaning for a point with $z = 0$, under action of the external load is indicated with u_z. The deformed center line is represented by the sum of these points with $z = 0$ and is referred to as the bending line $u_z(x)$.

In the case of a deformation in the x-z plane, it is important to precisely distinguish between the positive orientation of the internal reactions, the positive rotational angle, and the slope see Fig. 3.5. The internal reactions at a right-hand boundary are directed in the positive directions of the coordinate axes. Thus, a positive moment at a right-hand boundary is clockwise oriented (as the positive rotational angle), see Fig. 3.5. However, the slope is negative, see Fig. 3.5. This difference requires some careful derivations of the corresponding equations.

Only the center line of the deformed beam is considered in the following. Through the relation for an arbitrary point (x, u_z) on a circle with radius R around the center point (x_0, z_0), meaning

$$(x - x_0)^2 + (u_z(x) - z_0)^2 = R^2\,, \tag{3.1}$$

one obtains through differentiation with respect to the x-coordinate

$$2(x - x_0) + 2(u_z(x) - z_0)\frac{\mathrm{d}u_z(x)}{\mathrm{d}x} = 0\,, \tag{3.2}$$

alternatively after another differentiation:

$$2 + 2\frac{\mathrm{d}u_z}{\mathrm{d}x}\frac{\mathrm{d}u_z}{\mathrm{d}x} + 2(u_z(x) - z_0)\frac{\mathrm{d}^2u_z}{\mathrm{d}x^2} = 0\,. \tag{3.3}$$

Equation (3.3) provides the vertical distance between an arbitrary point on the center line of the beam and the center point of a circle as

$$(u_z - z_0) = -\frac{1 + \left(\dfrac{\mathrm{d}u_z}{\mathrm{d}x}\right)^2}{\dfrac{\mathrm{d}^2u_z}{\mathrm{d}x^2}}\,, \tag{3.4}$$

while the difference between the x-coordinates results from Eq. (3.2):

$$(x - x_0) = -(u_z - z_0)\frac{\mathrm{d}u_z}{\mathrm{d}x}. \tag{3.5}$$

If the expression according to Eq. (3.4) is used in Eq. (3.5) the following results:

$$(x - x_0) = \frac{\mathrm{d}u_z}{\mathrm{d}x} \frac{1 + \left(\frac{\mathrm{d}u_z}{\mathrm{d}x}\right)^2}{\frac{\mathrm{d}^2u_z}{\mathrm{d}x^2}}. \tag{3.6}$$

Inserting both expressions for the x- and z-coordinate differences according to Eqs. (3.6) and (3.4) in the circle equation according to (3.1) leads to:

$$\begin{aligned} R^2 &= (x - x_0)^2 + (u_z - z_0)^2 \\ &= \left(\frac{\mathrm{d}u_z}{\mathrm{d}x}\right)^2 \frac{\left(1 + \left(\frac{\mathrm{d}u_z}{\mathrm{d}x}\right)^2\right)^2}{\left(\frac{\mathrm{d}^2u_z}{\mathrm{d}x^2}\right)^2} + \frac{\left(1 + \left(\frac{\mathrm{d}u_z}{\mathrm{d}x}\right)^2\right)^2}{\left(\frac{\mathrm{d}^2u_z}{\mathrm{d}x^2}\right)^2} \\ &= \left(\left(\frac{\mathrm{d}^2u_z}{\mathrm{d}x^2}\right)^2 + 1\right) \frac{\left(1 + \left(\frac{\mathrm{d}u_z}{\mathrm{d}x}\right)^2\right)^2}{\left(\frac{\mathrm{d}^2u_z}{\mathrm{d}x^2}\right)^2} \end{aligned} \tag{3.7}$$

$$= \frac{\left(1 + \left(\frac{\mathrm{d}u_z}{\mathrm{d}x}\right)^2\right)^3}{\left(\frac{\mathrm{d}^2u_z}{\mathrm{d}x^2}\right)^2}. \tag{3.8}$$

Thus, the radius of curvature is obtained as:

$$|R| = \frac{\left(1 + \left(\frac{\mathrm{d}u_z}{\mathrm{d}x}\right)^2\right)^{3/2}}{\left|\frac{\mathrm{d}^2u_z}{\mathrm{d}x^2}\right|}. \tag{3.9}$$

To decide if the radius of curvature is positive or negative, let us have a look at Fig. 3.6 where a curve with its tangential and normal vectors is shown. Since the curve in

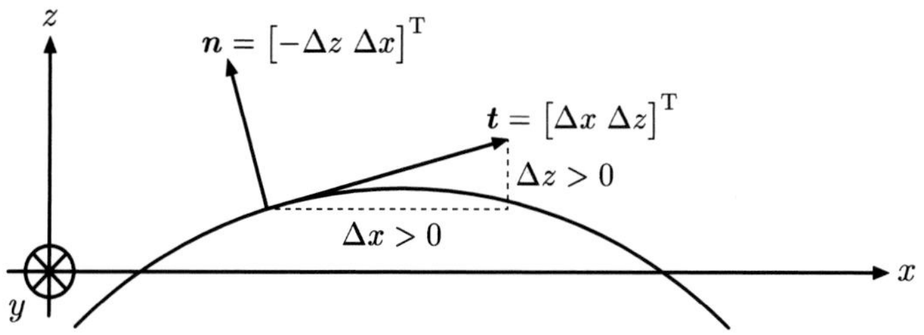

Fig. 3.6 On the definition of a negative curvature in the x-z plane

this configuration is bending away[10] from the normal vector $\boldsymbol{n}$, it holds that $\frac{\mathrm{d}^2u_z}{\mathrm{d}x^2} < 0$ and the radius of curvature is obtained for a positive bending moment as:

$$R = -\frac{\left(1+\left(\dfrac{\mathrm{d}u_z}{\mathrm{d}x}\right)^2\right)^{3/2}}{\dfrac{\mathrm{d}^2u_z}{\mathrm{d}x^2}}. \tag{3.10}$$

Note that the expression curvature, which results as a reciprocal value from the curvature radius, $\kappa = \frac{1}{R}$, is used as well.

For small bending deflections, meaning $u_z \ll L$, $\frac{\mathrm{d}u_z}{\mathrm{d}x} \ll 1$ results and Eq. (3.10) simplifies to:

$$R = -\frac{1}{\dfrac{\mathrm{d}^2u_z}{\mathrm{d}x^2}} \quad \text{or} \quad \kappa = \frac{1}{R} = -\frac{\mathrm{d}^2u_z}{\mathrm{d}x^2}. \tag{3.11}$$

For the determination of the strain, one refers to its general definition, meaning elongation referring to initial length. Relating to the configuration shown in Fig. 3.7, the longitudinal elongation of a fibre at distance z to the neutral fibre allows to express the strain as:

$$\varepsilon_x = \frac{\mathrm{d}s - \mathrm{d}x}{\mathrm{d}x}. \tag{3.12}$$

The lengths of the circular arcs $\mathrm{d}s$ and $\mathrm{d}x$ result from the corresponding radii and the enclosed angles in radian measure as:

$$\mathrm{d}x = R\mathrm{d}\varphi_y\,, \tag{3.13}$$

$$\mathrm{d}s = (R+z)\mathrm{d}\varphi_y\,. \tag{3.14}$$

[10] See Sect. A.14.2 for further details.

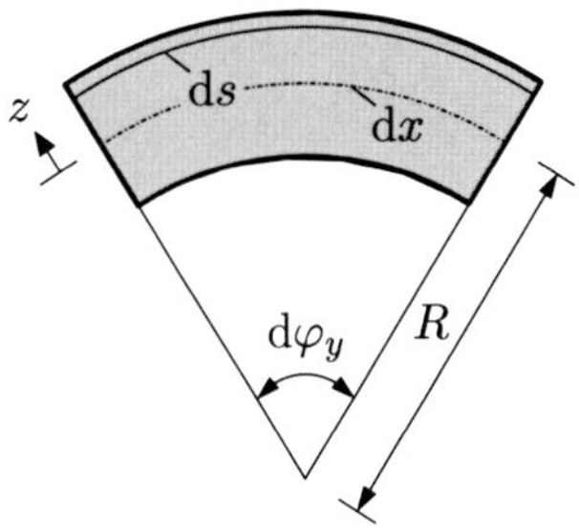

Fig. 3.7 Segment of a beam under pure bending in the x-z plane. Note that the deformation is exaggerated for better illustration

If these relations for the circular arcs are used in Eq. (3.12), the following results:

$$\varepsilon_x = \frac{(R+z)\mathrm{d}\varphi_y - R\mathrm{d}\varphi_y}{\mathrm{d}x} = z\frac{\mathrm{d}\varphi_y}{\mathrm{d}x}. \tag{3.15}$$

From Eq. (3.13) $\frac{\mathrm{d}\varphi_y}{\mathrm{d}x} = \frac{1}{R}$ results and together with relation (3.11) the strain can finally be expressed as follows:

$$\varepsilon_x(x, y) = z\,\frac{1}{R} \overset{(3.11)}{=} -z\frac{\mathrm{d}^2 u_z(x)}{\mathrm{d}x^2} \overset{(3.11)}{=} z\kappa\,. \tag{3.16}$$

An alternative derivation of the kinematics relation results from consideration of Fig. 3.8. From the relation of the right-angled triangle $0'1'2'$, this means[11] $\sin\varphi_y = \frac{u_x}{z}$, the following relation results for small angles ($\sin\varphi_y \approx \varphi_y$):

$$u_x = +z\varphi_y\,. \tag{3.17}$$

Furthermore, it holds that the rotation angle of the slope equals the center line for small angles:

$$\tan\varphi_y = \frac{-\,\mathrm{d}u_z(x)}{\mathrm{d}x} \approx \varphi_y\,. \tag{3.18}$$

If Eqs. (3.18) and (3.17) are combined, the following results:

$$u_x = -z\frac{\mathrm{d}u_z(x)}{\mathrm{d}x}. \tag{3.19}$$

The last relation equals $(\mathrm{d}s - \mathrm{d}x)$ in Eq. (3.12) and differentiation with respect to the x-coordinate leads directly to Eq. (3.16).

[11] Note that according to the assumptions of the BERNOULLI beam the lengths 01 and $0'1'$ remain unchanged.

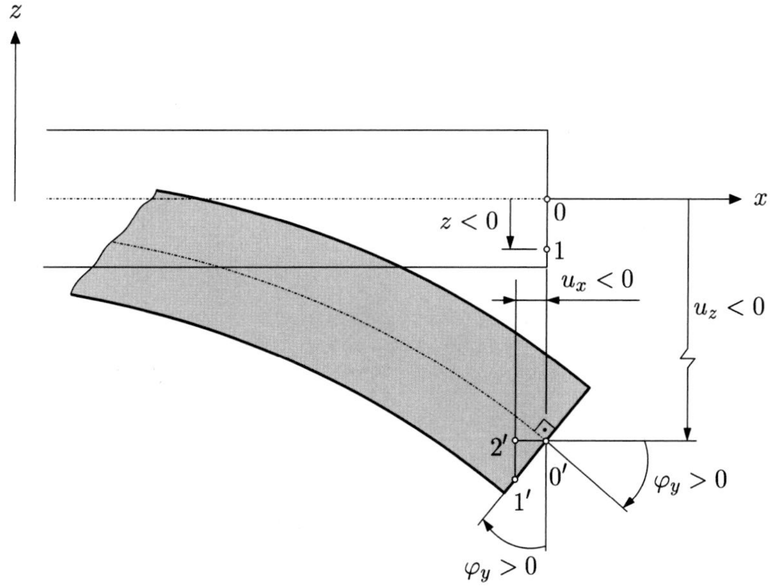

Fig. 3.8 Alternative configuration for the derivation of the kinematics relation. Note that the deformation is exaggerated for better illustration

3.2.2 Constitutive Equation

The one-dimensional HOOKE's law according to Eq. (2.3) can also be assumed in the case of the bending beam, since, according to the requirement, only normal stresses are regarded in this section:

$$\sigma_x = E\varepsilon_x \,. \tag{3.20}$$

Through the kinematics relation according to Eq. (3.16), the stress results as a function of deflection to:

$$\sigma_x(x, z) = -Ez\frac{\mathrm{d}^2 u_z(x)}{\mathrm{d}x^2}. \tag{3.21}$$

The stress distribution shown in Fig. 3.9a generates the internal moment, which acts in this cross section. To calculate this internal moment, the stress is multiplied by a surface element, so that the resulting force is obtained. Multiplication with the corresponding lever arm then gives the internal moment. Since the stress is linearly distributed over the height, the evaluation is done for an infinitesimally small surface element:

$$\mathrm{d}M_y = (+z)(+\sigma_x)\mathrm{d}A = z\sigma_x\mathrm{d}A \,. \tag{3.22}$$

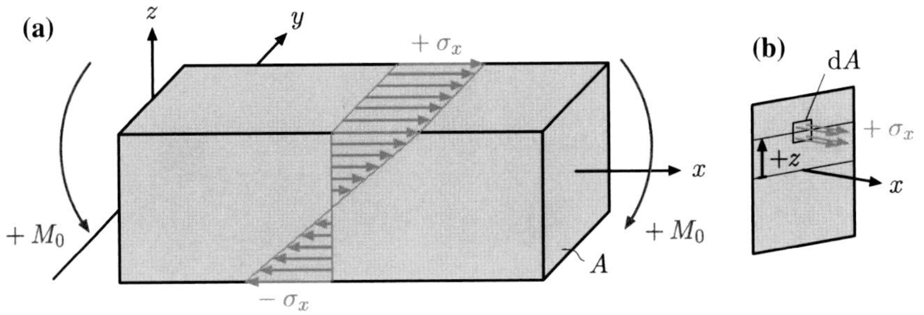

Fig. 3.9 **a** Schematic representation of the normal stress distribution $\sigma_x = \sigma_x(z)$ of a bending beam; **b** Definition and position of an infinitesimal surface element for the derivation of the resulting moment action due to the normal stress distribution

Therefore, the entire moment results via integration over the entire surface in:

$$M_y = \int_A z\sigma_x \mathrm{d}A \overset{(3.21)}{=} -\int_A zEz\frac{\mathrm{d}^2 u_z(x)}{\mathrm{d}x^2}\mathrm{d}A\,. \tag{3.23}$$

Assuming that the YOUNG's modulus is constant, the internal moment around the y-axis results in:

$$M_y = -E\frac{\mathrm{d}^2 u_z}{\mathrm{d}x^2}\underbrace{\int_A z^2 \mathrm{d}A}_{I_y} = \frac{I_y \sigma_z}{z}\,. \tag{3.24}$$

The integral in Eq. (3.24) is the so-called axial second moment of area or axial surface moment of 2nd order in the SI unit m^4. This factor is only dependent on the geometry of the cross section and is also a measure of the stiffness of a plane cross section against bending. The values of the axial second moment of area for simple geometric cross sections are collected in Table 3.3.

Consequently the internal moment can also be expressed as

$$M_y = -EI_y\frac{\mathrm{d}^2 u_z}{\mathrm{d}x^2} \overset{(3.11)}{=} \frac{EI_y}{R} = EI_y\kappa. \tag{3.25}$$

Equation (3.25) describes the bending line $u_z(x)$ as a function of the bending moment and is therefore also referred to as the bending line-moment relation. The product EI_y in Eq. (3.25) is also called the bending stiffness. If the result from Eq. (3.25) is used in the relation for the bending stress according to Eq. (3.21), the distribution of stress over the cross section results in:

$$\sigma_x(x, z) = +\frac{M_y(x)}{I_y}z(x)\,. \tag{3.26}$$

Table 3.3 Axial second moment of area around the y- and z-axis

Cross section	I_y	I_z
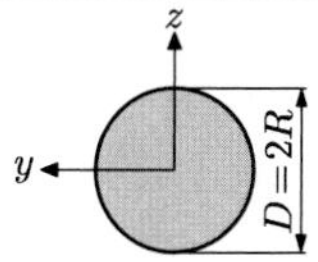	$\frac{\pi D^4}{64} = \frac{\pi R^4}{4}$	$\frac{\pi D^4}{64} = \frac{\pi R^4}{4}$
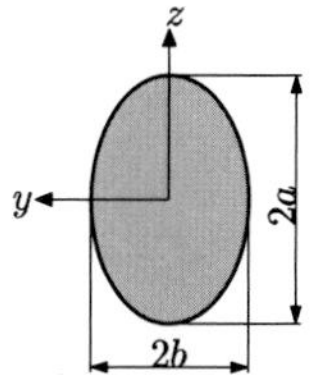	$\frac{\pi b a^3}{4}$	$\frac{\pi a b^3}{4}$
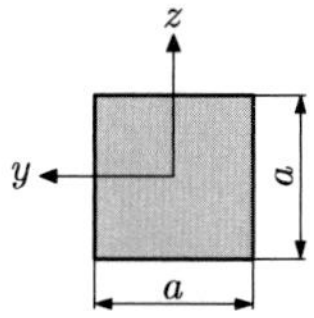	$\frac{a^4}{12}$	$\frac{a^4}{12}$
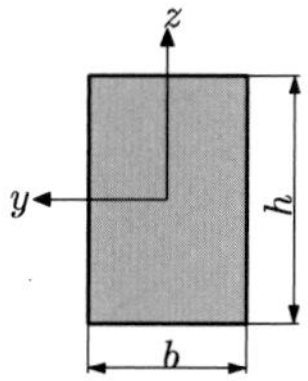	$\frac{bh^3}{12}$	$\frac{hb^3}{12}$
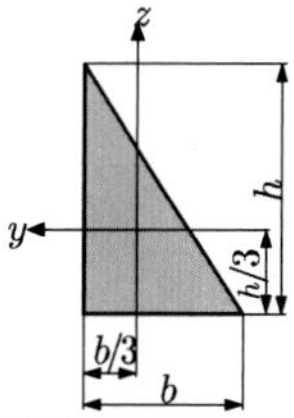	$\frac{bh^3}{36}$	$\frac{hb^3}{36}$
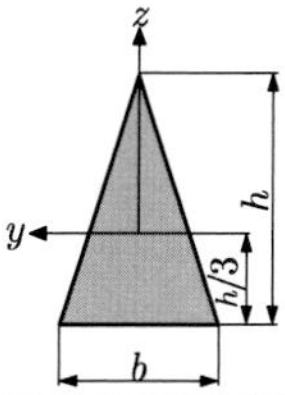	$\frac{bh^3}{36}$	$\frac{bh^3}{48}$

The plus sign in Eq. (3.26) causes that a positive bending moment (see Fig. 3.4) leads to a tensile stress in the upper beam half (meaning for $z > 0$). The corresponding equations for a deformation in the x-y plane can be found in [11].

In the case of plane bending with $M_y(x) \neq \text{const.}$, the bending line can be approximated in each case locally through a circle of curvature, see Fig. 3.10. Therefore,

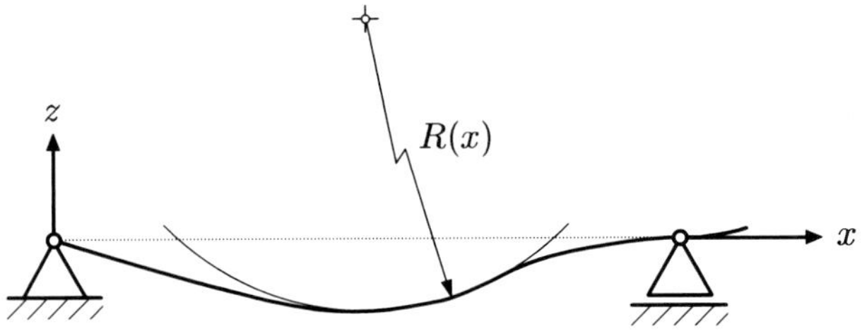

Fig. 3.10 Deformation of a beam in the x-z plane with $M_y(x) \neq$ const.

the result for *pure* bending according to Eq. (3.25) can be transferred to the case of plane bending as:

$$-EI_y \frac{\mathrm{d}^2 u_z(x)}{\mathrm{d}x^2} = M_y(x). \tag{3.27}$$

Let us note at the end of this section that HOOKE's law in the form of Eq. (3.20) is not so easy to apply[12] in the case of beams since the stress and strain is linearly changing over the height of the cross section, see Eq. (3.26) and Fig. 3.9. Thus, it might be easier to apply a so-called stress resultant or generalized stress, i.e. a simplified representation of the normal stress state[13] based on the acting bending moment:

$$M_y(x) = \iint z\sigma_x(x, z)\,\mathrm{d}A\,, \tag{3.28}$$

which was already introduced in Eq. (3.22). Using in addition the curvature[14] $\kappa = \kappa(x)$ (see Eq. (3.16)) instead of the strain $\varepsilon_x = \varepsilon_x(x, z)$, the constitutive equation can be easier expressed as shown in Fig. 3.11. The variables M_y and κ have both the advantage that they are constant for any location x of the beam.

3.2.3 Equilibrium

The equilibrium conditions are derived from an infinitesimal beam element of length $\mathrm{d}x$, which is loaded by a constant distributed load q_z, see Fig. 3.12. The internal reactions are drawn on both cut faces, i.e. at location x and $x + \mathrm{d}x$. One can see that a positive shear force is oriented in the positive z-direction at the right-hand face[15]

---

[12] However, this formulation works well in the case of rod elements since stress and strain are constant over the cross section, i.e. $\sigma_x = \sigma_x(x)$ and $\varepsilon_x = \varepsilon_x(x)$, see Fig. 2.4.

[13] A similar stress resultant can be stated for the shear stress based on the shear force: $Q_z(x) = \iint \tau_{xz}(x, z)\,\mathrm{d}A$.

[14] The curvature is then called a generalized strain.

[15] A positive cut face is defined by the surface normal on the cut plane which has the same orientation as the positive x-axis. It should be regarded that the surface normal is always directed outward.

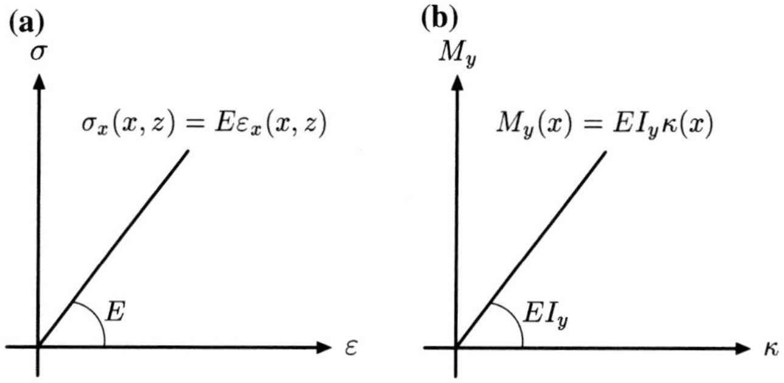

Fig. 3.11 Formulation of the constitutive law based on **a** stress and **b** stress resultant

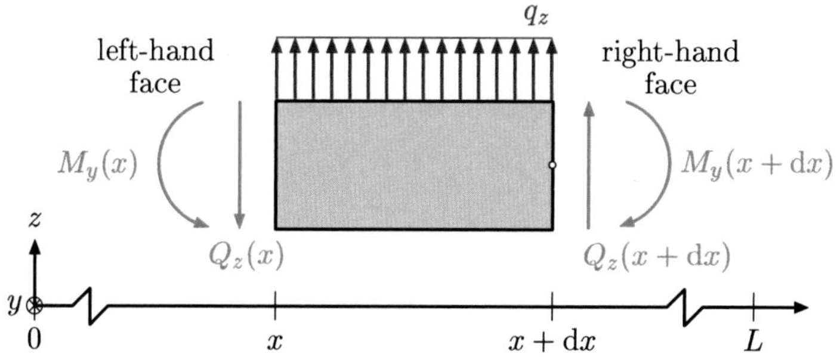

Fig. 3.12 Infinitesimal beam element in the x-z plane with internal reactions and constant distributed load

and that a positive bending moment has the same rotational direction as the positive y-axis (right-hand grip rule[16]). The orientation of shear force and bending moment is reversed at the left-hand face in order to cancel in sum the effect of the internal reactions at both faces. This convention for the direction of the internal directions is maintained in the following. Furthermore, it can be derived from Fig. 3.12 that an upwards directed *external* force or alternatively a mathematically positive oriented *external* moment at the right-hand face leads to a positive shear force or alternatively a positive internal moment. In a corresponding way, it results that a downwards directed *external* force or alternatively a mathematically negative oriented *external* moment at the left-hand face leads to a positive shear force or alternatively a positive internal moment.

[16]If the axis is grasped with the right hand in a way so that the spread out thumb points in the direction of the positive axis, the bent fingers then show the direction of the positive rotational direction.

The equilibrium condition will be determined in the following for the vertical forces. Assuming that forces in the direction of the positive z-axis are considered positive, the following results:

$$-Q_z(x)+Q_z(x+\mathrm{d}x)+q_z\mathrm{d}x=0\,. \tag{3.29}$$

If the shear force on the right-hand face is expanded in a TAYLOR's series of first order, meaning

$$Q(x+\mathrm{d}x)\approx Q(x)+\frac{\mathrm{d}Q(x)}{\mathrm{d}x}\mathrm{d}x\,, \tag{3.30}$$

Equation (3.29) results in

$$-Q(x)+Q(x)+\frac{\mathrm{d}Q(x)}{\mathrm{d}x}\mathrm{d}x+q_z\mathrm{d}x=0\,, \tag{3.31}$$

or alternatively after simplification finally to:

$$\frac{\mathrm{d}Q_z(x)}{\mathrm{d}x}=-q_z\,. \tag{3.32}$$

For the special case that no distributed load is acting ($q_z=0$), Eq. (3.32) simplifies to:

$$\frac{\mathrm{d}Q(x)}{\mathrm{d}x}=0\,. \tag{3.33}$$

The equilibrium of moments around the reference point at $x+\mathrm{d}x$ gives:

$$M_y(x+\mathrm{d}x)-M_y(x)-Q_z(x)\mathrm{d}x+\frac{1}{2}q_z\mathrm{d}x^2=0\,. \tag{3.34}$$

If the bending moment on the right-hand face is expanded into a TAYLOR's series of first order similar to Eq. (3.30) and consideration that the term $\frac{1}{2}\,q_z\mathrm{d}x^2$ as infinitesimal small size of higher order can be disregarded, finally the following results:

$$\frac{\mathrm{d}M_y(x)}{\mathrm{d}x}=Q_z(x)\,. \tag{3.35}$$

The combination of Eqs. (3.32) and (3.35) leads to the relation between the bending moment and the distributed load:

$$\frac{\mathrm{d}^2M_y(x)}{\mathrm{d}x^2}=\frac{\mathrm{d}Q_z(x)}{\mathrm{d}x}=-q_z(x)\,. \tag{3.36}$$

Finally, the elementary basic equations for the bending of a beam in the x-z plane for arbitrary moment loading $M_y(x)$ are summarized in Table 3.4.

Table 3.4 Elementary basic equations for the bending of a BERNOULLI beam in the x-z plane. The differential equations are given under the assumption of constant bending stiffness EI_y

Name	Equation
Kinematics	$\varepsilon_x(x,z) = -z\dfrac{\mathrm{d}^2 u_z(x)}{\mathrm{d}x^2}$
Equilibrium	$\dfrac{\mathrm{d}Q_z(x)}{\mathrm{d}x} = -q_z(x);\ \dfrac{\mathrm{d}M_y(x)}{\mathrm{d}x} = Q_z(x)$
Constitutive equation	$\sigma_x(x,z) = E\varepsilon_x(x,z)$
Stress	$\sigma_x(x,z) = \dfrac{M_y(x)}{I_y}z(x)$
Diff'equation	$EI_y\dfrac{\mathrm{d}^2 u_z(x)}{\mathrm{d}x^2} = -M_y(x)$
	$EI_y\dfrac{\mathrm{d}^3 u_z(x)}{\mathrm{d}x^3} = -Q_z(x)$
	$EI_y\dfrac{\mathrm{d}^4 u_z(x)}{\mathrm{d}x^4} = q_z(x)$

3.2.4 Differential Equation

Two-time differentiation of Eq. (3.25) and consideration of the relation between bending moment and distributed load according to Eq. (3.36) lead to the classical type of differential equation of the bending line,

$$\frac{\mathrm{d}^2}{\mathrm{d}x^2}\left(EI_y\frac{\mathrm{d}^2 u_z}{\mathrm{d}x^2}\right) = q_z\,, \tag{3.37}$$

which is also referred to as the bending line-distributed load relation. For a beam with constant bending stiffness EI_y along the beam axis, the following results:

$$EI_y\frac{\mathrm{d}^4 u_z}{\mathrm{d}x^4} = q_z\,. \tag{3.38}$$

The differential equation of the bending line can of course also be expressed through the bending moment or the shear force as

$$EI_y\frac{\mathrm{d}^2 u_z}{\mathrm{d}x^2} = -M_y \quad \text{or} \tag{3.39}$$

$$EI_y\frac{\mathrm{d}^3 u_z}{\mathrm{d}x^3} = -Q_z\,. \tag{3.40}$$

Equations (3.39) and (3.40) can be also written in the more general form for variable bending stiffness:

Table 3.5 Different formulations of the partial differential equation for a BERNOULLI beam in the x-z plane (x-axis: right facing; z-axis: upward facing)

Configuration	Partial differential equation
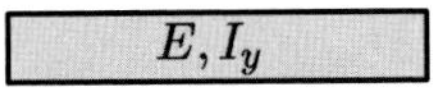	$EI_y \frac{d^4 u_z}{dx^4} = 0$
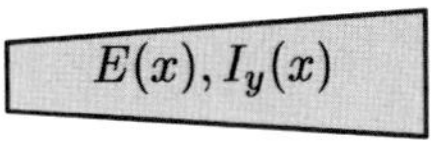	$\frac{d^2}{dx^2}\left(E(x)I_y(x)\frac{d^2 u_z}{dx^2}\right) = 0$
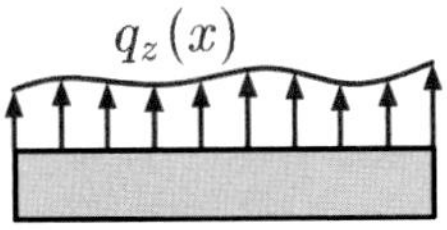	$EI_y \frac{d^4 u_z}{dx^4} = q_z(x)$
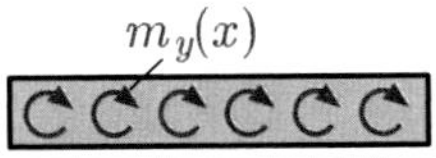	$EI_y \frac{d^4 u_z}{dx^4} = \frac{dm_y(x)}{dx}$
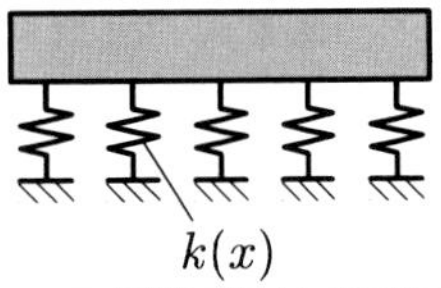	$EI_y \frac{d^4 u_z}{dx^4} = -k(x)u_z$

$$E(x)I_y(x)\frac{d^2 u_z}{dx^2} = -M_y(x)\,, \tag{3.41}$$

$$\frac{du_z}{dx}\left(E(x)I_y(x)\frac{d^2 u_z}{dx^2}\right) = -Q_z(x)\,. \tag{3.42}$$

Depending on the problem and the fact which distribution ($q_z(x)$, $M_y(x)$ or $Q_z(x)$) is easier to state, one may start from one of the three formulations to derive the displacement field $u_z(x)$.

Different formulations of the fourth order differential equation are collected in Table 3.5 where different types of loadings, geometry and bedding are differentiated. The last case in Table 3.5 refers to the elastic foundation of a beam which is also know in the literature as WINKLER[17] foundation [14]. The elastic foundation or WINKLER foundation modulus k has in the case of beams[18] the unit of force per unit area.

If we replace the common formulation of the second order derivative, i.e. $\frac{d^2 \ldots}{dx^2}$, by a formal operator symbol, i.e. $\mathcal{L}_2(\ldots)$, the basic equations can be stated in a more formal way as given in Table 3.6.

---

[17]Emil WINKLER (1835–1888), German engineer.

[18]In the general case, the unit of the elastic foundation modulus is force per unit area per unit length, i.e. $\frac{\text{N}}{\text{m}^2}/\text{m} = \frac{\text{N}}{\text{m}^3}$.

Table 3.6 Different formulations of the basic equations for an EULER–BERNOULLI beam (bending in the x-z plane; x-axis along the principal beam axis). E: YOUNG's modulus; I_y: second moment of area; q_x: length-specific distributed force; $\mathcal{L}_2 = \frac{\mathrm{d}^2(\ldots)}{\mathrm{d}x^2}$: second-order derivative

Specific formulation	General formulation
Kinematics	
$\varepsilon_x(x,z) = -z\dfrac{\mathrm{d}^2 u_z(x)}{\mathrm{d}x^2}$	$\varepsilon_x(x,z) = -z\mathcal{L}_2\left(u_z(x)\right)$
$\kappa = -\dfrac{\mathrm{d}^2 u_z(x)}{\mathrm{d}x^2}$	$\kappa = -\mathcal{L}_2\left(u_z(x)\right)$
Constitution	
$\sigma_x(x,z) = E\varepsilon_x(x,z)$	$\sigma_x(x,z) = C\varepsilon_x(x,z)$
$M_y(x) = EI_y\kappa(x)$	$M_y(x) = D\kappa(x)$
Equilibrium	
$\dfrac{\mathrm{d}^2 M_y(x)}{\mathrm{d}x^2} + q_z(x) = 0$	$\mathcal{L}_2^{\mathrm{T}}\left(M_y(x)\right) + q_z(x) = 0$
PDE	
$\dfrac{\mathrm{d}^2}{\mathrm{d}x^2}\left(EI_y\dfrac{\mathrm{d}^2 u_z(x)}{\mathrm{d}x^2}\right) - q_z(x) = 0$	$\mathcal{L}_2^{\mathrm{T}}\left(D\mathcal{L}_2\left(u_z(x)\right)\right) - q_z(x) = 0$

3.3 Finite Element Solution

3.3.1 Derivation of the Principal Finite Element Equation

Let us consider in the following the governing differential equation according to Eq. (3.38). This formulation assumes that the bending stiffness EI_y is constant and we obtain

$$EI_y\frac{\mathrm{d}^4 u_z^0(x)}{\mathrm{d}x^4} - q_z(x) = 0\,, \tag{3.43}$$

where $u_z^0(x)$ represents the exact solution of the problem. The last equation, which contains the exact solution of the problem, is fulfilled at each location x of the beam and is called the *strong formulation* of the problem. Replacing the exact solution in Eq. (3.43) by an approximate solution $u_z(x)$, a residual r is obtained:

$$r(x) = EI_y\frac{\mathrm{d}^4 u_z(x)}{\mathrm{d}x^4} - q_z(x) \neq 0\,. \tag{3.44}$$

As a consequence of the introduction of the approximate solution $u_z(x)$, it is in general no longer possible to satisfy the differential equation at each location x of the beam. In the scope of the weighted residual method, it is alternatively requested that the differential equation is fulfilled over a certain length (and no longer at each location x) and the following integral statement[19] is obtained

$$\int_0^L W^{\mathrm{T}}(x)\left(EI_y\frac{\mathrm{d}^4 u_z(x)}{\mathrm{d}x^4} - q_z(x)\right)\mathrm{d}x \overset{!}{=} 0\,, \tag{3.45}$$

which is called the *inner product*. The function $W(x)$ in Eq. (3.45) is called the weight function which distributes the error or the residual in the considered domain.

Integrating by parts[20] of the first expression in the parentheses of Eq. (3.45) gives:

$$\int_0^L \underbrace{W^{\mathrm{T}}}_{f} EI_y \underbrace{\frac{\mathrm{d}^4 u_z}{\mathrm{d}x^4}}_{g'}\mathrm{d}x = EI_y\left[W^{\mathrm{T}}\frac{\mathrm{d}^3 u_z}{\mathrm{d}x^3}\right]_0^L - EI_y\int_0^L \frac{\mathrm{d}W^{\mathrm{T}}}{\mathrm{d}x}\frac{\mathrm{d}^3 u_z}{\mathrm{d}x^3}\mathrm{d}x = 0\,. \tag{3.46}$$

Integrating by parts of the integral on the right-hand side of Eq. (3.46) results in:

$$EI_y\int_0^L \underbrace{\frac{\mathrm{d}W^{\mathrm{T}}}{\mathrm{d}x}}_{f}\underbrace{\frac{\mathrm{d}^3 u_z}{\mathrm{d}x^3}}_{g'}\mathrm{d}x = EI_y\left[\frac{\mathrm{d}W^{\mathrm{T}}}{\mathrm{d}x}\frac{\mathrm{d}^2 u_z}{\mathrm{d}x^2}\right]_0^L - EI_y\int_0^L \frac{\mathrm{d}^2 W^{\mathrm{T}}}{\mathrm{d}x^2}\frac{\mathrm{d}^2 u_z}{\mathrm{d}x^2}\mathrm{d}x\,. \tag{3.47}$$

Combination of Eqs. (3.46) and (3.47) gives under consideration of Eq. (3.45) the so-called *weak formulation* of the problem as:

$$EI_y\int_0^L \frac{\mathrm{d}^2 W^{\mathrm{T}}}{\mathrm{d}x^2}\frac{\mathrm{d}^2 u_z}{\mathrm{d}x^2}\mathrm{d}x = EI_y\left[-W^{\mathrm{T}}\frac{\mathrm{d}^3 u_z}{\mathrm{d}x^3} + \frac{\mathrm{d}W^{\mathrm{T}}}{\mathrm{d}x}\frac{\mathrm{d}^2 u_z}{\mathrm{d}x^2}\right]_0^L + \int_0^L W^{\mathrm{T}} q_z\,\mathrm{d}x\,. \tag{3.48}$$

Looking at the weak formulation, it can be seen that the integration by parts shifted two derivatives from the approximate solution to the weight function and a symmetrical formulation with respect to the derivatives is obtained. This symmetry with respect to the derivatives of the approximate solution and the weight function will again guarantee in the following—as in the case of the rod element—that a symmetric stiffness matrix is obtained for the beam element. In order to continue the derivation of the principal finite element equation, the displacement $u_z(x)$ and the weight function $W(x)$ must be expressed within an element (superscript 'e') in the form of the nodal approach. This nodal approach can be generally stated as

$$u_z^{\mathrm{e}}(x) = \boldsymbol{N}^{\mathrm{T}}(x)\,\boldsymbol{u}_{\mathrm{p}}\,, \tag{3.49}$$

$$W(x) = \boldsymbol{N}^{\mathrm{T}}\delta\boldsymbol{u}_{\mathrm{p}}(x)\,, \tag{3.50}$$

[19]The use of the transposed 'T' for the scalar weight function W is not obvious at the first glance. However, the following matrix operations will clarify this approach.

[20]A common representation of integration by parts of two functions $f(x)$ and $g(x)$ is: $\int fg'\,\mathrm{d}x = fg - \int f'g\,\mathrm{d}x$.

where $\boldsymbol{N}(x)$ is the column matrix of the interpolation functions, $\boldsymbol{u}_{\mathrm{p}}$ is the column matrix of the nodal unknowns and $\delta\boldsymbol{u}_{\mathrm{p}}$ is the column matrix of the virtual displacements.

The kinematics relation according to Eq. (3.16) allows to express the strain distribution within an element based on the nodal approach as:

$$\varepsilon_x^{\mathrm{e}}(x,z) = -z\frac{\mathrm{d}^2 u_z^{\mathrm{e}}(x)}{\mathrm{d}x^2} = -z\frac{\mathrm{d}^2}{\mathrm{d}x^2}\left(\boldsymbol{N}^{\mathrm{T}}(x)\,\boldsymbol{u}_{\mathrm{p}}\right) = -z\frac{\mathrm{d}^2\boldsymbol{N}^{\mathrm{T}}(x)}{\mathrm{d}x^2}\boldsymbol{u}_{\mathrm{p}}\,. \tag{3.51}$$

Analogous to the procedure in Sect. 2.3.1, one can introduce for the beam element a generalized $\boldsymbol{B}$-matrix. Thus, an equivalent description as in Eq. (2.147), i.e. $\varepsilon_x^{\mathrm{e}}(x) = \boldsymbol{B}^{\mathrm{T}}\boldsymbol{u}_{\mathrm{p}}$, is obtained with:

$$\boldsymbol{B}^{\mathrm{T}} = -z\frac{\mathrm{d}^2\boldsymbol{N}^{\mathrm{T}}(x)}{\mathrm{d}x^2}\,. \tag{3.52}$$

This definition of the $\boldsymbol{B}$-matrix allows a formal derivation (cf. [10]) of the stiffness matrix based on the general formulation as presented in Eq. (8.34), i.e. $\boldsymbol{K}^{\mathrm{e}} = \int_V \boldsymbol{B}\boldsymbol{C}\boldsymbol{B}^{\mathrm{T}}\mathrm{d}V$. If a formulation corresponding to Eq. (2.33), i.e. $\boldsymbol{K}^{\mathrm{e}} = EI_y\int_x \boldsymbol{B}\boldsymbol{B}^{\mathrm{T}}\mathrm{d}x$, is the goal of the derivation, the definition of the $\boldsymbol{B}$-matrix should be rather given as $\boldsymbol{B}^{\mathrm{T}} = \frac{\mathrm{d}^2\boldsymbol{N}^{\mathrm{T}}(x)}{\mathrm{d}x^2}$.

Let us introduce now the formulations for $u_z^{\mathrm{e}}(x)$ and $W(x)$ according to Eqs. (3.49) and (3.50) in the weak formulation (3.48):

$$EI_y\int_0^L \frac{\mathrm{d}^2}{\mathrm{d}x^2}\left(\boldsymbol{N}^{\mathrm{T}}(x)\delta\boldsymbol{u}_{\mathrm{p}}\right)^{\mathrm{T}}\frac{\mathrm{d}^2}{\mathrm{d}x^2}\left(\boldsymbol{N}^{\mathrm{T}}(x)\,\boldsymbol{u}_{\mathrm{p}}\right)\mathrm{d}x = EI_y\left[-\left(\boldsymbol{N}^{\mathrm{T}}(x)\delta\boldsymbol{u}_{\mathrm{p}}\right)^{\mathrm{T}}\frac{\mathrm{d}^3u_z}{\mathrm{d}x^3}+\right.$$

$$\left.+\frac{\mathrm{d}}{\mathrm{d}x}\left(\boldsymbol{N}^{\mathrm{T}}(x)\delta\boldsymbol{u}_{\mathrm{p}}\right)^{\mathrm{T}}\frac{\mathrm{d}^2u_z}{\mathrm{d}x^2}\right]_0^L + \int_0^L\left(\boldsymbol{N}^{\mathrm{T}}(x)\delta\boldsymbol{u}_{\mathrm{p}}\right)^{\mathrm{T}}q_z(x)\,\mathrm{d}x\,. \tag{3.53}$$

Since the virtual displacements $\delta\boldsymbol{u}_{\mathrm{p}}$ and the displacements $\boldsymbol{u}_{\mathrm{p}}$ are not a function of x, they can be considered as constants with respect to the integration and can be taken out of the integral on the left-hand side of Eq. (3.53). Furthermore, the virtual displacements $\delta\boldsymbol{u}_{\mathrm{p}}$ occur on both sides of Eq. (3.53) and can be 'canceled'. Considering $W^{\mathrm{T}}(x) = \left(\boldsymbol{N}(x)^{\mathrm{T}}\delta\boldsymbol{u}_{\mathrm{p}}\right)^{\mathrm{T}} = \delta\boldsymbol{u}_{\mathrm{p}}^{\mathrm{T}}\boldsymbol{N}(x)$, the weak formulation takes the following form:

$$\underbrace{EI_y\int_0^L\frac{\mathrm{d}^2}{\mathrm{d}x^2}\left(\boldsymbol{N}(x)\right)\frac{\mathrm{d}^2}{\mathrm{d}x^2}\left(\boldsymbol{N}^{\mathrm{T}}(x)\right)\mathrm{d}x}_{\boldsymbol{K}^{\mathrm{e}}}\;\boldsymbol{u}_{\mathrm{p}}^{\mathrm{e}} =$$

$$\underbrace{EI_y\left[-\boldsymbol{N}(x)\frac{\mathrm{d}^3u_z}{\mathrm{d}x^3}+\frac{\mathrm{d}}{\mathrm{d}x}\left(\boldsymbol{N}(x)\right)\frac{\mathrm{d}^2u_z}{\mathrm{d}x^2}\right]_0^L+\int_0^L\boldsymbol{N}(x)\,q_z(x)\,\mathrm{d}x}_{\boldsymbol{f}^{\mathrm{e}}}\,. \tag{3.54}$$

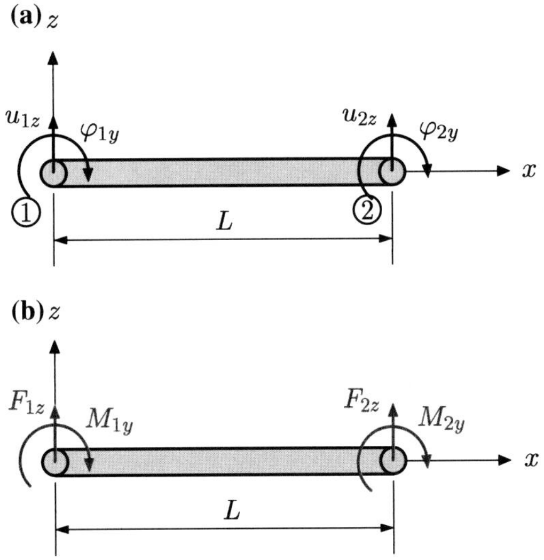

Fig. 3.13 Definition of the BERNOULLI beam element for deformation in the x-z-plane: **a** deformations; **b** external loads. The nodes are symbolized by two circles at the ends (◯)

In order to continue the derivation of the principal finite element equation, it is at this point now required that the column matrix of the interpolation functions $\boldsymbol{N}$ and the column matrix of the nodal unknowns $\boldsymbol{u}_{\mathrm{p}}^{\mathrm{e}}$ are further specified.

Let us consider in the following a BERNOULLI beam element which is composed of two nodes as schematically shown in Fig. 3.13. Each node has two degrees of freedom, i.e. a displacement u_z and a rotation $\varphi_y = -\frac{\mathrm{d}u_z}{\mathrm{d}x}$, and can be loaded by a single force F_z and a bending moment M_y. Since the BERNOULLI beam element has two nodes with in total four nodal unknowns, the nodal approaches given in Eqs. (3.49) and (3.50) can be stated as[21]

$$u_z^{\mathrm{e}}(x) = \boldsymbol{N}^{\mathrm{T}}(x)\boldsymbol{u}_{\mathrm{p}}^{\mathrm{e}} = \begin{bmatrix} N_{1u} & N_{1\varphi} & N_{2u} & N_{2\varphi} \end{bmatrix} \times \begin{bmatrix} u_{1z} \\ \varphi_{1y} \\ u_{2z} \\ \varphi_{2y} \end{bmatrix} \tag{3.55}$$

and

$$W(x) = \boldsymbol{N}^{\mathrm{T}}(x)\delta\boldsymbol{u}_{\mathrm{p}} = \begin{bmatrix} N_{1u} & N_{1\varphi} & N_{2u} & N_{2\varphi} \end{bmatrix} \times \begin{bmatrix} \delta u_{1z} \\ \delta\varphi_{1y} \\ \delta u_{2z} \\ \delta\varphi_{2y} \end{bmatrix}, \tag{3.56}$$

where N_u are the interpolation functions for the displacement field and N_φ for the rotational field.

[21] A detailed description of the derivation of the nodal approach and the respective interpolation functions for the BERNOULLI beam element is given in Sect. 3.3.2.

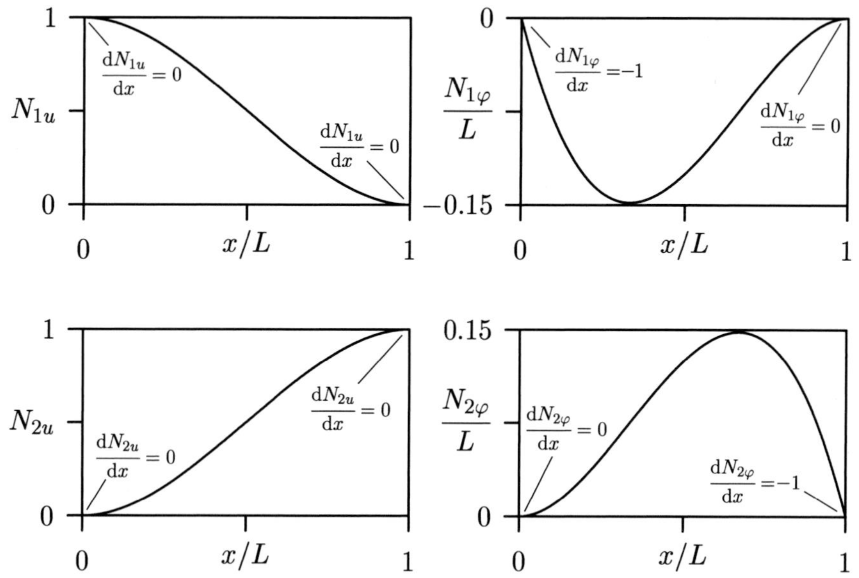

Fig. 3.14 Interpolation functions for the BERNOULLI beam element given in the physical coordinate for bending in the x-z plane

These four interpolation functions are commonly taken from the family of cubic HERMITE[22] interpolation functions as:

$$N_{1u}(x) = 1 - 3\left(\frac{x}{L}\right)^2 + 2\left(\frac{x}{L}\right)^3, \tag{3.57}$$

$$N_{1\varphi}(x) = -x + 2\frac{x^2}{L} - \frac{x^3}{L^2}, \tag{3.58}$$

$$N_{2u}(x) = 3\left(\frac{x}{L}\right)^2 - 2\left(\frac{x}{L}\right)^3, \tag{3.59}$$

$$N_{2\varphi}(x) = \frac{x^2}{L} - \frac{x^3}{L^2}. \tag{3.60}$$

The graphical representation of these cubic interpolation functions is given in Fig. 3.14.

Let us first consider in the following only the left-hand side of Eq. (3.54) in order to derive the expression for the elemental stiffness matrix $\boldsymbol{K}^{\mathrm{e}}$ of the BERNOULLI beam

[22]Charles HERMITE (1822–1901), French mathematician.

element. Introducing the components of the interpolation function column matrix gives

$$\boldsymbol{K}^{\mathrm{e}} = EI_y \int_0^L \begin{bmatrix} \frac{\mathrm{d}^2 N_{1u}}{\mathrm{d}x^2} \\ \frac{\mathrm{d}^2 N_{1\varphi}}{\mathrm{d}x^2} \\ \frac{\mathrm{d}^2 N_{2u}}{\mathrm{d}x^2} \\ \frac{\mathrm{d}^2 N_{2\varphi}}{\mathrm{d}x^2} \end{bmatrix} \begin{bmatrix} \frac{\mathrm{d}^2 N_{1u}}{\mathrm{d}x^2} & \frac{\mathrm{d}^2 N_{1\varphi}}{\mathrm{d}x^2} & \frac{\mathrm{d}^2 N_{2u}}{\mathrm{d}x^2} & \frac{\mathrm{d}^2 N_{2\varphi}}{\mathrm{d}x^2} \end{bmatrix} \mathrm{d}x\,, \tag{3.61}$$

or after performing the multiplication as:

$$\boldsymbol{K}^{\mathrm{e}} = EI_y \int_0^L \left[\begin{array}{c|c|c|c} \frac{\mathrm{d}^2 N_{1u}}{\mathrm{d}x^2}\frac{\mathrm{d}^2 N_{1u}}{\mathrm{d}x^2} & \frac{\mathrm{d}^2 N_{1u}}{\mathrm{d}x^2}\frac{\mathrm{d}^2 N_{1\varphi}}{\mathrm{d}x^2} & \frac{\mathrm{d}^2 N_{1u}}{\mathrm{d}x^2}\frac{\mathrm{d}^2 N_{2u}}{\mathrm{d}x^2} & \frac{\mathrm{d}^2 N_{1u}}{\mathrm{d}x^2}\frac{\mathrm{d}^2 N_{2\varphi}}{\mathrm{d}x^2} \\ \hline \frac{\mathrm{d}^2 N_{1\varphi}}{\mathrm{d}x^2}\frac{\mathrm{d}^2 N_{1u}}{\mathrm{d}x^2} & \frac{\mathrm{d}^2 N_{1\varphi}}{\mathrm{d}x^2}\frac{\mathrm{d}^2 N_{1\varphi}}{\mathrm{d}x^2} & \frac{\mathrm{d}^2 N_{1\varphi}}{\mathrm{d}x^2}\frac{\mathrm{d}^2 N_{2u}}{\mathrm{d}x^2} & \frac{\mathrm{d}^2 N_{1\varphi}}{\mathrm{d}x^2}\frac{\mathrm{d}^2 N_{2\varphi}}{\mathrm{d}x^2} \\ \hline \frac{\mathrm{d}^2 N_{2u}}{\mathrm{d}x^2}\frac{\mathrm{d}^2 N_{1u}}{\mathrm{d}x^2} & \frac{\mathrm{d}^2 N_{2u}}{\mathrm{d}x^2}\frac{\mathrm{d}^2 N_{1\varphi}}{\mathrm{d}x^2} & \frac{\mathrm{d}^2 N_{2u}}{\mathrm{d}x^2}\frac{\mathrm{d}^2 N_{2u}}{\mathrm{d}x^2} & \frac{\mathrm{d}^2 N_{2u}}{\mathrm{d}x^2}\frac{\mathrm{d}^2 N_{2\varphi}}{\mathrm{d}x^2} \\ \hline \frac{\mathrm{d}^2 N_{2\varphi}}{\mathrm{d}x^2}\frac{\mathrm{d}^2 N_{1u}}{\mathrm{d}x^2} & \frac{\mathrm{d}^2 N_{2\varphi}}{\mathrm{d}x^2}\frac{\mathrm{d}^2 N_{1\varphi}}{\mathrm{d}x^2} & \frac{\mathrm{d}^2 N_{2\varphi}}{\mathrm{d}x^2}\frac{\mathrm{d}^2 N_{2u}}{\mathrm{d}x^2} & \frac{\mathrm{d}^2 N_{2\varphi}}{\mathrm{d}x^2}\frac{\mathrm{d}^2 N_{2\varphi}}{\mathrm{d}x^2} \end{array}\right] \mathrm{d}x\,. \tag{3.62}$$

The derivatives of the interpolation functions according to Eq. (3.57) till (3.60) can be calculated as

$$\frac{\mathrm{d}N_{1u}(x)}{\mathrm{d}x} = -\frac{6x}{L^2} + \frac{6x^2}{L^3}\,, \tag{3.63}$$

$$\frac{\mathrm{d}N_{1\varphi}(x)}{\mathrm{d}x} = -1 + \frac{4x}{L} - \frac{3x^2}{L^2}\,, \tag{3.64}$$

$$\frac{\mathrm{d}N_{2u}(x)}{\mathrm{d}x} = \frac{6x}{L^2} - \frac{6x^2}{L^3}\,, \tag{3.65}$$

$$\frac{\mathrm{d}N_{2\varphi}(x)}{\mathrm{d}x} = \frac{2x}{L} - \frac{3x^2}{L^2}\,, \tag{3.66}$$

and accordingly the second-order derivatives as

$$\frac{\mathrm{d}^2 N_{1u}(x)}{\mathrm{d}x^2} = -\frac{6}{L^2} + \frac{12x}{L^3}\,, \tag{3.67}$$

$$\frac{\mathrm{d}^2 N_{1\varphi}(x)}{\mathrm{d}x^2} = \frac{4}{L} - \frac{6x}{L^2}\,, \tag{3.68}$$

$$\frac{\mathrm{d}^2 N_{2u}(x)}{\mathrm{d}x^2} = \frac{6}{L^2} - \frac{12x}{L^3}, \tag{3.69}$$

$$\frac{\mathrm{d}^2 N_{2\varphi}(x)}{\mathrm{d}x^2} = \frac{2}{L} - \frac{6x}{L^2}. \tag{3.70}$$

These derivatives introduced into Eq. (3.62) give after analytical integration of the polynomials the elemental stiffness matrix of the BERNOULLI beam element as:

$$\boldsymbol{K}^{\mathrm{e}} = \frac{EI_y}{L^3} \begin{bmatrix} 12 & -6L & -12 & -6L \\ -6L & 4L^2 & 6L & 2L^2 \\ -12 & 6L & 12 & 6L \\ -6L & 2L^2 & 6L & 4L^2 \end{bmatrix}. \tag{3.71}$$

It must be noted here that the analytical integration as performed to obtain Eq. (3.71) cannot be performed in commercial finite element codes since they are written in traditional programming languages such as FORTRAN. As in the case of the rod element, a numerical integration is performed (cf. Appendix A.9) by GAUSS–LEGENDRE quadrature. Transforming the interpolation functions, cf. Eqs. (3.57) till (3.60), from the Cartesian coordinate x to its natural coordinate ξ based on the transformation given in Table 2.4 gives:

$$N_{1u}(\xi) = \frac{1}{4}\left[2 - 3\xi + \xi^3\right], \tag{3.72}$$

$$N_{1\varphi}(\xi) = -\frac{1}{4}\left[1 - \xi - \xi^2 + \xi^3\right]\frac{L}{2}, \tag{3.73}$$

$$N_{2u}(\xi) = \frac{1}{4}\left[2 + 3\xi - \xi^3\right], \tag{3.74}$$

$$N_{1\varphi}(\xi) = -\frac{1}{4}\left[-1 - \xi + \xi^2 + \xi^3\right]\frac{L}{2}. \tag{3.75}$$

The second-order derivatives which are required for Eq. (3.62) can easily be derived as:

$$\frac{\mathrm{d}^2 N_{1u}(\xi)}{\mathrm{d}x^2} = \frac{4}{L^2}\frac{\mathrm{d}^2 N_{1u}(\xi)}{\mathrm{d}\xi^2} = \frac{6}{L^2}\xi, \tag{3.76}$$

$$\frac{\mathrm{d}^2 N_{1\varphi}(\xi)}{\mathrm{d}x^2} = \frac{4}{L^2}\frac{\mathrm{d}^2 N_{1\varphi}(\xi)}{\mathrm{d}\xi^2} = -\frac{1}{L}(-1 + 3\xi), \tag{3.77}$$

$$\frac{\mathrm{d}^2 N_{2u}(\xi)}{\mathrm{d}x^2} = \frac{4}{L^2}\frac{\mathrm{d}^2 N_{2u}(\xi)}{\mathrm{d}\xi^2} = -\frac{6}{L^2}\xi, \tag{3.78}$$

$$\frac{\mathrm{d}^2 N_{2\varphi}(\xi)}{\mathrm{d}x^2} = \frac{4}{L^2}\frac{\mathrm{d}^2 N_{2\varphi}(\xi)}{\mathrm{d}\xi^2} = -\frac{1}{L}(1 + 3\xi). \tag{3.79}$$

Introducing these relationships in Eq. (3.62), one receives

$$
EI_z \int_{-1}^{1} \left[\begin{array}{c|c|c|c}
\frac{36}{L^4}\xi^2 & -\frac{6}{L^3}(-\xi+3\xi^2) & -\frac{36}{L^4}\xi^2 & -\frac{6}{L^3}(\xi+3\xi^2) \\ \hline
-\frac{6}{L^3}(-\xi+3\xi^2) & \frac{1}{L^2}(-1+3\xi)^2 & \frac{6}{L^3}(-\xi+3\xi^2) & \frac{1}{L^2}(9\xi^2-1) \\ \hline
-\frac{36}{L^4}\xi^2 & \frac{6}{L^3}(-\xi+3\xi^2) & \frac{36}{L^4}\xi^2 & \frac{6}{L^3}(\xi+3\xi^2) \\ \hline
-\frac{6}{L^3}(\xi+3\xi^2) & \frac{1}{L^2}(9\xi^2-1) & \frac{6}{L^3}(\xi+3\xi^2) & \frac{1}{L^2}(1+3\xi)^2
\end{array}\right] \frac{L\mathrm{d}\xi}{2}. \tag{3.80}
$$

The polynomials included in Eq. (3.80) are of maximum order of three and thus, a two-point integration rule (cf. Table A.5) is sufficient to accurately integrate as:

$$
\begin{aligned}
\boldsymbol{K}^{\mathrm{e}} = EI_y \int_{-1}^{1} \left[\ \cdots\ \right] \frac{L}{2}\,\mathrm{d}\xi \approx \frac{EI_yL}{2}\left[\ \cdots\ \right]\Bigg|_{\xi=\frac{1}{\sqrt{3}}} \times 1 + \\
+ \frac{EI_yL}{2}\left[\ \cdots\ \right]\Bigg|_{\xi=-\frac{1}{\sqrt{3}}} \times 1\,,
\end{aligned} \tag{3.81}
$$

which gives the same result for the stiffness matrix as the analytical integration shown in Eq. (3.71).

The transformation between Cartesian (x) and natural coordinates (ξ) and the integration over the natural coordinate can be further generalized. Let us assume that the interpolation functions $N(\xi)$ as given in Eqs. (3.76)–(3.79) are known or derived. Then, the second-order derivative in Eq. (3.62) can be expressed, for example, for the first component as:

$$
\frac{\mathrm{d}^2 N_{1u}(\xi)}{\mathrm{d}x^2} = \frac{\mathrm{d}}{\mathrm{d}x}\left(\frac{\mathrm{d}N_{1u}(\xi)}{\mathrm{d}\xi}\frac{\mathrm{d}\xi}{\mathrm{d}x}\right), \tag{3.82}
$$

$$
\begin{aligned}
&= \frac{\mathrm{d}}{\mathrm{d}x}\left(\frac{\mathrm{d}N_{1u}(\xi)}{\mathrm{d}\xi}\right)\frac{\mathrm{d}\xi}{\mathrm{d}x} + \frac{\mathrm{d}N_{1u}(\xi)}{\mathrm{d}\xi}\frac{\mathrm{d}^2\xi}{\mathrm{d}x^2}, \\
&= \frac{\mathrm{d}^2 N_{1u}(\xi)}{\mathrm{d}\xi^2}\left(\frac{\mathrm{d}\xi}{\mathrm{d}x}\right)^2 + \frac{\mathrm{d}N_{1u}(\xi)}{\mathrm{d}\xi}\frac{\mathrm{d}^2\xi}{\mathrm{d}x^2}.
\end{aligned} \tag{3.83}
$$

The last equation requires the evaluation of geometrical derivatives and the same interpolation of the coordinate as in the case of the rod element can be applied, see

Eqs. (2.44) and (2.45):

$$\frac{\mathrm{d}x(\xi)}{\mathrm{d}\xi} = \frac{\mathrm{d}\overline{N}_1(\xi)}{\mathrm{d}\xi}x_1 + \frac{\mathrm{d}\overline{N}_2(\xi)}{\mathrm{d}\xi}x_2 = \frac{L}{2}. \tag{3.84}$$

Based on the results $\frac{\mathrm{d}\xi}{\mathrm{d}x} = \frac{2}{L}$ and $\frac{\mathrm{d}^2\xi}{\mathrm{d}x^2} = 0$, Eq. (3.83) gives the same result as Eq. (3.76). Considering that the shape functions $\overline{N}_i$ are different to the interpolation functions N_i, or more specifically $\deg(\overline{N}) < \deg(N)$, the above derivation is an example for a subparametric element formulation.

Let us now consider the right-hand side of Eq. (3.54) in order to derive the expression for the load column matrix $\boldsymbol{f}^{\mathrm{e}}$ of the beam element. The first part of the right-hand side, i.e.

$$EI_y\left[-N(x)\frac{\mathrm{d}^3u_z}{\mathrm{d}x^3} + \frac{\mathrm{d}}{\mathrm{d}x}(N(x))\frac{\mathrm{d}^2u_z}{\mathrm{d}x^2}\right]_0^L \tag{3.85}$$

results with the definition of the column matrix of interpolation functions according to Eq. (3.55) in

$$EI_y\left[-\begin{bmatrix} N_{1u} \\ N_{1\varphi} \\ N_{2u} \\ N_{2\varphi} \end{bmatrix}\frac{\mathrm{d}^3u_z}{\mathrm{d}x^3} + \frac{\mathrm{d}}{\mathrm{d}x}\begin{bmatrix} N_{1u} \\ N_{1\varphi} \\ N_{2u} \\ N_{2\varphi} \end{bmatrix}\frac{\mathrm{d}^2u_z}{\mathrm{d}x^2}\right]_0^L . \tag{3.86}$$

Equation (3.86) represents a system of four equations which are to be evaluated at the boundaries of the integration, i.e. $x = 0$ and $x = L$. The first row of Eq. (3.86) gives:

$$\left(-N_{1u}EI_y\frac{\mathrm{d}^3u_z}{\mathrm{d}x^3} + \frac{\mathrm{d}N_{1u}}{\mathrm{d}x}\frac{\mathrm{d}^2u_z}{\mathrm{d}x^2}\right)_{x=L} - \left(-N_{1u}EI_y\frac{\mathrm{d}^3u_z}{\mathrm{d}x^3} + \frac{\mathrm{d}N_{1u}}{\mathrm{d}x}\frac{\mathrm{d}^2u_z}{\mathrm{d}x^2}\right)_{x=0} . \tag{3.87}$$

Under consideration of the boundary values of the interpolation functions respectively their derivatives according to Fig. 3.14, i.e. $N_{1u}(L) = 0$, $\frac{\mathrm{d}N_{1u}}{\mathrm{d}x}(L) = \frac{\mathrm{d}N_{1u}}{\mathrm{d}x}(0) = 0$ and $N_{1u}(0) = 1$, one receives the following expression:

$$+EI_y\left.\frac{\mathrm{d}^3u_z}{\mathrm{d}x^3}\right|_{x=0} \overset{(3.40)}{=} -Q_z(0)\,. \tag{3.88}$$

Corresponding expressions can be derived for the other rows of Eq. (3.86):

$$\text{Row 2:} \qquad -EI_y\left.\frac{\mathrm{d}^2u_z}{\mathrm{d}x^2}\right|_{x=0} \overset{(3.39)}{=} +M_y(0)\,, \tag{3.89}$$

$$\text{Row 3:} \qquad -EI_y \left.\frac{\mathrm{d}^3 u_z}{\mathrm{d}x^3}\right|_{x=L} \overset{(3.40)}{=} +Q_z(L)\,, \tag{3.90}$$

$$\text{Row 4:} \qquad +EI_y \left.\frac{\mathrm{d}^2 u_z}{\mathrm{d}x^2}\right|_{x=L} \overset{(3.39)}{=} -M_y(L)\,. \tag{3.91}$$

It must be noted here that shear forces Q_z and bending moments M_y are the internal reactions according to Fig. 3.12. Furthermore, the minus sign in Eqs. (3.89) and (3.90) should be transferred to the right-hand side to correctly interpret the meaning of these equations (see the definitions in Eqs. 3.39 and 3.40). The external loads with their positive directions according to Fig. 3.13 can be obtained from the internal loads by inverting the sign at the left-hand boundary and by maintaining the positive direction of the internal reactions at the right-hand boundary, see Fig. 3.15.

Thus, the contribution to the load matrix due to single forces F_{iz} and moments M_{iy} at the nodes is expressed by:

$$\boldsymbol{f}^{\mathrm{e}}_{FM} = \begin{bmatrix} F_{1z} \\ M_{1y} \\ F_{2z} \\ M_{2y} \end{bmatrix}. \tag{3.92}$$

The second part of Eq. (3.54), i.e.

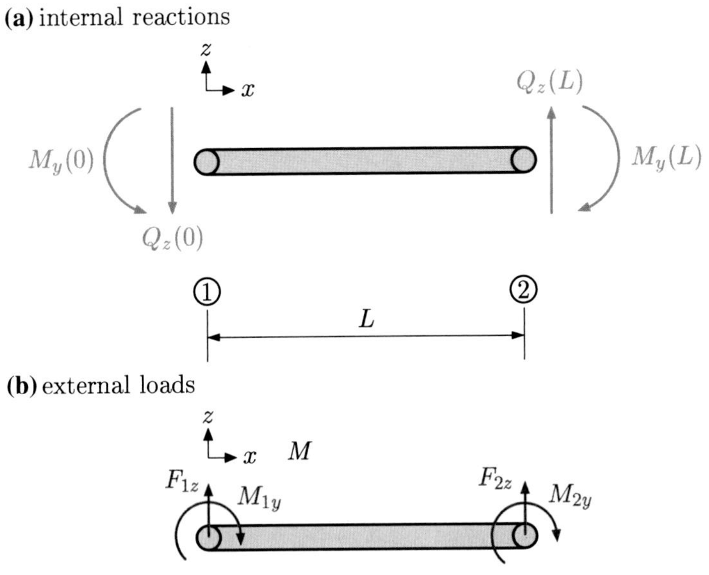

Fig. 3.15 Positive definition of **a** internal reactions and **b** external loads

$$\int_0^L \boldsymbol{N}(x)\, q_z(x)\, \mathrm{d}x \tag{3.93}$$

represents the general rule to determine equivalent nodal loads in the case of arbitrarily distributed loads $q_z(x)$. As an example, the evaluation of Eq. (3.93) for a constant load $+q_z$ results in the following load matrix:

$$\boldsymbol{f}_q^{\mathrm{e}} = q_z \int_0^L \begin{bmatrix} N_{1u}(x) \\ N_{1\varphi}(x) \\ N_{2u}(x) \\ N_{2\varphi}(x) \end{bmatrix} \mathrm{d}x = \frac{q_z L}{2} \begin{bmatrix} 1 \\ -\frac{L}{6} \\ 1 \\ +\frac{L}{6} \end{bmatrix}. \tag{3.94}$$

Further expressions for equivalent nodal loads can be taken from Table 3.7. Let us remind at this step that in the scope of the finite element method any type of loading can be only introduced at nodes into the discretized structure.

Based on the derived results, the principal finite element equation for a single BERNOULLI beam element with constant bending stiffness EI_y can be expressed in a general form as

$$\boldsymbol{K}^{\mathrm{e}} \boldsymbol{u}_{\mathrm{p}}^{\mathrm{e}} = \boldsymbol{f}^{\mathrm{e}}, \tag{3.95}$$

or in components as

$$\frac{EI_y}{L^3} \begin{bmatrix} 12 & -6L & -12 & -6L \\ -6L & 4L^2 & 6L & 2L^2 \\ -12 & 6L & 12 & 6L \\ -6L & 2L^2 & 6L & 4L^2 \end{bmatrix} \begin{bmatrix} u_{1z} \\ \varphi_{1y} \\ u_{2z} \\ \varphi_{2y} \end{bmatrix} = \begin{bmatrix} F_{1z} \\ M_{1y} \\ F_{2z} \\ M_{2y} \end{bmatrix} + \int_0^L q_z(x) \begin{bmatrix} N_{1u} \\ N_{1\varphi} \\ N_{2u} \\ N_{2\varphi} \end{bmatrix} \mathrm{d}x. \tag{3.96}$$

Let us summarize at the end of this section the major steps that were undertaken to transform the partial differential equation into the principal finite element equation, see Table 3.8.

The following description is related to a more formalized derivation of the principal finite element equation (this approach will be consistently use do derive the principal finite element equation for two- and three-dimensional elements). Based on the general formulation of the partial differential equation given in Table 3.6, the strong formulation can be written as[23]:

$$\mathcal{L}_2^{\mathrm{T}} \left(D \mathcal{L}_2 u_z^0(x) \right) - q_z(x) = 0. \tag{3.97}$$

Replacing the exact solution u_z^0 by an approximate solution u_z, a residual r is obtained:

$$r = \mathcal{L}_2^{\mathrm{T}} \left(D \mathcal{L}_2 u_z(x) \right) - q_z(x) \neq 0. \tag{3.98}$$

---

[23]The use of the transposed 'T' for the scalar operator $\mathcal{L}_2 = \frac{\mathrm{d}^2}{\mathrm{d}x^2}$ is not obvious at the first glance. However, the following matrix operations will clarify this approach.

Table 3.7 Equivalent nodal loadings for a BERNOULLI beam element (x-axis: right facing; z-axis: upward facing), partially adapted from [3]

Loading	Shear force	Bending moment
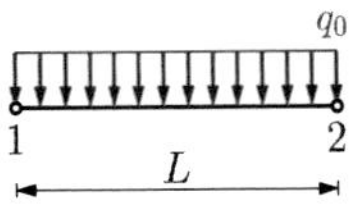	$F_{1z} = -\frac{q_0 L}{2}$	$M_{1y} = +\frac{q_0 L^2}{12}$
	$F_{2z} = -\frac{q_0 L}{2}$	$M_{2y} = -\frac{q_0 L^2}{12}$
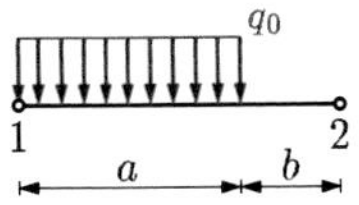	$F_{1z} = -\frac{q_0 a}{2L^3}(a^3 - 2a^2 L + 2L^3)$	$M_{1y} = +\frac{q_0 a^2}{12L^2}(3a^2 - 8aL + 6L^2)$
	$F_{2z} = -\frac{q_0 a^3}{2L^3}(2L - a)$	$M_{2y} = -\frac{q_0 a^3}{12L^2}(4L - 3a)$
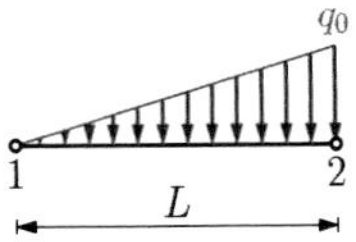	$F_{1z} = -\frac{3}{20} q_0 L$	$M_{1y} = +\frac{q_0 L^2}{30}$
	$F_{2z} = -\frac{7}{20} q_0 L$	$M_{2y} = -\frac{q_0 L^2}{20}$
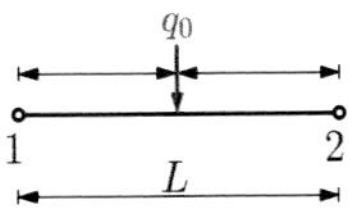	$F_{1z} = -\frac{1}{4} q_0 L$	$M_{1y} = +\frac{5 q_0 L^2}{96}$
	$F_{2z} = -\frac{1}{4} q_0 L$	$M_{2y} = -\frac{5 q_0 L^2}{96}$
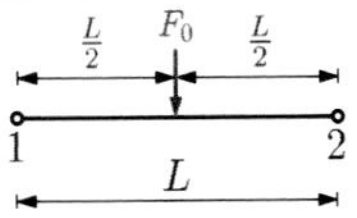	$F_{1z} = -\frac{F_0}{2}$	$M_{1y} = +\frac{F_0 L}{8}$
	$F_{2z} = -\frac{F_0}{2}$	$M_{2y} = -\frac{F_0 L}{8}$
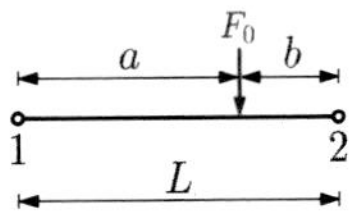	$F_{1z} = -\frac{F_0 b^2 (3a + b)}{L^3}$	$M_{1y} = +\frac{F_0 b^2 a}{L^2}$
	$F_{2z} = -\frac{F_0 a^2 (a + 3b)}{L^3}$	$M_{2y} = -\frac{F_0 a^2 b}{L^2}$
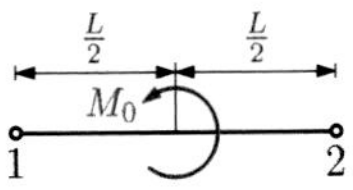	$F_{1z} = -\frac{3}{2}\frac{M_0}{L}$	$M_{1y} = +\frac{M_0}{4}$
	$F_{2z} = +\frac{3}{2}\frac{M_0}{L}$	$M_{2y} = +\frac{M_0}{4}$
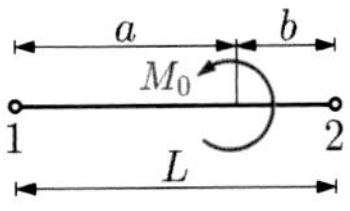	$F_{1z} = -6M_0 \frac{ab}{L^3}$	$M_{1y} = +M_0 \frac{b(2a - b)}{L^2}$
	$F_{2z} = +6M_0 \frac{ab}{L^3}$	$M_{2y} = +M_0 \frac{a(2b - a)}{L^2}$

Table 3.8 Summary: derivation of principal finite element equation for EULER–BERNOULLI beam elements (bending occurs in the x-z plane)

Strong formulation
$EI_y \dfrac{\mathrm{d}^4 u_z^0(x)}{\mathrm{d}x^4} - q_z(x) = 0$
Inner product
$\int_0^L W^{\mathrm{T}}(x) \left(EI_y \dfrac{\mathrm{d}^4 u_z(x)}{\mathrm{d}x^4} - q_z(x) \right) \mathrm{d}x \stackrel{!}{=} 0$
Weak formulation
$EI_y \int_0^L \dfrac{\mathrm{d}^2 W^{\mathrm{T}}}{\mathrm{d}x^2} \dfrac{\mathrm{d}^2 u_z}{\mathrm{d}x^2} \mathrm{d}x = EI_y \left[-W^{\mathrm{T}} \dfrac{\mathrm{d}^3 u_z}{\mathrm{d}x^3} + \dfrac{\mathrm{d}W^{\mathrm{T}}}{\mathrm{d}x} \dfrac{\mathrm{d}^2 u_z}{\mathrm{d}x^2} \right]_0^L + \int_0^L W^{\mathrm{T}} q_z \, \mathrm{d}x$
Principal finite element equation
$\dfrac{EI_y}{L^3} \begin{bmatrix} 12 & -6L & -12 & -6L \\ -6L & 4L^2 & 6L & 2L^2 \\ -12 & 6L & 12 & 6L \\ -6L & 2L^2 & 6L & 4L^2 \end{bmatrix} \begin{bmatrix} u_{1z} \\ \varphi_{1y} \\ u_{2z} \\ \varphi_{2y} \end{bmatrix} = \begin{bmatrix} F_{1z} \\ M_{1y} \\ F_{2z} \\ M_{2y} \end{bmatrix} + \int_0^L q_z(x) \begin{bmatrix} N_{1u} \\ N_{1\varphi} \\ N_{2u} \\ N_{2\varphi} \end{bmatrix} \mathrm{d}x$

The inner product is obtained by weighting the residual and integration as

$$\int_L W^{\mathrm{T}}(x) \left(\mathcal{L}_2^{\mathrm{T}} \left(D \mathcal{L}_2 u_z(x) \right) - q_z(x) \right) \mathrm{d}L \stackrel{!}{=} 0 \,, \tag{3.99}$$

where $W(x)$ is the scalar weight function. Any further development of Eq. (3.99) requires the following intermediate step:

$$\int_L W^{\mathrm{T}}(x) \left(\mathcal{L}_1^{\mathrm{T}} \left[\mathcal{L}_1^{\mathrm{T}} \left(D \mathcal{L}_2 u_z(x) \right) \right] - q_z(x) \right) \mathrm{d}L \stackrel{!}{=} 0 \,. \tag{3.100}$$

A first application of the GREEN–GAUSS theorem (cf. Sect. A.7) to the first expression in Eq. (3.100) gives:

$$\int_L W^{\mathrm{T}} \left(\mathcal{L}_1^{\mathrm{T}} \left[\mathcal{L}_1^{\mathrm{T}} \left(D \mathcal{L}_2 u_z(x) \right) \right] \right) \mathrm{d}L = - \int_L \left(\mathcal{L}_1 W \right)^{\mathrm{T}} \left[\mathcal{L}_1^{\mathrm{T}} \left(D \mathcal{L}_2 u_z(x) \right) \right] \mathrm{d}L + \int_s W^{\mathrm{T}} \left[\mathcal{L}_1^{\mathrm{T}} \left(D \mathcal{L}_2 u_z(x) \right) \right] n \mathrm{d}s \,, \tag{3.101}$$

where $\mathcal{L}_1 = \frac{\mathrm{d}}{\mathrm{d}x}$. Application of the GREEN–GAUSS theorem to the first expression on the right-hand side of Eq. (3.101) results in:

$$\int_L (\mathcal{L}_1 W)^{\mathrm{T}} \left[\mathcal{L}_1^{\mathrm{T}} \left(D\mathcal{L}_2 u_z(x)\right)\right] \mathrm{d}L = -\int_L (\mathcal{L}_1(\mathcal{L}_1 W))^{\mathrm{T}} \left(D\mathcal{L}_2 u_z(x)\right) \mathrm{d}L$$
$$+ \int_s (\mathcal{L}_1 W)^{\mathrm{T}} \left(D\mathcal{L}_2 u_z(x)\right) n\mathrm{d}s\,. \qquad (3.102)$$

Combining the last three equations, i.e. Eqs. (3.100)–(3.102), gives

$$\int_L (\mathcal{L}_1(\mathcal{L}_1 W))^{\mathrm{T}} \left(D\mathcal{L}_2 u_z(x)\right) \mathrm{d}L - \int_s (\mathcal{L}_1 W)^{\mathrm{T}} \underbrace{\left(D\mathcal{L}_2 u_z(x)\right)}_{-M_y(x)} n\mathrm{d}s +$$
$$+ \int_s W^{\mathrm{T}} \underbrace{\left[\mathcal{L}_1^{\mathrm{T}} \left(D\mathcal{L}_2 u_z(x)\right)\right]}_{-Q_z(x)} n\mathrm{d}s - \int_L W^{\mathrm{T}} q_z \mathrm{d}L = 0\,, \qquad (3.103)$$

from which the weak formulation can be obtained as:

$$\int_L (\mathcal{L}_2 W)^{\mathrm{T}} D\,(\mathcal{L}_2 u_z)\,\mathrm{d}L = \int_s W^{\mathrm{T}} Q_z n\mathrm{d}s - \int_s (\mathcal{L}_1 W)^{\mathrm{T}} M_y n\mathrm{d}s + \int_L W^{\mathrm{T}} q_z \mathrm{d}L\,. \qquad (3.104)$$

Any further development of Eq. (3.104) requires that the general expressions for the displacement and weight functions, i.e. u_z and W, are now approximated by some functional representations. With the nodal approaches for the displacements u_z (3.55) and the weight function W (3.56), the weak formulation reads:

$$\int_L \left((\mathcal{L}_2 \boldsymbol{N}^{\mathrm{T}})\delta \boldsymbol{u}_{\mathrm{p}}\right)^{\mathrm{T}} D \left(\mathcal{L}_2 \boldsymbol{N}^{\mathrm{T}} \boldsymbol{u}_{\mathrm{p}}\right) \mathrm{d}L = \int_s \delta \boldsymbol{u}_{\mathrm{p}}^{\mathrm{T}} \boldsymbol{N} Q_z n\mathrm{d}s$$
$$- \int_s ((\mathcal{L}_1 \boldsymbol{N}^{\mathrm{T}})\delta \boldsymbol{u}_{\mathrm{p}})^{\mathrm{T}} M_y n\mathrm{d}s + \int_L \delta \boldsymbol{u}_{\mathrm{p}}^{\mathrm{T}} \boldsymbol{N} q_z \mathrm{d}L\,, \qquad (3.105)$$

or finally after the elimination of $\delta \boldsymbol{u}_{\mathrm{p}}^{\mathrm{T}}$:

$$\int_L \left(\mathcal{L}_2 \boldsymbol{N}^{\mathrm{T}}\right)^{\mathrm{T}} D \left(\mathcal{L}_2 \boldsymbol{N}^{\mathrm{T}}\right) \mathrm{d}L \boldsymbol{u}_{\mathrm{p}}^{\mathrm{e}} = \int_s \boldsymbol{N} Q_z n\mathrm{d}s - \int_s (\mathcal{L}_1 \boldsymbol{N}^{\mathrm{T}})^{\mathrm{T}} M_y n\mathrm{d}s + \int_L \boldsymbol{N} q_z \mathrm{d}L\,. \qquad (3.106)$$

Let us summarize at the end of this section the major steps that were undertaken to transform the partial differential equation into the principal finite element equation, see Table 3.9.

Table 3.9 Summary: derivation of principal finite element equation for EULER–BERNOULLI beam elements (bending occurs in the x-z plane)

Strong formulation
$\mathcal{L}_2^{\mathrm{T}}\left(D\mathcal{L}_2 u_z^0(x)\right) - q_z = 0$
Inner product
$\int_L W^{\mathrm{T}}(x)\left(\mathcal{L}_2^{\mathrm{T}}\left(D\mathcal{L}_2 u_z(x)\right) - q_z\right)\mathrm{d}L = 0$
Weak formulation
$\int_L (\mathcal{L}_2 W)^{\mathrm{T}} D (\mathcal{L}_2 u_z)\,\mathrm{d}L = \int_s W^{\mathrm{T}} Q_z n \mathrm{d}s - \int_s (\mathcal{L}_1 W)^{\mathrm{T}} M_y n \mathrm{d}s + \int_L W^{\mathrm{T}} q_z \mathrm{d}L$
Principal finite element equation (line 2 with 4 DOF)
$\underbrace{\int_L \underbrace{\left(\mathcal{L}_2 N^{\mathrm{T}}\right)^{\mathrm{T}}}_{B} D \underbrace{\left(\mathcal{L}_2 N^{\mathrm{T}}\right)}_{B^{\mathrm{T}}} \mathrm{d}L}_{K^{\mathrm{e}}} \begin{bmatrix} u_{1z} \\ \varphi_{1y} \\ u_{2z} \\ \varphi_{2y} \end{bmatrix} = \begin{bmatrix} F_{1z} \\ M_{1y} \\ F_{2z} \\ M_{2y} \end{bmatrix} + \int_L N q_z \mathrm{d}L$

3.3.2 *Derivation of Interpolation Functions*

Looking at Fig. 3.13 which schematically shows the definition of the BERNOULLI beam element, one can see that four deformation quantities are the unknowns at the nodes. A nodal interpolation of the displacement and rotational field must therefore fulfill the following four conditions:

$$u_z(0) = u_{1z}\,,\ \varphi_y(0) = \varphi_{1y}\,,\ u_z(L) = u_{2z}\,,\ \varphi_y(L) = \varphi_{2y}\,. \tag{3.107}$$

Furthermore, the second-order derivative of the interpolation of the displacement field must be nonzero as can be concluded from the weak formulation given in Eq. (3.53). Thus, the following general third-order polynomial with four unknowns $(a_0, \ldots, a_3)$ can be introduced to describe the displacement field:

$$u_z^{\mathrm{e}}(x) = a_0 + a_1 x + a_2 x^2 + a_3 x^3 = \begin{bmatrix} 1 & x & x^2 & x^3 \end{bmatrix} \begin{bmatrix} a_0 \\ a_1 \\ a_2 \\ a_3 \end{bmatrix} = \boldsymbol{\chi}_u^{\mathrm{T}} \boldsymbol{a}\,. \tag{3.108}$$

Differentiation with respect to the x-coordinate gives the rotational field as:

$$\varphi_y^{\mathrm{e}}(x) = -\frac{\mathrm{d}u_z^{\mathrm{e}}(x)}{\mathrm{d}x} = -a_1 - 2a_2 x - 3a_3 x^2 = \begin{bmatrix} 0 & -1 & -2x & -3x^2 \end{bmatrix} \begin{bmatrix} a_0 \\ a_1 \\ a_2 \\ a_3 \end{bmatrix} = \boldsymbol{\chi}_\varphi^{\mathrm{T}} \boldsymbol{a}\,. \tag{3.109}$$

Equations (3.108) and (3.109) can be written in matrix form as:

$$\begin{bmatrix} u_z \\ \varphi_y \end{bmatrix} = \underbrace{\begin{bmatrix} 1 & x & x^2 & x^3 \\ 0 & -1 & -2x & -3x^2 \end{bmatrix}}_{\chi^{\mathrm{T}}} \underbrace{\begin{bmatrix} a_0 \\ a_1 \\ a_2 \\ a_3 \end{bmatrix}}_{\boldsymbol{a}} . \tag{3.110}$$

Evaluation of the functional expressions for the displacement, $u_z^{\mathrm{e}}(x)$, and rotational, $\varphi_y^{\mathrm{e}}(x)$, fields at both nodes, i.e. for $x = 0$ and $x = L$, gives:

$$\text{Node 1:} \quad u_{1z}^{\mathrm{e}}(0) = a_0 , \tag{3.111}$$

$$\varphi_{1y}^{\mathrm{e}}(0) = -a_1 , \tag{3.112}$$

$$\text{Node 2:} \quad u_{2z}^{\mathrm{e}}(L) = a_0 + a_1 L + a_2 L^2 + a_3 L^3 , \tag{3.113}$$

$$\varphi_{2y}^{\mathrm{e}}(L) = -a_1 - 2a_2 L - 3a_3 L^2 . \tag{3.114}$$

The last four equations can be expressed in matrix notation as:

$$\begin{bmatrix} u_{1y} \\ \varphi_{1z} \\ u_{2y} \\ \varphi_{2z} \end{bmatrix} = \underbrace{\begin{bmatrix} 1 & 0 & 0 & 0 \\ 0 & -1 & 0 & 0 \\ 1 & L & L^2 & L^3 \\ 0 & -1 & -2L & -3L^3 \end{bmatrix}}_{\boldsymbol{X}} \begin{bmatrix} a_0 \\ a_1 \\ a_2 \\ a_3 \end{bmatrix} . \tag{3.115}$$

Solving this system of equations for the unknown basis functions a_i gives

$$\begin{bmatrix} a_0 \\ a_1 \\ a_2 \\ a_3 \end{bmatrix} = \underbrace{\begin{bmatrix} 1 & 0 & 0 & 0 \\ 0 & -1 & 0 & 0 \\ -\frac{3}{L^2} & \frac{2}{L} & \frac{3}{L^2} & \frac{1}{L} \\ \frac{2}{L^3} & -\frac{1}{L^2} & -\frac{2}{L^3} & -\frac{1}{L^2} \end{bmatrix}}_{\boldsymbol{X}^{-1}=\boldsymbol{A}} \begin{bmatrix} u_{1y} \\ \varphi_{1z} \\ u_{2y} \\ \varphi_{2z} \end{bmatrix} , \tag{3.116}$$

and the matrix of the interpolation functions for the displacement field results according to Eq. (2.86), i.e. $\boldsymbol{N}^{\mathrm{T}} = \boldsymbol{\chi}^{\mathrm{T}} \boldsymbol{A}$, as:

$$\begin{bmatrix} N_{1u} & N_{1\varphi} & N_{2u} & N_{2\varphi} \end{bmatrix} = \begin{bmatrix} 1 & x & x^2 & x^3 \end{bmatrix} \begin{bmatrix} 1 & 0 & 0 & 0 \\ 0 & -1 & 0 & 0 \\ -\frac{3}{L^2} & \frac{2}{L} & \frac{3}{L^2} & \frac{1}{L} \\ \frac{2}{L^3} & -\frac{1}{L^2} & -\frac{2}{L^3} & -\frac{1}{L^2} \end{bmatrix} =$$

$$
\left[\left(1-\frac{3x^2}{L^2}+\frac{2x^3}{L^3}\right)\left(-x+\frac{2x^2}{L}-\frac{x^3}{L^2}\right)\left(\frac{3x^2}{L^2}-\frac{2x^3}{L^3}\right)\left(\frac{x^2}{L}-\frac{x^3}{L^2}\right)\right]. \quad (3.117)
$$

This gives the same results as in Eq. (3.57) till (3.60). In a similar way, the matrix of the interpolation functions for the rotational field results

$$
\left[N^*_{1u}\; N^*_{1\varphi}\; N^*_{2u}\; N^*_{2\varphi}\right] = \left[0\; -1\; -2x\; -3x^2\right]\begin{bmatrix} 1 & 0 & 0 & 0 \\ 0 & -1 & 0 & 0 \\ -\frac{3}{L^2} & \frac{2}{L} & \frac{3}{L^2} & \frac{1}{L} \\ \frac{2}{L^3} & -\frac{1}{L^2} & -\frac{2}{L^3} & -\frac{1}{L^2} \end{bmatrix} =
$$

$$
\left[\left(\frac{6x}{L^2}-\frac{6x^2}{L^3}\right)\left(1-\frac{4x}{L}+\frac{3x^2}{L^2}\right)\left(-\frac{6x}{L^2}+\frac{6x^2}{L^3}\right)\left(-\frac{2x}{L}+\frac{3x^2}{L^2}\right)\right]. \quad (3.118)
$$

The last two equations for the interpolation functions of the displacement and rotational field indicate that $N^* = -\frac{\mathrm{d}N}{\mathrm{d}x}$, which is a direct result of the relationship between rotation and displacement: $\varphi^{\mathrm{e}}_y(x) = -\frac{\mathrm{d}u^{\mathrm{e}}_z(x)}{\mathrm{d}x}$.

If the interpolation functions are required in the natural coordinate $-1 \leq \xi \leq +1$ (→ numerical integration), the transformation $\xi = \frac{2x}{L} - 1$ (see Table 2.4) or more appropriate $x = \frac{L}{2}(\xi + 1)$ can be used in expressions (3.117) and (3.118):

$$
\left[N_{1u}(\xi)\; N_{1\varphi}(\xi)\; N_{2u}(\xi)\; N_{2\varphi}(\xi)\right] =
$$
$$
\left[\left(\frac{1}{2}-\frac{3\xi}{4}+\frac{\xi^3}{4}\right)\left(-\frac{L}{8}+\frac{L\xi}{8}+\frac{L\xi^2}{8}-\frac{L\xi^3}{8}\right)\right.
$$
$$
\left.\left(\frac{1}{2}+\frac{3\xi}{4}-\frac{\xi^3}{4}\right)\left(\frac{L}{8}+\frac{L\xi}{8}-\frac{L\xi^2}{8}-\frac{L\xi^3}{8}\right)\right], \quad (3.119)
$$

$$
\left[N^*_{1u}(\xi)\; N^*_{1\varphi}(\xi)\; N^*_{2u}(\xi)\; N^*_{2\varphi}(\xi)\right] =
$$
$$
\left[\left(-\frac{3(\xi^2-1)}{2L}\right)\left(-\frac{1}{4}-\frac{\xi}{2}+\frac{3\xi^2}{4}\right)\right.
$$
$$
\left.\left(\frac{3(\xi^2-1)}{2L}\right)\left(-\frac{1}{4}+\frac{\xi}{2}+\frac{3\xi^2}{4}\right)\right]. \quad (3.120)
$$

Alternatively, the derivations could start immediately based on the natural coordinate $-1 \leq \xi \leq +1$. The displacement field reads then

$$
u^{\mathrm{e}}_z(\xi) = a_0 + a_1\xi + a_2\xi^2 + a_3\xi^3. \quad (3.121)
$$

Based on the relationship for the rotation, i.e. $\varphi^{\mathrm{e}}_y(x) = -\frac{\mathrm{d}u^{\mathrm{e}}_z(x)}{\mathrm{d}x}$, we can write:

$$\varphi_y^{\mathrm{e}}(\xi) = -\tfrac{\mathrm{d}u_z^{\mathrm{e}}(x)}{\mathrm{d}x} = -\frac{\mathrm{d}u_z^{\mathrm{e}}(\xi)}{\mathrm{d}\xi}\frac{\mathrm{d}\xi}{\mathrm{d}x} = -\left(0 + a_1 + 2a_2\xi + 3a_3\xi^2\right)\frac{2}{L}. \tag{3.122}$$

Equations (3.121) and (3.122) can be written in matrix notation as:

$$\begin{bmatrix} u_z \\ \varphi_y \end{bmatrix} = \underbrace{\begin{bmatrix} 1 & \xi & \xi^2 & \xi^3 \\ 0 & -\frac{2}{L} & -\frac{4\xi}{L} & -\frac{6\xi^2}{L} \end{bmatrix}}_{\chi^{\mathrm{T}}} \underbrace{\begin{bmatrix} a_0 \\ a_1 \\ a_2 \\ a_3 \end{bmatrix}}_{\boldsymbol{a}}. \tag{3.123}$$

Following the same way of reasoning results in the same interpolation functions.

In generalization of the procedure we may state that a thin beam element with n nodes requires for the representation of the displacement field n interpolation functions $(N_{1u}, \ldots, N_{nu})$ which relate to the displacement unknowns and n interpolation functions $(N_{1\varphi}, \ldots, N_{n\varphi})$ which relate to the rotation unknowns. Thus, the displacement field can be expressed as:

$$u_z^{\mathrm{e}}(\xi) = N_{1u}u_{1z} + N_{1\varphi}\varphi_{1y} + N_{2u}u_{2z} + N_{2\varphi}\varphi_{2y} + \cdots + N_{nu}u_{nz} + N_{n\varphi}\varphi_{ny}\,. \tag{3.124}$$

It should be noted that only the interpolation functions N_{nu} and $N_{n\varphi}$ are used to derive the elemental stiffness matrix.

In a similar way as Eq. (3.124), we may state the expression for the rotational field as:

$$\varphi_y^{\mathrm{e}}(\xi) = N_{1u}^{*}u_{1z} + N_{1\varphi}^{*}\varphi_{1y} + N_{2u}^{*}u_{2z} + N_{2\varphi}^{*}\varphi_{2y} + \cdots + N_{nu}^{*}u_{nz} + N_{n\varphi}^{*}\varphi_{ny}\,. \tag{3.125}$$

However, this is not an independent field since the following equation

$$\varphi_y^{\mathrm{e}}(x) = -\frac{\mathrm{d}u_z^{\mathrm{e}}(x)}{\mathrm{d}x} = -\frac{\mathrm{d}u_z^{\mathrm{e}}(\xi)}{\mathrm{d}\xi}\frac{\mathrm{d}\xi}{\mathrm{d}x}. \tag{3.126}$$

holds. For the interpolation functions of a thin beam, the following characteristics can be stated:

- At node i: $N_{iu} = 1$ and all other N_{ju} $(j = 1, \ldots, n \text{ and } j \neq i)$ and $N_{j\varphi}$ $(j = 1, \ldots, n)$ are zero.
- At node i: $N_{i\varphi}^{*} = 1$ and all other N_{ju}^{*} $(j = 1, \ldots, n)$ and $N_{j\varphi}^{*}$ $(j = 1, \ldots, n$ and $j \neq i)$ are zero.
- $\sum\limits_{i=1}^{n} N_{iu} = 1$.

To conclude this section, a more descriptive approach to derive the interpolation functions is presented in the following. To this end, let us consider the general

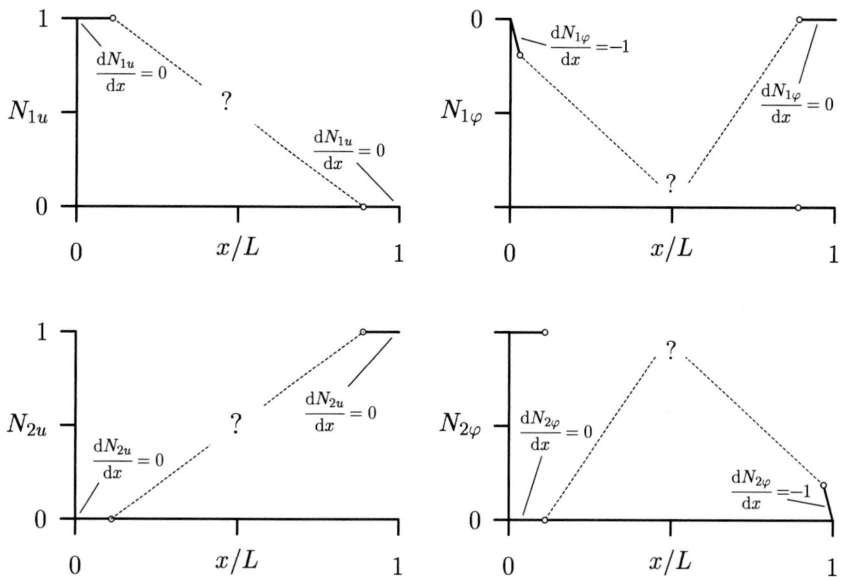

Fig. 3.16 Boundary conditions for the interpolation functions of a BERNOULLI beam element for bending in the x-z plane. Note that the regions for the given slopes are exaggerated for better illustration

characteristic of an interpolation function, i.e. that a function N_i takes the value 1 at node i and is zero at all other nodes.[24] Furthermore, it must be considered that the displacement and rotational fields are decoupled at the nodes in the case of a beam in bending. As a result, one obtains that an interpolation function for the displacement field takes at 'its' node a value of 1 whilst the slope must be equal to 0. At all other nodes j, the functional value and the slope are equal to zero:

$$N_{iu}(x_i) = 1\,, \tag{3.127}$$

$$N_{iu}(x_j) = 0\,, \tag{3.128}$$

$$\frac{\mathrm{d}N_{iu}(x_i)}{\mathrm{d}x} = 0\,, \tag{3.129}$$

$$\frac{\mathrm{d}N_{iu}(x_j)}{\mathrm{d}x} = 0\,. \tag{3.130}$$

In the same way, it is concluded that an interpolation function for the rotational field takes at 'its' node a value of -1 for the slope whilst the functional value must be equal to 0. At all other nodes, the functional values and slopes are identical zero. Thus, one obtains the in Fig. 3.16 represented boundary conditions for the four interpolation functions.

[24] A further characteristic of the interpolation functions is that their sum is unity.

Each interpolation function must change its curvature if there should be no geometrical discontinuities, i.e. kinks, in the course of the function. This can be achieved by a third-order polynomial whose curvature, i.e. the second-order derivative, is a linear function:

$$N(x) = a_0 + a_1 x + a_2 x^2 + a_3 x^3 \,. \tag{3.131}$$

Since a third-order polynomial contains in the general case four unknowns, $a_0, \ldots, a_3$, this approach allows via the four boundary conditions—two for the functional values and two for the slopes—to determine all unknowns. As an example, let us look at the first interpolation function for the displacement field. The boundary conditions for this case are as follows:

$$N_{1u}(0) = 1 \,, \tag{3.132}$$

$$\frac{\mathrm{d}N_{1u}}{\mathrm{d}x}(0) = \frac{\mathrm{d}N_{1u}}{\mathrm{d}x}(L) = 0 \,, \tag{3.133}$$

$$N_{1u}(L) = 0 \,. \tag{3.134}$$

Evaluation of these boundary conditions based on the formulation (3.131) gives:

$$1 = a_0 \,, \tag{3.135}$$

$$0 = a_1 \,, \tag{3.136}$$

$$0 = a_0 + a_1 L + a_2 L^2 + a_3 L^3 \,, \tag{3.137}$$

$$0 = a_1 + 2a_2 L + 3a_3 L^3 \,, \tag{3.138}$$

or in matrix notation:

$$\begin{bmatrix} 1 \\ 0 \\ 0 \\ 0 \end{bmatrix} = \begin{bmatrix} 1 & 0 & 0 & 0 \\ 0 & 1 & 0 & 0 \\ 1 & L & L^2 & L^3 \\ 0 & 1 & 2L & 3L^3 \end{bmatrix} \begin{bmatrix} a_0 \\ a_1 \\ a_2 \\ a_3 \end{bmatrix} . \tag{3.139}$$

The solution of this system of equations for the unknowns gives $\boldsymbol{a} = \begin{bmatrix} 1 & 0 & -\frac{3}{L^2} & \frac{2}{L^3} \end{bmatrix}^{\mathrm{T}}$. Based on these constants, the formulation of the first interpolation function as given in Eq. (3.57) is obtained. Finally it should be noted here that the HERMITE interpolation considers in addition to the nodal value also the slope in the considered node. Thus, the displacements and the rotations are continuous at the nodes.

3.3.3 Assembly of Elements and Consideration of Boundary Conditions

The assembly procedure of the single elemental finite element equations $\boldsymbol{K}^{\mathrm{e}}\boldsymbol{u}_{\mathrm{p}}^{\mathrm{e}} = \boldsymbol{f}^{\mathrm{e}}$ to a global system of equations, i.e. $\boldsymbol{K}\boldsymbol{u}_{\mathrm{p}} = \boldsymbol{f}$, where $\boldsymbol{K}$ is the global stiffness matrix,

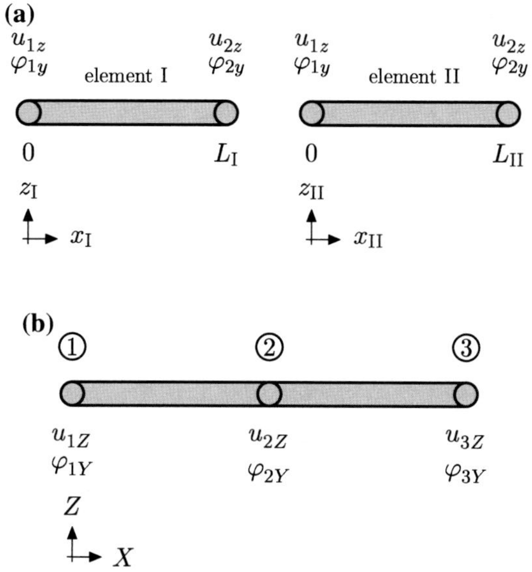

Fig. 3.17 Relationship between **a** elemental and **b** global nodes and deformations in a horizontal beam structure. The nodes are symbolized by two circles at the ends (○)

$\boldsymbol{u}_{\mathrm{p}}$ the global column matrix of unknowns and $\boldsymbol{f}$ the global column matrix of loads, can be treated as described in Sect. 2.3.3 for rod elements. Let us illustrate the process to assemble the global system of equations for BERNOULLI beam elements due to a two-element horizontal structure as shown in Fig. 3.17. As can be seen in Fig. 3.17a, each element has its own coordinate system (x_i, z_i) with $i = \mathrm{I}, \mathrm{II}$ and its own nodal deformations $(u^i_{1z}, \varphi^i_{1y})$ and $(u^i_{2z}, \varphi^i_{2y})$. In order to assemble the single elements to a connected structure as shown in Fig. 3.17b, it is useful to introduce a global coordinate system (X, Z) and global nodal deformations denoted by (u_{iZ}, φ_{iY}). Comparing the elemental and global nodal displacements shown in Fig. 3.17, the following mapping between the local and global displacements can be derived:

$$u_{1Z} = u^{\mathrm{I}}_{1z}\,, \tag{3.140}$$

$$u_{2Z} = u^{\mathrm{I}}_{2z} = u^{\mathrm{II}}_{1z}\,, \tag{3.141}$$

$$u_{3Z} = u^{\mathrm{II}}_{2z}\,, \tag{3.142}$$

and in the same way the relations between the rotations:

$$\varphi_{1Y} = \varphi^{\mathrm{I}}_{1y}\,, \tag{3.143}$$

$$\varphi_{2Y} = \varphi^{\mathrm{I}}_{2y} = \varphi^{\mathrm{II}}_{1y}\,, \tag{3.144}$$

$$\varphi_{3Y} = \varphi^{\mathrm{II}}_{2y}\,. \tag{3.145}$$

One possible way to assemble the elemental stiffness matrices to the global system will be illustrated in the following. In a first step, each single element is considered separately and its elemental stiffness matrix is written, for example, as given in Eq. (3.96). In addition, the corresponding *global* nodal deformations are written over the matrix and on the right-hand side which gives the following expressions:

$$\boldsymbol{K}^{\mathrm{e}}_{\mathrm{I}} = \frac{EI_Y}{L^3} \begin{array}{c} \begin{matrix} u_{1Z} & \varphi_{1Y} & u_{2Z} & \varphi_{2Y} \end{matrix} \\ \begin{bmatrix} 12 & -6L & -12 & -6L \\ -6L & 4L^2 & 6L & 2L^2 \\ -12 & 6L & 12 & 6L \\ -6L & 2L^2 & 6L & 4L^2 \end{bmatrix} \end{array} \begin{matrix} u_{1Z} \\ \varphi_{1Y} \\ u_{2Z} \\ \varphi_{2Y} \end{matrix}, \tag{3.146}$$

$$\boldsymbol{K}^{\mathrm{e}}_{\mathrm{II}} = \frac{EI_Y}{L^3} \begin{array}{c} \begin{matrix} u_{2Z} & \varphi_{2Y} & u_{3Z} & \varphi_{3Y} \end{matrix} \\ \begin{bmatrix} 12 & -6L & -12 & -6L \\ -6L & 4L^2 & 6L & 2L^2 \\ -12 & 6L & 12 & 6L \\ -6L & 2L^2 & 6L & 4L^2 \end{bmatrix} \end{array} \begin{matrix} u_{2Z} \\ \varphi_{2Y} \\ u_{3Z} \\ \varphi_{3Y} \end{matrix}. \tag{3.147}$$

By indicating the global unknowns in the described manner at each elemental stiffness matrix, it is easy to assign to each element in a matrix a unique index. For example, the upper right element of the stiffness matrix $\boldsymbol{K}^{\mathrm{e}}_{\mathrm{I}}$ has the index[25] (u_{1Z}, φ_{2Y}) and the value $\frac{-6EI_Y}{L^2}$. The next step consists in indicating the structure of the global stiffness matrix with its correct dimensions. To this end, the total number of global unknowns must be determined. This is given here by the number of nodes times degrees of freedom per node. It should be noted here that the determination of the unknowns at this step of the procedure is without any consideration of boundary conditions. For the problem shown in Fig. 3.17a, the number of nodes is three and two unknowns per node. Thus, the dimensions of the global stiffness matrix are given by (number global nodes × number global unknowns) or for our example as (3 × 2) and the structure can be written as a (6 × 6) matrix:

$$\boldsymbol{K} = \begin{array}{c} \begin{matrix} u_{1Z} & \varphi_{1Y} & u_{2Z} & \varphi_{2Y} & u_{3Z} & \varphi_{3Y} \end{matrix} \\ \left[\begin{array}{c|c|c|c|c|c} & & & & & \\ \hline & & & & & \\ \hline & & & & & \\ \hline & & & & & \\ \hline & & & & & \\ \hline & & & & & \end{array}\right] \end{array} \begin{matrix} u_{1Z} \\ \varphi_{1Y} \\ u_{2Z} \\ \varphi_{2Y} \\ u_{3Z} \\ \varphi_{3Y} \end{matrix}. \tag{3.148}$$

It is now required to indicate the global unknowns over the empty global stiffness matrix and on its right-hand side. Any order can be chosen but it is common for the problem under consideration to start with u_{1Z} and φ_{1Y} and simply move to the next node. The scheme for this consecutive use of the global unknowns from the lowest

[25]We follow here the convention where the first expression specifies the row and the second one the column: (row, column).

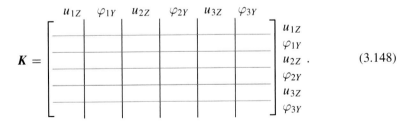

to the highest number is drawn on the matrix in Eq. (3.148). Each cell of the global stiffness matrix has now its unique index expressed by the global unknowns and each element of the elemental stiffness matrix must be placed in the global matrix based on this unique index scheme. As an example, the upper right element of the stiffness matrix $\boldsymbol{K}_{\text{I}}^{\text{e}}$ with the index (u_{1Z}, φ_{2Y}) must be placed in the global stiffness matrix in the first row and the fourth column. If each entry of the elemental stiffness matrices is inserted into the global matrix based on the described index scheme, the assembly of the global stiffness matrix is completed. As the elemental stiffness matrices, the global stiffness matrix is symmetric and commercial finite element codes store only half of the entries in order to reduce the requirements for data storage.

In order to complete the assembly of the global finite element equation, the global load vector $\boldsymbol{f}$ must be composed. Here, it is more advantageous to look from the beginning at the assembled structure and fill the external single loads F_{iZ} and moments M_{iY} which are acting at nodes in the proper order in the column matrix $\boldsymbol{f}$. A bit care must be taken if distributed loads were converted to equivalent nodal loads. For this case, components f_i from both elements must be summed up at inner nodes:

$$\boldsymbol{f} = \begin{bmatrix} F_{1Z} \\ M_{1Y} \\ F_{2Z} \\ M_{2Y} \\ F_{3Z} \\ M_{3Y} \end{bmatrix} + \begin{bmatrix} f_{1Z,\text{I}} \\ f_{1Y,\text{I}} \\ f_{2Z,\text{I}} + f_{2Z,\text{II}} \\ f_{2Y,\text{I}} + f_{2Y,\text{II}} \\ f_{3Z,\text{II}} \\ f_{3Y,\text{II}} \end{bmatrix} . \tag{3.149}$$

The global system of equations for the problem shown in Fig. 3.17b is finally obtained as:

$$\frac{EI_Y}{L^3}\begin{bmatrix} 12 & -6L & -12 & -6L & 0 & 0 \\ -6L & 4L^2 & 6L & 2L^2 & 0 & 0 \\ -12 & 6L & 24 & 0 & -12 & -6L \\ -6L & 2L^2 & 0 & 8L^2 & 6L & 2L^2 \\ 0 & 0 & -12 & 6L & 12 & 6L \\ 0 & 0 & -6L & 2L^2 & 6L & 4L^2 \end{bmatrix}\begin{bmatrix} u_{1Z} \\ \varphi_{1Y} \\ u_{2Z} \\ \varphi_{2Y} \\ u_{3Z} \\ \varphi_{3Y} \end{bmatrix} = \begin{bmatrix} \cdots \\ \cdots \\ \cdots \\ \cdots \\ \cdots \\ \cdots \end{bmatrix} , \tag{3.150}$$

where the right-hand side is not specified since nothing on the loading is indicated in Fig. 3.17. This system of equations without consideration of any boundary conditions is called the non-reduced system. For this system, the global stiffness matrix $\boldsymbol{K}$ is still singular and cannot be inverted in order to solve the global system of equations. Boundary conditions must be introduced in order to make this matrix regular und thus invertible.

For the beam elements under consideration, two types of boundary conditions must be distinguished. The DIRICHLET boundary condition specifies the displacement u_Z or a rotation φ_Y at a node while the NEUMANN boundary condition assigns a force F_Z or a moment M_Y at a node. The different ways to handle these different types of boundary conditions will be explained in the following based on the problem shown

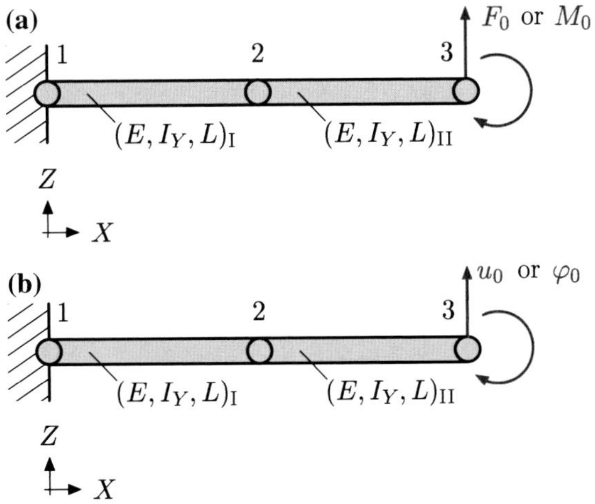

Fig. 3.18 Consideration of boundary conditions for a cantilever beam structure: **a** force or moment boundary condition; **b** displacement or rotation boundary condition at the right-hand boundary node. The nodes are symbolized by two circles at the ends (◯)

in Fig. 3.18 where a cantilever beam structure has different boundary conditions at its right-hand end node.

The consideration of the homogeneous DIRICHLET boundary condition at the fixed support, i.e. $u_{1Z} = u(X = 0) = 0$ and $\varphi_{1Y} = \varphi(X = 0) = 0$, is the simplest case. To incorporate these boundary conditions in the system (3.150), the first two rows and columns can be canceled to obtain a reduced systems as:

$$\frac{EI_Y}{L^3}\begin{bmatrix} 24 & 0 & -12 & -6L \\ 0 & 8L^2 & 6L & 2L^2 \\ -12 & 6L & 12 & 6L \\ -6L & 2L^2 & 6L & 4L^2 \end{bmatrix}\begin{bmatrix} u_{2Z} \\ \varphi_{2Y} \\ u_{3Z} \\ \varphi_{3Y} \end{bmatrix} = \begin{bmatrix} \cdots \\ \cdots \\ \cdots \\ \cdots \end{bmatrix}. \tag{3.151}$$

In general we can state that a homogenous DIRICHLET boundary condition at node n can be considered in the non-reduced system of equations by eliminating in the case of a displacement $u_{nZ} = 0$ the $(2n - 1)$th row and $(2n - 1)$th column of the system. In the case of a rotation $\varphi_{nY} = 0$, the $(2n)$th row and $(2n)$th column of the system can be eliminated. Let us consider next the case shown in Fig. 3.18a where the right-hand end node is subjected to an external load, i.e. a force F_0 or a moment M_0. This external force or moment can simply be specified on the right-hand side and since no other external forces are acting, the reduced system of equations is finally obtained as:

$$\frac{EI_Y}{L^3}\begin{bmatrix} 24 & 0 & -12 & -6L \\ 0 & 8L^2 & 6L & 2L^2 \\ -12 & 6L & 12 & 6L \\ -6L & 2L^2 & 6L & 4L^2 \end{bmatrix}\begin{bmatrix} u_{2Z} \\ \varphi_{2Y} \\ u_{3Z} \\ \varphi_{3Y} \end{bmatrix} = \begin{bmatrix} 0 \\ 0 \\ F_0 \\ 0 \end{bmatrix}, \tag{3.152}$$

or

$$\frac{EI_Y}{L^3}\begin{bmatrix} 24 & 0 & -12 & -6L \\ 0 & 8L^2 & 6L & 2L^2 \\ -12 & 6L & 12 & 6L \\ -6L & 2L^2 & 6L & 4L^2 \end{bmatrix}\begin{bmatrix} u_{2Z} \\ \varphi_{2Y} \\ u_{3Z} \\ \varphi_{3Y} \end{bmatrix} = \begin{bmatrix} 0 \\ 0 \\ 0 \\ M_0 \end{bmatrix}. \tag{3.153}$$

This system of equations can be solved, e.g. by inverting the reduced stiffness matrix and solving for the unknown nodal deformations in the form $\boldsymbol{u}_{\text{p}} = \boldsymbol{K}^{-1}\boldsymbol{f}$:

$$\begin{bmatrix} u_{2Z} \\ \varphi_{2Y} \\ u_{3Z} \\ \varphi_{3Y} \end{bmatrix} = \frac{L^3 F_0}{EI_Y}\begin{bmatrix} \frac{5}{6} \\ -\frac{3}{2L} \\ \frac{8}{3} \\ -\frac{2}{L} \end{bmatrix}, \tag{3.154}$$

or

$$\begin{bmatrix} u_{2Z} \\ \varphi_{2Y} \\ u_{3Z} \\ \varphi_{3Y} \end{bmatrix} = \frac{L^2 M_0}{EI_Y}\begin{bmatrix} -\frac{1}{2} \\ \frac{1}{L} \\ -\frac{2}{1} \\ \frac{2}{L} \end{bmatrix}. \tag{3.155}$$

To incorporate a non-homogeneous DIRICHLET boundary condition ($u_Z \neq 0 \vee \varphi_Y \neq 0$) as shown in Fig. 3.18b, three different strategies can be mentioned, see Sect. 2.3.3. We will illustrate only the first approach and the interested reader may refer to the mentioned section for alternative strategies. The first one modifies the reduced system shown in Eqs. (3.152) or (3.153) in such a way that the boundary condition is directly introduced:

$$\frac{EI_Y}{L^3}\begin{bmatrix} 24 & 0 & -12 & -6L \\ 0 & 8L^2 & 6L & 2L^2 \\ 0 & 0 & 1\times\frac{L^3}{EI_Y} & 0 \\ -6L & 2L^2 & 6L & 4L^2 \end{bmatrix}\begin{bmatrix} u_{2Z} \\ \varphi_{2Y} \\ u_{3Z} \\ \varphi_{3Y} \end{bmatrix} = \begin{bmatrix} 0 \\ 0 \\ u_0 \\ 0 \end{bmatrix}, \tag{3.156}$$

or

$$\frac{EI_Y}{L^3}\begin{bmatrix} 24 & 0 & -12 & -6L \\ 0 & 8L^2 & 6L & 2L^2 \\ -12 & 6L & 12 & 6L \\ 0 & 0 & 0 & 1\times\frac{L^3}{EI_Z} \end{bmatrix}\begin{bmatrix} u_{2Z} \\ \varphi_{2Y} \\ u_{3Z} \\ \varphi_{3Y} \end{bmatrix} = \begin{bmatrix} 0 \\ 0 \\ 0 \\ \varphi_0 \end{bmatrix}, \tag{3.157}$$

where the third equation of (3.156) and the fourth equation of (3.157) gives immediately the boundary condition as $u_{3Z} = u_0$ or $\varphi_{3Y} = \varphi_0$. The solution of the systems of equations given in Eqs. (3.156) and (3.157) can be obtained by inverting the coefficient matrices and multiplying them with the matrices on the right-hand sides as:

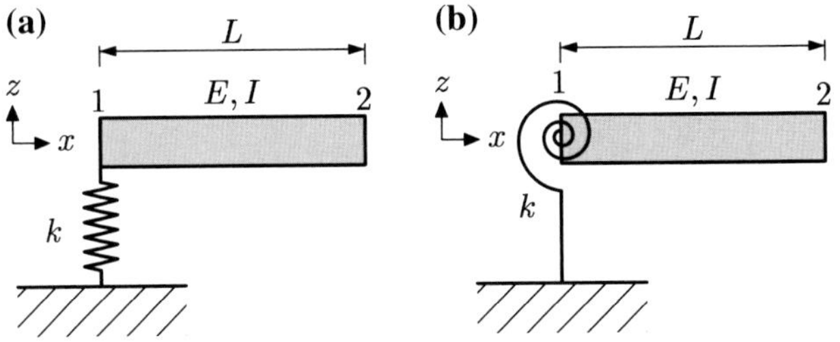

Fig. 3.19 Consideration of springs in beam elements: **a** tension/compression spring; **b** torsion spring

$$\begin{bmatrix} u_{2Z} \\ \varphi_{2Y} \\ u_{3Z} \\ \varphi_{3Y} \end{bmatrix} = u_0 \begin{bmatrix} \frac{5}{16} \\ -\frac{9}{16L} \\ 1 \\ -\frac{3}{4L} \end{bmatrix}, \text{ or } \begin{bmatrix} u_{2Z} \\ \varphi_{2Y} \\ u_{3Z} \\ \varphi_{3Y} \end{bmatrix} = \varphi_0 \begin{bmatrix} -\frac{L}{4} \\ \frac{1}{2} \\ -\frac{L}{1} \\ 1 \end{bmatrix}. \tag{3.158}$$

In general we can state that a non-homogeneous DIRICHLET boundary condition at node n can be introduced into the system of equations by the following modifications: In the case of a given displacement u_0: Modify the $(2n-1)$th row in such a way that at the position of the $(2n-1)$th column a '1' is obtained while all other entries of the $(2n-1)$th row are set to zero. On the right-hand side, the given displacement u_0 is introduced at the $(2n-1)$th position of the column matrix. In the case of a given rotation φ_0: Modify the $(2n)$th row in such a way that at the position of the $(2n)$th column a '1' is obtained while all other entries of the $(2n)$th row are set to zero. On the right-hand side, the given rotation φ_0 is introduced at the $(2n)$th position of the column matrix.

A special type of 'boundary condition' can be realized by attaching a spring to a beam element a shown in Fig. 3.19. The two different degrees of freedom at each node can be influenced by different types of springs: The translatory degree of freedom is addressed by a tension/compression spring (cf. Fig. 3.19a) while the rotatory degree of freedom is addressed by a torsion spring (cf. Fig. 3.19b).

Let us consider the configuration where the spring[26] is attached to node 1 as shown in Fig. 3.19. Then,[27] the elemental stiffness matrix in the case of the tension/compression spring is written as

$$\boldsymbol{k}^{\mathrm{e}} = \frac{EI_y}{L^3} \begin{bmatrix} 12 + \frac{L^3}{EI_y}k & -6L & -12 & -6L \\ -6L & 4L^2 & 6L & 2L^2 \\ -12 & 6L & 12 & 6L \\ -6L & 2L^2 & 6L & 4L^2 \end{bmatrix}, \tag{3.159}$$

[26] It is assumed here that the spring is in its unstrained state in the sketched configuration, i.e. without the application of any force or displacement boundary conditions at the nodes of the beam.

[27] See the comments in Sect. 2.3.3.

or in the case of the torsion spring as:

$$\boldsymbol{k}^{\mathrm{e}} = \frac{EI_y}{L^3}\begin{bmatrix} 12 & 6-L & -12 & -6L \\ -6L & 4L^2 + \frac{L^3}{EI_y}k & 6L & 2L^2 \\ -12 & 6L & 12 & 6L \\ -6L & 2L^2 & 6L & 4L^2 \end{bmatrix}. \tag{3.160}$$

3.3.4 Post-Computation: Determination of Strain, Stress and Further Quantities

The previous section explained how to compose the global system of equations from which the primary unknowns, i.e. the nodal displacements and rotations, can be obtained. After the solution for the nodal unknowns, the distributions of these deformations and the curvature within an element can be obtained based on the equations provided in Table 3.10. These distributions allow then the calculation of strain and stress values at any location of the beam (cf. Table 3.4) as:

$$\varepsilon_x^{\mathrm{e}}(x,z) = -z\frac{\mathrm{d}^2 u_z^{\mathrm{e}}(x)}{\mathrm{d}x^2} \quad \text{and} \quad \sigma_x^{\mathrm{e}}(x,z) = E\varepsilon_x^{\mathrm{e}}(x,z)\,. \tag{3.161}$$

It should be noted here that the bending and shear stress distribution (cf. Table 3.4) are at this stage defined based on the nodal displacements and rotations, see Table 3.11.

3.3.5 Solved Beam Problems

3.1 Sample: Beam loaded by end force or moment—approximation through one single finite element

Determine through one single finite element the displacement and the rotation of the right-hand end of the beam, which is illustrated in Fig. 3.20. Furthermore, determine the course of the bending line $u_z^{\mathrm{e}} = u_z^{\mathrm{e}}(x)$ and compare the finite element solution with the analytical solution.

3.1 Solution

(a) The finite element equation on element level according to Eq. (3.96) reduces for the illustrated load case to:

$$\frac{EI_y}{L^3}\begin{bmatrix} 12 & -6L & -12 & -6L \\ -6L & 4L^2 & 6L & 2L^2 \\ -12 & 6L & 12 & 6L \\ -6L & 2L^2 & 6L & 4L^2 \end{bmatrix}\begin{bmatrix} u_{1z} \\ \varphi_{1y} \\ u_{2z} \\ \varphi_{2y} \end{bmatrix} = \begin{bmatrix} 0 \\ 0 \\ -F_0 \\ 0 \end{bmatrix}. \tag{3.162}$$

Table 3.10 Displacement, rotation and curvature distribution for a BERNOULLI beam element given as being dependent on the nodal values as function of the physical coordinate $0 \leq x \leq L$ and natural coordinate $-1 \leq \xi \leq 1$. Bending occurs in the x-z plane

Vertical displacement (Deflection) u_z

$$u_z^{\mathrm{e}}(x) = \left[1 - 3\left(\frac{x}{L}\right)^2 + 2\left(\frac{x}{L}\right)^3\right] u_{1z} + \left[-x + \frac{2x^2}{L} - \frac{x^3}{L^2}\right] \varphi_{1y} + \left[3\left(\frac{x}{L}\right)^2 - 2\left(\frac{x}{L}\right)^3\right] u_{2z} + \left[+\frac{x^2}{L} - \frac{x^3}{L^2}\right] \varphi_{2y}$$

$$u_z^{\mathrm{e}}(\xi) = \frac{1}{4}\left[2 - 3\xi + \xi^3\right] u_{1z} - \frac{1}{4}\left[1 - \xi - \xi^2 + \xi^3\right] \frac{L}{2} \varphi_{1y} + \frac{1}{4}\left[2 + 3\xi - \xi^3\right] u_{2z} - \frac{1}{4}\left[-1 - \xi + \xi^2 + \xi^3\right] \frac{L}{2} \varphi_{2y}$$

Rotation (Slope) $\varphi_y = -\frac{\mathrm{d}u_z}{\mathrm{d}x} = -\frac{2}{L}\frac{\mathrm{d}u_z}{\mathrm{d}\xi}$

$$\varphi_y^{\mathrm{e}}(x) = \left[+\frac{6x}{L^2} - \frac{6x^2}{L^3}\right] u_{1z} + \left[1 - \frac{4x}{L} + \frac{3x^2}{L^2}\right] \varphi_{1y} + \left[-\frac{6x}{L^2} + \frac{6x^2}{L^3}\right] u_{2z} + \left[-\frac{2x}{L} + \frac{3x^2}{L^2}\right] \varphi_{2y}$$

$$\varphi_y^{\mathrm{e}}(\xi) = \frac{1}{2L}\left[+3 - 3\xi^2\right] u_{1z} + \frac{1}{4}\left[-1 - 2\xi + 3\xi^2\right] \varphi_{1y} + \frac{1}{2L}\left[-3 + 3\xi^2\right] u_{2z} + \frac{1}{4}\left[-1 + 2\xi + 3\xi^2\right] \varphi_{2y}$$

Curvature $\kappa_y = -\frac{\mathrm{d}^2 u_z}{\mathrm{d}x^2} = -\frac{4}{L^2}\frac{\mathrm{d}^2 u_z}{\mathrm{d}\xi^2}$

$$\kappa_y^{\mathrm{e}}(x) = \left[+\frac{6}{L^2} - \frac{12x}{L^3}\right] u_{1z} + \left[-\frac{4}{L} + \frac{6x}{L^2}\right] \varphi_{1y} + \left[-\frac{6}{L^2} + \frac{12x}{L^3}\right] u_{2z} + \left[-\frac{2}{L} + \frac{6x}{L^2}\right] \varphi_{2y}$$

$$\kappa_y^{\mathrm{e}}(\xi) = \frac{6}{L^2}[-\xi]\, u_{1z} + \frac{1}{L}[-1 + 3\xi]\, \varphi_{1y} + \frac{6}{L^2}[\xi]\, u_{2z} + \frac{1}{L}[1 + 3\xi]\, \varphi_{2y}$$

Table 3.11 Bending moment and shear stress distribution for a BERNOULLI beam element given as being dependent on the nodal values as function of the physical coordinate $0 \leq x \leq L$ and natural coordinate $-1 \leq \xi \leq 1$. Bending occurs in the x-z plane

Bending moment $M_y = -EI_y\frac{\mathrm{d}^2 u_z}{\mathrm{d}x^2} = -\frac{4}{L^2}EI_y\frac{\mathrm{d}^2 u_z}{\mathrm{d}\xi^2}$

$$M_y^{\mathrm{e}}(x) = EI_y\left(\left[+\frac{6}{L^2} - \frac{12x}{L^3}\right] u_{1z} + \left[-\frac{4}{L} + \frac{6x}{L^2}\right] \varphi_{1y} + \left[-\frac{6}{L^2} + \frac{12x}{L^3}\right] u_{2z} + \left[-\frac{2}{L} + \frac{6x}{L^2}\right] \varphi_{2y}\right)$$

$$M_y^{\mathrm{e}}(\xi) = EI_y\left(\frac{6}{L^2}[-\xi]\, u_{1z} + \frac{1}{L}[-1 + 3\xi]\, \varphi_{1y} + \frac{6}{L^2}[\xi]\, u_{2z} + \frac{1}{L}[1 + 3\xi]\, \varphi_{2y}\right)$$

Shear force $Q_z = -EI_y\frac{\mathrm{d}^3 u_z}{\mathrm{d}x^3} = -\frac{8}{L^3}EI_y\frac{\mathrm{d}^3 u_z}{\mathrm{d}\xi^3}$

$$Q_z^{\mathrm{e}}(x) = EI_y\left(\left[-\frac{12}{L^3}\right] u_{1z} + \left[+\frac{6}{L^2}\right] \varphi_{1y} + \left[+\frac{12}{L^3}\right] u_{2z} + \left[+\frac{6}{L^2}\right] \varphi_{2y}\right)$$

$$Q_z^{\mathrm{e}}(\xi) = EI_y\left(\frac{12}{L^3}[-1]\, u_{1z} + \frac{2}{L^2}[+3]\, \varphi_{1y} + \frac{12}{L^3}[1]\, u_{2z} + \frac{2}{L^2}[+3]\, \varphi_{2y}\right)$$

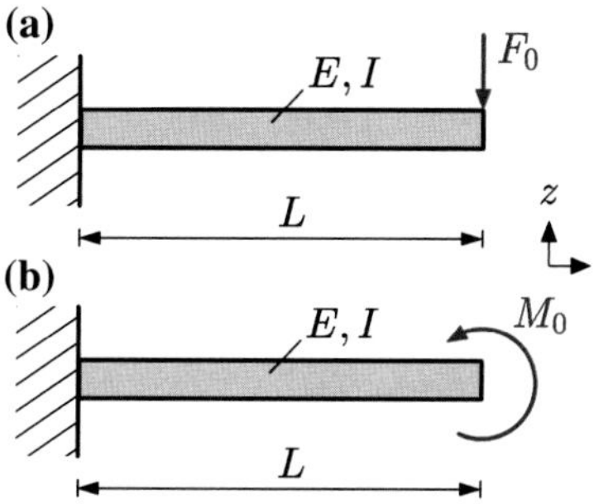

Fig. 3.20 Sample problem beam with end load: **a** single force; **b** single moment

Since the displacement and the rotation are zero on the left-hand boundary due to the fixed support, the first two rows and columns of the system of equations can be eliminated:

$$\frac{EI_y}{L^3}\begin{bmatrix} 12 & 6L \\ 6L & 4L^2 \end{bmatrix}\begin{bmatrix} u_{2z} \\ \varphi_{2y} \end{bmatrix} = \begin{bmatrix} -F_0 \\ 0 \end{bmatrix}. \tag{3.163}$$

Solving for the unknown deformations yields:

$$\begin{bmatrix} u_{2z} \\ \varphi_{2y} \end{bmatrix} = \frac{L^3}{EI_y}\begin{bmatrix} 12 & 6L \\ 6L & 4L^2 \end{bmatrix}^{-1}\begin{bmatrix} -F_0 \\ 0 \end{bmatrix} \tag{3.164}$$

$$= \frac{L^3}{EI_z(48L^2 - 36L^2)}\begin{bmatrix} 4L^2 & -6L \\ -6L & 12 \end{bmatrix}\begin{bmatrix} -F_0 \\ 0 \end{bmatrix} = \begin{bmatrix} -\frac{F_0 L^3}{3EI_y} \\ +\frac{F_0 L^2}{2EI_y} \end{bmatrix}. \tag{3.165}$$

According to Ref. [11], the analytical displacement results in:

$$u_z(x = L) = -\frac{F_0}{6EI_y}\left(3L^3 - L^3\right) = -\frac{F_0 L^3}{3EI_y}. \tag{3.166}$$

The analytical solution for the rotation results from differentiation of the general displacement distribution $u_z = u_z(x)$ according to Ref. [11] for $a = L$ to:

$$\varphi_y(x) = -\frac{\mathrm{d}u_z(x)}{\mathrm{d}x} = +\frac{F_0}{6EI_y} \times \left[6Lx - 3x^2\right], \tag{3.167}$$

or alternatively on the right-hand boundary:

$$\varphi_y(x = L) = +\frac{F_0}{6EI_y} \times \left[6L^2 - 3L^2\right] = +\frac{F_0 L^2}{2EI_y}. \tag{3.168}$$

The course of the bending line $u_z^{\mathrm{e}} = u_z^{\mathrm{e}}(x)$ results from Table 3.10 as:

$$u_z^{\mathrm{e}}(x) = N_{2u}(x)u_{2z} + N_{2\varphi}(x)\varphi_{2y}$$

$$= \left[3\left(\frac{x}{L}\right)^2 - 2\left(\frac{x}{L}\right)^3\right]\left(-\frac{F_0L^3}{3EI_z}\right) + \left[\frac{x^2}{L} - \frac{x^3}{L^2}\right]\left(+\frac{F_0L^2}{2EI_z}\right)$$

$$= \frac{F_0}{6EI_y}\left(x^3 - 3Lx^2\right). \tag{3.169}$$

According to Ref. [11], this course matches with the analytical solution.

Conclusion: Finite element solution and analytical solution are identical!

(b) The reduced system of equations in this case results in:

$$\frac{EI_y}{L^3}\begin{bmatrix} 12 & 6L \\ 6L & 4L^2 \end{bmatrix}\begin{bmatrix} u_{2z} \\ \varphi_{2y} \end{bmatrix} = \begin{bmatrix} 0 \\ -M_0 \end{bmatrix}. \tag{3.170}$$

Solving for the unknown deformations yields:

$$\begin{bmatrix} u_{2z} \\ \varphi_{2y} \end{bmatrix} = \frac{L^3}{12EI_yL^2}\begin{bmatrix} 4L^2 & -6L \\ -6L & 12 \end{bmatrix}\begin{bmatrix} 0 \\ -M_0 \end{bmatrix} = \begin{bmatrix} \frac{M_0L^2}{2EI_y} \\ -\frac{M_0L}{EI_y} \end{bmatrix}. \tag{3.171}$$

The analytical solution according to Ref. [11] is

$$u_y(x = L) = +\frac{M_0}{2EI_y}\left(L^2\right) = \frac{M_0L^2}{2EI_y}, \tag{3.172}$$

or alternatively the rotation in general for $a = L$:

$$\varphi_y(x) = -\frac{\mathrm{d}u_z(x)}{\mathrm{d}x} = -\frac{M_0}{2EI_y}(2x) \tag{3.173}$$

or only on the right-hand boundary:

$$\varphi_y(x = L) = -\frac{M_0}{2EI_y}(2L) = -\frac{M_0L}{EI_y}. \tag{3.174}$$

The course of the bending line $u_z^{\mathrm{e}} = u_z^{\mathrm{e}}(x)$ results from Table 3.10 as:

$$u_z^{\mathrm{e}}(x) = N_{2u}(x)u_{2z} + N_{2\varphi}(x)\varphi_{2y}$$

$$= \left[3\left(\frac{x}{L}\right)^2 - 2\left(\frac{x}{L}\right)^3\right]\left(\frac{M_0L^2}{2EI_z}\right) + \left[\frac{x^2}{L} - \frac{x^3}{L^2}\right]\left(-\frac{M_0L}{EI_z}\right)$$

$$= \frac{M_0x^2}{2EI_y}. \tag{3.175}$$

According to Ref. [11], this course matches with the analytical solution.

Conclusion: Finite element solution and analytical solution are identical!

3.2 Sample: Beam under constant distributed load—approximation through one single finite element

Determine through one single finite element the displacement and the rotation (a) of the right-hand boundary and (b) in the middle for the beam under constant distributed load, which is illustrated in Fig. 3.21. Furthermore, determine the course of the bending line $u_z^e = u_z^e(x)$ and compare the finite element solution with the analytical solution.

3.2 Solution

To solve the problem, the constant distributed load has to be converted into equivalent nodal loads. These equivalent nodal loads can be extracted from Table 3.7 for the considered case, and the finite element equation results to:

$$\frac{EI_y}{L^3}\begin{bmatrix} 12 & -6L & -12 & -6L \\ -6L & 4L^2 & 6L & 2L^2 \\ -12 & 6L & 12 & 6L \\ -6L & 2L^2 & 6L & 4L^2 \end{bmatrix}\begin{bmatrix} u_{1z} \\ \varphi_{1y} \\ u_{2z} \\ \varphi_{2y} \end{bmatrix} = \begin{bmatrix} -\frac{q_0 L}{2} \\ +\frac{q_0 L^2}{12} \\ -\frac{q_0 L}{2} \\ -\frac{q_0 L^2}{12} \end{bmatrix}. \tag{3.176}$$

(a) Consideration of the boundary conditions shown in Fig. 3.21a, meaning the fixed support on the left-hand boundary, and solving for the unknowns yields:

$$\begin{bmatrix} u_{2z} \\ \varphi_{2y} \end{bmatrix} = \frac{L}{12EI_z}\begin{bmatrix} 4L^2 & -6L \\ -6L & 12 \end{bmatrix}\begin{bmatrix} -\frac{q_0 L}{2} \\ -\frac{q_0 L^2}{12} \end{bmatrix} = \begin{bmatrix} -\frac{q_0 L^4}{8EI_y} \\ +\frac{q_0 L^3}{6EI_y} \end{bmatrix}. \tag{3.177}$$

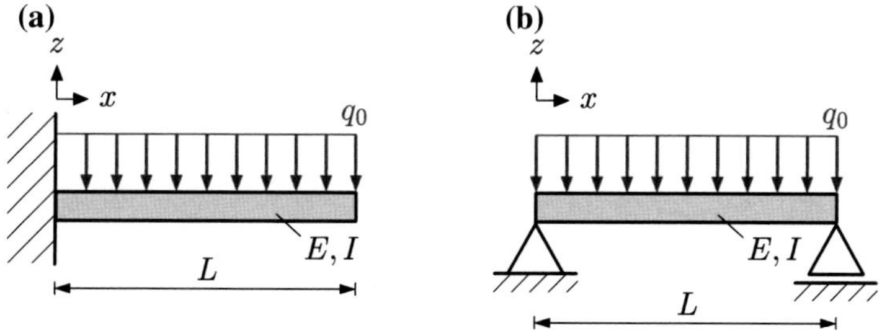

Fig. 3.21 Sample problem beam under constant distributed load and different boundary conditions: **a** cantilever and **b** simply supported beam

The analytical solution according to Ref. [11] yields

$$u_z(x = L) = -\frac{q_0}{24EI_y}\left(6L^4 - 4L^4 + L^4\right) = -\frac{q_0 L^4}{8EI_y}, \tag{3.178}$$

or alternatively the rotation based on $\varphi_y(x) = -\frac{\mathrm{d}u_z(x)}{\mathrm{d}x}$:

$$\varphi_y(x) = +\frac{q_0}{24EI_z}\left(12L^2x - 12Lx^2 + 4x^3\right), \tag{3.179}$$

or only at the right-hand boundary:

$$\varphi_y(x = L) = +\frac{q_0}{24EI_y}\left(12L^3 - 12L^3 + 4L^3\right) = +\frac{q_0 L^3}{6EI_y}. \tag{3.180}$$

The course of the bending line $u_z^{\mathrm{e}} = u_z^{\mathrm{e}}(x)$ results from Table 3.10 as:

$$\begin{aligned} u_z^{\mathrm{e}}(x) &= N_{2u}(x)u_{2z} + N_{2\varphi}(x)\varphi_{2y} \\ &= \left[3\left(\frac{x}{L}\right)^2 - 2\left(\frac{x}{L}\right)^3\right]\left(-\frac{q_0 L^4}{8EI_y}\right) + \left[\frac{x^2}{L} - \frac{x^3}{L^2}\right]\left(\frac{q_0 L^3}{6EI_y}\right) \\ &= -\frac{q_0}{24EI_y}\left(-2Lx^3 + 5L^2x^2\right), \end{aligned} \tag{3.181}$$

however the analytical course according to Ref. [11] results in $u_z(x) = -\frac{q_0}{24EI_y}\left(x^4 - 4Lx^3 + 6L^2x^2\right)$, meaning the analytical and therefore the exact course is not identical with the numerical solution between the nodes ($0 < x < L$), see Fig. 3.22. One can see that between the nodes a small difference between the two solutions arises. If a higher accuracy is demanded between those two nodes, the beam has to be divided into more elements.

Conclusion: Finite element solution and the analytical solution are only identical at the nodes!

(b) Consideration of the boundary conditions shown in Fig. 3.21b, meaning the simple support and the roller support, yields through the elimination of the first and third row and column of the system of Eqs. (3.176):

$$\frac{EI_y}{L^3}\begin{bmatrix} 4L^2 & 2L^2 \\ 2L^2 & 4L^2 \end{bmatrix}\begin{bmatrix} \varphi_{1y} \\ \varphi_{2y} \end{bmatrix} = \begin{bmatrix} +\frac{q_0 L^2}{12} \\ -\frac{q_0 L^2}{12} \end{bmatrix}. \tag{3.182}$$

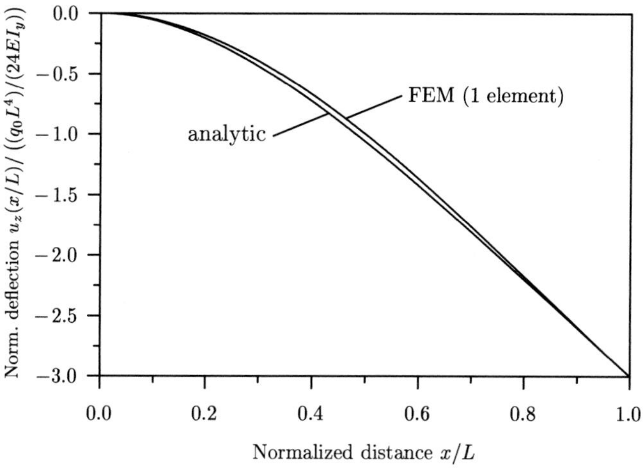

Fig. 3.22 Comparison of the analytical and the finite element solution for the beam according to Fig. 3.21a

Solving for the unknowns yields:

$$\begin{bmatrix} \varphi_{1y} \\ \varphi_{2y} \end{bmatrix} = \frac{1}{12EI_zL} \begin{bmatrix} 4L^2 & -2L^2 \\ -2L^2 & 4L^2 \end{bmatrix} \begin{bmatrix} +\frac{q_0L^2}{12} \\ -\frac{q_0L^2}{12} \end{bmatrix} = \begin{bmatrix} +\frac{q_0L^3}{24EI_y} \\ -\frac{q_0L^3}{24EI_y} \end{bmatrix}. \tag{3.183}$$

The course of the bending line $u_z^e = u_z^e(x)$ results from Table 3.10 as:

$$\begin{aligned} u_z^e(x) &= N_{1\varphi}(x)\varphi_{1y} + N_{2\varphi}(x)\varphi_{2y} \\ &= \left[-x + 2\frac{x^2}{L} - \frac{x^3}{L^2}\right]\left(+\frac{q_0L^3}{24EI_y}\right) + \left[+\frac{x^2}{L} - \frac{x^3}{L^2}\right]\left(-\frac{q_0L^3}{24EI_y}\right) \\ &= -\frac{q_0}{24EI_y}\left(-L^2x^2 + L^3x\right), \end{aligned} \tag{3.184}$$

however the analytical course according to Ref. [11] results in $u_z(x) = -\frac{q_0}{24EI_y}\left(x^4 - 2Lx^3 + L^3x\right)$, meaning the analytical and therefore exact course is also at this point not identical with the numerical solution between the nodes ($0 < x < L$), see Fig. 3.23.

The numerical solution for the deflection in the middle of the beam yields $u_z^e(x = \frac{1}{2}L) = \frac{-4q_0L^4}{384EI_y}$, however the exact solution is $u_z(x = \frac{1}{2}L) = \frac{-5q_0L^4}{384EI_y}$.

Conclusion: Finite element solution and analytical solution are only identical at the nodes!

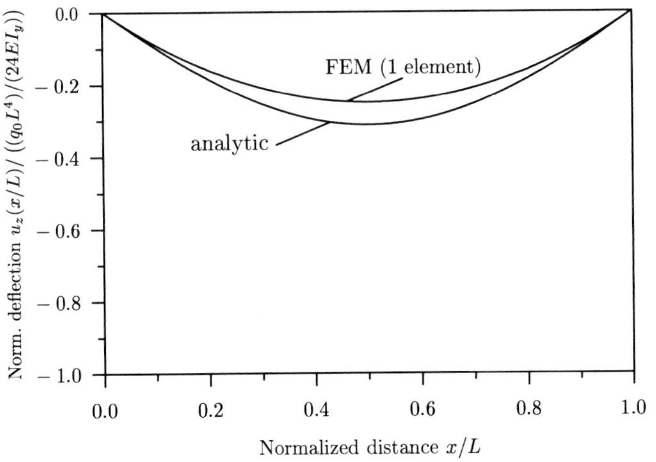

Fig. 3.23 Comparison of the analytical and the finite element solution for the beam according to Fig. 3.21b

3.3 Example: Beam with distributed load over half of the length

The following Fig. 3.24 shows a horizontal beam structure of length $2L$ which is fixed at both ends. The left-hand part of the structure ($0 \leq X \leq L$) is loaded by a constant distributed load q_0.

Use two EULER–BERNOULLI beam elements of equal length (see Fig. 3.24b) and:

- Assemble the global system of equations without consideration of the boundary conditions at the fixed supports.
- Obtain the reduced system of equations.
- Solve the system of equations for the unknowns at node 2.
- Calculate the reactions at node 1 and 3.

Fig. 3.24 Horizontal beam structure: **a** geometry and boundary conditions and **b** discretization

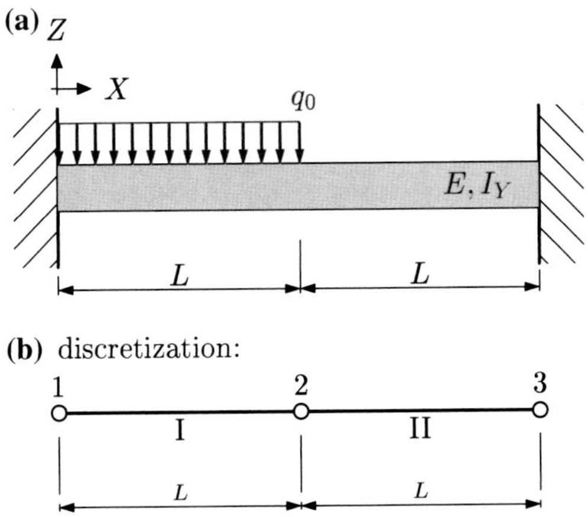

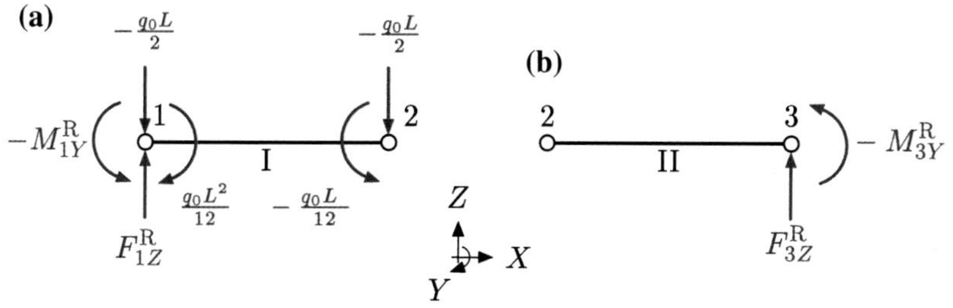

Fig. 3.25 Free body diagram of the beam structure problem

3.3 Solution

The free body diagram is shown in Fig. 3.25.

• Let us look in the following first separately at each element. The stiffness matrix for element I can be written as:

$$\boldsymbol{K}_{\mathrm{I}}^{\mathrm{e}} = \frac{EI_Y}{L^3}\begin{bmatrix} 12 & -6L & -12 & -6L \\ -6L & 4L^2 & 6L & 2L^2 \\ -12 & 6L & 12 & 6L \\ -6L & 2L^2 & 6L & 4L^2 \end{bmatrix}\begin{matrix} u_{1Z} \\ \varphi_{1Y} \\ u_{2Z} \\ \varphi_{2Y} \end{matrix}. \quad (3.185)$$

(column labels: u_{1Y} φ_{1Z} u_{2Y} φ_{2Z})

In the same way, the stiffness matrix for element II reads as:

$$\boldsymbol{K}_{\mathrm{II}}^{\mathrm{e}} = \frac{EI_Y}{L^3}\begin{bmatrix} 12 & -6L & -12 & -6L \\ -6L & 4L^2 & 6L & 2L^2 \\ -12 & 6L & 12 & 6L \\ -6L & 2L^2 & 6L & 4L^2 \end{bmatrix}\begin{matrix} u_{2Z} \\ \varphi_{2Y} \\ u_{3Z} \\ \varphi_{3Y} \end{matrix}. \quad (3.186)$$

(column labels: u_{2Y} φ_{2Z} u_{3Y} φ_{3Z})

The global system of equation without consideration of the boundary conditions is obtained as:

$$\frac{EI_Y}{L^3}\begin{bmatrix} 12 & -6L & -12 & -6L & 0 & 0 \\ -6L & 4L^2 & 6L & 2L^2 & 0 & 0 \\ -12 & 6L & 12+12 & 6L-6L & -12 & -6L \\ -6L & 2L^2 & 6L-6L & 4L^2+4L^2 & 6L & 2L^2 \\ 0 & 0 & -12 & 6L & 12 & 6L \\ 0 & 0 & -6L & 2L^2 & 6L & 4L^2 \end{bmatrix}\begin{bmatrix} u_{1Z} \\ \varphi_{1Y} \\ u_{2Z} \\ \varphi_{2Y} \\ u_{3Z} \\ \varphi_{3Y} \end{bmatrix} = \begin{bmatrix} F_{1Z}^{R} - \frac{q_0L}{2} \\ -M_{1Y}^{R} + \frac{q_0L^2}{12} \\ -\frac{q_0L}{2} \\ -\frac{q_0L^2}{12} \\ F_{3Z}^{R} \\ -M_{3Y}^{R} \end{bmatrix}. \quad (3.187)$$

• Introduction of the boundary conditions, i.e. $u_{1Z} = \varphi_{1Y} = 0$ and $u_{3Z} = \varphi_{3Y} = 0$, gives the following reduced system of equations:

$$\frac{EI_Y}{L^3}\begin{bmatrix} 24 & 0 \\ 0 & 8L^2 \end{bmatrix}\begin{bmatrix} u_{2Z} \\ \varphi_{2Y} \end{bmatrix} = \begin{bmatrix} -\frac{q_0 L}{2} \\ -\frac{q_0 L^2}{12} \end{bmatrix}. \tag{3.188}$$

• The solution of this system of equations can be obtained by calculating the inverse of the stiffness matrix to give:

$$\begin{bmatrix} u_{2Z} \\ \varphi_{2Y} \end{bmatrix} = \frac{L^3}{EI_Y}\begin{bmatrix} \frac{1}{24} & 0 \\ 0 & \frac{1}{8L^2} \end{bmatrix}\begin{bmatrix} -\frac{q_0 L}{2} \\ -\frac{q_0 L^2}{12} \end{bmatrix}, \tag{3.189}$$

or simplified as:

$$\begin{bmatrix} u_{2Z} \\ \varphi_{2Y} \end{bmatrix} = \frac{L^3}{EI_Y}\begin{bmatrix} -\frac{q_0 L}{48} \\ -\frac{q_0}{96} \end{bmatrix}. \tag{3.190}$$

• Reaction forces are obtained from the global non-reduced system under consideration of the known deformation matrix:

$$\frac{EI_Y}{L^3}(-12u_{2Z} - 6L\varphi_{2Y}) = F_{1Z}^{\mathrm{R}} - \frac{q_0 L}{2} \quad \Rightarrow \quad F_{1Z}^{\mathrm{R}} = \frac{13}{16} q_0 L. \tag{3.191}$$

$$\frac{EI_Y}{L^3}\left(6Lu_{2Z} + 2L^2\varphi_{2Y}\right) = -M_{1Y}^{\mathrm{R}} + \frac{q_0 L^2}{12} \quad \Rightarrow \quad M_{1Y}^{\mathrm{R}} = \frac{11}{48} q_0 L^2. \tag{3.192}$$

$$\frac{EI_Y}{L^3}(-12u_{2Z} + 6L\varphi_{2Y}) = F_{3Z}^{\mathrm{R}} \quad \Rightarrow \quad F_{3Z}^{\mathrm{R}} = \frac{3}{16} q_0 L. \tag{3.193}$$

$$\frac{EI_Y}{L^3}\left(-6Lu_{2Z} + 2L^2\varphi_{2Y}\right) = -M_{3Y}^{\mathrm{R}} \quad \Rightarrow \quad M_{3Y}^{\mathrm{R}} = -\frac{5}{48} q_0 L^2. \tag{3.194}$$

3.4 Example: Beam with linearly increasing distributed load

The following Fig. 3.26a shows a cantilever beam structure of length L which is loaded with a triangular shaped distributed load (maximum value of q_0 at $X = L$).

Use two beam elements of equal length $\frac{L}{2}$ (see Fig. 3.26b) and:

- Calculate for each element separately the vector of the equivalent nodal loads based on the general statement $\int \boldsymbol{N} q(x)\mathrm{d}x$ and analytical integration.
- Assemble the global system of equations without consideration of the boundary conditions at the fixed support.
- Obtain the reduced system of equations.
- Solve the linear system of equations.
- Additional question: Check the results for the equivalent nodal loads based on the natural coordinate, i.e. $\int \boldsymbol{N} q(\xi)\mathrm{d}\xi$ for a one-, two- and three-point numerical integration rule.

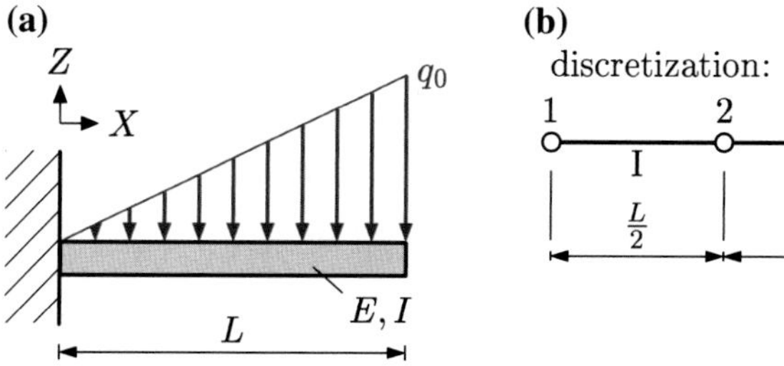

Fig. 3.26 Cantilever beam with triangular shaped distributed load: **a** geometry and boundary conditions and **b** discretization

3.4 Solution

The separated elements and the corresponding distributed loads are shown in Fig. 3.27.

Special consideration requires the elemental length $\frac{L}{2}$ since the interpolation functions and integrals are defined from $0 \ldots L$. Thus, let us calculate the equivalent nodal loads first for the length L and at the end we substitute $L := \frac{L}{2}$.

• Let us look in the following first separately at each element. The load vector for element I can be written as:

$$f_{\mathrm{I}} = \int_0^L \boldsymbol{N}(x) q(x)\,\mathrm{d}x = \int_0^L \begin{bmatrix} N_{1u}(x) \\ N_{1\varphi}(x) \\ N_{2u}(x) \\ N_{2\varphi}(x) \end{bmatrix} \left(-\frac{q_0}{2}\frac{x}{L} \right) \mathrm{d}x =$$

$$= -\frac{q_0}{2L} \int_0^L \begin{bmatrix} 1 - \frac{3x^2}{L^2} + \frac{2x^3}{L^3} \\ -x + \frac{2x^2}{L} - \frac{x^3}{L^2} \\ \frac{3x^2}{L^2} - \frac{2x^3}{L^3} \\ \frac{x^2}{L} - \frac{x^3}{L^2} \end{bmatrix} x\,\mathrm{d}x = -\frac{q_0 L}{2} \begin{bmatrix} \frac{3}{20} \\ -\frac{L}{30} \\ \frac{7}{20} \\ \frac{L}{20} \end{bmatrix}. \tag{3.195}$$

Fig. 3.27 Single elements and corresponding distributed loads: **a** element I and **b** element II

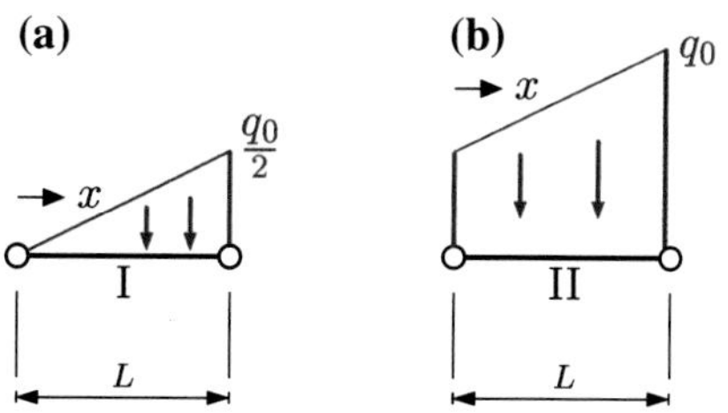

$$L := \frac{L}{2} \Rightarrow \boldsymbol{f}_{\mathrm{I}}^{\mathrm{e}} = -\frac{q_0 L}{4} \begin{bmatrix} \frac{3}{20} \\ -\frac{L}{60} \\ \frac{7}{20} \\ \frac{L}{40} \end{bmatrix} . \tag{3.196}$$

In a similar way we obtain for element II:

$$\boldsymbol{f}_{\mathrm{II}} = \int_0^L \boldsymbol{N}(x) q(x)\,\mathrm{d}x = \int_0^L \begin{bmatrix} N_{1u}(x) \\ N_{1\varphi}(x) \\ N_{2u}(x) \\ N_{2\varphi}(x) \end{bmatrix} \left(-\frac{q_0}{2}\left(1 + \frac{x}{L}\right)\right) \mathrm{d}x =$$

$$= -q_0 \int_0^L \begin{bmatrix} 1 - \frac{3x^2}{L^2} + \frac{2x^3}{L^3} \\ -x + \frac{2x^2}{L} - \frac{x^3}{L^2} \\ \frac{3x^2}{L^2} - \frac{2x^3}{L^3} \\ \frac{x^2}{L} - \frac{x^3}{L^2} \end{bmatrix} \left(\frac{1}{2} + \frac{x}{2L}\right) \mathrm{d}x = -q_0 L \begin{bmatrix} \frac{13}{40} \\ -\frac{7L}{120} \\ \frac{17}{40} \\ \frac{L}{15} \end{bmatrix} . \tag{3.197}$$

$$L := \frac{L}{2} \Rightarrow \boldsymbol{f}_{\mathrm{II}}^{\mathrm{e}} = -\frac{q_0 L}{4} \begin{bmatrix} \frac{26}{40} \\ -\frac{14L}{240} \\ \frac{34}{20} \\ \frac{2L}{30} \end{bmatrix} . \tag{3.198}$$

- The principal finite element equation for element I reads:

$$\boldsymbol{K}_{\mathrm{I}}^{\mathrm{e}} = \frac{8EI_Y}{L^3} \begin{bmatrix} 12 & -6\left(\frac{L}{2}\right) & -12 & -6\left(\frac{L}{2}\right) \\ -6\left(\frac{L}{2}\right) & 4\left(\frac{L}{2}\right)^2 & 6\left(\frac{L}{2}\right) & 2\left(\frac{L}{2}\right)^2 \\ -12 & 6\left(\frac{L}{2}\right) & 12 & 6\left(\frac{L}{2}\right) \\ -6\left(\frac{L}{2}\right) & 2\left(\frac{L}{2}\right)^2 & 6\left(\frac{L}{2}\right) & 4\left(\frac{L}{2}\right)^2 \end{bmatrix} \begin{matrix} u_{1Z} \\ \varphi_{1Y} \\ u_{2Z} \\ \varphi_{2Y} \end{matrix} , \tag{3.199}$$

(columns: u_{1Z}, φ_{1Y}, u_{2Z}, φ_{2Y})

and for element II:

$$
\boldsymbol{K}^{\mathrm{e}}_{\mathrm{II}} = \frac{8EI_Y}{L^3}
\begin{array}{c}
\begin{array}{cccc} u_{2Z} & \varphi_{2Y} & u_{3Z} & \varphi_{3Y} \end{array} \\
\begin{bmatrix}
12 & -6\left(\frac{L}{2}\right) & -12 & -6\left(\frac{L}{2}\right) \\
-6\left(\frac{L}{2}\right) & 4\left(\frac{L}{2}\right)^2 & 6\left(\frac{L}{2}\right) & 2\left(\frac{L}{2}\right)^2 \\
-12 & 6\left(\frac{L}{2}\right) & 12 & 6\left(\frac{L}{2}\right) \\
-6\left(\frac{L}{2}\right) & 2\left(\frac{L}{2}\right)^2 & 6\left(\frac{L}{2}\right) & 4\left(\frac{L}{2}\right)^2
\end{bmatrix}
\end{array}
\begin{array}{c} u_{2Z} \\ \varphi_{2Y} \\ u_{3Z} \\ \varphi_{3Y} \end{array} . \tag{3.200}
$$

The global stiffness matrix $\boldsymbol{K}$ without consideration of the boundary conditions at the fixed support is obtained by combining both elemental matrices as:

$$
\boldsymbol{K} = \frac{8EI_Y}{L^3}
\begin{array}{c}
\begin{array}{cccccc} u_{1Z} & \varphi_{1Y} & u_{2Z} & \varphi_{2Y} & u_{3Z} & \varphi_{3Y} \end{array} \\
\begin{bmatrix}
12 & -3L & -12 & -3L & 0 & 0 \\
-3L & L^2 & 3L & \frac{L^2}{2} & 0 & 0 \\
-12 & 3L & 12+12 & 3L-3L & -12 & -3L \\
-3L & \frac{L^2}{2} & 3L-3L & L^2+L^2 & 3L & \frac{L^2}{2} \\
0 & 0 & -12 & 3L & 12 & 3L \\
0 & 0 & -3L & \frac{L^2}{2} & 3L & L^2
\end{bmatrix}
\end{array}
\begin{array}{c} u_{1Z} \\ \varphi_{1Y} \\ u_{2Z} \\ \varphi_{2Y} \\ u_{3Z} \\ \varphi_{3Y} \end{array} . \tag{3.201}
$$

• Consideration of the boundary condition at the left-hand boundary, i.e. $u_{1Z} = 0$ and $\varphi_{1Y} = 0$, results in the following reduced system of equations where the global load matrix $\boldsymbol{f} = \boldsymbol{f}^{\mathrm{e}}_{\mathrm{I}} + \boldsymbol{f}^{\mathrm{e}}_{\mathrm{I}}$ was assembled based on Eqs. (3.196) and (3.198):

$$
\frac{8EI_Y}{L^3}
\begin{bmatrix}
12+12 & 3L-3L & -12 & -3L \\
3L-3L & L^2+L^2 & 3L & \frac{L^2}{2} \\
-12 & 3L & 12 & 3L \\
-3L & \frac{L^2}{2} & 3L & L^2
\end{bmatrix}
\begin{bmatrix} u_{2Z} \\ \varphi_{2Y} \\ u_{3Z} \\ \varphi_{3Y} \end{bmatrix}
= -\frac{qL}{4}
\begin{bmatrix} \frac{7}{20} + \frac{26}{40} \\ \frac{L}{40} - \frac{14L}{240} \\ \frac{34}{40} \\ \frac{2L}{30} \end{bmatrix} . \tag{3.202}
$$

• The solution of the linear system of equations can be obtained based on $\boldsymbol{u} = \boldsymbol{K}^{-1}\boldsymbol{f}$ as:

$$
\begin{bmatrix} u_{2Z} \\ \varphi_{2Y} \\ u_{3Z} \\ \varphi_{3Y} \end{bmatrix}
= \frac{q_0 L^4}{EI_Y}
\begin{bmatrix} -\frac{121}{3840} \\ \frac{41}{384L} \\ -\frac{11}{120} \\ \frac{1}{8L} \end{bmatrix} . \tag{3.203}
$$

This finite element solution is equal to the analytical solution.

• $\int \boldsymbol{N} q(\xi)\mathrm{d}\xi$ will be checked for the first component of $\boldsymbol{f}_{\mathrm{I}}^{\mathrm{e}}$:

$$f_{1Z} = \int_{-1}^{1} \frac{1}{4}(2 - 3\xi + \xi^3)\left(-\frac{q_0}{4}\right)(1+\xi)\,\frac{L}{4}\,\mathrm{d}\xi\,. \tag{3.204}$$

$$\text{one-point rule} \quad f_{1Z} = -\frac{q_0 L}{16}, \tag{3.205}$$

$$\text{two-point rule} \quad f_{1Z} = -\frac{5 q_0 L}{144}, \tag{3.206}$$

$$\text{three-point rule} \quad f_{1Z} = -\frac{3 q_0 L}{80}. \tag{3.207}$$

3.5 Example: Simply supported beam problems—comparison between finite element and analytical approach

Given is a simply supported EULER–BERNOULLI beam as shown in Fig. 3.28. The length of the beam is L and the bending stiffness is EI. Consider two different load cases in the following: (a) a single force F_0 acting in the middle of the beam. (b) A constant distributed load q_0.

Calculate based on (I) one single and (II) four beam finite elements of equal length the deformations at the nodes and in the middle of each element. In addition evaluate the maximum stress at the nodes and in the middle of the element. Compare your results with the analytical (exact) solution and calculate the relative error. The deformations and stresses will be a function of L, EI, F_0, q_0, or z_{max}.

3.5 Solution

▶ **Analytical Solutions**

• ***Load Case with Single Force***

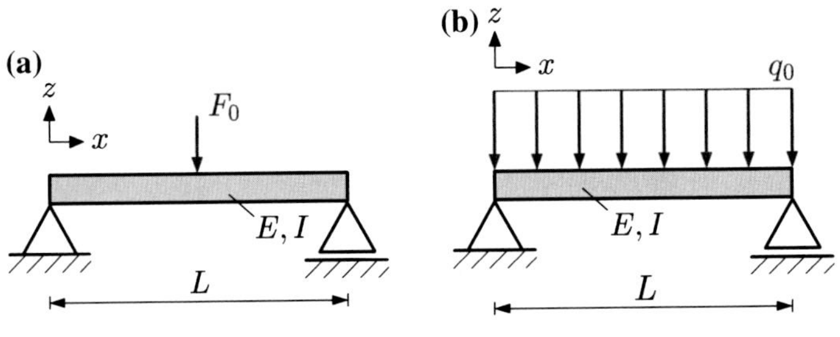

Fig. 3.28 Simply supported beam: **a** single force case; **b** distributed load case

For the range $0 \leq x \leq \frac{L}{2}$, the displacement $u_z(x)$ and rotation $\varphi_y(x)$ are given by:

$$u_z(x) = -\frac{F_0 x}{48EI}\left(3L^2 - 4x^2\right) , \tag{3.208}$$

$$\varphi_y(x) = -\frac{\mathrm{d}u_z(x)}{\mathrm{d}x} = \frac{F_0}{16EI}\left(L^2 - 4x^2\right) . \tag{3.209}$$

The stress $\sigma_x(x, z)$ can be obtained from:

$$\sigma_x(x, z) = \frac{M_y(x)}{I} \times z(x). \tag{3.210}$$

We assume in the following that the maximum stress at $z = z_{\text{max}}$ should be evaluated. The moment equilibrium gives the following distributions of the bending moment:

$$M_y(x) = -\frac{F_0 x}{2} \quad \text{for} \quad 0 \leq x \leq \tfrac{L}{2} , \tag{3.211}$$

$$M_y(x) = -\frac{F_0(L - x)}{2} \quad \text{for} \quad \tfrac{L}{2} \leq x \leq L . \tag{3.212}$$

Based on the above equations, the following Tables 3.12 and 3.13 can be generated to benchmark the finite element simulations based on one, four, and ten elements. Since the problem is symmetric, only the range $0 \leq x \leq \frac{L}{2}$ needs to be covered (note that the rotation may change the sign in the second part of the coordinate range which is not shown in the following).

Table 3.12 Analytical results for displacement, rotation, and stress for comparison with the one-element approach (single force case)

Coordinate $\frac{x}{L}$	$\frac{u_z(x)}{\frac{F_0 L^3}{EI}}$	$\frac{\varphi_y(x)}{\frac{F_0 L^2}{EI}}$	$\frac{\sigma_{\text{max}}}{\frac{F_0 L z_{\text{max}}}{I}}$
0	0	$\frac{1}{16}$	0
$\frac{1}{2}$	$-\frac{1}{48}$	0	$-\frac{1}{4}$

Table 3.13 Analytical results for displacement, rotation, and stress for comparison with the four-element approach (single force case)

Coordinate $\frac{x}{L}$	$\frac{u_z(x)}{\frac{F_0 L^3}{EI}}$	$\frac{\varphi_y(x)}{\frac{F_0 L^2}{EI}}$	$\frac{\sigma_{\text{max}}}{\frac{F_0 L z_{\text{max}}}{I}}$
0	0	$\frac{1}{16}$	0
$\frac{1}{8}$	$-\frac{47}{6144}$	$\frac{15}{256}$	$-\frac{1}{16}$
$\frac{1}{4}$	$-\frac{11}{768}$	$\frac{3}{64}$	$-\frac{1}{8}$
$\frac{3}{8}$	$-\frac{39}{2048}$	$\frac{7}{256}$	$-\frac{3}{16}$
$\frac{1}{2}$	$-\frac{1}{48}$	0	$-\frac{1}{4}$

• Load Case with Distributed Load

For the range $0 \leq x \leq L$, the displacement $u_z(x)$ and rotation $\varphi_y(x)$ are given by:

$$u_z(x) = -\frac{q_0 x}{24EI}\left(x^3 - 2Lx^2 + L^3\right) , \tag{3.213}$$

$$\varphi_y(x) = -\frac{du_z(x)}{dx} = \frac{q_0}{24EI}\left(4x^3 - 6Lx^2 + L^3\right) . \tag{3.214}$$

The stress $\sigma_x(x,z)$ can be obtained from:

$$\sigma_x(x,z) = \frac{M_y(x)}{I_y} \times z(x). \tag{3.215}$$

We assume in the following that the maximum stress at $z = z_{\text{max}}$ should be evaluated. The moment equilibrium gives the following distribution of the bending moment:

$$M_y(x) = \frac{q_0 x}{2}(x - L) . \tag{3.216}$$

Based on the above equations, the following Tables 3.14 and 3.15 can be generated to benchmark the finite element simulations based on one, four, and ten elements. Since the problem is symmetric, only the range $0 \leq x \leq \frac{L}{2}$ needs to be covered (note that the rotation may change the sign in the second part of the coordinate range which is not shown in the following).

Table 3.14 Analytical results for displacement, rotation, and stress for comparison with the one-element approach (distributed load case)

Coordinate $\frac{x}{L}$	$\frac{u_z(x)}{\frac{q_0 L^4}{EI}}$	$\frac{\varphi_y(x)}{\frac{q_0 L^3}{EI}}$	$\frac{\sigma_{\text{max}}}{\frac{q_0 L^2 z_{\text{max}}}{I}}$
0	0	$\frac{1}{24}$	0
$\frac{1}{2}$	$-\frac{5}{384}$	0	$-\frac{1}{8}$

Table 3.15 Analytical results for displacement, rotation, and stress for comparison with the four-element approach (distributed load case)

Coordinate $\frac{x}{L}$	$\frac{u_z(x)}{\frac{q_0 L^4}{EI}}$	$\frac{\varphi_y(x)}{\frac{q_0 L^3}{EI}}$	$\frac{\sigma_{\text{max}}}{\frac{q_0 L^2 z_{\text{max}}}{I}}$
0	0	$\frac{1}{24}$	0
$\frac{1}{8}$	$-\frac{497}{98304}$	$\frac{39}{1024}$	$-\frac{7}{128}$
$\frac{1}{4}$	$-\frac{19}{2048}$	$\frac{11}{384}$	$-\frac{3}{32}$
$\frac{3}{8}$	$-\frac{395}{32768}$	$\frac{47}{3072}$	$-\frac{15}{128}$
$\frac{1}{2}$	$-\frac{5}{384}$	0	$-\frac{1}{8}$

▶ Finite Element Solutions

• One Element Approach—Single Force Case

The global system of equations for one beam element without consideration of the boundary conditions can be written as:

$$\frac{EI}{L^3}\begin{bmatrix} 12 & -6L & -12 & -6L \\ -6L & 4L^2 & 6L & 2L^2 \\ -12 & 6L & 12 & 6L \\ -6L & 2L^2 & 6L & 4L^2 \end{bmatrix}\begin{bmatrix} u_{1z} \\ \varphi_{1y} \\ u_{2z} \\ \varphi_{2y} \end{bmatrix} = \begin{bmatrix} -\frac{F_0}{2} \\ \frac{F_0 L}{8} \\ -\frac{F_0}{2} \\ -\frac{F_0 L}{8} \end{bmatrix}. \tag{3.217}$$

Consideration of the boundary conditions, i.e. $u_{1z} = u_{2z} = 0$, gives the reduced system as:

$$\frac{EI}{L^3}\begin{bmatrix} 4L^2 & 2L^2 \\ 2L^2 & 4L^2 \end{bmatrix}\begin{bmatrix} \varphi_{1y} \\ \varphi_{2y} \end{bmatrix} = \begin{bmatrix} \frac{F_0 L}{8} \\ -\frac{F_0 L}{8} \end{bmatrix}. \tag{3.218}$$

Calculation of the inverse of the stiffness matrix allows the determination of the nodal unknowns:

$$\begin{bmatrix} \varphi_{1y} \\ \varphi_{2y} \end{bmatrix} = \frac{L^3}{EI}\frac{1}{16L^4 - 4L^4}\begin{bmatrix} 4L^2 & -2L^2 \\ -2L^2 & 4L^2 \end{bmatrix}\begin{bmatrix} \varphi_{1y} \\ \varphi_{2y} \end{bmatrix}\begin{bmatrix} \frac{F_0 L}{8} \\ -\frac{F_0 L}{8} \end{bmatrix} = \frac{F_0 L^2}{16EI}\begin{bmatrix} 1 \\ -1 \end{bmatrix}. \tag{3.219}$$

Comment: The obtained nodal rotations are equal to the analytical solution. Furthermore, the nodal displacements are equal to the analytical solution since they were imposed as boundary condition.

The calculation of the displacement distribution within a single element is based on the following nodal approach:

$$u_z^{\mathrm{e}}(x) = \begin{bmatrix} N_{1u} & N_{1\varphi} & N_{2u} & N_{2\varphi} \end{bmatrix}\begin{bmatrix} u_{1z} \\ \varphi_{1y} \\ u_{2z} \\ \varphi_{2y} \end{bmatrix}. \tag{3.220}$$

The displacement in the middle of the element is obtained for $x = \frac{L}{2}$ from Eq. (3.220) as:

$$u_z^{\mathrm{e}}\left(\tfrac{L}{2}\right) = -\frac{L}{8} \times \frac{F_0 L^2}{16EI} - \frac{L}{8} \times \frac{F_0 L^2}{16EI} = -\frac{1}{64}\frac{F_0 L^2}{EI}. \tag{3.221}$$

$$\text{relative error} = \left| \frac{-\frac{1}{64} + \frac{1}{48}}{-\frac{1}{48}} \right| \times 100 = 25\%\,. \tag{3.222}$$

The calculation of the rotation distribution within a single element is based on the following nodal approach:

$$\varphi_y^{\mathrm{e}}(x) = -\frac{\mathrm{d}u_z^{\mathrm{e}}(x)}{\mathrm{d}x} = -\begin{bmatrix} \frac{\mathrm{d}N_{1u}}{\mathrm{d}x} & \frac{\mathrm{d}N_{1\varphi}}{\mathrm{d}x} & \frac{\mathrm{d}N_{2u}}{\mathrm{d}x} & \frac{\mathrm{d}N_{2\varphi}}{\mathrm{d}x} \end{bmatrix} \begin{bmatrix} u_{1z} \\ \varphi_{1y} \\ u_{2z} \\ \varphi_{2y} \end{bmatrix} . \tag{3.223}$$

The rotation in the middle of the element is obtained for $x = \frac{L}{2}$ from Eq. (3.223) as:

$$\varphi_y^{\mathrm{e}}\left(\tfrac{L}{2}\right) = -\frac{1}{4} \times \frac{F_0 L^2}{16EI} + \frac{1}{4} \times \frac{F_0 L^2}{16EI} = 0 . \tag{3.224}$$

This result for the rotation is equal to the analytical solution.

The calculation of the maximum stress distribution, i.e. for $z = z_{\mathrm{max}}$, within a single element is based on the following nodal approach:

$$\sigma^{\mathrm{e}}(x, z_{\mathrm{max}}) = -E \frac{\mathrm{d}^2 u_z^{\mathrm{e}}(x)}{\mathrm{d}x^2} z_{\mathrm{max}} = -E z_{\mathrm{max}} \begin{bmatrix} \frac{\mathrm{d}^2 N_{1u}}{\mathrm{d}x^2} & \frac{\mathrm{d}^2 N_{1\varphi}}{\mathrm{d}x^2} & \frac{\mathrm{d}^2 N_{2u}}{\mathrm{d}x^2} & \frac{\mathrm{d}^2 N_{2\varphi}}{\mathrm{d}x^2} \end{bmatrix} \begin{bmatrix} u_{1z} \\ \varphi_{1y} \\ u_{2z} \\ \varphi_{2y} \end{bmatrix} . \tag{3.225}$$

Thus, the maximum stresses at the nodes and in the middle of the element are obtained from Eq. (3.225) as:

$$\sigma^{\mathrm{e}}(0, z_{\mathrm{max}}) = -E z_{\mathrm{max}} \left(\frac{4}{L} \frac{F_0 L^2}{16EI} - \frac{2}{L} \frac{F_0 L^2}{16EI} \right) = -\frac{1}{8} \frac{F_0 L z_{\mathrm{max}}}{I} , \tag{3.226}$$

$$\sigma^{\mathrm{e}}(\tfrac{L}{2}, z_{\mathrm{max}}) = -E z_{\mathrm{max}} \left(\frac{1}{L} \frac{F_0 L^2}{16EI} + \frac{1}{L} \frac{F_0 L^2}{16EI} \right) = -\frac{1}{8} \frac{F_0 L z_{\mathrm{max}}}{I} , \tag{3.227}$$

$$\sigma^{\mathrm{e}}(L, z_{\mathrm{max}}) = -E z_{\mathrm{max}} \left(-\frac{2}{L} \frac{F_0 L^2}{16EI} + \frac{4}{L} \frac{F_0 L^2}{16EI} \right) = -\frac{1}{8} \frac{F_0 L z_{\mathrm{max}}}{I} . \tag{3.228}$$

The corresponding relative errors are (∞),[28] 50% and (∞).

- **One Element Approach—Distributed Load Case**

The global system of equations without consideration of the boundary conditions can be written as:

[28] A division by zero occurs during the calculation of the relative error.

$$\frac{EI}{L^3}\begin{bmatrix} 12 & -6L & -12 & -6L \\ -6L & 4L^2 & 6L & 2L^2 \\ -12 & 6L & 12 & 6L \\ -6L & 2L^2 & 6L & 4L^2 \end{bmatrix}\begin{bmatrix} u_{1z} \\ \varphi_{1y} \\ u_{2z} \\ \varphi_{2y} \end{bmatrix} = \begin{bmatrix} -\frac{q_0L}{2} \\ \frac{q_0L^2}{12} \\ -\frac{q_0L}{2} \\ -\frac{q_0L^2}{12} \end{bmatrix}. \tag{3.229}$$

Consideration of the boundary conditions, i.e. $u_{1z} = u_{2z} = 0$, gives the reduced system as:

$$\frac{EI}{L^3}\begin{bmatrix} 4L^2 & 2L^2 \\ 2L^2 & 4L^2 \end{bmatrix}\begin{bmatrix} \varphi_{1y} \\ \varphi_{2y} \end{bmatrix} = \begin{bmatrix} \frac{q_0L^2}{12} \\ -\frac{q_0L^2}{12} \end{bmatrix}. \tag{3.230}$$

Comparing this expression with Eq. (3.218), it can be concluded that changing the load from the single force to the distributed load affects only the right-hand side of the principal finite element equation.

Calculation of the inverse of the stiffness matrix allows the determination of the nodal unknowns:

$$\begin{bmatrix} \varphi_{1y} \\ \varphi_{2y} \end{bmatrix} = \frac{L^3}{EI}\frac{1}{16L^4 - 4L^4}\begin{bmatrix} 4L^2 & -2L^2 \\ -2L^2 & 4L^2 \end{bmatrix}\begin{bmatrix} \varphi_{1y} \\ \varphi_{2y} \end{bmatrix}\begin{bmatrix} \frac{q_0L^2}{12} \\ -\frac{q_0L^2}{12} \end{bmatrix} = \frac{q_0L^3}{24EI}\begin{bmatrix} 1 \\ -1 \end{bmatrix}. \tag{3.231}$$

Comment: The obtained nodal rotations are equal to the analytical solution.

The displacement in the middle of the element is obtained for $x = \frac{L}{2}$ from Eq. (3.220) as:

$$u_z^{\mathrm{e}}\left(\tfrac{L}{2}\right) = -\frac{L}{8} \times \frac{q_0L^3}{24EI} - \frac{L}{8} \times \frac{q_0L^3}{24EI} = -\frac{1}{96}\frac{q_0L^4}{EI}. \tag{3.232}$$

$$\text{relative error} = \left|\frac{-\frac{1}{96} + \frac{5}{384}}{-\frac{5}{384}}\right| \times 100 = 20\%\,. \tag{3.233}$$

The rotation in the middle of the element is obtained for $x = \frac{L}{2}$ from Eq. (3.223) as:

$$\varphi_y^{\mathrm{e}}\left(\tfrac{L}{2}\right) = \frac{1}{4} \times \frac{q_0L^3}{24EI} - \frac{1}{4} \times \frac{q_0L^3}{24EI} = 0\,. \tag{3.234}$$

This result for the rotation is equal to the analytical solution.

The maximum stresses at the nodes and in the middle of the element are obtained from Eq. (3.225) as:

$$\sigma^{\mathrm{e}}(0, z_{\max}) = -Ez_{\max}\left(\frac{4}{L} \times \frac{q_0L^3}{24EI} - \frac{2}{L} \times \frac{q_0L^3}{24EI}\right) = -\frac{1}{12}\frac{q_0L^2z_{\max}}{I}, \tag{3.235}$$

$$\sigma^{e}(\tfrac{L}{2}, z_{max}) = -Ez_{max}\left(\frac{1}{L}\times\frac{q_0L^3}{24EI}+\frac{1}{L}\times\frac{q_0L^3}{24EI}\right) = -\frac{1}{12}\frac{q_0L^2z_{max}}{I}, \tag{3.236}$$

$$\sigma^{e}(L, z_{max}) = -Ez_{max}\left(-\frac{2}{L}\times\frac{q_0L^3}{24EI}+\frac{4}{L}\times\frac{q_0L^3}{24EI}\right) = -\frac{1}{12}\frac{q_0L^2z_{max}}{I}. \tag{3.237}$$

The corresponding relative errors are (∞),[29] 33% and (∞).

- **Four Element Approach—Single Force Case**

The reduced stiffness matrix under consideration of the symmetry of the problem (i.e. only two elements considered; symmetry condition $\varphi_{3y} = 0$) reads for this case as:

$$\begin{bmatrix} 16\frac{EJ}{L} & 96\frac{EJ}{L^2} & 8\frac{EJ}{L} & 0 \\ 96\frac{EJ}{L^2} & 1536\frac{EJ}{L^3} & 0 & -768\frac{EJ}{L^3} \\ 8\frac{EJ}{L} & 0 & 32\frac{EJ}{L} & 96\frac{EJ}{L^2} \\ 0 & -768\frac{EJ}{L^3} & 96\frac{EJ}{L^2} & 768\frac{EJ}{L^3} \end{bmatrix}, \tag{3.238}$$

where $L^{e} = \frac{L}{4}$. The inversion of this reduced stiffness matrix reads:

$$\begin{bmatrix} 1/2\frac{L}{EJ} & -\frac{3L^2}{32EJ} & 1/4\frac{L}{EJ} & -1/8\frac{L^2}{EJ} \\ -\frac{3L^2}{32EJ} & 1/48\frac{L^3}{EJ} & -1/16\frac{L^2}{EJ} & \frac{11L^3}{384EJ} \\ 1/4\frac{L}{EJ} & -1/16\frac{L^2}{EJ} & 1/4\frac{L}{EJ} & -\frac{3L^2}{32EJ} \\ -1/8\frac{L^2}{EJ} & \frac{11L^3}{384EJ} & -\frac{3L^2}{32EJ} & 1/24\frac{L^3}{EJ} \end{bmatrix}, \tag{3.239}$$

and the nodal unknowns are obtained by multiplying this inverse with the reduced load vector, i.e. $\left[0\ 0\ 0\ -\frac{F}{2}\right]$:

$$\begin{bmatrix} \varphi_{1y} \\ u_{2z} \\ \varphi_{2y} \\ u_{3z} \end{bmatrix} = \begin{bmatrix} \frac{1}{16}\frac{L^2F_0}{EI} \\ -\frac{11}{768}\frac{L^3F_0}{EI} \\ \frac{3}{64}\frac{L^2F_0}{EI} \\ -\frac{1}{48}\frac{L^3F_0}{EI} \end{bmatrix}. \tag{3.240}$$

These nodal values are equal to the analytical solution. The determination of the deformations in the middle of each element and the stress values follow the proce-

[29] A division by zero occurs during the calculation of the relative error.

dure for the single element approach (each element is separately considered). The following listing summarizes the nodal values and the values in the middle (index 'm'). The corresponding relative errors are given in brackets.

• Element I:

$$u^{\mathrm{e}}_{\mathrm{m}z} = -\frac{7L^3F_0}{768EI} \quad (19.15\%)\,, \tag{3.241}$$

$$\varphi^{\mathrm{e}}_{\mathrm{m}y} = \frac{15L^2F_0}{256EI} \quad (0\%)\,. \tag{3.242}$$

$$\sigma^{\mathrm{e}}_1(z_{\max}) = 0 \quad (0\%)\,, \tag{3.243}$$

$$\sigma^{\mathrm{e}}_{\mathrm{m}}(z_{\max}) = -\frac{LF_0z_{\max}}{16I} \quad (0\%)\,, \tag{3.244}$$

$$\sigma^{\mathrm{e}}_2(z_{\max}) = -\frac{LF_0z_{\max}}{8I} \quad (0\%)\,. \tag{3.245}$$

• Element II:

$$u^{\mathrm{e}}_{\mathrm{m}z} = -\frac{3L^3F_0}{128EI} \quad (23.08\%)\,, \tag{3.246}$$

$$\varphi^{\mathrm{e}}_{\mathrm{m}y} = \frac{7L^2F_0}{256EI} \quad (0\%)\,. \tag{3.247}$$

$$\sigma^{\mathrm{e}}_1(z_{\max}) = -\frac{LF_0z_{\max}}{8I} \quad (0\%)\,, \tag{3.248}$$

$$\sigma^{\mathrm{e}}_{\mathrm{m}}(z_{\max}) = -\frac{3LF_0z_{\max}}{16I} \quad (0\%)\,, \tag{3.249}$$

$$\sigma^{\mathrm{e}}_2(z_{\max}) = -\frac{LF_0z_{\max}}{4I} \quad (0\%)\,. \tag{3.250}$$

• **Four Element Approach—Distributed Load Case**

The reduced stiffness matrix and the corresponding inverse is the same as in the single force case. The reduced load vector changes to $\left[q(L^{\mathrm{e}})^2/12 \;\; -aL^{\mathrm{e}} \;\; 0 \;\; -aL^{\mathrm{e}}/2\right]$. Thus, the nodal deformations are obtained as:

$$\begin{bmatrix} \varphi_{1y} \\ u_{2z} \\ \varphi_{2y} \\ u_{3z} \end{bmatrix} = \begin{bmatrix} \frac{1}{24}\frac{L^3q_0}{EI} \\ -\frac{19}{2048}\frac{L^4q_0}{EI} \\ \frac{11}{384}\frac{L^3q_0}{EI} \\ -\frac{5}{384}\frac{l^4q_0}{EI} \end{bmatrix}. \tag{3.251}$$

These nodal values are equal to the analytical solution. The following listing summarizes the nodal values and the values in the middle (index 'm'). The corresponding relative errors are given in brackets.

- Element I:

$$u^{\mathrm{e}}_{\mathrm{m}z} = -\frac{77L^4 q_0}{12288EI} \quad (23.94\%)\,, \tag{3.252}$$

$$\varphi^{\mathrm{e}}_{\mathrm{m}y} = \frac{39L^3 q_0}{1024EI} \quad (0\%)\,. \tag{3.253}$$

$$\sigma^{\mathrm{e}}_1(z_{\mathrm{max}}) = -\frac{L^2 q_0 z_{\mathrm{max}}}{192I} \quad (\to \infty)\,, \tag{3.254}$$

$$\sigma^{\mathrm{e}}_{\mathrm{m}}(z_{\mathrm{max}}) = -\frac{5L^2 q_0 z_{\mathrm{max}}}{96I} \quad (4.76\%)\,, \tag{3.255}$$

$$\sigma^{\mathrm{e}}_2(z_{\mathrm{max}}) = -\frac{19L^2 q_0 z_{\mathrm{max}}}{192I} \quad (5.56\%)\,. \tag{3.256}$$

- Element II:

$$u^{\mathrm{e}}_{\mathrm{m}z} = -\frac{181L^4 q_0}{12288EI} \quad (22.19\%)\,, \tag{3.257}$$

$$\varphi^{\mathrm{e}}_{\mathrm{m}y} = \frac{47L^3 q_0}{3072EI} \quad (0\%)\,. \tag{3.258}$$

$$\sigma^{\mathrm{e}}_1(z_{\mathrm{max}}) = -\frac{19L^2 q_0 z_{\mathrm{max}}}{192I} \quad (5.56\%)\,, \tag{3.259}$$

$$\sigma^{\mathrm{e}}_{\mathrm{m}}(z_{\mathrm{max}}) = -\frac{11L^2 q_0 z_{\mathrm{max}}}{96I} \quad (2.22\%)\,, \tag{3.260}$$

$$\sigma^{\mathrm{e}}_2(z_{\mathrm{max}}) = -\frac{25L^2 q_0 z_{\mathrm{max}}}{192I} \quad (4.17\%)\,. \tag{3.261}$$

3.6 Example: Beam with variable cross section

The beam shown in Fig. 3.29 has along its x-axis a variable cross section. Derive for

(a) a circular cross section, and
(b) a square cross section

the element stiffness matrix for the case $d_1 = 2h$ and $d_2 = h$.

3.6 Solution

(a) Circular cross section:

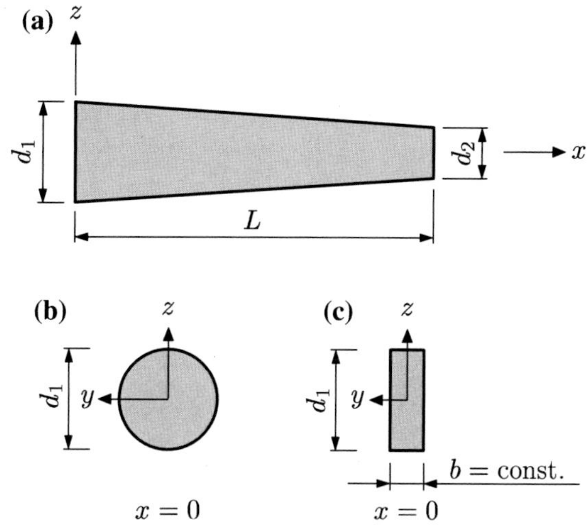

Fig. 3.29 Beam with variable cross section: **a** change along the x-axis; **b** circular cross section; **c** square cross section

The following expression can be used as an initial point for the derivation of the stiffness matrix:

$$\boldsymbol{K}^{\mathrm{e}} = E \int_x I_y(x) \frac{\mathrm{d}^2 \boldsymbol{N}(x)}{\mathrm{d}x^2} \frac{\mathrm{d}^2 \boldsymbol{N}^{\mathrm{T}}(x)}{\mathrm{d}x^2} \mathrm{d}x \,. \tag{3.262}$$

Since the axial second moment of area changes along the x-axis, a corresponding function has to be derived at first. An elegant method would be to use the polar second moment of area of the circle, since in this case the function of the radius along the x-axis can be used. Hereby the relation, that the polar second moment of area composes additively of the two axial second moments of area I_y and I_z, is being used:

$$I_{\mathrm{p}} = \int_A r^2 \mathrm{d}A = I_y + I_z \,. \tag{3.263}$$

Since the axial second moments of area of a circle are identical, the following expression can be derived for I_z:

$$I_y(x) = \frac{1}{2} I_{\mathrm{p}}(x) = \frac{1}{2} \int_A r^2 \mathrm{d}A = \frac{1}{2} \int_{\alpha=0}^{2\pi} \int_0^{r(x)} \hat{r}^2 \underbrace{\hat{r}\mathrm{d}\hat{r}\mathrm{d}\alpha}_{\mathrm{d}A} \tag{3.264}$$

$$= \pi \int_0^{r(x)} \hat{r}^3 \mathrm{d}\hat{r} = \pi \left[\frac{1}{4} \hat{r}^4 \right]_0^{r(x)} = \frac{\pi}{4} r(x)^4 \,. \tag{3.265}$$

The change of the radius along the x-axis can easily be derived from Fig. 3.29a:

$$r(x) = h \left(1 - \frac{x}{2L} \right) = \frac{h}{2} \left(2 - \frac{x}{L} \right) \,. \tag{3.266}$$

Therefore, the axial second moment of area results in

$$I_y(x) = \frac{\pi h^4}{64}\left(2 - \frac{x}{L}\right)^4 \tag{3.267}$$

and can be used in Eq. (3.262):

$$\boldsymbol{K}^{\mathrm{e}} = E\frac{\pi h^4}{64}\int\limits_L \left(2 - \frac{x}{L}\right)^4 \frac{\mathrm{d}^2\boldsymbol{N}(x)}{\mathrm{d}x^2}\frac{\mathrm{d}^2\boldsymbol{N}^{\mathrm{T}}(x)}{\mathrm{d}x^2}\mathrm{d}x\,. \tag{3.268}$$

The integration can be carried out through the second order derivatives of the interpolation function according to Eqs. (3.67) up to (3.70). As an example for the first component of the stiffness matrix

$$K_{11} = E\frac{\pi h^4}{64}\int\limits_L \left(2 - \frac{x}{L}\right)^4 \left(-\frac{6}{L^2} + \frac{12x}{L^3}\right)^2 \mathrm{d}x\,, \tag{3.269}$$

is being used and the entire stiffness matrix finally results after a short calculation:

$$\boldsymbol{K}^{\mathrm{e}}_{\text{circle}} = \frac{E\ \pi h^4}{L^3\ 64}\begin{bmatrix} \frac{2988}{35} & -\frac{1998}{35}L & -\frac{2988}{35} & -\frac{198}{7}L \\ -\frac{1998}{35}L & \frac{1468}{35}L^2 & \frac{1998}{35}L & \frac{106}{7}L^2 \\ -\frac{2988}{35} & \frac{1998}{35}L & \frac{2988}{35} & \frac{198}{7}L \\ -\frac{198}{7}L & \frac{106}{7}L^2 & \frac{198}{7}L & \frac{92}{7}L^2 \end{bmatrix}. \tag{3.270}$$

(b) Square cross section:

Regarding the square cross section, Eq. (3.262) serves as a basis as well. However, in this case it seems to be a good idea to go back to the definition of I_y:

$$I_y(x) = \int\limits_A z^2\mathrm{d}A = \int\limits_{-z(x)}^{z(x)} \hat{z}^2\,\underbrace{b\mathrm{d}\hat{z}}_{\mathrm{d}A} = b\left[\frac{1}{3}\hat{z}^3\right]_{-z(x)}^{z(x)} = \frac{2b}{3}z(x)^3\,. \tag{3.271}$$

The course of the function $z(x)$ of the cross section is identical with the radius in part (a) meaning $z(x) = h(1 - \frac{x}{2L})$ and the second moment of area in this case results in:

$$I_y(x) = \frac{2bh^3}{3}\left(1 - \frac{x}{2L}\right)^3 = \frac{bh^3}{12}\left(2 - \frac{x}{L}\right)^3. \tag{3.272}$$

Due to the special form of the second moment of area, the stiffness matrix therefore results in

$$\boldsymbol{K}^{\mathrm{e}} = E\frac{bh^3}{12}\int\limits_L \left(2 - \frac{x}{L}\right)^3 \frac{\mathrm{d}^2\boldsymbol{N}(x)}{\mathrm{d}x^2}\frac{\mathrm{d}^2\boldsymbol{N}^{\mathrm{T}}(x)}{\mathrm{d}x^2}\mathrm{d}x\,, \tag{3.273}$$

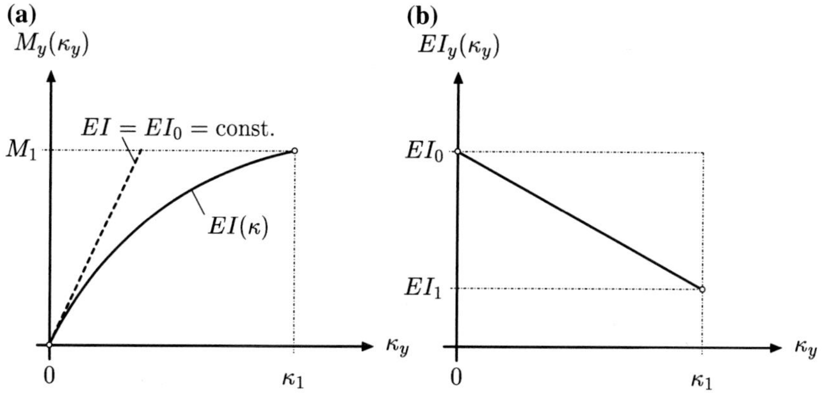

Fig. 3.30 **a** Nonlinear moment-curvature diagram; **b** curvature-dependent bending stiffness

or after the integration finally as:

$$\boldsymbol{K}^{\mathrm{e}}_{\text{square}} = \frac{E\,bh^3}{L^3\,12}\begin{bmatrix} \frac{243}{5} & -\frac{156}{5}L & -\frac{243}{5} & -\frac{87}{5}L \\ -\frac{156}{5}L & \frac{114}{5}L^2 & \frac{156}{5}L & \frac{42}{5}L^2 \\ -\frac{243}{5} & \frac{156}{5}L & \frac{243}{5} & \frac{87}{5}L \\ -\frac{87}{5}L & \frac{42}{5}L^2 & \frac{87}{5}L & 9\,L^2 \end{bmatrix}. \tag{3.274}$$

3.7 Advanced example: Beam element with nonlinear bending stiffness

Derive the elemental stiffness matrix $\boldsymbol{K}^{\mathrm{e}}$ for a BERNOULLI beam element with nonlinear bending stiffness[30] $EI = EI(\kappa)$, cf. Fig. 3.30. Consider the case that the bending stiffness changes linearly with the curvature κ as shown in Fig. 3.30b. The linear relationship of the bending stiffness should be defined by the two sampling points $EI(\kappa = 0) = IE_0$ and $EI(\kappa = \kappa_1) = IE_1$.

3.7 Solution

The relationship for the curvature-dependent bending stiffness can be derived based on the two given sampling points as:

$$EI(\kappa) = EI_0 - \frac{\kappa}{\kappa_1}(EI_0 - EI_1) = EI_0\left(1 - \kappa \times \underbrace{\frac{1 - EI_1/EI_0}{\kappa_1}}_{\alpha_{01}}\right)$$
$$= EI_0\,(1 - \kappa \times \alpha_{01})\,. \tag{3.275}$$

[30]The product of YOUNG's modulus E and moment of inertia I is considered in the following as a single variable: $(EI) \rightarrow EI$.

It should be noted here that the constant factor α_{01} has the unit of a length, e.g. meter, and the curvature κ one divided by length.

The describing partial differential equation can be derived from Table 3.5 as

$$\frac{d^2}{dx^2}\left(EI(\kappa)\frac{d^2u_z}{dx^2}\right) = 0\,. \tag{3.276}$$

Thus, the weighted residual statement (inner product) can be written as (cf. Eq. (3.45)):

$$\int_0^L W^{\mathrm{T}}(x)\frac{d^2}{dx^2}\left(EI(\kappa)\frac{d^2u_z}{dx^2}\right)dx \overset{!}{=} 0\,. \tag{3.277}$$

Twice integrating by parts results in the following weak form of the problem:

$$\int_0^L \frac{d^2W^{\mathrm{T}}}{dx^2}EI(\kappa)\frac{d^2u_z}{dx^2}dx = \left[-W^{\mathrm{T}}\frac{d}{dx}\left(EI(\kappa)\frac{d^2u_z}{dx^2}\right) + \frac{dW^{\mathrm{T}}}{dx}EI(\kappa)\frac{d^2u_z}{dx^2}\right]_0^L\,. \tag{3.278}$$

The left-hand side of Eq. (3.278) represents the elemental stiffness matrix $\boldsymbol{K}^{\mathrm{e}}$. Introducing the nodal approach for the displacement, i.e. $u_z^{\mathrm{e}}(x) = \boldsymbol{N}^{\mathrm{T}}(x)\boldsymbol{u}_{\mathrm{p}}$, and the weight function, i.e. $W(x) = \boldsymbol{N}^{\mathrm{T}}(x)\delta\boldsymbol{u}_{\mathrm{p}}$, the following statement for the stiffness matrix can be derived under consideration of Eq. (3.275):

$$\boldsymbol{K}^{\mathrm{e}} = \int_0^L \frac{d^2\boldsymbol{N}}{dx^2}\underbrace{EI_0\left(1-\kappa\times\alpha_{01}\right)}_{\text{scalar}}\frac{d^2\boldsymbol{N}^{\mathrm{T}}}{dx^2}dx\,. \tag{3.279}$$

Since the bending stiffness $EI(\kappa)$ is a scalar function, the last equation can be rearranged to obtain the following expression:

$$\begin{aligned}\boldsymbol{K}^{\mathrm{e}} &= \int_0^L EI_0\left(1-\kappa\times\alpha_{01}\right)\frac{d^2\boldsymbol{N}}{dx^2}\frac{d^2\boldsymbol{N}^{\mathrm{T}}}{dx^2}dx\\ &= EI_0\int_0^L \frac{d^2\boldsymbol{N}}{dx^2}\frac{d^2\boldsymbol{N}^{\mathrm{T}}}{dx^2}dx - EI_0\alpha_{01}\int_0^L \kappa\frac{d^2\boldsymbol{N}}{dx^2}\frac{d^2\boldsymbol{N}^{\mathrm{T}}}{dx^2}dx\,,\end{aligned} \tag{3.280}$$

where κ is given by the expression (cf. Table 3.10):

$$\kappa(x) = \frac{d^2N_{1u}(x)}{dx^2}u_{1z} + \frac{d^2N_{1\varphi}(x)}{dx^2}\varphi_{1y} + \frac{d^2N_{2u}(x)}{dx^2}u_{2z} + \frac{d^2N_{2\varphi}(x)}{dx^2}\varphi_{2y}\,. \tag{3.281}$$

It should be noted here that the scalar function $\kappa(x)$ must be multiplied by each cell of the 4×4 matrix $\frac{d^2\boldsymbol{N}}{dx^2}\frac{d^2\boldsymbol{N}^{\mathrm{T}}}{dx^2}$ in Eq. (3.280) and after performing the integration, the elemental stiffness matrix is obtained as shown in Eq. (3.282).

$$\boldsymbol{K}^{\mathrm{e}} = \frac{(EI_y)_0}{L^3}\begin{bmatrix} 12 & -6L & -12 & -6L \\ -6L & 4L^2 & 6L & 2L^2 \\ -12 & 6L & 12 & 6L \\ -6L & 2L^2 & 6L & 4L^2 \end{bmatrix} -$$

$$- \frac{(EI_y)_0 \times \alpha_{01}}{L^4}\begin{bmatrix} 12(\varphi_{1y} - \varphi_{2y}) & -12(-u_{1z} + \varphi_{1y}L + u_{2z}) & -12(\varphi_{1y} - \varphi_{2y}) & 12(-u_{1z} + \varphi_{2y}L + u_{2z}) \\ -12(-u_{1z} + \varphi_{1y}L + u_{2z}) & 2(-6u_{1z} + 5\varphi_{1y}L + \varphi_{2y}L + 6u_{2z})L & 12(-u_{1z} + \varphi_{1y}L + u_{2z}) & 2(\varphi_{1y} - \varphi_{2y})L^2 \\ -12(\varphi_{1y} - \varphi_{2y}) & 12(-u_{1z} + \varphi_{1y}L + u_{2z}) & 12(\varphi_{1y} - \varphi_{2y}) & -12(-u_{1z} + \varphi_{2y}L + u_{2z}) \\ 12(-u_{1z} + \varphi_{2y}L + u_{2z}) & 2(\varphi_{1y} - \varphi_{2y})L^2 & -12(-u_{1z} + \varphi_{2y}L + u_{2z}) & -2(-6u_{1z} + \varphi_{1y}L + 5\varphi_{2y}L + 6u_{2z})L \end{bmatrix} \tag{3.282}$$

It can be seen from Eq. (3.282) that the expression of the elemental stiffness matrix $\boldsymbol{K}^{\mathrm{e}}$ is composed of a part which is identical to the expression for a constant stiffness matrix and a second component which is dependent on the nodal unknowns. Because of this dependence on the nodal unknowns ($\boldsymbol{K}^{\mathrm{e}} = \boldsymbol{K}^{\mathrm{e}}(\boldsymbol{u}_{\mathrm{p}})$), the resulting system of equations is *nonlinear* and its solution requires the application of iteration techniques such as the NEWTON–RAPHSON[31,32] scheme.

3.4 Assembly of Elements to Plane Frame Structures

3.4.1 Rotation of a Beam Element

We consider in the following the planar rotation of a beam element. As a result, an angle α is obtained between the global (X, Z) and the local (x, z) coordinate system, see Fig. 3.31.

Every node in the global coordinate system now has two degrees of freedom, i.e. a displacement in the X- and a displacement in the Z-direction. These two displacements at a node can be used to determine the displacement perpendicular to the beam axis, meaning in the direction of the local z-axis. By means of the right-angled triangles illustrated in Fig. 3.31, the displacements in the local coordinate system based on the global displacement values result in[33]

$$u_{1z} = \underbrace{\sin\alpha}_{<0}\, \underbrace{u_{1X}}_{<0} + \underbrace{\cos\alpha}_{>0}\, \underbrace{u_{1Z}}_{>0}\,, \tag{3.283}$$

$$u_{2z} = \underbrace{\sin\alpha}_{<0}\, \underbrace{u_{2X}}_{<0} + \underbrace{\cos\alpha}_{>0}\, \underbrace{u_{2Z}}_{>0}\,. \tag{3.284}$$

Corresponding to the above, the global displacements can be calculated from the local displacements:

$$\underbrace{u_{1X}}_{<0} = \underbrace{u_{1z}}_{>0}\, \underbrace{\sin\alpha}_{<0}\,, \qquad \underbrace{u_{2X}}_{<0} = \underbrace{u_{2z}}_{>0}\, \underbrace{\sin\alpha}_{<0}\,, \tag{3.285}$$

$$\underbrace{u_{1Z}}_{>0} = \underbrace{u_{1z}}_{>0}\, \underbrace{\cos\alpha}_{>0}\,, \qquad \underbrace{u_{2Z}}_{>0} = \underbrace{u_{2z}}_{>0}\, \underbrace{\cos\alpha}_{>0}\,. \tag{3.286}$$

The last relations between global and local displacements can be written in matrix notation as:

$$\begin{bmatrix} u_{1X} \\ u_{1Z} \\ u_{2X} \\ u_{2Z} \end{bmatrix} = \begin{bmatrix} \sin\alpha & 0 \\ \cos\alpha & 0 \\ 0 & \sin\alpha \\ 0 & \cos\alpha \end{bmatrix} \begin{bmatrix} u_{1z} \\ u_{2z} \end{bmatrix}. \tag{3.287}$$

---

[31] Isaac NEWTON (1642–1727), English physicist and mathematician.

[32] Joseph RAPHSON (ca. 1648–1715), English mathematician.

[33] Consider that the rotational angle α is negative in Fig. 3.31.

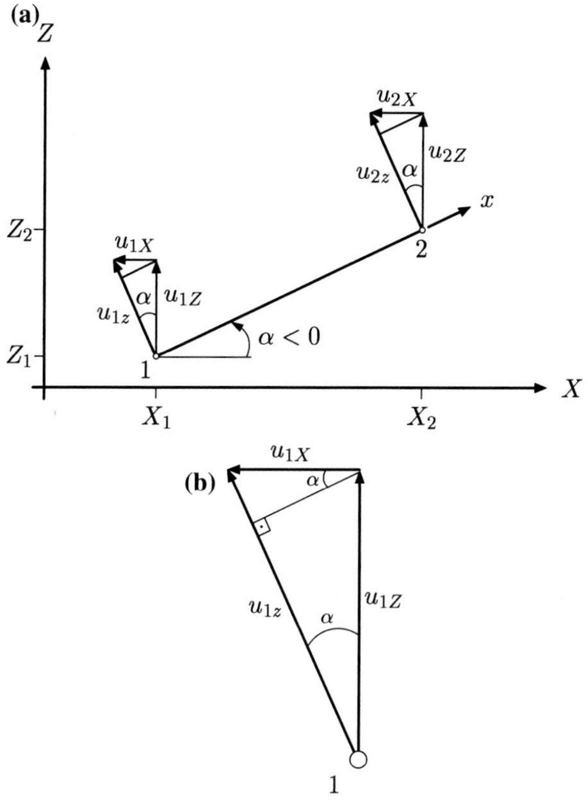

Fig. 3.31 Rotatory transformation of a beam element in the X-Z plane: **a** total view and **b** detail for node 1

The nodal rotations do not need a transformation and the general transformation rule for the calculation of the global parameters from the local deformations results in abbreviated notation in

$$\boldsymbol{u}_{XZ} = \boldsymbol{T}^{\mathrm{T}} \boldsymbol{u}_{xz} , \tag{3.288}$$

or alternatively in components:

$$\begin{bmatrix} u_{1X} \\ u_{1Z} \\ \varphi_{1Y} \\ u_{2X} \\ u_{2Z} \\ \varphi_{2Y} \end{bmatrix} = \begin{bmatrix} \sin\alpha & 0 & 0 & 0 \\ \cos\alpha & 0 & 0 & 0 \\ 0 & 1 & 0 & 0 \\ 0 & 0 & \sin\alpha & 0 \\ 0 & 0 & \cos\alpha & 0 \\ 0 & 0 & 0 & 1 \end{bmatrix} \begin{bmatrix} u_{1z} \\ \varphi_{1y} \\ u_{2z} \\ \varphi_{2y} \end{bmatrix} . \tag{3.289}$$

The last equation can also be solved for the deformations in the local coordinate system and through inversion[34] the transformation matrix results

[34]Since the transformation matrix $\boldsymbol{T}$ is an orthogonal matrix, the following applies: $\boldsymbol{T}^{\mathrm{T}} = \boldsymbol{T}^{-1}$.

$$\boldsymbol{u}_{xz} = \boldsymbol{T}\boldsymbol{u}_{XZ}\,, \tag{3.290}$$

or alternatively in components:

$$\begin{bmatrix} u_{1z} \\ \varphi_{1y} \\ u_{2z} \\ \varphi_{2y} \end{bmatrix} = \begin{bmatrix} \sin\alpha & \cos\alpha & 0 & 0 & 0 & 0 \\ 0 & 0 & 1 & 0 & 0 & 0 \\ 0 & 0 & 0 & \sin\alpha & \cos\alpha & 0 \\ 0 & 0 & 0 & 0 & 0 & 1 \end{bmatrix} \begin{bmatrix} u_{1X} \\ u_{1Z} \\ \varphi_{1Y} \\ u_{2X} \\ u_{2Z} \\ \varphi_{2Y} \end{bmatrix}. \tag{3.291}$$

The matrix of the external loads can be transformed in the same way:

$$\boldsymbol{f}_{XZ} = \boldsymbol{T}^{\mathrm{T}}\boldsymbol{f}_{xz}\,, \tag{3.292}$$

$$\boldsymbol{f}_{xz} = \boldsymbol{T}\boldsymbol{f}_{XZ}\,. \tag{3.293}$$

If the transformation of the local deformation into the global coordinate system is considered in the expression for the principal finite element equation according to Eq. (3.95), the transformation of the stiffness matrix into the global coordinate system results in

$$\boldsymbol{K}^{\mathrm{e}}_{XZ} = \boldsymbol{T}^{\mathrm{T}}\boldsymbol{K}^{\mathrm{e}}_{xz}\boldsymbol{T}\,, \tag{3.294}$$

or alternatively in components:

$$\boldsymbol{K}^{\mathrm{e}}_{XZ} = \frac{EI_y}{L^3}\begin{bmatrix} 12\mathrm{s}^2\alpha & 12\mathrm{s}\alpha\mathrm{c}\alpha & -6L\mathrm{s}\alpha & -12\mathrm{s}^2\alpha & -12\mathrm{s}\alpha\mathrm{c}\alpha & -6L\mathrm{s}\alpha \\ 12\mathrm{s}\alpha\mathrm{c}\alpha & 12\mathrm{c}^2\alpha & -6L\mathrm{c}\alpha & -12\mathrm{s}\alpha\mathrm{c}\alpha & -12\mathrm{c}^2\alpha & -6L\mathrm{c}\alpha \\ -6L\mathrm{s}\alpha & -6L\mathrm{c}\alpha & 4L^2 & 6L\mathrm{s}\alpha & 6L\mathrm{c}\alpha & 2L^2 \\ -12\mathrm{s}^2\alpha & -12\mathrm{s}\alpha\mathrm{c}\alpha & 6L\mathrm{s}\alpha & 12\mathrm{s}^2\alpha & 12\mathrm{s}\alpha\mathrm{c}\alpha & 6L\mathrm{s}\alpha \\ -12\mathrm{s}\alpha\mathrm{c}\alpha & -12\mathrm{c}^2\alpha & 6L\mathrm{c}\alpha & 12\mathrm{s}\alpha\mathrm{c}\alpha & 12\mathrm{c}^2\alpha & 6L\mathrm{c}\alpha \\ -6L\mathrm{s}\alpha & -6L\mathrm{c}\alpha & 2L^2 & 6L\mathrm{s}\alpha & 6L\mathrm{c}\alpha & 4L^2 \end{bmatrix}. \tag{3.295}$$

The sines and cosines values of the rotation angle α can be calculated through the global node coordinates via

$$\mathrm{s}\alpha \triangleq \sin\alpha = -\frac{Z_2 - Z_1}{L} \quad \text{or} \quad \mathrm{c}\alpha \triangleq \cos\alpha = \frac{X_2 - X_1}{L} \tag{3.296}$$

and

$$L = \sqrt{(X_2 - X_1)^2 + (Z_2 - Z_1)^2}. \tag{3.297}$$

It needs to be remarked at this point, that in a mathematical positive sense the angle α always should be plotted from the global to the local coordinate system. The mathematical positive direction of rotation and therefore the algebraic sign of α is illustrated in Fig. 3.32. However independent from the algebraic sign of α the calculation can always occur according to Eq. (3.296).

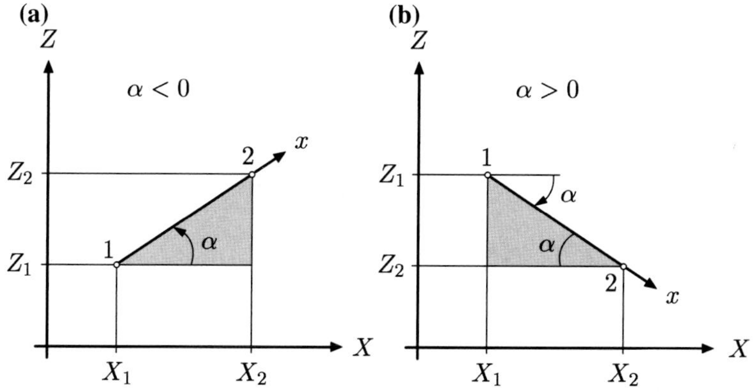

Fig. 3.32 Rotation angle in the XZ-plane: **a** α negative; **b** α positive

3.4.2 Generalized Beam Element

A generalized beam element, i.e. an element which can deform in direction and perpendicular to the principal beam axis, can be obtained by superposition of the stiffness matrix of a rod and a beam element as given in Eqs. (2.41) and (3.71) as:

$$\begin{bmatrix} \frac{EA}{L} & 0 & 0 & -\frac{EA}{L} & 0 & 0 \\ 0 & \frac{12EI}{L^3} & -\frac{6EI}{L^2} & 0 & -\frac{12EI}{L^3} & -\frac{6EI}{L^2} \\ 0 & -\frac{6EI}{L^2} & \frac{4EI}{L} & 0 & \frac{6EI}{L^2} & \frac{2EI}{L} \\ -\frac{EA}{L} & 0 & 0 & \frac{EA}{L} & 0 & 0 \\ 0 & -\frac{12EI}{L^3} & \frac{6EI}{L^2} & 0 & \frac{12EI}{L^3} & \frac{6EI}{L^2} \\ 0 & -\frac{6EI}{L^2} & \frac{2EI}{L} & 0 & \frac{6EI}{L^2} & \frac{4EI}{L} \end{bmatrix} \begin{bmatrix} u_{1x} \\ u_{1z} \\ \varphi_{1y} \\ u_{2x} \\ u_{2z} \\ \varphi_{2y} \end{bmatrix} = \begin{bmatrix} F_{1x} \\ F_{1z} \\ M_{1y} \\ F_{2x} \\ F_{2z} \\ M_{2y} \end{bmatrix} . \tag{3.298}$$

The equation of the stiffness matrix in a global coordinate system, i.e. after a rotation by an angle α, can be obtained by superposition of the deformations of a truss element (cf. Eqs. (2.204) and (2.205)) and a beam element (cf. Eqs. (3.285) and (3.286)) for the first node as

$$u_{1X} = \cos\alpha\, u_{1x} + \sin\alpha\, u_{1z} \,, \tag{3.299}$$

$$u_{1Z} = -\sin\alpha\, u_{1x} + \cos\alpha\, u_{1z} \,, \tag{3.300}$$

and for the second node in a similar way as:

$$u_{2X} = \cos\alpha\, u_{2x} + \sin\alpha\, u_{2z} \,, \tag{3.301}$$

$$u_{2Z} = -\sin\alpha\, u_{2x} + \cos\alpha\, u_{2x}\,. \tag{3.302}$$

These four transformation relationships can be arranged in matrix notation as:

$$\begin{bmatrix} u_{1X} \\ u_{1Z} \\ u_{2X} \\ u_{2Z} \end{bmatrix} = \begin{bmatrix} \cos\alpha & \sin\alpha & 0 & 0 \\ -\sin\alpha & \cos\alpha & 0 & 0 \\ 0 & 0 & \cos\alpha & \sin\alpha \\ 0 & 0 & -\sin\alpha & \cos\alpha \end{bmatrix} \begin{bmatrix} u_{1x} \\ u_{1z} \\ u_{2x} \\ u_{2z} \end{bmatrix}, \tag{3.303}$$

or in abbreviated form as:

$$\boldsymbol{u}_{XZ} = \boldsymbol{T}^{\mathrm{T}} \boldsymbol{u}_{xz}\,. \tag{3.304}$$

Since the rotational degrees of freedom do not need to be transformed, the final transformation for all six degrees of freedom can be obtained as:

$$\begin{bmatrix} u_{1X} \\ u_{1Z} \\ \varphi_{1Y} \\ \hline u_{2X} \\ u_{2Z} \\ \varphi_{2Y} \end{bmatrix} = \underbrace{\left[\begin{array}{ccc|ccc} \cos\alpha & \sin\alpha & 0 & 0 & 0 & 0 \\ -\sin\alpha & \cos\alpha & 0 & 0 & 0 & 0 \\ 0 & 0 & 1 & 0 & 0 & 0 \\ \hline 0 & 0 & 0 & \cos\alpha & \sin\alpha & 0 \\ 0 & 0 & 0 & -\sin\alpha & \cos\alpha & 0 \\ 0 & 0 & 0 & 0 & 0 & 1 \end{array}\right]}_{\boldsymbol{T}^{\mathrm{T}}} \begin{bmatrix} u_{1x} \\ u_{1z} \\ \varphi_{1y} \\ \hline u_{2x} \\ u_{2z} \\ \varphi_{2y} \end{bmatrix}. \tag{3.305}$$

The last equation can be rearranged for the unknowns in the local coordinate system as:

$$\begin{bmatrix} u_{1x} \\ u_{1z} \\ \varphi_{1y} \\ \hline u_{2x} \\ u_{2z} \\ \varphi_{2y} \end{bmatrix} = \underbrace{\left[\begin{array}{ccc|ccc} \cos\alpha & -\sin\alpha & 0 & 0 & 0 & 0 \\ \sin\alpha & \cos\alpha & 0 & 0 & 0 & 0 \\ 0 & 0 & 1 & 0 & 0 & 0 \\ \hline 0 & 0 & 0 & \cos\alpha & -\sin\alpha & 0 \\ 0 & 0 & 0 & \sin\alpha & \cos\alpha & 0 \\ 0 & 0 & 0 & 0 & 0 & 1 \end{array}\right]}_{\boldsymbol{T}} \begin{bmatrix} u_{1X} \\ u_{1Z} \\ \varphi_{1Y} \\ \hline u_{2X} \\ u_{2Z} \\ \varphi_{2Y} \end{bmatrix}, \tag{3.306}$$

or

$$\boldsymbol{u}_{xz} = \boldsymbol{T} \boldsymbol{u}_{XZ}\,, \tag{3.307}$$

where the transformation matrix $\boldsymbol{T}$ can be expressed as:

$$\boldsymbol{T} = \begin{bmatrix} \boldsymbol{T}_1 & \boldsymbol{0} \\ \boldsymbol{0} & \boldsymbol{T}_2 \end{bmatrix}. \tag{3.308}$$

It should be noted here that the submatrix $\boldsymbol{T}_1$ in Eq. (3.308) performs the transformation at node 1 while submatrix $\boldsymbol{T}_2$ handles everything at node 2. The global stiffness matrix, i.e. the stiffness matrix in the global coordinate system (X, Z), can be obtained by multiplication of the transformation matrix (3.308) with the beam stiffness matrix (3.71) according to relation (2.212), i.e. $\boldsymbol{K}^{\mathrm{e}}_{XZ} = \boldsymbol{T}^{\mathrm{T}} \boldsymbol{K}^{\mathrm{e}}_{xz} \boldsymbol{T}$. The result of this multiplication is shown in Eq. (3.309) where the principal finite element equation in the global coordinate system is presented.

To simplify the solution of simple frame structures, Tables 3.16 and 3.17 collect expressions for the global stiffness matrix for some common angles α.

Table 3.16 Elemental stiffness matrices for plane frame elements given for different rotation angles α in the X-Z plane, cf. Eq. (3.309)

0°

$$E\begin{bmatrix} \frac{A}{L} & 0 & 0 & -\frac{A}{L} & 0 & 0 \\ 0 & \frac{12I}{L^3} & -\frac{6I}{L^2} & 0 & -\frac{12I}{L^3} & -\frac{6I}{L^2} \\ 0 & -\frac{6I}{L^2} & \frac{4I}{L} & 0 & \frac{6I}{L^2} & \frac{2I}{L} \\ -\frac{A}{L} & 0 & 0 & \frac{A}{L} & 0 & 0 \\ 0 & -\frac{12I}{L^3} & \frac{6I}{L^2} & 0 & \frac{12I}{L^3} & \frac{6I}{L^2} \\ 0 & -\frac{6I}{L^2} & \frac{2I}{L} & 0 & \frac{6I}{L^2} & \frac{4I}{L} \end{bmatrix}$$

180°

$$E\begin{bmatrix} \frac{A}{L} & 0 & 0 & -\frac{A}{L} & 0 & 0 \\ 0 & \frac{12I}{L^3} & \frac{6I}{L^2} & 0 & -\frac{12I}{L^3} & -\frac{6I}{L^2} \\ 0 & \frac{6I}{L^2} & \frac{4I}{L} & 0 & -\frac{6I}{L^2} & \frac{2I}{L} \\ -\frac{A}{L} & 0 & 0 & \frac{A}{L} & 0 & 0 \\ 0 & -\frac{12I}{L^3} & -\frac{6J}{L^2} & 0 & \frac{12I}{L^3} & -\frac{6I}{L^2} \\ 0 & -\frac{6I}{L^2} & \frac{2I}{L} & 0 & -\frac{6I}{L^2} & \frac{4I}{L} \end{bmatrix}$$

−90°

$$E\begin{bmatrix} \frac{12I}{L^3} & 0 & \frac{6I}{L^2} & -\frac{12I}{L^3} & 0 & \frac{6I}{L^2} \\ 0 & \frac{A}{L} & 0 & 0 & -\frac{A}{L} & 0 \\ \frac{6I}{L^2} & 0 & \frac{4I}{L} & -\frac{6I}{L^2} & 0 & \frac{2I}{L} \\ -\frac{12I}{L^3} & 0 & -\frac{6I}{L^2} & \frac{12I}{L^3} & 0 & -\frac{6I}{L^2} \\ 0 & -\frac{A}{L} & 0 & 0 & \frac{A}{L} & 0 \\ \frac{6I}{L^2} & 0 & \frac{2I}{L} & -\frac{6I}{L^2} & 0 & \frac{4I}{L} \end{bmatrix}$$

90°

$$E\begin{bmatrix} \frac{12I}{L^3} & 0 & -\frac{6I}{L^2} & -\frac{12I}{L^3} & 0 & -\frac{6I}{L^2} \\ 0 & \frac{A}{L} & 0 & 0 & -\frac{A}{L} & 0 \\ -\frac{6I}{L^2} & 0 & \frac{4I}{L} & \frac{6I}{L^2} & 0 & \frac{2I}{L} \\ -\frac{12I}{L^3} & 0 & \frac{6I}{L^2} & \frac{12I}{L^3} & 0 & \frac{6I}{L^2} \\ 0 & -\frac{A}{L} & 0 & 0 & \frac{A}{L} & 0 \\ -\frac{6I}{L^2} & 0 & \frac{2I}{L} & \frac{6I}{L^2} & 0 & \frac{4I}{L} \end{bmatrix}$$

−45°

$$E\begin{bmatrix} \frac{6I}{L^3}+\frac{1}{2}\frac{A}{L} & -\frac{6I}{L^3}+\frac{1}{2}\frac{A}{L} & \frac{3I\sqrt{2}}{L^2} & -\frac{6I}{L^3}-\frac{1}{2}\frac{A}{L} & +\frac{6I}{L^3}-\frac{1}{2}\frac{A}{L} & \frac{3I\sqrt{2}}{L^2} \\ -\frac{6I}{L^3}+\frac{1}{2}\frac{A}{L} & \frac{6I}{L^3}+\frac{1}{2}\frac{A}{L} & -\frac{3I\sqrt{2}}{L^2} & +\frac{6I}{L^3}-\frac{1}{2}\frac{A}{L} & -\frac{6I}{L^3}-\frac{1}{2}\frac{A}{L} & -\frac{3I\sqrt{2}}{L^2} \\ \frac{3I\sqrt{2}}{L^2} & -\frac{3I\sqrt{2}}{L^2} & \frac{4I}{L} & -\frac{3I\sqrt{2}}{L^2} & +\frac{3I\sqrt{2}}{L^2} & \frac{2I}{L} \\ -\frac{6I}{L^3}-\frac{1}{2}\frac{A}{L} & \frac{6I}{L^3}-\frac{1}{2}\frac{A}{L} & -\frac{3I\sqrt{2}}{L^2} & \frac{6I}{L^3}+\frac{1}{2}\frac{A}{L} & -\frac{6I}{L^3}+\frac{1}{2}\frac{A}{L} & -\frac{3I\sqrt{2}}{L^2} \\ +\frac{6I}{L^3}-\frac{1}{2}\frac{A}{L} & -\frac{6I}{L^3}-\frac{1}{2}\frac{A}{L} & +\frac{3I\sqrt{2}}{L^2} & -\frac{6I}{L^3}+\frac{1}{2}\frac{A}{L} & \frac{6I}{L^3}+\frac{1}{2}\frac{A}{L} & \frac{3I\sqrt{2}}{L^2} \\ \frac{3I\sqrt{2}}{L^2} & -\frac{3I\sqrt{2}}{L^2} & \frac{2I}{L} & -\frac{3I\sqrt{2}}{L^2} & \frac{3I\sqrt{2}}{L^2} & \frac{4I}{L} \end{bmatrix}$$

45°

$$E\begin{bmatrix} \frac{6I}{L^3}+\frac{1}{2}\frac{A}{L} & \frac{6I}{L^3}-\frac{1}{2}\frac{A}{L} & -\frac{3I\sqrt{2}}{L^2} & -\frac{6I}{L^3}-\frac{1}{2}\frac{A}{L} & -\frac{6I}{L^3}+\frac{1}{2}\frac{A}{L} & -\frac{3I\sqrt{2}}{L^2} \\ \frac{6I}{L^3}-\frac{1}{2}\frac{A}{L} & \frac{6I}{L^3}+\frac{1}{2}\frac{A}{L} & -\frac{3I\sqrt{2}}{L^2} & -\frac{6I}{L^3}+\frac{1}{2}\frac{A}{L} & -\frac{6I}{L^3}-\frac{1}{2}\frac{A}{L} & -\frac{3I\sqrt{2}}{L^2} \\ -\frac{3I\sqrt{2}}{L^2} & -\frac{3I\sqrt{2}}{L^2} & \frac{4I}{L} & \frac{3I\sqrt{2}}{L^2} & \frac{3I\sqrt{2}}{L^2} & \frac{2I}{L} \\ -\frac{6I}{L^3}-\frac{1}{2}\frac{A}{L} & -\frac{6I}{L^3}+\frac{1}{2}\frac{A}{L} & \frac{3I\sqrt{2}}{L^2} & \frac{6I}{L^3}+\frac{1}{2}\frac{A}{L} & \frac{6I}{L^3}-\frac{1}{2}\frac{A}{L} & \frac{3I\sqrt{2}}{L^2} \\ -\frac{6I}{L^3}+\frac{1}{2}\frac{A}{L} & -\frac{6I}{L^3}-\frac{1}{2}\frac{A}{L} & \frac{3I\sqrt{2}}{L^2} & \frac{6I}{L^3}-\frac{1}{2}\frac{A}{L} & \frac{6I}{L^3}+\frac{1}{2}\frac{A}{L} & \frac{3I\sqrt{2}}{L^2} \\ -\frac{3I\sqrt{2}}{L^2} & -\frac{3I\sqrt{2}}{L^2} & \frac{2I}{L} & \frac{3I\sqrt{2}}{L^2} & \frac{3I\sqrt{2}}{L^2} & \frac{4I}{L} \end{bmatrix}$$

$$
E\begin{bmatrix}
\frac{12I}{L^3}\sin^2\alpha+\frac{A}{L}\cos^2\alpha & \left(\frac{12I}{L^3}-\frac{A}{L}\right)\sin\alpha\cos\alpha & -\frac{6I}{L^2}\sin\alpha & -\frac{12I}{L^3}\sin^2\alpha-\frac{A}{L}\cos^2\alpha & \left(-\frac{12I}{L^3}+\frac{A}{L}\right)\sin\alpha\cos\alpha & -\frac{6I}{L^2}\sin\alpha \\
\left(\frac{12I}{L^3}-\frac{A}{L}\right)\sin\alpha\cos\alpha & \frac{12I}{L^3}\cos^2\alpha+\frac{A}{L}\sin^2\alpha & -\frac{6I}{L^2}\cos\alpha & \left(-\frac{12I}{L^3}+\frac{A}{L}\right)\sin\alpha\cos\alpha & -\frac{12I}{L^3}\cos^2\alpha-\frac{A}{L}\sin^2\alpha & -\frac{6I}{L^2}\cos\alpha \\
-\frac{6I}{L^2}\sin\alpha & -\frac{6I}{L^2}\cos\alpha & \frac{4I}{L} & \frac{6I}{L^2}\sin\alpha & \frac{6I}{L^2}\cos\alpha & \frac{2I}{L} \\
-\frac{12I}{L^3}\sin^2\alpha-\frac{A}{L}\cos^2\alpha & \left(-\frac{12I}{L^3}+\frac{A}{L}\right)\sin\alpha\cos\alpha & \frac{6I}{L^2}\sin\alpha & \frac{12I}{L^3}\sin^2\alpha+\frac{A}{L}\cos^2\alpha & \left(\frac{12I}{L^3}-\frac{A}{L}\right)\sin\alpha\cos\alpha & \frac{6I}{L^2}\sin\alpha \\
\left(-\frac{12I}{L^3}+\frac{A}{L}\right)\sin\alpha\cos\alpha & -\frac{12I}{L^3}\cos^2\alpha-\frac{A}{L}\sin^2\alpha & \frac{6I}{L^2}\cos\alpha & \left(\frac{12I}{L^3}-\frac{A}{L}\right)\sin\alpha\cos\alpha & \frac{12I}{L^3}\cos^2\alpha+\frac{A}{L}\sin^2\alpha & \frac{6I}{L^2}\cos\alpha \\
-\frac{6I}{L^2}\sin\alpha & -\frac{6I}{L^2}\cos\alpha & \frac{2I}{L} & \frac{6I}{L^2}\sin\alpha & \frac{6I}{L^2}\cos\alpha & \frac{4I}{L}
\end{bmatrix}
\begin{bmatrix} u_{1X} \\ u_{1Z} \\ \varphi_{1Y} \\ u_{2X} \\ u_{2Z} \\ \varphi_{2Y} \end{bmatrix}
=
\begin{bmatrix} F_{1X} \\ F_{1Z} \\ M_{1Y} \\ F_{2X} \\ F_{2Z} \\ M_{2Y} \end{bmatrix}
\quad (3.309)
$$

Table 3.17 Elemental stiffness matrices for plane frame elements given for different rotation angles α in the X-Z plane, cf. Eq. (3.309)

$-30°$	$E\begin{bmatrix} \frac{3I}{L^3}+\frac{3}{4}\frac{A}{L} & \frac{\sqrt{3}}{4}\left(-\frac{12I}{L^3}+\frac{A}{L}\right) & \frac{3I}{L^2} & -\frac{3I}{L^3}-\frac{3}{4}\frac{A}{L} & \frac{\sqrt{3}}{4}\left(\frac{12I}{L^3}-\frac{A}{L}\right) & \frac{3I}{L^2} \\ \frac{\sqrt{3}}{4}\left(-\frac{12I}{L^3}+\frac{A}{L}\right) & \frac{9I}{L^3}+\frac{1}{4}\frac{A}{L} & -\frac{3I\sqrt{3}}{L^2} & \frac{\sqrt{3}}{4}\left(\frac{12I}{L^3}-\frac{A}{L}\right) & -\frac{9I}{L^3}-\frac{1}{4}\frac{A}{L} & -\frac{3I\sqrt{3}}{L^2} \\ \frac{3I}{L^2} & -\frac{3I\sqrt{3}}{L^2} & \frac{4I}{L} & -\frac{3I}{L^2} & \frac{3I\sqrt{3}}{L^2} & \frac{2I}{L} \\ -\frac{3I}{L^3}-\frac{3}{4}\frac{A}{L} & \frac{\sqrt{3}}{4}\left(\frac{12I}{L^3}-\frac{A}{L}\right) & -\frac{3I}{L^2} & \frac{3I}{L^3}+\frac{3}{4}\frac{A}{L} & \frac{\sqrt{3}}{4}\left(-\frac{12I}{L^3}+\frac{A}{L}\right) & -\frac{3I}{L^2} \\ \frac{\sqrt{3}}{4}\left(\frac{12I}{L^3}-\frac{A}{L}\right) & -\frac{9I}{L^3}-\frac{1}{4}\frac{A}{L} & \frac{3I\sqrt{3}}{L^2} & -\frac{\sqrt{3}}{4}\left(-\frac{12I}{L^3}+\frac{A}{L}\right) & \frac{9I}{L^3}+\frac{1}{4}\frac{A}{L} & \frac{3I\sqrt{3}}{L^2} \\ \frac{3I}{L^2} & -\frac{3I\sqrt{3}}{L^2} & \frac{2I}{L} & -\frac{3I}{L^2} & \frac{3I\sqrt{3}}{L^2} & \frac{4I}{L} \end{bmatrix}$
$30°$	$E\begin{bmatrix} \frac{3I}{L^3}+\frac{3}{4}\frac{A}{L} & -\frac{\sqrt{3}}{4}\left(-\frac{12I}{L^3}+\frac{A}{L}\right) & -\frac{3I}{L^2} & -\frac{3I}{L^3}-\frac{3}{4}\frac{A}{L} & -\frac{\sqrt{3}}{4}\left(\frac{12I}{L^3}-\frac{A}{L}\right) & -\frac{3I}{L^2} \\ -\frac{\sqrt{3}}{4}\left(-\frac{12I}{L^3}+\frac{A}{L}\right) & \frac{9I}{L^3}+\frac{1}{4}\frac{A}{L} & -\frac{3I\sqrt{3}}{L^2} & -\frac{\sqrt{3}}{4}\left(\frac{12I}{L^3}-\frac{A}{L}\right) & -\frac{9I}{L^3}-\frac{1}{4}\frac{A}{L} & -\frac{3I\sqrt{3}}{L^2} \\ -\frac{3I}{L^2} & -\frac{3I\sqrt{3}}{L^2} & \frac{4I}{L} & \frac{3I}{L^2} & \frac{3I\sqrt{3}}{L^2} & \frac{2I}{L} \\ -\frac{3I}{L^3}-\frac{3}{4}\frac{A}{L} & -\frac{\sqrt{3}}{4}\left(\frac{12I}{L^3}-\frac{A}{L}\right) & \frac{3I}{L^2} & \frac{3I}{L^3}+\frac{3}{4}\frac{A}{L} & -\frac{\sqrt{3}}{4}\left(-\frac{12I}{L^3}+\frac{A}{L}\right) & \frac{3I}{L^2} \\ -\frac{\sqrt{3}}{4}\left(\frac{12I}{L^3}-\frac{A}{L}\right) & -\frac{9I}{L^3}-\frac{1}{4}\frac{A}{L} & \frac{3I\sqrt{3}}{L^2} & \frac{\sqrt{3}}{4}\left(-\frac{12I}{L^3}+\frac{A}{L}\right) & \frac{9I}{L^3}+\frac{1}{4}\frac{A}{L} & \frac{3I\sqrt{3}}{L^2} \\ -\frac{3I}{L^2} & -\frac{3I\sqrt{3}}{L^2} & \frac{2I}{L} & \frac{3I}{L^2} & \frac{3I\sqrt{3}}{L^2} & \frac{4I}{L} \end{bmatrix}$

It might be required for certain problems to apply the transformation given in Eq. (3.308) only at one node of the element. This can be the case if a support is rotated only at one node as shown in Fig. 3.33. Let us have a look, for example, on the case shown in Fig. 3.33a where it would be quite difficult to describe the boundary condition in the local (x, z) system. However, the global (X, Z) system easily allows to specify the boundary conditions at the first node as: $u_{1Z} = 0 \wedge u_{1X} \neq 0$.

Thus, the transformation (3.308) can be individually applied at each node with a different transformation angle α_1 at node 1 and α_2 at node 2 as:

$$\boldsymbol{T} = \begin{bmatrix} \boldsymbol{T}_1(\alpha_1) & \boldsymbol{0} \\ \boldsymbol{0} & \boldsymbol{T}_2(\alpha_2) \end{bmatrix}. \tag{3.310}$$

The last equation implies that the global coordinate system can be differently chosen at each node. In the case that the rotation is only required at the first node as shown in Fig. 3.33a, Eq. (3.310) can be simplified to

$$\boldsymbol{T} = \begin{bmatrix} \boldsymbol{T}_1(\alpha_1) & \boldsymbol{0} \\ \boldsymbol{0} & \mathbf{I} \end{bmatrix}, \tag{3.311}$$

where $\mathbf{I}$ is the identity matrix. Thus, the elemental stiffness matrix in the global coordinate system can be obtained for this special case based on the following relationship:

$$\boldsymbol{K}^{\mathrm{e}}_{XZ} = \begin{bmatrix} \boldsymbol{T}_1(\alpha_1) & \boldsymbol{0} \\ \boldsymbol{0} & \mathbf{I} \end{bmatrix}^{\mathrm{T}} \boldsymbol{K}_{xz} \begin{bmatrix} \boldsymbol{T}_1(\alpha_1) & \boldsymbol{0} \\ \boldsymbol{0} & \mathbf{I} \end{bmatrix}. \tag{3.312}$$

In a similar way, the transformation can be only performed at node 2. The elemental stiffness matrices for these two special cases are summarized in Eqs. (3.313) and (3.314).

$$\boldsymbol{K}^{\mathrm{e}}_{XZ}(\alpha_1, \alpha_2 = 0) = E \times \begin{bmatrix} \frac{12I - 12I\cos^2\alpha_1 + AL^2\cos^2\alpha_1}{L^3} & -\frac{(-12I + AL^2)\cos\alpha_1\sin\alpha_1}{L^3} & -\frac{6I\sin\alpha_1}{L^2} & -\frac{A\cos\alpha_1}{L} & -\frac{12I\sin\alpha_1}{L^3} & -\frac{6I\sin\alpha_1}{L^2} \\ -\frac{(-12I + AL^2)\cos\alpha_1\sin\alpha_1}{L^3} & \frac{12I\cos^2\alpha_1 + AL^2 - AL^2\cos^2\alpha_1}{L^3} & -\frac{6I\cos\alpha_1}{L^2} & +\frac{A\sin\alpha_1}{L} & -\frac{12I\cos\alpha_1}{L^3} & -\frac{6I\cos\alpha_1}{L^2} \\ -\frac{6I\sin\alpha_1}{L^2} & -\frac{6I\cos\alpha_1}{L^2} & \frac{4I}{L} & 0 & \frac{6I}{L^2} & \frac{2I}{L} \\ -\frac{A\cos\alpha_1}{L} & \frac{A\sin\alpha_1}{L} & 0 & \frac{A}{L} & 0 & 0 \\ -\frac{12I\sin\alpha_1}{L^3} & -\frac{12I\cos\alpha_1}{L^3} & \frac{6I}{L^2} & 0 & \frac{12I}{L^3} & \frac{6I}{L^2} \\ -\frac{6I\sin\alpha_1}{L^2} & -\frac{6I\cos\alpha_1}{L^2} & \frac{2I}{L} & 0 & \frac{6I}{L^2} & \frac{4I}{L} \end{bmatrix} \tag{3.313}$$

$$\boldsymbol{K}^{\mathrm{e}}_{XZ}(\alpha_1 = 0, \alpha_2) = E \times \begin{bmatrix} \frac{A}{L} & 0 & 0 & -\frac{A\cos\alpha_2}{L} & \frac{A\sin\alpha_2}{L} & 0 \\ 0 & \frac{12I}{L^3} & -\frac{6I}{L^2} & -\frac{12I\sin\alpha_2}{L^3} & -\frac{12I\cos\alpha_2}{L^3} & -\frac{6I}{L^2} \\ 0 & -\frac{6I}{L^2} & \frac{4I}{L} & \frac{6I\sin\alpha_2}{L^2} & \frac{6I\cos\alpha_2}{L^2} & \frac{2I}{L} \\ -\frac{A\cos\alpha_2}{L} & -\frac{12I\sin\alpha_2}{L^3} & \frac{6I\sin\alpha_2}{L^2} & \frac{12I - 12I\cos^2\alpha_2 + AL^2\cos^2\alpha_2}{L^3} & -\frac{(-12I + AL^2)\cos\alpha_2\sin\alpha_2}{L^3} & \frac{6I\sin\alpha_2}{L^2} \\ \frac{A\sin\alpha_2}{L} & -\frac{12I\cos\alpha_2}{L^3} & +\frac{6I\cos\alpha_2}{L^2} & -\frac{(-12I + AL^2)\cos\alpha_2\sin\alpha_2}{L^3} & \frac{12I\cos^2\alpha_2 + AL^2 - AL^2\cos^2\alpha_2}{L^3} & \frac{6I\cos\alpha_2}{L^2} \\ 0 & -\frac{6I}{L^2} & \frac{2I}{L} & \frac{6I\sin\alpha_2}{L^2} & \frac{6I\cos\alpha_2}{L^2} & \frac{4I}{L} \end{bmatrix} \tag{3.314}$$

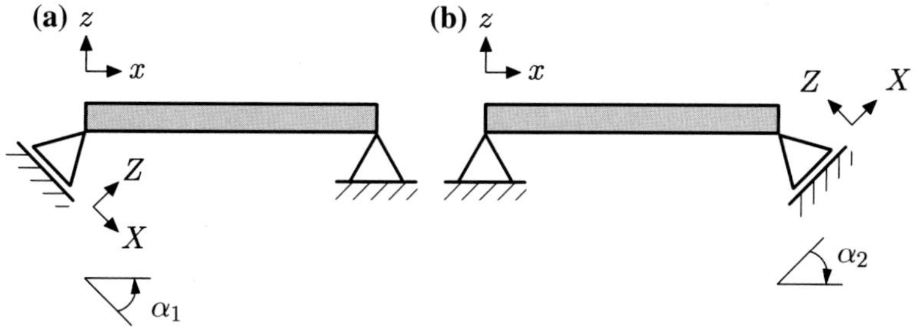

Fig. 3.33 Beam element with rotated support: **a** rotation at node 1; **b** rotation at node 2

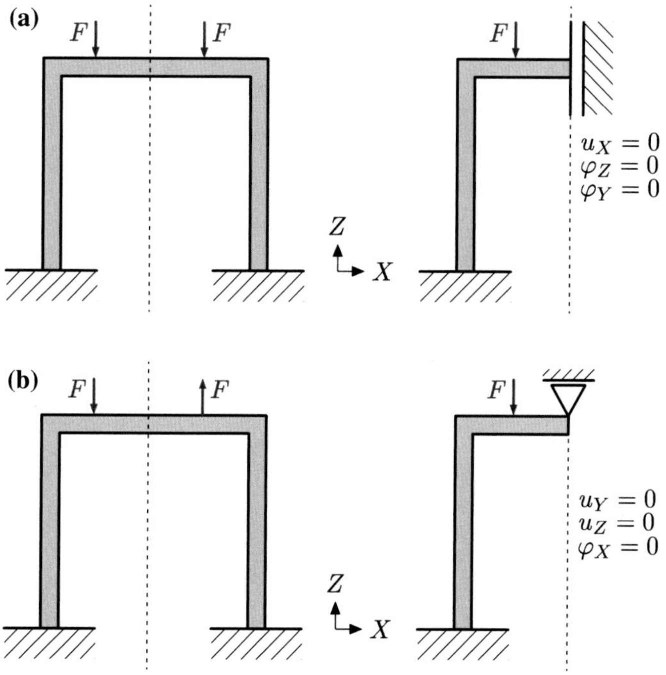

Fig. 3.34 Portal frame structure with **a** symmetry and **b** anti-symmetry

The size of some structures can be reduced by consideration of symmetry and anti-symmetry boundary conditions as shown in Fig. 3.34. The case of symmetry requires that the geometry *and* the load case is symmetric with respect to a certain plane while the case of anti-symmetry requires a symmetric geometry and anti-symmetric loading and results with respect to the same plane. A systematic summary of symmetric and anti-symmetric boundary conditions is given in Table 3.18.

Table 3.18 Conditions for symmetry and anti-symmetry

Case	u_X	u_Y	u_Z	φ_X	φ_Y	φ_Z
X-symmetry (resp. Y-Z plane)	0	free	free	free	0	0
Y-symmetry (resp. X-Y plane)	free	0	free	0	free	0
Z-symmetry (resp. X-Y plane)	free	free	0	0	0	free
X-anti-symmetry (resp. Y-Z plane)	free	0	0	0	free	free
Y-anti-symmetry (resp. X-Y plane)	0	free	0	free	0	free
Z-anti-symmetry (resp. X-Y plane)	0	0	free	free	free	0

3.4.3 *Solved Problems*

3.8 Example: Portal frame structure

The portal frame structure shown in Fig. 3.35 is loaded by a constant distributed load q_0 and a single horizontal force F_0. The three parts of the frame have the same length L, the same YOUNG's modulus E, the same cross-sectional area A, and the same second moment of area I. Determine

(a)

- the elemental stiffness matrices in the global X-Z system,
- the reduced system of equations,

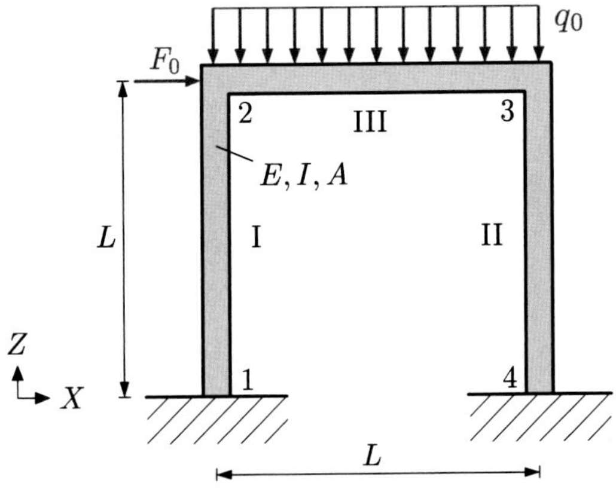

Fig. 3.35 Portal frame structure

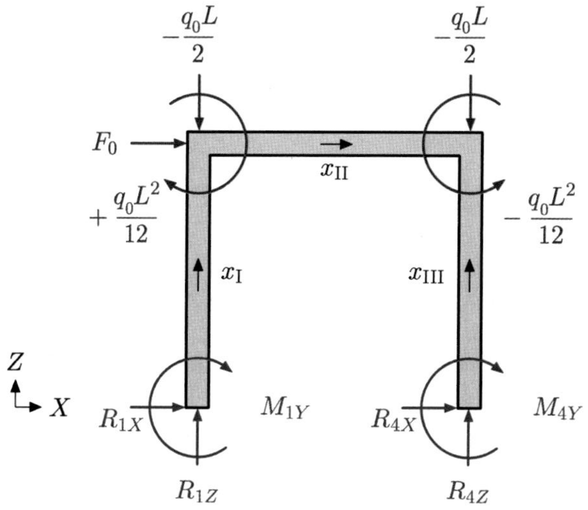

Fig. 3.36 Free body diagram of the portal frame structure, see Fig. 3.44

- all nodal displacements and rotations, and
- all reaction forces.

(b)
Consider now the case $F_0 = 0$, i.e. only the distributed load is acting on the frame. Develop a simpler model under consideration of the symmetry and determine the quantities as requested in part (a).

3.8 Solution

(a)

The free body diagram and the local coordinate axes of each element are shown in Fig. 3.36. In addition, the equivalent nodal loads resulting from the distributed load q_0 (cf. Table 3.7) are introduced at nodes 2 and 3. It should be noted that the unknown reactions at the supports are introduced in the positive direction of the respective coordinate axes. From this figure, the rotational angles from the global to the local coordinate system can be determined as $\alpha_{\mathrm{I}} = \alpha_{\mathrm{III}} = -90°$ and $\alpha_{\mathrm{II}} = 0°$.

The elemental stiffness matrices in the global X-Z system can be obtained from Eq. (3.309) as:

$$
\boldsymbol{K}_{\mathrm{I}}^{\mathrm{e}} = E
\begin{array}{c}
\begin{array}{cccccc} u_{1X} & u_{1Z} & \varphi_{1Y} & u_{2X} & u_{2Z} & \varphi_{2Y} \end{array} \\
\begin{bmatrix}
\frac{12I}{L^3} & 0 & \frac{6I}{L^2} & -\frac{12I}{L^3} & 0 & \frac{6I}{L^2} \\
0 & \frac{A}{L} & 0 & 0 & -\frac{A}{L} & 0 \\
\frac{6I}{L^2} & 0 & \frac{4I}{L} & -\frac{6I}{L^2} & 0 & \frac{2I}{L} \\
-\frac{12I}{L^3} & 0 & -\frac{6I}{L^2} & \frac{12I}{L^3} & 0 & -\frac{6I}{L^2} \\
0 & -\frac{A}{L} & 0 & 0 & \frac{A}{L} & 0 \\
\frac{6I}{L^2} & 0 & \frac{2I}{L} & -\frac{6I}{L^2} & 0 & \frac{4I}{L}
\end{bmatrix}
\end{array}
\begin{array}{c} u_{1X} \\ u_{1Z} \\ \varphi_{1Y} \\ u_{2X} \\ u_{2Z} \\ \varphi_{2Y} \end{array}, \tag{3.315}
$$

$$
\boldsymbol{K}_{\mathrm{II}}^{\mathrm{e}} = E
\begin{array}{c}
\begin{array}{cccccc} u_{2X} & u_{2Z} & \varphi_{2Y} & u_{3X} & u_{3Z} & \varphi_{3Y} \end{array} \\
\begin{bmatrix}
\frac{A}{L} & 0 & 0 & -\frac{A}{L} & 0 & 0 \\
0 & \frac{12I}{L^3} & -\frac{6I}{L^2} & 0 & -\frac{12I}{L^3} & -\frac{6I}{L^2} \\
0 & -\frac{6I}{L^2} & \frac{4I}{L} & 0 & \frac{6I}{L^2} & \frac{2I}{L} \\
-\frac{A}{L} & 0 & 0 & \frac{A}{L} & 0 & 0 \\
0 & -\frac{12I}{L^3} & \frac{6I}{L^2} & 0 & \frac{12I}{L^3} & \frac{6I}{L^2} \\
0 & -\frac{6I}{L^2} & \frac{2I}{L} & 0 & \frac{6I}{L^2} & \frac{4I}{L}
\end{bmatrix}
\end{array}
\begin{array}{c} u_{2X} \\ u_{2Y} \\ \varphi_{2Z} \\ u_{3X} \\ u_{3Y} \\ \varphi_{3Z} \end{array}. \tag{3.316}
$$

The third stiffness matrix is the same as given in Eq. (3.315), only the degrees of freedom are different:

$$
\boldsymbol{K}_{\mathrm{III}}^{\mathrm{e}} = E
\begin{array}{c}
\begin{array}{cccccc} u_{4X} & u_{4Z} & \varphi_{4Y} & u_{3X} & u_{3Z} & \varphi_{3Y} \end{array} \\
\begin{bmatrix}
\frac{12I}{L^3} & 0 & \frac{6I}{L^2} & -\frac{12I}{L^3} & 0 & \frac{6I}{L^2} \\
0 & \frac{A}{L} & 0 & 0 & -\frac{A}{L} & 0 \\
\frac{6I}{L^2} & 0 & \frac{4I}{L} & -\frac{6I}{L^2} & 0 & \frac{2I}{L} \\
-\frac{12I}{L^3} & 0 & -\frac{6I}{L^2} & \frac{12I}{L^3} & 0 & -\frac{6I}{L^2} \\
0 & -\frac{A}{L} & 0 & 0 & \frac{A}{L} & 0 \\
\frac{6I}{L^2} & 0 & \frac{2I}{L} & -\frac{6I}{L^2} & 0 & \frac{4I}{L}
\end{bmatrix}
\end{array}
\begin{array}{c} u_{4X} \\ u_{4Z} \\ \varphi_{4Y} \\ u_{3X} \\ u_{3Z} \\ \varphi_{3Y} \end{array}. \tag{3.317}
$$

The global system of equations, i.e. $\boldsymbol{K}\boldsymbol{u}_{\text{p}} = \boldsymbol{f}$, can be assembled based on the scheme of unknowns which is indicated at each elemental stiffness matrix. Introducing the boundary conditions, i.e. $u_{1X} = u_{1Z} = u_{4X} = u_{4Z} = 0$ and $\varphi_{1Y} = \varphi_{4Y} = 0$, gives the following reduced system of equations:

$$\begin{bmatrix} \frac{12I}{L^3}+\frac{A}{L} & 0+0 & -\frac{6I}{L^2}+0 & -\frac{A}{L} & 0 & 0 \\ 0+0 & \frac{A}{L}+\frac{12I}{L^3} & 0-\frac{6I}{L^2} & 0 & -\frac{12I}{L^3} & -\frac{6I}{L^2} \\ -\frac{6I}{L^2}+0 & 0-\frac{6I}{L^2} & \frac{4I}{L}+\frac{4I}{L} & 0 & \frac{6I}{L^2} & \frac{2I}{L} \\ -\frac{A}{L} & 0 & 0 & \frac{A}{L}+\frac{12I}{L^3} & 0+0 & 0-\frac{6I}{L^2} \\ 0 & -\frac{12I}{L^3} & \frac{6I}{L^2} & 0+0 & \frac{12I}{L^3}+\frac{A}{L} & \frac{6I}{L^2}+0 \\ 0 & -\frac{6I}{L^2} & \frac{2I}{L} & 0-\frac{6I}{L^2} & \frac{6I}{L^2}+0 & \frac{4I}{L}+\frac{4I}{L} \end{bmatrix} \begin{bmatrix} u_{2X} \\ u_{2Z} \\ \varphi_{2Y} \\ u_{3X} \\ u_{3Z} \\ \varphi_{3Y} \end{bmatrix} = \begin{bmatrix} F_0 \\ -\frac{q_0 L}{2} \\ +\frac{q_0 L^2}{12} \\ 0 \\ -\frac{q_0 L}{2} \\ \frac{q_0 L^2}{12} \\ - \end{bmatrix}. \tag{3.318}$$

The solution of this linear (6×6) system can be obtained, for example, by inverting the reduced stiffness matrix to give the reduced result vector $\left[u_{2X}\ u_{2Z}\ \varphi_{2Y}\ u_{3X}\ u_{3Z}\ \varphi_{3Y}\right]^{\text{T}}$ as:

$$\begin{bmatrix} \frac{L^3\left(10F_0L^4A^2+7q_0AIL^3+168F_0AIL^2+24q_0I^2L+432F_0I^2\right)}{24\left(7A^2L^4+45AL^2I+72I^2\right)EI} \\ \frac{L^2\left(6LF_0A-7\,q_0AL^2-24q_0I\right)}{2\left(7AL^2+24I\right)EA} \\ \frac{L^2\left(18F_0L^4A^2+612F_0AIL^2+1728F_0I^2+7q_0L^5A^2+66q_0AIL^3+144q_0I^2L\right)}{72\left(AL^2+3I\right)\left(7AL^2+24I\right)EI} \\ \frac{L^3\left(10F_0L^4A^2+84F_0AIL^2+144F_0I^2-7q_0AIL^3-24q_0I^2L\right)}{24\left(7A^2L^4+45AL^2I+72I^2\right)EI} \\ -\frac{L^2\left(6F_0LA+24q_0I+7q_0AL^2\right)}{2\left(7AL^2+24I\right)EA} \\ \frac{L^2\left(18F_0L^4A^2+360F_0AIL^2+864F_0I^2-7q_0L^5A^2-66q_0AIL^3-144q_0I^2L\right)}{72\left(7A^2L^4+45AL^2I+72I^2\right)EJ} \end{bmatrix}. \tag{3.319}$$

The reaction forces can be obtained by multiplying the global (non-reduced) stiffness matrix with the total displacement vector, i.e.

$$\boldsymbol{u}^{\text{T}} = [0\ \ 0\ \ 0\ \ u_{2X}\ \ u_{2Z}\ \ \varphi_{2Y}\ \ u_{3X}\ \ u_{3Z}\ \ \varphi_{3Y}\ \ 0\ \ 0\ \ 0]\,, \tag{3.320}$$

to give $\left[R_{1X}\ R_{1Z}\ M_{1Y}\ R_{4X}\ R_{4Z}\ M_{4Y}\right]^{\mathrm{T}}$ as:

$$\begin{bmatrix} -\dfrac{-q_0L^3A+6F_0L^2A+36F_0I}{12\left(AL^2+3I\right)} \\ -\dfrac{L\left(6LF_0A-7q_0AL^2-24q_0I\right)}{2\left(7AL^2+24I\right)} \\ -\dfrac{\left(-7q_0L^5A^2+72F_0L^4A^2-3q_0AIL^3+900F_0AIL^2+72q_0I^2L+2160F_0I^2\right)L}{36\left(AL^2+3I\right)\left(7AL^2+24I\right)} \\ -\dfrac{\left(q_0L+6F_0\right)AL^2}{12\left(AL^2+3I\right)} \\ \dfrac{L\left(6F_0LA+24q_0I+7q_0AL^2\right)}{2\left(7AL^2+24I\right)} \\ -\dfrac{L\left(72F_0L^4A^2+396F_0AIL^2+432F_0I^2+3q_0AIL^3-72q_0I^2L+7q_0L^5A^2\right)}{36\left(7A^2L^4+45AL^2I+72I^2\right)} \end{bmatrix}. \tag{3.321}$$

(b)

In the case of $F_0=0$, the symmetry in regards to the geometry *and* the load case can be used to create a simplified model under consideration of appropriate symmetry conditions. As can be seen in Fig. 3.37a, only half of the structure needs to be modeled if at the symmetry line ($X=\frac{L}{2}$) the condition $u_{3X}=\varphi_{3Y}=0$ is imposed. The free body diagram of this structure is shown in Fig. 3.37b where now only two finite elements are required to simulate the structure.

The elemental stiffness matrices can be taken from Eqs. (3.315) and (3.316) in which the transformation $L_{\mathrm{II}}=\frac{L}{2}$ must be applied. Assembling to the global system

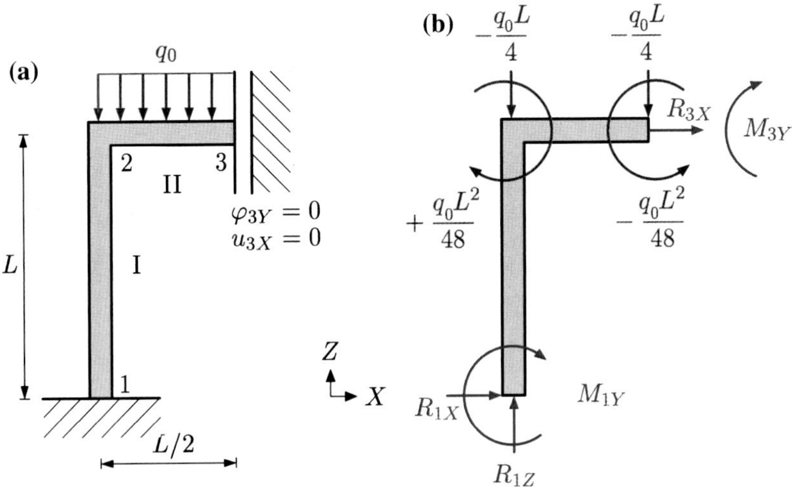

Fig. 3.37 Symmetrical portal frame structure: **a** half model; **b** free body diagram

of equations and consideration of the boundary condition, i.e. $u_{1X} = u_{1Z} = u_{3X} = 0$ and $\varphi_{1Y} = \varphi_{3Y} = 0$, gives the following reduced system of equations:

$$\begin{bmatrix} \frac{12I}{L^3}+\frac{2A}{L} & 0+0 & -\frac{6I}{L^2}+0 & 0 \\ 0+0 & \frac{A}{L}+\frac{96I}{L^3} & 0-\frac{24I}{L^2} & -\frac{96I}{L^3} \\ -\frac{6I}{L^2}+0 & 0-\frac{24I}{L^2} & \frac{4I}{L}+\frac{8I}{L} & \frac{24I}{L^2} \\ 0 & -\frac{96I}{L^3} & \frac{24I}{L^2} & \frac{96I}{L^3} \end{bmatrix} \begin{bmatrix} u_{2X} \\ u_{2Z} \\ \varphi_{2Y} \\ u_{3Z} \end{bmatrix} = \begin{bmatrix} 0 \\ -\frac{q_0 L}{4} \\ +\frac{q_0 L^2}{48} \\ -\frac{q_0 L}{4} \end{bmatrix}. \tag{3.322}$$

The solution of this systems of equations is obtained, for example, by inverting the reduced stiffness matrix as:

$$\begin{bmatrix} u_{2X} \\ u_{2Z} \\ \varphi_{2Y} \\ u_{3Z} \end{bmatrix} = \begin{bmatrix} \frac{q_0 L^4}{24\left(AL^2+3I\right)E} \\ -\frac{q_0 L^2}{2EA} \\ +\frac{q_0 L^3\left(6I+AL^2\right)}{72\left(AL^2+3I\right)EI} \\ -\frac{q_0 L^2\left(609IAL^2+1728I^2+7A^2L^4\right)}{1152\left(AL^2+3I\right)AEI} \end{bmatrix}. \tag{3.323}$$

Multiplying the global (non-reduced) stiffness matrix with the known vector of nodal displacements and rotations serves to calculate the reactions as:

$$\begin{bmatrix} R_{1X} \\ R_{1Z} \\ M_{1Y} \\ R_{3X} \\ M_{3Y} \end{bmatrix} = \begin{bmatrix} \frac{q_0 AL^3}{12\left(AL^2+3I\right)} \\ \frac{q_0 L}{2} \\ +\frac{q_0 L^2\left(-3I+AL^2\right)}{36\left(AL^2+3I\right)} \\ -\frac{q_0 AL^3}{12\left(AL^2+3I\right)} \\ -\frac{q_0 L^2\left(21I+5AL^2\right)}{72\left(AL^2+3I\right)} \end{bmatrix}. \tag{3.324}$$

3.9 Example: Plane frame structure with rotated support and spring

The plane frame structure shown in Fig. 3.38 is loaded by a single vertical force F_0 and a single moment $M_0 = F_0 L$ at point 1. The structure is composed of a generalized beam of length $3L$ between point 1 and 3 and a rod element between point 2 and 4.

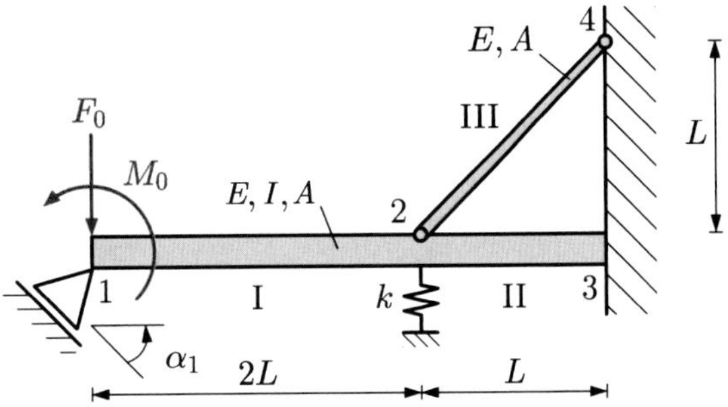

Fig. 3.38 Plane frame structure with rotated support and spring

The members of the frame have the same YOUNG's modulus E and the same cross-sectional area A. The generalized beam is in addition characterized by the second moment of area I. The structure is supported at point 2 by a spring with a spring constant of $k = \frac{EA}{L}$. The simple support at point 1 is rotated by $\alpha_1 = -45°$. Use three elements as indicated in the figure to model the problem. Determine

- the free body diagram,
- the elemental stiffness matrices of the three elements, and
- the reduced system of equations.

3.9 Solution

The free body diagram and the local coordinate systems are shown in Fig. 3.39. The unknown reactions at the supports are introduced in the positive direction of the respective global coordinate system. Let us have first a look at element I. At node 1, it is required to introduce a rotated global coordinate system in order to be able to impose the boundary condition resulting from the rotated support. Thus, the global coordinate system is rotated by $\alpha_1 = -45°$ as indicated in Fig. 3.39. At the second node, there is no transformation between the local and global coordinate system

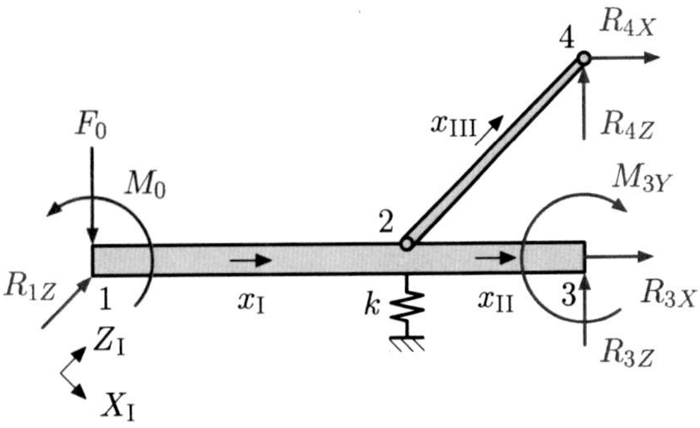

Fig. 3.39 Free body diagram of the plane frame structure with rotated support and spring, see Fig. 3.38

required ($\alpha_2 = 0$). Thus, the global elemental stiffness matrix of element I can be obtained from Eq. (3.313) with $\alpha_{\mathrm{I}} = -45°$ and $L_{\mathrm{I}} = 2L$ as:

$$\boldsymbol{K}_{\mathrm{I}}^{\mathrm{e}} = E \times \begin{bmatrix} \frac{AL^2+3I}{4L^3} & \frac{AL^2-3I}{4L^3} & \frac{3\sqrt{2}I}{4L^2} & -\frac{\sqrt{2}A}{4L} & \frac{3\sqrt{2}I}{4L^3} & \frac{3\sqrt{2}I}{4L^2} \\ \frac{AL^2-3I}{4L^3} & \frac{AL^2+3I}{4L^3} & -\frac{3\sqrt{2}I}{4L^2} & -\frac{\sqrt{2}A}{4L} & -\frac{3\sqrt{2}I}{4L^3} & -\frac{3\sqrt{2}I}{4L^2} \\ \frac{3\sqrt{2}I}{4L^2} & -\frac{3\sqrt{2}I}{4L^2} & \frac{2I}{L} & 0 & \frac{3I}{2L^2} & \frac{I}{L} \\ -\frac{\sqrt{2}A}{4L} & -\frac{\sqrt{2}A}{4L} & 0 & \frac{A}{2L} & 0 & 0 \\ \frac{3\sqrt{2}I}{4L^3} & -\frac{3\sqrt{2}I}{4L^3} & \frac{3I}{2L^2} & 0 & \frac{3I}{2L^3} & \frac{3I}{2L^2} \\ \frac{3\sqrt{2}I}{4L^2} & -\frac{3\sqrt{2}I}{4L^2} & \frac{I}{L} & 0 & \frac{3I}{2L^2} & \frac{2I}{L} \end{bmatrix}. \tag{3.325}$$

Considering the boundary condition at node 1, i.e. $u_{1Z} = 0$, the second column and second row can be canceled to form the global stiffness matrix. The second element does not require any transformation and the stiffness matrix in the global system is given by Eq. (3.298), i.e.

$$\boldsymbol{K}_{\mathrm{II}}^{\mathrm{e}} = \begin{bmatrix} \frac{EA}{L} & 0 & 0 & -\frac{EA}{L} & 0 & 0 \\ 0 & \frac{12EI}{L^3} & -\frac{6EI}{L^2} & 0 & -\frac{12EI}{L^3} & -\frac{6EI}{L^2} \\ 0 & -\frac{6EI}{L^2} & \frac{4EI}{L} & 0 & \frac{6EI}{L^2} & \frac{2EI}{L} \\ -\frac{EA}{L} & 0 & 0 & \frac{EA}{L} & 0 & 0 \\ 0 & -\frac{12EI}{L^3} & \frac{6EI}{L^2} & 0 & \frac{12EI}{L^3} & \frac{6EI}{L^2} \\ 0 & -\frac{6EI}{L^2} & \frac{2EI}{L} & 0 & \frac{6EI}{L^2} & \frac{4EI}{L} \end{bmatrix}. \tag{3.326}$$

The fixed support at node 3 allows to cancel the last three columns and rows to form the global stiffness matrix. The rod element is rotated by $-45°$ and the stiffness matrix in the global coordinate system can be calculated from Eq. (2.228) with $L_{\mathrm{III}} = \sqrt{2}L$ as:

$$\boldsymbol{K}_{\mathrm{III}}^{\mathrm{e}} = \frac{EA}{\sqrt{2}L} \begin{bmatrix} \frac{1}{2} & \frac{1}{2} & -\frac{1}{2} & -\frac{1}{2} \\ \frac{1}{2} & \frac{1}{2} & -\frac{1}{2} & -\frac{1}{2} \\ -\frac{1}{2} & -\frac{1}{2} & \frac{1}{2} & \frac{1}{2} \\ -\frac{1}{2} & -\frac{1}{2} & \frac{1}{2} & \frac{1}{2} \end{bmatrix}. \tag{3.327}$$

Considering the support at node 4, the last two columns and rows can be canceled to form the global matrix. Assembling the three matrices and considering the boundary conditions, the following reduced global system of equations can be obtained:

$$\begin{bmatrix} \frac{(3I+AL^2)E}{4L^3} & \frac{3EI\sqrt{2}}{4L^2} & -\frac{EA\sqrt{2}}{4L} & \frac{3EI\sqrt{2}}{4L^3} & \frac{3EI\sqrt{2}}{4L^2} \\ \frac{3EI\sqrt{2}}{4L^2} & \frac{2EI}{L} & 0 & \frac{3EI}{2L^2} & \frac{EI}{L} \\ -\frac{EA\sqrt{2}}{4L} & 0 & \frac{3EA}{2L}+\frac{EA\sqrt{2}}{4L} & \frac{EA\sqrt{2}}{4L} & 0 \\ \frac{3EI\sqrt{2}}{4L^3} & \frac{3EI}{2L^2} & \frac{EA\sqrt{2}}{4L} & \frac{27EI}{2L^3}+\frac{EA\sqrt{2}}{4L}+\frac{EA}{L} & -\frac{9EI}{2L^2} \\ \frac{3EI\sqrt{2}}{4L^2} & \frac{EI}{L} & 0 & -\frac{9EI}{2L^2} & \frac{6EI}{L} \end{bmatrix} \begin{bmatrix} u_{1X} \\ \varphi_{1Y} \\ u_{2X} \\ u_{2Z} \\ \varphi_{2Y} \end{bmatrix} = \begin{bmatrix} \frac{1}{2}\sqrt{2}F_0 & -M_0 & 0 & 0 & 0 \end{bmatrix}^{\mathrm{T}} . \tag{3.328}$$

It should be noted here that the spring constant $k = \frac{EA}{L}$ was added in the cell (u_{2Z}, u_{2Z}). The solution of this system of equations is obtained, for example, by inverting the reduced stiffness matrix.

3.5 Supplementary Problems

3.10 Knowledge questions on beams and frames

- State the major assumptions for the EULER–BERNOULLI beam theory.
- State the major assumptions for the TIMOSHENKO beam theory.
- Name the primary unknown in the partial differential equation of an EULER–BERNOULLI beam.
- Sketch (a) the normal and (b) the shear stress distribution in a square cross section of an EULER–BERNOULLI beam under pure bending load.

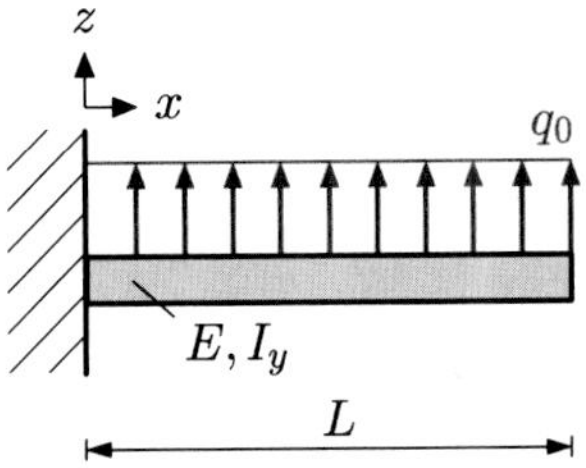

Fig. 3.40 Cantilever beam with constant distributed load

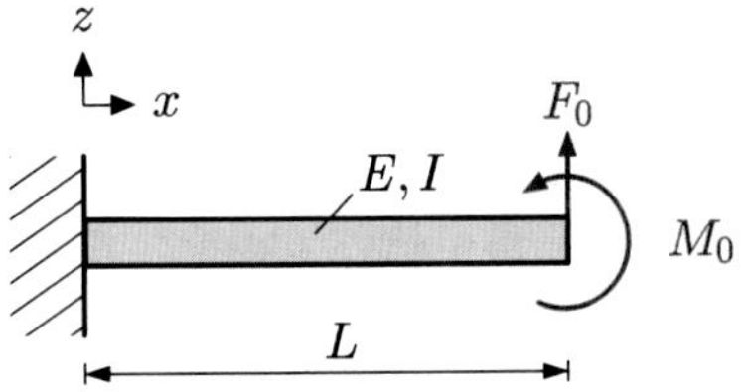

Fig. 3.41 Cantilever beam with point loads

- Sketch (a) the normal and (b) the shear stress distribution of a TIMOSHENKO beam under pure bending load.
- Consider a beam bending problem. To which internal reaction are the (a) normal stress and (b) the shear stress related?
- The following Fig. 3.40 shows a cantilever beam which is loaded by a constant distributed load q_0.
  Sketch schematically (without calculation) the distribution of the internal shear force $Q_z(x)$ and bending moment $M_y(x)$ (based on the theory of continuum mechanics).
- The following Fig. 3.41 shows a cantilever beam which is simultaneously loaded by point loads F_0 and M_0.
  Sketch schematically (without calculation) the distribution of the internal shear force $Q_z(x)$ and bending moment $M_y(x)$ (based on the theory of continuum mechanics).
- The following Fig. 3.42 shows a generalized cantilever beam which is simultaneously loaded at its right-hand end by a horizontal force F_0 and a moment M_0.
  Sketch schematically (without calculation) the distribution of the internal shear force $Q_z(x)$, bending moment $M_y(x)$, and normal force $N_x(x)$ (based on the theory of continuum mechanics).
- State the required (a) geometrical parameters and (b) material parameters to define an EULER–BERNOULLI beam element.
- Sketch the interpolation functions $N_i(\xi)$ of an EULER–BERNOULLI beam element.
- Sketch the shape functions $\overline{N}_i(\xi)$ of an EULER–BERNOULLI beam element.
- The following equation shows the elemental stiffness matrix for an EULER–BERNOULLI beam element.

$$\boldsymbol{K}^{\mathrm{e}} = \frac{EI_y}{L^3}\begin{bmatrix} 12 & -6L & -12 & -6L \\ -6L & 4L^2 & 6L & 2L^2 \\ -12 & 6L & 12 & 6L \\ -6L & 2L^2 & 6L & 4L^2 \end{bmatrix}.$$

  State three (3) assumptions for the derivation.
- Explain (in words) the difference between an EULER–BERNOULLI beam element and a generalized beam element in regards to the nodal unknowns.

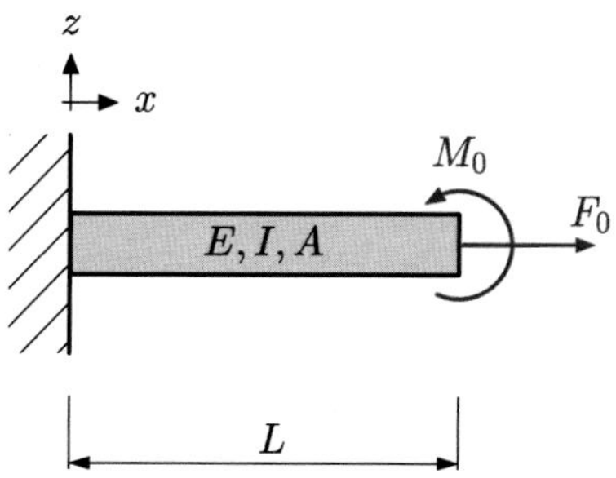

Fig. 3.42 Cantilever beam with point loads

- Explain (in words) the difference between a rod element and a generalized beam element in regards to the nodal unknowns.
- State the DOF per node for a generalized beam element in a plane (2D) problem.
- State the DOF per node for a generalized beam element in a 3D problem.
- State the required (a) geometrical parameters and (b) material parameters to define a generalized beam element.
- Consider a generalized beam element in a finite element code. Which fundamental modes of deformation can be applied to such an element?
- The following Fig. 3.43a shows schematically a cantilever EULER–BERNOULLI beam. In a finite element approach, such a beam can be modeled based on one-dimensional beam elements (Fig. 3.43b), two-dimensional plane elasticity elements (Fig. 3.43c), or three-dimensional solid elements (Fig. 3.43d).
  State for each approach *one* advantage.
- The following Fig. 3.44 shows a *plane* frame structure which should be modeled with three generalized beam (I, II, III) elements.
  State the dimensions of the stiffness matrix of the *non-reduced* system of equations, i.e. without consideration of the boundary conditions. What are the dimensions of the stiffness matrix of the *reduced* system of equations, i.e. under consideration of the boundary conditions?
- Given is a generalized beam as shown in Fig. 3.45.
  Sketch the distributions of the internal bending moment, shear force and normal force along the principal axis (X) and indicate the maximum values as a function of E, I, A, L, and F_0.
- Consider a generalized beam (cross section: hollow tube with inner radius r_i and outer radius r_a) which is simultaneously loaded by an axial tensile force, axial

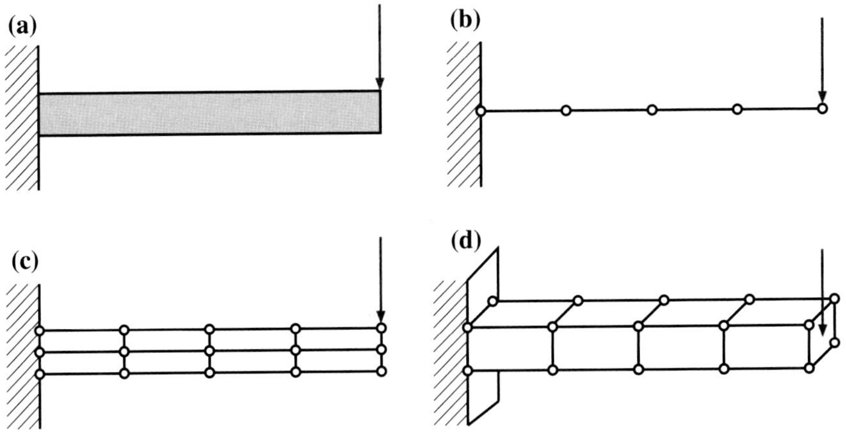

Fig. 3.43 Different modelling approaches for a bending problem: **a** Problem sketch; **b** 1D beam elements; **c** 2D plane elasticity elements; **d** 3D solid elements

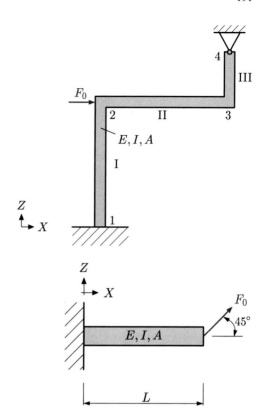

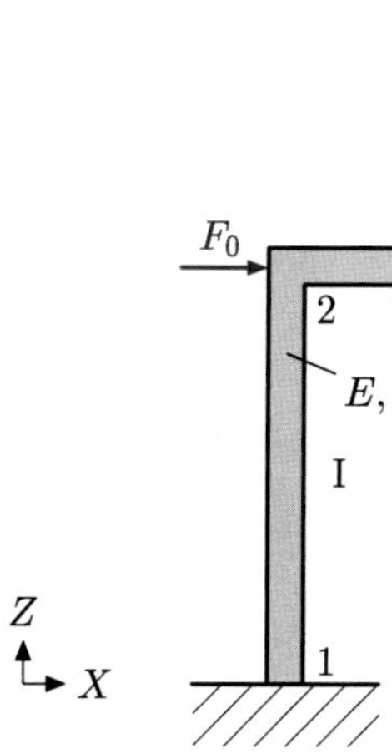

Fig. 3.44 Plane frame structure composed of generalized beam elements

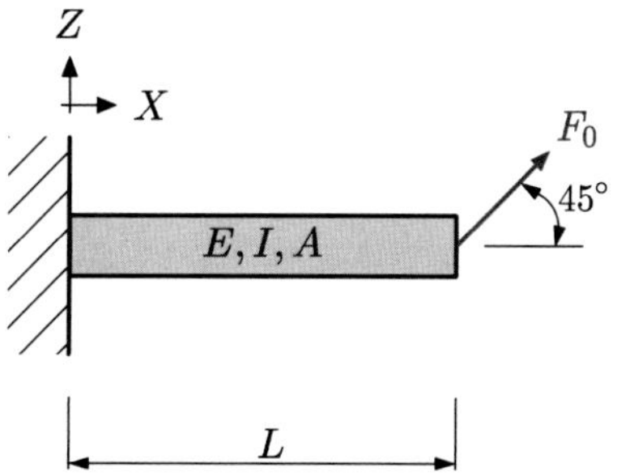

Fig. 3.45 Cantilever beam with constant distributed load

torque, and bending moment. Describe in words where we can expect the maximum (a) normal stress and (b) shear stress.

3.11 Cantilever beam with a distributed load: analytical solution

Calculate the analytical solution for the deflection $u_z(x)$ and rotation $\varphi_y(x)$ of the cantilever beam shown in Fig. 3.46. Start your derivation from the fourth order differential equation. It can be assumed for this exercise that the bending stiffness EI_y is constant.

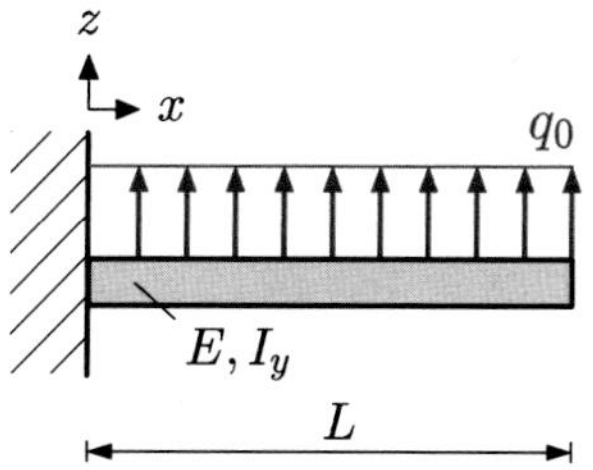

Fig. 3.46 Cantilever beam loaded by a distributed load

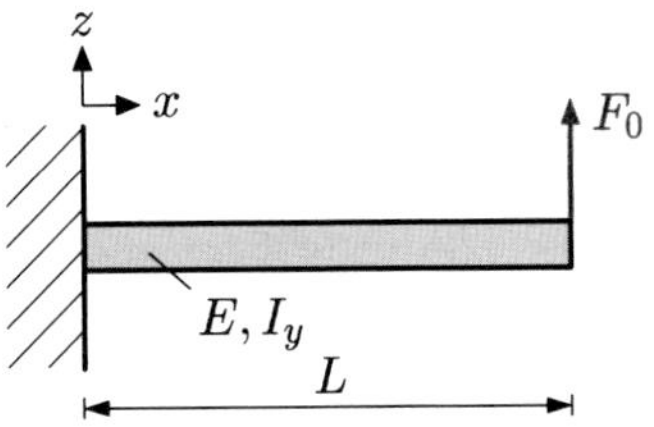

Fig. 3.47 Cantilever beam loaded by a point load F_0

3.12 Cantilever beam with a point load: analytical solution

Calculate the analytical solution for the deflection $u_z(x)$ of the cantilever beam shown in Fig. 3.47. Start your derivation from the (a) fourth order differential equation. It can be assumed for this exercise that the bending stiffness EI_y is constant. As an alternative solution procedure, start your derivation from the (b) moment distribution $M_y(x)$.

3.13 Cantilever beam with different end loads and deformations: analytical solution

Calculate the analytical solution for the deflection $u_z(x)$ and rotation $\varphi_y(x)$ of the cantilever beams shown in Fig. 3.48. Start your derivation from the fourth order differential equation. It can be assumed for this exercise that the bending stiffness EI_y is constant. Calculate in addition for all four cases the reactions at the fixed support and the distributions of the bending moment and shear force.

3.14 Simply supported beam with centered single force: analytical solution

Calculate the analytical solution for the deflection $u_z(x)$ and $u_z\left(\frac{L}{2}\right)$ of the simply supported BERNOULLI beam shown in Fig. 3.49 based on the fourth order differential equation given in Table 3.5. It can be assumed for this exercise that the bending stiffness EI_y is constant.

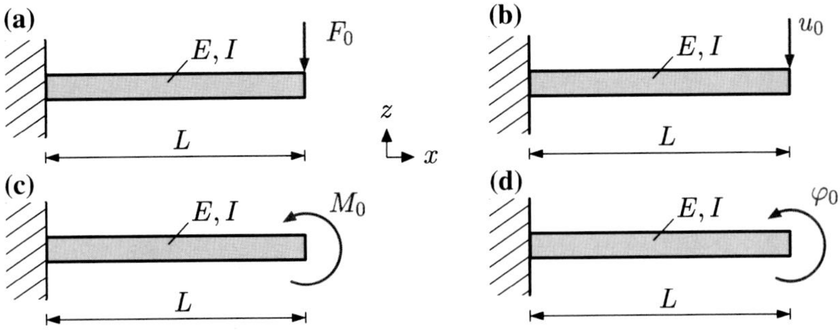

Fig. 3.48 Cantilever beam with different end loads and deformations: **a** single force; **b** single moment; **c** displacement; **d** rotation

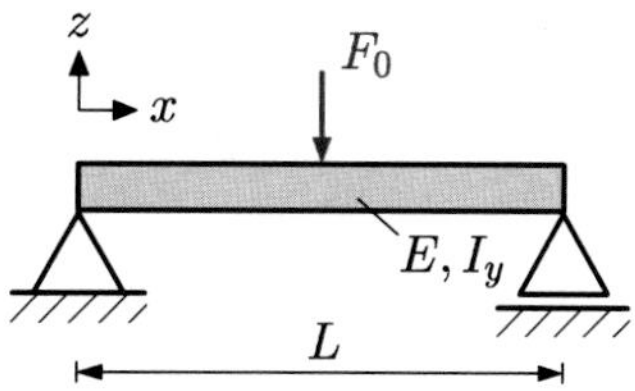

Fig. 3.49 Simply supported BERNOULLI beam with centered single force

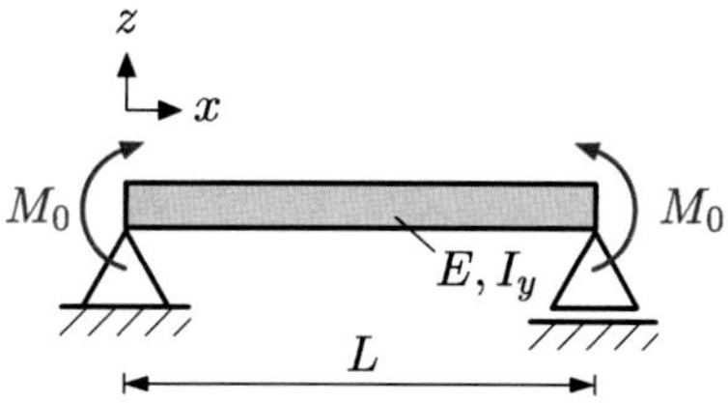

Fig. 3.50 Simply supported BERNOULLI beam under pure bending load

3.15 Simply supported beam under pure bending load: analytical solution

Calculate the analytical solution for the deflection $u_z(x)$ and $u_z\left(\frac{L}{2}\right)$ of the simply supported BERNOULLI beam shown in Fig. 3.50 based on the second order differential equation for the bending moment distribution given in Eq. (3.39). It can be assumed for this exercise that the bending stiffness EI_y is constant.

3.16 Bernoulli beam fixed at both ends: analytical solution

Calculate the analytical solution for the deflection $u_z(x)$ and slope $\varphi_y(x)$ of the BERNOULLI beams shown in Fig. 3.51 based on the fourth order differential equation given in Table 3.5. Determine in addition the maximum deflection and slope. It can be assumed for this exercise that the bending stiffness EI_y is constant.

3.17 Cantilever Bernoulli beam with triangular shaped distributed load: analytical solution

Calculate the analytical solution for the deflection $u_z(x)$ of the BERNOULLI beams shown in Fig. 3.52 based on the fourth order, third order and second order differential equation given in Table 3.5. It can be assumed for this exercise that the bending stiffness EI_y is constant.

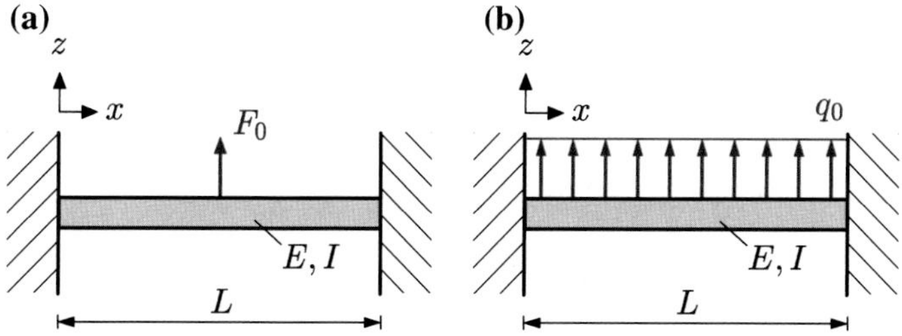

Fig. 3.51 BERNOULLI beam fixed at both ends: **a** single force case; **b** distributed load case

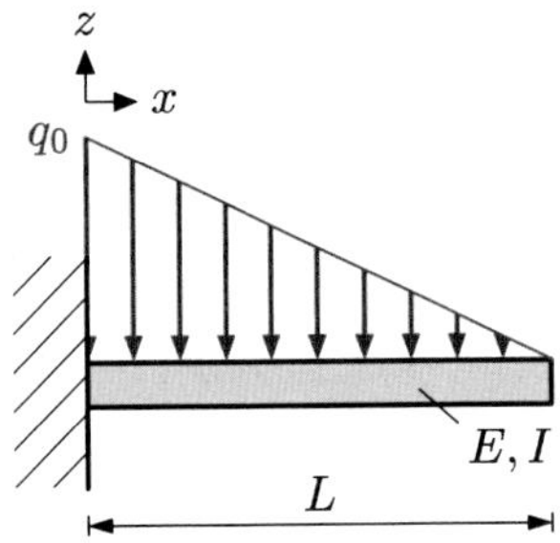

Fig. 3.52 Cantilever BERNOULLI beam with triangular shaped distributed load

3.18 Weighted residual method based on general formulation of partial differential equation

The partial differential equation for a BERNOULLI beam, i.e.

$$\frac{\mathrm{d}^2}{\mathrm{d}x^2}\left(EI_y\frac{\mathrm{d}^2u_z(x)}{\mathrm{d}x^2}\right) - q_z = 0\,, \tag{3.329}$$

can be generalized as

$$\mathcal{L}_1^{\mathrm{T}}\left[\mathcal{L}_1^{\mathrm{T}}\left(EI_y\mathcal{L}_2(u_z(x))\right)\right] - q_z = 0\,, \tag{3.330}$$

where $\mathcal{L}_1 = \frac{\mathrm{d}}{\mathrm{d}x}$ and $\mathcal{L}_2 = \frac{\mathrm{d}^2}{\mathrm{d}x^2}$. Simplify the GREEN–GAUSS theorem as given in Eq. (A.27) to derive the the weak formulation based on Eq. (3.330).

3.19 Weighted residual method with arbitrary distributed load for a beam

Derive the principal finite element equation for a BERNOULLI beam element based on the weighted residual method. Starting point should be the partial differential equation with an arbitrary distributed load $q_z(x)$. In addition, it can be assumed that the bending stiffness EI_y is constant.

3.20 Stiffness matrix for bending in the x-y plane

Derive the stiffness matrix for a BERNOULLI beam element for bending in the x-y plane.

3.21 Investigation of displacement and slope consistency along boundaries

Investigate the interelement continuity of the displacement and slope for a BERNOULLI beam element.

3.22 Bending moment distribution for a cantilever beam

Given is a cantilever beam of length L and constant bending stiffness given by EI_y as shown in Fig. 3.53. At the left-hand side there is a fixed support. The right-hand side of the beam is loaded by a point load F_0 in negative z-direction.

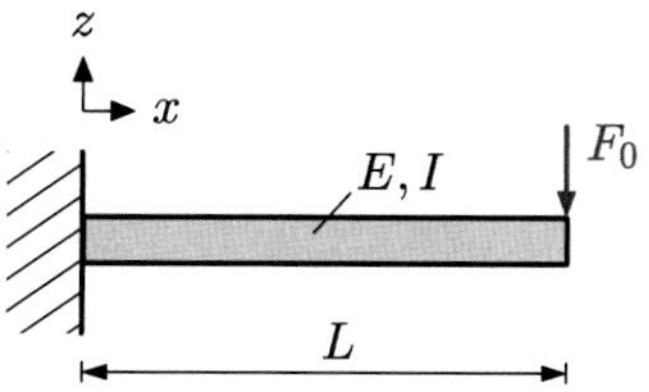

Fig. 3.53 Cantilever beam loaded by a point load

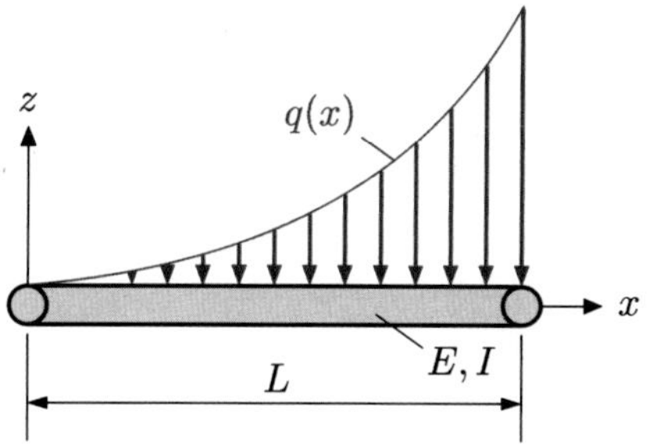

Fig. 3.54 Quadratic distributed load

Use a single EULER–BERNOULLI beam element to determine the moment distribution $M_y^e(\xi)$ and evaluate the numerical values at the 3 integration points for the values $E = 200000$, $L = 10$, $I_y = \frac{1}{192}$, and $F = -50$. Consistent units can be assumed.

3.23 Beam with variable cross-sectional area

Solve Example 3.6 for arbitrary values of d_1 and d_2!

3.24 Equivalent nodal loads for quadratic distributed load

Determine the equivalent nodal loads for a BERNOULLI beam, cf. Fig. 3.54, for the cases:

(a) $q(x) = q_0 x^2$,

(b) $q(x) = q_0 \left(\frac{x}{L}\right)^2$.

3.25 Beam with variable cross section loaded by a single force

Calculate for the BERNOULLI beam shown in Fig. 3.55 the vertical displacement of the right-hand boundary for $d_1 = 2h$ and $d_2 = h$. Use one single finite element and compare the numerical solution with the analytical solution. Advice: The stiffness matrix can be taken from problem 3.6.

3.26 Beam on elastic foundation: stiffness matrix

A beam on an elastic foundation is schematically shown in Fig. 3.56. Derive the elemental stiffness matrix for a BERNOULLI beam element under the assumption that the elastic foundation modulus k is constant. The describing partial differential equation can be taken from Table 3.5.

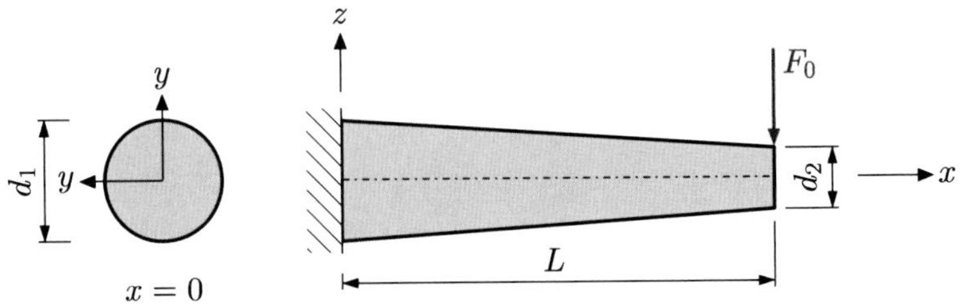

Fig. 3.55 Beam with variable cross-sectional area loaded by a single force

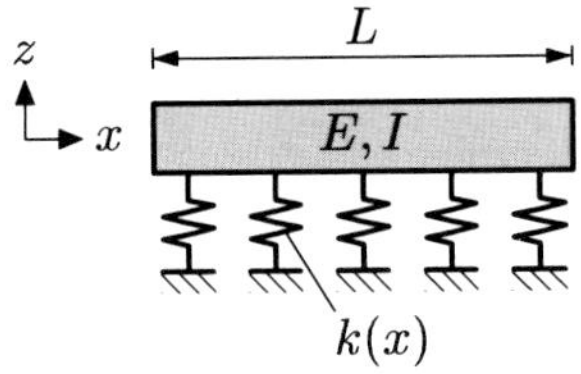

Fig. 3.56 Beam on elastic foundation

3.27 Beam on elastic foundation: single force case

A cantilever beam on an elastic foundation is loaded by a single force F_0 as shown in Fig. 3.57. Assume that the elastic foundation modulus k and the bending stiffness EI_y are constant. Use one single BERNOULLI beam element to determine:

- the reduced system of equations,
- the deflection and rotation of the beam at $x = L$,
- simplify your result for the special case $k = 0$,
- simplify your result for the special case $EI_y = 0$.
- Compare the finite element solution with the analytical solution for the case $k = 4$, $EI_y = 1$ and $L = 1$ (assume consistent units).

3.28 Beam on nonlinear elastic foundation: stiffness matrix

Derive the elemental stiffness matrix $\boldsymbol{K}^{\text{e}}$ for a BERNOULLI beam element with constant bending stiffness EI_y and nonlinear elastic foundation modulus $k = k(u_z)$, cf.

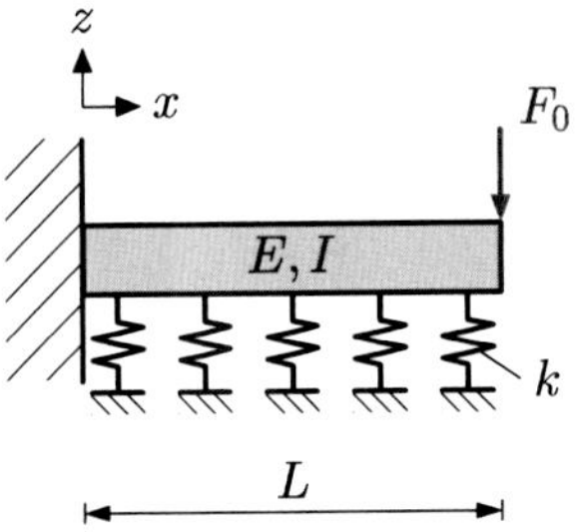

Fig. 3.57 Beam on elastic foundation loaded by a single force

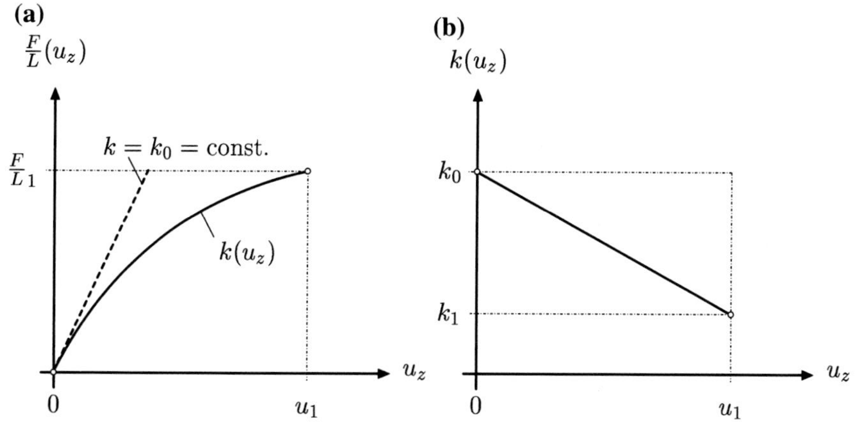

Fig. 3.58 **a** Nonlinear load-displacement diagram; **b** displacement-dependent elastic foundation modulus

Fig. 3.56. Consider the case that the elastic foundation modulus changes linearly with the vertical displacement u_z as shown in Fig. 3.58b. The linear relationship of the elastic foundation modulus should be defined by two sampling points $k(u_z = 0) = k_0$ and $k(u_z = u_1) = k_1$.

3.29 Cantilever beam with triangular shaped distributed load

Given is a cantilever beam of length L and constant bending stiffness EI, cf. Fig. 3.59. The beam is loaded with a triangular shaped distributed load with a maximum value of q_0. Use one single finite element to calculate

- the maximum deflection of the beam,
- the distribution of the deflection between the nodes: $u_z = u_z(x)$.
- Calculate based on the appropriate partial differential equation the analytical solution for the deflection $u_z = u_z(x)$ and compare this result with the finite element solution at $x = L$.
- Sketch the analytical and finite element solution, i.e. $u_z = u_z(x)$, in the range $0 \leq x \leq L$.

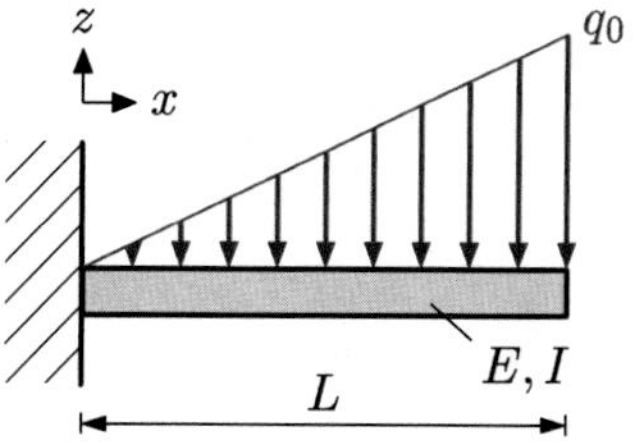

Fig. 3.59 Cantilever beam with triangular shaped distributed load

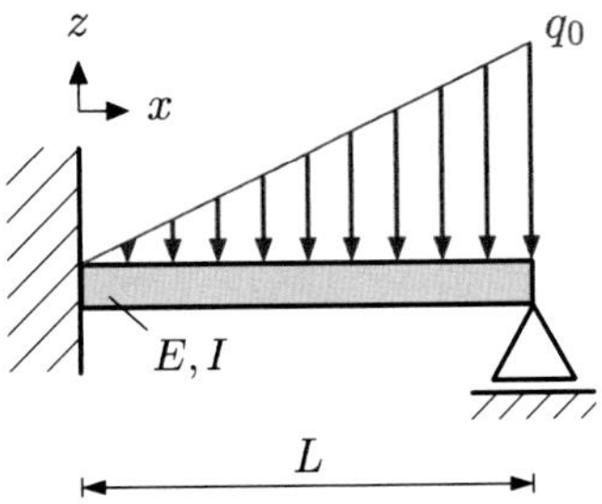

Fig. 3.60 Cantilever beam with triangular shaped distributed load and roller support

3.30 Cantilever beam with triangular shaped distributed load and roller support

Given is a cantilever beam of length L and constant bending stiffness EI, cf. Fig. 3.60. The beam is loaded with a triangular shaped distributed load with a maximum value of q_0. Use the analytical approach to derive

- the bending line, and
- the bending moment distribution.
- Use now a single finite element to calculate the rotation at the right-hand end and compare your result with the analytical approach.

3.31 Finite element approximation with a single beam element

Given is a beam with different supports as shown in Fig. 3.61. The bending stiffness EI is constant and the length is equal to $L = a + b$. The beam is loaded by a single force at location $x = a$. Derive the finite element solution based on one single beam element and compare the displacements $u_z(0)$, $u_z(L)$ and $u_z(a)$ with the analytical solution.

3.32 Cantilever beam: moment curvature relationship

The cantilever beam shown in Fig. 3.62 is loaded by a single force F_0 in the negative Z-direction at its right-hand end. The total length of the beam is $2L$ and the bending stiffness is EI. Use two elements of length L to determine:

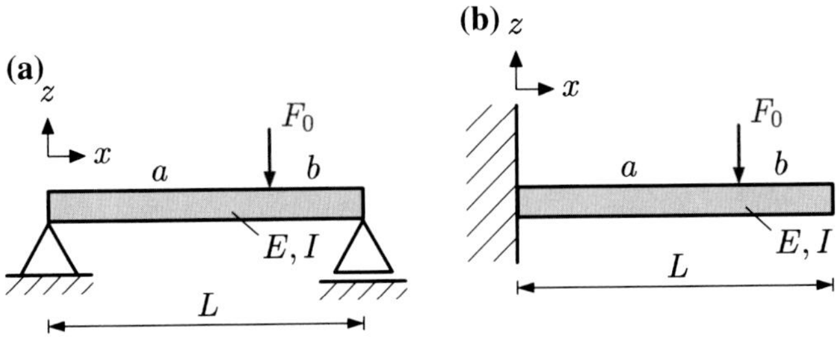

Fig. 3.61 Finite element approximation with a single beam element: **a** simply supported beam; **b** cantilever beam

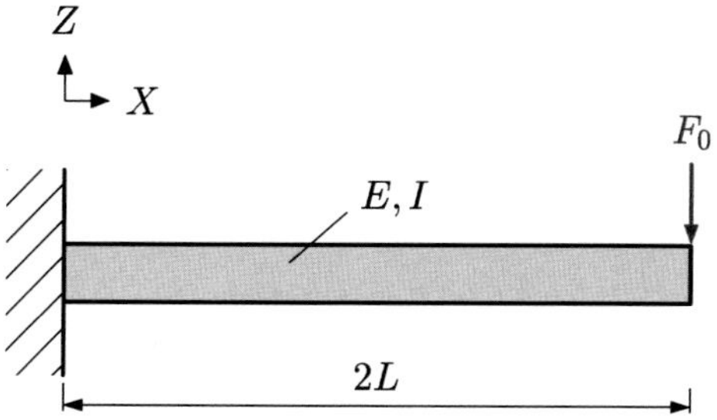

Fig. 3.62 Cantilever beam: Moment curvature relationship

- the displacements and rotations at each node,
- the curvature at each node,
- the bending moment at each node,
- check at each node the moment curvature relationship $\frac{M_y}{\kappa_y} = EI_y$.
- Compare the results with the analytical solution.

3.33 Fixed-end beam with distributed load and displacement boundary condition

The beam shown in Fig. 3.63 is loaded by a triangular shaped distributed load (maximum value q_0) and a vertical displacement u_0 in the middle ($X = L$). The bending stiffness EI is constant and the total length of the beam is equal to $2L$. The beam is supported at both ends by fixed supports.

Determine for two finite elements of equal length (I and II):

- the free body diagram of the discretized structure,
- the stiffness matrix for each element,
- the global system of equations without consideration of any support conditions,
- the reduced system of equations.
- Solve the reduced system of equations for the nodal unknowns.
- Determine the reaction force F^{R}_{1Z} and the reaction moment M^{R}_{1Y} at the left-hand end.

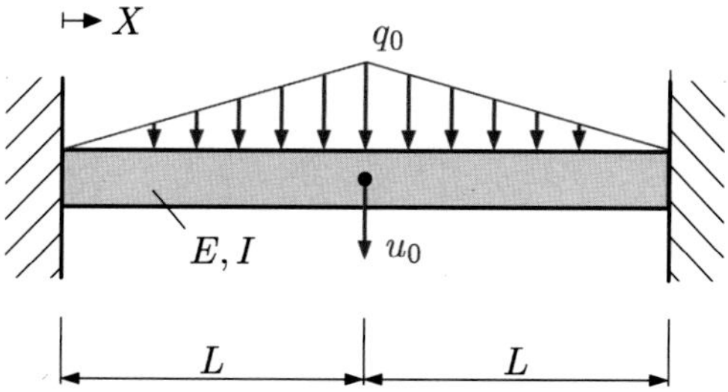

Fig. 3.63 Fixed-end beam with distributed load and displacement boundary condition

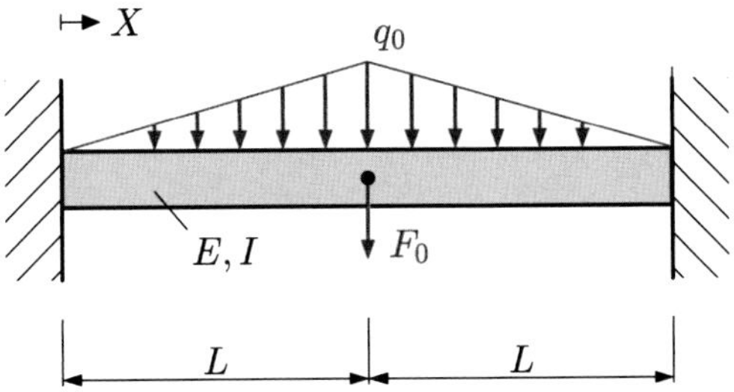

Fig. 3.64 Fixed-end beam with distributed load and single force load

3.34 Fixed-end beam with distributed load and single force load

The beam shown in Fig. 3.64 is loaded by a triangular shaped distributed load (maximum value q_0) and a vertical force F_0 in the middle ($X = L$). The bending stiffness EI is constant and the total length of the beam is equal to $2L$. The beam is supported at both ends by fixed supports.

Determine for two finite elements of equal length (I and II):

- the free body diagram of the discretized structure,
- the stiffness matrix for each element,
- the global system of equations without consideration of any support conditions,
- the reduced system of equations.
- Solve the reduced system of equations for the nodal unknowns.
- Determine the bending line $u_z(x_{\mathrm{I}})$ of the first element.
- Determine the reaction force F_{1Z}^{R} and the reaction moment M_{1Y}^{R} at the left-hand end.

3.35 Cantilever stepped beam with two sections

The cantilever beam shown in Fig. 3.65 is loaded by a single force F_0 in negative Z-direction at its right-hand end. The beam is divided in two sections of length L_{I} and L_{II}. The geometrical and material properties of the structure are given as

(a) $L_{\mathrm{I}} = L_{\mathrm{II}} = L$, $E_{\mathrm{I}} = E_{\mathrm{II}} = E$, $I_{\mathrm{I}} = 2I$ and $I_{\mathrm{II}} = I$,
(b) $L_{\mathrm{I}} = L_{\mathrm{II}} = L$, $E_{\mathrm{I}} = E_{\mathrm{II}} = E$, $I_{\mathrm{I}} = I_{\mathrm{II}} = I$.

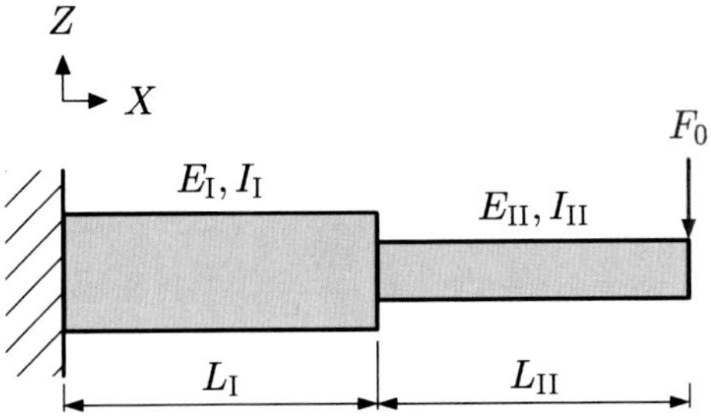

Fig. 3.65 Cantilever stepped beam with two sections

Determine for both cases:

- the global system of equations,
- the reduced system of equations,
- all nodal unknowns,
- all reactions at the support,
- the bending moment at the nodes and its distribution in each element,
- the shear force in each element,
- the stress distribution in Z-direction at locations $X = 0, \frac{L}{2}, L, \frac{3L}{2}, L$ which results from the bending moment.

3.36 Cantilever stepped beam with three sections

The cantilever beam shown in Fig. 3.66 is loaded by a single force F_0 in negative Z-direction at its right-hand end. The beam is divided into three sections of length $L_{\mathrm{I}} = L_{\mathrm{II}} = L_{\mathrm{III}} = \frac{L}{3}$. The material of each section is the same, i.e. $E_{\mathrm{I}} = E_{\mathrm{II}} = E_{\mathrm{III}} = E$, but each cross section is different. Determine the deformations at each node.

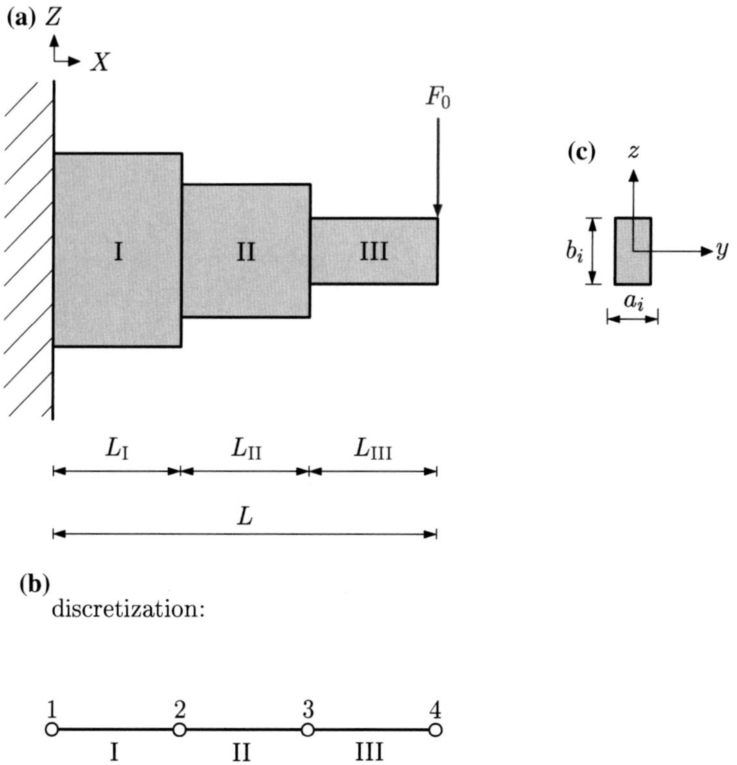

Fig. 3.66 Cantilever stepped beam with three sections: **a** general configuration, **b** discretization, and **c** cross section of element i

3.37 Simply supported stepped beam with four sections

The simply supported beam shown in Fig. 3.67 is loaded by a single force F_0 in negative Z-direction at its middle. The beam is divided into four sections of length $L_{\text{I}} = \cdots = L_{\text{IV}} = \frac{L}{4}$. The material of each section is the same, i.e. $E_{\text{I}} = \cdots = E_{\text{IV}} = E$, but each cross section is different. Determine the deformations at each node under the assumption that the beam is symmetric, i.e $I_{\text{IV}} = I_{\text{I}}$, and $I_{\text{III}} = I_{\text{II}}$.

3.38 Overhang beam with distributed load and single force

The beam shown in Fig. 3.68 is loaded by a constant distributed load q_0 and a single force F_0. The bending stiffness EI is constant and the total length of the beam is equal to $(L_{\text{I}} + L_{\text{II}})$. Model the beam with two elements to determine:

- the unknown rotations and displacements at the nodes,
- the reaction forces at the supports,
- the vertical deflection in the middle of the section with the distributed load, i.e. $X = \frac{1}{2}L_{\text{I}}$,

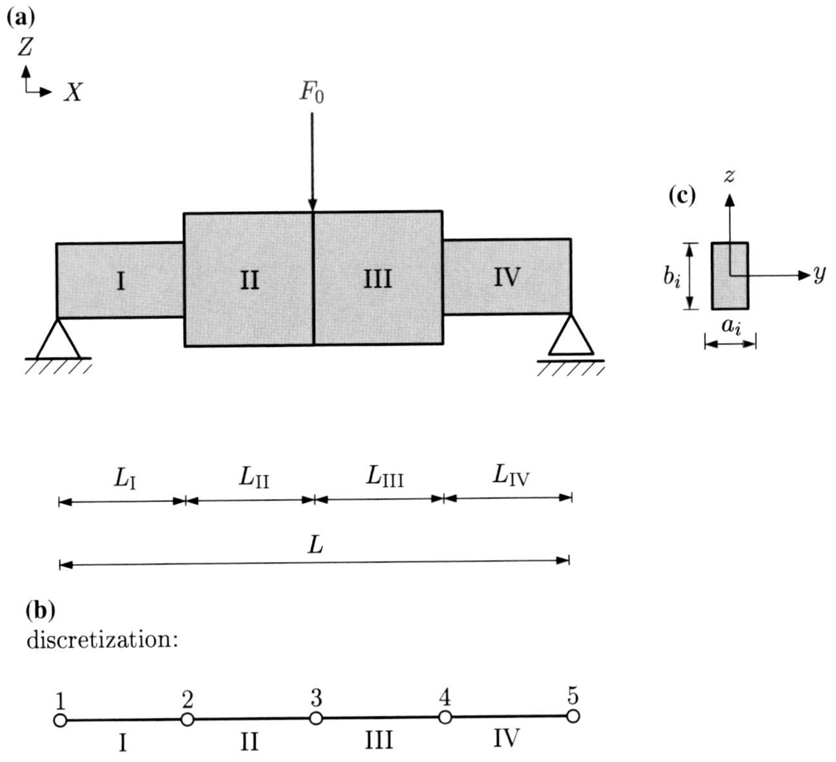

Fig. 3.67 Simply supported stepped beam with four sections: **a** general configuration, **b** discretization, and **c** cross section of element i

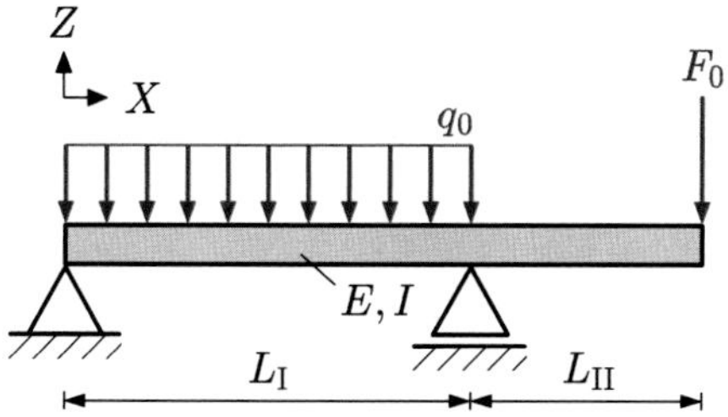

Fig. 3.68 Overhang beam with different types of vertical loads

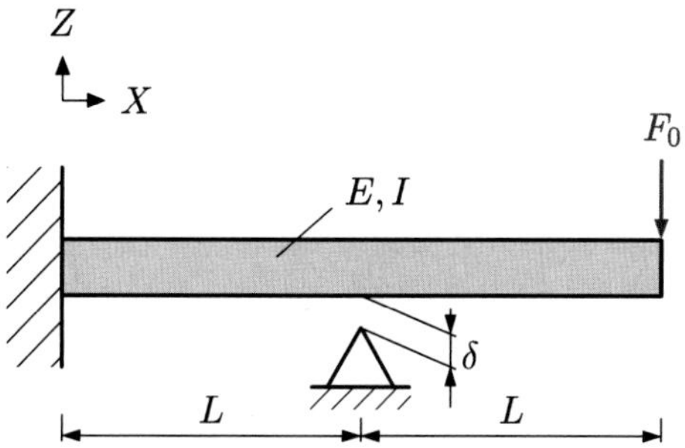

Fig. 3.69 Beam structure with a gap in the middle and vertical point load at the right end

- the bending moment and shear force at the midpoint of the section with the distributed load, i.e. $X = \frac{1}{2}L_{\mathrm{I}}$.
- Improve the approximate solution for $u_Z(X = \frac{1}{2}L_{\mathrm{I}})$ by subdividing the section with the distributed load in two elements of equal length.

3.39 Beam structure with a gap

The beam shown in Fig. 3.69 is loaded by a single vertical force F_0 at its right-hand end. The bending stiffness EI is constant and the total length of the beam is equal to $2L$. In the middle of the entire structure, there is a pap of length δ between the beam and a simple support. Model the beam with two elements to determine:

- the force F_0^* to close the gap,
- the deflection and rotation at the free end, i.e. $X = 2L$, as a function of the increasing force F,
- the maximum normalized stress $\frac{\sigma_x(z_{\max})}{z_{\max}}$ at the nodes for the situation $u_{2z}(X = L) < \delta$ and $u_{2z}(X = L) \geq \delta$.

3.40 Advanced example: beam element with nonlinear bending stiffness

Derive the elemental stiffness matrix $\boldsymbol{K}^{\mathrm{e}}$ for a BERNOULLI beam element with nonlinear bending stiffness $EI = EI(\kappa)$. Consider the case that the bending stiffness changes quadratically with the curvature κ as shown in Fig. 3.70. The quadratic relationship of the bending stiffness should be defined by the three sampling points $EI(\kappa = 0) = IE_0$, $EI(\kappa = \frac{1}{2}\kappa_1) = \beta_{05}IE_0$ and $EI(\kappa = \kappa_1) = \beta_1 IE_0$. To simplify the notation, the product of YOUNG's modulus E and moment of inertia I is considered as a single variable: $(EI) \rightarrow EI$.

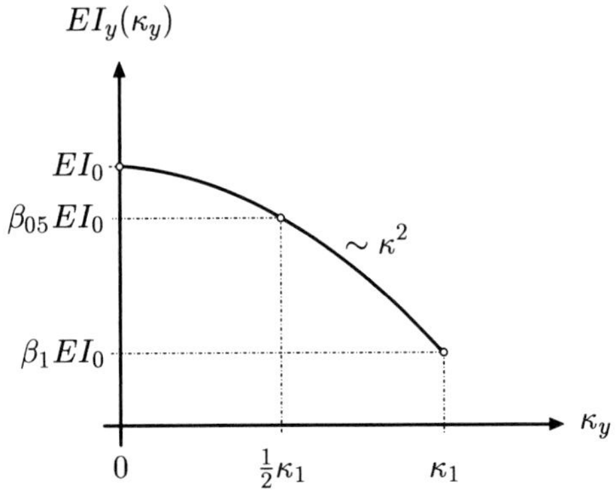

Fig. 3.70 Curvature-dependent bending stiffness

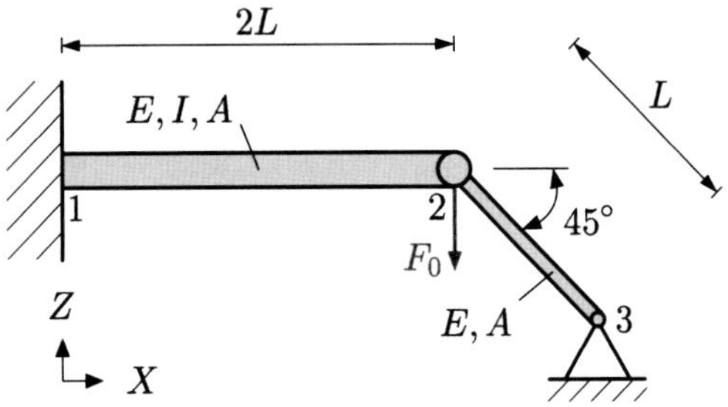

Fig. 3.71 Plane beam rod structure

3.41 Plane beam-rod structure

Given is the two-dimensional beam-rod structure as shown in Fig. 3.71. The generalized beam element of length $2L$ has the bending stiffness EI and the tensile stiffness EA. The rod of length L has the tensile stiffness EA. The structure is loaded by a vertical force F_0 at node 2.

Determine

- the global system of equations,
- the reduced system of equations,
- all nodal displacements and rotations,
- all reaction forces and reaction moments.
- Validate your results by checking the global equilibrium of forces and moments.

3.42 Plane beam-rod structure with distributed load

Figure 3.72 shows a horizontal EULER–BERNOULLI beam element (1–2) which is at point (2) supported by a vertical rod element. Both elements have the same length L and the beam is loaded by a vertical distributed load q_0.

Determine

- based on a finite element approach the general solution for the unknown deformations at point (2).
- Simplify your general solution for the special case that the rod is absent.
- Simplify your general solution for the special case that the beam is absent.

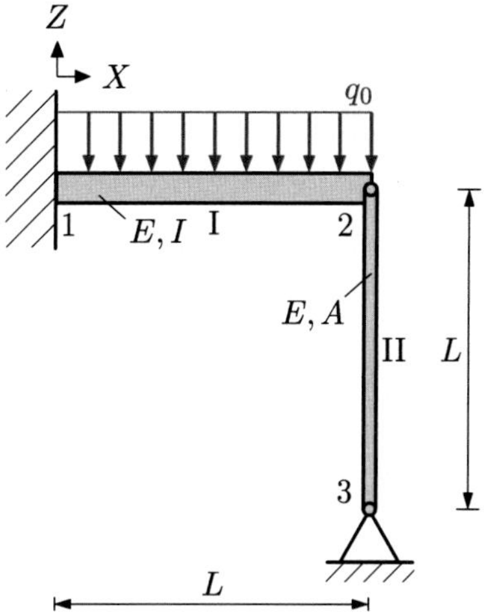

Fig. 3.72 Plane beam rod structure with distributed load

3.43 Plane beam-rod structure with a triangular shaped distributed load

Figure 3.73 shows a horizontal EULER–BERNOULLI beam element (1–3) of length L which is in the middle (at point 2) supported by a vertical rod element of length $\frac{L}{2}$. The beam is loaded by a vertical distributed triangular shaped load (maximum value: q_0).

Determine based on a finite element approach with 3 elements of length $\frac{L}{2}$:

- The global system of equations without the consideration of the displacement boundary conditions.
- The reduced system of equations.
- The unknown deformations at node 2.

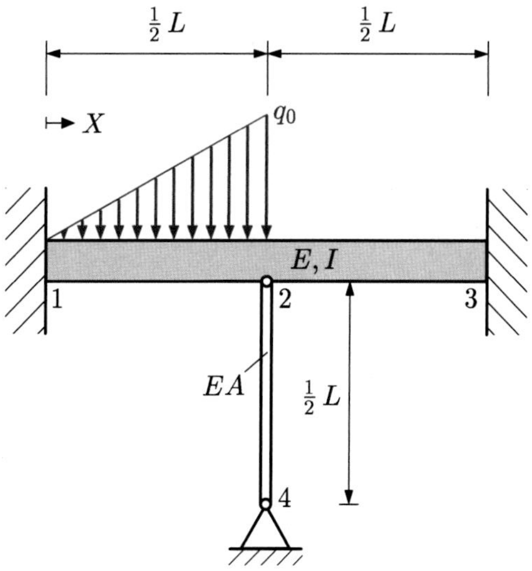

Fig. 3.73 Plane beam-rod structure with a triangular shaped distributed load

- The bending line of the beam, i.e. $u_Z(X)$.
- Sketch the normalized bending line for the special case $\frac{AL^2}{I} = 20$.
- The vertical reaction force at node 4.
- Simplify your general solution for the unknown deformations at node 2 for the special case that the rod is absent.

3.44 Plane generalized beam-rod structure with different distributed loads

Figure 3.74 shows a horizontal generalized EULER–BERNOULLI beam element (1–3) of length $2L$ which is in the middle (at point 2) supported by a vertical rod element of length $\frac{L}{2}$. The beam is loaded in the range $0 \leq X \leq L$ by a vertical distributed triangular shaped load (maximum value: q_0) and in the range $L \leq X \leq 2L$ by a horizontal distributed constant load of magnitude p_0. Both members are made of the same material (E).

Determine based on a finite element approach with 3 elements, i.e. two generalized beams of length L and one rod element of length $\frac{L}{2}$:

- The global system of equations without the consideration of the displacement and rotation boundary conditions.
- The reduced system of equations.
- The unknown deformations at node 2.
- All the reactions at the supports.
- Check the results based on the global force and moment equilibria.
- The normal stress distribution in each element.
- Use in the following the simplification $2AL^2 := 4I$ to determine the horizontal ($u_X(x_i)$) and vertical displacement ($u_Z(x_i)$) of the generalized beam ($i = \text{I, II}$). Sketch in addition the normalized distributions along the beam axis.

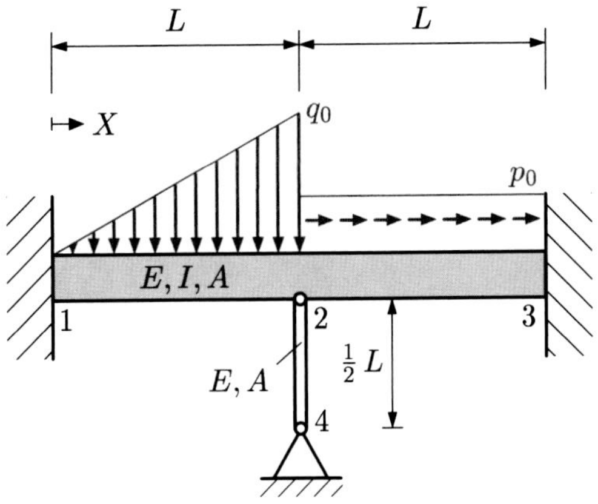

Fig. 3.74 Plane generalized beam-rod structure with different distributed loads

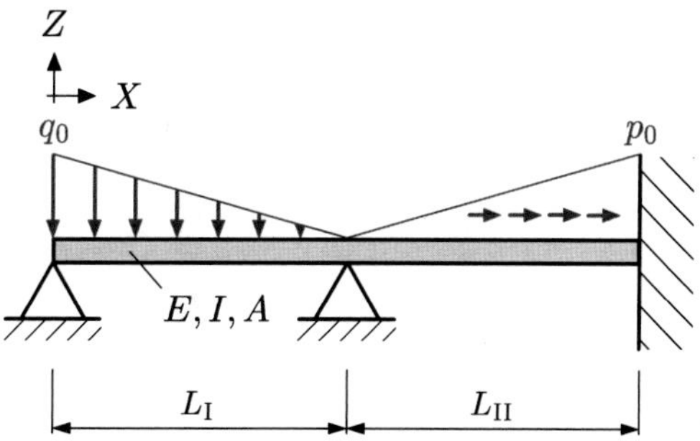

Fig. 3.75 Plane generalized beam structure with different distributed loads

3.45 Plane generalized beam structure with different distributed loads

The generalized beam shown in Fig. 3.75 is loaded by a linear vertical distributed load (maximum value: q_0) in the range $0 \leq X \leq L_{\mathrm{I}}$ and a linear horizontal load (maximum value: p_0) in the range $L_{\mathrm{I}} \leq X \leq L_{\mathrm{I}} + L_{\mathrm{II}}$. The material constant (E) and the geometrical properties (I, A) are constant and the total length of the beam is equal to $L_{\mathrm{I}} + L_{\mathrm{II}}$. Model the member with two generalized beam finite elements of length L_{I} and L_{II} to determine:

- The unknowns at the nodes.
- The horizontal reaction force at node 1, i.e. for $X = 0$.
- Simplify the nodal unknowns for the special case $L_{\mathrm{I}} = L_{\mathrm{II}} = L$.
- Which nodal deformations are obtained for the special case $L_{\mathrm{I}} = L$ and $L_{\mathrm{II}} \to 0$.
- Calculate and schematically sketch the bending line $u_Z(X)$ in the range $0 \leq X \leq L_{\mathrm{I}} + L_{\mathrm{II}}$ for the special case $L_{\mathrm{I}} = L_{\mathrm{II}} = L$.

3.46 Stiffness matrix for a generalized beam element for different rotation angles at the nodes

Derive the stiffness matrix $\boldsymbol{K}^{\mathrm{e}}_{XY}(\alpha_1, \alpha_1)$ for a generalized beam (BERNOULLI) element which can deform in the global X-Y plane. Consider that the rotation angle between the local and global coordinate system is different at both nodes.

3.47 Mechanical properties of a square frame structure

Given is a square frame structure made of generalized beam elements with side length L as shown in Fig. 3.76. Two different orientations, i.e. flat or angled orientation, should be considered in the following. Calculate the displacement (BC: F_0) or reaction force (BC: u_0) of the point of load application and estimate the macroscopic stiffness E_{struct} of the frame structure. Simplify your results for the macroscopic stiffness for the special case $A = \frac{\pi d^2}{4}$ and $I = \frac{\pi d^4}{64}$, i.e. a circular cross section of the beam elements. The derivation should be performed first for the full model, then for the half model and finally for the quarter model.

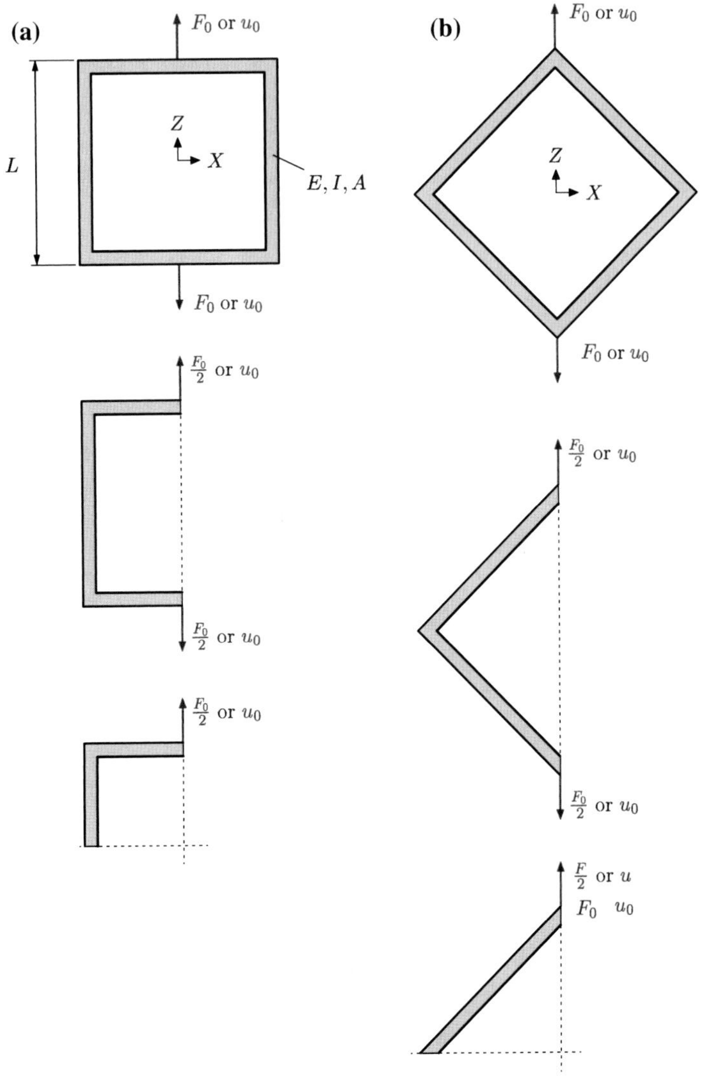

Fig. 3.76 Square frame structure in **a** flat and **b** angled orientation. Top: full model; middle: half model and bottom: quarter model

3.48 Square frame structure: different ways of load application

Given is a square frame structure made of generalized beam elements with side length L as shown in Fig. 3.77. Two different load cases, i.e. central and on the sides, should be compared. Calculate the displacement of the point of load application and estimate the macroscopic stiffness E_{struct} of the frame structure. Simplify your results for the macroscopic stiffness for the special case $A = \frac{\pi d^2}{4}$ and $I = \frac{\pi d^4}{64}$, i.e. a circular cross

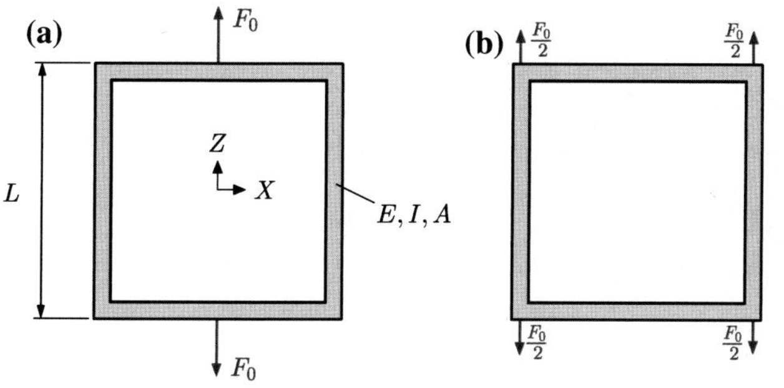

Fig. 3.77 Square frame structure: **a** central load application and **b** loads on the sides

section of the beam elements. The derivation should be performed for the quarter model under consideration of the double symmetry.

3.49 Mechanical properties of idealized honeycomb structure

A honeycomb structure should be idealized by a regular hexagon as shown in Fig. 3.78. Such a regular hexagon has all sides of the same length L, and all internal angles are 120°. Assume for this simplified approach that the honeycomb structure is represented by a single cell, either in flat or angled orientation. Furthermore, a two-dimensional approach based on a frame structure made of generalized beam elements is to consider. Calculate the displacement of the point of load application and estimate the macroscopic stiffness E_{struct} of the idealized honeycomb structure. Simplify your results for the macroscopic stiffness for the special case $A = \frac{\pi d^2}{4}$ and $I = \frac{\pi d^4}{64}$, i.e. a circular cross section of the beam elements. To facilitate the finite element approach, exploit the symmetry of the problem.

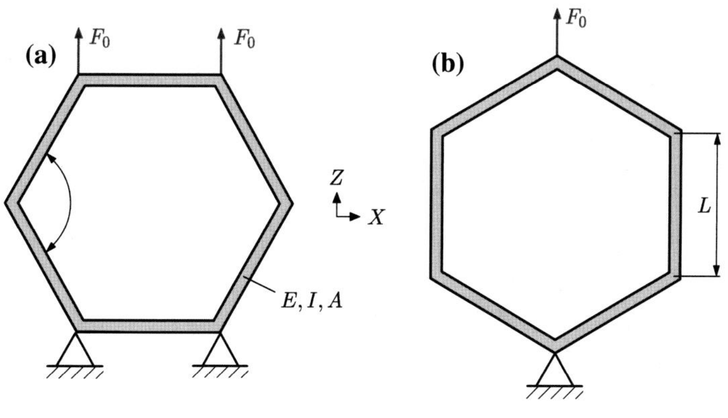

Fig. 3.78 Honeycomb structure approximated by a regular hexagon: **a** flat orientation and **b** pointy or angled orientation

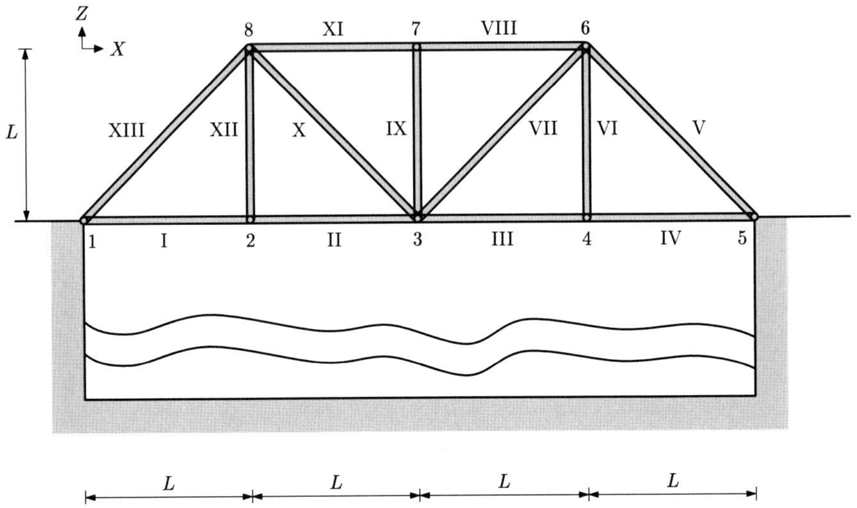

Fig. 3.79 Plane bridge structure over a valley

3.50 Bridge structure (computational problem)

Given is a simplified plane bridge structure over a valley as shown in Fig. 3.79. The bridge structure is idealized in the X-Z plane based on thirteen line elements (I, ..., XIII) which are connected at eight nodes (1, ..., 8). Consider the following numerical values for the geometrical and material parameters (assume consistent units): $L = 4000\,\text{mm}$; $E = 200000\,\text{MPa}$; $I = 80\,\text{mm}^4$ and $A = 10\,\text{mm}^2$.

Consider the following cases that

(a) all elements are generalized beams (i.e. with EA and EI),
(b) only elements I, II, III, IV are generalized beams, all other elements are rod elements (only EA),
(c) all elements are rod elements,
(d) the structure is only composed of the beam elements I, II, III, IV,

to calculate the unknowns at the nodes. Sketch in addition the vertical deformation of elements I, II, III, IV and use the interpolation functions to interpolate between the nodes. Consider the following load cases:

(1) a vertical force of magnitude $-F_0$ at node 3 (the force F_0 is given in general, i.e. not as a numerical value),
(2) a vertical force of magnitude $-F_0$ at node 2,
(3) a vertical force of magnitude $-F_0$ in the middle of element I, i.e. for $X = \frac{L}{2}$ (only case (a), (b) and (d)),
(4) a vertical force of magnitude $-F_0$ in the middle of element II, i.e. for $X = \frac{3L}{2}$ (only case (a), (b) and (d)).

At the supports, i.e. at node 1 and 5, the horizontal and vertical displacements are zero but the beam element can rotate ($\varphi_Y \neq 0$). In the cases (c) and (d), it is possible to derive general expressions for the unknowns, i.e. as a general function of L, E, (A), (I) and F_0, which are not too complicated.

References

1. Altenbach H, Altenbach J, Naumenko K (1998) Ebene Flächentragwerke: Grundlagen der Modellierung und Berechnung von Scheiben und Platten. Springer, Berlin
2. Boresi AP, Schmidt RJ (2003) Advanced mechanics of materials. Wiley, New York
3. Buchanan GR (1995) Schaum's outline of theory and problems of finite element analysis. McGraw-Hill, New York
4. Budynas RG (1999) Advanced strength and applied stress analysis. McGraw-Hill Book, Singapore
5. Gould PL (1988) Analysis of shells and plates. Springer-Verlag, New York
6. Gross D, Hauger W, Schröder J, Wall WA (2009) Technische Mechanik 2: Elastostatik. Springer, Berlin
7. Hartmann F, Katz C (2007) Structural analysis with finite elements. Springer, Berlin
8. Heyman J (1998) Structural analysis: a historical approach. Cambridge University Press, Cambridge
9. Hibbeler RC (2008) Mechanics of materials. Prentice Hall, Singapore
10. Öchsner A, Merkel M (2013) One-dimensional finite elements: an introduction to the FE method. Springer, Berlin
11. Öchsner A (2014) Elasto-plasticity of frame structure elements: modeling and simulation of rods and beams. Springer, Berlin
12. Szabó I (2003) Einführung in die Technische Mechanik: Nach Vorlesungen István Szabó. Springer, Berlin
13. Timoshenko S, Woinowsky-Krieger S (1959) Theory of plates and shells. McGraw-Hill Book Company, New York
14. Winkler E (1867) Die Lehre von der Elasticität und Festigkeit mit besonderer Rücksicht auf ihre Anwendung in der Technik. H. Dominicus, Prag

Chapter 4
Timoshenko Beams

Abstract This chapter starts with the analytical description of beam members under the additional influence of shear stresses. Based on the three basic equations of continuum mechanics, i.e., the kinematics relationship, the constitutive law and the equilibrium equation, the partial differential equations, which describe the physical problem, are derived. The weighted residual method is then used to derive the principal finite element equation for TIMOSHENKO beam elements. In addition to linear interpolation functions, a general concept for arbitrary polynomials of interpolation functions is introduced.

4.1 Introduction

The general difference regarding the deformation of a beam with and without shear influence has already been discussed in Sect. 3.1. In this section, the shear influence on the deformation is considered with the help of the TIMOSHENKO beam theory [21, 22]. Within the framework of the following remarks, the definition of the shear strain and the relation between shear force and shear stress will first be covered.

For the derivation of the equation for the shear strain in the x-z plane, the infinitesimal rectangular beam element $ABCD$, shown in Fig. 4.1, is considered, which deforms under the influence of a pure shear stress. Here, a change of the angle of the original right angles as well as a change in the lengths of the edges occurs.

The deformation of the point A can be described via the displacement fields $u_x(x, z)$ and $u_z(x, z)$. These two functions of *two* variables can be expanded in TAYLOR's series[1] of first order around point A to approximately calculate the deformations of the points B and D:

[1]For a function $f(x, z)$ of two variables usually a TAYLOR's series expansion of first order is formulated around the point (x_0, z_0) as follows: $f(x, z) = f(x_0 + \mathrm{d}x, z_0 + \mathrm{d}z) \approx f(x_0, z_0) + \left(\frac{\partial f}{\partial x}\right)_{x_0, z_0} \times (x - x_0) + \left(\frac{\partial f}{\partial z}\right)_{x_0, z_0} \times (z - z_0)$.

A. Öchsner, *Computational Statics and Dynamics*,
https://doi.org/10.1007/978-981-15-1278-0_4

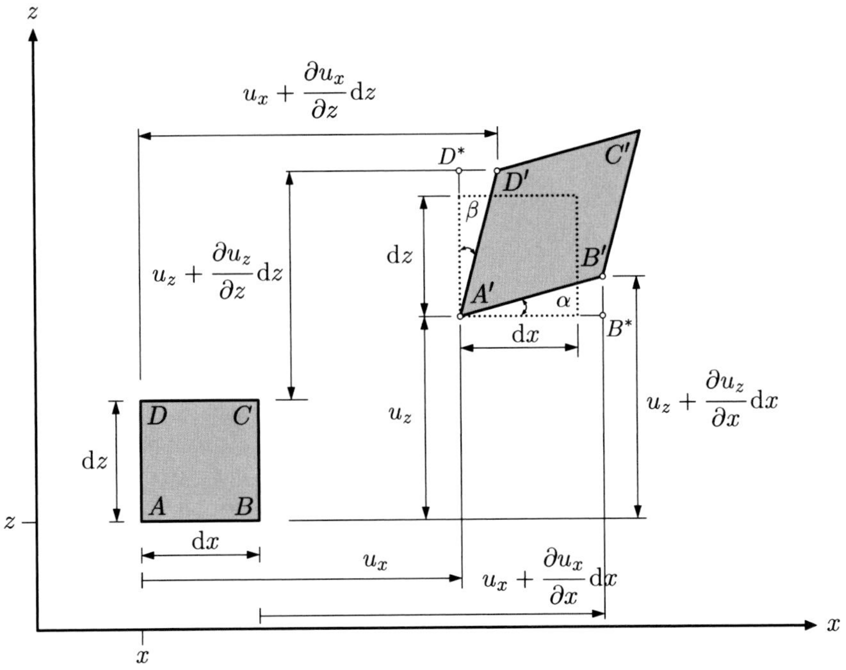

Fig. 4.1 Definition of the shear strain γ_{xz} in the x-z plane at an infinitesimal beam element

$$u_{x,B} = u_x(x + \mathrm{d}x, z) = u_x(x, z) + \frac{\partial u_x}{\partial x}\mathrm{d}x + \frac{\partial u_x}{\partial z}\mathrm{d}z\,, \tag{4.1}$$

$$u_{z,B} = u_z(x + \mathrm{d}x, z) = u_z(x, z) + \frac{\partial u_z}{\partial x}\mathrm{d}x + \frac{\partial u_z}{\partial z}\mathrm{d}z\,, \tag{4.2}$$

or alternatively

$$u_{x,D} = u_x(x, z + \mathrm{d}z) = u_x(x, z) + \frac{\partial u_x}{\partial x}\mathrm{d}x + \frac{\partial u_x}{\partial z}\mathrm{d}z\,, \tag{4.3}$$

$$u_{z,D} = u_z(x, z + \mathrm{d}z) = u_z(x, z) + \frac{\partial u_z}{\partial x}\mathrm{d}x + \frac{\partial u_z}{\partial z}\mathrm{d}z\,. \tag{4.4}$$

In Eqs. (4.1) up to (4.4), $u_x(x, z)$ and $u_z(x, z)$ represent the so-called rigid-body displacements, which do not cause a deformation. If one considers that point B has the coordinates $(x + \mathrm{d}x, z)$ and D the coordinates $(x, z + \mathrm{d}z)$, the following results:

$$u_{x,B} = u_x(x,z) + \frac{\partial u_x}{\partial x}\mathrm{d}x\,, \tag{4.5}$$

$$u_{z,B} = u_z(x,z) + \frac{\partial u_z}{\partial x}\mathrm{d}x\,, \tag{4.6}$$

or alternatively

$$u_{x,D} = u_x(x,z) + \frac{\partial u_x}{\partial z}\mathrm{d}z\,, \tag{4.7}$$

$$u_{z,D} = u_z(x,z) + \frac{\partial u_z}{\partial z}\mathrm{d}z\,. \tag{4.8}$$

The total shear strain γ_{xz} of the deformed beam element $A'B'C'D'$ results, according to Fig. 4.1, from the sum of the angles α and β. The two angles can be identified in the rectangle, which is deformed to a rhombus. Under consideration of the two right-angled triangles $A'D^*D'$ and $A'B^*B'$, these two angles can be expressed as:

$$\tan\alpha = \frac{\frac{\partial u_z}{\partial x}\mathrm{d}x}{\mathrm{d}x + \frac{\partial u_x}{\partial x}\mathrm{d}x} \quad \text{and} \quad \tan\beta = \frac{\frac{\partial u_x}{\partial z}\mathrm{d}z}{\mathrm{d}z + \frac{\partial u_z}{\partial z}\mathrm{d}z}\,. \tag{4.9}$$

It holds approximately for small deformations that $\tan\alpha \approx \alpha$ and $\tan\beta \approx \beta$ or alternatively $\frac{\partial u_x}{\partial x} \ll 1$ and $\frac{\partial u_z}{\partial z} \ll 1$, so that the following expression results for the shear strain:

$$\gamma_{xz} = \alpha + \beta \approx \frac{\partial u_z}{\partial x} + \frac{\partial u_x}{\partial z}\,. \tag{4.10}$$

This total change of the angle is also called the engineering shear strain. In contrast to this, the expression $\varepsilon_{xz} = \frac{1}{2}\gamma_{xz} = \frac{1}{2}(\frac{\partial u_z}{\partial x} + \frac{\partial u_x}{\partial z})$ is known as the tensorial definition (tensor shear strain) in the literature [25]. Due to the symmetry of the strain tensor, the identity $\gamma_{ij} = \gamma_{ji}$ applies to the tensor elements outside the main diagonal.

The algebraic sign of the shear strain needs to be explained in the following with the help of Fig. 4.2 for the special case that only one shear force acts in parallel to the z-axis. If a shear force acts in the direction of the positive z-axis at the right-hand face—hence a positive shear force distribution is being assumed at this point—, according to Fig. 4.2a under consideration of Eq. (4.10) a positive shear strain results. In a similar way, a negative shear force distribution leads to a negative shear strain according to Fig. 4.2b.

It has already been mentioned in Sect. 3.1 that the shear stress distribution is variable over the cross-section. As an example, the parabolic shear stress distribution was illustrated over a rectangular cross section in Fig. 3.3. Based on HOOKE's law for a one-dimensional shear stress state, it can be derived that the shear strain has to exhibit a corresponding parabolic course. From the shear stress distribution in the

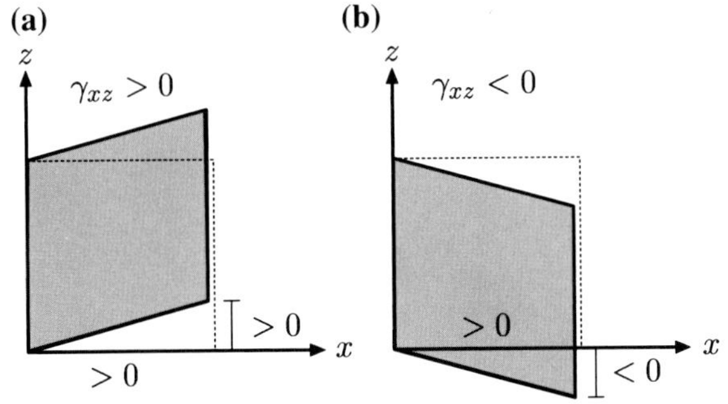

Fig. 4.2 Definition of a **a** positive and **b** negative shear strain in the x-z plane

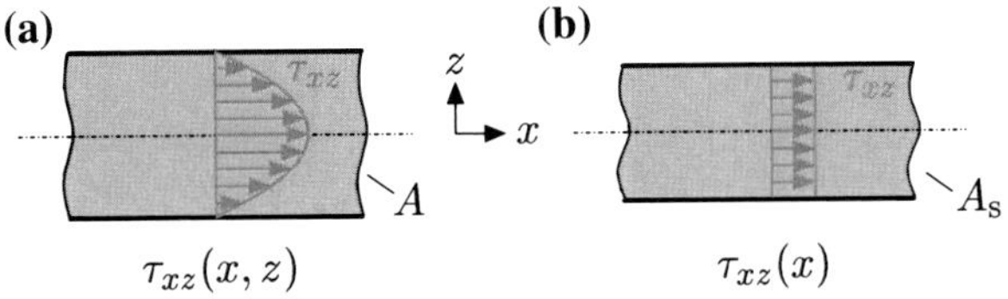

Fig. 4.3 Shear stress distribution: **a** real distribution for a rectangular cross section and **b** TIMOSHENKO's approximation

cross-sectional area at location x of the beam,[2] one receives the acting shear force through integration as:

$$Q_z = \int_A \tau_{xz}(y, z)\,\mathrm{d}A\,. \tag{4.11}$$

However, to simplify the problem, it is assumed for the TIMOSHENKO beam that an equivalent *constant* shear stress and strain act, see Fig. 4.3:

$$\tau_{xz}(y, z) \to \tau_{xz}\,. \tag{4.12}$$

This constant shear stress results from the shear force, which acts in an equivalent cross-sectional area, the so-called shear area A_s:

$$\tau_{xz} = \frac{Q_z}{A_s}\,, \tag{4.13}$$

---

[2] A closer analysis of the shear stress distribution in the cross-sectional area shows that the shear stress does not just alter over the height of the beam but also through the width of the beam. If the width of the beam is small when compared to the height, only a small change along the width occurs and one can assume in the first approximation a constant shear stress throughout the width: $\tau_{xz}(y, z) \to \tau_{xz}(z)$. See for example [2, 24].

Table 4.1 Comparison of shear correction factor values for a rectangular cross section based on different approaches

k_s	Comment	Reference
$\frac{2}{3}$	–	[21, 23]
$0.833\left(=\frac{5}{6}\right)$	$\nu = 0.0$	[4]
0.850	$\nu = 0.3$	
0.870	$\nu = 0.5$	

whereupon the relation between the shear area A_s and the actual cross-sectional area A is referred to as the shear correction factor k_s:

$$k_s = \frac{A_s}{A}. \tag{4.14}$$

Different assumptions can be made to calculate the shear correction factor [4]. As an example, it can be demanded [1] that the elastic strain energy of the equivalent shear stress has to be identical with the energy, which results from the acting shear stress distribution in the actual cross-sectional-area. A comparison for a rectangular cross section is presented in Table 4.1.

Different geometric characteristics of simple geometric cross-sections—including the shear correction factor[3]—are collected in Table 4.2 [5, 28]. Further details regarding the shear correction factor for arbitrary cross-sections can be taken from [6].

It is obvious that the equivalent constant shear stress can alter along the center line of the beam, in case the shear force along the center line of the beam changes. The attribute 'constant' thus just refers to the cross-sectional area at location x and the equivalent constant shear stress is therefore in general a function of the coordinate of length for the TIMOSHENKO beam:

$$\tau_{xz} = \tau_{xz}(x). \tag{4.15}$$

The so-called TIMOSHENKO beam can be generated by superposing a shear deformation on a BERNOULLI beam according to Fig. 4.4.

One can see that the BERNOULLI hypothesis is partly no longer fulfilled for the TIMOSHENKO beam: Plane cross sections remain plane after the deformation. However, a cross section which stood at right angles on the beam axis before the deformation is not at right angles on the beam axis after the deformation. If the demand for planeness of the cross sections is also given up, one reaches theories of higher-order [9, 14, 15], at which, for example, a parabolic course of the shear strain and stress in the displacement field are considered, see Fig. 4.5. Therefore, a shear correction factor is not required for these theories of higher-order.

[3]It should be noted that the so-called form factor for shear is also known in the literature. This results as the reciprocal of the shear correction factor.

Table 4.2 Characteristics of different cross sections in the y-z plane. I_y and I_z: axial second moment of area; A: cross-sectional area; k_s: shear correction factor. Adapted from [28]

Cross-section	I_y	I_z	A	k_s
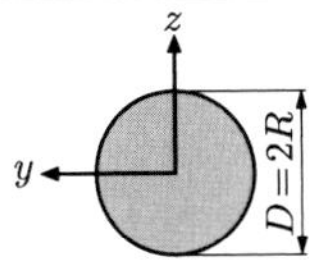	$\frac{\pi R^4}{4}$	$\frac{\pi R^4}{4}$	πR^2	$\frac{9}{10}$
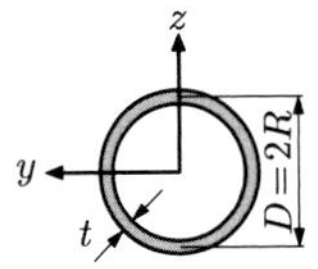	$\pi R^3 t$	$\pi R^3 t$	$2\pi Rt$	0.5
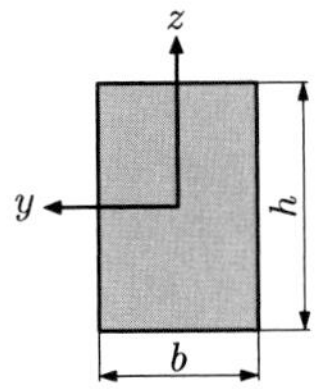	$\frac{bh^3}{12}$	$\frac{hb^3}{12}$	hb	$\frac{5}{6}$
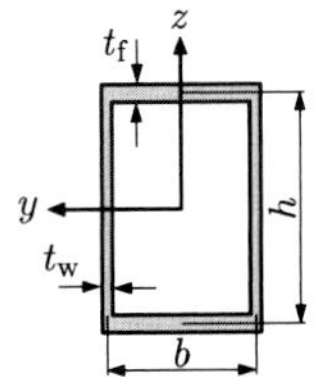	$\frac{h^2}{6}(ht_w + 3bt_f)$	$\frac{b^2}{6}(bt_f + 3ht_w)$	$2(bt_f + ht_w)$	$\frac{2ht_w}{A}$
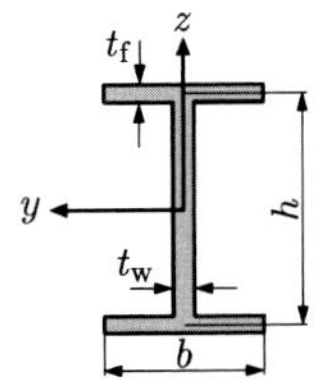	$\frac{h^2}{12}(ht_w + 6bt_f)$	$\frac{b^3 t_f}{6}$	$ht_w + 2bt_f$	$\frac{ht_w}{A}$

4.2 Derivation of the Governing Differential Equation

4.2.1 Kinematics

According to the alternative derivation in Sect. 3.2.1, the kinematics relation can also be derived for the beam with shear action, by considering the angle ϕ_y instead of the angle φ_y, see Figs. 4.4c and 4.6.

Following an equivalent procedure as in Sect. 3.2.1, the corresponding relationships are obtained:

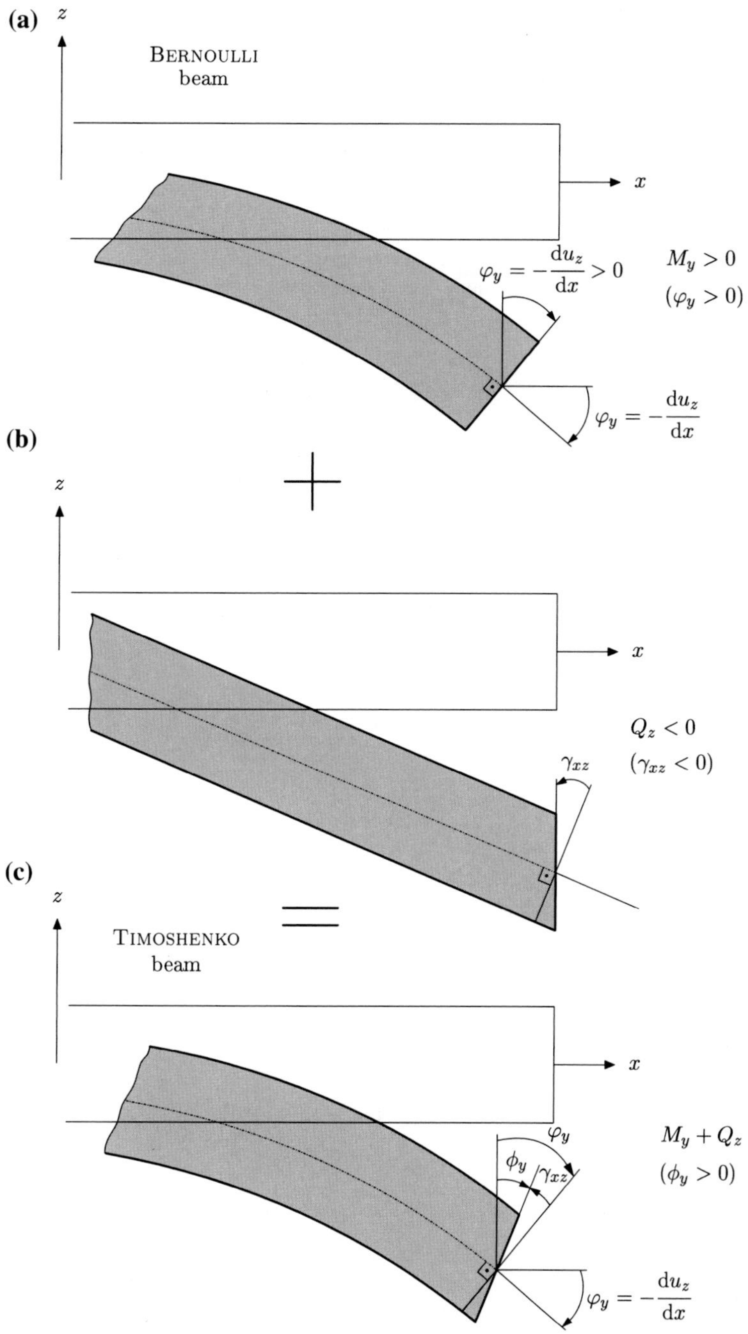

Fig. 4.4 Superposition of the BERNOULLI beam **(a)** and the shear deformation **(b)** to the TIMOSHENKO beam **(c)** in the x-z plane. Note that the deformation is exaggerated for better illustration

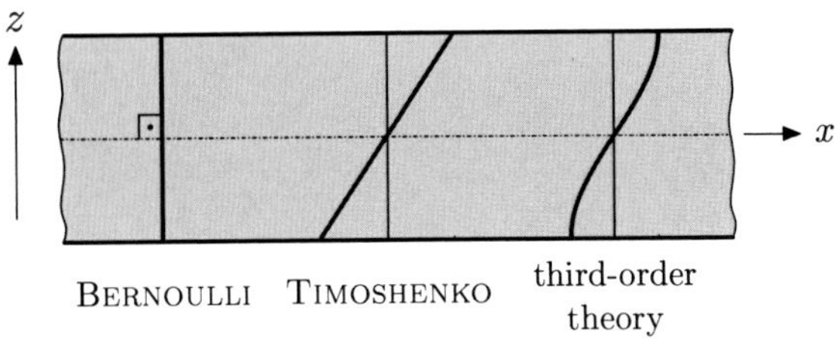

Fig. 4.5 Deformation of originally plane cross sections for the BERNOULLI beam (left), the TIMOSHENKO beam (middle) and a higher-order theory (right) [16]

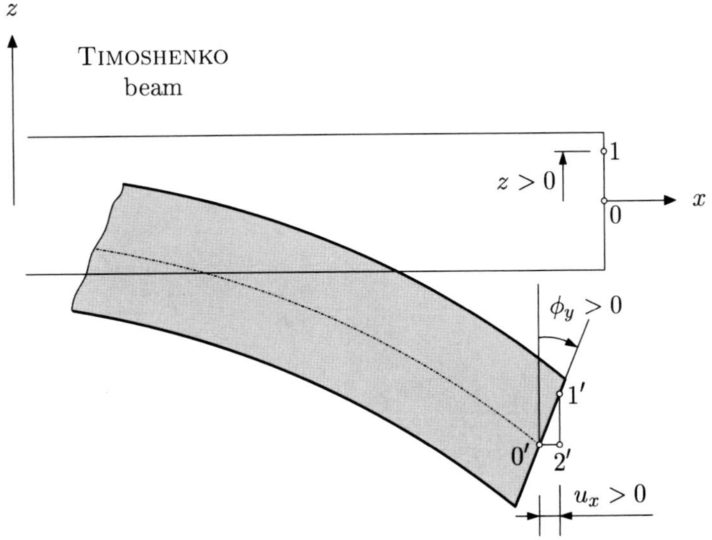

Fig. 4.6 Derivation of the kinematics relation. Note that the deformation is exaggerated for better illustration

$$\sin\phi_y = \frac{u_x}{z} \approx \phi_y \quad \text{or} \quad u_x = +z\phi_y\,, \tag{4.16}$$

wherefrom, via the general relation for the strain, meaning $\varepsilon_x = \mathrm{d}u_x/\mathrm{d}x$, the kinematics relation results through differentiation with respect to the x-coordinate:

$$\varepsilon_x = +z\frac{\mathrm{d}\phi_y}{\mathrm{d}x}\,. \tag{4.17}$$

Note that $\phi_y \rightarrow \varphi_y = -\frac{\mathrm{d}u_z}{\mathrm{d}x}$ results from neglecting the shear deformation and a relation according to Eq. (3.16) results as a special case. Furthermore, the following relation between the angles can be derived from Fig. 4.4c

$$\phi_y = \varphi_y + \gamma_{xz} = -\frac{\mathrm{d}u_z}{\mathrm{d}x} + \gamma_{xz}\,, \tag{4.18}$$

which complements the set of the kinematics relations. It needs to be remarked that at this point the so-called bending line was considered. Therefore, the displacement field u_z is only a function of *one* variable: $u_z = u_z(x)$.

4.2.2 Equilibrium

The derivation of the equilibrium condition for the TIMOSHENKO beam is identical with the derivation for the BERNOULLI beam according to Sect. 3.2.3:

$$\frac{\mathrm{d}Q_z(x)}{\mathrm{d}x} = -q_z(x)\,, \tag{4.19}$$

$$\frac{\mathrm{d}M_y(x)}{\mathrm{d}x} = +Q_z(x)\,. \tag{4.20}$$

4.2.3 Constitutive Equation

For the consideration of the constitutive relation, HOOKE's law for a one-dimensional normal stress state and for a one-dimensional shear stress state is used:

$$\sigma_x = E\varepsilon_x\,, \tag{4.21}$$

$$\tau_{xz} = G\gamma_{xz}\,, \tag{4.22}$$

whereupon the shear modulus G can be calculated based on the YOUNG's modulus E and the POISSON's ratio ν as:

$$G = \frac{E}{2(1+\nu)}\,. \tag{4.23}$$

According to the equilibrium configuration of Fig. 3.9 and Eq. (3.22), the relation between the internal moment and the bending stress can be used for the TIMOSHENKO beam as follows:

$$\mathrm{d}M_y = (+z)(+\sigma_x)\mathrm{d}A\,, \tag{4.24}$$

or alternatively after integration under the consideration of the constitutive Eq. (4.21) and the kinematics relation (4.17):

$$M_y(x) = +EI_y\frac{\mathrm{d}\phi_y(x)}{\mathrm{d}x}\,. \tag{4.25}$$

Table 4.3 Elementary basic equations for the bending of a TIMOSHENKO beam in the x-z plane (x-axis: right facing; z-axis: upward facing)

Relation	Equation
Kinematics	$\varepsilon_x(x,z) = +z\dfrac{\mathrm{d}\phi_y(x)}{\mathrm{d}x}$ and $\phi_y(x) = -\dfrac{\mathrm{d}u_z(x)}{\mathrm{d}x} + \gamma_{xz}(x)$
Equilibrium	$\dfrac{\mathrm{d}Q_z(x)}{\mathrm{d}x} = -q_z(x)$ $\dfrac{\mathrm{d}M_y(x)}{\mathrm{d}x} = +Q_z(x)$
Constitution	$\sigma_x(x,z) = E\varepsilon_x(x,z)$ and $\tau_{xz}(x) = G\gamma_{xz}(x)$

The relation between shear force and cross-sectional rotation results from the equilibrium Eq. (4.20) as:

$$Q_z(x) = +\frac{\mathrm{d}M_y(x)}{\mathrm{d}x} = +EI_y\frac{\mathrm{d}^2\phi_y(x)}{\mathrm{d}x^2}. \tag{4.26}$$

Before looking in more detail at the differential equations of the bending line, let us summarize the basic equations for the TIMOSHENKO beam in Table 4.3. Note that the normal stress and normal strain are functions of both spatial coordinates, i.e. x and z. However, the shear stress and shear strain are only dependent on the x-coordinate, since an equivalent *constant* shear stress has been introduced over the cross section as an approximation of the TIMOSHENKO beam theory.

4.2.4 Differential Equation

Within the previous section, the relation between the internal moment and the cross-sectional rotation was derived from the normal stress distribution with the help of HOOKE's law, see Eq. (4.25). Differentiation of this relation with respect to the x-coordinate leads to the following expression

$$\frac{\mathrm{d}M_y}{\mathrm{d}x} = \frac{\mathrm{d}}{\mathrm{d}x}\left(EI_y\frac{\mathrm{d}\phi_y}{\mathrm{d}x}\right), \tag{4.27}$$

which can be transformed with the help of the equilibrium relation (4.20), the constitutive Eq. (4.22), and the relation for the shear stress according to (4.13) and (4.14) to

$$\frac{\mathrm{d}}{\mathrm{d}x}\left(EI_y\frac{\mathrm{d}\phi_y}{\mathrm{d}x}\right) = +k_sGA\gamma_{xz}. \tag{4.28}$$

If the kinematics relation (4.18) is considered in the last equation, the so-called bending differential equation results in:

$$\frac{\mathrm{d}}{\mathrm{d}x}\left(EI_y\frac{\mathrm{d}\phi_y}{\mathrm{d}x}\right) - k_s GA\left(\frac{\mathrm{d}u_z}{\mathrm{d}x} + \phi_y\right) = 0\,. \tag{4.29}$$

Considering the shear stress according to Eqs. (4.13) and (4.14) in the expression of HOOKE's law according to (4.22), one obtains

$$Q_z = k_s AG\gamma_{xz}\,. \tag{4.30}$$

Introducing the equilibrium relation (4.20) and the kinematics relation (4.18) in the last equation gives:

$$\frac{\mathrm{d}M_y}{\mathrm{d}x} = +k_s AG\left(\frac{\mathrm{d}u_z}{\mathrm{d}x} + \phi_y\right)\,. \tag{4.31}$$

After differentiation and the consideration of the equilibrium relations according to Eqs. (4.19) and (4.20), the so-called shear differential equation results finally in:

$$\frac{\mathrm{d}}{\mathrm{d}x}\left[k_s AG\left(\frac{\mathrm{d}u_z}{\mathrm{d}x} + \phi_y\right)\right] = -q_z(x)\,. \tag{4.32}$$

Therefore, the shear flexible TIMOSHENKO beam is described through the following two coupled differential equations of second order:

$$\frac{\mathrm{d}}{\mathrm{d}x}\left(EI_y\frac{\mathrm{d}\phi_y}{\mathrm{d}x}\right) - k_s AG\left(\frac{\mathrm{d}u_z}{\mathrm{d}x} + \phi_y\right) = 0\,, \tag{4.33}$$

$$\frac{\mathrm{d}}{\mathrm{d}x}\left[k_s AG\left(\frac{\mathrm{d}u_z}{\mathrm{d}x} + \phi_y\right)\right] = -q_z(x)\,. \tag{4.34}$$

This system contains two unknown functions, namely the deflection $u_z(x)$ and the cross-sectional rotation $\phi_y(x)$. Boundary conditions must be formulated for both functions to be able to solve the system of differential equations for a specific problem.

Different formulations of these coupled differential equations are collected in Table 4.4 where different types of loadings, geometry and bedding are differentiated. The last case in Table 4.4 refers again to the elastic or WINKLER foundation of a beam, [29]. The elastic foundation modulus k has in the case of beams the unit of force per unit area.

Table 4.4 Different formulations of the partial differential equation for a TIMOSHENKO beam in the x-z plane (x-axis: right facing; z-axis: upward facing)

Configuration	Partial Differential Equation
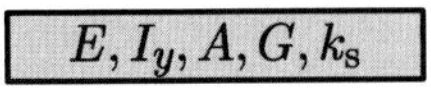	$EI_y \frac{\mathrm{d}^2\phi_y}{\mathrm{d}x^2} - k_s GA\left(\frac{\mathrm{d}u_z}{\mathrm{d}x} + \phi_y\right) = 0$ $k_s GA\left(\frac{\mathrm{d}^2 u_z}{\mathrm{d}x^2} + \frac{\mathrm{d}\phi_y}{\mathrm{d}x}\right) = 0$
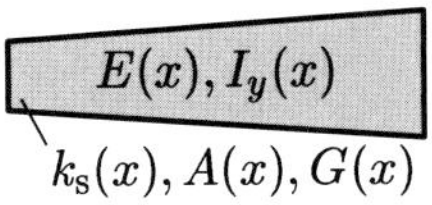	$\frac{\mathrm{d}}{\mathrm{d}x}\left(E(x)I_y(x)\frac{\mathrm{d}\phi_y}{\mathrm{d}x}\right) - k_s(x)G(x)A(x)\left(\frac{\mathrm{d}u_z}{\mathrm{d}x} + \phi_y\right) = 0$ $\frac{\mathrm{d}}{\mathrm{d}x}\left[k_s(x)G(x)A(x)\left(\frac{\mathrm{d}u_z}{\mathrm{d}x} + \phi_y\right)\right] = 0$
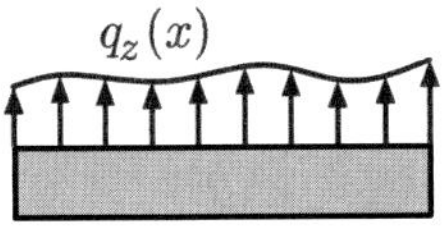	$EI_y \frac{\mathrm{d}^2\phi_y}{\mathrm{d}x^2} - k_s GA\left(\frac{\mathrm{d}u_z}{\mathrm{d}x} + \phi_y\right) = 0$ $k_s GA\left(\frac{\mathrm{d}^2 u_z}{\mathrm{d}x^2} + \frac{\mathrm{d}\phi_y}{\mathrm{d}x}\right) = -q_z(x)$
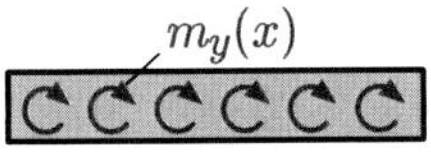	$EI_y \frac{\mathrm{d}^2\phi_y}{\mathrm{d}x^2} - k_s GA\left(\frac{\mathrm{d}u_z}{\mathrm{d}x} + \phi_y\right) = -m_y(x)$ $k_s GA\left(\frac{\mathrm{d}^2 u_z}{\mathrm{d}x^2} + \frac{\mathrm{d}\phi_y}{\mathrm{d}x}\right) = 0$
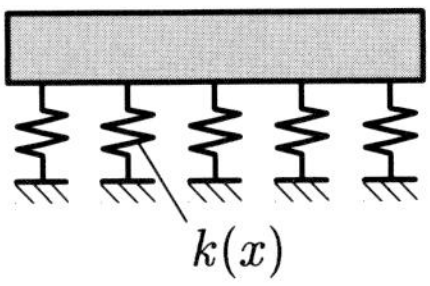	$EI_y \frac{\mathrm{d}^2\phi_y}{\mathrm{d}x^2} - k_s GA\left(\frac{\mathrm{d}u_z}{\mathrm{d}x} + \phi_y\right) = 0$ $k_s GA\left(\frac{\mathrm{d}^2 u_z}{\mathrm{d}x^2} + \frac{\mathrm{d}\phi_y}{\mathrm{d}x}\right) = k(x)u_z$

A single-equation description for the TIMOSHENKO beam can be obtained under the assumption of constant material (E, G) and geometrical (I_z, A, k_s) properties: Rearranging and two-times differentiation of Eq. (4.34) gives:

$$\frac{\mathrm{d}\phi_y}{\mathrm{d}x} = -\frac{\mathrm{d}^2 u_z}{\mathrm{d}x^2} - \frac{q_z}{k_s GA}, \tag{4.35}$$

$$\frac{\mathrm{d}^3\phi_y}{\mathrm{d}x^3} = -\frac{\mathrm{d}^4 u_z}{\mathrm{d}x^4} - \frac{\mathrm{d}^2 q_z}{k_s GA \mathrm{d}x^2}. \tag{4.36}$$

Fig. 4.7 Analytical solution of a cantilever TIMOSHENKO beam under constant distributed load

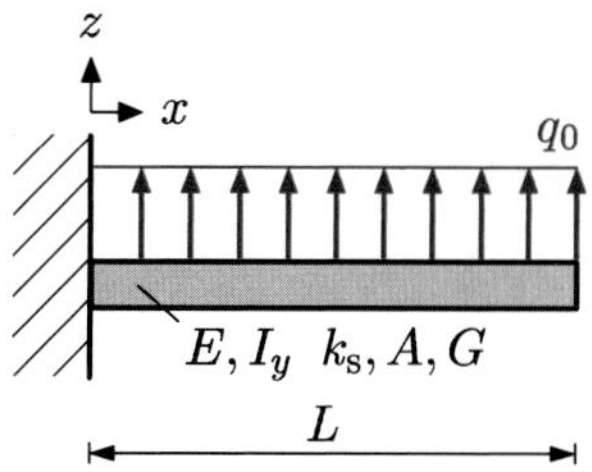

One-time differentiation of Eq. (4.33) gives:

$$EI_y \frac{d^3\phi_y}{dx^3} - k_s AG \left(\frac{d^2 u_z}{dx^2} + \frac{d\phi_y}{dx} \right) = 0\,. \tag{4.37}$$

Inserting Eqs. (4.35) into (4.37) and consideration of (4.36) gives finally the following expression:

$$EI_y \frac{d^4 u_z(x)}{dx^4} = q_z(x) - \frac{EI_y}{k_s AG} \frac{d^2 q_z(x)}{dx^2}\,. \tag{4.38}$$

The last equation reduces for shear-rigid beams, i.e. $k_s AG \to \infty$, to the classical BERNOULLI formulation as given in Table 3.5.

For the derivation of analytical solutions, the system of coupled differential equations as summarized in Table 4.4 has to be solved. Through the use of computer algebra systems (CAS) for the symbolic calculation of mathematical expressions,[4] the general solution of the system as given in Eqs. (4.33) and (4.34) results for constant EI_y, AG, and $q_z = q_0$ in:

$$u_z(x) = \frac{1}{EI_y} \left(\frac{q_0 x^4}{24} + c_1 \frac{x^3}{6} + c_2 \frac{x^2}{2} + c_3 x + c_4 \right), \tag{4.39}$$

$$\phi_y(x) = -\frac{1}{EI_y} \left(\frac{q_0 x^3}{6} + c_1 \frac{x^2}{2} + c_2 x + c_3 \right) - \frac{q_0 x}{k_s AG} - \frac{c_1}{k_s AG}\,. \tag{4.40}$$

The constants of integration $c_1, \ldots, c_4$ must be defined through appropriate boundary conditions to calculate the specific solution of a given problem, meaning under consideration of the support and load conditions.

Consider the TIMOSHENKO beam, which is illustrated in Fig. 4.7, as an example in the following. A constant distributed load q_0 acts in the positive z-direction. Determine the bending line $u_z(x)$ and compare the result to the classical EULER-BERNOULLI beam solution.

[4]Maple®, Mathematica® and Matlab® can be mentioned as commercial examples of computer algebra systems.

The boundary conditions are given as follows for this example:

$$u_z(x=0)=0\ ,\quad \phi_y(x=0)=0\,, \tag{4.41}$$

$$M_y(x=0)=-\frac{q_0L^2}{2}\ ,\quad M_y(x=L)=0\,. \tag{4.42}$$

The application of the boundary condition $(4.41)_1$ in the general analytical solution for the deflection according to Eq. (4.39) immediately yields $c_4=0$. With the second boundary condition in Eq. (4.41) and the general analytical solution for the rotation according to Eq. (4.40), the third constant of integration is obtained as $c_3=-c_1\frac{EI_y}{k_sAG}$.

The further determination of the constants of integration demands that the bending moment is expressed with the help of the deformation. Application of Eq. (4.25), i.e. $M_y=EI_y\frac{\mathrm{d}\phi_y}{\mathrm{d}x}$, gives the moment distribution as

$$M_y(x)=EI_y\frac{\mathrm{d}\phi_y}{\mathrm{d}x}=-\left(\frac{3q_0x^2}{6}+c_1x+c_2\right)-\frac{q_0EI_y}{k_sAG}, \tag{4.43}$$

and the consideration of boundary conditions $(4.42)_1$ yields $c_2=\frac{q_0L^2}{2}-\frac{q_0EI_y}{k_sAG}$. In a similar way, consideration of the second boundary condition in Eq. (4.42) yields the first constant of integration to $c_1=-q_0L$ and finally $c_3=\frac{q_0LEI_y}{k_sAG}$.

Therefore, the bending line results in

$$u_z(x)=\frac{1}{EI_y}\left(\frac{q_0x^4}{24}-q_0L\frac{x^3}{6}+\left[\frac{q_0L^2}{2}-\frac{q_0EI_y}{k_sAG}\right]\frac{x^2}{2}+\frac{q_0LEI_y}{k_sAG}x\right), \tag{4.44}$$

or alternatively the maximal deflection on the right-hand boundary of the beam, meaning for $x=L$, to:

$$u_z(x=L)=\frac{q_0L^4}{8EI_y}+\frac{q_0L^2}{2k_sAG}. \tag{4.45}$$

Through comparison with the analytical solutions in Sect. 3.5 it becomes obvious that the analytical solutions for the maximal deflection compose additively from the classical solution for the BERNOULLI beam and an additional shear part.

To highlight the influence of the shear contribution, the maximum deflection is presented in the following as a function of the fraction between beam height and beam length. As an example three different loading and boundary conditions for a rectangular cross section with the width b and the height h are presented in Fig. 4.8. It becomes obvious that the difference between the BERNOULLI and the TIMOSHENKO beam becomes smaller and smaller for a decreasing slenderness ratio, meaning for beams at which the length L is significantly larger compared to the height h.

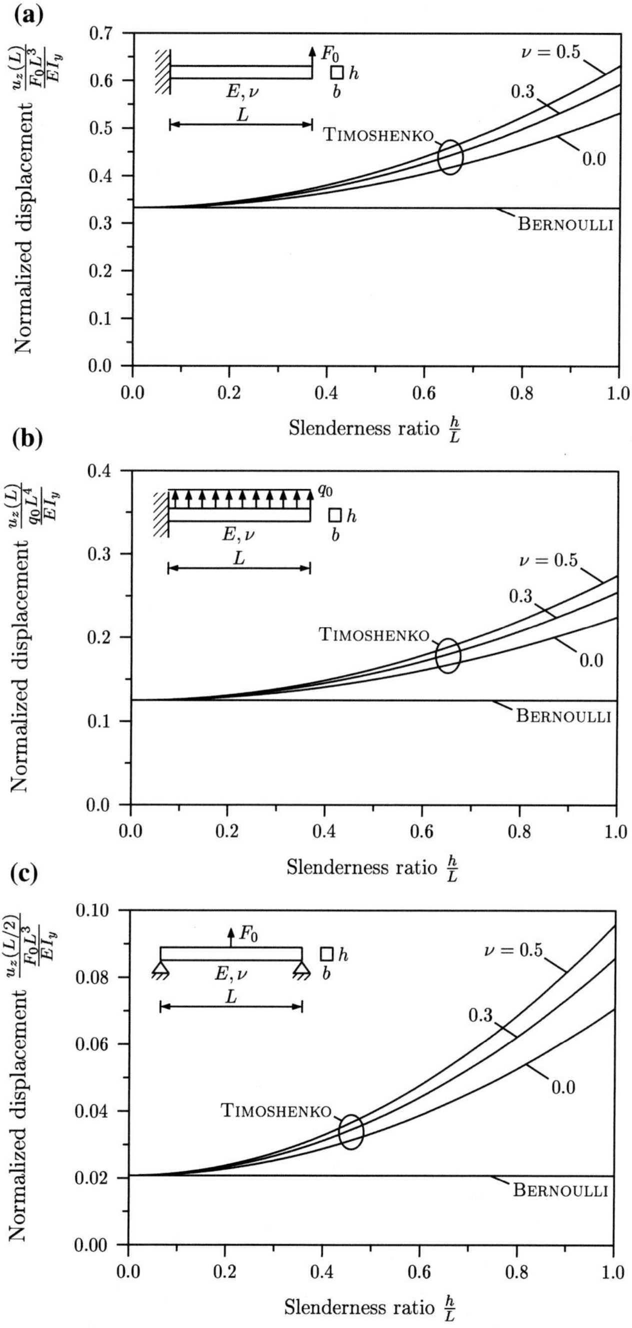

Fig. 4.8 Comparison of the analytical solutions for the BERNOULLI and TIMOSHENKO beam for different loading and boundary conditions: **a** Cantilever beam with end load; **b** cantilever beam with distributed load; **c** simply supported beam with point load

Table 4.5 Different formulations of the basic equations for a TIMOSHENKO beam (bending in the x-z plane; x-axis along the principal beam axis). E: YOUNG's modulus; G: shear modulus; A: cross-sectional area; I_y: second moment of area; k_s: shear correction factor; q_z: length-specific distributed force; m_z: length-specific distributed moment; $\boldsymbol{e}$: generalized strains; $\boldsymbol{s}$: generalized stresses

Specific Formulation	General Formulation
Kinematics	
$\begin{bmatrix} \frac{du_z}{dx} + \phi_y \\ \frac{d\phi_y}{dx} \end{bmatrix} = \begin{bmatrix} \frac{d}{dx} & 1 \\ 0 & \frac{d}{dx} \end{bmatrix} \begin{bmatrix} u_z \\ \phi_y \end{bmatrix}$	$\boldsymbol{e} = \boldsymbol{\mathcal{L}}_1 \boldsymbol{u}$
Constitution	
$\begin{bmatrix} -Q_z \\ M_y \end{bmatrix} = \begin{bmatrix} -k_s AG & 0 \\ 0 & EI_y \end{bmatrix} \begin{bmatrix} \frac{du_z}{dx} + \phi_y \\ \frac{d\phi_y}{dx} \end{bmatrix}$	$\boldsymbol{s} = \boldsymbol{D}\boldsymbol{e}$
Equilibrium	
$\begin{bmatrix} \frac{d}{dx} & 0 \\ 1 & \frac{d}{dx} \end{bmatrix} \begin{bmatrix} -Q_z \\ M_y \end{bmatrix} + \begin{bmatrix} -q_z \\ +m_z \end{bmatrix} = \begin{bmatrix} 0 \\ 0 \end{bmatrix}$	$\boldsymbol{\mathcal{L}}_1^{\mathrm{T}} \boldsymbol{s} + \boldsymbol{b} = \boldsymbol{0}$
PDE	
$-\frac{d}{dx}\left[k_s GA\left(\frac{du_z}{dx} + \phi_y\right)\right] - q_z = 0$	
$\frac{d}{dx}\left(EI_y \frac{d\phi_y}{dx}\right) - k_s GA\left(\frac{du_z}{dx} + \phi_y\right) + m_y = 0\,,$	$\boldsymbol{\mathcal{L}}_1^{\mathrm{T}} \boldsymbol{D} \boldsymbol{\mathcal{L}}_1 \boldsymbol{u} + \boldsymbol{b} = \boldsymbol{0}$

The relative difference between the BERNOULLI and the TIMOSHENKO solutions, for example for a POISSON's ratio of 0.3 and a slenderness ratio of 0.1—meaning for a beam, for which the length is ten times larger than the height—depending on the loading and boundary conditions is: 0.77% for the cantilever beam with point load, 1.03% for the cantilever beam with distributed load and 3.03% for the simply supported beam. Further analytical solutions for the TIMOSHENKO beam can be obtained, for example, from [13, 26].

If we replace in the previous formulations the first order derivative, i.e. $\frac{d\ldots}{dx}$, by a formal operator symbol, i.e. the $\boldsymbol{\mathcal{L}}_1$–matrix, then the basic equations of the TIMOSHENKO beam can be stated in a more formal way as given in Table 4.5.

It should be noted here that the formulation of the partial differential equation in Table 4.5 is slightly different to Eqs. (4.33) and (4.34): First of all, we considered here a distributed moment m_z. Furthermore, the first PDE in Table 4.5 is multiplied by -1 compared to Eq. (4.34). This must be carefully considered as soon as finite element derivations based on both approaches are compared.

Similar to the end of Sect. 3.2.2, it is more advantageous in a general approach to work with the generalized stresses $\boldsymbol{s} = \left[-Q_z,\, M_y\right]^{\mathrm{T}}$ and generalized strains $\boldsymbol{e} = \left[\frac{du_z}{dx} + \phi_y,\, \frac{d\phi_y}{dx}\right]^{\mathrm{T}} = \left[\gamma_{xz},\, \kappa_y\right]^{\mathrm{T}}$ since these quantities do not depend on the vertical coordinate z. Classical stress and strain values are changing along the vertical

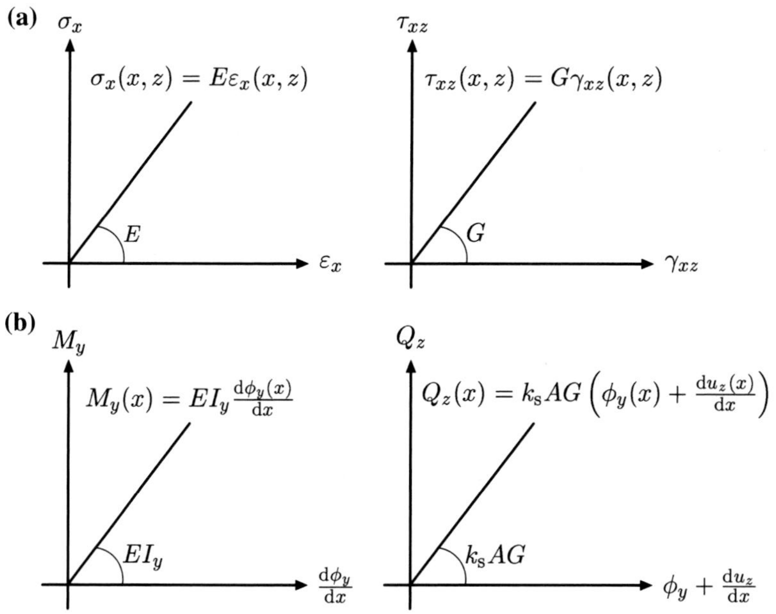

Fig. 4.9 Formulation of the constitutive law based on **a** classical stress-strain and **b** generalized-stress-generalized-strain relations

Table 4.6 Elementary basic equations for bending of a TIMOSHENKO beam in the x-y plane (x-axis: right facing; y-axis: upward facing)

Relation	Equation
Kinematics	$\varepsilon_x(x,y) = -y\frac{d\phi_z(x)}{dx}$ and $\phi_z(x) = \frac{du_z(x)}{dx} - \gamma_{xz}(x)$
Equilibrium	$\frac{dQ_y(x)}{dx} = -q_y(x)\,;\ \frac{dM_z(x)}{dx} = -Q_y(x)$
Constitution	$\sigma_x(x,y) = E\varepsilon_x(x,y)$ and $\tau_{xy}(x) = G\gamma_{xy}(x)$
Diff. Equation	$\frac{d}{dx}\left(EI_z\frac{d\phi_z}{dx}\right) + k_s AG\left(\frac{du_y}{dx} - \phi_z\right) = -m_z(x)$ $\frac{d}{dx}\left[k_s AG\left(\frac{du_y}{dx} - \phi_z\right)\right] = -q_y(x)$

coordinate. The representation of the constitutive relationship based on the classical stress-strain quantities and the corresponding generalized quantities is represented in Fig. 4.9.

Let us mention at the end of this section that for bending in the x-y plane slightly modified equations occur compared to Tables 4.3 and 4.4. The corresponding equations for bending in the x-y plane with shear contribution are summarized in Table 4.6.

4.3 Finite Element Solution

4.3.1 Derivation of the Principal Finite Element Equation

The TIMOSHENKO element is here defined as a prismatic body with the center line x and the z-axis orthogonally to the center line. Nodes, at which displacements and rotations or alternatively forces and moments, as drafted in Fig. 4.10, are defined, will be introduced at both ends of the beam element. The deformation and load parameters are drafted in their positive orientation.

The two unknowns, meaning the deflection $u_z(x)$ and the herefrom independent cross-sectional rotation $\phi_y(x)$ are approximated with the help of the following nodal approaches:

$$u_z^{\mathrm{e}}(x) = N_{1u}(x)u_{1z} + N_{2u}(x)u_{2z}\,, \tag{4.46}$$

$$\phi_y^{\mathrm{e}}(x) = N_{1\phi}(x)\phi_{1y} + N_{2\phi}(x)\phi_{2y}\,, \tag{4.47}$$

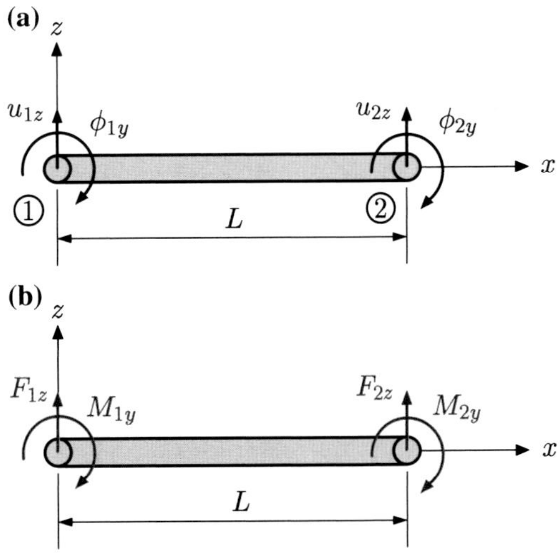

Fig. 4.10 Definition of the TIMOSHENKO beam element for deformation in the x-z plane: **a** deformations; **b** external loads. The nodes are symbolized by two circles at the end (◯)

or alternatively in matrix notation as

$$u_z^{\mathrm{e}}(x) = \begin{bmatrix} N_{1u}(x) \; 0 \; N_{2u}(x) \; 0 \end{bmatrix} \begin{bmatrix} u_{1z} \\ \phi_{1y} \\ u_{2z} \\ \phi_{2y} \end{bmatrix} = \boldsymbol{N}_u^{\mathrm{T}} \boldsymbol{u}_{\mathrm{p}} \,, \tag{4.48}$$

$$\phi_y^{\mathrm{e}}(x) = \begin{bmatrix} 0 \; N_{1\phi} \; 0 \; N_{2\phi}(x) \end{bmatrix} \begin{bmatrix} u_{1z} \\ \phi_{1y} \\ u_{2z} \\ \phi_{2y} \end{bmatrix} = \boldsymbol{N}_\phi^{\mathrm{T}} \boldsymbol{u}_{\mathrm{p}} \,. \tag{4.49}$$

With these relations, the derivative of the cross-sectional rotation in the coupled differential Eqs. (4.33) and (4.34) results in

$$\frac{\mathrm{d}\phi_y^{\mathrm{e}}(x)}{\mathrm{d}x} = \frac{\mathrm{d}N_{1\phi}(x)}{\mathrm{d}x}\phi_{1y} + \frac{\mathrm{d}N_{2\phi}(x)}{\mathrm{d}x}\phi_{2y} = \frac{\mathrm{d}\boldsymbol{N}_\phi^{\mathrm{T}}}{\mathrm{d}x}\boldsymbol{u}_{\mathrm{p}} \,. \tag{4.50}$$

In the following, the shear differential Eq. (4.34) is first considered, which is multiplied by a deflection weight function $W_u(x)$ to obtain the following inner product:

$$\int_0^L W_u^{\mathrm{T}}(x) \left\{ k_{\mathrm{s}} AG \left(\frac{\mathrm{d}^2 u_z}{\mathrm{d}x^2} + \frac{\mathrm{d}\phi_y}{\mathrm{d}x} \right) + q_z(x) \right\} \mathrm{d}x \overset{!}{=} 0 \,. \tag{4.51}$$

Partial integrating of both expressions in the parentheses yields:

$$\int_0^L W_u^{\mathrm{T}} k_{\mathrm{s}} AG \frac{\mathrm{d}^2 u_z}{\mathrm{d}x^2} \mathrm{d}x = \left[W_u^{\mathrm{T}} k_{\mathrm{s}} AG \frac{\mathrm{d}u_z}{\mathrm{d}x} \right]_0^L - \int_0^L \frac{\mathrm{d}W_u^{\mathrm{T}}}{\mathrm{d}x} k_{\mathrm{s}} AG \frac{\mathrm{d}u_z}{\mathrm{d}x} \mathrm{d}x \,, \tag{4.52}$$

$$\int_0^L W_u^{\mathrm{T}} k_{\mathrm{s}} AG \frac{\mathrm{d}\phi_y}{\mathrm{d}x} \mathrm{d}x = \left[W_u^{\mathrm{T}} k_{\mathrm{s}} AG \phi_y \right]_0^L - \int_0^L \frac{\mathrm{d}W_u^{\mathrm{T}}}{\mathrm{d}x} k_{\mathrm{s}} AG \phi_y \, \mathrm{d}x \,. \tag{4.53}$$

Next, the bending differential Eq. (4.33) is multiplied by a rotation weight function $W_\phi(x)$ and is transformed in the inner product:

$$\int_0^L W_\phi^{\mathrm{T}}(x) \left\{ \frac{\mathrm{d}}{\mathrm{d}x} \left(EI_y \frac{\mathrm{d}\phi_y}{\mathrm{d}x} \right) - k_{\mathrm{s}} AG \left(\frac{\mathrm{d}u_z}{\mathrm{d}x} + \phi_y \right) \right\} \mathrm{d}x \overset{!}{=} 0 \,. \tag{4.54}$$

Partial integrating of the first expression[5] yields

$$\int_0^L W_\phi^{\mathrm{T}} EI_y \frac{\mathrm{d}^2\phi_y}{\mathrm{d}x^2}\,\mathrm{d}x = \left[W_\phi^{\mathrm{T}} EI_y \frac{\mathrm{d}\phi_y}{\mathrm{d}x}\right]_0^L - \int_0^L \frac{\mathrm{d}W_\phi^{\mathrm{T}}}{\mathrm{d}x} EI_y \frac{\mathrm{d}\phi_y}{\mathrm{d}x}\,\mathrm{d}x \tag{4.55}$$

and the bending differential equations results in:

$$\left[W_\phi^{\mathrm{T}} EI_y \frac{\mathrm{d}\phi_y}{\mathrm{d}x}\right]_0^L - \int_0^L \frac{\mathrm{d}W_\phi^{\mathrm{T}}}{\mathrm{d}x} EI_y \frac{\mathrm{d}\phi_y}{\mathrm{d}x}\,\mathrm{d}x - \int_0^L W_\phi^{\mathrm{T}}(x)\, k_{\mathrm{s}}AG\left(\frac{\mathrm{d}u_z}{\mathrm{d}x} + \phi_y\right)\mathrm{d}x = 0\,. \tag{4.56}$$

Adding of the two converted differential equations yields

$$\begin{aligned}
&\left[W_u^{\mathrm{T}} k_{\mathrm{s}}AG \frac{\mathrm{d}u_z}{\mathrm{d}x}\right]_0^L - \int_0^L \frac{\mathrm{d}W_u^{\mathrm{T}}}{\mathrm{d}x} k_{\mathrm{s}}AG \frac{\mathrm{d}u_z}{\mathrm{d}x}\,\mathrm{d}x + \left[W_u^{\mathrm{T}} k_{\mathrm{s}}AG\phi_y\right]_0^L \\
&- \int_0^L \frac{\mathrm{d}W_u^{\mathrm{T}}}{\mathrm{d}x} k_{\mathrm{s}}AG\phi_y\,\mathrm{d}x + \int_0^L W_u^{\mathrm{T}} q_z \mathrm{d}x - \int_0^L W_\phi^{\mathrm{T}}(x) k_{\mathrm{s}}AG\left(\frac{\mathrm{d}u_z}{\mathrm{d}x} + \phi_y\right)\mathrm{d}x \\
&- \int_0^L \frac{\mathrm{d}W_\phi^{\mathrm{T}}}{\mathrm{d}x} EI_y \frac{\mathrm{d}\phi_y}{\mathrm{d}x}\,\mathrm{d}x + \left[W_\phi^{\mathrm{T}} EI_y \frac{\mathrm{d}\phi_y}{\mathrm{d}x}\right]_0^L = 0\,,
\end{aligned} \tag{4.57}$$

or alternatively after a short conversion the weak form of the shear flexible bending beam:

$$\begin{aligned}
&\int_0^L \frac{\mathrm{d}W_\phi^{\mathrm{T}}}{\mathrm{d}x} EI_y \frac{\mathrm{d}\phi_y}{\mathrm{d}x}\,\mathrm{d}x + \int_0^L \underbrace{\left(\frac{\mathrm{d}W_u^{\mathrm{T}}}{\mathrm{d}x} + W_\phi^{\mathrm{T}}\right)}_{\delta\gamma_{xz}} k_{\mathrm{s}}AG \underbrace{\left(\frac{\mathrm{d}u_z}{\mathrm{d}x} + \phi_y\right)}_{\gamma_{xz}}\mathrm{d}x \\
&= \int_0^L W_u^{\mathrm{T}} q_z \mathrm{d}x + \left[W_u^{\mathrm{T}} k_{\mathrm{s}}AG\left(\frac{\mathrm{d}u_z}{\mathrm{d}x} + \phi_y\right)\right]_0^L + \left[W_\phi^{\mathrm{T}} EI_y \frac{\mathrm{d}\phi_y}{\mathrm{d}x}\right]_0^L.
\end{aligned} \tag{4.58}$$

[5]The second expression is proportional to γ_{xz} and does not need any integration by parts since no derivative is applied to the angle.

One can see that the first part of the left-hand half represents the bending part and the second half the shear part. The right-hand side results from the external loads of the beam. In the following, the left-hand half of the weak form will be first considered to derive the stiffness matrix:

$$\int_0^L \frac{\mathrm{d}W_\phi^{\mathrm{T}}}{\mathrm{d}x} EI_y \frac{\mathrm{d}\phi_y}{\mathrm{d}x} \mathrm{d}x + \int_0^L \left(\frac{\mathrm{d}W_u^{\mathrm{T}}}{\mathrm{d}x} + W_\phi^{\mathrm{T}}\right) k_{\mathrm{s}} AG \left(\frac{\mathrm{d}u_z}{\mathrm{d}x} + \phi_y\right) \mathrm{d}x\,. \tag{4.59}$$

In the next step, the approaches for the deflection and rotation of the nodes or alternatively their derivatives according to Eqs. (4.48) and (4.49), meaning

$$u_z^{\mathrm{e}}(x) = \boldsymbol{N}_u^{\mathrm{T}}(x)\boldsymbol{u}_{\mathrm{p}}\,, \quad \frac{\mathrm{d}u_z(x)}{\mathrm{d}x} = \frac{\mathrm{d}\boldsymbol{N}_u^{\mathrm{T}}(x)}{\mathrm{d}x}\boldsymbol{u}_{\mathrm{p}}\,, \tag{4.60}$$

$$\phi_y^{\mathrm{e}}(x) = \boldsymbol{N}_\phi^{\mathrm{T}}(x)\boldsymbol{u}_{\mathrm{p}}\,, \quad \frac{\mathrm{d}\phi_y(x)}{\mathrm{d}x} = \frac{\mathrm{d}\boldsymbol{N}_\phi^{\mathrm{T}}(x)}{\mathrm{d}x}\boldsymbol{u}_{\mathrm{p}}\,, \tag{4.61}$$

have to be considered. The approaches for the weight functions are chosen analogous to the approaches for the unknowns:

$$W_u(x) = \boldsymbol{N}_u^{\mathrm{T}}(x)\delta\boldsymbol{u}_{\mathrm{p}}\,, \tag{4.62}$$

$$W_\phi(x) = \boldsymbol{N}_\phi^{\mathrm{T}}(x)\delta\boldsymbol{u}_{\mathrm{p}}\,, \tag{4.63}$$

and accordingly for the transposed expressions

$$W(x)^{\mathrm{T}} = \left(N^{\mathrm{T}}(x)\,\delta\boldsymbol{u}_{\mathrm{p}}\right)^{\mathrm{T}} = \delta\boldsymbol{u}_{\mathrm{p}}^{\mathrm{T}} N(x)\,, \tag{4.64}$$

or alternatively for the derivatives:

$$\frac{W_u(x)}{\mathrm{d}x} = \frac{\boldsymbol{N}_u^{\mathrm{T}}(x)}{\mathrm{d}x}\delta\boldsymbol{u}_{\mathrm{p}}\,, \tag{4.65}$$

$$\frac{W_\phi(x)}{\mathrm{d}x} = \frac{\boldsymbol{N}_\phi^{\mathrm{T}}(x)}{\mathrm{d}x}\delta\boldsymbol{u}_{\mathrm{p}}\,. \tag{4.66}$$

Therefore the left-hand side of Eq. (4.58)—under consideration that the nodal deformations ($\boldsymbol{u}_{\mathrm{p}}$) or alternatively the virtual deformations ($\delta\boldsymbol{u}_{\mathrm{p}}^{\mathrm{T}}$) can be considered as constant with respect to the integration over x—results in:

$$\delta \boldsymbol{u}_{\mathrm{p}}^{\mathrm{T}} \int_{0}^{L} EI_y \frac{\mathrm{d}\boldsymbol{N}_{\phi}}{\mathrm{d}x}\frac{\mathrm{d}\boldsymbol{N}_{\phi}^{\mathrm{T}}}{\mathrm{d}x}\mathrm{d}x\, \boldsymbol{u}_{\mathrm{p}}+$$

$$+\,\delta \boldsymbol{u}_{\mathrm{p}}^{\mathrm{T}} \int_{0}^{L} k_{\mathrm{s}}AG\left(\frac{\mathrm{d}\boldsymbol{N}_{u}}{\mathrm{d}x}+\boldsymbol{N}_{\phi}\right)\left(\frac{\mathrm{d}\boldsymbol{N}_{u}^{\mathrm{T}}}{\mathrm{d}x}+\boldsymbol{N}_{\phi}^{\mathrm{T}}\right)\mathrm{d}x\, \boldsymbol{u}_{\mathrm{p}}\,. \tag{4.67}$$

In the following, it remains to be seen that the virtual deformations $\delta\boldsymbol{u}^{\mathrm{T}}$ can be 'canceled' with a corresponding expression on the right-hand side of Eq. (4.58). Therefore, on the left-hand side there remains

$$\underbrace{\int_{0}^{L} EI_y \frac{\mathrm{d}\boldsymbol{N}_{\phi}}{\mathrm{d}x}\frac{\mathrm{d}\boldsymbol{N}_{\phi}^{\mathrm{T}}}{\mathrm{d}x}\mathrm{d}x\, \boldsymbol{u}_{\mathrm{p}}}_{\boldsymbol{k}_{\mathrm{b}}^{\mathrm{e}}}+\underbrace{\int_{0}^{L} k_{\mathrm{s}}AG\left(\frac{\mathrm{d}\boldsymbol{N}_{u}}{\mathrm{d}x}+\boldsymbol{N}_{\phi}\right)\left(\frac{\mathrm{d}\boldsymbol{N}_{u}^{\mathrm{T}}}{\mathrm{d}x}+\boldsymbol{N}_{\phi}^{\mathrm{T}}\right)\mathrm{d}x\, \boldsymbol{u}_{\mathrm{p}}}_{\boldsymbol{k}_{\mathrm{s}}^{\mathrm{e}}}\,, \tag{4.68}$$

and the bending or alternatively the shear stiffness matrix can be identified. The element stiffness matrices for constant bending stiffness EI_y and constant shear stiffness GA results herefrom in components to:

$$\boldsymbol{K}_{\mathrm{b}}^{\mathrm{e}} = EI_y \int_{0}^{L} \begin{bmatrix} 0 & 0 & 0 & 0 \\ 0 & \frac{\mathrm{d}N_{1\phi}}{\mathrm{d}x}\frac{\mathrm{d}N_{1\phi}}{\mathrm{d}x} & 0 & \frac{\mathrm{d}N_{1\phi}}{\mathrm{d}x}\frac{\mathrm{d}N_{2\phi}}{\mathrm{d}x} \\ 0 & 0 & 0 & 0 \\ 0 & \frac{\mathrm{d}N_{2\phi}}{\mathrm{d}x}\frac{\mathrm{d}N_{1\phi}}{\mathrm{d}x} & 0 & \frac{\mathrm{d}N_{2\phi}}{\mathrm{d}x}\frac{\mathrm{d}N_{2\phi}}{\mathrm{d}x} \end{bmatrix} \mathrm{d}x\,, \tag{4.69}$$

$$\boldsymbol{K}_{\mathrm{s}}^{\mathrm{e}} = k_{\mathrm{s}}GA \int_{0}^{L} \begin{bmatrix} \frac{\mathrm{d}N_{1u}}{\mathrm{d}x}\frac{\mathrm{d}N_{1u}}{\mathrm{d}x} & \frac{\mathrm{d}N_{1u}}{\mathrm{d}x}N_{1\phi} & \frac{\mathrm{d}N_{1u}}{\mathrm{d}x}\frac{\mathrm{d}N_{2u}}{\mathrm{d}x} & \frac{\mathrm{d}N_{1u}}{\mathrm{d}x}N_{2\phi} \\ N_{1\phi}\frac{\mathrm{d}N_{1u}}{\mathrm{d}x} & N_{1\phi}N_{1\phi} & N_{1\phi}\frac{\mathrm{d}N_{2u}}{\mathrm{d}x} & N_{1\phi}N_{2\phi} \\ \frac{\mathrm{d}N_{2u}}{\mathrm{d}x}\frac{\mathrm{d}N_{1u}}{\mathrm{d}x} & \frac{\mathrm{d}N_{2u}}{\mathrm{d}x}N_{1\phi} & \frac{\mathrm{d}N_{2u}}{\mathrm{d}x}\frac{\mathrm{d}N_{2u}}{\mathrm{d}x} & \frac{\mathrm{d}N_{2u}}{\mathrm{d}x}N_{2\phi} \\ N_{2\phi}\frac{\mathrm{d}N_{1u}}{\mathrm{d}x} & N_{2\phi}N_{1\phi} & N_{2\phi}\frac{\mathrm{d}N_{2u}}{\mathrm{d}x} & N_{2\phi}N_{2\phi} \end{bmatrix} \mathrm{d}x\,. \tag{4.70}$$

The two expressions for the bending and shear parts of the element stiffness matrix according to Eqs. (4.69) and (4.70) can be superposed for the principal finite element equation of the TIMOSHENKO beam on the element level

$$\boldsymbol{K}^{\mathrm{e}}\boldsymbol{u}_{\mathrm{p}}^{\mathrm{e}} = \boldsymbol{f}^{\mathrm{e}}\,, \tag{4.71}$$

whereupon the total stiffness matrix according to Eq. (4.72) is given.

$$
\underbrace{\left[\begin{array}{c|c|c|c}
k_sGA\int\limits_0^L \frac{dN_{1u}}{dx}\frac{dN_{1u}}{dx}\,dx & k_sGA\int\limits_0^L \frac{dN_{1u}}{dx}N_{1\phi}\,dx & k_sGA\int\limits_0^L \frac{dN_{1u}}{dx}\frac{dN_{2u}}{dx}\,dx & k_sGA\int\limits_0^L \frac{dN_{1u}}{dx}N_{2\phi}\,dx \\
k_sGA\int\limits_0^L N_{1\phi}\frac{dN_{1u}}{dx}\,dx & k_sGA\int\limits_0^L N_{1\phi}N_{1\phi}\,dx + EI_y\int\limits_0^L \frac{dN_{1\phi}}{dx}\frac{dN_{1\phi}}{dx}\,dx & k_sGA\int\limits_0^L N_{1\phi}\frac{dN_{2u}}{dx}\,dx & k_sGA\int\limits_0^L N_{1\phi}N_{2\phi}\,dx + EI_y\int\limits_0^L \frac{dN_{2\phi}}{dx}\frac{dN_{1\phi}}{dx}\,dx \\
k_sGA\int\limits_0^L \frac{dN_{2u}}{dx}\frac{dN_{1u}}{dx}\,dx & k_sGA\int\limits_0^L \frac{dN_{2u}}{dx}N_{1\phi}\,dx & k_sGA\int\limits_0^L \frac{dN_{2u}}{dx}\frac{dN_{2u}}{dx}\,dx & k_sGA\int\limits_0^L \frac{dN_{2u}}{dx}N_{2\phi}\,dx \\
k_sGA\int\limits_0^L N_{2\phi}\frac{dN_{1u}}{dx}\,dx & k_sGA\int\limits_0^L N_{2\phi}N_{1\phi}\,dx + EI_y\int\limits_0^L \frac{dN_{1\phi}}{dx}\frac{dN_{2\phi}}{dx}\,dx & k_sGA\int\limits_0^L N_{2\phi}\frac{dN_{2u}}{dx}\,dx & k_sGA\int\limits_0^L N_{2\phi}N_{2\phi}\,dx + EI_y\int\limits_0^L \frac{dN_{2\phi}}{dx}\frac{dN_{2\phi}}{dx}\,dx
\end{array}\right]}_{\boldsymbol{K}^e}
\tag{4.72}
$$

Any further evaluation of these stiffness matrices requires the introduction of the interpolation functions N_i.

Finally, the right-hand side of the weak form according to Eq. (4.58) is considered:

$$\int_0^L W_u^{\mathrm{T}} q_z \mathrm{d}x + \left[W_u^{\mathrm{T}} k_{\mathrm{s}} AG \left(\frac{\mathrm{d}u_z}{\mathrm{d}x} + \phi_y \right) \right]_0^L + \left[W_\phi^{\mathrm{T}} EI_y \frac{\mathrm{d}\phi_y}{\mathrm{d}x} \right]_0^L . \tag{4.73}$$

Consideration of the relations for the shear force and the internal moment according to Eqs. (4.30) and (4.25) in the right-hand side of the weak form yields

$$\int_0^L W_u^{\mathrm{T}} q_z \mathrm{d}x + \left[W_u(x)^{\mathrm{T}} Q_z(x) \right]_0^L + \left[W_\phi(x)^{\mathrm{T}} M_y(x) \right]_0^L , \tag{4.74}$$

or alternatively after the introduction of the approaches for the weight functions for the displacements and rotations according to Eqs. (4.62) and (4.63):

$$\delta \boldsymbol{u}_{\mathrm{p}}^{\mathrm{T}} \int_0^L q_z \boldsymbol{N}_u \mathrm{d}x + \delta \boldsymbol{u}_{\mathrm{p}}^{\mathrm{T}} \left[Q_z(x) \boldsymbol{N}_u(x) \right]_0^L + \delta \boldsymbol{u}_{\mathrm{p}}^{\mathrm{T}} \left[M_y(x) \boldsymbol{N}_\phi(x) \right]_0^L . \tag{4.75}$$

$\delta \boldsymbol{u}_{\mathrm{p}}^{\mathrm{T}}$ can be eliminated with a corresponding expression in Eq. (4.67) and the following remains

$$\int_0^L q_z \boldsymbol{N}_u \mathrm{d}x + \left[Q_z(x) \boldsymbol{N}_u(x) \right]_0^L + \left[M_y(x) \boldsymbol{N}_\phi(x) \right]_0^L , \tag{4.76}$$

or alternatively in components:

$$\int_0^L q_z(x) \begin{bmatrix} N_{1u} \\ 0 \\ N_{2u} \\ 0 \end{bmatrix} \mathrm{d}x + \begin{bmatrix} -Q_z(0) \\ 0 \\ +Q_z(L) \\ 0 \end{bmatrix} + \begin{bmatrix} 0 \\ -M_y(0) \\ 0 \\ +M_y(L) \end{bmatrix} . \tag{4.77}$$

One notes that the general characteristics of the interpolation functions have been used during the evaluation of the boundary integrals:

$$1^{\mathrm{st}}\text{row:} \quad Q_z(L) \underbrace{N_{1u}(L)}_{0} - Q_z(0) \underbrace{N_{1u}(0)}_{1} , \tag{4.78}$$

$$2^{\mathrm{nd}}\text{row:} \quad M_y(L) \underbrace{N_{1\phi}(L)}_{0} - M_y(0) \underbrace{N_{1\phi}(0)}_{1} , \tag{4.79}$$

3rd row: $$Q_z(L)\underbrace{N_{2u}(L)}_{1} - Q_z(0)\underbrace{N_{2u}(0)}_{0}, \tag{4.80}$$

4th row: $$M_y(L)\underbrace{N_{2\phi}(L)}_{1} - M_y(0)\underbrace{N_{2\phi}(0)}_{0}. \tag{4.81}$$

The following description is related to a more formalized derivation of the principal finite element equation (see Sect. 3.3.1 for a similar approach in the case of the EULER–BERNOULLI beam). Based on the general formulation of the partial differential equation given in Table 4.5, the strong formulation can be written as:

$$\boldsymbol{\mathcal{L}}_1^{\mathrm{T}} \boldsymbol{D} \boldsymbol{\mathcal{L}}_1 \boldsymbol{u}^0 + \boldsymbol{b} = \boldsymbol{0}. \tag{4.82}$$

Replacing the exact solution $\boldsymbol{u}^0$ by an approximate solution $\boldsymbol{u}$, a residual $\boldsymbol{r}$ is obtained:

$$\boldsymbol{r} = \boldsymbol{\mathcal{L}}_1^{\mathrm{T}} \boldsymbol{D} \boldsymbol{\mathcal{L}}_1 \boldsymbol{u} + \boldsymbol{b} \neq \boldsymbol{0}. \tag{4.83}$$

The inner product is obtained by weighting the residual and integration as

$$\int_L \boldsymbol{W}^{\mathrm{T}} \left(\boldsymbol{\mathcal{L}}_1^{\mathrm{T}} \boldsymbol{D} \boldsymbol{\mathcal{L}}_1 \boldsymbol{u} + \boldsymbol{b}\right) \mathrm{d}L = 0, \tag{4.84}$$

where $\boldsymbol{W}(x) = \begin{bmatrix} W_\phi & W_u \end{bmatrix}^{\mathrm{T}}$ is the column matrix of weight functions. Application of the GREEN–GAUSS theorem (cf. Sect. A.7) would shift the derivative to the weight functions $\boldsymbol{W}$. However, the matrix of differential operators, $\boldsymbol{\mathcal{L}}_1^{\mathrm{T}}$, contains in addition to derivatives a constant value '1' and it is therefore appropriate to split the matrix into a part which contains all the derivatives, $\boldsymbol{\mathcal{L}}_{1,a}^{\mathrm{T}}$, and a part with the constant value, $\boldsymbol{\mathcal{L}}_{1,b}^{\mathrm{T}}$:

$$\underbrace{\begin{bmatrix} \frac{\mathrm{d}}{\mathrm{d}x} & 0 \\ 1 & \frac{\mathrm{d}}{\mathrm{d}x} \end{bmatrix}}_{\boldsymbol{\mathcal{L}}_1^{\mathrm{T}}} = \underbrace{\begin{bmatrix} \frac{\mathrm{d}}{\mathrm{d}x} & 0 \\ 0 & \frac{\mathrm{d}}{\mathrm{d}x} \end{bmatrix}}_{\boldsymbol{\mathcal{L}}_{1,a}^{\mathrm{T}}} + \underbrace{\begin{bmatrix} 0 & 0 \\ 1 & 0 \end{bmatrix}}_{\boldsymbol{\mathcal{L}}_{1,b}^{\mathrm{T}}}. \tag{4.85}$$

Thus, we can write the inner product as:

$$\int_L \boldsymbol{W}^{\mathrm{T}} \left[\left(\boldsymbol{\mathcal{L}}_{1,a}^{\mathrm{T}} + \boldsymbol{\mathcal{L}}_{1,b}^{\mathrm{T}}\right) \boldsymbol{D} \boldsymbol{\mathcal{L}}_1 \boldsymbol{u} + \boldsymbol{b}\right] \mathrm{d}L = 0, \tag{4.86}$$

or

$$\int_L \boldsymbol{W}^{\mathrm{T}} \boldsymbol{\mathcal{L}}_{1,a}^{\mathrm{T}} \boldsymbol{D} \boldsymbol{\mathcal{L}}_1 \boldsymbol{u}\, \mathrm{d}L + \int_L \boldsymbol{W}^{\mathrm{T}} \boldsymbol{\mathcal{L}}_{1,b}^{\mathrm{T}} \boldsymbol{D} \boldsymbol{\mathcal{L}}_1 \boldsymbol{u}\, \mathrm{d}L + \int_L \boldsymbol{W}^{\mathrm{T}} \boldsymbol{b}\, \mathrm{d}L = 0. \tag{4.87}$$

Application of the GREEN–GAUSS theorem to the first integral[6] gives:

$$\int_L \boldsymbol{W}^{\mathrm{T}} \boldsymbol{\mathcal{L}}_{1,a}^{\mathrm{T}} \boldsymbol{D} \boldsymbol{\mathcal{L}}_1 \boldsymbol{u} \, \mathrm{d}L = -\int_L \left(\boldsymbol{\mathcal{L}}_{1,a} \boldsymbol{W}\right)^{\mathrm{T}} \boldsymbol{D} \left(\boldsymbol{\mathcal{L}}_1 \boldsymbol{u}\right) \mathrm{d}L + \int_s \boldsymbol{W}^{\mathrm{T}} \underbrace{(\boldsymbol{D} \boldsymbol{\mathcal{L}}_1 \boldsymbol{u})^{\mathrm{T}}}_{\boldsymbol{s}^{\mathrm{T}}} \boldsymbol{n} \mathrm{d}s \,, \tag{4.88}$$

and the weak formulation can be obtained as:

$$\int_L \left(\boldsymbol{\mathcal{L}}_{1,a} \boldsymbol{W}\right)^{\mathrm{T}} \boldsymbol{D} \left(\boldsymbol{\mathcal{L}}_1 \boldsymbol{u}\right) \mathrm{d}L - \int_L \boldsymbol{W}^{\mathrm{T}} \boldsymbol{\mathcal{L}}_{1,b}^{\mathrm{T}} \boldsymbol{D} \boldsymbol{\mathcal{L}}_1 \boldsymbol{u} \, \mathrm{d}L = \int_s \boldsymbol{W}^{\mathrm{T}} \underbrace{(\boldsymbol{D} \boldsymbol{\mathcal{L}}_1 \boldsymbol{u})^{\mathrm{T}}}_{\boldsymbol{s}^{\mathrm{T}}} \boldsymbol{n} \mathrm{d}s + \int_L \boldsymbol{W}^{\mathrm{T}} \boldsymbol{b} \, \mathrm{d}L \,. \tag{4.89}$$

Summarizing the equations for the nodal unknowns as given in Eqs. (4.48) and (4.49) in a single matrix equation gives:

$$\boldsymbol{u}^{\mathrm{e}} = \boldsymbol{N}^{\mathrm{T}} \boldsymbol{u}_{\mathrm{p}}^{\mathrm{e}} = \left[\begin{array}{cc|cc} N_{1u} & 0 & N_{2u} & 0 \\ 0 & N_{1\phi_y} & 0 & N_{2\phi_y} \end{array}\right] \left[\begin{array}{c} u_{1z} \\ \phi_{1y} \\ \hline u_{2z} \\ \phi_{2y} \end{array}\right] . \tag{4.90}$$

The same approach is adopted for the weight functions:

$$\boldsymbol{W} = \boldsymbol{N}^{\mathrm{T}} \delta \boldsymbol{u}_{\mathrm{p}} = \left[\begin{array}{cc|cc} N_{1u} & 0 & N_{2u} & 0 \\ 0 & N_{1\phi_y} & 0 & N_{2\phi_y} \end{array}\right] \left[\begin{array}{c} \delta u_{1z} \\ \delta\phi_{1y} \\ \hline \delta u_{2z} \\ \delta\phi_{2y} \end{array}\right] , \tag{4.91}$$

or for its transposed:

$$(\boldsymbol{W})^{\mathrm{T}} = \left(\boldsymbol{N}^{\mathrm{T}} \delta \boldsymbol{u}_{\mathrm{p}}\right)^{\mathrm{T}} = \delta \boldsymbol{u}_{\mathrm{p}}^{\mathrm{T}} \boldsymbol{N} \,. \tag{4.92}$$

With these nodal approaches, the first expression on the left-hand side of Eq. (4.89) can be expressed as:

$$\begin{aligned} \int_L \left(\boldsymbol{\mathcal{L}}_{1,a} \boldsymbol{W}\right)^{\mathrm{T}} \boldsymbol{D} \left(\boldsymbol{\mathcal{L}}_1 \boldsymbol{u}\right) \mathrm{d}L &= \int_L \boldsymbol{W}^{\mathrm{T}} \boldsymbol{\mathcal{L}}_{1,a}^{\mathrm{T}} \boldsymbol{D} \boldsymbol{\mathcal{L}}_1 \boldsymbol{u} \, \mathrm{d}L \\ &= \delta \boldsymbol{u}_{\mathrm{p}}^{\mathrm{T}} \int_L (\boldsymbol{N} \boldsymbol{\mathcal{L}}_{1,a}^{\mathrm{T}}) \boldsymbol{D} (\boldsymbol{\mathcal{L}}_1 \boldsymbol{N}^{\mathrm{T}}) \mathrm{d}L \boldsymbol{u}_{\mathrm{p}} \,. \end{aligned} \tag{4.93}$$

[6]This is the same approach as in the case of Eq. (4.54).

The virtual deformations can be eliminated from this statement and the first expression of the left-hand side of Eq. (4.93) reads:

$$\int_L (\boldsymbol{\mathcal{L}}_{1,a}\boldsymbol{N}^{\mathrm{T}})^{\mathrm{T}}\boldsymbol{D}(\boldsymbol{\mathcal{L}}_1\boldsymbol{N}^{\mathrm{T}})\mathrm{d}L\boldsymbol{u}_{\mathrm{p}}\,. \tag{4.94}$$

Let us shave a closer look on the components of this matrix equation. The integrand of the first integral can be written as:

$$\boldsymbol{\mathcal{L}}_{1,a}\boldsymbol{N}^{\mathrm{T}} = \begin{bmatrix} \frac{\mathrm{d}}{\mathrm{d}x} & 0 \\ 0 & \frac{\mathrm{d}}{\mathrm{d}x} \end{bmatrix} \left[\begin{array}{cc|cc} N_{1u} & 0 & N_{2u} & 0 \\ 0 & N_{1\phi_y} & 0 & N_{2\phi_y} \end{array}\right] = \left[\begin{array}{cc|cc} \frac{\mathrm{d}N_{1u}}{\mathrm{d}x} & 0 & \frac{\mathrm{d}N_{2u}}{\mathrm{d}x} & 0 \\ 0 & \frac{\mathrm{d}N_{1\phi_y}}{\mathrm{d}x} & 0 & \frac{\mathrm{d}N_{2\phi_y}}{\mathrm{d}x} \end{array}\right]$$

$$(\boldsymbol{\mathcal{L}}_{1,a}\boldsymbol{N}^{\mathrm{T}})^{\mathrm{T}}\boldsymbol{D} = \begin{bmatrix} \frac{\mathrm{d}N_{1u}}{\mathrm{d}x} & 0 \\ 0 & \frac{\mathrm{d}N_{1\phi_y}}{\mathrm{d}x} \\ \frac{\mathrm{d}N_{2u}}{\mathrm{d}x} & 0 \\ 0 & \frac{\mathrm{d}N_{2\phi_y}}{\mathrm{d}x} \end{bmatrix} \begin{bmatrix} -k_{\mathrm{s}}AG & 0 \\ 0 & EI_y \end{bmatrix} = \begin{bmatrix} -k_{\mathrm{s}}AG\frac{\mathrm{d}N_{1u}}{\mathrm{d}x} & 0 \\ 0 & EI_y\frac{\mathrm{d}N_{1\phi_y}}{\mathrm{d}x} \\ -k_{\mathrm{s}}AG\frac{\mathrm{d}N_{2u}}{\mathrm{d}x} & 0 \\ 0 & EI_y\frac{\mathrm{d}N_{2\phi_y}}{\mathrm{d}x} \end{bmatrix}$$

$$(\boldsymbol{\mathcal{L}}_{1,a}\boldsymbol{N}^{\mathrm{T}})^{\mathrm{T}}\boldsymbol{D}(\boldsymbol{\mathcal{L}}_1\boldsymbol{N}^{\mathrm{T}}) = \begin{bmatrix} -k_{\mathrm{s}}AG\frac{\mathrm{d}N_{1u}}{\mathrm{d}x} & 0 \\ 0 & EI_y\frac{\mathrm{d}N_{1\phi_y}}{\mathrm{d}x} \\ -k_{\mathrm{s}}AG\frac{\mathrm{d}N_{2u}}{\mathrm{d}x} & 0 \\ 0 & EI_y\frac{\mathrm{d}N_{2\phi_y}}{\mathrm{d}x} \end{bmatrix} \left[\begin{array}{cc|cc} \frac{\mathrm{d}N_{1u}}{\mathrm{d}x} & N_{1\phi_y} & \frac{\mathrm{d}N_{2u}}{\mathrm{d}x} & \mathrm{d}N_{2\phi_y} \\ 0 & \frac{\mathrm{d}N_{1\phi_y}}{\mathrm{d}x} & 0 & \frac{\mathrm{d}N_{2\phi_y}}{\mathrm{d}x} \end{array}\right]$$

$$= \begin{bmatrix} -k_{\mathrm{s}}AG\frac{\mathrm{d}N_{1u}}{\mathrm{d}x}\frac{\mathrm{d}N_{1u}}{\mathrm{d}x} & -k_{\mathrm{s}}AG\frac{\mathrm{d}N_{1u}}{\mathrm{d}x}N_{1\phi_y} & -k_{\mathrm{s}}AG\frac{\mathrm{d}N_{1u}}{\mathrm{d}x}\frac{\mathrm{d}N_{2u}}{\mathrm{d}x} & -k_{\mathrm{s}}AG\frac{\mathrm{d}N_{1u}}{\mathrm{d}x}N_{2\phi_y} \\ 0 & EI_y\frac{\mathrm{d}N_{1\phi_y}}{\mathrm{d}x}\frac{\mathrm{d}N_{1\phi_y}}{\mathrm{d}x} & 0 & EI_y\frac{\mathrm{d}N_{1\phi_y}}{\mathrm{d}x}\frac{\mathrm{d}N_{2\phi_y}}{\mathrm{d}x} \\ -k_{\mathrm{s}}AG\frac{\mathrm{d}N_{2u}}{\mathrm{d}x}\frac{\mathrm{d}N_{1u}}{\mathrm{d}x} & -k_{\mathrm{s}}AG\frac{\mathrm{d}N_{2u}}{\mathrm{d}x}N_{1\phi_y} & -k_{\mathrm{s}}AG\frac{\mathrm{d}N_{2u}}{\mathrm{d}x}\frac{\mathrm{d}N_{2u}}{\mathrm{d}x} & -k_{\mathrm{s}}AG\frac{\mathrm{d}N_{2u}}{\mathrm{d}x}N_{2\phi_y} \\ 0 & EI_y\frac{\mathrm{d}N_{1\phi_y}}{\mathrm{d}x}\frac{\mathrm{d}N_{2\phi_y}}{\mathrm{d}x} & 0 & EI_y\frac{\mathrm{d}N_{2\phi_y}}{\mathrm{d}x}\frac{\mathrm{d}N_{2\phi_y}}{\mathrm{d}x} \end{bmatrix}. \tag{4.95}$$

Considering the nodal approaches, the first expression on the right-hand side of the weak formulation, i.e. Eq. (4.89), can be expressed as:

$$\int_s \boldsymbol{W}^{\mathrm{T}}\boldsymbol{s}^{\mathrm{T}}\boldsymbol{n}\mathrm{d}s = \int_s \delta\boldsymbol{u}_{\mathrm{p}}^{\mathrm{T}}\boldsymbol{N}\boldsymbol{s}^{\mathrm{T}}\boldsymbol{n}\mathrm{d}s = \delta\boldsymbol{u}_{\mathrm{p}}^{\mathrm{T}}\int_s \boldsymbol{N}\boldsymbol{s}^{\mathrm{T}}\boldsymbol{n}\mathrm{d}s\,, \tag{4.96}$$

or after elimination of the virtual displacements:

$$\int_s \boldsymbol{N}\boldsymbol{s}^{\mathrm{T}}\boldsymbol{n}\mathrm{d}s\,, \tag{4.97}$$

where the expression $(\boldsymbol{s})^{\mathrm{T}} n$ can be understood as the 'tractions' $\boldsymbol{t}$ as in the three-dimensional case [8].

The second expression on the left-hand side of Eq. (4.89) can be expressed as:

$$\int_L \boldsymbol{W}^{\mathrm{T}} \boldsymbol{\mathcal{L}}_{1,b}^{\mathrm{T}} \boldsymbol{D} \boldsymbol{\mathcal{L}}_1 \boldsymbol{u}\, \mathrm{d}L = \delta \boldsymbol{u}_{\mathrm{p}}^{\mathrm{T}} \int_L (\boldsymbol{N} \boldsymbol{\mathcal{L}}_{1,b}^{\mathrm{T}}) \boldsymbol{D} (\boldsymbol{\mathcal{L}}_1 \boldsymbol{N}^{\mathrm{T}}) \mathrm{d}L \boldsymbol{u}_{\mathrm{p}}\,. \tag{4.98}$$

The virtual displacements can be eliminated and the integrand reads:

$$\boldsymbol{N} \boldsymbol{\mathcal{L}}_{1,b}^{\mathrm{T}} = (\boldsymbol{\mathcal{L}}_{1,b} \boldsymbol{N}^{\mathrm{T}})^{\mathrm{T}} = \left(\begin{bmatrix} 0 & 1 \\ 0 & 0 \end{bmatrix} \left[\begin{array}{cc|cc} N_{1u} & 0 & N_{2u} & 0 \\ 0 & N_{1\phi_y} & 0 & N_{2\phi_y} \end{array} \right] \right)^{\mathrm{T}}$$

$$= \left[\begin{array}{cc|cc} 0 & N_{1\phi_y} & 0 & N_{2\phi_y} \\ 0 & 0 & 0 & 0 \end{array} \right]^{\mathrm{T}} = \begin{bmatrix} 0 & 0 \\ N_{1\phi_y} & 0 \\ 0 & \\ N_{2\phi_y} & 0 \end{bmatrix}$$

$$(\boldsymbol{\mathcal{L}}_{1,b} \boldsymbol{N}^{\mathrm{T}})^{\mathrm{T}} \boldsymbol{D} = \begin{bmatrix} 0 & 0 \\ N_{1\phi_y} & 0 \\ 0 & 0 \\ N_{2\phi_y} & 0 \end{bmatrix} \begin{bmatrix} -k_{\mathrm{s}} AG & 0 \\ 0 & EI_y \end{bmatrix} = \begin{bmatrix} 0 & 0 \\ -k_{\mathrm{s}} AG N_{1\phi_y} & 0 \\ 0 & 0 \\ -k_{\mathrm{s}} AG N_{2\phi_y} & 0 \end{bmatrix}$$

$$(\boldsymbol{\mathcal{L}}_{1,b} \boldsymbol{N}^{\mathrm{T}})^{\mathrm{T}} \boldsymbol{D} (\boldsymbol{\mathcal{L}}_1 \boldsymbol{N}^{\mathrm{T}}) = \begin{bmatrix} 0 & 0 \\ -k_{\mathrm{s}} AG N_{1\phi_y} & 0 \\ 0 & 0 \\ -k_{\mathrm{s}} AG N_{2\phi_y} & 0 \end{bmatrix} \left[\begin{array}{cc|cc} \frac{\mathrm{d}N_{1u}}{\mathrm{d}x} & N_{1\phi_y} & \frac{\mathrm{d}N_{2u}}{\mathrm{d}x} & \mathrm{d}N_{2\phi_y} \\ 0 & \frac{\mathrm{d}N_{1\phi_y}}{\mathrm{d}x} & 0 & \frac{\mathrm{d}N_{2\phi_y}}{\mathrm{d}x} \end{array} \right]$$

$$= \begin{bmatrix} 0 & 0 & 0 & 0 \\ -k_{\mathrm{s}} AG N_{1\phi_y} \frac{\mathrm{d}N_{1u}}{\mathrm{d}x} & -k_{\mathrm{s}} AG N_{1\phi_y} N_{1\phi_y} & -k_{\mathrm{s}} AG N_{1\phi_y} \frac{\mathrm{d}N_{2u}}{\mathrm{d}x} & -k_{\mathrm{s}} AG N_{1\phi_y} N_{2\phi_y} \\ 0 & 0 & 0 & 0 \\ -k_{\mathrm{s}} AG N_{2\phi_y} \frac{\mathrm{d}N_{1u}}{\mathrm{d}x} & -k_{\mathrm{s}} AG N_{2\phi_y} N_{1\phi_y} & -k_{\mathrm{s}} AG N_{2\phi_y} \frac{\mathrm{d}N_{2u}}{\mathrm{d}x} & -k_{\mathrm{s}} AG N_{2\phi_y} N_{2\phi_y} \end{bmatrix}. \tag{4.99}$$

The integral over the body forces in the weak formulation, cf. Eq. (4.89), can be expressed as:

$$\int_L \boldsymbol{W}^{\mathrm{T}} \boldsymbol{b}\, \mathrm{d}L = \delta \boldsymbol{u}_{\mathrm{p}}^{\mathrm{T}} \int_L \boldsymbol{N} \boldsymbol{b}\, \mathrm{d}L\,. \tag{4.100}$$

The virtual displacements can be eliminated and the integrand reads:

$$\boldsymbol{N}\boldsymbol{b} = \begin{bmatrix} N_{1u} & 0 \\ 0 & N_{1\phi_y} \\ N_{2u} & 0 \\ 0 & N_{2\phi_y} \end{bmatrix} \begin{bmatrix} -q_z \\ m_z \end{bmatrix} = \begin{bmatrix} -N_{1u}q_z \\ N_{1\phi_y}m_z \\ -N_{2u}q_z \\ N_{2\phi_y}m_z \end{bmatrix}. \tag{4.101}$$

Summing up, one can state that the weak formulation can be expressed after the consideration of the nodal approaches in matrix form as:

$$\int_L (\boldsymbol{\mathcal{L}}_{1,a}\boldsymbol{N}^{\mathrm{T}})^{\mathrm{T}}\boldsymbol{D}(\boldsymbol{\mathcal{L}}_1\boldsymbol{N}^{\mathrm{T}})\mathrm{d}L\boldsymbol{u}_{\mathrm{p}}^{\mathrm{e}} - \int_L (\boldsymbol{\mathcal{L}}_{1,b}\boldsymbol{N}^{\mathrm{T}})^{\mathrm{T}}\boldsymbol{D}(\boldsymbol{\mathcal{L}}_1\boldsymbol{N}^{\mathrm{T}})\mathrm{d}L\boldsymbol{u}_{\mathrm{p}}^{\mathrm{e}}$$
$$= \int_s \boldsymbol{N}\boldsymbol{s}^{\mathrm{T}}\boldsymbol{n}\mathrm{d}s + \int_L \boldsymbol{N}\boldsymbol{b}\,\mathrm{d}L. \tag{4.102}$$

Introducing Eqs. (4.95), (4.99) and (4.101) in the weak formulation according to Eq. (4.102) gives:

$$\int_L \begin{bmatrix} -k_sAG\frac{\mathrm{d}N_{1u}}{\mathrm{d}x}\frac{\mathrm{d}N_{1u}}{\mathrm{d}x} & -k_sAG\frac{\mathrm{d}N_{1u}}{\mathrm{d}x}N_{1\phi_y} & -k_sAG\frac{\mathrm{d}N_{1u}}{\mathrm{d}x}\frac{\mathrm{d}N_{2u}}{\mathrm{d}x} & -k_sAG\frac{\mathrm{d}N_{1u}}{\mathrm{d}x}N_{2\phi_y} \\ 0 & EI_y\frac{\mathrm{d}N_{1\phi_y}}{\mathrm{d}x}\frac{\mathrm{d}N_{1\phi_y}}{\mathrm{d}x} & 0 & EI_y\frac{\mathrm{d}N_{1\phi_y}}{\mathrm{d}x}\frac{\mathrm{d}N_{2\phi_y}}{\mathrm{d}x} \\ -k_sAG\frac{\mathrm{d}N_{2u}}{\mathrm{d}x}\frac{\mathrm{d}N_{1u}}{\mathrm{d}x} & -k_sAG\frac{\mathrm{d}N_{2u}}{\mathrm{d}x}N_{1\phi_y} & -k_sAG\frac{\mathrm{d}N_{2u}}{\mathrm{d}x}\frac{\mathrm{d}N_{2u}}{\mathrm{d}x} & -k_sAG\frac{\mathrm{d}N_{2u}}{\mathrm{d}x}N_{2\phi_y} \\ 0 & EI_y\frac{\mathrm{d}N_{1\phi_y}}{\mathrm{d}x}\frac{\mathrm{d}N_{2\phi_y}}{\mathrm{d}x} & 0 & EI_y\frac{\mathrm{d}N_{2\phi_y}}{\mathrm{d}x}\frac{\mathrm{d}N_{2\phi_y}}{\mathrm{d}x} \end{bmatrix} \mathrm{d}L\boldsymbol{u}_{\mathrm{p}}$$
$$-\int_L \begin{bmatrix} 0 & 0 & 0 & 0 \\ -k_sAGN_{1\phi_y}\frac{\mathrm{d}N_{1u}}{\mathrm{d}x} & -k_sAGN_{1\phi_y}N_{1\phi_y} & -k_sAGN_{1\phi_y}\frac{\mathrm{d}N_{2u}}{\mathrm{d}x} & -k_sAGN_{1\phi_y}N_{2\phi_y} \\ 0 & 0 & 0 & 0 \\ -k_sAGN_{2\phi_y}\frac{\mathrm{d}N_{1u}}{\mathrm{d}x} & -k_sAGN_{2\phi_y}N_{1\phi_y} & -k_sAGN_{2\phi_y}\frac{\mathrm{d}N_{2u}}{\mathrm{d}x} & -k_sAGN_{2\phi_y}N_{2\phi_y} \end{bmatrix} \mathrm{d}L\boldsymbol{u}_{\mathrm{p}}$$
$$= \int_s \boldsymbol{N}\boldsymbol{t}\mathrm{d}s + \int_L \begin{bmatrix} -N_{1u}q_z \\ N_{1\phi_y}m_z \\ -N_{2u}q_z \\ N_{2\phi_y}m_z \end{bmatrix} \mathrm{d}L. \tag{4.103}$$

To obtain the same formulation as in the first part of this section, it is required to change the sign of each first and third row. This can be achieved by multiplying each matrix with the diagonal matrix $\lceil -1\ 1\ -1\ 1 \rfloor$. Rearranging in the classical form of $\boldsymbol{k}^{\mathrm{e}}\boldsymbol{u}_{\mathrm{p}}^{\mathrm{e}} = \boldsymbol{f}^{\mathrm{e}}$, we get:

$$\int_L \begin{bmatrix} k_s AG \frac{dN_{1u}}{dx}\frac{dN_{1u}}{dx} & k_s AG \frac{dN_{1u}}{dx} N_{1\phi_y} & k_s AG \frac{dN_{1u}}{dx}\frac{dN_{2u}}{dx} & k_s AG \frac{dN_{1u}}{dx} N_{2\phi_y} \\ \hline k_s AG N_{1\phi_y}\frac{dN_{1u}}{dx} & \begin{array}{c} EI_y \frac{dN_{1\phi_y}}{dx}\frac{dN_{1\phi_y}}{dx} \\ +k_s AG N_{1\phi_y} N_{1\phi_y} \end{array} & k_s AG N_{1\phi_y}\frac{dN_{2u}}{dx} & \begin{array}{c} EI_y \frac{dN_{1\phi_y}}{dx}\frac{dN_{2\phi_y}}{dx} \\ +k_s AG N_{1\phi_y} N_{2\phi_y} \end{array} \\ \hline k_s AG \frac{dN_{2u}}{dx}\frac{dN_{1u}}{dx} & k_s AG \frac{dN_{2u}}{dx} N_{1\phi_y} & k_s AG \frac{dN_{2u}}{dx}\frac{dN_{2u}}{dx} & k_s AG \frac{dN_{2u}}{dx} N_{2\phi_y} \\ \hline k_s AG N_{2\phi_y}\frac{dN_{1u}}{dx} & \begin{array}{c} EI_y \frac{dN_{2\phi_y}}{dx}\frac{dN_{1\phi_y}}{dx} \\ +k_s AG N_{2\phi_y} N_{1\phi_y} \end{array} & k_s AG N_{2\phi_y}\frac{dN_{2u}}{dx} & \begin{array}{c} EI_y \frac{dN_{2\phi_y}}{dx}\frac{dN_{2\phi_y}}{dx} \\ +k_s AG N_{2\phi_y} N_{2\phi_y} \end{array} \end{bmatrix} dL\, \boldsymbol{u}_p$$

$$= \int_s \boldsymbol{N}\boldsymbol{t}\,ds + \int_L \begin{bmatrix} +N_{1u}q_z \\ N_{1\phi_y}m_z \\ +N_{2u}q_z \\ N_{2\phi_y}m_z \end{bmatrix} dL\,. \tag{4.104}$$

The general formulations of the basic equations for a TIMOSHENKO beam as given in Table 4.5 can be slightly modified to avoid some esthetic appeals. The representations of generalized stresses, generalized stiffness matrix, and distributed loads contain a minus sign. Let us start with the constitutive equation, i.e.,

$$\begin{bmatrix} -Q_z \\ M_y \end{bmatrix} = \begin{bmatrix} -k_s AG & 0 \\ 0 & EI_y \end{bmatrix} \begin{bmatrix} \frac{du_z}{dx} + \phi_y \\ \frac{d\phi_y}{dx} \end{bmatrix}, \tag{4.105}$$

or under elimination of the minus sign in the mentioned matrices:

$$\begin{bmatrix} -1 & 0 \\ 0 & 1 \end{bmatrix} \begin{bmatrix} Q_z \\ M_y \end{bmatrix} = \begin{bmatrix} -1 & 0 \\ 0 & 1 \end{bmatrix} \begin{bmatrix} k_s AG & 0 \\ 0 & EI_y \end{bmatrix} \begin{bmatrix} \frac{du_z}{dx} + \phi_y \\ \frac{d\phi_y}{dx} \end{bmatrix}. \tag{4.106}$$

The diagonal matrix $\lceil -1 \;\; 1 \rfloor$ can be eliminated from the last equation to obtain the modified constitutive law in matrix form:

$$\begin{bmatrix} Q_z \\ M_y \end{bmatrix} = \begin{bmatrix} k_s AG & 0 \\ 0 & EI_y \end{bmatrix} \begin{bmatrix} \frac{du_z}{dx} + \phi_y \\ \frac{d\phi_y}{dx} \end{bmatrix}. \tag{4.107}$$

The next step is to have a closer look on the equilibrium equation, i.e.,

$$\begin{bmatrix} \frac{d}{dx} & 0 \\ 1 & \frac{d}{dx} \end{bmatrix} \begin{bmatrix} -Q_z \\ M_y \end{bmatrix} + \begin{bmatrix} -q_z \\ +m_z \end{bmatrix} = \begin{bmatrix} 0 \\ 0 \end{bmatrix}, \tag{4.108}$$

or again re-written based on the diagonal matrix:

$$\begin{bmatrix} \frac{\mathrm{d}}{\mathrm{d}x} & 0 \\ 1 & \frac{\mathrm{d}}{\mathrm{d}x} \end{bmatrix} \begin{bmatrix} -1 & 0 \\ 0 & 1 \end{bmatrix} \begin{bmatrix} Q_z \\ M_y \end{bmatrix} + \begin{bmatrix} -1 & 0 \\ 0 & 1 \end{bmatrix} \begin{bmatrix} q_z \\ m_z \end{bmatrix} = \begin{bmatrix} 0 \\ 0 \end{bmatrix}, \tag{4.109}$$

$$\begin{bmatrix} -\frac{\mathrm{d}}{\mathrm{d}x} & 0 \\ -1 & \frac{\mathrm{d}}{\mathrm{d}x} \end{bmatrix} \begin{bmatrix} Q_z \\ M_y \end{bmatrix} + \begin{bmatrix} -1 & 0 \\ 0 & 1 \end{bmatrix} \begin{bmatrix} q_z \\ m_z \end{bmatrix} = \begin{bmatrix} 0 \\ 0 \end{bmatrix}. \tag{4.110}$$

Let us now multiply the last equation with the diagonal matrix from the left-hand side:

$$\begin{bmatrix} -1 & 0 \\ 0 & 1 \end{bmatrix} \begin{bmatrix} -\frac{\mathrm{d}}{\mathrm{d}x} & 0 \\ -1 & \frac{\mathrm{d}}{\mathrm{d}x} \end{bmatrix} \begin{bmatrix} Q_z \\ M_y \end{bmatrix} + \underbrace{\begin{bmatrix} -1 & 0 \\ 0 & 1 \end{bmatrix} \begin{bmatrix} -1 & 0 \\ 0 & 1 \end{bmatrix}}_{\mathbf{I}} \begin{bmatrix} q_z \\ m_z \end{bmatrix} = \begin{bmatrix} -1 & 0 \\ 0 & 1 \end{bmatrix} \begin{bmatrix} 0 \\ 0 \end{bmatrix}, \tag{4.111}$$

Or finally as the modified expression of the equilibrium equation:

$$\begin{bmatrix} \frac{\mathrm{d}}{\mathrm{d}x} & 0 \\ -1 & \frac{\mathrm{d}}{\mathrm{d}x} \end{bmatrix} \begin{bmatrix} Q_z \\ M_y \end{bmatrix} + \begin{bmatrix} q_z \\ m_z \end{bmatrix} = \begin{bmatrix} 0 \\ 0 \end{bmatrix}. \tag{4.112}$$

The modified basic equations, i.e., 'without the minus sign', are summarized in Table 4.7.

Let us now refer to Eq. (4.82) and the following procedure to derive again the principal finite element equation based on the modified formulations given in Table 4.7. The strong formulation can be written as:

$$\mathcal{L}_{1*}^{\mathrm{T}} \boldsymbol{D}^* \mathcal{L}_1 \boldsymbol{u}^0 + \boldsymbol{b}^* = \boldsymbol{0}. \tag{4.113}$$

Replacing the exact solution by an approximate solution, a residual is obtained and the following integration gives the inner product as

$$\int_L \boldsymbol{W}^{\mathrm{T}} \left(\mathcal{L}_{1*}^{\mathrm{T}} \boldsymbol{D}^* \mathcal{L}_1 \boldsymbol{u} + \boldsymbol{b}^* \right) \mathrm{d}L = 0. \tag{4.114}$$

Application of the GREEN–GAUSS theorem (cf. Sect. A.7) would shift the derivative to the weight functions $\boldsymbol{W}$. However, the matrix of differential operators, $\mathcal{L}_{1*}^{\mathrm{T}}$, contains in addition to derivatives a constant value '-1' and it is therefore appropriate to split the matrix into a part which contains all the derivatives, $\mathcal{L}_{1*,a}^{\mathrm{T}}$, and a part with the constant value, $\mathcal{L}_{1*,b}^{\mathrm{T}}$:

$$\underbrace{\begin{bmatrix} \frac{\mathrm{d}}{\mathrm{d}x} & 0 \\ -1 & \frac{\mathrm{d}}{\mathrm{d}x} \end{bmatrix}}_{\mathcal{L}_{1*}^{\mathrm{T}}} = \underbrace{\begin{bmatrix} \frac{\mathrm{d}}{\mathrm{d}x} & 0 \\ 0 & \frac{\mathrm{d}}{\mathrm{d}x} \end{bmatrix}}_{\mathcal{L}_{1*,a}^{\mathrm{T}}} + \underbrace{\begin{bmatrix} 0 & 0 \\ -1 & 0 \end{bmatrix}}_{\mathcal{L}_{1*,b}^{\mathrm{T}}}. \tag{4.115}$$

Table 4.7 Alternative formulations of the basic equations for a TIMOSHENKO beam (bending occurs in the x-z plane)

Specific Formulation	General Formulation
Kinematics	
$\begin{bmatrix} \frac{du_z}{dx} + \phi_y \\ \frac{d\phi_y}{dx} \end{bmatrix} = \begin{bmatrix} \frac{d}{dx} & 1 \\ 0 & \frac{d}{dx} \end{bmatrix} \begin{bmatrix} u_z \\ \phi_y \end{bmatrix}$	$\boldsymbol{e} = \boldsymbol{\mathcal{L}}_1 \boldsymbol{u}$
Constitution	
$\begin{bmatrix} Q_z \\ M_y \end{bmatrix} = \begin{bmatrix} k_s AG & 0 \\ 0 & EI_y \end{bmatrix} \begin{bmatrix} \frac{du_z}{dx} + \phi_y \\ \frac{d\phi_y}{dx} \end{bmatrix}$	$\boldsymbol{s}^* = \boldsymbol{D}^* \boldsymbol{e}$
Equilibrium	
$\begin{bmatrix} \frac{d}{dx} & 0 \\ -1 & \frac{d}{dx} \end{bmatrix} \begin{bmatrix} Q_z \\ M_y \end{bmatrix} + \begin{bmatrix} q_z \\ +m_z \end{bmatrix} = \begin{bmatrix} 0 \\ 0 \end{bmatrix}$	$\boldsymbol{\mathcal{L}}_{1*}^{\mathrm{T}} \boldsymbol{s}^* + \boldsymbol{b}^* = \boldsymbol{0}$
PDE	
$\frac{d}{dx}\left[k_s GA \left(\frac{du_z}{dx} + \phi_y\right)\right] + q_z = 0$ $\frac{d}{dx}\left(EI_y \frac{d\phi_y}{dx}\right) - k_s GA \left(\frac{du_z}{dx} + \phi_y\right) + m_y = 0,$	$\boldsymbol{\mathcal{L}}_{1*}^{\mathrm{T}} \boldsymbol{D}^* \boldsymbol{\mathcal{L}}_1 \boldsymbol{u} + \boldsymbol{b}^* = \boldsymbol{0}$

Thus, we can write the inner product as:

$$\int_L \boldsymbol{W}^{\mathrm{T}} \left[\left(\boldsymbol{\mathcal{L}}_{1^*,a}^{\mathrm{T}} + \boldsymbol{\mathcal{L}}_{1^*,b}^{\mathrm{T}} \right) \boldsymbol{D}^* \boldsymbol{\mathcal{L}}_1 \boldsymbol{u} + \boldsymbol{b}^* \right] \mathrm{d}L = 0 \,, \tag{4.116}$$

or

$$\int_L \boldsymbol{W}^{\mathrm{T}} \boldsymbol{\mathcal{L}}_{1^*,a}^{\mathrm{T}} \boldsymbol{D}^* \boldsymbol{\mathcal{L}}_1 \boldsymbol{u} \, \mathrm{d}L + \int_L \boldsymbol{W}^{\mathrm{T}} \boldsymbol{\mathcal{L}}_{1^*,b}^{\mathrm{T}} \boldsymbol{D}^* \boldsymbol{\mathcal{L}}_1 \boldsymbol{u} \, \mathrm{d}L + \int_L \boldsymbol{W}^{\mathrm{T}} \boldsymbol{b}^* \, \mathrm{d}L = 0 \,. \tag{4.117}$$

Application of the GREEN–GAUSS theorem to the first integral[7] gives:

$$\int_L \boldsymbol{W}^{\mathrm{T}} \boldsymbol{\mathcal{L}}_{1^*,a}^{\mathrm{T}} \boldsymbol{D}^* \boldsymbol{\mathcal{L}}_1 \boldsymbol{u} \, \mathrm{d}L = - \int_L \left(\boldsymbol{\mathcal{L}}_{1^*,a} \boldsymbol{W} \right)^{\mathrm{T}} \boldsymbol{D}^* \left(\boldsymbol{\mathcal{L}}_1 \boldsymbol{u} \right) \mathrm{d}L + \int_s \boldsymbol{W}^{\mathrm{T}} \underbrace{\left(\boldsymbol{D}^* \boldsymbol{\mathcal{L}}_1 \boldsymbol{u} \right)^{\mathrm{T}}}_{(\boldsymbol{s}^*)^{\mathrm{T}}} \boldsymbol{n} \mathrm{d}s \,, \tag{4.118}$$

[7] This is the same approach as in the case of Eq. (4.54).

and the weak formulation can be obtained as:

$$\int_L (\boldsymbol{\mathcal{L}}_{1^*,a} \boldsymbol{W})^{\mathrm{T}} \boldsymbol{D}^* (\boldsymbol{\mathcal{L}}_1 \boldsymbol{u})\,\mathrm{d}L - \int_L \boldsymbol{W}^{\mathrm{T}} \boldsymbol{\mathcal{L}}_{1^*,b}^{\mathrm{T}} \boldsymbol{D}^* \boldsymbol{\mathcal{L}}_1 \boldsymbol{u}\,\mathrm{d}L =$$

$$\int_s \boldsymbol{W}^{\mathrm{T}} \underbrace{(\boldsymbol{D}^* \boldsymbol{\mathcal{L}}_1 \boldsymbol{u})^{\mathrm{T}}}_{(\boldsymbol{s}^*)^{\mathrm{T}}} n \mathrm{d}s + \int_L \boldsymbol{W}^{\mathrm{T}} \boldsymbol{b}^*\,\mathrm{d}L\,. \tag{4.119}$$

The last equation can be simplified (see the distributive law in the Appendix A.11.1) to

$$\int_L \left(\boldsymbol{W}^{\mathrm{T}} \boldsymbol{\mathcal{L}}_{1^*,a}^{\mathrm{T}} - \boldsymbol{W}^{\mathrm{T}} \boldsymbol{\mathcal{L}}_{1^*,b}^{\mathrm{T}}\right) \boldsymbol{D}^* (\boldsymbol{\mathcal{L}}_1 \boldsymbol{u})\,\mathrm{d}L = \int_s \boldsymbol{W}^{\mathrm{T}} (\boldsymbol{s}^*)^{\mathrm{T}} n \mathrm{d}s + \int_L \boldsymbol{W}^{\mathrm{T}} \boldsymbol{b}^*\,\mathrm{d}L\,,$$

$$\int_L \boldsymbol{W}^{\mathrm{T}} \underbrace{\left(\boldsymbol{\mathcal{L}}_{1^*,a}^{\mathrm{T}} - \boldsymbol{\mathcal{L}}_{1^*,b}^{\mathrm{T}}\right)}_{\boldsymbol{\mathcal{L}}_1^{\mathrm{T}}} \boldsymbol{D}^* (\boldsymbol{\mathcal{L}}_1 \boldsymbol{u})\,\mathrm{d}L = \int_s \boldsymbol{W}^{\mathrm{T}} (\boldsymbol{s}^*)^{\mathrm{T}} n \mathrm{d}s + \int_L \boldsymbol{W}^{\mathrm{T}} \boldsymbol{b}^*\,\mathrm{d}L\,,$$

$$\tag{4.120}$$

or finally

$$\int_L (\boldsymbol{\mathcal{L}}_1 \boldsymbol{W})^{\mathrm{T}} \boldsymbol{D}^* (\boldsymbol{\mathcal{L}}_1 \boldsymbol{u})\,\mathrm{d}L = \int_s \boldsymbol{W}^{\mathrm{T}} (\boldsymbol{s}^*)^{\mathrm{T}} n \mathrm{d}s + \int_L \boldsymbol{W}^{\mathrm{T}} \boldsymbol{b}^*\,\mathrm{d}L\,. \tag{4.121}$$

Introducing the nodal approaches for the deformations and weight functions as given in Eqs. (4.90)–(4.92), the following formulation of the weak formulation on the element level is obtained:

$$\int_L (\boldsymbol{\mathcal{L}}_1 \boldsymbol{N}^{\mathrm{T}})^{\mathrm{T}} \boldsymbol{D}^* (\boldsymbol{\mathcal{L}}_1 \boldsymbol{N}^{\mathrm{T}})\,\mathrm{d}L \boldsymbol{u}_{\mathrm{p}} = \int_s \boldsymbol{N} (\boldsymbol{s}^*)^{\mathrm{T}} n \mathrm{d}s + \int_L \boldsymbol{N} \boldsymbol{b}^*\,\mathrm{d}L\,. \tag{4.122}$$

Thus, we can identify the following three element matrices from the principal finite element equation:

$$\text{Stiffness matrix:}\ \boldsymbol{k}^{\mathrm{e}} = \int_L (\boldsymbol{\mathcal{L}}_1 \boldsymbol{N}^{\mathrm{T}})^{\mathrm{T}} \boldsymbol{D}^* (\boldsymbol{\mathcal{L}}_1 \boldsymbol{N}^{\mathrm{T}})\,\mathrm{d}L\,, \tag{4.123}$$

$$\text{Boundary force matrix:}\ \boldsymbol{f}_{\boldsymbol{t}}^{\mathrm{e}} = \int_s \boldsymbol{N} (\boldsymbol{s}^*)^{\mathrm{T}} n \mathrm{d}s\,, \tag{4.124}$$

$$\text{Body force matrix:}\ \boldsymbol{f}_{\boldsymbol{b}}^{\mathrm{e}} = \int_L \boldsymbol{N} \boldsymbol{b}^*\,\mathrm{d}L\,. \tag{4.125}$$

The evaluation of these equations results finally in the expression given in Eq. (4.104).

4.3.2 Linear Interpolation Functions for the Displacement and Rotational Field

Only the first derivatives of the interpolation functions appear in the element stiffness matrices $\boldsymbol{K}_{\mathrm{b}}^{\mathrm{e}}$ and $\boldsymbol{K}_{\mathrm{s}}^{\mathrm{e}}$ according to Eqs. (4.69) and (4.70). This demand on the differentiability of the interpolation functions leads to polynomials of minimum first order (linear functions) for the displacement and rotational field, so that in the approaches according to Eqs. (4.46) and (4.47) the following linear interpolation functions can be used:

$$N_{1u}(x) = N_{1\phi}(x) = 1 - \frac{x}{L}, \tag{4.126}$$

$$N_{2u}(x) = N_{2\phi}(x) = \frac{x}{L}. \tag{4.127}$$

The necessary derivatives result in:

$$\frac{\mathrm{d}N_{1u}}{\mathrm{d}x} = \frac{\mathrm{d}N_{1\phi}}{\mathrm{d}x} = -\frac{1}{L}, \tag{4.128}$$

$$\frac{\mathrm{d}N_{2u}}{\mathrm{d}x} = \frac{\mathrm{d}N_{2\phi}}{\mathrm{d}x} = \frac{1}{L}. \tag{4.129}$$

A graphical illustration of the interpolation function is given in Fig. 4.11. Additionally, the interpolation functions in the natural coordinate $\xi \in [-1, 1]$ are given. This formulation is more beneficial for the numerical integration of the stiffness matrices.

The integrals of the element stiffness matrices $\boldsymbol{K}_{\mathrm{b}}^{\mathrm{e}}$ and $\boldsymbol{K}_{\mathrm{s}}^{\mathrm{e}}$ according to Eqs. (4.69) and (4.70) are analytically calculated in the following. Using the linear approaches for the interpolation functions, the following results for the bending stiffness matrix:

$$\boldsymbol{K}_{\mathrm{b}}^{\mathrm{e}} = EI_y \int_0^L \begin{bmatrix} 0 & 0 & 0 & 0 \\ 0 & \frac{1}{L^2} & 0 & -\frac{1}{L^2} \\ 0 & 0 & 0 & 0 \\ 0 & -\frac{1}{L^2} & 0 & \frac{1}{L^2} \end{bmatrix} \mathrm{d}x = EI_y \begin{bmatrix} 0 & 0 & 0 & 0 \\ 0 & \frac{x}{L^2} & 0 & -\frac{x}{L^2} \\ 0 & 0 & 0 & 0 \\ 0 & -\frac{x}{L^2} & 0 & \frac{x}{L^2} \end{bmatrix}_0^L, \tag{4.130}$$

or alternatively under consideration of the integration boundaries:

$$\boldsymbol{K}_{\mathrm{b}}^{\mathrm{e}} = EI_y \begin{bmatrix} 0 & 0 & 0 & 0 \\ 0 & \frac{1}{L} & 0 & -\frac{1}{L} \\ 0 & 0 & 0 & 0 \\ 0 & -\frac{1}{L} & 0 & \frac{1}{L} \end{bmatrix}. \tag{4.131}$$

Using the linear approaches for the interpolation functions, the following results for the shear stiffness matrix:

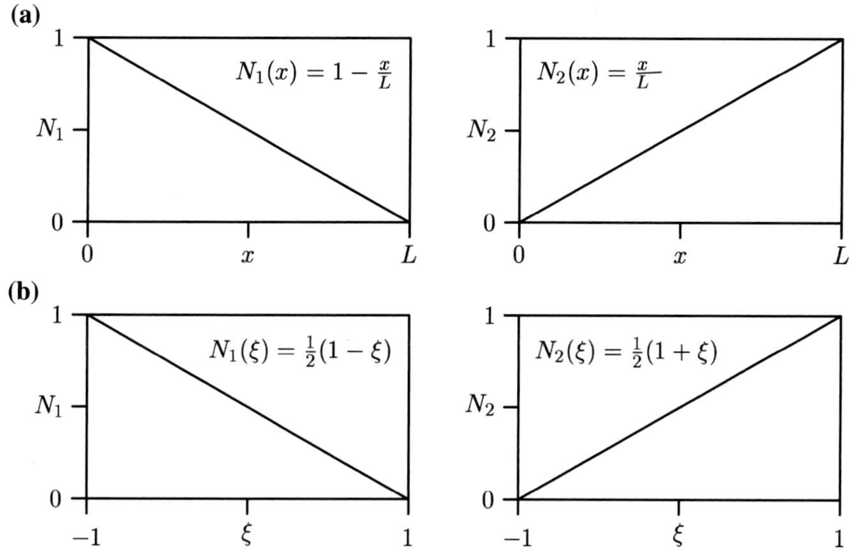

Fig. 4.11 Linear interpolation functions $N_1 = N_{1u} = N_{1\phi}$ and $N_2 = N_{2u} = N_{2\phi}$ for the TIMOSHENKO element: **a** physical coordinate (x); **b** natural coordinate (ξ)

$$
\boldsymbol{K}_{\mathrm{s}}^{\mathrm{e}} = k_{\mathrm{s}} AG \int_0^L \begin{bmatrix} \frac{1}{L^2} & \left(1-\frac{x}{L}\right)\left(-\frac{1}{L}\right) & -\frac{1}{L^2} & -\frac{x}{L^2} \\ \left(1-\frac{x}{L}\right)\left(-\frac{1}{L}\right) & \left(1-\frac{x}{L}\right)^2 & \left(1-\frac{x}{L}\right)\frac{1}{L} & \left(1-\frac{x}{L}\right)\frac{x}{L} \\ -\frac{1}{L^2} & \left(1-\frac{x}{L}\right)\frac{1}{L} & \frac{1}{L^2} & \frac{x}{L^2} \\ -\frac{x}{L^2} & \left(1-\frac{x}{L}\right)\frac{x}{L} & \frac{x}{L^2} & \frac{x^2}{L^2} \end{bmatrix} \mathrm{d}x \tag{4.132}
$$

$$
= k_{\mathrm{s}} AG \begin{bmatrix} \frac{x}{L^2} & \frac{x(-2L+x)}{2L^2} & -\frac{x}{L^2} & -\frac{x^2}{2L^2} \\ \frac{x(-2L+x)}{2L^2} & \frac{(-L+x)^3}{3L^2} & \frac{x(2L-x)}{2L^2} & \frac{x^2(3L-2x)}{6L^2} \\ -\frac{x}{L^2} & \frac{x(2L-x)}{2L^2} & \frac{x}{L^2} & \frac{x^2}{2L^2} \\ -\frac{x^2}{2L^2} & \frac{x^2(3L-2x)}{6L^2} & \frac{x^2}{2L^2} & \frac{x^3}{3L^2} \end{bmatrix}_0^L \tag{4.133}
$$

and finally after considering the constants of integration:

$$\boldsymbol{K}_{\mathrm{s}}^{\mathrm{e}} = k_{\mathrm{s}}AG \begin{bmatrix} +\frac{1}{L} & -\frac{1}{2} & -\frac{1}{L} & -\frac{1}{2} \\ -\frac{1}{2} & +\frac{L}{3} & +\frac{1}{2} & +\frac{L}{6} \\ -\frac{1}{L} & +\frac{1}{2} & +\frac{1}{L} & +\frac{1}{2} \\ -\frac{1}{2} & +\frac{L}{6} & +\frac{1}{2} & +\frac{L}{3} \end{bmatrix}. \tag{4.134}$$

The two stiffness matrices according to Eqs. (4.131) and (4.134) can be summarized additively to the total stiffness matrix of the TIMOSHENKO beam:

$$\boldsymbol{K}^{\mathrm{e}} = \begin{bmatrix} +\frac{k_{\mathrm{s}}AG}{L} & -\frac{k_{\mathrm{s}}AG}{2} & -\frac{k_{\mathrm{s}}AG}{L} & -\frac{k_{\mathrm{s}}AG}{2} \\ -\frac{k_{\mathrm{s}}AG}{2} & +\frac{k_{\mathrm{s}}AGL}{3}+\frac{EI_y}{L} & +\frac{k_{\mathrm{s}}AG}{2} & +\frac{k_{\mathrm{s}}AGL}{6}-\frac{EI_y}{L} \\ -\frac{k_{\mathrm{s}}AG}{L} & +\frac{k_{\mathrm{s}}AG}{2} & +\frac{k_{\mathrm{s}}AG}{L} & +\frac{k_{\mathrm{s}}AG}{2} \\ -\frac{k_{\mathrm{s}}AG}{2} & +\frac{k_{\mathrm{s}}AGL}{6}-\frac{EI_y}{L} & +\frac{k_{\mathrm{s}}AG}{2} & +\frac{k_{\mathrm{s}}AGL}{3}+\frac{EI_y}{L} \end{bmatrix}, \tag{4.135}$$

or alternatively via the abbreviation $\alpha = \frac{4EI_y}{k_{\mathrm{s}}AG}$

$$\boldsymbol{K}^{\mathrm{e}} = \frac{k_{\mathrm{s}}AG}{4L} \begin{bmatrix} 4 & -2L & -4 & -2L \\ -2L & \frac{4}{3}L^2+\alpha & 2L & \frac{4}{6}L^2-\alpha \\ -4 & 2L & 4 & 2L \\ -2L & \frac{4}{6}L^2-\alpha & 2L & \frac{4}{3}L^2+\alpha \end{bmatrix}, \tag{4.136}$$

or alternatively via the abbreviation $\Lambda = \dfrac{EI_z}{k_{\mathrm{s}}AGL^2}$ [17, 27]:

$$\boldsymbol{K}^{\mathrm{e}} = \frac{EI_z}{6\Lambda L^3} \begin{bmatrix} 6 & -3L & -6 & -3L \\ -3L & L^2(2+6\Lambda) & 3L & L^2(1-6\Lambda) \\ -6 & 3L & 6 & 3L \\ -3L & L^2(1-6\Lambda) & 3L & L^2(2+6\Lambda) \end{bmatrix}. \tag{4.137}$$

Thus, the principal finite element equation for the linear TIMOSHENKO beam element can be written as:

$$\frac{k_{\mathrm{s}}AG}{4L} \begin{bmatrix} 4 & -2L & -4 & -2L \\ -2L & \frac{4}{3}L^2+\alpha & 2L & \frac{4}{6}L^2-\alpha \\ -4 & 2L & 4 & 2L \\ -2L & \frac{4}{6}L^2-\alpha & 2L & \frac{4}{3}L^2+\alpha \end{bmatrix} \begin{bmatrix} u_{1z} \\ \phi_{1y} \\ u_{2z} \\ \phi_{2y} \end{bmatrix} = \begin{bmatrix} F_{1z} \\ M_{1y} \\ F_{2z} \\ M_{2y} \end{bmatrix} + \int_0^L q_z(x) \begin{bmatrix} N_{1u} \\ 0 \\ N_{2u} \\ 0 \end{bmatrix} \mathrm{d}x\,. \tag{4.138}$$

Table 4.8 Summary: Derivation of principal finite element equation for linear TIMOSHENKO beam elements

Strong Formulation

$$k_{\mathrm{s}}GA\left(\frac{\mathrm{d}^2u_z^0}{\mathrm{d}x^2}+\frac{\mathrm{d}\phi_y^0}{\mathrm{d}x}\right)+q_z(x)=0\ \text{(shear)}$$

$$EI_y\frac{\mathrm{d}^2\phi_y^0}{\mathrm{d}x^2}-k_{\mathrm{s}}GA\left(\frac{\mathrm{d}u_z^0}{\mathrm{d}x}+\phi_y^0\right)=0\ \text{(bending)}$$

Inner Product

$$\int_0^L\left\{k_{\mathrm{s}}AG\left(\frac{\mathrm{d}^2u_z}{\mathrm{d}x^2}+\frac{\mathrm{d}\phi_y}{\mathrm{d}x}\right)+q_z(x)\right\}W_u(x)\mathrm{d}x\overset{!}{=}0\ \text{(shear)}$$

$$\int_0^L\left\{\frac{\mathrm{d}}{\mathrm{d}x}\left(EI_y\frac{\mathrm{d}\phi_y}{\mathrm{d}x}\right)-k_{\mathrm{s}}AG\left(\frac{\mathrm{d}u_z}{\mathrm{d}x}+\phi_y\right)\right\}W_\phi(x)\mathrm{d}x\overset{!}{=}0\ \text{(bending)}$$

Weak Formulation

$$\int_0^L EI_y\frac{\mathrm{d}\phi_y}{\mathrm{d}x}\frac{\mathrm{d}W_\phi}{\mathrm{d}x}\mathrm{d}x+\int_0^L k_{\mathrm{s}}AG\left(\frac{\mathrm{d}u_z}{\mathrm{d}x}+\phi_y\right)\left(\frac{\mathrm{d}W_u}{\mathrm{d}x}+W_\phi\right)\mathrm{d}x$$
$$=\int_0^L q_zW_u\mathrm{d}x+\left[k_{\mathrm{s}}AG\left(\frac{\mathrm{d}u_z}{\mathrm{d}x}+\phi_y\right)W_u\right]_0^L+\left[EI_y\frac{\mathrm{d}\phi_y}{\mathrm{d}x}W_\phi\right]_0^L$$

Principal Finite Element Equation ($\alpha=\frac{4EI_y}{k_{\mathrm{s}}AG}$)

$$\frac{k_{\mathrm{s}}AG}{4L}\begin{bmatrix}4 & -2L & -4 & -2L\\ -2L & \frac{4}{3}L^2+\alpha & 2L & \frac{4}{6}L^2-\alpha\\ -4 & 2L & 4 & 2L\\ -2L & \frac{4}{6}L^2-\alpha & 2L & \frac{4}{3}L^2+\alpha\end{bmatrix}\begin{bmatrix}u_{1z}\\ \phi_{1y}\\ u_{2z}\\ \phi_{2y}\end{bmatrix}=\begin{bmatrix}F_{1z}\\ M_{1y}\\ F_{2z}\\ M_{2y}\end{bmatrix}+\int_0^L q_z(x)\begin{bmatrix}N_{1u}\\ 0\\ N_{2u}\\ 0\end{bmatrix}\mathrm{d}x$$

Let us summarize at the end of this derivation the major steps that were undertaken to transform the partial differential equations into the principal finite element equation, see Table 4.8.

Based on the general formulation of the partial differential equation given in Table 4.5 and the general derivations presented in second part of Sect. 4.3.1, the major steps to transform the partial differential equation into the principal finite element equation are summarized in Table 4.9. Alternatively, Table 4.10 summarizes the major steps based on the modified equations from Table 4.7.

In the following, the deformation behavior of this analytically integrated[8] TIMOSHENKO element needs to be analyzed. For this, the configuration in Fig. 4.12 needs

[8] A numerical GAUSS integration with two integration points yields the same results as the exact analytical integration.

Table 4.9 Summary: Derivation of principal finite element equation for linear TIMOSHENKO beam elements; general approach (see Table 4.5)

Strong Formulation

$$\boldsymbol{\mathcal{L}}_1^{\mathrm{T}} \boldsymbol{D} \boldsymbol{\mathcal{L}}_1 \boldsymbol{u}^0 + \boldsymbol{b} = \boldsymbol{0}$$

Inner Product

$$\int_L \boldsymbol{W}^{\mathrm{T}} \left(\boldsymbol{\mathcal{L}}_1^{\mathrm{T}} \boldsymbol{D} \boldsymbol{\mathcal{L}}_1 \boldsymbol{u} + \boldsymbol{b} \right) \mathrm{d}L = 0 \text{ with } \boldsymbol{\mathcal{L}}_1^{\mathrm{T}} = \boldsymbol{\mathcal{L}}_{1,a}^{\mathrm{T}} + \boldsymbol{\mathcal{L}}_{1,b}^{\mathrm{T}}$$

Weak Formulation

$$\int_L \left(\boldsymbol{\mathcal{L}}_{1,a} \boldsymbol{W} \right)^{\mathrm{T}} \boldsymbol{D} \left(\boldsymbol{\mathcal{L}}_1 \boldsymbol{u} \right) \mathrm{d}L - \int_L \boldsymbol{W}^{\mathrm{T}} \boldsymbol{\mathcal{L}}_{1,b}^{\mathrm{T}} \boldsymbol{D} \boldsymbol{\mathcal{L}}_1 \boldsymbol{u} \, \mathrm{d}L$$

$$= \int_s \boldsymbol{W}^{\mathrm{T}} \boldsymbol{s}^{\mathrm{T}} \boldsymbol{n} \mathrm{d}s + \int_L \boldsymbol{W}^{\mathrm{T}} \boldsymbol{b} \, \mathrm{d}L$$

Principal Finite Element Equation (line 2)

$$\underbrace{\left(\int_L (\boldsymbol{\mathcal{L}}_{1,a} \boldsymbol{N}^{\mathrm{T}})^{\mathrm{T}} \boldsymbol{D} (\boldsymbol{\mathcal{L}}_1 \boldsymbol{N}^{\mathrm{T}}) \mathrm{d}L - \int_L (\boldsymbol{\mathcal{L}}_{1,b} \boldsymbol{N}^{\mathrm{T}})^{\mathrm{T}} \boldsymbol{D} (\boldsymbol{\mathcal{L}}_1 \boldsymbol{N}^{\mathrm{T}}) \mathrm{d}L \right)}_{\boldsymbol{k}^{\mathrm{e}}} \begin{bmatrix} u_{1z} \\ \phi_{1y} \\ u_{2z} \\ \phi_{2y} \end{bmatrix}$$

$$= \begin{bmatrix} F_{1z} \\ M_{1y} \\ F_{2z} \\ M_{2y} \end{bmatrix} + \int_L \boldsymbol{N} \boldsymbol{b} \mathrm{d}L$$

Comment: Multiplied with $\lceil -1 \;\; 1 \;\; -1 \;\; -1 \rfloor$ to obtain the same formulation as in Eqs. (4.72) and (4.77).

to be considered for which the beam has a fixed support on the left-hand side and a point load on the right-hand side. The displacement of the loading point has to be analyzed.

Through the stiffness matrix according to Eq. (4.136), the principal finite element equation for a single element results in

$$\frac{k_s AG}{4L} \begin{bmatrix} 4 & -2L & -4 & -2L \\ -2L & \frac{4}{3}L^2 + \alpha & 2L & \frac{4}{6}L^2 - \alpha \\ -4 & 2L & 4 & 2L \\ -2L & \frac{4}{6}L^2 - \alpha & 2L & \frac{4}{3}L^2 + \alpha \end{bmatrix} \begin{bmatrix} u_{1z} \\ \phi_{1y} \\ u_{2z} \\ \phi_{2y} \end{bmatrix} = \begin{bmatrix} \cdots \\ \cdots \\ F_0 \\ 0 \end{bmatrix}, \tag{4.139}$$

Table 4.10 Summary: Derivation of principal finite element equation for linear TIMOSHENKO beam elements; general approach based on modified equations (see Table 4.7)

Strong Formulation
$\boldsymbol{\mathcal{L}}_{1*}^{\mathrm{T}}\boldsymbol{D}^*\boldsymbol{\mathcal{L}}_1\boldsymbol{u}^0 + \boldsymbol{b}^* = \boldsymbol{0}$
Inner Product
$\int_L \boldsymbol{W}^{\mathrm{T}}\left(\boldsymbol{\mathcal{L}}_{1*}^{\mathrm{T}}\boldsymbol{D}^*\boldsymbol{\mathcal{L}}_1\boldsymbol{u} + \boldsymbol{b}^*\right)\mathrm{d}L = 0$ with $\boldsymbol{\mathcal{L}}_{1*}^{\mathrm{T}} = \boldsymbol{\mathcal{L}}_{1*,a}^{\mathrm{T}} + \boldsymbol{\mathcal{L}}_{1*,b}^{\mathrm{T}}$
Weak Formulation
$\int_L (\boldsymbol{\mathcal{L}}_1\boldsymbol{W})^{\mathrm{T}}\boldsymbol{D}^*(\boldsymbol{\mathcal{L}}_1\boldsymbol{u})\,\mathrm{d}L = \int_s \boldsymbol{W}^{\mathrm{T}}(\boldsymbol{s}^*)^{\mathrm{T}}\boldsymbol{n}\mathrm{d}s + \int_L \boldsymbol{W}^{\mathrm{T}}\boldsymbol{b}^*\,\mathrm{d}L$
Principal Finite Element Equation (line 2)
$\underbrace{\int_L (\boldsymbol{\mathcal{L}}_1\boldsymbol{N}^{\mathrm{T}})^{\mathrm{T}}\boldsymbol{D}^*(\boldsymbol{\mathcal{L}}_1\boldsymbol{N}^{\mathrm{T}})\mathrm{d}L}_{\boldsymbol{k}^{\mathrm{e}}}\begin{bmatrix} u_{1z} \\ \phi_{1y} \\ u_{2z} \\ \phi_{2y}\end{bmatrix} = \begin{bmatrix} F_{1z} \\ M_{1y} \\ F_{2z} \\ M_{2y}\end{bmatrix} + \int_L \boldsymbol{N}\boldsymbol{b}^*\mathrm{d}L$

or alternatively after considering the fixed support ($u_{1y} = 0, \phi_{1z} = 0$) of the left-hand side:

$$\frac{k_{\mathrm{s}}AG}{4L}\begin{bmatrix} 4 & 2L \\ 2L & \frac{4}{3}L^2 + \alpha \end{bmatrix}\begin{bmatrix} u_{2z} \\ \phi_{2y}\end{bmatrix} = \begin{bmatrix} F_0 \\ 0 \end{bmatrix}. \tag{4.140}$$

Solving this 2×2 system of equations for the unknown parameters on the right-hand boundary yields:

$$\begin{bmatrix} u_{2z} \\ \phi_{2y}\end{bmatrix} = \frac{4L}{k_{\mathrm{s}}AG} \times \frac{1}{4(\frac{4}{3}L^2 + \alpha) - (2L)(2L)}\begin{bmatrix} \frac{4}{3}L^2 + \alpha & -2L \\ -2L & 4 \end{bmatrix}\begin{bmatrix} F_0 \\ 0 \end{bmatrix}, \tag{4.141}$$

or alternatively solved for the unknown displacement at the right-hand boundary:

$$u_{2z}(L) = \frac{12EI_y + 4k_{\mathrm{s}}AGL^2}{12EI_y + k_{\mathrm{s}}AGL^2} \times \left(\frac{F_0 L}{k_{\mathrm{s}}AG}\right). \tag{4.142}$$

Considering the rectangular cross section, illustrated in Fig. 4.12, meaning $A = hb$, $I_y = \frac{1}{12}bh^3$, and $k_{\mathrm{s}} = \frac{5}{6}$ and furthermore the relation for the shear modulus according to Eq. (4.23), after a short calculation the displacement on the right-hand end results:

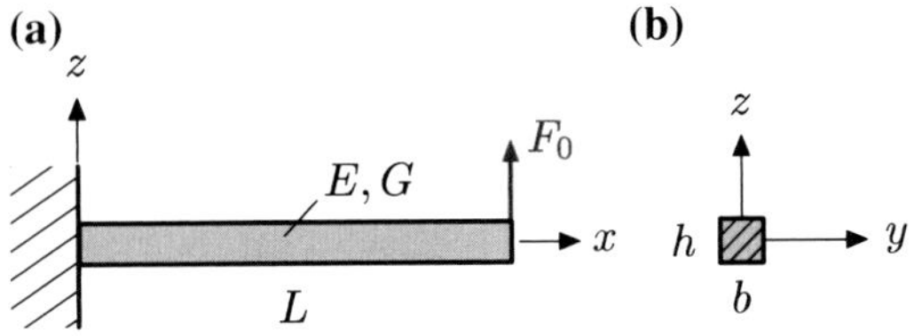

Fig. 4.12 Analysis of a TIMOSHENKO element under point load

$$u_{2z}(L) = \frac{12(1+\nu)\left(\frac{h}{L}\right)^2 + 20}{60 + 25\left(\frac{L}{h}\right)^2 \dfrac{1}{1+\nu}} \times \left(\frac{F_0 L^3}{E I_y}\right). \tag{4.143}$$

For very thick beams, meaning $h \gg L$, $\frac{L}{h} \to 0$ results and Eq. (4.143) converges against the analytical solution.[9] For very slender beams, however, meaning $h \ll L$, a boundary value[10] of $\frac{4F_0L}{k_sAG} = \frac{4}{5}(1+\nu)(\frac{h}{L})^2 \frac{F_0L^3}{EI_y}$ results from Eq. (4.142).

This boundary value only contains the shear part without bending and runs against a wrong solution. This phenomenon is called *shear locking*. A graphical illustration of this behavior is given in Fig. 4.13 via the normalized deflection with the BERNOULLI solution. One can clearly see the different convergence behaviors for different domains of the slenderness ratio, meaning for slender and compact beams.

For the improvement of the convergence behavior, the literature suggests [3, 18] to conduct the integration via numerical GAUSS integration with only one integration point. Therefore, the arguments and the integration boundaries in the formulations of the element stiffness matrices for $\boldsymbol{K}_\mathrm{b}^\mathrm{e}$ and $\boldsymbol{K}_\mathrm{s}^\mathrm{e}$ according to Eqs. (4.69) and (4.70) have to be transformed into the natural coordinate $-1 \leq \xi \leq 1$. Furthermore, the interpolation functions need to be used according to Fig. 4.11. Via the transformation of the derivative to the new coordinate, meaning $\frac{\mathrm{d}N}{\mathrm{d}x} = \frac{\mathrm{d}N}{\mathrm{d}\xi}\frac{\mathrm{d}\xi}{\mathrm{d}x}$ and the transformation of the coordinate $\xi = -1 + 2\frac{x}{L}$ or alternatively $\mathrm{d}\xi = \frac{2}{L}\mathrm{d}x$, the bending stiffness matrix results in:

$$\boldsymbol{K}_\mathrm{b}^\mathrm{e} = EI_y \int_{-1}^{1} \frac{4}{L^2} \begin{bmatrix} 0 & 0 & 0 & 0 \\ 0 & \frac{\mathrm{d}N_{1\phi}}{\mathrm{d}\xi}\frac{\mathrm{d}N_{1\phi}}{\mathrm{d}\xi} & 0 & \frac{\mathrm{d}N_{1\phi}}{\mathrm{d}\xi}\frac{\mathrm{d}N_{2\phi}}{\mathrm{d}\xi} \\ 0 & 0 & 0 & 0 \\ 0 & \frac{\mathrm{d}N_{2\phi}}{\mathrm{d}\xi}\frac{\mathrm{d}N_{1\phi}}{\mathrm{d}\xi} & 0 & \frac{\mathrm{d}N_{2\phi}}{\mathrm{d}\xi}\frac{\mathrm{d}N_{2\phi}}{\mathrm{d}\xi} \end{bmatrix} \frac{L}{2}\mathrm{d}\xi\,, \tag{4.144}$$

$$\boldsymbol{K}_\mathrm{b}^\mathrm{e} = \frac{2EI_y}{L}\int_{-1}^{1} \frac{4}{L^2} \begin{bmatrix} 0 & 0 & 0 & 0 \\ 0 & \frac{1}{4} & 0 & -\frac{1}{4} \\ 0 & 0 & 0 & 0 \\ 0 & -\frac{1}{4} & 0 & \frac{1}{4} \end{bmatrix} \mathrm{d}\xi = \frac{EI_y}{2L}\sum_{i=1}^{1} \begin{bmatrix} 0 & 0 & 0 & 0 \\ 0 & 1 & 0 & -1 \\ 0 & 0 & 0 & 0 \\ 0 & -1 & 0 & 1 \end{bmatrix} \times 2 \tag{4.145}$$

[9]For this see Fig. 4.8 and the supplementary Problem 4.11.

[10]One considers the definition of I_y and A in Eq. (4.142). Factor our L^2 and divide the fraction by h.

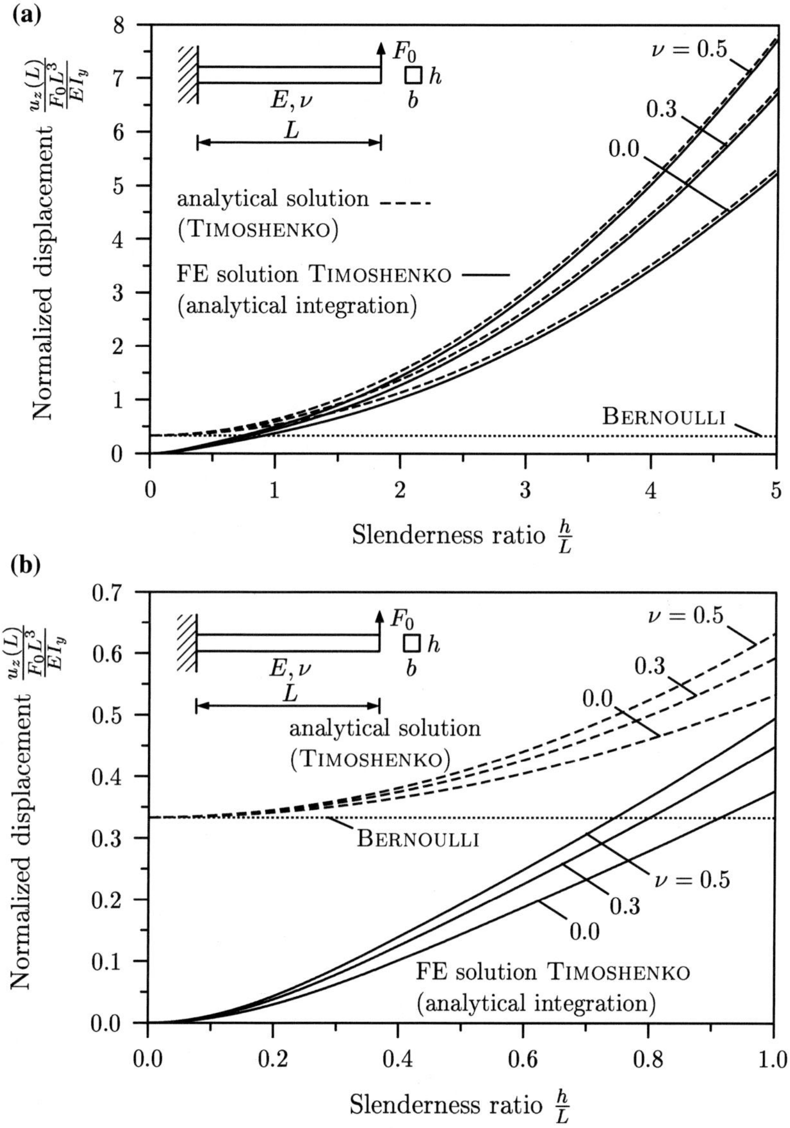

Fig. 4.13 Comparison of the analytical solution for a TIMOSHENKO beam and the corresponding discretization via one single finite element with analytical integration of the stiffness matrix: **a** general view and **b** magnification for small slenderness ratios

and after all in the final formulation in:

$$\boldsymbol{K}_{\mathrm{b}}^{\mathrm{e}} = EI_y \begin{bmatrix} 0 & 0 & 0 & 0 \\ 0 & \frac{1}{L} & 0 & -\frac{1}{L} \\ 0 & 0 & 0 & 0 \\ 0 & -\frac{1}{L} & 0 & \frac{1}{L} \end{bmatrix} . \tag{4.146}$$

One can see that the same result for the bending stiffness matrix results as for the analytical integration. In the case of the bending stiffness matrix therefore the GAUSS integration with just one integration point is accurate.

The following expression results for the shear stiffness matrix under the use of the natural coordinate:

$$\frac{2k_{\mathrm{s}}GA}{L}\int_{-1}^{1}\begin{bmatrix} \frac{\mathrm{d}N_{1u}}{\mathrm{d}\xi}\frac{\mathrm{d}N_{1u}}{\mathrm{d}\xi} & \frac{L}{2}\frac{\mathrm{d}N_{1u}}{\mathrm{d}\xi}(N_{1\phi}) & \frac{\mathrm{d}N_{1u}}{\mathrm{d}\xi}\frac{\mathrm{d}N_{2u}}{\mathrm{d}\xi} & \frac{L}{2}\frac{\mathrm{d}N_{1u}}{\mathrm{d}\xi}(N_{2\phi}) \\ \frac{L}{2}(N_{1\phi})\frac{\mathrm{d}N_{1u}}{\mathrm{d}\xi} & \frac{L^2}{4}(N_{1\phi})(N_{1\phi}) & \frac{L}{2}(N_{1\phi})\frac{\mathrm{d}N_{2u}}{\mathrm{d}\xi} & \frac{L^2}{4}(N_{1\phi})(N_{2\phi}) \\ \frac{\mathrm{d}N_{2u}}{\mathrm{d}\xi}\frac{\mathrm{d}N_{1u}}{\mathrm{d}\xi} & \frac{L}{2}\frac{\mathrm{d}N_{2u}}{\mathrm{d}\xi}(N_{1\phi}) & \frac{\mathrm{d}N_{2u}}{\mathrm{d}\xi}\frac{\mathrm{d}N_{2u}}{\mathrm{d}\xi} & \frac{L}{2}\frac{\mathrm{d}N_{2u}}{\mathrm{d}\xi}(N_{2\phi}) \\ \frac{L}{2}(N_{2\phi})\frac{\mathrm{d}N_{1u}}{\mathrm{d}\xi} & \frac{L^2}{4}(N_{2\phi})(N_{1\phi}) & \frac{L}{2}(N_{2\phi})\frac{\mathrm{d}N_{2u}}{\mathrm{d}\xi} & \frac{L^2}{4}(N_{2\phi})(N_{2\phi}) \end{bmatrix}\mathrm{d}\xi\,, \tag{4.147}$$

or alternatively after the introduction of the interpolation functions

$$\frac{2k_{\mathrm{s}}GA}{L}\int_{-1}^{1}\begin{bmatrix} \frac{1}{4} & \frac{L}{2}\left(-\frac{1}{4}+\frac{\xi}{4}\right) & -\frac{1}{4} & \frac{L}{2}\left(-\frac{1}{4}-\frac{\xi}{4}\right) \\ \frac{L}{2}\left(-\frac{1}{4}+\frac{\xi}{4}\right) & \frac{L^2}{4}\left(\frac{(1-\xi)^2}{4}\right) & \frac{L}{2}\left(\frac{1}{4}-\frac{\xi}{4}\right) & \frac{L^2}{4}\left(\frac{1}{4}-\frac{\xi^2}{4}\right) \\ -\frac{1}{4} & \frac{L}{2}\left(\frac{1}{4}-\frac{\xi}{4}\right) & \frac{1}{4} & \frac{L}{2}\left(\frac{1}{4}+\frac{\xi}{4}\right) \\ \frac{L}{2}\left(-\frac{1}{4}-\frac{\xi}{4}\right) & \frac{L^2}{4}\left(\frac{1}{4}-\frac{\xi^2}{4}\right) & \frac{L}{2}\left(\frac{1}{4}+\frac{\xi}{4}\right) & \frac{L^2}{4}\left(\frac{(1+\xi)^2}{4}\right) \end{bmatrix}\mathrm{d}\xi\,, \tag{4.148}$$

or after the transition to the numerical integration

$$\frac{2k_{\mathrm{s}}GA}{L}\begin{bmatrix} \frac{1}{4} & -\frac{L}{2}\frac{1}{4} & -\frac{1}{4} & -\frac{L}{2}\frac{1}{4} \\ -\frac{L}{2}\frac{1}{4} & \frac{L^2}{4}\frac{1}{4} & \frac{L}{2}\frac{1}{4} & \frac{L^2}{4}\frac{1}{4} \\ -\frac{1}{4} & \frac{L}{2}\frac{1}{4} & \frac{1}{4} & \frac{L}{2}\frac{1}{4} \\ -\frac{L}{2}\frac{1}{4} & \frac{L^2}{4}\frac{1}{4} & \frac{L}{2}\frac{1}{4} & \frac{L^2}{4}\frac{1}{4} \end{bmatrix}_{\xi_i=0} \times 2 \tag{4.149}$$

and after all in the final formulation as:

$$\boldsymbol{K}_{\mathrm{s}}^{\mathrm{e}} = k_{\mathrm{s}}AG\begin{bmatrix} \frac{1}{L} & -\frac{1}{2} & -\frac{1}{L} & -\frac{1}{2} \\ -\frac{1}{2} & \frac{L}{4} & \frac{1}{2} & \frac{L}{4} \\ -\frac{1}{L} & \frac{1}{2} & \frac{1}{L} & \frac{1}{2} \\ -\frac{1}{2} & \frac{L}{4} & \frac{1}{2} & \frac{L}{4} \end{bmatrix} . \tag{4.150}$$

The two stiffness matrices according to Eqs. (4.146) and (4.150) can be summarized additively to the total stiffness matrix of the TIMOSHENKO beam and with the abbreviation $\alpha = \frac{4EI_y}{k_s AG}$ the following results:

$$\boldsymbol{K}^{\mathrm{e}} = \frac{k_s AG}{4L} \begin{bmatrix} 4 & -2L & -4 & -2L \\ -2L & L^2 + \alpha & 2L & L^2 - \alpha \\ -4 & 2L & 4 & 2L \\ -2L & L^2 - \alpha & 2L & L^2 + \alpha \end{bmatrix}, \tag{4.151}$$

or alternatively via the abbreviation $\Lambda = \dfrac{EI_z}{k_s AGL^2}$:

$$\boldsymbol{K}^{\mathrm{e}} = \frac{EI_y}{6\Lambda L^3} \begin{bmatrix} 6 & -3L & -6 & -3L \\ -3L & L^2(1.5 + 6\Lambda) & 3L & L^2(1.5 - 6\Lambda) \\ -6 & 3L & 6 & 3L \\ -3L & L^2(1.5 - 6\Lambda) & 3L & L^2(1.5 + 6\Lambda) \end{bmatrix}. \tag{4.152}$$

With the help of this formulation for the stiffness matrix the example according to Fig. 4.12 needs to be analyzed once again in the following to investigate the differences to the analytical integration. Via the stiffness matrix according to Eq. (4.151) the principal finite element equation for a single element under consideration of the fixed support ($u_{1z} = 0$, $\phi_{1y} = 0$) on the left-hand side results in:

$$\frac{k_s AG}{4L} \begin{bmatrix} 4 & 2L \\ 2L & L^2 + \alpha \end{bmatrix} \begin{bmatrix} u_{2z} \\ \phi_{2y} \end{bmatrix} = \begin{bmatrix} F_0 \\ 0 \end{bmatrix}. \tag{4.153}$$

Solving of this 2×2 system of equations for the unknown displacement on the right-hand side yields:

$$u_{2z}(L) = \left(1 + \frac{4EI_y}{k_s AGL^2}\right) \times \frac{F_0 L^3}{4EI_y}. \tag{4.154}$$

If the illustrated rectangular cross section in Fig. 4.12 is being considered at this point as well, after a short calculation the displacement on the right-hand side, via $A = hb$, $I_y = \frac{1}{12}bh^3$, $k_s = \frac{5}{6}$ and the relation for the shear modulus according to Eq. (4.23) results in:

$$u_{2z}(L) = \left(\frac{1}{4} + \frac{1}{5}(1 + \nu)\left(\frac{h}{L}\right)^2\right) \times \left(\frac{F_0 L^3}{EI_y}\right). \tag{4.155}$$

For very thick beams, meaning $h \gg L$, the solution converges against the analytical solution.[11] For very slender beams, however, meaning $h \ll L$, a boundary value

[11] For this see Fig. 4.8 and the supplementary Problem 4.11.

of $\frac{F_0 L^3}{4EI_y}$ results from Eq. (4.155), whereupon the analytical solution yields a value of $\frac{F_0 L^3}{3EI_y}$. However, the phenomenon of shear locking does not occur and therefore, compared to the stiffness matrix based on the analytical integration, an improvement of the element formulation has been achieved.

A graphical illustration of this behavior via the normalized deflection is given in Fig. 4.14. One can clearly see the improved convergence behavior for small slenderness ratios. For larger slenderness ratios the behavior remains according to the result of the analytical integration, since both approaches converge against the analytical solution.

When the differential equations according to (4.33) and (4.34) are considered, it becomes obvious that the derivative $\frac{\mathrm{d}u_z}{\mathrm{d}x}$ and the function ϕ_y itself are contained there. If linear interpolation functions are being used for u_z and ϕ_y, the degree for polynomials for $\frac{\mathrm{d}u_z}{\mathrm{d}x}$ and ϕ_y is different. In the limiting case of slender beams, however, the relation $\phi \approx \frac{\mathrm{d}u_z}{\mathrm{d}x}$ has to be fulfilled and the consistency of the polynomials for $\frac{\mathrm{d}u_z}{\mathrm{d}x}$ and ϕ_y is of importance. The linear approach for u_z yields for $\frac{\mathrm{d}u_z}{\mathrm{d}x}$ a constant function and therefore also for ϕ_y a constant would be desirable. However, it needs to be considered at this point that the demand for the differentiability of ϕ_y at least results in a linear function. The one-point integration[12] in the case of the shear stiffness matrix with the expressions $N_{i\phi}N_{j\phi}$ causes however that the linear approach for ϕ_y is treated as a constant term, since two integration points would have to be used for an exact integration. A one-point integration can at most integrate a polynomial of first order exactly, meaning proportional to x^1, and therefore the following point of view results $(N_{i\phi}N_{j\phi}) \sim x^1$. However, this means that at most $N_{i\phi} \sim x^{0.5}$ or alternatively $N_{j\phi} \sim x^{0.5}$ holds. Since the polynomial approach solely allows integer values for the exponent of x, $N_{i\phi} \sim x^0$ or alternatively $N_{j\phi} \sim x^0$ results and the rotation needs to be seen as a constant term. This is consistent with the demand that the shear strain $\gamma_{xz} = \frac{\mathrm{d}u_z}{\mathrm{d}x} + \phi_y$ has to be constant in an element for constant bending stiffness EI_y. Therefore, *shear locking* does not occur in this case.

As another option for the improvement of the convergence behavior of linear Timoshenko elements with numerical one-point integrations, [3, 11] suggests to correct the shear stiffness $k_s AG$ according to the analytical correct solution.[13] To this end, the elastic strain energy is being regarded (see [12]), which results as follows:

$$\Pi_{\text{int}} = \frac{1}{2}\int\limits_{\Omega} \boldsymbol{\varepsilon}^{\mathrm{T}}\boldsymbol{\sigma}\mathrm{d}\Omega = \frac{1}{2}\int\limits_{\Omega} \varepsilon_x \sigma_x \mathrm{d}A\mathrm{d}x + \frac{1}{2}\int\limits_{\Omega} \gamma_{xz}\tau_{xz}\mathrm{d}A\mathrm{d}x\,, \tag{4.156}$$

$$\Pi_{\text{int}} = \frac{1}{2}\int\limits_{0}^{L} EI_y \left(\frac{\mathrm{d}\phi_y(x)}{\mathrm{d}x}\right)^2 \mathrm{d}x + \frac{1}{2}\int\limits_{0}^{L} k_s AG \left(\frac{\mathrm{d}u_z(x)}{\mathrm{d}x} + \phi_y(x)\right)^2 \mathrm{d}x\,. \tag{4.157}$$

[12]The numerical integration according to the Gauss-Legendre method with n integration points integrates a polynomial, which degree is at most $2n-1$, exactly.

[13]MacNeal herefore uses the expression 'residual bending flexibility' [10, 19].

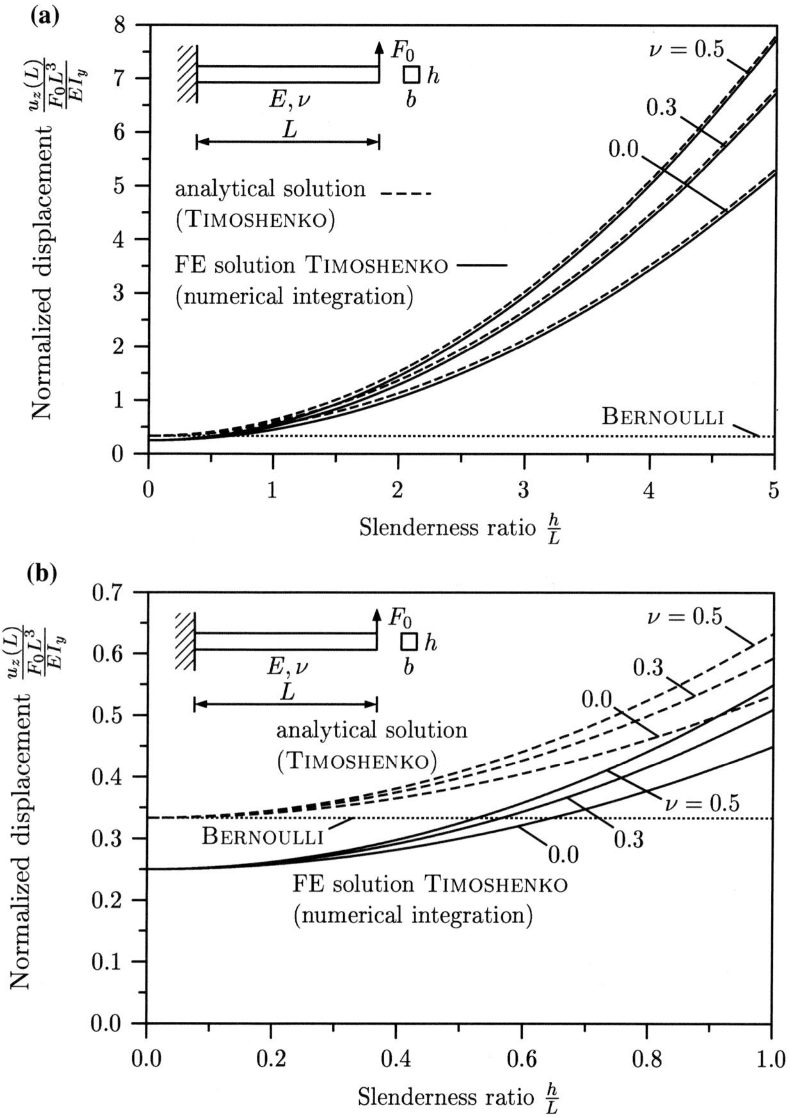

Fig. 4.14 Comparison of the analytical solution for a TIMOSHENKO beam and the appropriate discretization via one single finite element with one-point numerical integration of the corresponding matrix: **a** general view and **b** magnification for small slenderness ratios

It is now demanded that the strain energy for the analytical solution and the finite element solution under the use of the corrected shear stiffness $(k_s AG)^*$ are identical. The analytical solution[14] for the problem in Fig. 4.12 results in

[14]For this see the supplementary Problem 4.9.

$$u_z(x) = \frac{1}{EI_y}\left(-F_0\frac{x^3}{6} + F_0L\frac{x^2}{2} + \frac{EI_yF_0}{k_sAG}x\right), \tag{4.158}$$

$$\phi_y(x) = -\frac{1}{EI_y}\left(-F_0\frac{x^2}{2} + F_0Lx\right), \tag{4.159}$$

and the elastic strain energy for the analytical solution therefore results in:

$$\Pi_{\text{int}} = \frac{F_0^2}{2EI_y}\int_0^L (L-x)^2\text{d}x + \frac{F_0^2(EI_y)^2}{2k_sAG}\int_0^L \text{d}x = \frac{F_0^2L^3}{6EI_y} + \frac{F_0^2L}{2k_sAG}. \tag{4.160}$$

Via Eq. (4.153) the finite element solution of the displacement and rotation distributions can be obtained as

$$u_z(x) = 0 + N_{2u}u_{2z} = \frac{x}{L}\times\left(1+\frac{4EI_y}{k_sAGL^2}\right)\times\frac{F_0L^2}{4EI_y}, \tag{4.161}$$

$$\phi_y(x) = 0 + N_{2\phi}\phi_{2y} = -\frac{x}{L}\times\frac{F_0L^2}{2EI_y}, \tag{4.162}$$

and the elastic strain energy results in:

$$\begin{aligned}\Pi_{\text{int}} =& \frac{EI_y}{2}\int_0^L\left(\frac{F_0L}{2EI_y}\right)^2\text{d}x\\ &+\frac{(k_sAG)^*}{2}\int_0^L\left(\left(1+\frac{4EI_y}{(k_sAG)^*L^2}\right)\frac{F_0L^2}{4EI_y}-\frac{F_0Lx}{2EI_y}\right)^2\text{d}x\,.\end{aligned} \tag{4.163}$$

This integral has to be evaluated numerically with a one-point integration rule and it is therefore necessary to introduce the natural coordinate via the transformations $x = \frac{L}{2}(\xi+1)$ and $\text{d}x = \frac{L}{2}\text{d}\xi$:

$$\begin{aligned}\Pi_{\text{int}} =& \frac{F_0^2L^3}{8EI_y}\\ &+\frac{(k_sAG)^*}{2}\int_{-1}^{1}\left(\left(1+\frac{4EI_y}{(k_sAG)^*L^2}\right)\frac{F_0L^2}{4EI_y}-\frac{F_0L}{2EI_y}(\xi+1)\frac{L}{2}\right)^2\frac{L}{2}\text{d}\xi\\ =& \frac{F_0^2L^3}{8EI_y}+\frac{(k_sAG)^*}{2}\left(\frac{4EI_y}{(k_sAG)^*L^2}\times\frac{F_0L^2}{4EI_y}\right)^2\frac{L}{2}2\end{aligned} \tag{4.164}$$

and finally

$$\Pi_{\text{int}} = \frac{F_0^2 L^3}{8EI_z} + \frac{F_0^2 L}{2(k_s AG)^*}. \tag{4.165}$$

Equalizing of the two energy expressions according to Eqs. (4.160) and (4.165) finally yields the corrected shear stiffness:

$$(k_s AG)^* = \left(\frac{L^2}{12EI_y} + \frac{1}{k_s AG}\right)^{-1}. \tag{4.166}$$

By inserting this—with the 'residual bending flexibility' $\frac{L^2}{12EI_y}$ corrected—shear stiffness into the finite element solution according to Eq. (4.154), the analytically exact solution results. The same result is derived in [11], starting from the general—meaning without considering a certain support of the beam—solution for the beam deflection, and in [3] the derivation for the equality of the deflection on the loading point according to the analytical and the corrected finite element solution takes place. It is to be considered that the derived corrected shear stiffness is not just valid for the cantilever beam under point load, but yields the same value for arbitrary supports and loads on the ends of the beam. However, the derivation of the corrected shear stiffness for nonhomogeneous, anisotropic and non-linear materials appears problematic [3].

4.3.2.1 Post-computation: Determination of Strain, Stress and Further Quantities

The solution of the principal finite element equation, i.e. a linear system of equations according to Tables 4.8 and 4.9, results in the nodal displacements (u_{1z}, u_{2z}) and the nodal rotations (ϕ_{1y}, ϕ_{2y}).

Based on the nodal values and the corresponding distributions given in Table 4.11, the values for the strains and stresses (see Table 4.3) can be obtained. For the normal components

$$\varepsilon_x(x, z) = +\frac{\mathrm{d}\phi_y(x)}{\mathrm{d}x} \quad \text{and} \quad \sigma_{x,z} = E\varepsilon_x(x, z), \tag{4.167}$$

while the shear stress reads:

$$\tau_{xy} = \frac{Q_z(x)}{k_s A} = G\gamma_{xz}(x). \tag{4.168}$$

Table 4.11 Displacement, rotation, curvature, shear strain, shear force and bending moment distribution for a linear TIMOSHENKO beam element given as a function of the nodal values. Bending occurs in the x-z plane

Vertical Displacement (Deflection) u_z
$u_z^{\mathrm{e}}(x) = \left[1 - \frac{x}{L}\right]u_{1z} + \left[\frac{x}{L}\right]u_{2z}$ $u_z^{\mathrm{e}}(\xi) = \left[\frac{1}{2}(1-\xi)\right]u_{1z} + \left[\frac{1}{2}(1+\xi)\right]u_{2z}$
Rotation ϕ_y
$\phi_y^{\mathrm{e}}(x) = \left[1 - \frac{x}{L}\right]\phi_{1y} + \left[\frac{x}{L}\right]\phi_{2y}$ $\phi_y^{\mathrm{e}}(\xi) = \left[\frac{1}{2}(1-\xi)\right]\phi_{1y} + \left[\frac{1}{2}(1+\xi)\right]\phi_{2y}$
Curvature $\kappa_y = \frac{\mathrm{d}\phi_y}{\mathrm{d}x} = \frac{\mathrm{d}\phi_y}{\mathrm{d}\xi}\frac{\mathrm{d}\xi}{\mathrm{d}x} = \frac{2}{L}\frac{\mathrm{d}\phi_y}{\mathrm{d}\xi}$
$\kappa_y^{\mathrm{e}}(x) = \left[-\frac{1}{L}\right]\phi_{1y} + \left[\frac{1}{L}\right]\phi_{2y}$ $\kappa_y^{\mathrm{e}}(\xi) = \left[-\frac{1}{L}\right]\phi_{1y} + \left[\frac{1}{L}\right]\phi_{2y}$
Shear Strain $\gamma_{xz} = \frac{\mathrm{d}u_z}{\mathrm{d}x} + \phi_y = \frac{\mathrm{d}\xi}{\mathrm{d}x}\frac{\mathrm{d}u_z}{\mathrm{d}\xi} + \phi_y$
$\gamma_{xz}^{\mathrm{e}}(x) = \left[-\frac{1}{L}\right]u_{1z} + \left[\frac{1}{L}\right]u_{2z} + \left[1 - \frac{x}{L}\right]\phi_{1y} + \left[\frac{x}{L}\right]\phi_{2y}$ $\gamma_{xz}^{\mathrm{e}}(\xi) = \left[-\frac{1}{L}\right]u_{1z} + \left[\frac{1}{L}\right]u_{2z} + \left[\frac{1}{2}(1-\xi)\right]\phi_{1y} + \left[\frac{1}{2}(1+\xi)\right]\phi_{2y}$
Shear Force $Q_z = k_{\mathrm{s}}AG\gamma_{xz} = k_{\mathrm{s}}AG\left(\frac{\mathrm{d}u_z}{\mathrm{d}x} + \phi_y\right) = \frac{\mathrm{d}M_y}{\mathrm{d}x}$
$Q_z^{\mathrm{e}}(x) = k_{\mathrm{s}}AG\left(\left[-\frac{1}{L}\right]u_{1z} + \left[\frac{1}{L}\right]u_{2z} + \left[1 - \frac{x}{L}\right]\phi_{1y} + \left[\frac{x}{L}\right]\phi_{2y}\right)$ $Q_z^{\mathrm{e}}(\xi) = k_{\mathrm{s}}AG\left(\left[-\frac{1}{L}\right]u_{1z} + \left[\frac{1}{L}\right]u_{2z} + \left[\frac{1}{2}(1-\xi)\right]\phi_{1y} + \left[\frac{1}{2}(1+\xi)\right]\phi_{2y}\right)$
Bending Moment $M_y = +EI_y\kappa_y = EI_y\frac{\mathrm{d}\phi_y}{\mathrm{d}x} = EI_y\frac{\mathrm{d}\phi_y}{\mathrm{d}\xi}\frac{\mathrm{d}\xi}{\mathrm{d}x}$
$M_y^{\mathrm{e}}(x) = EI_y\left(\left[-\frac{1}{L}\right]\phi_{1y} + \left[\frac{1}{L}\right]\phi_{2y}\right)$ $M_y^{\mathrm{e}}(\xi) = EI_y\left(\left[-\frac{1}{L}\right]\phi_{1y} + \left[\frac{1}{L}\right]\phi_{2y}\right)$

4.3.3 *Higher-Order Interpolation Functions for the Beam with Shear Contribution*

This subsection follows the derivations presented in [18] and derives first a general approach for a TIMOSHENKO element with an arbitrary number of nodes. In generalization of Eqs. (4.46) and (4.47), the following approach results for the unknowns at the nodes:

$$u_z^{\mathrm{e}}(x) = \sum_{i=1}^{m} N_{iu}(x)u_{iz}\,, \tag{4.169}$$

$$\phi_y^{\mathrm{e}}(x) = \sum_{i=1}^{n} N_{i\phi}(x)\phi_{iy}\,, \tag{4.170}$$

or alternatively in matrix notation as

$$u_z^{\mathrm{e}}(x) = \begin{bmatrix} N_{1u} \ldots N_{mu} \; 0 \ldots 0 \end{bmatrix} \begin{bmatrix} u_{1z} \\ \vdots \\ u_{mz} \\ \phi_{1y} \\ \vdots \\ \phi_{ny} \end{bmatrix} = \boldsymbol{N}_u^{\mathrm{T}} \boldsymbol{u}_{\mathrm{p}}\,, \tag{4.171}$$

$$\phi_y^{\mathrm{e}}(x) = \begin{bmatrix} 0 \ldots 0 \; N_{1\phi} \ldots N_{n\phi} \end{bmatrix} \begin{bmatrix} u_{1z} \\ \vdots \\ u_{mz} \\ \phi_{1y} \\ \vdots \\ \phi_{ny} \end{bmatrix} = \boldsymbol{N}_\phi^{\mathrm{T}} \boldsymbol{u}_{\mathrm{p}}\,. \tag{4.172}$$

With this generalized approach, the deflection can be evaluated at m nodes and the rotation at n nodes. For the interpolation functions N_i usually LAGRANGE polynomials[15] are used, which in general are calculated in the case of the deflection as follows:

$$\begin{aligned} N_i &= \prod_{j=1 \wedge j \neq i}^{m} \frac{x_j - x}{x_j - x_i} \\ &= \frac{(x_1 - x)(x_2 - x) \cdots [x_i - x] \cdots (x_m - x)}{(x_1 - x_i)(x_2 - x_i) \cdots [x_i - x_i] \cdots (x_m - x_i)}\,, \end{aligned} \tag{4.173}$$

whereupon the expressions in the square brackets for the ith interpolation function remains unconsidered. The abscissa values $x_1, \ldots, x_m$ represent the x-coordinates of the m nodes. In the case of the rotation, the variable m has to be replaced by n in Eq. (4.173).

For the derivation of the general stiffness matrix, the weighted residual method is considered. One can use the new approaches (4.171) and (4.172) in Eq. (4.68). Execution of the multiplication for the bending stiffness matrix yields

[15] In the case of the so-called LAGRANGE interpolation, m points are approximated via the ordinate values with the help of a polynomial of the order $m - 1$. In the case of the HERMITE interpolation, the slope of the regarded points is considered in addition to the ordinate value.

$$\boldsymbol{K}_{\mathrm{b}}^{\mathrm{e}} = \int_0^L EI_y \left[\begin{array}{ccc|ccc} 0 & \cdots & 0 & 0 & \cdots & 0 \\ \vdots & (m \times m) & \vdots & \vdots & (m \times n) & \vdots \\ 0 & \cdots & 0 & 0 & \cdots & 0 \\ \hline 0 & \cdots & 0 & \frac{\mathrm{d}N_{1\phi}}{\mathrm{d}x}\frac{\mathrm{d}N_{1\phi}}{\mathrm{d}x} & \cdots & \frac{\mathrm{d}N_{1\phi}}{\mathrm{d}x}\frac{\mathrm{d}N_{n\phi}}{\mathrm{d}x} \\ \vdots & (n \times m) & \vdots & \vdots & (n \times n) & \vdots \\ 0 & \cdots & 0 & \frac{\mathrm{d}N_{n\phi}}{\mathrm{d}x}\frac{\mathrm{d}N_{1\phi}}{\mathrm{d}x} & \cdots & \frac{\mathrm{d}N_{n\phi}}{\mathrm{d}x}\frac{\mathrm{d}N_{n\phi}}{\mathrm{d}x} \end{array}\right] \mathrm{d}x \tag{4.174}$$

and a corresponding execution of the multiplication for the shear stiffness matrix $\boldsymbol{K}_{\mathrm{s}}^{\mathrm{e}}$ yields

$$\int_0^L k_{\mathrm{s}}AG \left[\begin{array}{ccc|ccc} \frac{\mathrm{d}N_{1u}}{\mathrm{d}x}\frac{\mathrm{d}N_{1u}}{\mathrm{d}x} & \cdots & \frac{\mathrm{d}N_{1u}}{\mathrm{d}x}\frac{\mathrm{d}N_{mu}}{\mathrm{d}x} & \frac{\mathrm{d}N_{1u}}{\mathrm{d}x}N_{1\phi} & \cdots & \frac{\mathrm{d}N_{1u}}{\mathrm{d}x}N_{n\phi} \\ \vdots & (m\times m) & \vdots & \vdots & (m\times n) & \vdots \\ \frac{\mathrm{d}N_{mu}}{\mathrm{d}x}\frac{\mathrm{d}N_{1u}}{\mathrm{d}x} & \cdots & \frac{\mathrm{d}N_{mu}}{\mathrm{d}x}\frac{\mathrm{d}N_{mu}}{\mathrm{d}x} & \frac{\mathrm{d}N_{mu}}{\mathrm{d}x}N_{1\phi} & \cdots & \frac{\mathrm{d}N_{mu}}{\mathrm{d}x}N_{n\phi} \\ \hline N_{1\phi}\frac{\mathrm{d}N_{1u}}{\mathrm{d}x} & \cdots & N_{1\phi}\frac{\mathrm{d}N_{mu}}{\mathrm{d}x} & N_{1\phi}N_{1\phi} & \cdots & N_{1\phi}N_{n\phi} \\ \vdots & (n\times m) & \vdots & \vdots & (n\times n) & \vdots \\ N_{n\phi}\frac{\mathrm{d}N_{1u}}{\mathrm{d}x} & \cdots & N_{n\phi}\frac{\mathrm{d}N_{mu}}{\mathrm{d}x} & N_{n\phi}N_{1\phi} & \cdots & N_{n\phi}N_{n\phi} \end{array}\right] \mathrm{d}x\,. \tag{4.175}$$

These two stiffness matrices can be superposed additively at this point and the following general structure for the total stiffness matrix is obtained:

$$\boldsymbol{K}^{\mathrm{e}} = \begin{bmatrix} \boldsymbol{K}^{11} & \boldsymbol{K}^{12} \\ \boldsymbol{K}^{21} & \boldsymbol{K}^{22} \end{bmatrix}, \tag{4.176}$$

with

$$\boldsymbol{K}_{kl}^{11} = \int_0^L k_{\mathrm{s}}AG \frac{\mathrm{d}N_{ku}}{\mathrm{d}x}\frac{\mathrm{d}N_{lu}}{\mathrm{d}x}\mathrm{d}x\,, \tag{4.177}$$

$$\boldsymbol{K}_{kl}^{12} = \int_0^L k_{\mathrm{s}}AG \frac{\mathrm{d}N_{ku}}{\mathrm{d}x}N_{l\phi}\mathrm{d}x\,, \tag{4.178}$$

$$\boldsymbol{K}_{kl}^{21} = \boldsymbol{k}_{kl}^{12,\mathrm{T}} = \int_0^L k_s AG N_{k\phi} \frac{\mathrm{d}N_{lu}}{\mathrm{d}x} \mathrm{d}x \,, \tag{4.179}$$

$$\boldsymbol{K}_{kl}^{22} = \int_0^L \left(k_s AG N_{k\phi} N_{l\phi} + EI_y \frac{\mathrm{d}N_{k\phi}}{\mathrm{d}x} \frac{\mathrm{d}N_{l\phi}}{\mathrm{d}x} \right) \mathrm{d}x \,. \tag{4.180}$$

The derivation of the right-hand side can be performed according to Eq. (4.77) and the following column matrix of the loads results:

$$\boldsymbol{f}^{\mathrm{e}} = \int_0^L q_z(x) \begin{bmatrix} N_{1u} \\ \vdots \\ N_{mu} \\ 0 \\ \vdots \\ 0 \end{bmatrix} \mathrm{d}x + \begin{bmatrix} F_{1z} \\ \vdots \\ F_{mz} \\ M_{1y} \\ \vdots \\ M_{ny} \end{bmatrix} . \tag{4.181}$$

In the following, a quadratic interpolation for $u_y(x)$ as well as a linear interpolation for $\phi_z(x)$ are chosen [18]. Therefore, for $\frac{\mathrm{d}u_y(x)}{\mathrm{d}x}$ and $\phi_z(x)$ functions of the same order result and the phenomenon of *shear locking* can be avoided. Quadratic interpolation for the deflection means that the deflection will be evaluated at three nodes. The linear approach for the rotation means that the unknowns will be evaluated at only two nodes. Therefore, the illustrated configuration in Fig. 4.15 for this TIMOSHENKO element results.

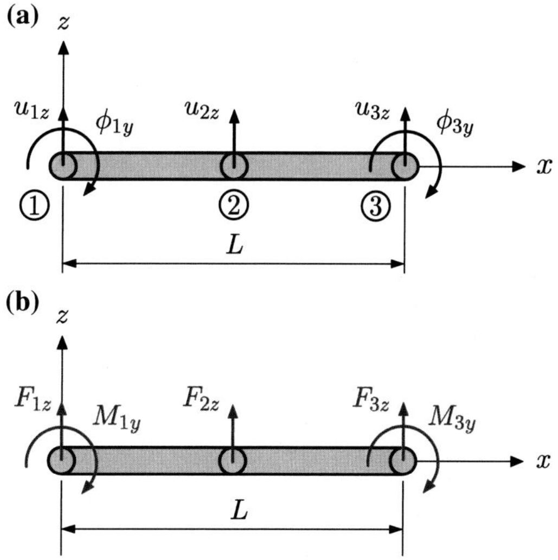

Fig. 4.15 TIMOSHENKO beam element with quadratic interpolation functions for the deflection and linear interpolation functions for the rotation: **a** deformation parameters; **b** load parameters

Evaluation of the general LAGRANGE polynomial according to Eq. (4.173) for the deflection, meaning under consideration of three nodes, yields

$$N_{1u} = \frac{(x_2 - x)(x_3 - x)}{(x_2 - x_1)(x_3 - x_1)} = 1 - 3\frac{x}{L} + 2\left(\frac{x}{L}\right)^2 , \tag{4.182}$$

$$N_{2u} = \frac{(x_1 - x)(x_3 - x)}{(x_1 - x_2)(x_3 - x_2)} = 4\frac{x}{L} - 4\left(\frac{x}{L}\right)^2 , \tag{4.183}$$

$$N_{3u} = \frac{(x_1 - x)(x_2 - x)}{(x_1 - x_3)(x_2 - x_3)} = -\frac{x}{L} + 2\left(\frac{x}{L}\right)^2 , \tag{4.184}$$

or alternatively for both nodes for the rotation:

$$N_{1\phi} = \frac{(x_2 - x)}{(x_2 - x_1)} = 1 - \frac{x}{L}, \tag{4.185}$$

$$N_{2\phi} = \frac{(x_1 - x)}{(x_1 - x_2)} = \frac{x}{L}. \tag{4.186}$$

A graphical illustration of the interpolation functions is given in Fig. 4.16. One can see that the typical characteristics for interpolation functions, meaning $N_i(x_i) = 1 \wedge N_i(x_j) = 0$ and $\sum_i N_i = 1$ are fulfilled.

With these interpolation functions the submatrices $\boldsymbol{K}^{11}, \ldots, \boldsymbol{K}^{22}$ in Eq. (4.176) result in the following via analytical integration as:

$$\boldsymbol{K}^{11} = \frac{k_s AG}{3L}\begin{bmatrix} 7 & -8 & 1 \\ -8 & 16 & -8 \\ 1 & -8 & 7 \end{bmatrix} , \tag{4.187}$$

$$\boldsymbol{K}^{12} = \frac{k_s AG}{6}\begin{bmatrix} -5 & -1 \\ 4 & -4 \\ 1 & 5 \end{bmatrix} = (\boldsymbol{K}^{21})^{\mathrm{T}} , \tag{4.188}$$

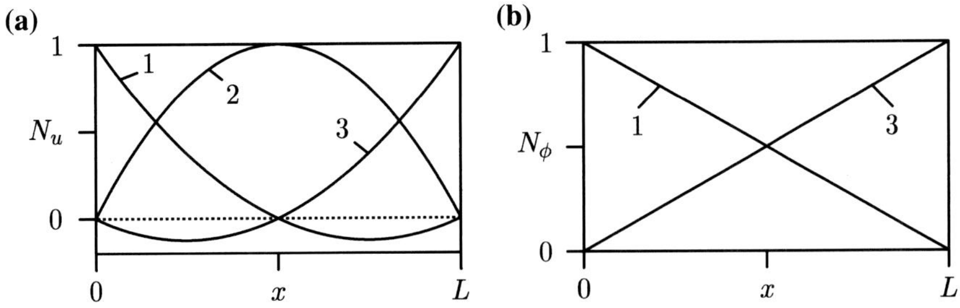

Fig. 4.16 Interpolation functions for a TIMOSHENKO beam element with **a** quadratic approach for the deflection and **b** linear approach for the rotation

$$\boldsymbol{K}^{22} = \frac{k_s AGL}{6}\begin{bmatrix} 2 & 1 \\ 1 & 2 \end{bmatrix} + \frac{EI_y}{L}\begin{bmatrix} 1 & -1 \\ -1 & 1 \end{bmatrix}, \tag{4.189}$$

which can be assembled to the principal finite element equation by making use of the abbreviation $\Lambda = \frac{EI_y}{k_s AGL^2}$:

$$\frac{k_s AG}{6L}\begin{bmatrix} 14 & -16 & 2 & -5L & -1L \\ -16 & 32 & -16 & 4L & -4L \\ 2 & -16 & 14 & 1L & 5L \\ -5L & 4L & 1L & 2L^2(1+3\Lambda) & L^2(1-6\Lambda) \\ -1L & -4L & 5L & L^2(1-6\Lambda) & 2L^2(1+3\Lambda) \end{bmatrix}\begin{bmatrix} u_{1z} \\ u_{2z} \\ u_{3z} \\ \phi_{1y} \\ \phi_{3y} \end{bmatrix} = \begin{bmatrix} F_{1z} \\ F_{2z} \\ F_{3z} \\ M_{1y} \\ M_{3y} \end{bmatrix}. \tag{4.190}$$

Since only a displacement is evaluated at the middle node, the number of unknowns is not the same at each node. This circumstance complicates the creation of the global system of equations for several of these elements. However, the degree of freedom u_{2z} can be expressed via the remaining unknowns and therefore the possibility exists to eliminate this node from the system of equations. For this, the second Eq. (4.190) has[16] to be evaluated:

$$\frac{k_s AG}{6L}\left(-16u_{1z} + 32u_{2z} - 16u_{3z} + 4L\phi_{1y} - 4L\phi_{3y}\right) = F_{2z}\,, \tag{4.191}$$

$$u_{2z} = \frac{6L}{32k_s AG}F_{2z} + \frac{u_{1z} + u_{3z}}{2} + \frac{-\phi_{1y} + \phi_{3y}}{8}L\,. \tag{4.192}$$

Furthermore, it can be demanded that no external force should act at the middle node, so that the relation between the deflection at the middle node and the other unknowns yields as follows:

$$u_{2z} = \frac{u_{1z} + u_{3z}}{2} + \frac{-\phi_{1z} + \phi_{3z}}{8}L\,. \tag{4.193}$$

This relation can be introduced into the system of Eq. (4.190) to eliminate the degree of freedom u_{2u}. Finally, after a new arrangement of the unknowns, the following principal finite element equation results, which is reduced by one column and one row:

$$\frac{EI_y}{6\Lambda L^3}\begin{bmatrix} 6 & -3L & -6 & -3L \\ -3L & L^2(1.5+6\Lambda) & 3L & L^2(1.5-6\Lambda) \\ -6 & 3L & 6 & 3L \\ -3L & L^2(1.5-6\Lambda) & 3L & L^2(1.5+6\Lambda) \end{bmatrix} = \begin{bmatrix} u_{1z} \\ \phi_{1y} \\ u_{3z} \\ \phi_{3y} \end{bmatrix} = \begin{bmatrix} F_{1z} \\ M_{1y} \\ F_{3z} \\ M_{3y} \end{bmatrix}. \tag{4.194}$$

---

[16]It needs to be remarked that the influence of distributed loads is disregarded in this derivation. If distributed loads occur, the equivalent nodal loads have to be distributed on the remaining nodes.

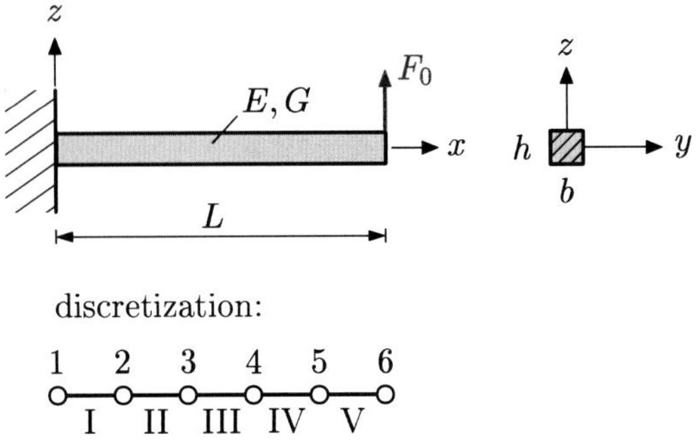

Fig. 4.17 Discretization of a beam structure with elements under consideration of the shear contribution

This element formulation is identical with Eq. (4.137), which was derived for linear interpolation functions and numerical one-point integration. However, it should be considered that the interpolation between the nodes during the use of (4.194) takes place with quadratic functions.

Further details and formulations regarding the TIMOSHENKO beam element can be found in the scientific papers [16, 17].

4.3.4 Solved Problems

4.1 Discretization of a beam with five linear elements with shear contribution

Discretize equally with five linear TIMOSHENKO elements the beam[17] which is illustrated in Fig. 4.17. Discuss the displacement of the loading point as a function of the slenderness ratio and the POISSON's ratio. Consider the case of (a) an analytically and (b) a numerically (one integration point) integrated stiffness matrix.

4.1 Solution

(a) Stiffness matrix via analytical integration

The element stiffness matrix according to Eq. (4.136) can be used for each of the five elements, whereupon it has to be considered that the single element length is $\frac{L}{5}$. The resulting total stiffness matrix has the dimensions 12×12, which reduces to a 10×10 matrix due to the consideration of the fixed support on the left-hand boundary ($u_{1z} = 0$, $\phi_{1y} = 0$). Through inversion of the stiffness matrix, the reduced system of equations can be solved via $\boldsymbol{u} = \boldsymbol{K}^{-1}\boldsymbol{F}$. The following extract shows the most important entries in this system of equations:

[17] A similar example is presented in [20].

$$\begin{bmatrix} u_{2z} \\ \vdots \\ u_{6z} \\ \phi_{6y} \end{bmatrix} = \frac{\frac{4}{5}L}{k_s AG} \underbrace{\begin{bmatrix} \text{x} & \cdots & \text{x} & \text{x} \\ \vdots & & \vdots & \vdots \\ \text{x} & \cdots & \frac{125(3\alpha+4L^2)}{4(75\alpha+L^2)} & \text{x} \\ \text{x} & \cdots & \text{x} & \text{x} \end{bmatrix}}_{10\times 10 \text{ matrix}} \begin{bmatrix} 0 \\ \vdots \\ F_0 \\ 0 \end{bmatrix}. \tag{4.195}$$

Multiplication of the 9th row of the matrix with the load matrix yields the displacement of the loading point to:

$$u_{6z} = \frac{25(3\alpha + 4L^2)}{75\alpha + L^2} \times \frac{F_0 L}{k_s AG}, \tag{4.196}$$

or alternatively via $A = hb$, $k_s = \frac{5}{6}$ and the relation for the shear modulus according to Eq. (4.23) after a short calculation:

$$u_{6z} = \frac{12(1+\nu)\left(\frac{h}{L}\right)^2 + 20}{60 + \left(\frac{L}{h}\right)^2 \frac{1}{1+\nu}} \times \frac{F_0 L^3}{EI_y}. \tag{4.197}$$

A graphical illustration of the displacement distribution dependent on the slenderness ratio can be seen in Fig. 4.18. A comparison with Fig. 4.13 shows that the convergence behavior in the lower domain of the slenderness ratio for $0.2 < \frac{h}{L} < 1.0$ has significantly improved through the fine discretization. However, the phenomenon of *shear locking* for $\frac{h}{L} \to 0$ still occurs.

(b) Stiffness matrix via numerical integration with one integration point

According to the procedure in part (a) of this problem, the following 10×10 system of equations results at this point via the stiffness matrix according to Eq. (4.151)

$$\begin{bmatrix} u_{2z} \\ \vdots \\ u_{6z} \\ \phi_{6y} \end{bmatrix} = \frac{\frac{4}{5}L}{k_s AG} \underbrace{\begin{bmatrix} \text{x} & \cdots & \text{x} & \text{x} \\ \vdots & & \vdots & \vdots \\ \text{x} & \cdots & \frac{25\alpha+33L^2}{20\alpha} & \text{x} \\ \text{x} & \cdots & \text{x} & \text{x} \end{bmatrix}}_{10\times 10 \text{ matrix}} \begin{bmatrix} 0 \\ \vdots \\ F_0 \\ 0 \end{bmatrix}, \tag{4.198}$$

from which the displacement on the right-hand boundary can be defined as the following

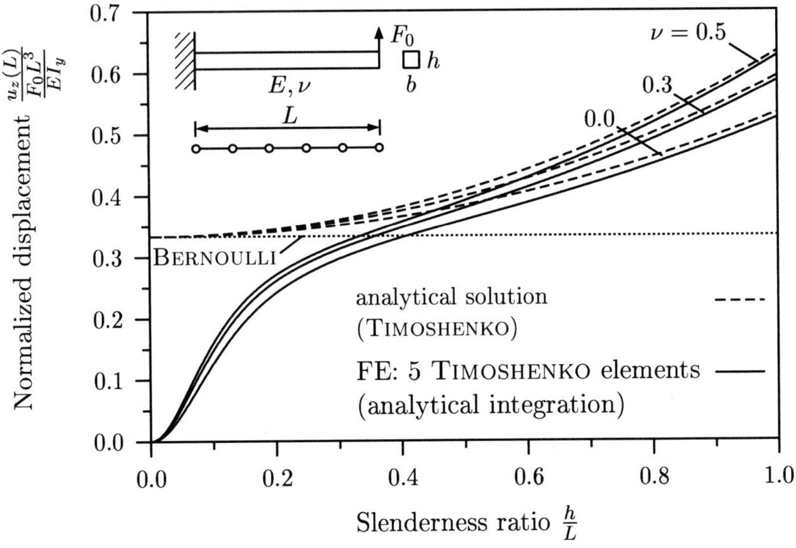

Fig. 4.18 Discretization of a beam with five linear TIMOSHENKO elements and analytical integration of the stiffness matrix

$$u_{6z} = \frac{4}{5}\left(\frac{5}{4} + \frac{33L^2}{20\alpha}\right) \times \frac{F_0 L}{k_s AG}. \tag{4.199}$$

With the use of $A = hb$, $k_s = \frac{5}{6}$ and the relation for the shear modulus according to Eq. (4.23), the following results after a short calculation:

$$u_{6z} = \left(\frac{33}{100} + \frac{1}{5}(1+\nu)\left(\frac{h}{L}\right)^2\right) \times \frac{F_0 L^3}{EI_y}. \tag{4.200}$$

The graphical illustration of the displacement distribution in Fig. 4.19 shows that an excellent conformity with the analytical solution throughout the entire domain of the slenderness ratio results through the mesh refinement. Therefore, the accuracy of a TIMOSHENKO element with linear interpolation functions and reduced numerical integration can be increased considerably through mesh refinement.

4.2 Timoshenko bending element with quadratic interpolation functions for the deflection and the rotation

Derive the stiffness matrix and the principal finite element equation $\boldsymbol{K}^e \boldsymbol{u}_p = \boldsymbol{f}^e$ for the illustrated TIMOSHENKO beam element in Fig. 4.20 with quadratic interpolation functions. Distinguish in the derivation between the analytical and numerical integration. Analyze the convergence behavior of an element for the illustrated configuration in Fig. 4.12.

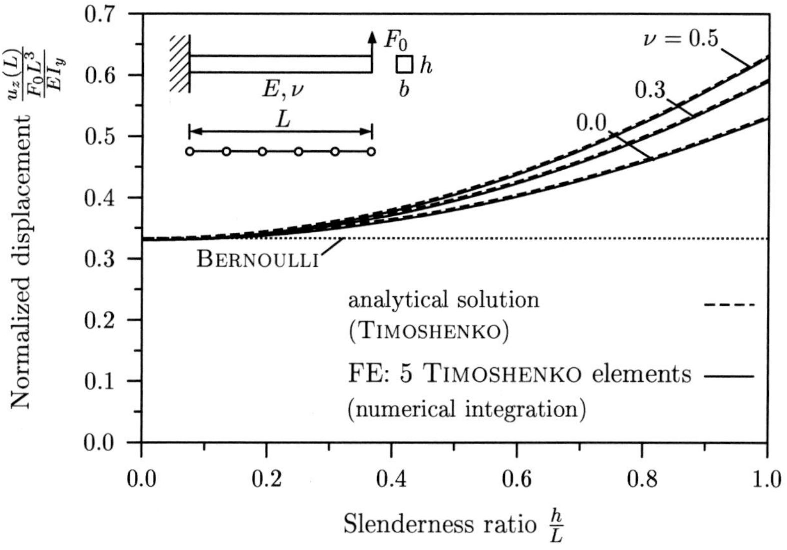

Fig. 4.19 Discretization of a beam via five linear TIMOSHENKO elements and numerical integration of the stiffness matrix with one integration point

Fig. 4.20 TIMOSHENKO beam element with quadratic interpolation functions for the deflection and the rotation: **a** deformation parameters; **b** load parameters

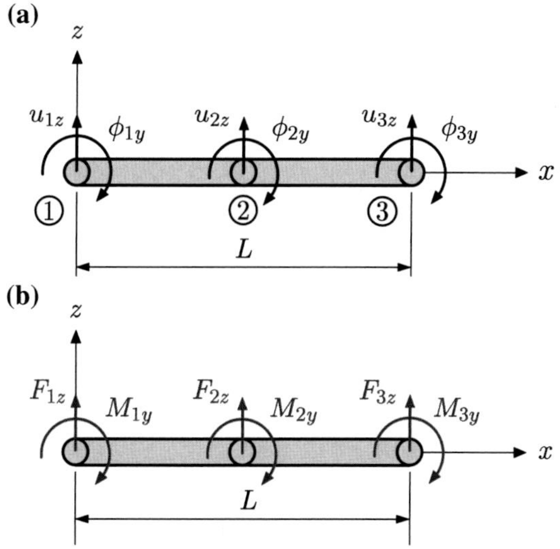

4.2 Solution

Evaluation of the general LAGRANGE polynomial according to Eq. (4.173) under consideration of three nodes yields the following interpolation functions for the deflection and the rotation:

$$N_{1u} = N_{1\phi} = \frac{(x_2 - x)(x_3 - x)}{(x_2 - x_1)(x_3 - x_1)} = 1 - 3\frac{x}{L} + 2\left(\frac{x}{L}\right)^2, \tag{4.201}$$

$$N_{2u} = N_{2\phi} = \frac{(x_1 - x)(x_3 - x)}{(x_1 - x_2)(x_3 - x_2)} = 4\frac{x}{L} - 4\left(\frac{x}{L}\right)^2, \tag{4.202}$$

$$N_{3u} = N_{3\phi} = \frac{(x_1 - x)(x_2 - x)}{(x_1 - x_3)(x_2 - x_3)} = -\frac{x}{L} + 2\left(\frac{x}{L}\right)^2. \tag{4.203}$$

With these interpolation functions the submatrices $\boldsymbol{K}^{11}, \ldots, \boldsymbol{K}^{22}$ in Eq. (4.176) result as follows through *analytical* integration:

$$\boldsymbol{K}^{11} = \frac{k_s AG}{6L}\begin{bmatrix} 14 & -16 & 2 \\ -16 & 32 & -16 \\ 2 & -16 & 14 \end{bmatrix}, \tag{4.204}$$

$$\boldsymbol{K}^{12} = \frac{k_s AG}{6L}\begin{bmatrix} -3L & -4L & 1L \\ 4L & 0 & -4L \\ -1L & 4L & 3L \end{bmatrix} = (\boldsymbol{K}^{21})^{\mathrm{T}}, \tag{4.205}$$

$$\boldsymbol{K}^{22} = \frac{k_s AGL}{30}\begin{bmatrix} 4 & 2 & -1 \\ 2 & 16 & 2 \\ -1 & 2 & 4 \end{bmatrix} + \frac{EI_y}{3L}\begin{bmatrix} 7 & -8 & 1 \\ -8 & 16 & -8 \\ 1 & -8 & 7 \end{bmatrix}, \tag{4.206}$$

which can be composed to the stiffness matrix $\boldsymbol{K}^{\mathrm{e}}$ via the use of the abbreviation $\Lambda = \frac{EI_y}{k_s AGL^2}$:

$$\frac{k_s AG}{6L}\left[\begin{array}{ccc|ccc} 14 & -16 & 2 & -3L & -4L & 1L \\ -16 & 32 & -16 & 4L & 0 & -4L \\ 2 & -16 & 14 & -1L & 4L & 3L \\ \hline -3L & 4L & -1L & L^2(\frac{4}{5} + 14\Lambda) & L^2(\frac{2}{5} - 16\Lambda) & L^2(-\frac{1}{5} + 2\Lambda) \\ -4L & 0 & 4L & L^2(\frac{2}{5} - 16\Lambda) & L^2(\frac{16}{5} + 32\Lambda) & L^2(\frac{2}{5} - 16\Lambda) \\ 1L & -4L & 3L & L^2(-\frac{1}{5} + 2\Lambda) & L^2(\frac{2}{5} - 16\Lambda) & L^2(\frac{4}{5} + 14\Lambda) \end{array}\right]. \tag{4.207}$$

With this stiffness matrix the principal finite element equation results in $\boldsymbol{K}^{\mathrm{e}}\boldsymbol{u}_{\mathrm{p}} = \boldsymbol{f}^{\mathrm{e}}$, at which the deformation and load vector contains the following components:

$$\boldsymbol{u}_{\text{p}} = \left[u_{1z}\ u_{2z}\ u_{3z}\ \phi_{1y}\ \phi_{2y}\ \phi_{3y}\right]^{\text{T}}, \tag{4.208}$$

$$\boldsymbol{f}^{\text{e}} = \left[F_{1z}\ F_{2z}\ F_{3z}\ M_{1y}\ M_{2y}\ M_{3y}\right]^{\text{T}}. \tag{4.209}$$

For the analysis of the convergence behavior of an element for the illustrated beam in Fig. 4.12 with point load, the columns and rows for the entries u_{1y} and ϕ_{1z} in Eq. (4.207) can be canceled due to the fixed support at this node. This reduced 4×4 stiffness matrix can be inverted and the following system of equations for the definition of the unknown degrees of freedom results:

$$\begin{bmatrix} u_{2z} \\ u_{3z} \\ \vdots \\ \phi_{3y} \end{bmatrix} = \frac{6L}{k_{\text{s}}AG} \underbrace{\begin{bmatrix} \times & \cdots & \cdots\times \\ \times & \frac{-3+340\Lambda+1200\Lambda^2}{8(-1-45\Lambda+900\Lambda^2)} & \cdots\times \\ \vdots & & \vdots \\ \times & & \cdots\times \end{bmatrix}}_{4\times 4 \text{ matrix}} \begin{bmatrix} 0 \\ F_0 \\ \vdots \\ 0 \end{bmatrix}, \tag{4.210}$$

from which, through evaluation of the second row, the displacement at the right-hand boundary can be defined as:

$$u_{3z} = \underbrace{\frac{6L}{k_{\text{s}}AG}}_{\frac{6\Lambda L^3}{EI_y}} \times \frac{-3+340\Lambda+1200\Lambda^2}{8(-1-45\Lambda+900\Lambda^2)} \times F_0\,. \tag{4.211}$$

For a rectangular cross section $\Lambda = \frac{1}{5}(1+\nu)\left(\frac{h}{L}\right)^2$ results, and one can see that *shear locking* occurs also at this point for slender beams with $L \gg h$, since in the limit case $u_{3z} \rightarrow 0$ occurs.

In the following, the reduced numerical integration of the stiffness matrix needs to be analyzed. For the definition of a reasonable amount of integration points one takes into account the following consideration:

If quadratic interpolation functions are used for u_z and ϕ_y, the degree of the polynomials for $\frac{\text{d}u_z}{\text{d}x}$ and ϕ_y differs. The quadratic approach for u_z yields for $\frac{\text{d}u_z}{\text{d}x}$ a linear function and thus a linear function would also be desirable for ϕ_y. The two-point integration, however, determines that the quadratic approach for ϕ_y is treated as a linear function. A two-point integration can exactly integrate a polynomial of third order, meaning proportional to x^3, at most and therefore the following view results: $(N_{i\phi}N_{j\phi}) \sim x^3$. This however means that $N_{i\phi} \sim x^{1.5}$ or alternatively $N_{j\phi} \sim x^{1.5}$ applies at most. Since the polynomial approach only allows integer values for the exponent, $N_{i\phi} \sim x^1$ or alternatively $N_{j\phi} \sim x^1$ results and the rotation needs to be considered as a linear function.

The integration via numerical GAUSS integration with two integration points demands that the arguments and the integration boundaries in the formulations of the submatrices $\boldsymbol{K}^{11}, \ldots, \boldsymbol{K}^{22}$ in Eq. (4.176) have to be transformed to the natural coordinate $-1 \leq \xi \leq 1$. Via the transformation of the derivative to the new coordinate, meaning $\frac{\text{d}N}{\text{d}x} = \frac{\text{d}N}{\text{d}\xi}\frac{\text{d}\xi}{\text{d}x}$ and the transformation of the coordinate $\xi = -1 + 2\frac{x}{L}$

or alternatively $\mathrm{d}\xi = \frac{2}{L}\mathrm{d}x$, the numerical approximation of the submatrices for two integration points $\xi_{1,2} = \pm\frac{1}{\sqrt{3}}$ results in:

$$\boldsymbol{K}^{11} = \sum_{i=1}^{2} \frac{2k_{\mathrm{s}}AG}{L} \begin{bmatrix} \frac{\mathrm{d}N_{1u}}{\mathrm{d}\xi}\frac{\mathrm{d}N_{1u}}{\mathrm{d}\xi} & \frac{\mathrm{d}N_{1u}}{\mathrm{d}\xi}\frac{\mathrm{d}N_{2u}}{\mathrm{d}\xi} & \frac{\mathrm{d}N_{1u}}{\mathrm{d}\xi}\frac{\mathrm{d}N_{3u}}{\mathrm{d}\xi} \\ \frac{\mathrm{d}N_{2u}}{\mathrm{d}\xi}\frac{\mathrm{d}N_{1u}}{\mathrm{d}\xi} & \frac{\mathrm{d}N_{2u}}{\mathrm{d}\xi}\frac{\mathrm{d}N_{2u}}{\mathrm{d}\xi} & \frac{\mathrm{d}N_{2u}}{\mathrm{d}\xi}\frac{\mathrm{d}N_{3u}}{\mathrm{d}\xi} \\ \frac{\mathrm{d}N_{3u}}{\mathrm{d}\xi}\frac{\mathrm{d}N_{1u}}{\mathrm{d}\xi} & \frac{\mathrm{d}N_{3u}}{\mathrm{d}\xi}\frac{\mathrm{d}N_{2u}}{\mathrm{d}\xi} & \frac{\mathrm{d}N_{3u}}{\mathrm{d}\xi}\frac{\mathrm{d}N_{3u}}{\mathrm{d}\xi} \end{bmatrix} \times 1\,, \tag{4.212}$$

$$\boldsymbol{K}^{12} = \sum_{i=1}^{2} k_{\mathrm{s}}AG \begin{bmatrix} \frac{\mathrm{d}N_{1u}}{\mathrm{d}\xi}(N_{1\phi}) & \frac{\mathrm{d}N_{1u}}{\mathrm{d}\xi}(N_{2\phi}) & \frac{\mathrm{d}N_{1u}}{\mathrm{d}\xi}(N_{3\phi}) \\ \frac{\mathrm{d}N_{2u}}{\mathrm{d}\xi}(N_{1\phi}) & \frac{\mathrm{d}N_{2u}}{\mathrm{d}\xi}(N_{2\phi}) & \frac{\mathrm{d}N_{2u}}{\mathrm{d}\xi}(N_{3\phi}) \\ \frac{\mathrm{d}N_{3u}}{\mathrm{d}\xi}(N_{1\phi}) & \frac{\mathrm{d}N_{3u}}{\mathrm{d}\xi}(N_{2\phi}) & \frac{\mathrm{d}N_{3u}}{\mathrm{d}\xi}(N_{3\phi}) \end{bmatrix} \times 1\,, \tag{4.213}$$

$$\boldsymbol{K}^{22} = \sum_{i=1}^{2} \frac{k_{\mathrm{s}}AGL}{2} \begin{bmatrix} N_{1\phi}N_{1\phi} & N_{1\phi}N_{2\phi} & N_{1\phi}N_{3\phi} \\ N_{2\phi}N_{1\phi} & N_{2\phi}N_{2\phi} & N_{2\phi}N_{3\phi} \\ N_{3\phi}N_{1\phi} & N_{3\phi}N_{2\phi} & N_{3\phi}N_{3\phi} \end{bmatrix} \times 1 \tag{4.214}$$

$$+ \sum_{i=1}^{2} \frac{2EI_y}{L} \begin{bmatrix} \frac{\mathrm{d}N_{1\phi}}{\mathrm{d}\xi}\frac{\mathrm{d}N_{1\phi}}{\mathrm{d}\xi} & \frac{\mathrm{d}N_{1\phi}}{\mathrm{d}\xi}\frac{\mathrm{d}N_{2\phi}}{\mathrm{d}\xi} & \frac{\mathrm{d}N_{1\phi}}{\mathrm{d}\xi}\frac{\mathrm{d}N_{3\phi}}{\mathrm{d}\xi} \\ \frac{\mathrm{d}N_{2\phi}}{\mathrm{d}\xi}\frac{\mathrm{d}N_{1\phi}}{\mathrm{d}\xi} & \frac{\mathrm{d}N_{2\phi}}{\mathrm{d}\xi}\frac{\mathrm{d}N_{2\phi}}{\mathrm{d}\xi} & \frac{\mathrm{d}N_{2\phi}}{\mathrm{d}\xi}\frac{\mathrm{d}N_{3\phi}}{\mathrm{d}\xi} \\ \frac{\mathrm{d}N_{3\phi}}{\mathrm{d}\xi}\frac{\mathrm{d}N_{1\phi}}{\mathrm{d}\xi} & \frac{\mathrm{d}N_{3\phi}}{\mathrm{d}\xi}\frac{\mathrm{d}N_{2\phi}}{\mathrm{d}\xi} & \frac{\mathrm{d}N_{3\phi}}{\mathrm{d}\xi}\frac{\mathrm{d}N_{3\phi}}{\mathrm{d}\xi} \end{bmatrix} \times 1\,. \tag{4.215}$$

The quadratic interpolation functions, which have already been introduced into Eqs. (4.182) up to (4.184), still have to be transformed to the new coordinates via the transformation $x = (\xi + 1)\frac{L}{2}$. Therefore for the interpolation functions or alternatively their derivatives the following results:

$$N_1(\xi) = -\frac{1}{2}(\xi - \xi^2)\,, \qquad \frac{\mathrm{d}N_1}{\mathrm{d}\xi} = -\frac{1}{2}(1 - 2\xi)\,, \tag{4.216}$$

$$N_2(\xi) = 1 - \xi^2\,, \qquad \frac{\mathrm{d}N_2}{\mathrm{d}\xi} = -2\xi\,, \tag{4.217}$$

$$N_3(\xi) = \frac{1}{2}(\xi + \xi^2)\,, \qquad \frac{\mathrm{d}N_3}{\mathrm{d}\xi} = \frac{1}{2}(1 + 2\xi)\,. \tag{4.218}$$

The use of these interpolation functions or alternatively their derivatives finally leads to the following submatrices

$$\boldsymbol{K}^{11} = \frac{k_s AG}{6L} \begin{bmatrix} 14 & -16 & 2 \\ -16 & 32 & -16 \\ 2 & -16 & 14 \end{bmatrix}, \tag{4.219}$$

$$\boldsymbol{K}^{12} = \frac{k_s AG}{6L} \begin{bmatrix} -3L & -4L & L \\ 4L & 0 & -4L \\ -L & 4L & 3L \end{bmatrix}, \tag{4.220}$$

$$\boldsymbol{K}^{22} = \frac{k_s AG}{6L} \begin{bmatrix} \frac{2}{3}L^2 & \frac{2}{3}L^2 & -\frac{1}{3}L^2 \\ \frac{2}{3}L^2 & \frac{8}{3}L^2 & \frac{2}{3}L^2 \\ -\frac{1}{3}L^2 & \frac{2}{3}L^2 & \frac{2}{3}L^2 \end{bmatrix} + \frac{EI_y}{L^3} \begin{bmatrix} \frac{7}{3}L^2 & -\frac{8}{3}L^2 & \frac{1}{3}L^2 \\ -\frac{8}{3}L^2 & \frac{16}{3}L^2 & -\frac{8}{3}L^2 \\ \frac{1}{3}L^2 & -\frac{8}{3}L^2 & \frac{7}{3}L^2 \end{bmatrix}, \tag{4.221}$$

which can be put together to the stiffness matrix $\boldsymbol{K}^{\mathrm{e}}$ under the use of the abbreviation $\Lambda = \frac{EI_z}{k_s AGL^2}$:

$$\frac{k_s AG}{6L} \left[\begin{array}{ccc|ccc} 14 & -16 & 2 & -3L & -4L & 1L \\ -16 & 32 & -16 & 4L & 0 & -4L \\ 2 & -16 & 14 & -1L & 4L & 3L \\ \hline -3L & 4L & -1L & L^2(\frac{2}{3}+14\Lambda) & L^2(\frac{2}{3}-16\Lambda) & L^2(-\frac{1}{3}+2\Lambda) \\ -4L & 0 & 4L & L^2(\frac{2}{3}-16\Lambda) & L^2(\frac{8}{3}+32\Lambda) & L^2(\frac{2}{3}-16\Lambda) \\ 1L & -4L & 3L & L^2(-\frac{1}{3}+2\Lambda) & L^2(\frac{2}{3}-16\Lambda) & L^2(\frac{2}{3}+14\Lambda) \end{array} \right], \tag{4.222}$$

whereupon the deformation and load matrix also contains the following components at this point:

$$\boldsymbol{u}_{\mathrm{p}} = \left[u_{1z} \; u_{2z} \; u_{3z} \; \phi_{1y} \; \phi_{2y} \; \phi_{3y} \right]^{\mathrm{T}}, \tag{4.223}$$

$$\boldsymbol{f}^{\mathrm{e}} = \left[F_{1z} \; F_{2z} \; F_{3z} \; M_{1y} \; M_{2y} \; M_{3y} \right]^{\mathrm{T}}. \tag{4.224}$$

For the analysis of the convergence behavior for the beam according to Fig. 4.12 the columns and rows for the entries u_{1z} and ϕ_{1y} in the present system of equations can be canceled. The inverted 4×4 stiffness matrix can be used for the definition of the unknown degrees of freedom:

$$\begin{bmatrix} u_{2z} \\ u_{3z} \\ \vdots \\ \phi_{3y} \end{bmatrix} = \frac{6L}{k_s AG} \underbrace{\begin{bmatrix} \mathsf{X} & \cdots & & \cdots \mathsf{X} \\ \mathsf{X} & \frac{1+3\Lambda}{18\Lambda} & & \cdots \mathsf{X} \\ \vdots & & & \vdots \\ \mathsf{X} & & & \cdots \mathsf{X} \end{bmatrix}}_{4\times 4 \text{ matrix}} \begin{bmatrix} 0 \\ F_0 \\ \vdots \\ 0 \end{bmatrix}, \tag{4.225}$$

from which, through evaluation of the second row, the deformation on the right-hand boundary can be defined as:

$$u_{3z} = \underbrace{\frac{6L}{k_s AG}}_{\frac{6\Lambda L^3}{EI_y}} \times \frac{1+3\Lambda}{18\Lambda} \times F = \left(\frac{1}{3} + \Lambda\right) \frac{F_0 L^3}{EI_y}. \tag{4.226}$$

For a rectangular cross section $\Lambda = \frac{1}{5}(1+\nu)\left(\frac{h}{L}\right)^2$ results and one receives the exact solution[18] of the problem as:

$$u_{3z} = \left(\frac{1}{3} + \frac{1+\nu}{5}\left(\frac{h}{L}\right)^2\right) \times \frac{F_0 L^3}{EI_y}. \tag{4.227}$$

According to the procedure for the TIMOSHENKO element with quadratic-linear interpolation functions in Sect. 4.3.3, the middle node can be eliminated. Under the assumption that no forces or moments should have an effect on the middle node, the 2nd and 5th row of Eq. (4.222) yields the following relations for the unknowns at the middle node:

$$u_{2z} = \frac{1}{2}u_{1z} + \frac{1}{2}u_{3z} - \frac{1}{8}L\phi_{1y} + \frac{1}{8}L\phi_{3y}, \tag{4.228}$$

$$\phi_{2y} = \frac{+4u_{1z}}{L\left(\frac{8}{3}+32\Lambda\right)} + \frac{-4u_{3z}}{L\left(\frac{8}{3}+32\Lambda\right)} - \frac{\left(\frac{2}{3}-16\lambda\right)\phi_{1y}}{\left(\frac{8}{3}+32\Lambda\right)} - \frac{\left(\frac{2}{3}-16\lambda\right)\phi_{3y}}{\left(\frac{8}{3}+32\Lambda\right)}. \tag{4.229}$$

These two relations can be considered in Eq. (4.222) so that the following principal finite element equation results after a short conversion:

$$\frac{2EI_y}{L^3(1+12\Lambda)} \begin{bmatrix} 6 & -3L & -6 & -3L \\ -3L & 2L^2(1+3\Lambda) & 3L & L^2(1-6\Lambda) \\ -6 & 3L & 6 & 3L \\ -3L & L^2(1-6\Lambda) & 3L & 2L^2(1+3\Lambda) \end{bmatrix} \begin{bmatrix} u_{1z} \\ \phi_{1y} \\ u_{3z} \\ \phi_{3y} \end{bmatrix} = \begin{bmatrix} F_{1z} \\ M_{1y} \\ F_{3z} \\ M_{3y} \end{bmatrix}. \tag{4.230}$$

With this formulation the one-beam problem according to Fig. 4.12 can be solved a little bit faster since after the consideration of the boundary conditions only a 2×2 matrix needs to be inverted. In this case, for the definition of the unknown the following results:

[18] For this see the supplementary Problem 4.11.

$$\frac{L^3(1+12\Lambda)}{2EI_y}\begin{bmatrix} \frac{2(1+3\Lambda)}{3(1+12\Lambda)} & \frac{-1}{L(1+12\Lambda)} \\ \frac{-1}{L(1+12\Lambda)} & \frac{2}{L^2(1+12\Lambda)} \end{bmatrix}\begin{bmatrix} F_0 \\ 0 \end{bmatrix} = \begin{bmatrix} u_{3z} \\ \phi_{3y} \end{bmatrix}, \tag{4.231}$$

which results from the exact solution for the deflection according to Eq. (4.227).

4.4 Supplementary Problems

4.3 Knowledge Questions on Timoshenko Beams

- Name the primary unknowns in the partial differential equations of a TIMOSHENKO beam.
- State the one-dimensional HOOKE's law for a pure shear state in common variables. Which material parameter is involved?
- State for isotropic materials the relationship between the shear modulus G, YOUNG's modulus E, and POISSON's ratio ν.
- Explain in words the major difference between the EULER–BERNOULLI and TIMOSHENKO beam theory.
- Consider a beam bending problem which is described based on the EULER–BERNOULLI and TIMOSHENKO beam theories. Which theory gives the larger deflection and why?
- Sketch (a) the normal and (b) the shear stress distribution of a TIMOSHENKO beam under bending load.
- State the required (a) geometrical parameters and (b) material parameters to define a TIMOSHENKO beam element for finite element applications.
- Sketch the interpolation functions $N_1(x)$ and $N_2(x)$ of a linear TIMOSHENKO element.
- State the rule of thumb which allows the user of a finite element code to select between an EULER–BERNOULLI and TIMOSHENKO based on geometrical properties.

4.4 Calculation of the shear stress distribution in a rectangular cross section

Given is a beam with rectangular cross section of width b and height h. Calculate the distribution of the shear stress τ_{zx} over the cross section under the influence of a shear force $Q_z(x)$. Assume that the shear stress is constant along the width.

4.5 Calculation of the shear stress distribution in a circular cross section

Given is a beam with circular cross section of radius R. Calculate the distribution of the shear stress τ_{zx} over the cross section under the influence of a shear force $Q_z(x)$. Consider the distribution in the middle of the section, i.e. for $y = 0$.

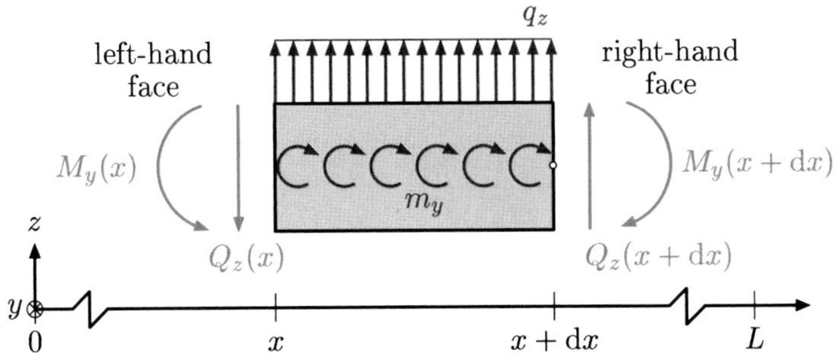

Fig. 4.21 Infinitesimal beam element with internal reactions and distributed load

4.6 Calculation of the shear correction factor for rectangular cross section

For a rectangular cross section with width b and height h, the shear stress distribution is given as follows [7]:

$$\tau_{zx}(z) = \frac{6Q_z}{bh^3}\left(\frac{h^2}{4} - z^2\right) \quad \text{with} \quad -\frac{h}{2} \leq z \leq \frac{h}{2}. \tag{4.232}$$

Compute the shear correction factor k_s under the assumption that the constant—in the surface A_s acting—equivalent shear stress $\tau_{xz} = Q_z/A_s$ yields the same shear strain energy as the actual shear stress distribution $\tau_{xz}(z)$, which acts in the actual cross-sectional area A of the beam.

4.7 Differential equation under consideration of distributed moment

For the derivation of the equilibrium condition, the infinitesimal beam element, illustrated in Fig. 4.21 needs to be considered, which is additionally loaded with a constant 'distributed moment' $m_y = \frac{\text{moment}}{\text{length}}$. Derive the differential equation for the TIMOSHENKO beam under consideration of a general moment distribution $m_y(x)$.

4.8 Differential equations for Timoshenko beam

Derive the general solution for the TIMOSHENKO differential equations in the following formulation:

$$EI_y \frac{d^2\phi_y}{dx^2} - k_s GA\left(\frac{du_z}{dx} + \phi_y\right) = -m_y\,, \tag{4.233}$$

$$k_s GA\left(\frac{d^2u_z}{dx^2} + \frac{d\phi_y}{dx}\right) = -q_z\,. \tag{4.234}$$

The distributed load q_z and the distributed moment m_y are constant in this case.

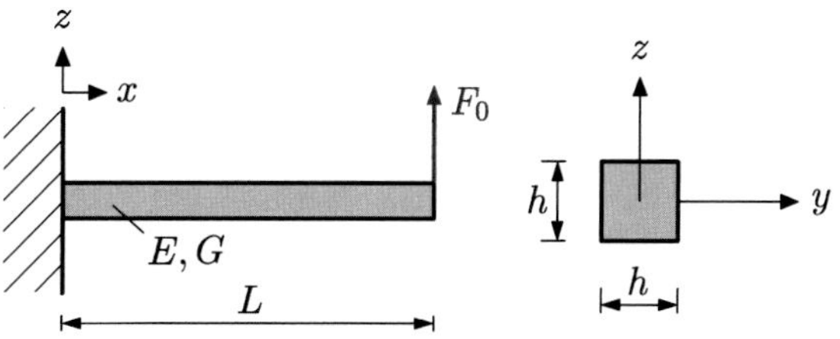

Fig. 4.22 Cantilever TIMOSHENKO beam under point load

4.9 Analytical calculation of the distribution of the deflection and rotation for a cantilever beam under point load

For a cantilever beam, which is loaded with a point load F_0 at the right-hand end in positive z-direction, calculate the distribution of the deflection $u_z(x)$ and the rotation $\phi_y(x)$ under consideration of the shear influence. Subsequently, the maximal deflection and the rotation at the loading point needs to be determined. Furthermore, the boundary value of the deflection at the loading point for slender ($h \ll L$) and compact ($h \gg L$) beams has to be determined.

4.10 Analytical calculation of various quantities for a cantilever beam under point load

Consider the cantilever beam as shown in Fig. 4.22 which is loaded by a single force F_0 at its right-hand end. Calculate based on the analytical approach for the TIMOSHENKO beam the following quantities:

- the deflection $u_z(x)$,
- the rotation $\phi_y(x)$,
- the bending moment distribution $M_y(x)$,
- the shear force distribution $Q_z(x)$,
- the absolute maximum normal strain $|\varepsilon_{x,\max}(x)|$,
- the curvature $\kappa_y(x)$,
- the absolute maximum normal stress $|\sigma_{x,\max}(x)|$,
- the absolute maximum shear stress $|\tau_{xz,\max}(x)|$, and
- the absolute maximum shear strain $|\gamma_{xz,\max}(x)|$.

Simplify your general solutions for the numerical values (assume consistent units) $h = 0.5$, $E = 200000$, $\nu = 0.3$, $F_0 = 100$, $L = 2h$ and calculate all values at $x = 0$ and $x = L$.

4.11 Analytical calculation of the normalized deflection for beams with shear contribution

For the illustrated courses of the maximal normalized deflection $u_{y,\,\text{norm}}$ in Fig. 4.8 as a function of the slenderness ratio, derive the corresponding equations.

4.12 Cantilever beam loaded by a single force

Calculate the analytical solution for the deflection $u_z(x)$ of a cantilever TIMOSHENKO beam shown in Fig. 4.23 based on the general solution given in Eqs. (4.39) and (4.40).

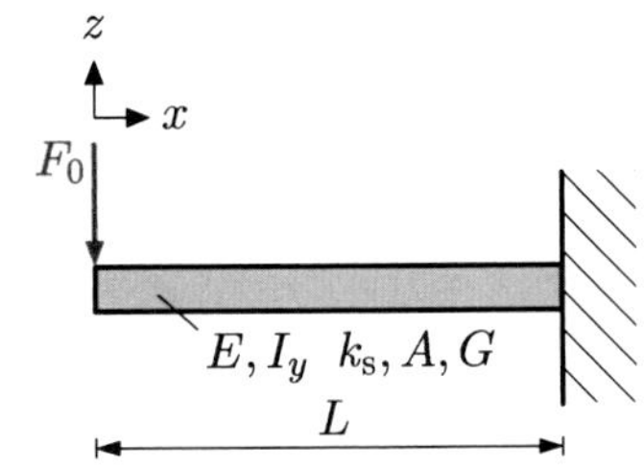

Fig. 4.23 Cantilever beam loaded by a single force

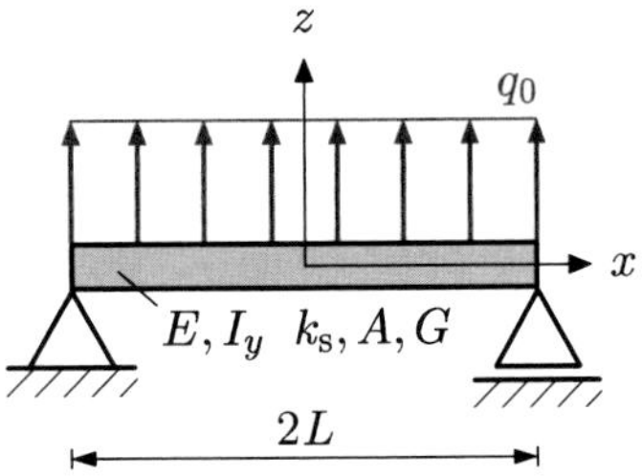

Fig. 4.24 Simply supported TIMOSHENKO beam in the elastic range loaded by a distributed load

It can be assumed for this exercise that the bending stiffness EI_y and the shear stiffness GA are constant. The rectangular cross-sectional area is equal to $A = bh$.

4.13 Simply supported beam in the elastic range loaded by a distributed load

Given is a simply supported TIMOSHENKO beam which is loaded by a constant distributed load of magnitude q_0 as shown in Fig. 4.24. The cross section of the beam can be assumed rectangular (width b and height h). Calculate the deflection $u_z(x)$ in the pure elastic range under the assumption that the bending stiffness EI_y and the shear stiffness GA are constant.

4.14 Timoshenko beam element with quadratic interpolation functions for the deflection and linear interpolation functions for the rotation

For a TIMOSHENKO beam element with quadratic interpolation functions for the deflection and linear interpolation functions for the rotation, the stiffness matrix, after elimination of the middle node according to Eq. (4.194), is given. Derive the additional load vector on the right-hand side of the principal finite element equation which results from a distributed load $q_z(x)$ in positive z-direction. Subsequently simplify the result for a constant load q_0.

4.15 Timoshenko beam element with cubic interpolation functions for the deflection and quadratic interpolation functions for the rotation

Derive the stiffness matrix and the principal finite element equation $\boldsymbol{K}^{\mathrm{e}}\boldsymbol{u}_{\mathrm{p}} = \boldsymbol{f}^{\mathrm{e}}$ for a TIMOSHENKO beam element with cubic interpolation functions for the deflection and quadratic interpolation functions for the rotation. Use the exact solution for the integration. Subsequently analyze the convergence behavior of an element configuration, which is illustrated in Fig. 4.12. The element deforms in the x-z plane. How

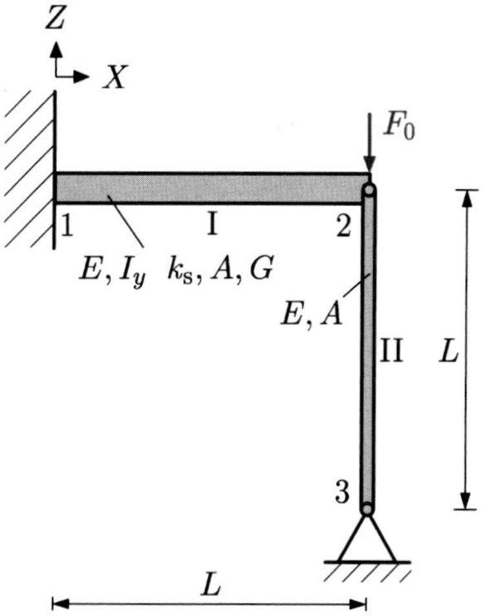

Fig. 4.25 Plane beam-rod structure with TIMOSHENKO element

does the principal finite element equation change, when the deformation occurs in the x-y plane?

4.16 Plane beam-rod structure with Timoshenko element

A horizontal TIMOSHENKO beam element 1-2 which is at point 2 supported by a vertical rod element is shown in Fig. 4.25. Both elements have the same length L and the frame is loaded by a vertical force F_0 at node 2.

Consider a linear TIMOSHENKO element with numerical integration to:

- Calculate based on a finite element approach the unknown deformations at point 2.
- Simplify your general solution for the special case that the rod is absent.
- Simplify your general solution for the special case that the beam is absent.

References

1. Bathe K-J (1996) Finite element procedures. Prentice-Hall, Upper Saddle River
2. Beer FP, Johnston ER Jr, DeWolf JT, Mazurek DF (2009) Mechanics of materials. McGraw-Hill, New York
3. Cook RD, Malkus DS, Plesha ME, Witt RJ (2002) Concepts and applications of finite element analysis. John Wiley & Sons, New York
4. Cowper GR (1966) The shear coefficient in Timoshenko's beam theory. J Appl Mech 33:335–340
5. Gere JM, Timoshenko SP (1991) Mechanics of materials. PWS-KENT Publishing Company, Boston
6. Gruttmann F, Wagner W (2001) Shear correction factors in Timoshenko's beam theory for arbitrary shaped cross-sections. Comput Mech 27:199–207
7. Hibbeler RC (2008) Mechanics of materials. Prentice Hall, Singapore

8. Hjelmstad DK (2005) Fundamentals of structural mechanics. Springer, New York
9. Levinson M (1981) A new rectangular beam theory. J Sound Vib 74:81–87
10. MacNeal RH (1978) A simple quadrilateral shell element. Comput Struct 8:175–183
11. MacNeal RH (1994) Finite elements: their design and performance. Marcel Dekker, New York
12. Öchsner A, Merkel M (2013) One-dimensional finite elements: an introduction to the FE method. Springer, Berlin
13. Öchsner A (2014) Elasto-plasticity of frame structure elements: modeling and simulation of rods and beams. Springer, Berlin
14. Reddy JN (1984) A simple higher-order theory for laminated composite plate. J Appl Mech 51:745–752
15. Reddy JN (1997) Mechanics of laminated composite plates: theory and analysis. CRC Press, Boca Raton
16. Reddy JN (1997) On locking-free shear deformable beam finite elements. Comput Method Appl Mech Eng 149:113–132
17. Reddy JN (1999) On the dynamic behaviour of the Timoshenko beam finite elements. Sadhana Acad Proc Eng Sci 24:175–198
18. Reddy JN (2006) An introduction to the finite element method. McGraw Hill, Singapore
19. Russel WT, MacNeal RH (1953) An improved electrical analogy for the analysis of beams in bending. J Appl Mech 20:349–354
20. Steinke P (2010) Finite-elemente-methode—rechnergestützte einführung. Springer, Berlin
21. Timoshenko SP (1921) On the correction for shear of the differential equation for transverse vibrations of prismatic bars. Philos Mag 41:744–746
22. Timoshenko SP (1922) On the transverse vibrations of bars of uniform cross-section. Philos Mag 43:125–131
23. Timoshenko S (1940) Strength of materials—part I elementary theory and problems. D. Van Nostrand Company, New York
24. Timoshenko SP, Goodier JN (1970) Theory of elasticity. McGraw-Hill, New York
25. Twiss RJ, Moores EM (1992) Structural geology. WH Freeman & Co, New York
26. Wang CM (1995) Timoshenko beam-bending solutions in terms of Euler-Bernoulli solutions. J Eng Mech ASCE 121:763–765
27. Wang CM, Reddy JN, Lee KH (2000) Shear deformable beams and plates: relationships with classical solution. Elsevier, Oxford
28. Weaver W Jr, Gere JM (1980) Matrix analysis of framed structures. Van Nostrand Reinhold Company, New York
29. Winkler E (1867) Die Lehre von der elasticität und Festigkeit mit besonderer Rücksicht auf ihre Anwendung in der Technik. H. Dominicus, Prag

Chapter 5
Plane Elements

Abstract This chapter starts with the analytical description of plane elasticity members. Based on the three basic equations of continuum mechanics, i.e., the kinematics relationship, the constitutive law and the equilibrium equation, the partial differential equation, which describes the physical problem, is derived. The weighted residual method is then used to derive the principal finite element equation for plane elements. Emphasis is given to the two plane elasticity cases, i.e., the plane stress and the plane strain case. The chapter exemplarily treats a four-node bilinear quadrilateral (quad 4) element.

5.1 Introduction

A plane elasticity element is defined as a thin two-dimensional member, as schematically shown in Fig. 5.1, with a much smaller thickness t than the planar dimensions. It can be seen as a two-dimensional extension or generalization of the rod. The following derivations are restricted to some simplifications:

- the thickness t is constant and much smaller than the planer dimensions a and b,
- the undeformed member shape is planar,
- the material is isotropic, homogenous and linear-elastic according to HOOKE's law for a plane stress or plane strain state,
- external forces act only at the boundary parallel to the plane of the member,
- external forces are distributed uniformly over the thickness,
- only rectangular members are considered.

The analogies between the rod and plane elasticity theories are summarized in Table 5.1.

A. Öchsner, *Computational Statics and Dynamics*,
https://doi.org/10.1007/978-981-15-1278-0_5

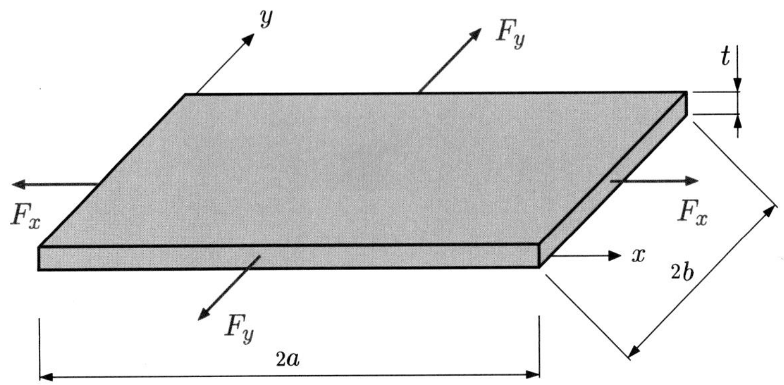

Fig. 5.1 General configuration for a plane elasticity problem

Table 5.1 Difference between rod, beam and plane element

Rod	Beam	Plane element
1D	1D	2D
Deformation along principal axis	Deformation perpendicular to principal axis	In-plane deformation
u_x	u_z, φ_y	u_x, u_y

5.2 Derivation of the Governing Differential Equation

5.2.1 Kinematics

The kinematics or strain-displacement relations extract the strain field contained in a displacement field. Using engineering definitions of strain, the following relations can be obtained [1, 2]:

$$\varepsilon_x = \frac{\partial u_x}{\partial x}; \ \varepsilon_y = \frac{\partial u_y}{\partial y}; \ \gamma_{xy} = 2\varepsilon_{xy} = \frac{\partial u_x}{\partial y} + \frac{\partial u_y}{\partial x}. \tag{5.1}$$

In matrix notation, these three relationships can be written as

$$\begin{bmatrix} \varepsilon_x \\ \varepsilon_y \\ 2\varepsilon_{xy} \end{bmatrix} = \begin{bmatrix} \frac{\partial}{\partial x} & 0 \\ 0 & \frac{\partial}{\partial y} \\ \frac{\partial}{\partial y} & \frac{\partial}{\partial x} \end{bmatrix} \begin{bmatrix} u_x \\ u_y \end{bmatrix}, \tag{5.2}$$

or symbolically as

$$\boldsymbol{\varepsilon} = \boldsymbol{\mathcal{L}}_1 \boldsymbol{u}, \tag{5.3}$$

where $\boldsymbol{\mathcal{L}}_1$ is the differential operator matrix.

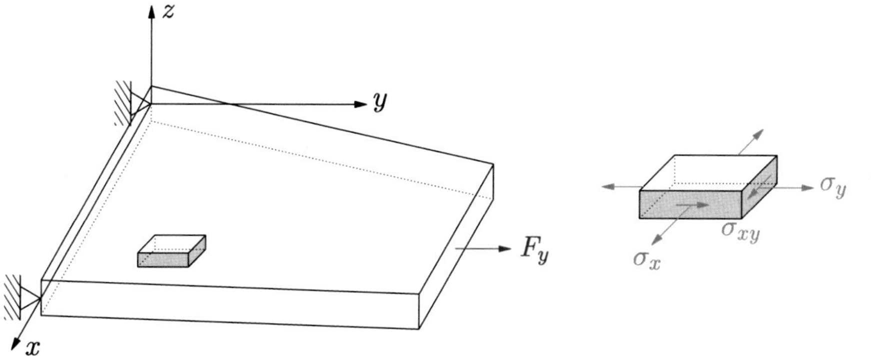

Fig. 5.2 Two-dimensional problem: plane stress case

5.2.2 Constitutive Equation

5.2.2.1 Plane Stress Case

The two-dimensional plane stress case ($\sigma_z = \sigma_{yz} = \sigma_{xz} = 0$) shown in Fig. 5.2 is commonly used for the analysis of thin, flat plates loaded in the plane of the plate (x-y plane).

It should be noted here that the normal thickness stress is zero ($\sigma_z = 0$) whereas the thickness normal strain is present ($\varepsilon_z \neq 0$).

The plane stress HOOKE's law for a linear-elastic isotropic material based on the YOUNG's modulus E and POISSON's ratio ν can be written for a constant temperature as

$$\begin{bmatrix} \sigma_x \\ \sigma_y \\ \sigma_{xy} \end{bmatrix} = \frac{E}{1-\nu^2} \begin{bmatrix} 1 & \nu & 0 \\ \nu & 1 & 0 \\ 0 & 0 & \frac{1-\nu}{2} \end{bmatrix} \begin{bmatrix} \varepsilon_x \\ \varepsilon_y \\ 2\,\varepsilon_{xy} \end{bmatrix}, \tag{5.4}$$

or in matrix notation as

$$\boldsymbol{\sigma} = \boldsymbol{C}\boldsymbol{\varepsilon}\,, \tag{5.5}$$

where $\boldsymbol{C}$ is the so-called elasticity matrix. It should be noted here that the engineering shear strain $\gamma_{xy} = 2\varepsilon_{xy}$ is used in the formulation of Eq. (5.4).

Rearranging the elastic stiffness form given in Eq. (5.4) for the strains gives the elastic compliance form

$$\begin{bmatrix} \varepsilon_x \\ \varepsilon_y \\ 2\,\varepsilon_{xy} \end{bmatrix} = \frac{1}{E} \begin{bmatrix} 1 & -\nu & 0 \\ -\nu & 1 & 0 \\ 0 & 0 & 2(\nu+1) \end{bmatrix} \begin{bmatrix} \sigma_x \\ \sigma_y \\ \sigma_{xy} \end{bmatrix}, \tag{5.6}$$

or in matrix notation as

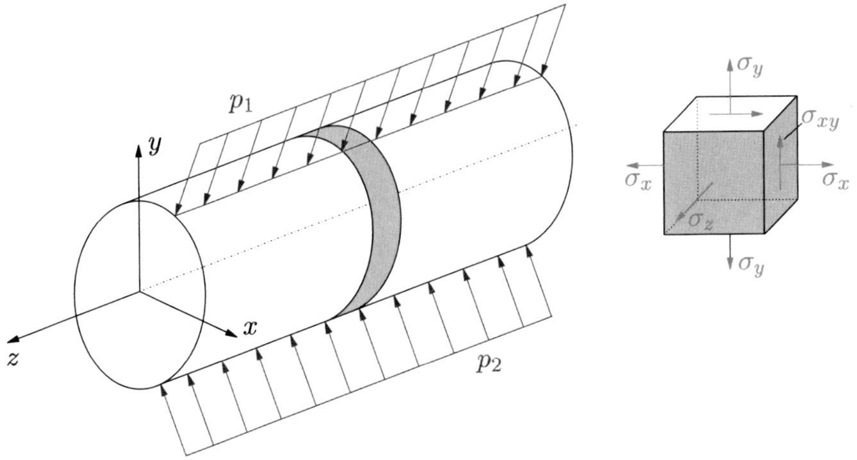

Fig. 5.3 Two dimensional problem: plane strain case

$$\varepsilon = \boldsymbol{D}\boldsymbol{\sigma}\,, \tag{5.7}$$

where $\boldsymbol{D} = \boldsymbol{C}^{-1}$ is the so-called elastic compliance matrix. The general characteristic of a plane HOOKE's law in the form of Eqs. (5.5) and (5.6) is that two independent material parameters are used.

It should be finally noted that the thickness strain ε_z can be obtained based on the two in-plane normal strains ε_x and ε_y as:

$$\varepsilon_z = -\frac{\nu}{1-\nu}\cdot\left(\varepsilon_x + \varepsilon_y\right). \tag{5.8}$$

The last equation can be derived from the tree-dimensional formulation, see Sect. 8.1.2.

5.2.2.2 Plane Strain Case

The two-dimensional plane strain case ($\varepsilon_z = \varepsilon_{yz} = \varepsilon_{xz} = 0$) shown in Fig. 5.3 is commonly used for the analysis of elongated prismatic bodies of uniform cross section subjected to uniform loading along their longitudinal axis but without any component in direction of the z-axis (e. g. pressure p_1 and p_2), such as in the case of tunnels, soil slopes, and retaining walls. It should be noted here that the normal thickness strain is zero ($\varepsilon_z = 0$) whereas the thickness normal stress is present ($\sigma_z \neq 0$).

The plane strain HOOKE's law for a linear-elastic isotropic material based on the YOUNG's modulus E and POISSON's ratio ν can be written for a constant temperature as

$$\begin{bmatrix} \sigma_x \\ \sigma_y \\ \sigma_{xy} \end{bmatrix} = \frac{E}{(1+\nu)(1-2\nu)} \begin{bmatrix} 1-\nu & \nu & 0 \\ \nu & 1-\nu & 0 \\ 0 & 0 & \frac{1-2\nu}{2} \end{bmatrix} \cdot \begin{bmatrix} \varepsilon_x \\ \varepsilon_y \\ 2\,\varepsilon_{xy} \end{bmatrix}, \tag{5.9}$$

or in matrix notation as

$$\boldsymbol{\sigma} = \boldsymbol{C}\boldsymbol{\varepsilon}\,, \tag{5.10}$$

where $\boldsymbol{C}$ is the so-called elasticity matrix.

Rearranging the elastic stiffness form given in Eq. (5.9) for the strains gives the elastic compliance form

$$\begin{bmatrix} \varepsilon_x \\ \varepsilon_y \\ 2\,\varepsilon_{xy} \end{bmatrix} = \frac{1-\nu^2}{E} \begin{bmatrix} 1 & -\frac{\nu}{1-\nu} & 0 \\ -\frac{\nu}{1-\nu} & 1 & 0 \\ 0 & 0 & \frac{2}{1-\nu} \end{bmatrix} \begin{bmatrix} \sigma_x \\ \sigma_y \\ \sigma_{xy} \end{bmatrix}, \tag{5.11}$$

or in matrix notation as

$$\boldsymbol{\varepsilon} = \boldsymbol{D}\boldsymbol{\sigma}\,, \tag{5.12}$$

where $\boldsymbol{D} = \boldsymbol{C}^{-1}$ is the so-called elastic compliance matrix. The general characteristic of a plane strain HOOKE's law in the form of Eqs. (5.9) and (5.11) is that two independent material parameters are used.

It should be finally noted that the thickness stress σ_z can be obtained based on the two in-plane normal stresses σ_x and σ_y as:

$$\sigma_z = \nu(\sigma_x + \sigma_y)\,. \tag{5.13}$$

The last equation can be derived from the tree-dimensional formulation, see Sect. 8.1.2.

5.2.3 Equilibrium

Figure 5.4 shows the normal and shear stresses which are acting on a differential volume element in the x-direction. All forces are drawn in their positive direction at each cut face. A positive cut face is obtained if the outward surface normal is directed in the positive direction of the corresponding coordinate axis. This means that the right-hand face in Fig. 5.4 is positive and the force $(\sigma_x + \frac{\partial \sigma_x}{\partial x}\mathrm{d}x)\mathrm{d}y\mathrm{d}z$ is oriented in the positive x-direction. In a similar way, the top face is positive, i.e. the outward surface normal is directed in the positive y-direction, and the shear force[1] is oriented in the positive x-direction. Since the volume element is assumed to be in equilibrium,

[1] In the case of a shear force σ_{ij}, the first index i indicates that the stress acts on a plane normal to the i-axis and the second index j denotes the direction in which the stress acts.

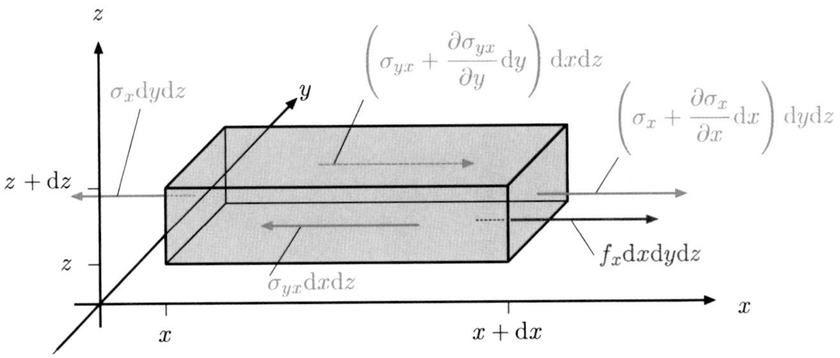

Fig. 5.4 Stress and body forces which act on a plane differential volume element in x-direction (note that the three directions $\mathrm{d}x$, $\mathrm{d}y$ and $\mathrm{d}z$ are differently sketched to indicate the plane problem)

forces resulting from stresses on the sides of the cuboid and from the body forces f_i $(i = x, y, z)$ must be balanced. These body forces are defined as forces per unit volume which can be produced by gravity,[2] acceleration, magnetic fields, and so on.

The static equilibrium of forces in the x-direction based on the five force components—two normal forces, two shear forces and one body force—indicated in Fig. 5.4 gives

$$\begin{aligned}\left(\sigma_x + \frac{\partial \sigma_x}{\partial x}\right)\mathrm{d}y\mathrm{d}z - \sigma_x \mathrm{d}y\mathrm{d}z + \left(\frac{\partial \sigma_{yx}}{\partial y}\right)\mathrm{d}x\mathrm{d}z \\ - \sigma_{yx}\mathrm{d}x\mathrm{d}z + f_x \mathrm{d}x\mathrm{d}y\mathrm{d}z = 0\,,\end{aligned} \tag{5.14}$$

or after simplification and canceling with $\mathrm{d}V = \mathrm{d}x\mathrm{d}y\mathrm{d}z$:

$$\frac{\partial \sigma_x}{\partial x} + \frac{\partial \sigma_{yx}}{\partial y} + f_x = 0\,. \tag{5.15}$$

Based on the same approach, a similar equation can be specified in the y-direction:

$$\frac{\partial \sigma_y}{\partial y} + \frac{\partial \sigma_{yx}}{\partial x} + f_y = 0\,. \tag{5.16}$$

These two balance equations can be written in matrix notation as

[2]If gravity is acting, the body force f results as the product of density times standard gravity: $f = \frac{F}{V} = \frac{mg}{V} = \frac{m}{V}g = \varrho g$. The units can be checked by consideration of $1\,\mathrm{N} = 1\,\frac{\mathrm{mkg}}{\mathrm{s}^2}$.

Table 5.2 Fundamental governing equations of a continuum in the plane elasticity case

Expression	Matrix notation	Tensor notation
Kinematics	$\varepsilon = \boldsymbol{\mathcal{L}}_1 \boldsymbol{u}$	$\varepsilon_{ij} = \frac{1}{2}\left(u_{i,j} + u_{j,i}\right)$
Constitution	$\boldsymbol{\sigma} = \boldsymbol{C}\boldsymbol{\varepsilon}$	$\sigma_{ij} = C_{ijkl}\varepsilon_{kl}$
Equilibrium	$\boldsymbol{\mathcal{L}}_1^{\mathrm{T}}\, \boldsymbol{\sigma} + \boldsymbol{b} = \boldsymbol{0}$	$\sigma_{ij,i} + b_j = 0$

$$\begin{bmatrix} \frac{\partial}{\partial x} & 0 & \frac{\partial}{\partial y} \\ 0 & \frac{\partial}{\partial y} & \frac{\partial}{\partial x} \end{bmatrix} \begin{bmatrix} \sigma_x \\ \sigma_y \\ \sigma_{xy} \end{bmatrix} + \begin{bmatrix} f_x \\ f_y \end{bmatrix} = \begin{bmatrix} 0 \\ 0 \end{bmatrix}, \tag{5.17}$$

or in symbolic notation:

$$\boldsymbol{\mathcal{L}}_1^{\mathrm{T}}\boldsymbol{\sigma} + \boldsymbol{b} = \boldsymbol{0}, \tag{5.18}$$

where $\boldsymbol{\mathcal{L}}_1$ is the differential operator matrix and $\boldsymbol{b}$ the column matrix of body forces.

5.2.4 *Differential Equation*

The basic equations introduced in the previous three sections, i.e., the kinematics, the constitutive, and the equilibrium equation, are summarized in the following Table 5.2 where in addition the tensor notation[3] is given.

For the solution of the eight unknown spatial functions (2 components of the displacement vector, 3 components of the symmetric strain tensor and 3 components of the symmetric stress tensor), a set of eight scalar field equations is available:

- Equilibrium: 2,
- Constitution: 3,
- Kinematics: 3.

Furthermore, the boundary conditions are given:

$$\boldsymbol{u} \text{ on } \Gamma_u, \tag{5.19}$$

$$\boldsymbol{t} \text{ on } \Gamma_t, \tag{5.20}$$

where Γ_u is the part of the boundary where a displacement boundary condition is prescribed and Γ_t is the part of the boundary where a traction boundary condition,

[3] A differentiation is there indicated by the use of a comma: The first index refers to the component and the comma indicates the partial derivative with respect to the second subscript corresponding to the relevant coordinate axis [1].

i.e. external force per unit area, is prescribed with $t_j = \sigma_{ij} n_j$, where n_j are the components of the normal vector.

The eight scalar field equations can be combined to eliminate the stress and strain fields. As a result, two scalar field equations for the three scalar displacement fields are obtained. These equations are called the LAMÉ-NAVIER[4] equations and can be derived as follows:

Introducing the constitutive equation according to (5.5) in the equilibrium equation (8.13) gives:

$$\mathcal{L}_1^{\mathrm{T}} \boldsymbol{C} \boldsymbol{\varepsilon} + \boldsymbol{b} = \boldsymbol{0}\,. \tag{5.21}$$

Introducing the kinematics relations in the last equation according to (5.3) finally gives the LAMÉ–NAVIER equations:

$$\mathcal{L}_1^{\mathrm{T}} \boldsymbol{C} \mathcal{L}_1 \boldsymbol{u} + \boldsymbol{b} = \boldsymbol{0}\,. \tag{5.22}$$

Alternatively, the displacements may be substituted and the differential equations are obtained in terms of stresses. This formulation is known as the BELTRAMI–MICHELL[5] equations. If the body forces vanish ($\boldsymbol{b} = \boldsymbol{0}$), the partial differential equations in terms of stresses are called the BELTRAMI equations.

Table 5.3 summarizes the different formulations of the basic equations for plane elasticity, once in their specific form and once in symbolic notation.

The following Table 5.4 shows a comparison between the basic equations for a rod and plane elasticity problem. It can be seen that the use of the differential operator $\mathcal{L}_1\{\ldots\}$ allows to depict a simple analogy between both sets of equations.

5.3 Finite Element Solution

5.3.1 Derivation of the Principal Finite Element Equation

Let us assume in the following that the elasticity matrix in Eq. (5.5) is constant and that the exact solution is given by $\boldsymbol{u}^0$. Thus, the differential equation in terms of displacements can be written as:

$$\mathcal{L}_1^{\mathrm{T}} \boldsymbol{C} \mathcal{L}_1 \boldsymbol{u}^0 + \boldsymbol{b} = \boldsymbol{0}\,. \tag{5.23}$$

[4]Gabriel Léon Jean Baptiste LAMÉ (1795–1870), French mathematician.
Claude-Louis NAVIER (1785–1836), French engineer and physicist.

[5]Eugenio BELTRAMI (1835–1900), Italian mathematician.
John Henry MICHELL (1863–1940), Australian mathematician.

Table 5.3 Different formulations of the basic equations for plane elasticity (deformation in the x-y plane) E: YOUNG's modulus; ν: POISSON's ratio; f_x volume-specific force in x-direction; f_y volume-specific force in y-direction

Specific formulation	General formulation
Kinematics	
$\begin{bmatrix} \varepsilon_x \\ \varepsilon_y \\ 2\varepsilon_{xy} \end{bmatrix} = \begin{bmatrix} \frac{\partial}{\partial x} & 0 \\ 0 & \frac{\partial}{\partial y} \\ \frac{\partial}{\partial y} & \frac{\partial}{\partial x} \end{bmatrix} \begin{bmatrix} u_x \\ u_y \end{bmatrix}$	$\boldsymbol{\varepsilon} = \boldsymbol{\mathcal{L}}_1 \boldsymbol{u}$
Constitution	
$\begin{bmatrix} \sigma_x \\ \sigma_y \\ \sigma_{xy} \end{bmatrix} = \frac{E'}{1-\nu'^2} \begin{bmatrix} 1 & \nu' & 0 \\ \nu' & 1 & 0 \\ 0 & 0 & \frac{1-\nu'}{2} \end{bmatrix} \begin{bmatrix} \varepsilon_x \\ \varepsilon_y \\ 2\,\varepsilon_{xy} \end{bmatrix}$ with $E' = E$ and $\nu' = \nu$ for plane stress and $E' = \frac{E}{1-\left(\frac{\nu}{1-\nu}\right)^2}$ and $\nu' = \frac{\nu}{1-\nu}$ for plane strain	$\boldsymbol{\sigma} = \boldsymbol{C}\boldsymbol{\varepsilon}$
Equilibrium	
$\begin{bmatrix} \frac{\partial}{\partial x} & 0 & \frac{\partial}{\partial y} \\ 0 & \frac{\partial}{\partial y} & \frac{\partial}{\partial x} \end{bmatrix} \begin{bmatrix} \sigma_x \\ \sigma_y \\ \sigma_{xy} \end{bmatrix} + \begin{bmatrix} f_x \\ f_y \end{bmatrix} = \begin{bmatrix} 0 \\ 0 \end{bmatrix}$	$\boldsymbol{\mathcal{L}}_1^{\mathrm{T}} \boldsymbol{\sigma} + \boldsymbol{b} = \boldsymbol{0}$
PDE	
$\frac{E'}{1-\nu'^2} \begin{bmatrix} \frac{\partial^2}{\partial x^2} + \frac{1-\nu'}{2}\frac{\partial^2}{\partial y^2} & \nu'\frac{\partial^2}{\partial x \partial y} + \frac{1-\nu'}{2}\frac{\partial^2}{\partial x \partial y} \\ \nu'\frac{\partial^2}{\partial x \partial y} + \frac{1-\nu'}{2}\frac{\partial^2}{\partial x \partial y} & \frac{\partial^2}{\partial y^2} + \frac{1-\nu'}{2}\frac{\partial^2}{\partial x^2} \end{bmatrix} \begin{bmatrix} u_x \\ u_y \end{bmatrix} + \begin{bmatrix} f_x \\ f_y \end{bmatrix} = \begin{bmatrix} 0 \\ 0 \end{bmatrix}$	$\boldsymbol{\mathcal{L}}_1^{\mathrm{T}} \boldsymbol{C} \boldsymbol{\mathcal{L}}_1 \, \boldsymbol{u} + \boldsymbol{b} = \boldsymbol{0}$

Replacing the exact solution by an approximate solution $\boldsymbol{u}$, a residual $\boldsymbol{r}$ is obtained:

$$\boldsymbol{r} = \boldsymbol{\mathcal{L}}_1^{\mathrm{T}} \boldsymbol{C} \boldsymbol{\mathcal{L}}_1 \boldsymbol{u} + \boldsymbol{b} \neq \boldsymbol{0}\,. \tag{5.24}$$

The inner product is obtained by weighting the residual and integrating over the volume V as

$$\int\limits_V \boldsymbol{W}^{\mathrm{T}} \left(\boldsymbol{\mathcal{L}}_1^{\mathrm{T}} \boldsymbol{C} \boldsymbol{\mathcal{L}}_1 \boldsymbol{u} + \boldsymbol{b}\right) \mathrm{d}V = 0\,, \tag{5.25}$$

where $\boldsymbol{W}(\boldsymbol{x}) = \begin{bmatrix} W_x & W_y \end{bmatrix}^{\mathrm{T}}$ is the column matrix of weight functions and $\boldsymbol{x} = \begin{bmatrix} x & y \end{bmatrix}^{\mathrm{T}}$ is the column matrix of Cartesian coordinates. Application of the GREEN-GAUSS theorem (cf. Sect. A.7) gives the weak formulation as:

Table 5.4 Comparison of basic equations for rod and plane elasticity

Rod	Plane elasticity
Kinematics	
$\varepsilon_x(x) = \mathcal{L}_1\left(u_x(x)\right)$	$\boldsymbol{\varepsilon} = \boldsymbol{\mathcal{L}}_1 \boldsymbol{u}$
Constitution	
$\sigma_x(x) = C\varepsilon_x(x)$	$\boldsymbol{\sigma} = \boldsymbol{C}\boldsymbol{\varepsilon}$
Equilibrium	
$\mathcal{L}_1^{\mathrm{T}}\left(\sigma_x(x)\right) + b = 0$	$\boldsymbol{\mathcal{L}}_1^{\mathrm{T}}\,\boldsymbol{\sigma} + \boldsymbol{b} = \boldsymbol{0}$
PDE	
$\mathcal{L}_1^{\mathrm{T}}\left(C\mathcal{L}_1\left(u_x(x)\right)\right) + b = 0$	$\boldsymbol{\mathcal{L}}_1^{\mathrm{T}}\,\boldsymbol{C}\,\boldsymbol{\mathcal{L}}_1\,\boldsymbol{u} + \boldsymbol{b} = \boldsymbol{0}$

$$\int\limits_V (\boldsymbol{\mathcal{L}}_1 \boldsymbol{W})^{\mathrm{T}} \boldsymbol{C}\, (\boldsymbol{\mathcal{L}}_1 \boldsymbol{u})\, \mathrm{d}V = \int\limits_A \boldsymbol{W}^{\mathrm{T}} \boldsymbol{t}\, \mathrm{d}A + \int\limits_V \boldsymbol{W}^{\mathrm{T}} \boldsymbol{b}\, \mathrm{d}V\,, \tag{5.26}$$

where the column matrix of traction forces $\boldsymbol{t} = \begin{bmatrix} t_x & t_y \end{bmatrix}^{\mathrm{T}}$ can be understood as the expression[6] $(\boldsymbol{C}\boldsymbol{\mathcal{L}}_1\boldsymbol{u})^{\mathrm{T}}\boldsymbol{n} = \boldsymbol{\sigma}^{\mathrm{T}}\boldsymbol{n}$.

Any further development of Eq. (5.26) requires that the general expressions for the displacement and weight function, i.e. $\boldsymbol{u}$ and $\boldsymbol{W}$, are now approximated by some functional representations. The nodal approach for the displacements[7] can be generally written for a two-dimensional element with n nodes as:

$$u_x^{\mathrm{e}}(\boldsymbol{x}) = N_1 u_{1x} + N_2 u_{2x} + N_3 u_{3x} + \cdots + N_n u_{nx}\,, \tag{5.27}$$

$$u_y^{\mathrm{e}}(\boldsymbol{x}) = N_1 u_{1y} + N_2 u_{2y} + N_3 u_{3y} + \cdots + N_n u_{ny}\,, \tag{5.28}$$

or in matrix notation as:

$$\boldsymbol{u}^{\mathrm{e}} = \begin{bmatrix} u_x^{\mathrm{e}}(x,y) \\ u_y^{\mathrm{e}}(x,y) \end{bmatrix} = \left[\begin{array}{cc|cc|c|cc} N_1 & 0 & N_2 & 0 & \cdots & N_n & 0 \\ 0 & N_1 & 0 & N_2 & \cdots & 0 & N_n \end{array}\right] \left[\begin{array}{c} u_{1x} \\ u_{1y} \\ \hline u_{2x} \\ u_{2y} \\ \hline \vdots \\ \hline u_{nx} \\ u_{ny} \end{array}\right]. \tag{5.29}$$

Introducing the notations

[6] Strictly speaking, the traction forces must be calculated based on the stress *tensor* as $t_i = \sigma_{ji} n_j$ and not based on the column matrix of stress components.

[7] The following derivations are written under the simplification that each node reveals only displacement DOF and no rotations.

$$N_i\mathbf{I} = \begin{bmatrix} N_i & 0 \\ 0 & N_i \end{bmatrix} \quad \text{and} \quad \boldsymbol{u}_{\mathrm{p}i} = \begin{bmatrix} u_{ix} \\ u_{iy} \end{bmatrix}, \tag{5.30}$$

Equation (5.29) can be written as

$$\begin{bmatrix} u_x \\ u_y \end{bmatrix} = \begin{bmatrix} N_1\mathbf{I} & N_2\mathbf{I} & \cdots & N_n\mathbf{I} \end{bmatrix} \begin{bmatrix} \boldsymbol{u}_{\mathrm{p}1} \\ \boldsymbol{u}_{\mathrm{p}2} \\ \vdots \\ \boldsymbol{u}_{\mathrm{p}n} \end{bmatrix}, \tag{5.31}$$

or with $\boldsymbol{N}_i = N_i\mathbf{I}$ as

$$\begin{bmatrix} u_x \\ u_y \end{bmatrix} = \begin{bmatrix} \boldsymbol{N}_1 & \boldsymbol{N}_2 & \cdots & \boldsymbol{u}_n \end{bmatrix} \begin{bmatrix} \boldsymbol{u}_{\mathrm{p}1} \\ \boldsymbol{u}_{\mathrm{p}2} \\ \vdots \\ \boldsymbol{u}_{\mathrm{p}n} \end{bmatrix}. \tag{5.32}$$

The last equation can be written in abbreviated form as:

$$\boldsymbol{u}^{\mathrm{e}}(\boldsymbol{x}) = \boldsymbol{N}^{\mathrm{T}}(\boldsymbol{x})\boldsymbol{u}_{\mathrm{p}}^{\mathrm{p}}, \tag{5.33}$$

which is the same structure as in the case of the one-dimensional elements. The column matrix of the weight functions in Eq. (5.26) is approximated in a similar way as the unknown displacements:

$$\boldsymbol{W}(\boldsymbol{x}) = \boldsymbol{N}^{\mathrm{T}}(\boldsymbol{x})\delta\boldsymbol{u}_{\mathrm{p}}. \tag{5.34}$$

Introducing the approximations for $\boldsymbol{u}^{\mathrm{e}}$ and $\boldsymbol{W}$ according to Eqs. (5.33) and (5.34) in the weak formulation gives:

$$\int_V \left(\boldsymbol{\mathcal{L}}_1\boldsymbol{N}^{\mathrm{T}}\delta\boldsymbol{u}_{\mathrm{p}}\right)^{\mathrm{T}} \boldsymbol{C}\,(\boldsymbol{\mathcal{L}}_1\boldsymbol{N}^{\mathrm{T}}\boldsymbol{u}_{\mathrm{p}})\,\mathrm{d}V = \int_A (\boldsymbol{N}^{\mathrm{T}}\delta\boldsymbol{u}_{\mathrm{p}})^{\mathrm{T}}\boldsymbol{t}\,\mathrm{d}A + \int_V (\boldsymbol{N}^{\mathrm{T}}\delta\boldsymbol{u}_{\mathrm{p}})^{\mathrm{T}}\boldsymbol{b}\,\mathrm{d}V, \tag{5.35}$$

which we can write under the consideration that the matrix of displacements and virtual displacements are not affected by the integration as:

$$\delta\boldsymbol{u}_{\mathrm{p}}^{\mathrm{T}} \int_V (\boldsymbol{\mathcal{L}}_1\boldsymbol{N}^{\mathrm{T}})^{\mathrm{T}} \boldsymbol{C}\,(\boldsymbol{\mathcal{L}}_1\boldsymbol{N}^{\mathrm{T}})\,\mathrm{d}V\boldsymbol{u}_{\mathrm{p}}^{\mathrm{e}} = \delta\boldsymbol{u}_{\mathrm{p}}^{\mathrm{T}} \int_A \boldsymbol{N}\boldsymbol{t}\,\mathrm{d}A + \delta\boldsymbol{u}_{\mathrm{p}}^{\mathrm{T}} \int_V \boldsymbol{N}\boldsymbol{b}\,\mathrm{d}V, \tag{5.36}$$

which gives after elimination of $\delta\boldsymbol{u}_{\mathrm{p}}^{\mathrm{T}}$ the following statement for the principal finite element equation on the element level as:

$$\int_V (\boldsymbol{\mathcal{L}}_1\boldsymbol{N}^{\mathrm{T}})^{\mathrm{T}} \boldsymbol{C}\,(\boldsymbol{\mathcal{L}}_1\boldsymbol{N}^{\mathrm{T}})\,\mathrm{d}V\boldsymbol{u}_{\mathrm{p}}^{\mathrm{e}} = \int_A \boldsymbol{N}\boldsymbol{t}\,\mathrm{d}A + \int_V \boldsymbol{N}\boldsymbol{b}\,\mathrm{d}V. \tag{5.37}$$

Thus, we can identify the following three element matrices from the principal finite element equation:

$$\text{Stiffness matrix } (2n \times 2n)\text{:}\ \ \boldsymbol{K}^{\text{e}} = \int_V \underbrace{\left(\boldsymbol{\mathcal{L}}_1 \boldsymbol{N}^{\text{T}}\right)^{\text{T}}}_{\boldsymbol{B}} \boldsymbol{C} \underbrace{\left(\boldsymbol{\mathcal{L}}_1 \boldsymbol{N}^{\text{T}}\right)}_{\boldsymbol{B}^{\text{T}}} \mathrm{d}V\,, \tag{5.38}$$

$$\text{Boundary force matrix } (2n \times 1)\text{:}\ \ \boldsymbol{f}_t^{\text{e}} = \int_A \boldsymbol{N}\boldsymbol{t}\,\mathrm{d}A\,, \tag{5.39}$$

$$\text{Body force matrix } (2n \times 1)\text{:}\ \ \boldsymbol{f}_b^{\text{e}} = \int_V \boldsymbol{N}\boldsymbol{b}\,\mathrm{d}V\,. \tag{5.40}$$

Based on these abbreviations, the principal finite element equation for a single element can be written as:

$$\boldsymbol{K}^{\text{e}}\boldsymbol{u}_{\text{p}}^{\text{e}} = \boldsymbol{f}_t^{\text{e}} + \boldsymbol{f}_b^{\text{e}}\,. \tag{5.41}$$

In the following, let us look at the $\boldsymbol{B}$-matrix, i.e. the matrix which contains the derivatives of the interpolation functions. Application of the matrix of differential operators according to Eq. (5.2) to the matrix of interpolation functions gives:

$$\boldsymbol{\mathcal{L}}_1\boldsymbol{N}^{\text{T}} = \begin{bmatrix} \frac{\partial}{\partial x} & 0 \\ 0 & \frac{\partial}{\partial y} \\ \frac{\partial}{\partial y} & \frac{\partial}{\partial x} \end{bmatrix} \left[\begin{array}{cc|cc|c|cc} N_1 & 0 & N_2 & 0 & \cdots & N_n & 0 \\ 0 & N_1 & 0 & N_2 & \cdots & 0 & N_n \end{array}\right] \tag{5.42}$$

$$= \left[\begin{array}{cc|cc|c|cc} \frac{\partial N_1}{\partial x} & 0 & \frac{\partial N_2}{\partial x} & 0 & & \frac{\partial N_n}{\partial x} & 0 \\ 0 & \frac{\partial N_1}{\partial y} & 0 & \frac{\partial N_2}{\partial y} & \cdots & 0 & \frac{\partial N_n}{\partial y} \\ \frac{\partial N_1}{\partial y} & \frac{\partial N_1}{\partial x} & \frac{\partial N_2}{\partial y} & \frac{\partial N_2}{\partial x} & & \frac{\partial N_n}{\partial y} & \frac{\partial N_n}{\partial x} \end{array}\right] = \boldsymbol{B}^{\text{T}}\,, \tag{5.43}$$

which is a $(3 \times 2n)$-matrix. The transposed, i.e. $(\boldsymbol{\mathcal{L}}_1\boldsymbol{N}^{\text{T}})^{\text{T}}$, is thus a $(2n \times 3)$-matrix. Multiplication with the elasticity matrix, i.e. a (3×3)-matrix, results in $(\boldsymbol{\mathcal{L}}_1\boldsymbol{N}^{\text{T}})^{\text{T}}\boldsymbol{C}$, which is a $(2n \times 6)$-matrix. The final multiplication, i.e. $(\boldsymbol{\mathcal{L}}_1\boldsymbol{N}^{\text{T}})^{\text{T}}\boldsymbol{C}(\boldsymbol{\mathcal{L}}_1\boldsymbol{N}^{\text{T}})$ gives after integration the stiffness matrix with a dimension of $(2n \times 2n)$.

The integrations for the element matrices given in Eqs. (5.38) till (5.40) are approximated by numerical integration. To this end, the coordinates (x, y) are transformed to the natural coordinates (unit space: ξ, η) where each coordinate ranges from -1 to 1. In the scope of the coordinate transformation, attention must be paid to the derivatives. For example, the derivative of the interpolation functions with respect to the x-coordinate is transformed in the following way:

$$\frac{\partial N_i}{\partial x} \rightarrow \frac{\partial N_i}{\partial \xi}\frac{\partial \xi}{\partial x} + \frac{\partial N_i}{\partial \eta}\frac{\partial \eta}{\partial x}\,. \tag{5.44}$$

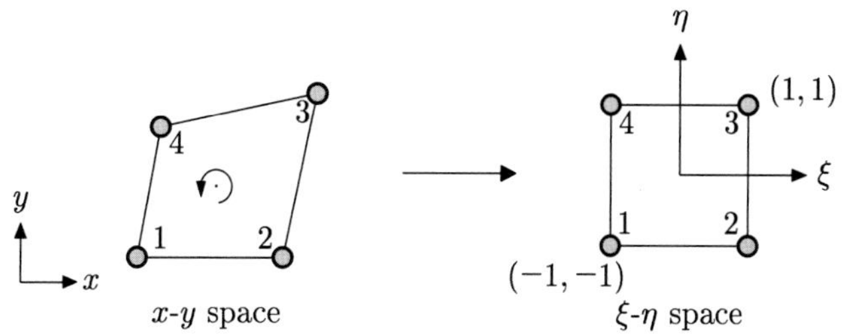

Fig. 5.5 Four-node planar element in the Cartesian (left) and parametric (right) space

Furthermore, the coordinate transformation requires that $\mathrm{d}V = \mathrm{d}x\mathrm{d}y \rightarrow \mathrm{d}V' = J\mathrm{d}\xi\mathrm{d}\eta$, where J is the Jacobian as given in the Appendix A.8.

5.3.2 *Four-Node Planar Element*

A simple representative of a two-dimensional finite element is a four-node planar bilinear quadrilateral (quad 4)[8] as shown in Fig. 5.5. This element uses bilinear interpolation functions and the node numbering must follow the right-hand convention as indicated in the figure.

The evaluation of the element stiffness matrix (5.38) requires the integration over derivatives of the *four* interpolation functions. Since numerical integration is applied, the unit space (ξ, η) requires some attention in regards to the derivatives, see Eq. (5.44):

$$\frac{\partial N_1(\xi,\eta)}{\partial x} = \frac{\partial N_1}{\partial \xi}\frac{\partial \xi}{\partial x} + \frac{\partial N_1}{\partial \eta}\frac{\partial \eta}{\partial x}, \tag{5.45}$$

$$\frac{\partial N_1(\xi,\eta)}{\partial y} = \frac{\partial N_1}{\partial \xi}\frac{\partial \xi}{\partial y} + \frac{\partial N_1}{\partial \eta}\frac{\partial \eta}{\partial y}. \tag{5.46}$$

Interpolation Functions and Derivatives

Let us assume in the following a linear displacement field in parametric $\xi{-}\eta$ space (demonstrated for the x-component here)

$$u_x^{\mathrm{e}}(\xi,\eta) = a_1 + a_2\xi + a_3\eta + a_4\xi\eta\,, \tag{5.47}$$

or in vector notation

[8]The derivation for a common linear three-node element is given in the Appendix D.

$$u_x^{\mathrm{e}}(\xi,\eta) = \boldsymbol{\chi}^{\mathrm{T}}\boldsymbol{a} = \begin{bmatrix} 1 & \xi & \eta & \xi\eta \end{bmatrix} \begin{bmatrix} a_1 \\ a_2 \\ a_3 \\ a_4 \end{bmatrix} . \tag{5.48}$$

Evaluating Eq. (5.47) for all four nodes of the quadrilateral element gives

$$\text{Node 1: } u_{1x} = u_x^{\mathrm{e}}(\xi=-1,\eta=-1) = a_1 - a_2 - a_3 + a_4 , \tag{5.49}$$

$$\text{Node 2: } u_{2x} = u_x^{\mathrm{e}}(\xi=1,\eta=-1) = a_1 + a_2 - a_3 - a_4 , \tag{5.50}$$

$$\text{Node 3: } u_{3x} = u_x^{\mathrm{e}}(\xi=1,\eta=1) = a_1 + a_2 + a_3 + a_4 , \tag{5.51}$$

$$\text{Node 4: } u_{4x} = u_x^{\mathrm{e}}(\xi=-1,\eta=1) = a_1 - a_2 + a_3 - a_4 , \tag{5.52}$$

or in matrix notation:

$$\begin{bmatrix} u_{1x} \\ u_{2x} \\ u_{3x} \\ u_{4x} \end{bmatrix} = \underbrace{\begin{bmatrix} 1 & -1 & -1 & 1 \\ 1 & 1 & -1 & -1 \\ 1 & 1 & 1 & 1 \\ 1 & -1 & 1 & -1 \end{bmatrix}}_{\boldsymbol{X}} \begin{bmatrix} a_1 \\ a_2 \\ a_3 \\ a_4 \end{bmatrix} . \tag{5.53}$$

Solving for $\boldsymbol{a}$ gives:

$$\begin{bmatrix} a_1 \\ a_2 \\ a_3 \\ a_4 \end{bmatrix} = \frac{1}{4} \begin{bmatrix} 1 & 1 & 1 & 1 \\ -1 & 1 & 1 & -1 \\ -1 & -1 & 1 & 1 \\ 1 & -1 & 1 & -1 \end{bmatrix} \begin{bmatrix} u_{1x} \\ u_{2x} \\ u_{3x} \\ u_{4x} \end{bmatrix} \tag{5.54}$$

or

$$\boldsymbol{a} = \boldsymbol{A}\boldsymbol{u}_{\mathrm{p}x} = \boldsymbol{X}^{-1}\boldsymbol{u}_{\mathrm{p},x} . \tag{5.55}$$

The matrix of interpolation functions results as:

$$\boldsymbol{N}_{\mathrm{e}}^{\mathrm{T}} = \boldsymbol{\chi}^{\mathrm{T}}\boldsymbol{A} = \begin{bmatrix} 1 & \xi & \eta & \xi\eta \end{bmatrix} \frac{1}{4} \begin{bmatrix} 1 & 1 & 1 & 1 \\ -1 & 1 & 1 & -1 \\ -1 & -1 & 1 & 1 \\ 1 & -1 & 1 & -1 \end{bmatrix} , \tag{5.56}$$

or

$$N_1(\xi,\eta) = \frac{1}{4}(1 - \xi - \eta + \xi\eta) = \frac{1}{4}(1-\xi)(1-\eta) , \tag{5.57}$$

$$N_2(\xi,\eta) = \frac{1}{4}(1 + \xi - \eta - \xi\eta) = \frac{1}{4}(1+\xi)(1-\eta) , \tag{5.58}$$

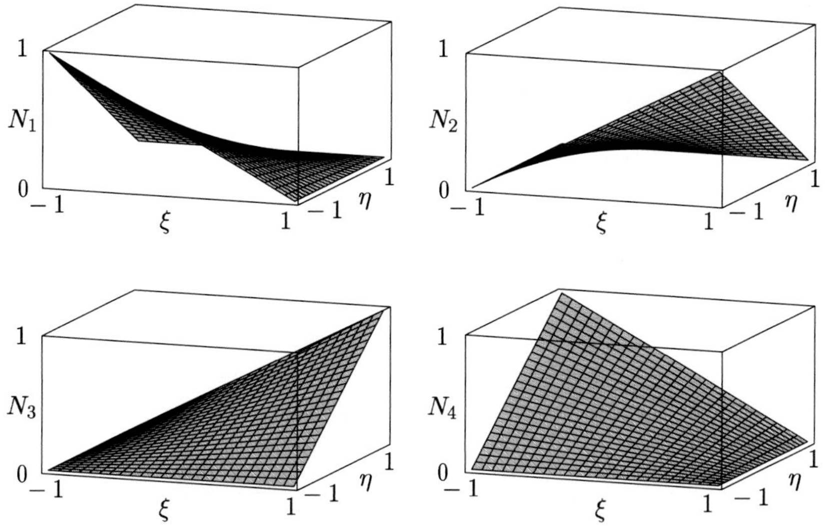

Fig. 5.6 Interpolation functions N_i $(i = 1, \dots, 4)$ for a four-node planar element in parametric ξ-η space

$$N_3(\xi, \eta) = \frac{1}{4}(1 + \xi + \eta + \xi\eta) = \frac{1}{4}(1 + \xi)(1 + \eta) , \tag{5.59}$$

$$N_4(\xi, \eta) = \frac{1}{4}(1 - \xi + \eta - \xi\eta) = \frac{1}{4}(1 - \xi)(1 + \eta) . \tag{5.60}$$

One may note that each N_i $(i = 1, 2, 3, 4)$ is unity when ξ and η assume coordinates of node i, but zero when ξ and η assume the coordinates of any other node. The graphical representation of the linear interpolation functions is shown in Fig. 5.6.

The derivatives with respect to the parametric coordinates can easily be obtained as:

$$\frac{\partial N_1(\xi, \eta)}{\partial \xi} = \frac{1}{4}(-1 + \eta) \quad ; \quad \frac{\partial N_1(\xi, \eta)}{\partial \eta} = \frac{1}{4}(-1 + \xi) , \tag{5.61}$$

$$\frac{\partial N_2(\xi, \eta)}{\partial \xi} = \frac{1}{4}(+1 - \eta) \quad ; \quad \frac{\partial N_2(\xi, \eta)}{\partial \eta} = \frac{1}{4}(-1 - \xi) , \tag{5.62}$$

$$\frac{\partial N_3(\xi, \eta)}{\partial \xi} = \frac{1}{4}(+1 + \eta) \quad ; \quad \frac{\partial N_3(\xi, \eta)}{\partial \eta} = \frac{1}{4}(+1 + \xi) , \tag{5.63}$$

$$\frac{\partial N_4(\xi, \eta)}{\partial \xi} = \frac{1}{4}(-1 - \eta) \quad ; \quad \frac{\partial N_4(\xi, \eta)}{\partial \eta} = \frac{1}{4}(+1 - \xi) . \tag{5.64}$$

Geometrical Derivatives

Let us assume the same interpolation for the global x- and y-coordinate as for the displacement (*isoparametric* element formulation), i.e. $\overline{N}_i = N_i$:

$$x(\xi,\eta) = \overline{N}_1(\xi,\eta) \times x_1 + \overline{N}_2(\xi,\eta) \times x_2 + \overline{N}_3(\xi,\eta) \times x_3 + \overline{N}_4(\xi,\eta) \times x_4\,, \tag{5.65}$$

$$y(\xi,\eta) = \overline{N}_1(\xi,\eta) \times y_1 + \overline{N}_2(\xi,\eta) \times y_2 + \overline{N}_3(\xi,\eta) \times y_3 + \overline{N}_4(\xi,\eta) \times y_4\,. \tag{5.66}$$

Remark: the *global* coordinates of the nodes $1, \ldots, 4$ can be used for $x_1, \ldots, x_4$ and $y_1, \ldots, y_4$.

Thus, the geometrical derivatives can easily be obtained as:

$$\frac{\partial x}{\partial \xi} = \frac{1}{4}\Big((-1+\eta)x_1 + (1-\eta)x_2 + (1+\eta)x_3 + (-1-\eta)x_4\Big)\,, \tag{5.67}$$

$$\frac{\partial y}{\partial \xi} = \frac{1}{4}\Big((-1+\eta)y_1 + (1-\eta)y_2 + (1+\eta)y_3 + (-1-\eta)y_4\Big)\,, \tag{5.68}$$

$$\frac{\partial x}{\partial \eta} = \frac{1}{4}\Big((-1+\xi)x_1 + (-1-\xi)x_2 + (1+\xi)x_3 + (1-\xi)x_4\Big)\,, \tag{5.69}$$

$$\frac{\partial y}{\partial \eta} = \frac{1}{4}\Big((-1+\xi)y_1 + (-1-\xi)y_2 + (1+\xi)y_3 + (1-\xi)y_4\Big)\,. \tag{5.70}$$

The calculation of the derivatives of the interpolation functions (see Eq. (5.44)) requires, however, the geometrical derivatives of the natural coordinates (ξ, η) with respect to the physical coordinates (x, y). These relations can be easily obtained from Eqs. (5.67)–(5.70) under consideration of the relationships provided in Sect. A.8:

$$\frac{\partial \xi}{\partial x} = +\frac{1}{\frac{\partial x}{\partial \xi}\frac{\partial y}{\partial \eta} - \frac{\partial x}{\partial \eta}\frac{\partial y}{\partial \xi}} \times \frac{\partial y}{\partial \eta}\,, \tag{5.71}$$

$$\frac{\partial \xi}{\partial y} = -\frac{1}{\frac{\partial x}{\partial \xi}\frac{\partial y}{\partial \eta} - \frac{\partial x}{\partial \eta}\frac{\partial y}{\partial \xi}} \times \frac{\partial x}{\partial \eta}\,, \tag{5.72}$$

$$\frac{\partial \eta}{\partial x} = -\frac{1}{\frac{\partial x}{\partial \xi}\frac{\partial y}{\partial \eta} - \frac{\partial x}{\partial \eta}\frac{\partial y}{\partial \xi}} \times \frac{\partial y}{\partial \xi}\,, \tag{5.73}$$

$$\frac{\partial \eta}{\partial y} = +\frac{1}{\frac{\partial x}{\partial \xi}\frac{\partial y}{\partial \eta} - \frac{\partial x}{\partial \eta}\frac{\partial y}{\partial \xi}} \times \frac{\partial x}{\partial \xi}\,. \tag{5.74}$$

Based on the derived equations, the triple matrix product $\boldsymbol{B}\boldsymbol{C}\boldsymbol{B}^{\mathrm{T}}$ (see Eq. (5.38)) can be numerically calculated to obtain the stiffness matrix.

Numerical Integration

The integration is performed as in the case of the one-dimensional integrals based on GAUSS–LEGENDRE quadrature. For the domain integrals, one can write that

Table 5.5 Integration rules for plane elasticity elements [3]

Points	ξ_i	η_i	Weight w_i	Error
1	0	0	4	$O(\xi^2)$
4	$\pm 1/\sqrt{3}$	$\pm 1/\sqrt{3}$	1	$O(\xi^4)$

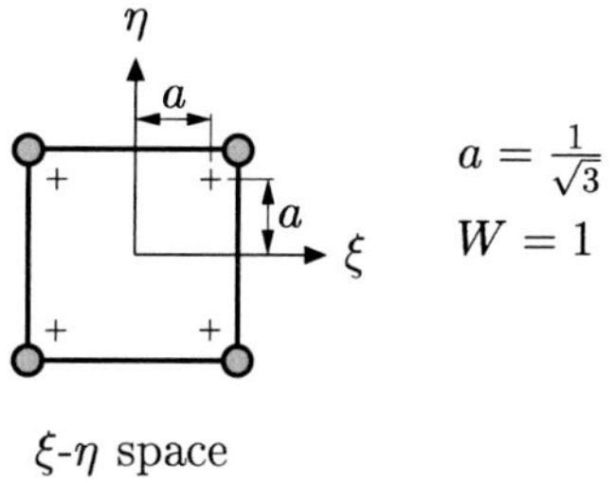

Fig. 5.7 Representation of a 2×2 integration for a plane elasticity element

$$\int\limits_{V_e} f(x,y)\mathrm{d}V = \int\limits_{V_e'} f'(\xi,\eta)J\mathrm{d}V' = \int\limits_{-1}^{1}\int\limits_{-1}^{1} tf'(\xi,\eta)J\mathrm{d}\xi\mathrm{d}\eta$$
$$= \sum_{i=1}^{n} tf'(\xi,\eta)_i J_i w_i\,, \tag{5.75}$$

where the Jacobian is $J = \frac{\partial x}{\partial \xi}\frac{\partial y}{\partial \eta} - \frac{\partial x}{\partial \eta}\frac{\partial y}{\partial \xi}$ (see Sect. A.8), $(\xi,\eta)_i$ are the coordinates of the GAUSS points and w_i are the corresponding weight factors. The location of the integration points and values of associated weights are given in Table 5.5.

Furthermore, it should be highlighted that the thickness t in Eq. (5.75) is assumed constant.

Thus, we can write for the 2×2 integration indicated in Fig. 5.7:

$$\begin{aligned}\boldsymbol{K}^{\mathrm{e}} = \int\limits_{V} (\boldsymbol{BCB}^{\mathrm{T}})\mathrm{d}V &= \boldsymbol{BCB}^{\mathrm{T}}J \times 1 \times t\Big|_{\left(-\frac{1}{\sqrt{3}},-\frac{1}{\sqrt{3}}\right)}\\ &+ \boldsymbol{BCB}^{\mathrm{T}}J \times 1 \times t\Big|_{\left(\frac{1}{\sqrt{3}},-\frac{1}{\sqrt{3}}\right)}\\ &+ \boldsymbol{BCB}^{\mathrm{T}}J \times 1 \times t\Big|_{\left(\frac{1}{\sqrt{3}},\frac{1}{\sqrt{3}}\right)}\\ &+ \boldsymbol{BCB}^{\mathrm{T}}J \times 1 \times t\Big|_{\left(-\frac{1}{\sqrt{3}},\frac{1}{\sqrt{3}}\right)}.\end{aligned} \tag{5.76}$$

Let us summarize here the major steps which are required to calculate the elemental stiffness matrix.

❶ Introduce an elemental coordinate system (x, y).
❷ Express the coordinates (x_i, y_i) of the corner nodes i $(i = 1, \cdots, 4)$ in this elemental coordinate system.
❸ Calculate the partial derivatives of the old Cartesian (x, y) coordinates with respect to the new natural (ξ, η) coordinates, see Eqs. (5.67)–(5.70):

$$\frac{\partial x}{\partial \xi} = x_\xi = \frac{1}{4}\Big((-1+\eta)x_1 + (1-\eta)x_2 + (1+\eta)x_3 + (-1-\eta)x_4\Big),$$

$$\vdots$$

$$\frac{\partial y}{\partial \eta} = y_\eta = \frac{1}{4}\Big((-1+\xi)y_1 + (-1-\xi)y_2 + (1+\xi)y_3 + (1-\xi)y_4\Big).$$

❹ Calculate the partial derivatives of the new natural (ξ, η) coordinates with respect to the old Cartesian (x, y) coordinates, see Eqs. (5.71)–(5.74):

$$\frac{\partial \xi}{\partial x} = +\frac{1}{\frac{\partial x}{\partial \xi}\frac{\partial y}{\partial \eta} - \frac{\partial x}{\partial \eta}\frac{\partial y}{\partial \xi}} \times \frac{\partial y}{\partial \eta}, \quad \frac{\partial \xi}{\partial y} = -\frac{1}{\frac{\partial x}{\partial \xi}\frac{\partial y}{\partial \eta} - \frac{\partial x}{\partial \eta}\frac{\partial y}{\partial \xi}} \times \frac{\partial x}{\partial \eta},$$

$$\frac{\partial \eta}{\partial x} = -\frac{1}{\frac{\partial x}{\partial \xi}\frac{\partial y}{\partial \eta} - \frac{\partial x}{\partial \eta}\frac{\partial y}{\partial \xi}} \times \frac{\partial y}{\partial \xi}, \quad \frac{\partial \eta}{\partial y} = +\frac{1}{\frac{\partial x}{\partial \xi}\frac{\partial y}{\partial \eta} - \frac{\partial x}{\partial \eta}\frac{\partial y}{\partial \xi}} \times \frac{\partial x}{\partial \xi}.$$

❺ Calculate the $\boldsymbol{B}$-matrix and its transposed, see Eq. (5.43):

$$\boldsymbol{B}^{\mathrm{T}} = \left[\begin{array}{cc|cc|c|cc} \frac{\partial N_1}{\partial x} & 0 & \frac{\partial N_2}{\partial x} & 0 & & \frac{\partial N_4}{\partial x} & 0 \\ 0 & \frac{\partial N_1}{\partial y} & 0 & \frac{\partial N_2}{\partial y} & \cdots & 0 & \frac{\partial N_4}{\partial y} \\ \frac{\partial N_1}{\partial y} & \frac{\partial N_1}{\partial x} & \frac{\partial N_2}{\partial y} & \frac{\partial N_2}{\partial x} & & \frac{\partial N_4}{\partial y} & \frac{\partial N_4}{\partial x} \end{array}\right],$$

where the partial derivatives are $\frac{\partial N_1(\xi,\eta)}{\partial x} = \frac{\partial N_1}{\partial \xi}\frac{\partial \xi}{\partial x} + \frac{\partial N_1}{\partial \eta}\frac{\partial \eta}{\partial x}, \ldots$ and the derivatives of the interpolation functions are given in Eqs. (5.61)–(5.64), i.e., $\frac{\partial N_1}{\partial \xi} = \frac{1}{4}(-1+\eta)\ , \ldots$
❻ Calculate the triple matrix product $\boldsymbol{B}\boldsymbol{C}^{\mathrm{T}}\boldsymbol{B}$, where the elasticity matrix $\boldsymbol{C}$ is given by Eqs. (5.4) and (5.9).
❼ Perform the numerical integration based on a 2×2 integration rule:

$$\int_V (\boldsymbol{B}\boldsymbol{C}\boldsymbol{B}^{\mathrm{T}})\mathrm{d}V = \boldsymbol{B}\boldsymbol{C}\boldsymbol{B}^{\mathrm{T}}J \times 1 \times t\Big|_{\left(-\frac{1}{\sqrt{3}},-\frac{1}{\sqrt{3}}\right)}$$
$$+ \boldsymbol{B}\boldsymbol{C}\boldsymbol{B}^{\mathrm{T}}J \times 1 \times t\Big|_{\left(\frac{1}{\sqrt{3}},-\frac{1}{\sqrt{3}}\right)} + \boldsymbol{B}\boldsymbol{C}\boldsymbol{B}^{\mathrm{T}}J \times 1 \times t\Big|_{\left(\frac{1}{\sqrt{3}},\frac{1}{\sqrt{3}}\right)}$$
$$+ \boldsymbol{B}\boldsymbol{C}\boldsymbol{B}^{\mathrm{T}}J \times 1 \times t\Big|_{\left(-\frac{1}{\sqrt{3}},\frac{1}{\sqrt{3}}\right)}.$$

❽ $\boldsymbol{K}$ obtained.

Let us summarize at the end of this section the major steps that were undertaken to transform the partial differential equation into the principal finite element equation, see Table 5.6.

Table 5.6 Summary: derivation of principal finite element equation for plane elements

Strong formulation
$\mathcal{L}_1^{\mathrm{T}} \boldsymbol{C} \mathcal{L}_1 \boldsymbol{u}^0 + \boldsymbol{b} = \boldsymbol{0}$
Inner product
$\int_V \boldsymbol{W}^{\mathrm{T}}(\boldsymbol{x}) \left(\mathcal{L}_1^{\mathrm{T}} \boldsymbol{C} \mathcal{L}_1 \boldsymbol{u} + \boldsymbol{b}\right) \mathrm{d}V = 0$
Weak formulation
$\int_V (\mathcal{L}_1 \boldsymbol{W})^{\mathrm{T}} \boldsymbol{C} (\mathcal{L}_1 \boldsymbol{u}) \mathrm{d}V = \int_A \boldsymbol{W}^{\mathrm{T}} \boldsymbol{t} \, \mathrm{d}A + \int_V \boldsymbol{W}^{\mathrm{T}} \boldsymbol{b} \, \mathrm{d}V$
Principal finite element equation (quad 4)
$\underbrace{\int_V \left(\mathcal{L}_1 \boldsymbol{N}^{\mathrm{T}}\right)^{\mathrm{T}} \boldsymbol{C} \left(\mathcal{L}_1 \boldsymbol{N}^{\mathrm{T}}\right) \mathrm{d}V}_{\boldsymbol{K}^{\mathrm{e}}} \begin{bmatrix} u_{1x} \\ u_{1y} \\ \vdots \\ u_{4x} \\ u_{4y} \end{bmatrix} = \begin{bmatrix} F_{1x} \\ F_{1y} \\ \vdots \\ F_{4x} \\ F_{4y} \end{bmatrix} + \int_V \boldsymbol{N} \boldsymbol{b} \, \mathrm{d}V$

Table 5.7 Coordinates of the four nodes for different locations of the elemental coordinate system, see Fig. 5.8

(a)	(b)	(c)
$1(-a, -a)$	$1(0, 0)$	$1(x_0, y_0)$
$2(a, -a)$	$2(2a, 0)$	$2(x_0 + 2a, y_0)$
$3(a, a)$	$3(2a, 2a)$	$3(x_0 + 2a, y_0 + 2a)$
$4(-a, a)$	$4(0, 2a)$	$4(x_0, y_0 + 2a)$

5.3.3 Solved Plane Elasticity Problems

5.1 Example: Influence of the coordinate system's origin on the geometrical derivatives

Given is a square two-dimensional element as shown in Fig. 5.8. Calculate the geometrical derivatives of the natural coordinates (ξ, η) with respect to the physical coordinates (x, y) and consider the different locations of the elemental coordinate system as shown in Figs. 5.8a–c. Comment: The parametric $\xi\eta$-space as shown in Fig. 5.8d would not allow this flexibility since $-1 \leq \xi \leq 1$ and $-1 \leq \eta \leq 1$ must hold.

This problem relates to steps ❶ to ❹ as given on p. 298.

5.1 Solution

The coordinates of the four corner nodes in the different xy-systems are collected in Table 5.7.

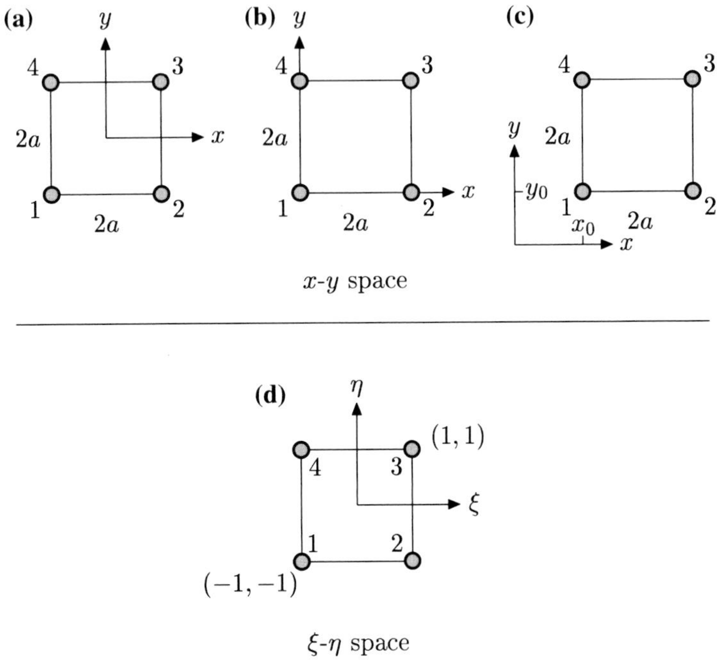

Fig. 5.8 Influence of the coordinate system's origin

The next step is to calculate the partial derivatives of the physical (x, y) coordinates with respect to the parametric (ξ, η) ones, see Eqs. (5.67)–(5.70).

For case **(a)**, this evaluation gives:

$$\frac{\partial x}{\partial \xi} = \frac{1}{4}\Big((-1+\eta)(-a) + (1-\eta)(a) + (1+\eta)(a) + (-1-\eta)(-a)\Big),$$

$$\frac{\partial y}{\partial \xi} = \frac{1}{4}\Big((-1+\eta)(-a) + (1-\eta)(-a) + (1+\eta)(a) + (-1-\eta)(a)\Big),$$

$$\frac{\partial x}{\partial \eta} = \frac{1}{4}\Big((-1+\xi)(-a) + (-1-\xi)(a) + (1+\xi)(a) + (1-\xi)(-a)\Big),$$

$$\frac{\partial y}{\partial \eta} = \frac{1}{4}\Big((-1+\xi)(-a) + (-1-\xi)(-a) + (1+\xi)(a) + (1-\xi)(a)\Big).$$

This finally gives after simplification:

$$\frac{\partial x}{\partial \xi} = a\ ;\ \frac{\partial y}{\partial \xi} = 0\ ;\ \frac{\partial x}{\partial \eta} = 0\ ;\ \frac{\partial y}{\partial \eta} = a\,.$$

Table 5.8 Coordinates of the four nodes for different element shapes, see Fig. 5.9

(a)	(b)	(c)	(d)
$1(-a, -a)$	$1(-a, -b)$	$1(-(a+d), -b)$	$1(-(a+d), -a)$
$2(a, -a)$	$2(a, -b)$	$2(a-d, -b)$	$2(a, -a)$
$3(a, a)$	$3(a, b)$	$3(a+d, b)$	$3(a, a)$
$4(-a, a)$	$4(-a, b)$	$4(-(a-d), b)$	$4(-a, a)$

Now we need to calculate the partial derivatives of the parametric (ξ, η) coordinates with respect to the physical (x, y) coordinates, see Eqs. (5.71)–(5.74).

For case **(a)**, this evaluation gives:

$$\frac{\partial \xi}{\partial x} = \frac{1}{a^2} \times a = \frac{1}{a},$$
$$\frac{\partial \xi}{\partial y} = \frac{1}{a^2} \times (-0) = 0,$$
$$\frac{\partial \eta}{\partial x} = \frac{1}{a^2} \times (-0) = 0,$$
$$\frac{\partial \eta}{\partial y} = \frac{1}{a^2} \times (a) = \frac{1}{a}.$$

It should be noted that cases **(a)**, **(b)** and **(c)** give the same results for the geometrical derivatives and the Jacobian.

5.2 Example: Influence of the shape regularity on the geometrical derivatives

Given are two-dimensional elements as shown in Fig. 5.9. Calculate the geometrical derivatives of the natural coordinates (ξ, η) with respect to the physical coordinates (x, y) and consider the different shapes as shown in Fig. 5.9a–d. This problem relates to steps ❶ to ❹ as given on p. 298.

5.2 Solution

The xy-coordinates of the four corner nodes for the different shapes are collected in Table 5.8

Case **(a)**: $J = a^2$

$$\frac{\partial x}{\partial \xi} = a\,; \; \frac{\partial y}{\partial \xi} = 0\,; \; \frac{\partial x}{\partial \eta} = 0\,; \; \frac{\partial y}{\partial \eta} = a\,.$$

$$\frac{\partial \xi}{\partial x} = \frac{1}{a}\,; \; \frac{\partial \xi}{\partial y} = 0\,; \; \frac{\partial \eta}{\partial x} = 0\,; \; \frac{\partial \eta}{\partial y} = \frac{1}{a}\,.$$

Case **(b)**: $J = ab$

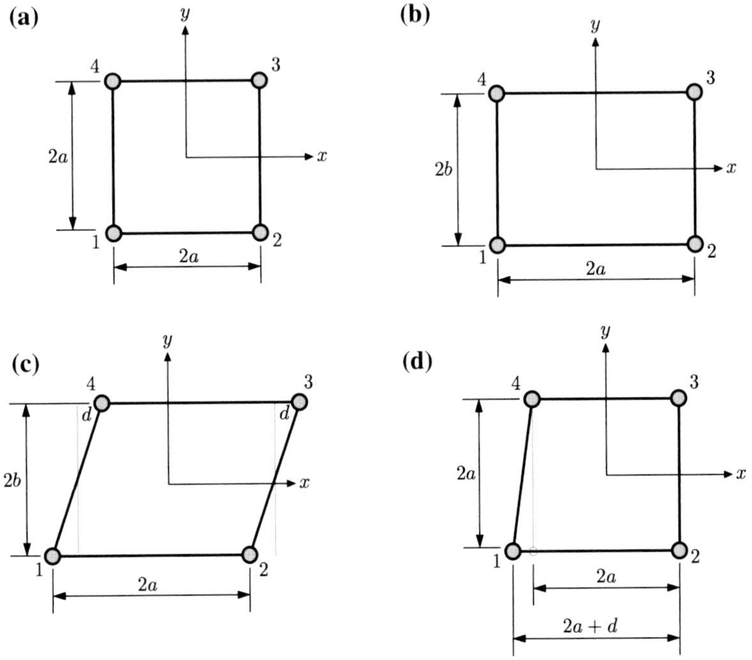

Fig. 5.9 Influence of the element regularity

$$\frac{\partial x}{\partial \xi} = a\ ;\ \frac{\partial y}{\partial \xi} = 0\ ;\ \frac{\partial x}{\partial \eta} = 0\ ;\ \frac{\partial y}{\partial \eta} = b\,.$$

$$\frac{\partial \xi}{\partial x} = \frac{1}{a}\ ;\ \frac{\partial \xi}{\partial y} = 0\ ;\ \frac{\partial \eta}{\partial x} = 0\ ;\ \frac{\partial \eta}{\partial y} = \frac{1}{b}.$$

Case (**c**): $J = ab$

$$\frac{\partial x}{\partial \xi} = a\ ;\ \frac{\partial y}{\partial \xi} = 0\ ;\ \frac{\partial x}{\partial \eta} = d\ ;\ \frac{\partial y}{\partial \eta} = b\,.$$

$$\frac{\partial \xi}{\partial x} = \frac{1}{a}\ ;\ \frac{\partial \xi}{\partial y} = -\frac{d}{ab}\ ;\ \frac{\partial \eta}{\partial x} = 0\ ;\ \frac{\partial \eta}{\partial y} = \frac{1}{b}.$$

Case (**d**): $J = a^2 + \frac{1}{4}ad - \frac{1}{4}\eta ad$

$$\frac{\partial x}{\partial \xi} = a + \tfrac{1}{4}d - \tfrac{1}{4}\eta d\ ;\ \frac{\partial y}{\partial \xi} = 0\ ;\ \frac{\partial x}{\partial \eta} = \tfrac{1}{4}d - \tfrac{1}{4}\xi d\ ;\ \frac{\partial y}{\partial \eta} = a\,.$$

Fig. 5.10 Distorted two-dimensional element

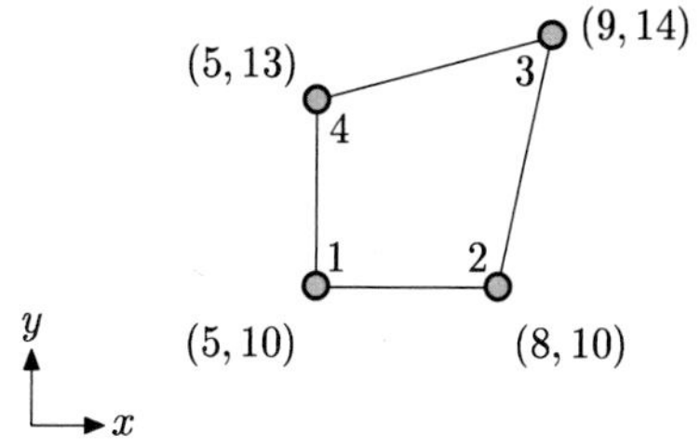

$$\frac{\partial \xi}{\partial x} = \frac{4}{4a+d-\eta d}\,;\ \frac{\partial \xi}{\partial y} = \frac{d(-1+\xi)}{a(-\eta d+4a+d)}\,;\ \frac{\partial \eta}{\partial x} = 0\,;\ \frac{\partial \eta}{\partial y} = \frac{1}{a}\,.$$

5.3 Example: Distorted two-dimensional element

Given is a distorted two-dimensional element as shown in Fig. 5.10. Calculate the geometrical derivatives of the natural coordinates (ξ, η) with respect to the physical coordinates (x, y). This problem relates to steps ❶ to ❹ as given on p. 298.

5.3 Solution

Application of Eqs. (5.71)–(5.74) under consideration of the Cartesian coordinates given in Fig. 5.10 gives:

$$\frac{\partial \xi}{\partial x} = \frac{7/4 + 1/4\,\xi}{(7/4 + 1/4\,\eta)\,(7/4 + 1/4\,\xi) - (1/4 + 1/4\,\xi)\,(1/4 + 1/4\,\eta)}\,, \tag{5.77}$$

$$\frac{\partial \xi}{\partial y} = -\frac{1/4 + 1/4\,\xi}{(7/4 + 1/4\,\eta)\,(7/4 + 1/4\,\xi) - (1/4 + 1/4\,\xi)\,(1/4 + 1/4\,\eta)}\,, \tag{5.78}$$

$$\frac{\partial \eta}{\partial x} = -\frac{1/4 + 1/4\,\eta}{(7/4 + 1/4\,\eta)\,(7/4 + 1/4\,\xi) - (1/4 + 1/4\,\xi)\,(1/4 + 1/4\,\eta)}\,, \tag{5.79}$$

$$\frac{\partial \eta}{\partial y} = \frac{7/4 + 1/4\,\eta}{(7/4 + 1/4\,\eta)\,(7/4 + 1/4\,\xi) - (1/4 + 1/4\,\xi)\,(1/4 + 1/4\,\eta)}\,. \tag{5.80}$$

5.4 Example: Plate under tensile load

Given is a regular two-dimensional element as shown in Fig. 5.11. The left-hand nodes are fixed and the right-hand nodes are loaded by horizontal point loads F_0.

Use a single plane elasticity element to:

- Derive the general expression for the stiffness matrix under plane stress condition.
- Calculate the nodal displacements for $a = 0.75$, $b = 0.5$ and $\nu = 0.2$ as a function of F_0, t, and E. Assume consistent units.

This problem relates to steps ❶ to ❽ as given on p. 298.

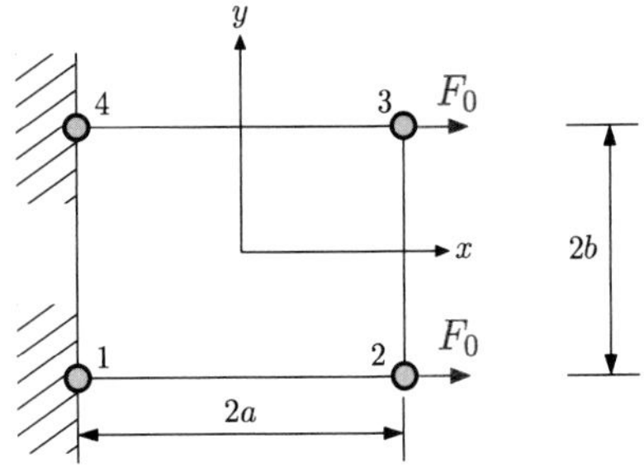

Fig. 5.11 Two-dimensional element under tensile load

5.4 Solution

The general expression for the stiffness matrix is given in Eq. (5.38):

$$(8\times 8):\ \boldsymbol{K}^{\text{e}} = \int\limits_V \underbrace{(\boldsymbol{\mathcal{L}}\boldsymbol{N}^{\text{T}})^{\text{T}}}_{\boldsymbol{B}} \boldsymbol{C} \underbrace{(\boldsymbol{\mathcal{L}}\boldsymbol{N}^{\text{T}})}_{\boldsymbol{B}^{\text{T}}} \text{d}V\,.$$

The calculation of the geometrical derivatives was covered in the preceding examples and the transposed of the $\boldsymbol{B}$-matrix (see Eq. (5.43)) is obtained as:

$$\boldsymbol{B}^{\text{T}} = \frac{1}{4}\times \begin{bmatrix} \frac{-1+\eta}{a} & 0 & -\frac{-1+\eta}{a} & 0 & \frac{1+\eta}{a} & 0 & -\frac{1+\eta}{a} & 0 \\ 0 & \frac{-1+\xi}{b} & 0 & -\frac{1+\xi}{b} & 0 & \frac{1+\xi}{b} & 0 & -\frac{-1+\xi}{b} \\ \frac{-1+\xi}{b} & \frac{-1+\eta}{a} & -\frac{1+\xi}{b} & -\frac{-1+\eta}{a} & \frac{1+\xi}{b} & \frac{1+\eta}{a} & -\frac{-1+\xi}{b} & -\frac{1+\eta}{a} \end{bmatrix}. \tag{5.81}$$

This matrix must be multiplied with the elasticity matrix $\boldsymbol{C}$ (see Eq. (5.4)), i.e.

$$\frac{E}{1-\nu^2}\begin{bmatrix} 1 & \nu & 0 \\ \nu & 1 & 0 \\ 0 & 0 & \frac{1-\nu}{2} \end{bmatrix},$$

and the result again multiplied with the $\boldsymbol{B}$-matrix. The components of the (8×8) $\boldsymbol{BCB}^{\text{T}}$ matrix are abbreviated with $\text{d}K_{ij}$ and a few components are stated here:

$$\text{d}K_{11} = \frac{1}{32}\frac{E\left(a^2\nu\xi^2 - 2a^2\nu\xi - a^2\xi^2 - 2b^2\eta^2 + a^2\nu + 2a^2\xi + 4b^2\eta - a^2 - 2b^2\right)}{a^2\left(\nu^2-1\right)b^2},$$

$$\text{d}K_{12} = -\frac{1}{32}\frac{(-1+\xi)(-1+\eta)E}{(\nu-1)ab},$$

$$\text{d}K_{13} = -\frac{1}{32}\frac{E\left(a^2\nu\xi^2 - a^2\xi^2 - 2b^2\eta^2 - a^2\nu + 4b^2\eta + a^2 - 2b^2\right)}{a^2\left(\nu^2-1\right)b^2},$$

$$\text{d}K_{14} = \frac{1}{32}\frac{(-1+\eta)E(\nu\xi + 3\nu + \xi - 1)}{ba\left(\nu^2-1\right)}.$$

In order to determine the stiffness matrix $\boldsymbol{K} = \int \boldsymbol{BCB}^{\mathrm{T}} \mathrm{d}V$, the integration over the components $\mathrm{d}K_{ij}$ must be performed. This is done by numerical 2×2 integration[9] as outlined in Eq. (5.76). The components of the (8×8) elemental stiffness matrix are abbreviated with K_{ij} and a few components should be stated here:

$$K_{11} = \frac{Et\left(a^2\nu - a^2 - 2b^2\right)}{6ba\left(\nu^2 - 1\right)},$$

$$K_{12} = -\frac{Et}{8(\nu - 1)},$$

$$K_{13} = \frac{Et\left(a^2\nu - a^2 + 4b^2\right)}{12ba\left(\nu^2 - 1\right)},$$

$$K_{14} = -\frac{Et\,(3\nu - 1)}{8(\nu^2 - 1)}.$$

The reduced stiffness matrix, i.e. under consideration of $u_{1x} = u_{1y} = u_{4x} = u_{4y}$, reads as:

$$\begin{bmatrix} \frac{Et\left(a^2\nu-a^2-2b^2\right)}{6ba\left(\nu^2-1\right)} & \frac{Et}{8(-1+\nu)} & -\frac{Et\left(a^2\nu-a^2+b^2\right)}{6ba\left(\nu^2-1\right)} & -\frac{Et(3\nu-1)}{8(\nu^2-1)} \\ \frac{Et}{8(-1+\nu)} & -\frac{Et\left(-b^2\nu+2a^2+b^2\right)}{6ba\left(\nu^2-1\right)} & \frac{Et(3\nu-1)}{8(\nu^2-1)} & \frac{Et\left(b^2\nu+4a^2-b^2\right)}{12ba\left(\nu^2-1\right)} \\ -\frac{Et\left(a^2\nu-a^2+b^2\right)}{6ba\left(\nu^2-1\right)} & \frac{Et(3\nu-1)}{8(\nu^2-1)} & \frac{Et\left(a^2\nu-a^2-2b^2\right)}{6ba\left(\nu^2-1\right)} & -\frac{Et}{8(-1+\nu)} \\ -\frac{Et(3\nu-1)}{8(\nu^2-1)} & \frac{Et\left(b^2\nu+4a^2-b^2\right)}{12ba\left(\nu^2-1\right)} & -\frac{Et}{8(-1+\nu)} & -\frac{Et\left(-b^2\nu+2a^2+b^2\right)}{6ba\left(\nu^2-1\right)} \end{bmatrix}.$$

The solution of the system of equations requires the calculation of the inverse of the stiffness matrix and the unknown displacements are obtained as $\boldsymbol{K}^{\mathrm{T}}\boldsymbol{f}$:

$$u_{2x} = 2.9652\,\frac{F_0}{tE},$$

$$u_{2y} = 0.2839\,\frac{F_0}{tE},$$

$$u_{3x} = 2.9652\,\frac{F_0}{tE},$$

$$u_{3y} = -0.2839\,\frac{F_0}{tE}.$$

[9]It should be noted here that the 2×2 numerical integration gives the same result as the analytical integration.

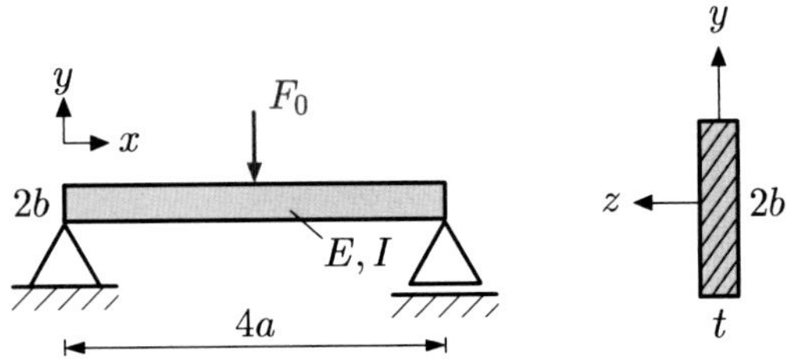

Fig. 5.12 Simply supported beam with rectangular cross section

5.5 Advanced Example: Different plane modeling approaches of a simply supported beam

Given is a simply supported Euler-Bernoulli beam as indicated Fig. 5.12. The length of the beam is $4a$ and the rectangular cross section has the dimensions $t \times 2b$. The beam is loaded by a single force F_0 acting in the middle of the beam. Note that the problem is *not* symmetric.

Use plane elasticity two-dimensional elements in the following to model the problem and to calculate the nodal unknowns under the assumption of a plane *stress* case. The modelling approach is based on two elements with nodes 1, ..., 6, see Fig. 5.13.

Case (a) Calculate the symbolic solution for all nodal unknowns (displacements) under the assumption that the force F_0 is acting at node 3.

Case (b) Calculate the symbolic solution for all nodal unknowns (displacements) under the assumption that the force F_0 is acting at node 2.

Case (c) Calculate the symbolic solution for all nodal unknowns (displacements) under the assumption that the force $\frac{1}{2}F_0$ is acting at node 3 and $\frac{1}{2}F$ is acting at node 2.

Compare the results for the vertical displacement at node 2 and 3 for the special case of $a = 0.75$, $b = 0.5$ and $\nu = 0.2$.

Case (d) Calculate the numerical solution ($a = 0.75$, $b = 0.5$, $\nu = 0.2$ and $d = 0.2a$) for all nodal unknowns (displacements) under the assumption that the force F_0 is acting at node 2. Pay attention to the fact that the elements are no longer rectangular.

5.5 Solution

The cases (a)–(c) can be solved with the same global stiffness matrix. Only the right-hand load matrix is different. Both elements have the same dimensions. Thus, it is sufficient to derive the stiffness matrix only for one element and then to assemble the global system of equations. Let us assume in the following that the elemental coordinate system is located in the center of the element, see Fig. 5.14.

The geometrical derivatives of the Cartesian coordinates (x, y) with respect to the natural coordinates (ξ, η) are obtained as:

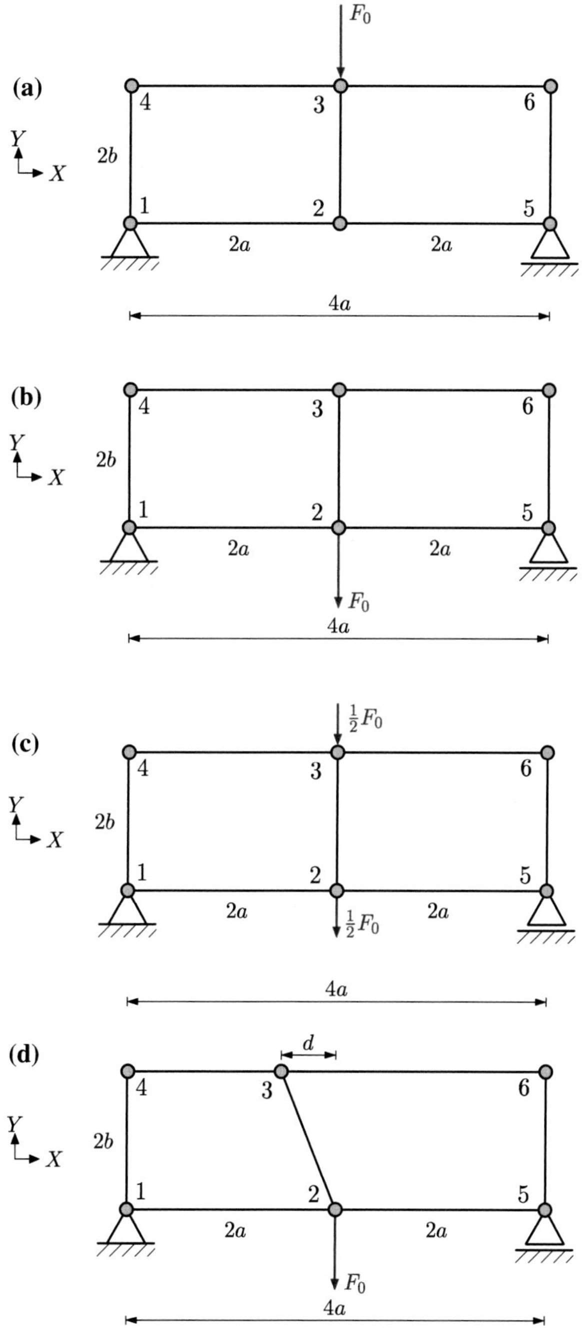

Fig. 5.13 Different modeling approaches for the beam shown in Fig. 5.12

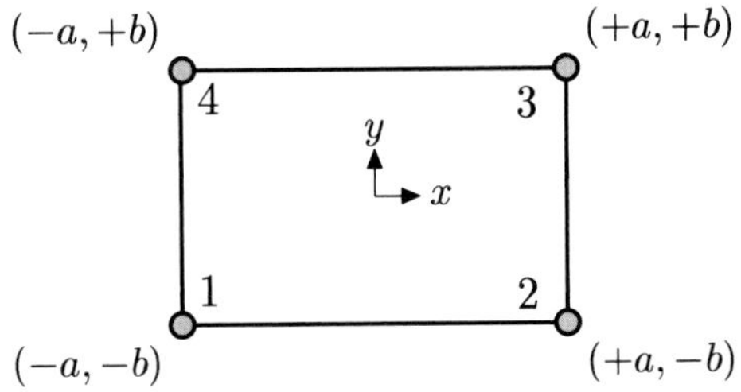

Fig. 5.14 Regular shaped plane elasticity element with local coordinate system

$$\frac{dx}{d\eta} = -\frac{1}{4}(-1+\xi)a + \frac{1}{4}(-1-\xi)a + \frac{1}{4}(1+\xi)a - \frac{1}{4}(1-\xi)a\,, \tag{5.82}$$

$$\frac{dx}{d\xi} = -\frac{1}{4}(-1+\eta)a + \frac{1}{4}(1-\eta)a + \frac{1}{4}(1+\eta)a - \frac{1}{4}(-1-\eta)a\,, \tag{5.83}$$

$$\frac{dy}{d\eta} = -\frac{1}{4}(-1+\xi)b - \frac{1}{4}(-1-\xi)b + \frac{1}{4}(1+\xi)b + \frac{1}{4}(1-\xi)b\,, \tag{5.84}$$

$$\frac{dy}{d\xi} = -\frac{1}{4}(-1+\eta)b - \frac{1}{4}(1-\eta)b + \frac{1}{4}(1+\eta)b + \frac{1}{4}(-1-\eta)b\,. \tag{5.85}$$

The derivatives of the natural coordinates (ξ, η) with respect to the Cartesian coordinates (x, y) are:

$$\frac{d\xi}{dx} = \frac{1}{a}\,, \quad \frac{d\xi}{dy} = 0\,, \tag{5.86}$$

$$\frac{d\eta}{dx} = 0\,, \quad \frac{d\eta}{dy} = \frac{1}{a}\,. \tag{5.87}$$

The Jacobian J is given by:

$$J = ab\,. \tag{5.88}$$

The derivatives of the interpolation functions N_i with respect to the Cartesian coordinate (x) are:

$$\frac{dN_1}{dx} = \frac{1}{4} \times \frac{-1+\eta}{a}\,, \tag{5.89}$$

$$\frac{dN_2}{dx} = -\frac{1}{4} \times \frac{-1+\eta}{a}\,, \tag{5.90}$$

$$\frac{dN_3}{dx} = \frac{1}{4} \times \frac{1+\eta}{a}\,, \tag{5.91}$$

$$\frac{dN_4}{dx} = -\frac{1}{4} \times \frac{1+\eta}{a}\,, \tag{5.92}$$

and with respect to the Cartesian coordinate (y):

$$\frac{\mathrm{d}N_1}{\mathrm{d}y} = \frac{1}{4} \times \frac{-1+\xi}{b}, \tag{5.93}$$

$$\frac{\mathrm{d}N_2}{\mathrm{d}y} = -\frac{1}{4} \times \frac{1+\xi}{b}, \tag{5.94}$$

$$\frac{\mathrm{d}N_3}{\mathrm{d}y} = \frac{1}{4} \times \frac{1+\xi}{b}, \tag{5.95}$$

$$\frac{\mathrm{d}N_4}{\mathrm{d}y} = -\frac{1}{4} \times \frac{-1+\xi}{b}. \tag{5.96}$$

Thus, the transposed of the $\boldsymbol{B}$-matrix is obtained as:

$$\boldsymbol{B}^{\mathrm{T}} = \frac{1}{4}\times \begin{bmatrix} \frac{-1+\eta}{a} & 0 & -\frac{-1+\eta}{a} & 0 & \frac{1+\eta}{a} & 0 & -\frac{1+\eta}{a} & 0 \\ 0 & \frac{-1+\xi}{b} & 0 & -\frac{1+\xi}{b} & 0 & \frac{1+\xi}{b} & 0 & -\frac{-1+\xi}{b} \\ \frac{-1+\xi}{b} & \frac{-1+\eta}{a} & -\frac{1+\xi}{b} & -\frac{-1+\eta}{a} & \frac{1+\xi}{b} & \frac{1+\eta}{a} & -\frac{-1+\xi}{b} & -\frac{1+\eta}{a} \end{bmatrix}. \tag{5.97}$$

This matrix must be multiplied with the elasticity matrix $\boldsymbol{C}$ for the plane stress case, i.e.

$$\frac{E}{1-\nu^2}\begin{bmatrix} 1 & \nu & 0 \\ \nu & 1 & 0 \\ 0 & 0 & \frac{1-\nu}{2} \end{bmatrix}, \tag{5.98}$$

and the result again multiplied with the $\boldsymbol{B}$-matrix. Integration results finally in the elemental stiffness matrix $\boldsymbol{K}$ as shown in Eq. (5.103). It should be noted here that Eq. (5.103) is valid for both elements.

Both elements can be assembled to the global stiffness matrix which can be written under consideration of the boundary conditions, i.e. $u_{1x} = u_{1y} = u_{5y} = 0$, as shown in Eq. (5.104).

The vector of unknowns is given by:

$$\boldsymbol{u} = \begin{bmatrix} u_{2x} \; u_{2y} \; u_{3x} \; u_{3y} \; u_{4x} \; u_{4y} \; u_{5x} \; u_{6x} \; u_{6y} \end{bmatrix}^{\mathrm{T}}. \tag{5.99}$$

The load vectors $\boldsymbol{f}_i$ for the three subcases (i = a, b, c) read as follows:

$$\boldsymbol{f}_{\mathrm{a}} = \begin{bmatrix} 0\;0\;0\;-F_0\;0\;0\;0\;0\;0 \end{bmatrix}^{\mathrm{T}}, \tag{5.100}$$

$$\boldsymbol{f}_{\mathrm{b}} = \begin{bmatrix} 0\;-F_0\;0\;0\;0\;0\;0\;0\;0 \end{bmatrix}^{\mathrm{T}}, \tag{5.101}$$

$$\boldsymbol{f}_{\mathrm{c}} = \begin{bmatrix} 0\;-\frac{F_0}{2}\;0\;-\frac{F_0}{2}\;0\;0\;0\;0\;0 \end{bmatrix}^{\mathrm{T}}. \tag{5.102}$$

$$
\frac{Et}{\nu^2-1}\begin{bmatrix}
\frac{a^2\nu-a^2-2b^2}{6ab} & -\frac{\nu+1}{8(\nu-1)} & \frac{a^2\nu-a^2+4b^2}{12ab} & -\frac{3\nu-1}{8} & -\frac{a^2\nu-a^2-2b^2}{12ab} & \frac{\nu+1}{8(\nu-1)} & -\frac{a^2\nu-a^2+b^2}{6ab} & \frac{3\nu-1}{8} \\
-\frac{\nu+1}{8(\nu-1)} & -\frac{-b^2\nu+2a^2+b^2}{6ab} & \frac{3\nu-1}{8} & -\frac{b^2\nu+a^2-b^2}{6ab} & \frac{\nu+1}{8(\nu-1)} & \frac{-b^2\nu+2a^2+b^2}{12ab} & -\frac{3\nu-1}{8} & \frac{b^2\nu+4a^2-b^2}{12ab} \\
\frac{a^2\nu-a^2+4b^2}{12ab} & \frac{3\nu-1}{8} & \frac{a^2\nu-a^2-2b^2}{6ab} & \frac{\nu+1}{8(\nu-1)} & -\frac{a^2\nu-a^2+b^2}{6ab} & -\frac{3\nu-1}{8} & -\frac{a^2\nu-a^2-2b^2}{12ab} & -\frac{\nu+1}{8(\nu-1)} \\
-\frac{3\nu-1}{8} & -\frac{b^2\nu+a^2-b^2}{6ab} & \frac{\nu+1}{8(\nu-1)} & -\frac{-b^2\nu+2a^2+b^2}{6ab} & \frac{3\nu-1}{8} & \frac{b^2\nu+4a^2-b^2}{12ab} & -\frac{\nu+1}{8(\nu-1)} & \frac{-b^2\nu+2a^2+b^2}{12ab} \\
-\frac{a^2\nu-a^2-2b^2}{12ab} & \frac{\nu+1}{8(\nu-1)} & -\frac{a^2\nu-a^2+b^2}{6ab} & \frac{3\nu-1}{8} & \frac{a^2\nu-a^2-2b^2}{6ab} & -\frac{\nu+1}{8(\nu-1)} & \frac{a^2\nu-a^2+4b^2}{12ab} & -\frac{3\nu-1}{8} \\
\frac{\nu+1}{8(\nu-1)} & \frac{-b^2\nu+2a^2+b^2}{12ab} & -\frac{3\nu-1}{8} & \frac{b^2\nu+4a^2-b^2}{12ab} & -\frac{\nu+1}{8(\nu-1)} & -\frac{-b^2\nu+2a^2+b^2}{6ab} & \frac{3\nu-1}{8} & -\frac{b^2\nu+a^2-b^2}{6ab} \\
-\frac{a^2\nu-a^2+b^2}{6ab} & -\frac{3\nu-1}{8} & -\frac{a^2\nu-a^2-2b^2}{12ab} & -\frac{\nu+1}{8(\nu-1)} & \frac{a^2\nu-a^2+4b^2}{12ab} & \frac{3\nu-1}{8} & \frac{a^2\nu-a^2-2b^2}{6ab} & \frac{\nu+1}{8(\nu-1)} \\
\frac{3\nu-1}{8} & \frac{b^2\nu+4a^2-b^2}{12ab} & -\frac{\nu+1}{8(\nu-1)} & \frac{-b^2\nu+2a^2+b^2}{12ab} & -\frac{3\nu-1}{8} & -\frac{b^2\nu+a^2-b^2}{6ab} & \frac{\nu+1}{8(\nu-1)} & -\frac{-b^2\nu+2a^2+b^2}{6ab}
\end{bmatrix}
\tag{5.103}
$$

$$
\frac{Et}{\nu^2-1}\begin{bmatrix}
\frac{a^2\nu-a^2-2b^2}{3ab} & 0 & -\frac{a^2\nu-a^2+b^2}{3ab} & 0 & -\frac{a^2\nu-a^2-2b^2}{12ab} & -\frac{\nu+1}{8} & \frac{a^2\nu-a^2+4b^2}{12ab} & -\frac{a^2\nu-a^2-2b^2}{12ab} & \frac{\nu+1}{8} \\
0 & -\frac{-b^2\nu+2a^2+b^2}{3ab} & 0 & \frac{b^2\nu+4a^2-b^2}{6ab} & -\frac{\nu+1}{8} & \frac{-b^2\nu+2a^2+b^2}{12ab} & \frac{3\nu-1}{8} & \frac{\nu+1}{8} & \frac{-b^2\nu+2a^2+b^2}{12ab} \\
-\frac{a^2\nu-a^2+b^2}{3ab} & 0 & \frac{a^2\nu-a^2-2b^2}{3ab} & 0 & \frac{a^2\nu-a^2+4b^2}{12ab} & \frac{-3\nu+1}{8} & -\frac{a^2\nu-a^2-2b^2}{12ab} & \frac{a^2\nu-12a^2+4b^2}{ab} & \frac{3\nu-1}{8} \\
0 & \frac{b^2\nu+4a^2-b^2}{6ab} & 0 & \frac{-b^2\nu+2a^2+b^2}{3ab} & \frac{3\nu-1}{8} & -\frac{b^2\nu+a^2-b^2}{6ab} & -\frac{\nu+1}{8} & \frac{-3\nu+1}{8} & -\frac{b^2\nu+a^2-b^2}{6ab} \\
-\frac{a^2\nu-a^2-2b^2}{12ab} & -\frac{\nu+1}{8} & \frac{a^2\nu-a^2+4b^2}{12ab} & \frac{3\nu-1}{8} & \frac{a^2\nu-a^2-2b^2}{6ab} & \frac{\nu+1}{8} & 0 & 0 & 0 \\
-\frac{\nu+1}{8} & \frac{-b^2\nu+2a^2+b^2}{12ab} & \frac{-3\nu+1}{8} & -\frac{b^2\nu+a^2-b^2}{6ab} & \frac{\nu+1}{8} & -\frac{-b^2\nu+2a^2+b^2}{6ab} & 0 & 0 & 0 \\
\frac{a^2\nu-a^2+4b^2}{12ab} & \frac{3\nu-1}{8} & -\frac{a^2\nu-a^2-2b^2}{12ab} & -\frac{\nu+1}{8} & 0 & 0 & \frac{a^2\nu-a^2-2b^2}{6ab} & -\frac{a^2\nu-a^2+b^2}{6ab} & \frac{-3\nu+1}{8} \\
-\frac{a^2\nu-a^2-2b^2}{12ab} & \frac{\nu+1}{8} & \frac{a^2\nu-a^2+4b^2}{12ab} & \frac{-3\nu+1}{8} & 0 & 0 & -\frac{a^2\nu-a^2+b^2}{6ab} & \frac{a^2\nu-a^2-2b^2}{6ab} & -\frac{\nu+1}{8} \\
\frac{\nu+1}{8} & \frac{-b^2\nu+2a^2+b^2}{12ab} & \frac{3\nu-1}{8} & -\frac{b^2\nu+a^2-b^2}{6ab} & 0 & 0 & \frac{-3\nu+1}{8} & -\frac{\nu+1}{8} & -\frac{-b^2\nu+2a^2+b^2}{6ab}
\end{bmatrix}. \tag{5.104}
$$

Table 5.9 Summary of the general and numerical solution of the vertical displacement at node 3 for the first three cases, i.e. for regular shaped elements

u_{3y}	$u_{3y}/\frac{F_0}{Et}$
Case (a)	
$-\dfrac{(8a^4\nu^2 - 3a^2b^2\nu - 8a^4 - 5a^2b^2 - 2b^4)F_0}{2atEb(a^2\nu - a^2 - 2b^2)}$	−4.691
Case (b)	
$-\dfrac{2a(2a^2\nu^2 - b^2\nu - 2a^2 - b^2)F_0}{bEt(a^2\nu - a^2 - 2b^2)}$	−4.358
Case (c)	
$\dfrac{(-4a^4\nu^2 + 1.75a^2b^2\nu + 4a^4 + 2.25a^2b^2 + 0.5b^4)F_0}{tEba(a^2\nu - a^2 - 2b^2)}$	−4.525

Table 5.10 Summary of the general and numerical solution of the vertical displacement at node 2 for the first three cases, i.e. for regular shaped elements

u_{2y}	$u_{2y}/\frac{F_0}{Et}$
Case (a)	
$-\dfrac{2(\nu+1)(2a^2\nu - 2a^2 - b^2)aF_0}{bEt(a^2\nu - a^2 - 2b^2)}$	−4.358
Case (b)	
$-\dfrac{(-7a^2b^2\nu^2 + 8a^4\nu + 14a^2b^2\nu + 8b^4\nu - 8a^4 - 11a^2b^2 - 8b^4)(\nu+1)aF_0}{btE(-b^2\nu + 2 * a^2 + b^2)(a^2\nu - a^2 - 2b^2)}$	−5.173
Case (c)	
$-\dfrac{F_0a(\nu+1)(-5.5a^2b^2\nu^2 + 8a^4\nu + 11a^2b^2\nu + 5b^4\nu - 8a^4 - 9.5a^2b^2 - 5b^4)}{Etb(a^2\nu - a^2 - 2b^2)(-b^2\nu + 2a^2 + b^2)}$	−4.765

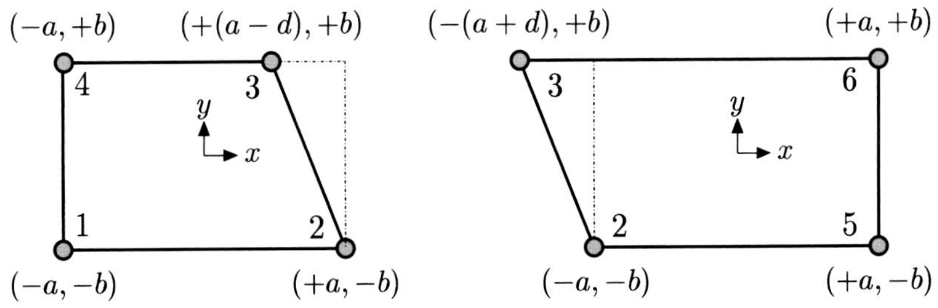

Fig. 5.15 Deformed plane elasticity elements with local coordinate systems

The solution of the system of equations, i.e. $\boldsymbol{u} = \boldsymbol{K}^{-1}\boldsymbol{f}$, is summarized in Tables 5.9 and 5.10 for the vertical displacements at nodes 2 and 3.

Let us assume for subtask (d) that the elemental coordinate systems are located in the original centers of the undeformed elements, see Fig. 5.15.

The geometrical derivatives of the Cartesian coordinates (x, y) with respect to the natural coordinates (ξ, η) are obtained as:

- Element I:

$$\frac{\mathrm{d}x}{\mathrm{d}\eta} = -0.0375 - 0.0375\xi\,, \tag{5.105}$$

$$\frac{\mathrm{d}x}{\mathrm{d}\xi} = 0.7125 - 0.0375\eta\,, \tag{5.106}$$

$$\frac{\mathrm{d}y}{\mathrm{d}\eta} = 0.5\,, \tag{5.107}$$

$$\frac{\mathrm{d}y}{\mathrm{d}\xi} = 0.0\,. \tag{5.108}$$

- Element II:

$$\frac{\mathrm{d}x}{\mathrm{d}\eta} = -0.0375 + 0.0375\xi\,, \tag{5.109}$$

$$\frac{\mathrm{d}x}{\mathrm{d}\xi} = 0.7875 + 0.0375\eta\,, \tag{5.110}$$

$$\frac{\mathrm{d}y}{\mathrm{d}\eta} = 0.5\,, \tag{5.111}$$

$$\frac{\mathrm{d}y}{\mathrm{d}\xi} = 0.0\,. \tag{5.112}$$

The derivatives of the natural coordinates (ξ, η) with respect to the Cartesian coordinates (x, y) are:

- Element I:

$$\frac{\mathrm{d}\xi}{\mathrm{d}x} = -\frac{26.\bar{6}}{-19+\eta}\,, \quad \frac{2(-1-\xi)}{-19+\eta} = 0\,, \tag{5.113}$$

$$\frac{\mathrm{d}\eta}{\mathrm{d}x} = 0\,, \quad \frac{\mathrm{d}\eta}{\mathrm{d}y} = 2\,. \tag{5.114}$$

- Element II:

$$\frac{\mathrm{d}\xi}{\mathrm{d}x} = \frac{26.\bar{6}}{21+\eta}\,, \quad \frac{\mathrm{d}\xi}{\mathrm{d}y} = \frac{2(1-\xi)}{21+\eta}\,, \tag{5.115}$$

$$\frac{\mathrm{d}\eta}{\mathrm{d}x} = 0\,, \quad \frac{\mathrm{d}\eta}{\mathrm{d}y} = 2\,. \tag{5.116}$$

The Jacobian J is given by:

- Element I:

$$J = 0.35625 - 0.01875\eta\,. \tag{5.117}$$

- Element II:

$$J = 0.39375 + 0.01875\eta\,. \tag{5.118}$$

The derivatives of the interpolation functions with respect to the Cartesian coordinates (x, y) are:
- Element I:

$$\frac{\mathrm{d}N_1}{\mathrm{d}x} = -\frac{6.\bar{6}(-1+\eta)}{-19+\eta}, \qquad \frac{\mathrm{d}N_1}{\mathrm{d}y} = \frac{10(1-0.1\eta-0.9\xi)}{-19+\eta}, \tag{5.119}$$

$$\frac{\mathrm{d}N_2}{\mathrm{d}x} = \frac{6.\bar{6}(-1+\eta)}{-19+\eta}, \qquad \frac{\mathrm{d}N_2}{\mathrm{d}y} = \frac{9(1+\xi)}{-19+\eta}, \tag{5.120}$$

$$\frac{\mathrm{d}N_3}{\mathrm{d}x} = -\frac{6.\bar{6}(1+\eta)}{-19+\eta}, \qquad \frac{\mathrm{d}N_3}{\mathrm{d}y} = 10(-1-\xi) - 19. + eta\,, \tag{5.121}$$

$$\frac{\mathrm{d}N_4}{\mathrm{d}x} = \frac{6.\bar{6}(1+\eta)}{-19+\eta}, \qquad \frac{\mathrm{d}N_4}{\mathrm{d}y} = \frac{10(-0.9+\xi+0.1\eta)}{-19+\eta}. \tag{5.122}$$

- Element II:

$$\frac{\mathrm{d}N_1}{\mathrm{d}x} = \frac{6.\bar{6}(-1+\eta)}{21+\eta}, \qquad \frac{\mathrm{d}N_1}{\mathrm{d}y} = \frac{11(-1+\xi)}{21+\eta}, \tag{5.123}$$

$$\frac{\mathrm{d}N_2}{\mathrm{d}x} = -\frac{6.\bar{6}(-1+\eta)}{21+\eta}, \qquad \frac{\mathrm{d}N_2}{\mathrm{d}y} = \frac{10(-1-0.1\eta-1.1\xi)}{21+\eta}, \tag{5.124}$$

$$\frac{\mathrm{d}N_3}{\mathrm{d}x} = \frac{6.\bar{6}(1+\eta)}{21+\eta}, \qquad \frac{\mathrm{d}N_3}{\mathrm{d}y} = \frac{10(1.1+0.1\eta+\xi)}{21+\eta}, \tag{5.125}$$

$$\frac{\mathrm{d}N_4}{\mathrm{d}x} = -\frac{6.\bar{6}(1+\eta)}{21+\eta}, \qquad \frac{\mathrm{d}N_4}{\mathrm{d}y} = \frac{10(1-\xi)}{21+\eta}. \tag{5.126}$$

The stiffness matrices for both elements can be obtained as indicated in Eqs. (5.127) and (5.128):

$$
\boldsymbol{K}_{\mathrm{I}} = Et \begin{bmatrix}
0.4465581500 & 0.1669380526 & -0.1340581499 & -0.0627713859 & -0.2368489693 & -0.1617354972 & -0.0756510308 & 0.0575688305 \\
0.1669380526 & 0.6176195876 & 0.0413952808 & 0.1636304127 & -0.1617354972 & -0.3361325572 & -0.0465978362 & -0.4451174431 \\
-0.1340581499 & 0.0413952808 & 0.4153081500 & -0.1455619475 & -0.0756510308 & -0.0465978362 & -0.2055989693 & 0.1507645029 \\
-0.0627713859 & 0.1636304127 & -0.1455619475 & 0.5394945875 & 0.0575688305 & -0.4451174431 & 0.1507645029 & -0.2580075572 \\
-0.2368489693 & -0.1617354972 & -0.0756510308 & 0.0575688305 & 0.4698591700 & 0.1675161144 & -0.1573591700 & -0.0633494477 \\
-0.1617354972 & -0.3361325572 & -0.0659783623 & -0.4451174431 & 0.1675161144 & 0.6488959242 & 0.04081721903 & 0.1323540761 \\
-0.0756510308 & -0.0465978362 & -0.2055989693 & 0.1507645029 & -0.1573591700 & 0.0408172190 & 0.4386091700 & -0.1449838857 \\
0.0575688305 & -0.4451174431 & 0.1507645029 & -0.2580075572 & -0.0633494477 & 0.1323540761 & -0.1449838857 & 0.5707709242
\end{bmatrix}.
\tag{5.127}
$$

$$
\boldsymbol{K}_{\mathrm{II}} = Et \begin{bmatrix}
0.4661760530 & 0.1664161739 & -0.122426053 & -0.0622495072 & -0.2356101550 & -0.1612125693 & -0.1081398450 & 0.0570459026 \\
0.1664161739 & 0.6910186855 & 0.0419171595 & 0.1683563147 & -0.1612125693 & -0.3574388216 & -0.0471207641 & -0.5019361787 \\
-0.1224260530 & 0.0419171595 & 0.4349260530 & -0.1460838262 & -0.1081398450 & -0.0471207641 & -0.2043601550 & 0.1512874308 \\
-0.0622495072 & 0.1683563147 & -0.1460838262 & 0.6128936855 & 0.0570459026 & -0.5019361787 & 0.1512874308 & -0.2793138216 \\
-0.2356101550 & -0.1612125693 & -0.1081398450 & 0.0570459026 & 0.4452155170 & 0.1659431189 & -0.1014655170 & -0.0617764522 \\
-0.1612125693 & -0.3574388216 & -0.0471207641 & -0.5019361787 & 0.1659431189 & 0.6606932433 & 0.0423902145 & 0.1986817570 \\
-0.1081398450 & -0.0471207641 & -0.2043601550 & 0.1512874308 & -0.1014655170 & 0.0423902145 & 0.4139655170 & -0.1465568812 \\
0.0570459026 & -0.5019361787 & 0.1512874308 & -0.2793138216 & -0.0617764522 & 0.1986817570 & -0.1465568812 & 0.5825682433
\end{bmatrix}.
\tag{5.128}
$$

Combining both elements to the global system of equations and consideration of the boundary conditions, i.e. $u_{1x} = u_{1y} = u_{5y} = 0$, gives the reduced stiffness matrix as shown in Eq. (5.129).

$$
\boldsymbol{K}_{\text{red}} = Et \times \begin{bmatrix}
0.8814842030 & 0.0208542264 & -0.1837908758 & 0.01044806635 & -0.2055989693 & 0.1507645029 & -0.122426053 & -0.235610155 & -0.1612125693 \\
0.0208542264 & 1.230513273 & 0.01044806635 & -0.9470536218 & 0.1507645029 & -0.2580075572 & 0.04191715950 & -0.1612125693 & -0.3574388216 \\
-0.1837908758 & 0.01044806635 & 0.883824687 & 0.0209592332 & -0.1573359170 & -0.06334944767 & -0.204360155 & -0.101465517 & 0.04239021447 \\
0.01044806635 & -0.9470536218 & 0.0209592332 & 1.231464168 & 0.04081721903 & 0.1323540761 & 0.1512874308 & -0.06177645223 & 0.1986817570 \\
-0.2055989693 & 0.1507645029 & -0.1573359170 & 0.04081721903 & 0.438609170 & -0.1449838857 & 0 & 0 & 0 \\
0.1507645029 & -0.2580075572 & -0.06334944767 & 0.1323540761 & -0.1449838857 & 0.5707709242 & 0 & 0 & 0 \\
-0.122426053 & 0.04191715950 & -0.204360155 & 0.1512874308 & 0 & 0 & 0.434926053 & -0.108139845 & -0.04712076412 \\
-0.235610155 & -0.1612125693 & -0.101465517 & -0.06177645223 & 0 & 0 & -0.108139845 & 0.445215517 & 0.1659431189 \\
-0.1612125693 & -0.3574388216 & 0.04239021447 & 0.1986817570 & 0 & 0 & -0.04712076412 & 0.1659431189 & 0.6606932433
\end{bmatrix}
\quad (5.129)
$$

Table 5.11 Summary of the numerical solution of all nodal displacements for the last cases, i.e. for irregular shaped elements

Node	$u_x / \frac{F_0}{Et}$	$u_y / \frac{F_0}{Et}$
2	1.795237262	−4.996792830
3	1.903135260	−4.172604435
4	3.402115127	−0.689931294
5	3.265376368	–
6	0.136738761	−0.934035113

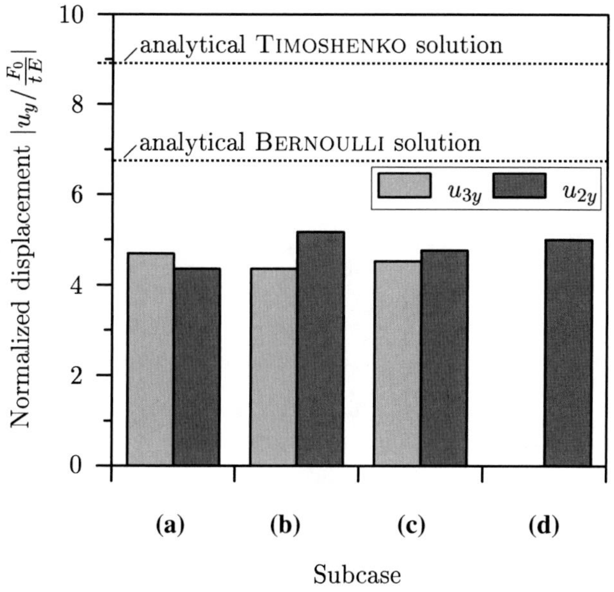

Fig. 5.16 Comparison of the numerical results for the vertical displacements of node 3 and 2

The solution of the system of equations, i.e. $\boldsymbol{u} = \boldsymbol{K}^{-1}\boldsymbol{f}$, is summarized in Table 5.11 for all displacements.

Figure 5.16 summarizes the numerical solutions for the vertical displacements of node 3 and 2. Furthermore, a comparison with the analytical solution [4] of a BERNOULLI and TIMOSHENKO beam is given.

5.4 Supplementary Problems

5.6 Knowledge questions on plane elements

- How many material parameters are required for the two-dimensional HOOKE's law? Name possible material parameters.
- Explain (in words) the difference between a plane stress and a plane strain state.
- State all the stress and strain components, which we distinguished in the case of a plane stress state.
- Consider a plane stress state. On which quantities (output values) does the thickness strain depend?
- Consider a plane strain state. On which quantities (output values) does the thickness stress depend?
- Describe (in words) the characteristics of the shape functions which we used for a linear plane stress finite element (quad 4).
- State the required (a) geometrical parameters and (b) material parameters to define a plane elasticity element.
- State the DOF per node for a plane elasticity element (Quad 4).
- State possible advantages to model a beam bending problem with plane elasticity elements and not with 1D beam elements.

5.7 Uniform representation of plane stress and plane strain

The constitutive representations for plane stress and plane strain as given in Eqs. (5.4) and (5.9) can be written in a single equation as

$$\begin{bmatrix} \sigma_x \\ \sigma_y \\ \sigma_{xy} \end{bmatrix} = \frac{E'}{1-\nu'^2} \begin{bmatrix} 1 & \nu' & 0 \\ \nu' & 1 & 0 \\ 0 & 0 & \frac{1-\nu'}{2} \end{bmatrix} \begin{bmatrix} \varepsilon_x \\ \varepsilon_y \\ 2\varepsilon_{xy} \end{bmatrix}, \tag{5.130}$$

where $E' = E$ and $\nu' = \nu$ for plane stress and $E' = \frac{E'}{1-\nu^2}$ and $\nu' = \frac{\nu}{1-\nu}$ for plane strain. Proof this uniform description.

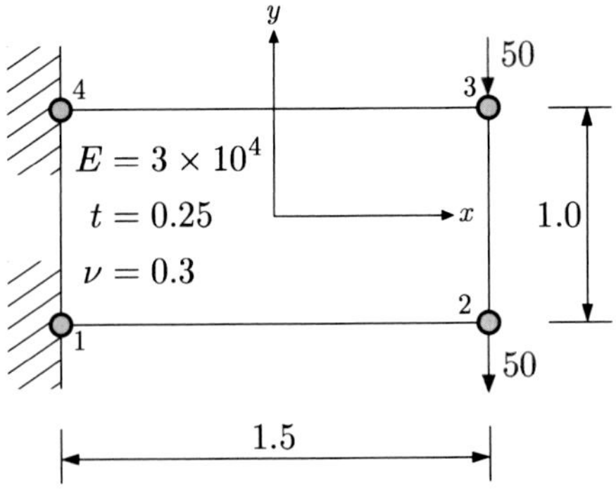

Fig. 5.17 Plane elasticity bending problem

5.8 Plane elasticity bending problem

Given is a rectangular two-dimensional element (dimensions $1.5 \times 1.0 \times 0.25$) as shown in Fig. 5.17. The left-hand nodes are fixed and the right-hand nodes are loaded by a vertical point of load $F_0 = -50$. Use a single plane elasticity element to calculate the nodal displacements for a plane stress state with $E = 3 \times 10^4$ and $\nu = 0.3$. Assume consistent units.

References

1. Chen WF, Han DJ (1988) Plasticity for structural engineers. Springer, New York
2. Eschenauer H, Olhoff N, Schnell W (1997) Applied structural mechanics: fundamentals of elasticity, load-bearing structures, structural optimization. Springer, Berlin
3. MacNeal RH (1994) Finite elements: their design and performance. Marcel Dekker, New York
4. Öchsner A (2014) Elasto-plasticity of frame structure elements: modelling and simulation of rods and beams. Springer, Berlin

Chapter 6
Classical Plate Elements

Abstract This chapter starts with the analytical description of classical plate members. Classical plates are thin plates where the contribution of the shear force on the deformations is neglected. Based on the three basic equations of continuum mechanics, i.e., the kinematics relationship, the constitutive law, and the equilibrium equation, the partial differential equation, which describes the physical problem, is derived. The weighted residual method is then used to derive the principal finite element equation for classical plate elements. The chapter exemplarily treats a four-node bilinear quadrilateral (quad 4) bending element.

6.1 Introduction

A classical plate is defined as a thin structural member, as schematically shown in Fig. 6.1, with a much smaller thickness h than the planar dimensions ($2a$ and $2b$). It can be seen as a two-dimensional extension or generalization of the EULER–BERNOULLI beam (see Chap. 3). The following derivations are restricted to some simplifications:

- the thickness h is constant and much smaller than the planer dimensions a and b: $\frac{h}{a}$ and $\frac{h}{b} < 0.1$,
- the thickness h is constant ($\rightarrow \varepsilon_z = 0$) and the undeformed plate shape is planar,
- the displacement $u_z(x, y)$ is small compared to the thickness dimension h: $u_z < 0.2h$,
- the material is isotropic, homogenous and linear-elastic according to HOOKE's law for a plane stress state ($\sigma_z = \tau_{xz} = \tau_{yz} = 0$),
- BERNOULLI's hypothesis is valid, i.e. a cross-sectional plane stays plane and unwrapped in the deformed state. This means that the shear strains γ_{yz} and γ_{xz} due to the distributed shear forces q_x and q_y are neglected,
- external forces act only perpendicular to the xy-plane, the vector of external moments lies within the xy-plane, and
- only rectangular plates are considered.

A. Öchsner, *Computational Statics and Dynamics*,
https://doi.org/10.1007/978-981-15-1278-0_6

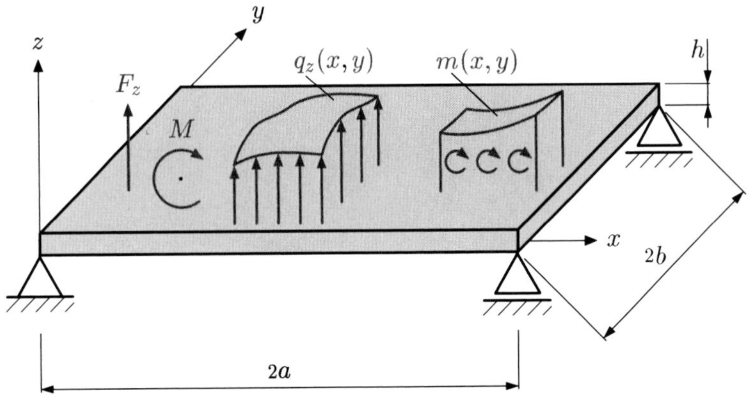

Fig. 6.1 General configuration for a classical plate problem

Table 6.1 Analogies between the classical beam and plate theories

Classical beam	Classical plate
1D	2D
Deformation perpendicular to principal x-axis	Deformation perpendicular to xy-plane
u_z, φ_y	u_z, φ_x, φ_y

The external loads, which are considered within this chapter, are single forces F_z, single moments M_x and M_y, area distributed forces $q_z(x, y)$, and area distributed moments $m_x(x, y)$ and $m_y(x, y)$.

The classical theories of plate bending distinguish between shear-rigid and shear-flexible models. The shear rigid-plate, also called the classical or KIRCHHOFF plate, neglects the shear deformation from the shear forces. This theory corresponds to the classical EULER–BERNOULLI beam theory (see Chap. 3). The consideration of the shear deformation leads to the REISSNER–MINDLIN plate (see Chap. 7) which corresponds to the TIMOSHENKO beam (see Chap. 4).

The analogies between the classical beam and plate theories are summarized in Table 6.1.

6.2 Derivation of the Governing Differential Equation

6.2.1 Kinematics

The kinematics or strain-displacement relations extract the strain field contained in a displacement field. Let us first derive a kinematics relation which relates the variation of u_x across the plate thickness in terms of the displacement u_z. For this purpose, let

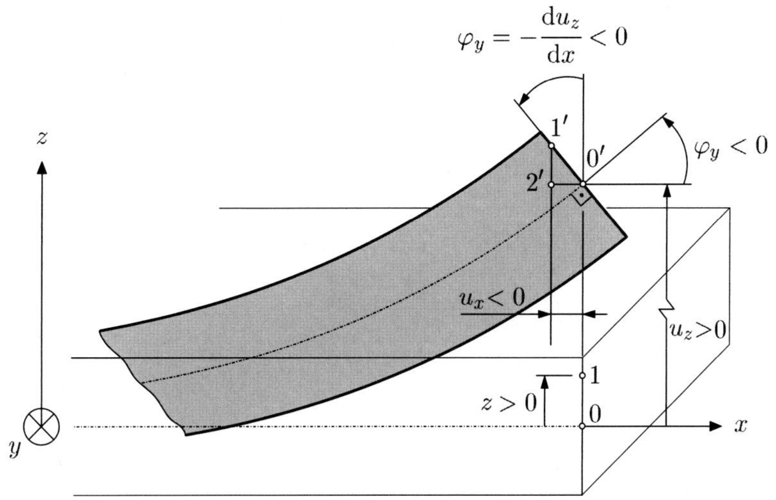

Fig. 6.2 Configuration for the derivation of kinematics relations in the x-z plane. Note that the deformation is exaggerated for better illustration

us imagine that a plate element is bent around the y-axis, see Fig. 6.2. We assume in the following the same definition of the rotational angle φ_y as in Chap. 3 for the EULER–BERNOULLI beam. This means that the angle φ_y is positive if the vector of the rotational direction is pointing in positive y-axis.

Looking at the right-angled triangle $0'1'2'$, we can state that[1]

$$\sin(-\varphi_y) = \frac{\overline{2'0'}}{\overline{0'1'}} = \frac{-u_x}{z}, \tag{6.1}$$

which results for small angles ($\sin(-\varphi_y) \approx -\varphi_y$) in:

$$u_x = +z\varphi_y\,. \tag{6.2}$$

Looking at the curved center line in Fig. 6.2, it holds that the slope of the tangent line at $0'$ equals:

$$\tan(-\varphi_y) = +\frac{du_z}{dx} \approx -\varphi_y\,. \tag{6.3}$$

If Eqs. (6.2) and (6.3) are combined, the following results:

$$u_x = -z\frac{du_z}{dx}. \tag{6.4}$$

[1] Note that according to the assumptions of the classical *thin* plate theory the lengths $\overline{01}$ and $\overline{0'1'}$ remain unchanged.

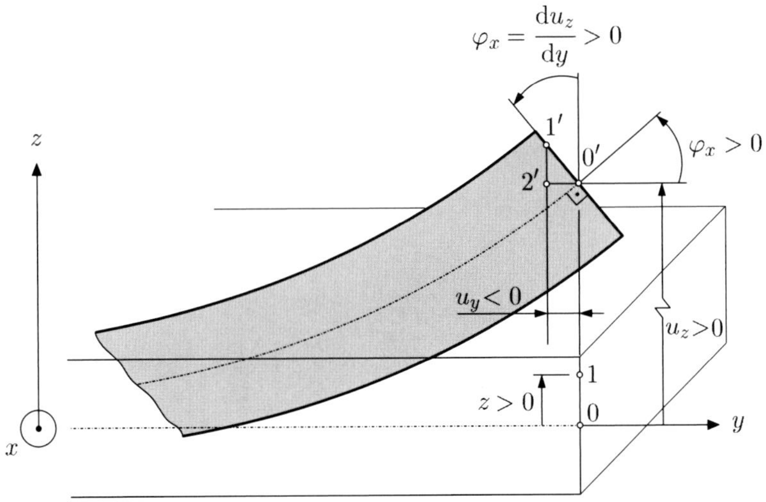

Fig. 6.3 Configuration for the derivation of kinematics relations in the y-z plane. Note that the deformation is exaggerated for better illustration

Considering a plate which is bent around the x-axis (see Fig. 6.3) and following the same line of reasoning (the angle φ_x is assumed positive if the vector of the rotational direction is pointing in positive x-axis.), similar equations can be derived for u_y:

$$\varphi_x \approx \frac{du_z}{dy}, \tag{6.5}$$

$$u_y = -z\varphi_x, \tag{6.6}$$

$$u_y = -z\frac{du_z}{dy}. \tag{6.7}$$

One may find in the scholarly literature other definitions of the rotational angles [1, 6, 8, 9]. The angle φ_y is introduced in the xz-plane (see Fig. 6.2) whereas φ_x is introduced in the yz-plane (see Fig. 6.3). These definitions are closer to the classical definitions of the angles in the scope of finite elements but not conform with the definitions of the stress resultants (see M_x^{n} and M_y^{n} in Fig. 6.5). Other definitions assume, for example, that the rotational angle φ_x (now defined in the x-z plane) is positive if it leads to a positive displacement u_x at the positive z-side of the neutral axis. The same definition holds for the angle φ_y (now defined in the y-z plane), see Suppl. Problem 6.4.

Using classical engineering definitions of strain, the following relations can be obtained [1, 7]:

$$\varepsilon_x = \frac{\partial u_x}{\partial x} \overset{(6.4)}{=} \frac{\partial}{\partial x}\left(-z\frac{\partial u_z}{\partial x}\right) = -z\frac{\partial^2 u_z}{\partial x^2} = z\kappa_x\,, \tag{6.8}$$

$$\varepsilon_y = \frac{\partial u_y}{\partial y} \overset{(6.7)}{=} \frac{\partial}{\partial y}\left(-z\frac{\partial u_z}{\partial y}\right) = -z\frac{\partial^2 u_z}{\partial y^2} = z\kappa_y\,, \tag{6.9}$$

$$\gamma_{xy} = \frac{\partial u_x}{\partial y} + \frac{\partial u_y}{\partial x} \overset{(6.4),(6.7)}{=} -2z\frac{\partial^2 u_z}{\partial x \partial y} = z\kappa_{xy}\,. \tag{6.10}$$

In matrix notation, these three relationships can be written as

$$\begin{bmatrix} \varepsilon_x \\ \varepsilon_y \\ \gamma_{xy} \end{bmatrix} = -z \begin{bmatrix} \frac{\partial^2}{\partial x^2} \\ \frac{\partial^2}{\partial y^2} \\ \frac{2\partial^2}{\partial x \partial y} \end{bmatrix} u_z = z \begin{bmatrix} \kappa_x \\ \kappa_y \\ \kappa_{xy} \end{bmatrix}, \tag{6.11}$$

or symbolically as

$$\boldsymbol{\varepsilon} = -z\boldsymbol{\mathcal{L}}_2 u_z = z\boldsymbol{\kappa}\,. \tag{6.12}$$

Let us recall here that we obtained in Chap. 3 the following kinematics relationship for the EULER–BERNOULLI beam, see Eq. (3.16):

$$\varepsilon_x(x,y) = -z\frac{\mathrm{d}^2 u_z(x)}{\mathrm{d}x^2} = z\kappa\,. \tag{6.13}$$

This last relationship corresponds exactly to Eq. (6.8).

6.2.2 Constitutive Equation

As stated in the introduction of this chapter, the classical plate theory assumes a plane stress state and the constitutive equation can be taken from Sect. 5.2.2 as:

$$\begin{bmatrix} \sigma_x \\ \sigma_y \\ \sigma_{xy} \end{bmatrix} = \frac{E}{1-\nu^2} \begin{bmatrix} 1 & \nu & 0 \\ \nu & 1 & 0 \\ 0 & 0 & \frac{1-\nu}{2} \end{bmatrix} \begin{bmatrix} \varepsilon_x \\ \varepsilon_y \\ \gamma_{xy} \end{bmatrix}, \tag{6.14}$$

or rearranged for the elastic compliance form:

$$\begin{bmatrix} \varepsilon_x \\ \varepsilon_y \\ \gamma_{xy} \end{bmatrix} = \frac{1}{E} \begin{bmatrix} 1 & -\nu & 0 \\ -\nu & 1 & 0 \\ 0 & 0 & 2(\nu+1) \end{bmatrix} \begin{bmatrix} \sigma_x \\ \sigma_y \\ \sigma_{xy} \end{bmatrix}. \tag{6.15}$$

The last two equations can be written in matrix notation as

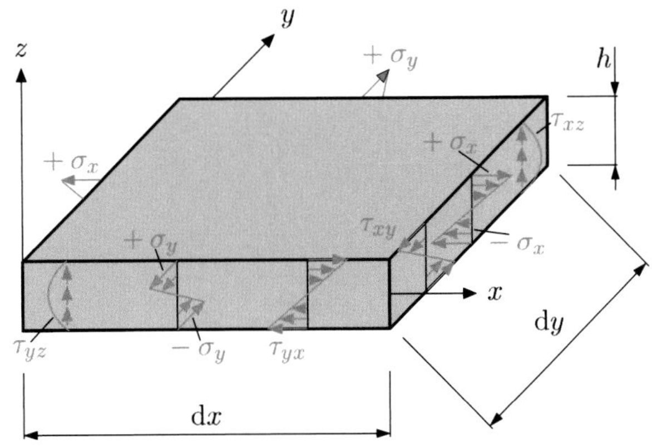

Fig. 6.4 Stresses acting on a classical plate element

$$\boldsymbol{\sigma} = \boldsymbol{C}\boldsymbol{\varepsilon}\,, \tag{6.16}$$

or

$$\boldsymbol{\varepsilon} = \boldsymbol{D}\boldsymbol{\sigma}\,, \tag{6.17}$$

where $\boldsymbol{C}$ is the elasticity matrix and $\boldsymbol{D} = \boldsymbol{C}^{-1}$ is the elastic compliance matrix.

Let us recall here that in Chap. 3 we obtained the following constitutive relationship for the EULER–BERNOULLI beam, see Eq. (3.20):

$$\sigma_x = E\varepsilon_x\,. \tag{6.18}$$

This last relationship corresponds to Eq. (6.14).

6.2.3 Equilibrium

Let us first look at the stress distributions through the thickness of a classical plate element $\mathrm{d}x\mathrm{d}yh$ as shown in Fig. 6.4. Linear distributed normal stresses (σ_x, σ_y), linear distributed shear stresses (τ_{yx}, τ_{xy}), and parabolic distributed shear stresses (τ_{yz}, τ_{xz}) can be identified. These stresses can be expressed by the so-called stress resultants, i.e. bending moments and shear forces as shown in Fig. 6.5. These stress resultants are taken to be positive if they cause a tensile stress (positive) at a point with positive z-coordinate.

These stress resultants are obtained as in the case of beams[2] by integrating over the stress distributions. In the case of plates, however, the integration is only performed over the thickness, i.e. the moments and forces are given per unit length (normalized

[2]See Eq. (3.23 for the EURLER–BERNOULLI beam or Eq. (F.364) for the TIMOSHENKO beam.

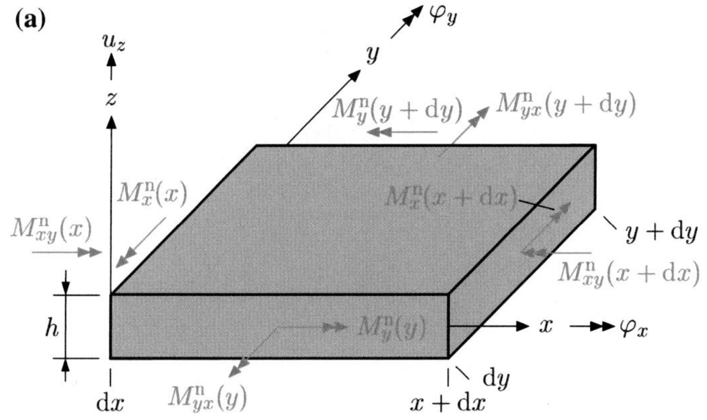

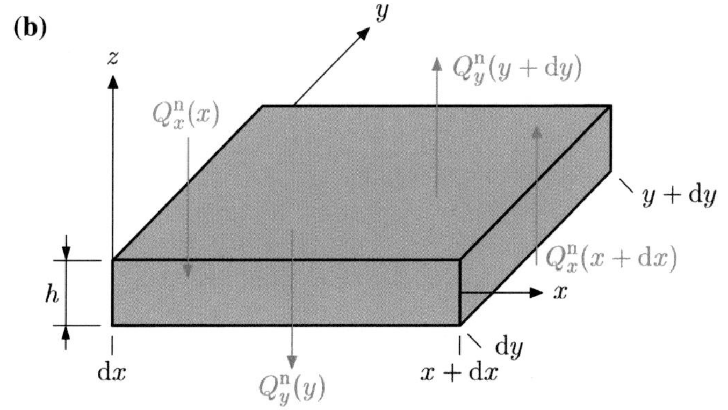

Fig. 6.5 Stress resultants acting on a classical plate element: **a** bending and twisting moments and **b** shear forces. Positive directions are drawn

with the corresponding side length of the plate element). The normalized (superscript 'n') bending moments are obtained as:

$$M_x^{\rm n} = \frac{M_x}{{\rm d}y} = \int_{-h/2}^{h/2} z\sigma_x {\rm d}z\,, \tag{6.19}$$

$$M_y^{\rm n} = \frac{M_y}{{\rm d}x} = \int_{-h/2}^{h/2} z\sigma_y {\rm d}z\,. \tag{6.20}$$

The twisting moment per unit length reads:

$$M_{xy}^{\mathrm{n}} = M_{yx}^{\mathrm{n}} = \frac{M_{xy}}{\mathrm{d}y} = \frac{M_{yx}}{\mathrm{d}x} = \int_{-h/2}^{h/2} z\tau_{xy}\mathrm{d}z\,. \tag{6.21}$$

Furthermore, the shear forces per unit length are calculated in the following way:

$$Q_x^{\mathrm{n}} = \frac{Q_x}{\mathrm{d}y} = \int_{-h/2}^{h/2} \tau_{xz}\mathrm{d}z\,, \tag{6.22}$$

$$Q_y^{\mathrm{n}} = \frac{Q_y}{\mathrm{d}x} = \int_{-h/2}^{h/2} \tau_{yz}\mathrm{d}z\,. \tag{6.23}$$

It should be noted that a slightly different notation when compared to the beam problems is used here. The bending moment around the y-axis is now called M_x^{n} (which directly corresponds to the causing stress σ_x) while in the beam notation it was M_y, see Fig. 3.12. Nevertheless, the orientation remains the same. The shear force, which was in the case of the beams given as Q_z is now either Q_x^{n} or Q_y^{n}. Thus, in the case of this plate notation, the index refers rather to the plane (check the surface normal vector) in which the corresponding resultant (vector) is located.

The equilibrium condition will be determined in the following for the vertical forces. Assuming that the distributed force is constant ($q_z(x, y) \rightarrow q_z$) and that forces in the direction of the positive z-axis are considered positive, the following results:

$$-Q_x^{\mathrm{n}}(x)\mathrm{d}y - Q_y^{\mathrm{n}}(y)\mathrm{d}x + Q_x^{\mathrm{n}}(x+\mathrm{d}x)\mathrm{d}y + Q_y^{\mathrm{n}}(y+\mathrm{d}y)\mathrm{d}x + q_z\mathrm{d}x\mathrm{d}y = 0\,. \tag{6.24}$$

Evaluating the shear forces at $x + \mathrm{d}x$ and $y + \mathrm{d}y$ in a TAYLOR's series of first order, meaning

$$Q_x^{\mathrm{n}}(x+\mathrm{d}x) \approx Q_x^{\mathrm{n}}(x) + \frac{\partial Q_x^{\mathrm{n}}}{\partial x}\mathrm{d}x\,, \tag{6.25}$$

$$Q_y^{\mathrm{n}}(y+\mathrm{d}y) \approx Q_y^{\mathrm{n}}(y) + \frac{\partial Q_y^{\mathrm{n}}}{\partial y}\mathrm{d}y\,, \tag{6.26}$$

Equation (6.24) results in

$$\frac{\partial Q_x^{\mathrm{n}}}{\partial x}\mathrm{d}x\mathrm{d}y + \frac{\partial Q_y^{\mathrm{n}}}{\partial y}\mathrm{d}y\mathrm{d}x + q_z\mathrm{d}x\mathrm{d}y = 0\,, \tag{6.27}$$

or alternatively after simplification to:

$$\frac{\partial Q_x^{\mathrm{n}}}{\partial x} + \frac{\partial Q_y^{\mathrm{n}}}{\partial y} + q_z = 0\,. \tag{6.28}$$

The equilibrium of moments around the reference axis at $x + \mathrm{d}x$ (positive if the moment vector is pointing in positive y-axis) gives:

$$\begin{aligned} &M_x^{\mathrm{n}}(x+\mathrm{d}x)\mathrm{d}y - M_x^{\mathrm{n}}(x)\mathrm{d}y + M_{yx}^{\mathrm{n}}(y+\mathrm{d}y)\mathrm{d}x - M_{yx}^{\mathrm{n}}\mathrm{d}x \\ &- Q_y^{\mathrm{n}}(y)\mathrm{d}x\tfrac{\mathrm{d}x}{2} + Q_y^{\mathrm{n}}(y+\mathrm{d}y)\mathrm{d}x\tfrac{\mathrm{d}x}{2} - Q_x^{\mathrm{n}}(x)\mathrm{d}y\mathrm{d}x + q_z\mathrm{d}x\mathrm{d}y\tfrac{\mathrm{d}x}{2} = 0\,. \end{aligned} \tag{6.29}$$

Expanding the stress resultants at $x + \mathrm{d}x$ and $y + \mathrm{d}y$ into a TAYLOR's series of first order, meaning

$$M_x^{\mathrm{n}}(x+\mathrm{d}x) = M_x^{\mathrm{n}}(x) + \frac{\partial M_x^{\mathrm{n}}}{\partial x}\mathrm{d}x\,, \tag{6.30}$$

$$M_{yx}^{\mathrm{n}}(y+\mathrm{d}y) = M_{yx}^{\mathrm{n}}(y) + \frac{\partial M_{yx}^{\mathrm{n}}}{\partial y}\mathrm{d}y\,, \tag{6.31}$$

$$Q_y^{\mathrm{n}}(y+\mathrm{d}y) = Q_y^{\mathrm{n}}(y) + \frac{\partial Q_y^{\mathrm{n}}}{\partial y}\mathrm{d}y\,, \tag{6.32}$$

Equation (6.29) results in

$$\frac{\partial M_x^{\mathrm{n}}}{\partial x}\mathrm{d}x\mathrm{d}y + \frac{\partial M_{yx}^{\mathrm{n}}}{\partial y}\mathrm{d}y\mathrm{d}x + \frac{\partial Q_y^{\mathrm{n}}}{\partial y}\mathrm{d}y\mathrm{d}x\frac{\mathrm{d}x}{2} - Q_x^{\mathrm{n}}(x)\mathrm{d}y\mathrm{d}x + q_z\mathrm{d}x\mathrm{d}y\frac{\mathrm{d}x}{2} = 0\,. \tag{6.33}$$

Seeing that the terms of third order ($\mathrm{d}x\mathrm{d}y\mathrm{d}x$) are considered as infinitesimally small and because of $M_{yx}^{\mathrm{n}} = M_{xy}^{\mathrm{n}}$, finally the following results:

$$\frac{\partial M_x^{\mathrm{n}}}{\partial x} + \frac{\partial M_{xy}^{\mathrm{n}}}{\partial y} - Q_x^{\mathrm{n}} = 0\,. \tag{6.34}$$

In a similar way, the equilibrium of moments around the reference axis at $y + \mathrm{d}y$ finally gives:

$$\frac{\partial M_y^{\mathrm{n}}}{\partial y} + \frac{\partial M_{xy}^{\mathrm{n}}}{\partial x} - Q_y^{\mathrm{n}} = 0\,. \tag{6.35}$$

Thus, the three equilibrium equations can be summarized as follows:

$$\frac{\partial Q_x^{\mathrm{n}}}{\partial x} + \frac{\partial Q_y^{\mathrm{n}}}{\partial y} + q_z = 0\,, \tag{6.36}$$

$$\frac{\partial M_x^{\mathrm{n}}}{\partial x} + \frac{\partial M_{xy}^{\mathrm{n}}}{\partial y} - Q_x^{\mathrm{n}} = 0\,, \tag{6.37}$$

$$\frac{\partial M_y^{\rm n}}{\partial y}+\frac{\partial M_{xy}^{\rm n}}{\partial x}-Q_y^{\rm n}=0\,. \tag{6.38}$$

Let us recall here that we obtained in Chap. 3 the following equilibrium equations for the EULER–BERNOULLI beam, see Eqs. (3.35) and (3.36):

$$\frac{\mathrm{d}M_y(x)}{\mathrm{d}x}=Q_z(x)\,,\quad \frac{\mathrm{d}^2M_y(x)}{\mathrm{d}x^2}=\frac{\mathrm{d}Q_z(x)}{\mathrm{d}x}=-q_z\,. \tag{6.39}$$

Rearranging Eqs. (6.37) and (6.38) for $Q^{\rm n}$ and introducing in Eq. (6.36) finally gives the combined equilibrium equation as:

$$\frac{\partial^2 M_x^{\rm n}}{\partial x^2}+2\frac{\partial^2 M_{xy}}{\partial x\partial y}+\frac{\partial^2 M_y^{\rm n}}{\partial y^2}+q_z=0\,. \tag{6.40}$$

The last equation can be written in matrix notation as

$$\begin{bmatrix}\frac{\partial^2}{\partial x^2} & \frac{\partial^2}{\partial y^2} & \frac{2\partial^2}{\partial x\partial y}\end{bmatrix}\begin{bmatrix}M_x^{\rm n}\\ M_y^{\rm n}\\ M_{xy}^{\rm n}\end{bmatrix}+q_z=0\,, \tag{6.41}$$

or symbolically as

$$\boldsymbol{\mathcal{L}}_2^{\rm T}\boldsymbol{M}^{\rm n}+q_z=0\,. \tag{6.42}$$

Equations (6.37) and (6.38) can be rearranged to obtain a relationship between the moments and shear forces similar to Eq. $(6.39)_1$:

$$\boldsymbol{\mathcal{L}}_1^{\rm T}\boldsymbol{M}^{\rm n}=\boldsymbol{Q}^{\rm n}\,, \tag{6.43}$$

where the first-order differential operator matrix $\boldsymbol{\mathcal{L}}_1$ is given by Eqs. (5.17) and (5.18).

6.2.4 Differential Equation

Let us combine the three equations for the resulting moments according to Eqs. (6.19)–(6.21) in matrix notation as

$$\boldsymbol{M}^{\rm n}=\begin{bmatrix}M_x^{\rm n}\\ M_y^{\rm n}\\ M_{xy}^{\rm n}\end{bmatrix}=\int\limits_{-h/2}^{h/2} z\begin{bmatrix}\sigma_x\\ \sigma_y\\ \tau_{xy}\end{bmatrix}\mathrm{d}z=\int\limits_{-h/2}^{h/2} z\boldsymbol{\sigma}\mathrm{d}z\,. \tag{6.44}$$

Introducing HOOKE's law (6.16) and the kinematics relation (6.12) gives for a constant elasticity matrix $\boldsymbol{C}$

$$\boldsymbol{M}^{\mathrm{n}} = -\int_{-h/2}^{h/2} z^2 \boldsymbol{C}\boldsymbol{\mathcal{L}}_2 u_z \mathrm{d}z = -\boldsymbol{C}\boldsymbol{\mathcal{L}}_2 u_z \underbrace{\int_{-h/2}^{h/2} z^2 \mathrm{d}z}_{\frac{h^3}{12}} = -\underbrace{\frac{h^3}{12}\boldsymbol{C}}_{\boldsymbol{D}}\, \boldsymbol{\mathcal{L}}_2 u_z \,, \tag{6.45}$$

where the plate elasticity matrix $\boldsymbol{D}$ is given by

$$\boldsymbol{D} = \frac{h^3}{12}\boldsymbol{C} = \underbrace{\frac{Eh^3}{12(1-\nu^2)}}_{D} \begin{bmatrix} 1 & \nu & 0 \\ \nu & 1 & 0 \\ 0 & 0 & \frac{1-\nu}{2} \end{bmatrix}, \tag{6.46}$$

and $D = \frac{Eh^3}{12(1-\nu^2)}$ is the bending rigidity of the plate. Using the kinematics relation in the curvature form (see Eq. (6.12)), it can be stated that

$$\boldsymbol{M}^{\mathrm{n}} = \boldsymbol{D}\boldsymbol{\kappa} \,. \tag{6.47}$$

Introducing the moment-displacement relation (6.45) in the equilibrium equation (6.42) results in the plate bending differential equation in the form:

$$\boldsymbol{\mathcal{L}}_2^{\mathrm{T}} \left(\boldsymbol{D}\boldsymbol{\mathcal{L}}_2 u_z\right) - q_z = 0 \,. \tag{6.48}$$

Using the definitions for $\boldsymbol{\mathcal{L}}_2$ and $\boldsymbol{D}$ given in Eqs. (6.41) and (6.46), the following classical form of the plate bending differential equation can be obtained:

$$\frac{Eh^3}{12(1-\nu^2)} \left(\frac{\partial^4 u_z}{\partial x^4} + 2\frac{\partial^4 u_z}{\partial x^2 \partial y^2} + \frac{\partial^4 u_z}{\partial y^4} \right) = q_z \,. \tag{6.49}$$

Let us recall here that in Chap. 3 we obtained the following partial differential equation for the EULER–BERNOULLI beam, see Table 3.5:

$$EI_y \frac{\mathrm{d}^4 u_z(x)}{\mathrm{d}x^4} = q_z(x) \,. \tag{6.50}$$

Table 6.2 summarizes the different formulations of the basic equations for a classical plate and Table 6.3 compares the general formulations with the relations for the EULER–BERNOULLI beam.

Table 6.2 Different formulations of the basic equations for a classical plate (bending perpendicular to the x-y plane). E: YOUNG's modulus; ν: POISSON's ratio; q_z: area-specific distributed force; h plate thickness; M^{n}: length-specific moment; Q^{n}: length-specific shear force

Specific formulation	General formulation
Kinematics	
$\begin{bmatrix}\varepsilon_x\\ \varepsilon_y\\ \gamma_{xy}\end{bmatrix} = -z\begin{bmatrix}\frac{\partial^2}{\partial x^2}\\ \frac{\partial^2}{\partial y^2}\\ \frac{2\partial^2}{\partial x \partial y}\end{bmatrix} u_z = z\begin{bmatrix}\kappa_x\\ \kappa_y\\ \kappa_{xy}\end{bmatrix}$	$\boldsymbol{\varepsilon}(x, y, z) = -z\boldsymbol{\mathcal{L}}_2 u_z = z\boldsymbol{\kappa}$
Constitution	
$\begin{bmatrix}\sigma_x\\ \sigma_y\\ \sigma_{xy}\end{bmatrix} = \frac{E}{1-\nu^2}\begin{bmatrix}1 & \nu & 0\\ \nu & 1 & 0\\ 0 & 0 & \frac{1-\nu}{2}\end{bmatrix}\begin{bmatrix}\varepsilon_x\\ \varepsilon_y\\ \gamma_{xy}\end{bmatrix}$	$\boldsymbol{\sigma} = \boldsymbol{C}\boldsymbol{\varepsilon}$
$\begin{bmatrix}M_x^{\mathrm{n}}\\ M_y^{\mathrm{n}}\\ M_{xy}^{\mathrm{n}}\end{bmatrix} = \frac{Eh^3}{12(1-\nu^2)}\begin{bmatrix}1 & \nu & 0\\ \nu & 1 & 0\\ 0 & 0 & \frac{1-\nu}{2}\end{bmatrix}\begin{bmatrix}\kappa_x\\ \kappa_y\\ \kappa_{xy}\end{bmatrix}$	$\boldsymbol{M}^{\mathrm{n}} = \boldsymbol{D}\boldsymbol{\kappa}$
Equilibrium	
$\frac{\partial^2 M_x^{\mathrm{n}}}{\partial x^2} + 2\frac{\partial^2 M_{xy}^{\mathrm{n}}}{\partial x \partial y} + \frac{\partial^2 M_y^{\mathrm{n}}}{\partial y^2} + q_z = 0$	$\boldsymbol{\mathcal{L}}_2^{\mathrm{T}}\boldsymbol{M}^{\mathrm{n}} + q_z = 0$
$\begin{bmatrix}\frac{\partial}{\partial x} & 0 & \frac{\partial}{\partial y}\\ 0 & \frac{\partial}{\partial y} & \frac{\partial}{\partial x}\end{bmatrix}\begin{bmatrix}M_x^{\mathrm{n}}\\ M_y^{\mathrm{n}}\\ M_{xy}^{\mathrm{n}}\end{bmatrix} = \begin{bmatrix}Q_x^{\mathrm{n}}\\ Q_y^{\mathrm{n}}\end{bmatrix}$	$\boldsymbol{\mathcal{L}}_1^{\mathrm{T}}\boldsymbol{M}^{\mathrm{n}} = \boldsymbol{Q}^{\mathrm{n}}$
PDE	
$\frac{Eh^3}{12(1-\nu^2)}\left(\frac{\partial^4 u_z}{\partial x^4} + 2\frac{\partial^4 u_z}{\partial x^2 \partial y^2} + \frac{\partial^4 u_z}{\partial y^4}\right) = q_z$	$\boldsymbol{\mathcal{L}}_2^{\mathrm{T}}(\boldsymbol{D}\boldsymbol{\mathcal{L}}_2 u_z) - q_z = 0$

Table 6.3 Comparison of basic equations for an EULER–BERNOULLI beam and a KIRCHHOFF plate (bending in z-direction)

EULER–BERNOULLI beam	KIRCHHOFF plate
Kinematics	
$\varepsilon_x(x, z) = -z\mathcal{L}_2(u_z(x))$	$\boldsymbol{\varepsilon}(x, y, z) = -z\boldsymbol{\mathcal{L}}_2 u_z(x, y)$
$\kappa(x) = -\mathcal{L}u_z(x)$	$\boldsymbol{\kappa}(x, y) = -\boldsymbol{\mathcal{L}}_2 u_z(x, y)$
Constitution	
$\sigma_x(x, z) = C\varepsilon_x(x, z)$	$\boldsymbol{\sigma}(x, y, z) = \boldsymbol{C}\boldsymbol{\varepsilon}(x, y, z)$
$M_y(x) = D\kappa(x)$	$\boldsymbol{M}^{\mathrm{n}}(x, y) = \boldsymbol{D}\boldsymbol{\kappa}(x, y)$
Equilibrium	
$\mathcal{L}_2^{\mathrm{T}}\left(M_y(x)\right) + q_z(x) = 0$	$\boldsymbol{\mathcal{L}}_2^{\mathrm{T}}\boldsymbol{M}^{\mathrm{n}}(x, y) + q_z(x, y) = 0$
PDE	
$\mathcal{L}_2^{\mathrm{T}}(D\mathcal{L}_2(u_z(x))) - q_z(x) = 0$	$\boldsymbol{\mathcal{L}}_2^{\mathrm{T}}(\boldsymbol{D}\boldsymbol{\mathcal{L}}_2 u_z(x, y)) - q_z(x, y) = 0$

6.3 Finite Element Solution

6.3.1 Derivation of the Principal Finite Element Equation

Let us consider in the following the governing differential equation according to Eq. (6.48). This formulation assumes that the plate elasticity matrix $\boldsymbol{D}$ is constant and we obtain

$$\boldsymbol{\mathcal{L}}_2^{\mathrm{T}}\left(\boldsymbol{D}\boldsymbol{\mathcal{L}}_2 u_z^0(x, y)\right) - q_z = 0\,, \tag{6.51}$$

where $u_z^0(x, y)$ represents the exact solution of the problem. The last equation which contains the exact solution of the problem is fulfilled at any location (x, y) of the plate and is called the *strong formulation* of the problem. Replacing the exact solution in Eq. (6.51) by an approximate solution $u_z(x, y)$, a residual r is obtained:

$$r(x, y) = \boldsymbol{\mathcal{L}}_2^{\mathrm{T}}\left(\boldsymbol{D}\boldsymbol{\mathcal{L}}_2 u_z(x, y)\right) - q_z \neq 0\,. \tag{6.52}$$

As a consequence of the introduction of the approximate solution $u_z(x, y)$, it is in general no longer possible to satisfy the differential equation at each location (x, y) of the plate. In the scope of the weighted residual method, it is alternatively requested that the differential equation is fulfilled over a certain area (and no longer at any location (x, y)) and the following integral statement is obtained

$$\int\limits_A W^{\mathrm{T}}(x, y)\left(\boldsymbol{\mathcal{L}}_2^{\mathrm{T}}\left(\boldsymbol{D}\boldsymbol{\mathcal{L}}_2 u_z(x, y)\right) - q_z\right)\mathrm{d}A \stackrel{!}{=} 0\,, \tag{6.53}$$

which is called the *inner product*.[3] The scalar function $W(\boldsymbol{x})$ in Eq. (6.53) is called the weight function which distributes the error or the residual in the considered domain and $\boldsymbol{x} = \begin{bmatrix} x & y \end{bmatrix}^{\mathrm{T}}$ is the column matrix of Cartesian coordinates. Applying the GREEN–GAUSS theorem[4] twice (cf. Sect. A.7), via the intermediate step[5]

$$\int\limits_A W^{\mathrm{T}}(x, y)\left(\boldsymbol{\mathcal{L}}_{1*}^{\mathrm{T}}\left[\boldsymbol{\mathcal{L}}_1^{\mathrm{T}}\left(\boldsymbol{D}\boldsymbol{\mathcal{L}}_2 u_z(x, y)\right)\right] - q_z\right)\mathrm{d}A \stackrel{!}{=} 0\,, \tag{6.54}$$

gives the weak formulation as:

---

[3]The general formulation of the inner product states the integration over the volume V, see Eq. (8.20). For this integration, the strong form (6.51) must be written as $\boldsymbol{\mathcal{L}}_2^{\mathrm{T}}\left(\frac{\boldsymbol{D}}{h}\boldsymbol{\mathcal{L}}_2 u_z^0(x, y)\right) - \frac{q_z}{h} = 0$ at which the distributed load is now given as force per unit volume.

[4]Consider for this purpose the formulation $\boldsymbol{\mathcal{L}}_{1*}^{\mathrm{T}}\left[\boldsymbol{\mathcal{L}}_1^{\mathrm{T}}\left(\boldsymbol{D}\boldsymbol{\mathcal{L}}_2 u_z(x, y)\right)\right] - q_z$, where $\boldsymbol{\mathcal{L}}_{1*}^{\mathrm{T}} = \begin{bmatrix} \frac{\partial}{\partial x} & \frac{\partial}{\partial y} \end{bmatrix}$, $\boldsymbol{\mathcal{L}}_1$ as given by Eqs. (5.17) and (5.18), and $\boldsymbol{\mathcal{L}}_2$ given by Eq. (6.41).

[5]Pay attention to the transposed of the scalar function W, cf. Eq. (6.58).

$$\int_A (\boldsymbol{\mathcal{L}}_2 W)^{\mathrm{T}} \boldsymbol{D} (\boldsymbol{\mathcal{L}}_2 u_z)\, \mathrm{d}A = \int_s W^{\mathrm{T}} \left(\boldsymbol{Q}^{\mathrm{n}}\right)^{\mathrm{T}} \boldsymbol{n} \mathrm{d}s - \int_s (\boldsymbol{\mathcal{L}}_{1*} W)^{\mathrm{T}} (\boldsymbol{M}^{\mathrm{n}})^{\mathrm{T}} \boldsymbol{n} \mathrm{d}s + \int_A W^{\mathrm{T}} q_z \mathrm{d}A\,. \tag{6.55}$$

Any further development of Eq. (6.55) requires that the general expressions for the displacement and weight function, i.e. u_z and W, are now approximated by some functional representations. The nodal approach for the displacements can be generally written for a two-dimensional element as:

$$u_z^{\mathrm{e}}(x, y) = \boldsymbol{N}^{\mathrm{T}}(x, y) \boldsymbol{u}_{\mathrm{p}}^{\mathrm{e}}\,, \tag{6.56}$$

which is the same structure as in the case of the one-dimensional elements, cf. Eq. (2.21). The weight function in Eq. (6.55) is approximated in a similar way as the unknown displacement:

$$W(x, y) = \boldsymbol{N}^{\mathrm{T}}(x, y) \delta \boldsymbol{u}_{\mathrm{p}}\,, \tag{6.57}$$

$$W^{\mathrm{T}}(x, y) = (\boldsymbol{N}^{\mathrm{T}}(x, y) \delta \boldsymbol{u}_{\mathrm{p}})^{\mathrm{T}} = \delta \boldsymbol{u}_{\mathrm{p}}^{\mathrm{T}} \boldsymbol{N}\,. \tag{6.58}$$

Introducing the approximations for u_z^{e} and W according to Eqs. (6.56) and (6.57) in the weak formulation (6.55) gives:

$$\delta \boldsymbol{u}_{\mathrm{p}}^{\mathrm{T}} \int_A \left(\boldsymbol{\mathcal{L}}_2 \boldsymbol{N}^{\mathrm{T}}\right)^{\mathrm{T}} \boldsymbol{D} \left(\boldsymbol{\mathcal{L}}_2 \boldsymbol{N}^{\mathrm{T}}\right) \mathrm{d}A \boldsymbol{u}_{\mathrm{p}} = \delta \boldsymbol{u}_{\mathrm{p}}^{\mathrm{T}} \int_s \boldsymbol{N} \left(\boldsymbol{Q}^{\mathrm{n}}\right)^{\mathrm{T}} \boldsymbol{n} \mathrm{d}s - \delta \boldsymbol{u}_{\mathrm{p}}^{\mathrm{T}} \int_s \boldsymbol{N} \boldsymbol{\mathcal{L}}_{1*}^{\mathrm{T}} (\boldsymbol{M}^{\mathrm{n}})^{\mathrm{T}} \boldsymbol{n} \mathrm{d}s + \delta \boldsymbol{u}_{\mathrm{p}}^{\mathrm{T}} \int_A \boldsymbol{N} q_z \mathrm{d}A\,. \tag{6.59}$$

The virtual deformations can be eliminated from both sides of the last equation and the general form of the principal finite element equation for a classical plate is obtained:

$$\int_A \left(\boldsymbol{\mathcal{L}}_2 \boldsymbol{N}^{\mathrm{T}}\right)^{\mathrm{T}} \boldsymbol{D} \left(\boldsymbol{\mathcal{L}}_2 \boldsymbol{N}^{\mathrm{T}}\right) \mathrm{d}A \boldsymbol{u}_{\mathrm{p}}^{\mathrm{e}} = \int_s \boldsymbol{N} \left(\boldsymbol{Q}^{\mathrm{n}}\right)^{\mathrm{T}} \boldsymbol{n} \mathrm{d}s - \int_s (\boldsymbol{\mathcal{L}}_{1*} \boldsymbol{N}^{\mathrm{T}})^{\mathrm{T}} (\boldsymbol{M}^{\mathrm{n}})^{\mathrm{T}} \boldsymbol{n} \mathrm{d}s + \int_A \boldsymbol{N} q_z \mathrm{d}A\,. \tag{6.60}$$

Thus, we can identify the following three element matrices from the principal finite element equation:

$$\text{Stiffness matrix: } \boldsymbol{K}^{\mathrm{e}} = \int_A \underbrace{\left(\boldsymbol{\mathcal{L}}_2 \boldsymbol{N}^{\mathrm{T}}\right)^{\mathrm{T}}}_{\boldsymbol{B}} \boldsymbol{D} \underbrace{\left(\boldsymbol{\mathcal{L}}_2 \boldsymbol{N}^{\mathrm{T}}\right)}_{\boldsymbol{B}^{\mathrm{T}}} \mathrm{d}A\,, \tag{6.61}$$

$$\text{Boundary load matrix: } \boldsymbol{f}_t^{\mathrm{e}} = \int_s \boldsymbol{N} \left(\boldsymbol{Q}^{\mathrm{n}}\right)^{\mathrm{T}} \boldsymbol{n} \mathrm{d}s - \int_s (\boldsymbol{\mathcal{L}}_{1*} \boldsymbol{N}^{\mathrm{T}})^{\mathrm{T}} (\boldsymbol{M}^{\mathrm{n}})^{\mathrm{T}} \boldsymbol{n} \mathrm{d}s\,, \tag{6.62}$$

$$\text{Body force matrix: } \boldsymbol{f}_b^{\mathrm{e}} = \int_A \boldsymbol{N} q_z \mathrm{d}A\,. \tag{6.63}$$

Based on these abbreviations, the principal finite element equation for a single element can be written as:

$$\boldsymbol{K}^{\mathrm{e}} \boldsymbol{u}_{\mathrm{p}}^{\mathrm{e}} = \boldsymbol{f}_t^{\mathrm{e}} + \boldsymbol{f}_b^{\mathrm{e}}\,. \tag{6.64}$$

In the following, let us look at the $\boldsymbol{B}$-matrix, i.e. the matrix which contains the second-order derivatives of the interpolation functions. Application of the matrix of differential operators, i.e., $\boldsymbol{\mathcal{L}}_2$, according to Eq. (6.12) to the matrix of interpolation functions gives:

$$\boldsymbol{B}^{\mathrm{T}} = \boldsymbol{\mathcal{L}}_2 \boldsymbol{N}^{\mathrm{T}} = \begin{bmatrix} \frac{\partial^2}{\partial x^2} \\ \frac{\partial^2}{\partial y^2} \\ 2\frac{\partial^2}{\partial x \partial y} \end{bmatrix} \begin{bmatrix} N_1 & N_2 & \cdots & N_n \end{bmatrix} = \tag{6.65}$$

$$= \begin{bmatrix} \frac{\partial^2 N_1}{\partial x^2} & \frac{\partial^2 N_2}{\partial x^2} & \cdots & \frac{\partial^2 N_n}{\partial x^2} \\ \frac{\partial^2 N_1}{\partial y^2} & \frac{\partial^2 N_2}{\partial y^2} & \cdots & \frac{\partial^2 N_n}{\partial y^2} \\ 2\frac{\partial^2 N_1}{\partial x \partial y} & 2\frac{\partial^2 N_2}{\partial x \partial y} & \cdots & 2\frac{\partial^2 N_n}{\partial x \partial y} \end{bmatrix}, \tag{6.66}$$

which is a $(3 \times n)$-matrix. The last relation can be also written directly for the $\boldsymbol{B}$-matrix as:

$$\boldsymbol{B} = \left(\boldsymbol{\mathcal{L}}_2 \boldsymbol{N}^{\mathrm{T}}\right)^{\mathrm{T}} = \boldsymbol{N} \boldsymbol{\mathcal{L}}_2^{\mathrm{T}} = \begin{bmatrix} N_1 \\ N_2 \\ \vdots \\ N_n \end{bmatrix} \begin{bmatrix} \frac{\partial^2}{\partial x^2} & \frac{\partial^2}{\partial y^2} & 2\frac{\partial^2}{\partial x \partial y} \end{bmatrix} = \tag{6.67}$$

$$= \begin{bmatrix} \frac{\partial^2 N_1}{\partial x^2} & \frac{\partial^2 N_1}{\partial y^2} & 2\frac{\partial^2 N_1}{\partial x \partial y} \\ \frac{\partial^2 N_2}{\partial x^2} & \frac{\partial^2 N_2}{\partial y^2} & 2\frac{\partial^2 N_2}{\partial x \partial y} \\ \vdots & \vdots & \vdots \\ \frac{\partial^2 N_n}{\partial x^2} & \frac{\partial^2 N_n}{\partial y^2} & 2\frac{\partial^2 N_n}{\partial x \partial y} \end{bmatrix}. \tag{6.68}$$

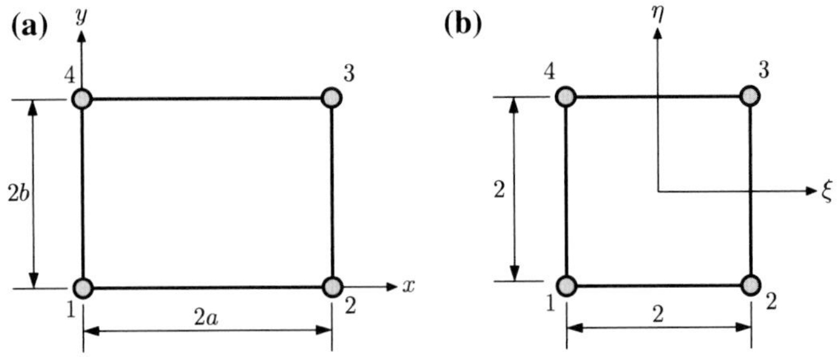

Fig. 6.6 Rectangular four-node plate element: **a** Cartesian and **b** parametric space

6.3.2 Rectangular Four-Node Plate Element

A simple representative of a two-dimensional plate[6] is a rectangular four-node element (also called 'quad 4') as shown in Fig. 6.6, see [2, 4, 5]. The node numbering must follow the right-hand convention as indicated in the figure.

Interpolation Functions and Derivatives

Let us assume in the following a fourth-order polynomial for the displacement field $u_z^e(\xi, \eta)$ in the parametric ξ-η space:

$$\begin{aligned} u_z^e(\xi, \eta) &= a_1 + a_2\xi + a_3\eta + a_4\xi^2 + a_5\xi\eta + a_6\eta^2 + a_7\xi^3 + a_8\xi^2\eta + \\ &\quad + a_9\xi\eta^2 + a_{10}\eta^3 + a_{11}\xi^3\eta + a_{12}\xi\eta^3 \,, \end{aligned} \tag{6.69}$$

or in vector notation

$$u_z^e(\xi, \eta) = \boldsymbol{\chi}^T \boldsymbol{a} = \begin{bmatrix} 1 \; \xi \; \eta \; \xi^2 \; \xi\eta \; \eta^2 \; \xi^3 \; \xi^2\eta \; \xi\eta^2 \; \eta^3 \; \xi^3\eta \; \xi\eta^3 \end{bmatrix} \begin{bmatrix} a_1 \\ a_2 \\ \vdots \\ a_{11} \\ a_{12} \end{bmatrix} . \tag{6.70}$$

Differentiation with respect to the y- and x-coordinate gives the rotational fields as (see Fig. 6.6):

$$\varphi_x^e(\xi, \eta) = \frac{\partial u_z^e}{\partial y} = \frac{\partial u_z^e(\xi, \eta)}{\partial \eta} \frac{\partial \eta}{\partial y} = \frac{1}{b} \frac{\partial u_z^e(\xi, \eta)}{\partial \eta} = \tag{6.71}$$

[6] An excellent review of classical plate elements is given in [3].

$$= \frac{1}{b}[0\ 0\ 1\ 0\ \xi\ 2\eta\ 0\ \xi^2\ 2\xi\eta\ 3\eta^2\ \xi^3\ 3\xi\eta^2] \begin{bmatrix} a_1 \\ a_2 \\ \vdots \\ a_{11} \\ a_{12} \end{bmatrix}, \tag{6.72}$$

or

$$\varphi_y^{\mathrm{e}}(\xi,\eta) = -\frac{\partial u_z^{\mathrm{e}}}{\partial x} = -\frac{\partial u_z^{\mathrm{e}}(\xi,\eta)}{\partial \xi}\frac{\partial \xi}{\partial x} = -\frac{1}{a}\frac{\partial u_z^{\mathrm{e}}(\xi,\eta)}{\partial \xi} = \tag{6.73}$$

$$= \frac{1}{a}[0\ -1\ 0\ -2\xi\ -\eta\ 0\ -3\xi^2\ -2\xi\eta\ -\eta^2\ 0\ -3\xi^2\eta\ -\eta^3] \begin{bmatrix} a_1 \\ a_2 \\ \vdots \\ a_{11} \\ a_{12} \end{bmatrix}. \tag{6.74}$$

Equations (6.70), (6.72) and (6.74) can be written in matrix notation for all four nodes as:

$$\underbrace{\begin{bmatrix} u_{1z}^{\mathrm{e}} \\ \varphi_{1x}^{\mathrm{e}} \\ \varphi_{1y}^{\mathrm{e}} \\ \hline u_{2z}^{\mathrm{e}} \\ \varphi_{2x}^{\mathrm{e}} \\ \varphi_{2y}^{\mathrm{e}} \\ \hline u_{3z}^{\mathrm{e}} \\ \varphi_{3x}^{\mathrm{e}} \\ \varphi_{3y}^{\mathrm{e}} \\ \hline u_{4z}^{\mathrm{e}} \\ \varphi_{4x}^{\mathrm{e}} \\ \varphi_{4y}^{\mathrm{e}} \end{bmatrix}}_{\boldsymbol{u}_{\mathrm{p}}} = \underbrace{\left[\begin{array}{l} 1\ \ \xi_1\ \ \eta_1\ \ \xi_1^2\ \ \xi_1\eta_1\ \ \eta_1^2\ \ \xi_1^3\ \ \xi_1^2\eta_1\ \ \xi_1\eta_1^2\ \ \eta_1^3\ \ \xi_1^3\eta_1\ \ \xi_1\eta_1^3 \\ \frac{1}{b}\left(0\ \ 0\ \ 1\ \ 0\ \ \xi_1\ \ 2\eta_1\ \ 0\ \ \xi_1^2\ \ 2\xi_1\eta_1\ \ 3\eta_1^2\ \ \xi_1^3\ \ 3\xi_1\eta_1^2\right) \\ \frac{1}{a}\left(0\ -1\ \ 0\ -2\xi_1\ -\eta_1\ \ 0\ -3\xi_1^2\ -2\xi_1\eta_1\ -\eta_1^2\ \ 0\ -3\xi_1^2\eta_1\ -\eta_1^3\right) \\ \hline 1\ \ \xi_2\ \ \eta_2\ \ \xi_2^2\ \ \xi_2\eta_2\ \ \eta_2^2\ \ \xi_2^3\ \ \xi_2^2\eta_2\ \ \xi_2\eta_2^2\ \ \eta_2^3\ \ \xi_2^3\eta_2\ \ \xi_2\eta_2^3 \\ \frac{1}{b}\left(0\ \ 0\ \ 1\ \ 0\ \ \xi_2\ \ 2\eta_2\ \ 0\ \ \xi_2^2\ \ 2\xi_2\eta_2\ \ 3\eta_2^2\ \ \xi_2^3\ \ 3\xi_2\eta_2^2\right) \\ \frac{1}{a}\left(0\ -1\ \ 0\ -2\xi_2\ -\eta_2\ \ 0\ -3\xi_2^2\ -2\xi_2\eta_2\ -\eta_2^2\ \ 0\ -3\xi_2^2\eta_2\ -\eta_2^3\right) \\ \hline 1\ \ \xi_3\ \ \eta_3\ \ \xi_3^2\ \ \xi_3\eta_3\ \ \eta_3^2\ \ \xi_3^3\ \ \xi_3^2\eta_3\ \ \xi_3\eta_3^2\ \ \eta_3^3\ \ \xi_3^3\eta_3\ \ \xi_3\eta_3^3 \\ \frac{1}{b}\left(0\ \ 0\ \ 1\ \ 0\ \ \xi_3\ \ 2\eta_3\ \ 0\ \ \xi_3^2\ \ 2\xi_3\eta_3\ \ 3\eta_3^2\ \ \xi_3^3\ \ 3\xi_3\eta_3^2\right) \\ \frac{1}{a}\left(0\ -1\ \ 0\ -2\xi_3\ -\eta_3\ \ 0\ -3\xi_3^2\ -2\xi_3\eta_3\ -\eta_3^2\ \ 0\ -3\xi_3^2\eta_3\ -\eta_3^3\right) \\ \hline 1\ \ \xi_4\ \ \eta_4\ \ \xi_4^2\ \ \xi_4\eta_4\ \ \eta_4^2\ \ \xi_4^3\ \ \xi_4^2\eta_4\ \ \xi_4\eta_4^2\ \ \eta_4^3\ \ \xi_4^3\eta_4\ \ \xi_4\eta_4^3 \\ \frac{1}{b}\left(0\ \ 0\ \ 1\ \ 0\ \ \xi_4\ \ 2\eta_4\ \ 0\ \ \xi_4^2\ \ 2\xi_4\eta_4\ \ 3\eta_4^2\ \ \xi_4^3\ \ 3\xi_4\eta_4^2\right) \\ \frac{1}{a}\left(0\ -1\ \ 0\ -2\xi_4\ -\eta_4\ \ 0\ -3\xi_4^2\ -2\xi_4\eta_4\ -\eta_4^2\ \ 0\ -3\xi_4^2\eta_4\ -\eta_4^3\right) \end{array}\right]}_{\boldsymbol{X}} \underbrace{\begin{bmatrix} a_1 \\ a_2 \\ a_3 \\ a_4 \\ a_5 \\ a_6 \\ a_7 \\ a_8 \\ a_9 \\ a_{10} \\ a_{11} \\ a_{12} \end{bmatrix}}_{\boldsymbol{a}}, \tag{6.75}$$

where ξ_i and η_i are the nodal coordinates in the ξ-η space, see Fig. 6.6b. Solving for $\boldsymbol{a}$ under consideration of these coordinates gives:

$$\begin{bmatrix} a_1 \\ a_2 \\ a_3 \\ a_4 \\ a_5 \\ a_6 \\ a_7 \\ a_8 \\ a_9 \\ a_{10} \\ a_{11} \\ a_{12} \end{bmatrix} = \begin{bmatrix} \frac{1}{4} & \frac{b}{8} & -\frac{a}{8} & \frac{1}{4} & \frac{b}{8} & \frac{a}{8} & \frac{1}{4} & -\frac{b}{8} & \frac{a}{8} & \frac{1}{4} & -\frac{b}{8} & -\frac{a}{8} \\ -\frac{3}{8} & -\frac{b}{8} & \frac{a}{8} & \frac{3}{8} & \frac{b}{8} & \frac{a}{8} & \frac{3}{8} & -\frac{b}{8} & \frac{a}{8} & -\frac{3}{8} & \frac{b}{8} & \frac{a}{8} \\ -\frac{3}{8} & -\frac{b}{8} & \frac{a}{8} & -\frac{3}{8} & -\frac{b}{8} & -\frac{a}{8} & \frac{3}{8} & -\frac{b}{8} & \frac{a}{8} & \frac{3}{8} & -\frac{b}{8} & -\frac{a}{8} \\ 0 & 0 & \frac{a}{8} & 0 & 0 & -\frac{a}{8} & 0 & 0 & -\frac{a}{8} & 0 & 0 & \frac{a}{8} \\ \frac{1}{2} & \frac{b}{8} & -\frac{a}{8} & -\frac{1}{2} & -\frac{b}{8} & -\frac{a}{8} & \frac{1}{2} & -\frac{b}{8} & \frac{a}{8} & -\frac{1}{2} & \frac{b}{8} & \frac{a}{8} \\ 0 & -\frac{b}{8} & 0 & 0 & -\frac{b}{8} & 0 & 0 & \frac{b}{8} & 0 & 0 & \frac{b}{8} & 0 \\ \frac{1}{8} & 0 & -\frac{a}{8} & -\frac{1}{8} & 0 & -\frac{a}{8} & -\frac{1}{8} & 0 & -\frac{a}{8} & \frac{1}{8} & 0 & -\frac{a}{8} \\ 0 & 0 & -\frac{a}{8} & 0 & 0 & \frac{a}{8} & 0 & 0 & -\frac{a}{8} & 0 & 0 & \frac{a}{8} \\ 0 & \frac{b}{8} & 0 & 0 & -\frac{b}{8} & 0 & 0 & \frac{b}{8} & 0 & 0 & -\frac{b}{8} & 0 \\ \frac{1}{8} & \frac{b}{8} & 0 & \frac{1}{8} & \frac{b}{8} & 0 & -\frac{1}{8} & \frac{b}{8} & 0 & -\frac{1}{8} & \frac{b}{8} & 0 \\ -\frac{1}{8} & 0 & \frac{a}{8} & \frac{1}{8} & 0 & \frac{a}{8} & -\frac{1}{8} & 0 & -\frac{a}{8} & \frac{1}{8} & 0 & -\frac{a}{8} \\ -\frac{1}{8} & -\frac{b}{8} & 0 & \frac{1}{8} & \frac{b}{8} & 0 & -\frac{1}{8} & \frac{b}{8} & 0 & \frac{1}{8} & -\frac{b}{8} & 0 \end{bmatrix} \begin{bmatrix} u_{1z}^{e} \\ \varphi_{1x}^{e} \\ \varphi_{1y}^{e} \\ u_{2z}^{e} \\ \varphi_{2x}^{e} \\ \varphi_{2y}^{e} \\ u_{3z}^{e} \\ \varphi_{3x}^{e} \\ \varphi_{3y}^{e} \\ u_{4z}^{e} \\ \varphi_{4x}^{e} \\ \varphi_{4y}^{e} \end{bmatrix}, \tag{6.76}$$

or

$$\boldsymbol{a} = \boldsymbol{A}\boldsymbol{u}_{\text{p}} = \boldsymbol{X}^{-1}\boldsymbol{u}_{\text{p}} \,. \tag{6.77}$$

The matrix of interpolation functions results as:

$$\boldsymbol{N}_{\text{e}}^{\text{T}} = \boldsymbol{\chi}^{\text{T}}\boldsymbol{A} = \left[1 \; \xi \; \eta \; \xi^2 \; \xi\eta \; \eta^2 \; \xi^3 \; \xi^2\eta \; \xi\eta^2 \; \eta^3 \; \xi^3\eta \; \xi\eta^3\right]\boldsymbol{A} \,, \tag{6.78}$$

or expressed in its components

$$N_{1u} = N_1 = -\frac{(\eta-1)\,(\xi-1)\,\left(\xi^2+\eta^2+\xi+\eta-2\right)}{8}, \tag{6.79}$$

$$N_{1\varphi_x} = N_2 = -\frac{b\,(\eta+1)\,(\eta-1)^2\,(\xi-1)}{8}, \tag{6.80}$$

$$N_{1\varphi_y} = N_3 = \frac{a\,(\xi+1)\,(\xi-1)^2\,(\eta-1)}{8}, \tag{6.81}$$

$$N_{2u} = N_4 = \frac{(\eta-1)\,(\xi+1)\,\left(\xi^2+\eta^2-\xi+\eta-2\right)}{8}, \tag{6.82}$$

$$N_{2\varphi_x} = N_5 = \frac{b\,(\eta+1)\,(\eta-1)^2\,(\xi+1)}{8}, \tag{6.83}$$

$$N_{2\varphi_y} = N_6 = \frac{a\,(\xi-1)\,(\xi+1)^2\,(\eta-1)}{8}, \tag{6.84}$$

$$N_{3u} = N_7 = -\frac{(\eta+1)\,(\xi+1)\,\left(\xi^2+\eta^2-\xi-\eta-2\right)}{8}, \tag{6.85}$$

$$N_{3\varphi_x} = N_8 = \frac{b\,(\eta - 1)\,(\eta + 1)^2\,(\xi + 1)}{8}, \tag{6.86}$$

$$N_{3\varphi_y} = N_9 = -\frac{a\,(\xi - 1)\,(\xi + 1)^2\,(\eta + 1)}{8}, \tag{6.87}$$

$$N_{4u} = N_{10} = \frac{(\eta + 1)\,(\xi - 1)\,\left(\xi^2 + \eta^2 + \xi - \eta - 2\right)}{8}, \tag{6.88}$$

$$N_{4\varphi_x} = N_{11} = -\frac{b\,(\eta - 1)\,(\eta + 1)^2\,(\xi - 1)}{8}, \tag{6.89}$$

$$N_{4\varphi_y} = N_{12} = -\frac{a\,(\xi + 1)\,(\xi - 1)^2\,(\eta + 1)}{8}. \tag{6.90}$$

It should be noted here that these twelve interpolation functions can be written in compact form ($i = 1, \ldots, 4$) as follows [6]:

$$N_{iu} = \frac{1}{8} \times (1 + \xi\xi_i)(1 + \eta\eta_i)(2 + \xi\xi_i + \eta\eta_i - \xi^2 - \eta^2), \tag{6.91}$$

$$N_{i\varphi_x} = \frac{b}{8} \times \eta_i(1 + \xi\xi_i)(\eta\eta_i - 1)(1 + \eta\eta_i)^2, \tag{6.92}$$

$$N_{i\varphi_y} = -\frac{a}{8} \times \xi_i(\xi\xi_i - 1)(1 + \eta\eta_i)(1 + \xi\xi_i)^2. \tag{6.93}$$

The corresponding interpolation functions for node 1 are shown in Fig. 6.7. Comparing with the characteristics of the cubic interpolation functions for a beam element (see Fig. 3.14), the same characteristics can be identified: The interpolation function for the displacement takes at its own node a value of one and is at all other nodes equal to zero. The interpolation functions for the rotations are at all nodes equal to zero but the slope takes absolute values of either one or zero. To illustrate the values of the slope, one may take the interpolation function $N_{1\varphi_y}$ according to Eq. (6.81). Assigning $\eta = -1$, i.e.

$$N_{1\varphi_y}\Big|_{\eta=-1} = \frac{a}{8}(\xi + 1)\,(\xi - 1)^2\,(-2)\,, \tag{6.94}$$

results after a short calculation in $N_{1\varphi}(\xi)$ according to Eq. (3.73).

The evaluation of the element stiffness matrix (6.61) requires the evaluation of the second-order derivatives of the interpolation functions. These second-order derivatives are to be evaluated with respect to the Cartesian coordinates (x, y) but the interpolation functions are given in the unit space (ξ, η). Thus, these derivations require some attention to correctly account for the different coordinate systems. The first-order derivatives can be stated under consideration of the product rule in the following manner:

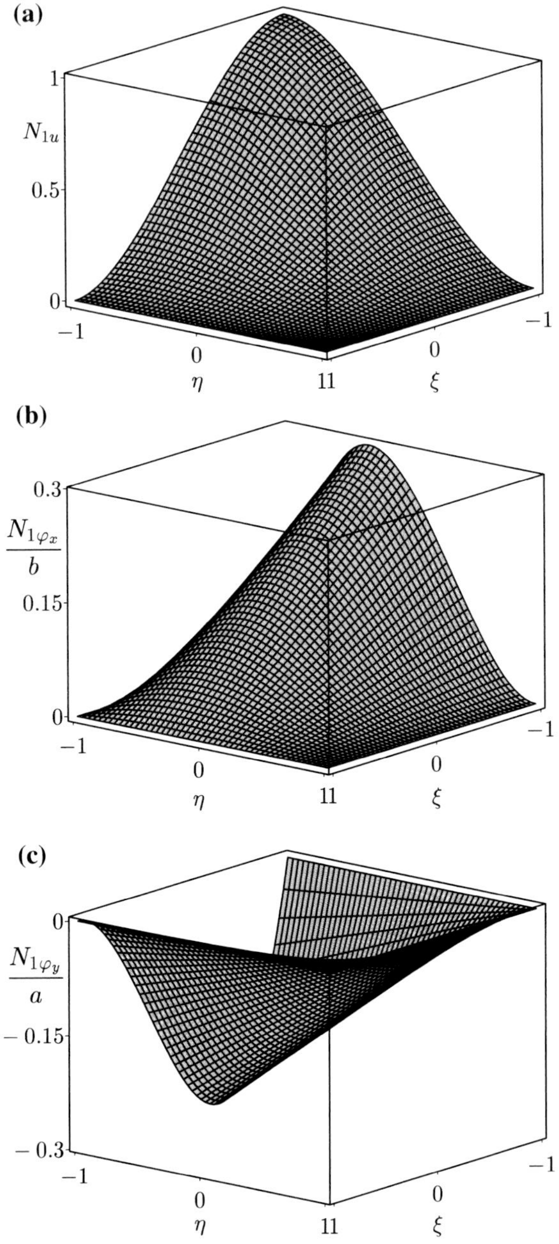

Fig. 6.7 Interpolation functions at node 1 ($\xi = -1, \eta = -1$) of a rectangular plate element (three DOF per node): **a** displacement; **b** and **c** rotations

$$\frac{\partial N(\xi,\eta)}{\partial x} = \frac{\partial N}{\partial \xi}\frac{\partial \xi}{\partial x} + \frac{\partial N}{\partial \eta}\frac{\partial \eta}{\partial x}, \tag{6.95}$$

$$\frac{\partial N(\xi,\eta)}{\partial y} = \frac{\partial N}{\partial \xi}\frac{\partial \xi}{\partial y} + \frac{\partial N}{\partial \eta}\frac{\partial \eta}{\partial y}. \tag{6.96}$$

These results for the first-order derivatives can be used to derive the calculation rules for the second-order derivatives under consideration of the product rule and different coordinate systems:

$$\begin{aligned}
\frac{\partial^2 N(\xi,\eta)}{\partial x^2} &= \frac{\partial}{\partial x}\left(\frac{\partial N(\xi,\eta)}{\partial x}\right) = \frac{\partial}{\partial x}\left(\frac{\partial N}{\partial \xi}\frac{\partial \xi}{\partial x} + \frac{\partial N}{\partial \eta}\frac{\partial \eta}{\partial x}\right), \\
&= \frac{\partial\left(\frac{\partial N}{\partial \xi}\right)}{\partial \xi}\frac{\partial \xi}{\partial x}\frac{\partial \xi}{\partial x} + \frac{\partial\left(\frac{\partial N}{\partial \xi}\right)}{\partial \eta}\frac{\partial \eta}{\partial x}\frac{\partial \xi}{\partial x} + \frac{\partial N}{\partial \xi}\frac{\partial^2 \xi}{\partial x^2} + \\
&\quad + \frac{\partial\left(\frac{\partial N}{\partial \eta}\right)}{\partial \xi}\frac{\partial \xi}{\partial x}\frac{\partial \eta}{\partial x} + \frac{\partial\left(\frac{\partial N}{\partial \eta}\right)}{\partial \eta}\frac{\partial \eta}{\partial x}\frac{\partial \eta}{\partial x} + \frac{\partial N}{\partial \eta}\frac{\partial^2 \eta}{\partial x^2} \\
&= \frac{\partial^2 N}{\partial \xi^2}\left(\frac{\partial \xi}{\partial x}\right)^2 + \frac{\partial^2 N}{\partial \xi \partial \eta}\frac{\partial \eta}{\partial x}\frac{\partial \xi}{\partial x} + \frac{\partial N}{\partial \xi}\frac{\partial^2 \xi}{\partial x^2} + \\
&\quad + \frac{\partial^2 N}{\partial \eta \partial \xi}\frac{\partial \xi}{\partial x}\frac{\partial \eta}{\partial x} + \frac{\partial^2 N}{\partial \eta^2}\left(\frac{\partial \eta}{\partial x}\right)^2 + \frac{\partial N}{\partial \eta}\frac{\partial^2 \eta}{\partial x^2},
\end{aligned} \tag{6.97}$$

or finally as:

$$\begin{aligned}
\frac{\partial^2 N(\xi,\eta)}{\partial x^2} &= \frac{\partial^2 N}{\partial \xi^2}\left(\frac{\partial \xi}{\partial x}\right)^2 + \frac{\partial^2 N}{\partial \eta^2}\left(\frac{\partial \eta}{\partial x}\right)^2 + 2\frac{\partial^2 N}{\partial \xi \partial \eta}\frac{\partial \xi}{\partial x}\frac{\partial \eta}{\partial x} + \\
&\quad + \frac{\partial N}{\partial \xi}\frac{\partial^2 \xi}{\partial x^2} + \frac{\partial N}{\partial \eta}\frac{\partial^2 \eta}{\partial x^2}.
\end{aligned} \tag{6.98}$$

The second-order derivative of an interpolation function with respect to the y-coordinate can be obtained from Eq. (6.98) by substituting x by y. The second-order mixed derivative is obtained in the following way:

$$\begin{aligned}
\frac{\partial^2 N(\xi,\eta)}{\partial x \partial y} &= \frac{\partial}{\partial y}\left(\frac{\partial N(\xi,\eta)}{\partial x}\right) = \frac{\partial}{\partial y}\left(\frac{\partial N}{\partial \xi}\frac{\partial \xi}{\partial x} + \frac{\partial N}{\partial \eta}\frac{\partial \eta}{\partial x}\right), \\
&= \frac{\partial\left(\frac{\partial N}{\partial \xi}\right)}{\partial y}\frac{\partial \xi}{\partial x} + \frac{\partial N}{\partial \xi}\frac{\partial^2 \xi}{\partial x \partial y} + \frac{\partial\left(\frac{\partial N}{\partial \eta}\right)}{\partial y}\frac{\partial \eta}{\partial x} + \frac{\partial N}{\partial \eta}\frac{\partial^2 \eta}{\partial x \partial y} \\
&= \frac{\partial^2 N}{\partial \xi^2}\frac{\partial \xi}{\partial y}\frac{\partial \xi}{\partial x} + \frac{\partial^2 N}{\partial \xi \partial \eta}\frac{\partial \eta}{\partial y}\frac{\partial \xi}{\partial x} + \frac{\partial N}{\partial \xi}\frac{\partial^2 \xi}{\partial x \partial y} +
\end{aligned} \tag{6.99}$$

$$+\frac{\partial^2 N}{\partial\eta\partial\xi}\frac{\partial\xi}{\partial y}\frac{\partial\eta}{\partial x}+\frac{\partial^2 N}{\partial\eta^2}\frac{\partial\eta}{\partial y}\frac{\partial\eta}{\partial x}+\frac{\partial N}{\partial\eta}\frac{\partial^2\eta}{\partial x\partial y},$$

or finally as:

$$\frac{\partial^2 N(\xi,\eta)}{\partial x\partial y}=\frac{\partial^2 N}{\partial\xi^2}\frac{\partial\xi}{\partial x}\frac{\partial\xi}{\partial y}+\frac{\partial^2 N}{\partial\eta^2}\frac{\partial\eta}{\partial x}\frac{\partial\eta}{\partial y}+\frac{\partial^2 N}{\partial\xi\partial\eta}\left(\frac{\partial\xi}{\partial x}\frac{\partial\eta}{\partial y}+\frac{\partial\xi}{\partial y}\frac{\partial\eta}{\partial x}\right)+ \\ +\frac{\partial N}{\partial\xi}\frac{\partial^2\xi}{\partial x\partial y}+\frac{\partial N}{\partial\eta}\frac{\partial^2\eta}{\partial x\partial y}. \tag{6.100}$$

The relationships for the second-order derivatives can be combined in matrix notation in the following way:

$$\begin{bmatrix}\frac{\partial^2 N}{\partial x^2}\\ \frac{\partial^2 N}{\partial y^2}\\ \frac{\partial^2 N}{\partial x\partial y}\end{bmatrix}=\begin{bmatrix}\left(\frac{\partial\xi}{\partial x}\right)^2 & \left(\frac{\partial\eta}{\partial x}\right)^2 & 2\frac{\partial\xi}{\partial x}\frac{\partial\eta}{\partial x}\\ \left(\frac{\partial\xi}{\partial y}\right)^2 & \left(\frac{\partial\eta}{\partial y}\right)^2 & 2\frac{\partial\xi}{\partial y}\frac{\partial\eta}{\partial y}\\ \frac{\partial\xi}{\partial x}\frac{\partial\xi}{\partial y} & \frac{\partial\eta}{\partial x}\frac{\partial\eta}{\partial y} & \frac{\partial\xi}{\partial x}\frac{\partial\eta}{\partial y}+\frac{\partial\xi}{\partial y}\frac{\partial\eta}{\partial x}\end{bmatrix}\begin{bmatrix}\frac{\partial^2 N}{\partial\xi^2}\\ \frac{\partial^2 N}{\partial\eta^2}\\ \frac{\partial^2 N}{\partial\xi\partial\eta}\end{bmatrix}+\begin{bmatrix}\frac{\partial^2\xi}{\partial x^2} & \frac{\partial^2\eta}{\partial x^2}\\ \frac{\partial^2\xi}{\partial y^2} & \frac{\partial^2\eta}{\partial y^2}\\ \frac{\partial^2\xi}{\partial x\partial y} & \frac{\partial^2\eta}{\partial x\partial y}\end{bmatrix}\begin{bmatrix}\frac{\partial N}{\partial\xi}\\ \frac{\partial N}{\partial\eta}\end{bmatrix}. \tag{6.101}$$

The evaluation of the second-order derivatives with respect to the Cartesian coordinates (x, y) as indicated in Eq. (6.101) requires the calculation of the first- and second-order derivatives with respect to the natural coordinates (ξ, η). These derivatives can be obtained from Eqs. (6.91)–(6.93) in the following way:

$$\frac{\partial N_{iu}}{\partial\xi}=-\tfrac{1}{8}(\eta\eta_i+1)(\eta^2\xi_i-\eta\eta_i\xi_i+3\xi^2\xi_i-2\xi\xi_i^2+2\xi-3\xi_i), \tag{6.102}$$

$$\frac{\partial N_{iu}}{\partial\eta}=-\tfrac{1}{8}(\xi\xi_i+1)(3\eta^2\eta_i-2\eta\eta_i^2+\eta_i\xi^2-\eta_i\xi\xi_i+2\eta-3\eta_i), \tag{6.103}$$

$$\frac{\partial^2 N_{iu}}{\partial\xi^2}=-\tfrac{1}{4}(\eta\eta_i+1)(3\xi\xi_i-\xi_i^2+1), \tag{6.104}$$

$$\frac{\partial^2 N_{iu}}{\partial\eta^2}=-\tfrac{1}{4}(\xi\xi_i+1)(3\eta\eta_i-\eta_i^2+1), \tag{6.105}$$

$$\frac{\partial^2 N_{iu}}{\partial\xi\partial\eta}=-\tfrac{1}{8}\left(\xi_i(3\eta^2\eta_i-2\eta\eta_i^2+\eta_i\xi^2-\eta_i\xi\xi_i+2\eta-3\eta_i)+(\xi\xi_i+1)(2\eta_i\xi-\eta_i\xi_i)\right). \tag{6.106}$$

$$\frac{\partial N_{i\varphi_x}}{\partial\xi}=\tfrac{1}{8}b\eta_i\xi_i(\eta\eta_i-1)(\eta\eta_i+1)^2, \tag{6.107}$$

$$\frac{\partial N_{i\varphi_x}}{\partial\eta}=\tfrac{1}{8}b\eta_1^2(\xi\xi_i+1)(\eta\eta_i+1)(3\eta\eta_i-1), \tag{6.108}$$

$$\frac{\partial^2 N_{i\varphi_x}}{\partial \xi^2} = 0\,, \tag{6.109}$$

$$\frac{\partial^2 N_{i\varphi_x}}{\partial \eta^2} = \tfrac{1}{4} b\eta_i^3(\xi\xi_i + 1)(3\eta\eta_i + 1)\,, \tag{6.110}$$

$$\frac{\partial^2 N_{i\varphi_x}}{\partial \xi \partial \eta} = \tfrac{1}{8} b\eta_i^2 \xi_1(\eta\eta_i + 1)(3\eta\eta_i - 1)\,. \tag{6.111}$$

$$\frac{\partial N_{i\varphi_y}}{\partial \xi} = -\tfrac{1}{8} a\xi_i^2(\eta\eta_i + 1)(\xi\xi_i + 1)(3\xi\xi_i - 1)\,, \tag{6.112}$$

$$\frac{\partial N_{i\varphi_y}}{\partial \eta} = -\tfrac{1}{8} a\xi_i(\xi\xi_i - 1)\eta_i(\xi\xi_i + 1)^2\,, \tag{6.113}$$

$$\frac{\partial^2 N_{i\varphi_y}}{\partial \xi^2} = -\tfrac{1}{4} a\xi_i^3(\eta\eta_i + 1)(3\xi\xi_i + 1)\,, \tag{6.114}$$

$$\frac{\partial^2 N_{i\varphi_y}}{\partial \eta^2} = 0\,, \tag{6.115}$$

$$\frac{\partial^2 N_{i\varphi_y}}{\partial \xi \partial \eta} = -\tfrac{1}{8} a\xi_i^2 \eta_i(\xi\xi_i + 1)(3\xi\xi_1 - 1)\,. \tag{6.116}$$

Geometrical Derivatives

Let us assume the same interpolation for the global x- and y-coordinate as for a four-node plane elasticity element (see Sect. 5.3.2)[7]:

$$x(\xi, \eta) = \overline{N}_1(\xi, \eta) \times x_1 + \overline{N}_2(\xi, \eta) \times x_2 + \overline{N}_3(\xi, \eta) \times x_3 + \overline{N}_4(\xi, \eta) \times x_4\,, \tag{6.117}$$

$$y(\xi, \eta) = \overline{N}_1(\xi, \eta) \times y_1 + \overline{N}_2(\xi, \eta) \times y_2 + \overline{N}_3(\xi, \eta) \times y_3 + \overline{N}_4(\xi, \eta) \times y_4\,, \tag{6.118}$$

where the *linear* shape functions $\overline{N}_i$ are given by Eqs. (5.57)–(5.60) and the *global* coordinates of the nodes $1, \ldots, 4$ can be used for $x_1, \ldots, x_4$ and $y_1, \ldots, y_4$.

Thus, the geometrical derivatives can easily be obtained as:

$$\frac{\partial x}{\partial \xi} = \frac{1}{4}\Big((-1+\eta)x_1 + (1-\eta)x_2 + (1+\eta)x_3 + (-1-\eta)x_4\Big)\,, \tag{6.119}$$

$$\frac{\partial y}{\partial \xi} = \frac{1}{4}\Big((-1+\eta)y_1 + (1-\eta)y_2 + (1+\eta)y_3 + (-1-\eta)y_4\Big)\,, \tag{6.120}$$

$$\frac{\partial x}{\partial \eta} = \frac{1}{4}\Big((-1+\xi)x_1 + (-1-\xi)x_2 + (1+\xi)x_3 + (1-\xi)x_4\Big)\,, \tag{6.121}$$

[7]Now we have the case $\overline{N}_i < N_i$ and a subparametric element formulation is obtained.

$$\frac{\partial y}{\partial \eta} = \frac{1}{4}\Big((-1+\xi)y_1 + (-1-\xi)y_2 + (1+\xi)y_3 + (1-\xi)y_4\Big). \tag{6.122}$$

The evaluation of Eq. (6.101) requires, however, the geometrical derivatives of the natural coordinates (ξ, η) with respect to the physical coordinates (x, y). These relations can be easily obtained from Eqs. (6.119)–(6.122) under consideration of the relationships provided in Sect. A.8. The first-order derivatives are

$$\frac{\partial \xi}{\partial x} = +\frac{1}{\frac{\partial x}{\partial \xi}\frac{\partial y}{\partial \eta} - \frac{\partial x}{\partial \eta}\frac{\partial y}{\partial \xi}} \times \frac{\partial y}{\partial \eta}, \tag{6.123}$$

$$\frac{\partial \xi}{\partial y} = -\frac{1}{\frac{\partial x}{\partial \xi}\frac{\partial y}{\partial \eta} - \frac{\partial x}{\partial \eta}\frac{\partial y}{\partial \xi}} \times \frac{\partial x}{\partial \eta}, \tag{6.124}$$

$$\frac{\partial \eta}{\partial x} = -\frac{1}{\frac{\partial x}{\partial \xi}\frac{\partial y}{\partial \eta} - \frac{\partial x}{\partial \eta}\frac{\partial y}{\partial \xi}} \times \frac{\partial y}{\partial \xi}, \tag{6.125}$$

$$\frac{\partial \eta}{\partial y} = +\frac{1}{\frac{\partial x}{\partial \xi}\frac{\partial y}{\partial \eta} - \frac{\partial x}{\partial \eta}\frac{\partial y}{\partial \xi}} \times \frac{\partial x}{\partial \xi}, \tag{6.126}$$

whereas the second-order derivatives follow as:

$$\frac{\partial^2 \xi}{\partial x^2} = \frac{\partial}{\partial x}\left(+\frac{1}{\frac{\partial x}{\partial \xi}\frac{\partial y}{\partial \eta} - \frac{\partial x}{\partial \eta}\frac{\partial y}{\partial \xi}} \times \frac{\partial y}{\partial \eta}\right), \tag{6.127}$$

$$\frac{\partial^2 \xi}{\partial y^2} = \frac{\partial}{\partial y}\left(-\frac{1}{\frac{\partial x}{\partial \xi}\frac{\partial y}{\partial \eta} - \frac{\partial x}{\partial \eta}\frac{\partial y}{\partial \xi}} \times \frac{\partial x}{\partial \eta}\right), \tag{6.128}$$

$$\frac{\partial^2 \xi}{\partial x \partial y} = \frac{\partial}{\partial y}\left(+\frac{1}{\frac{\partial x}{\partial \xi}\frac{\partial y}{\partial \eta} - \frac{\partial x}{\partial \eta}\frac{\partial y}{\partial \xi}} \times \frac{\partial y}{\partial \eta}\right), \tag{6.129}$$

$$\frac{\partial^2 \eta}{\partial x^2} = \frac{\partial}{\partial x}\left(-\frac{1}{\frac{\partial x}{\partial \xi}\frac{\partial y}{\partial \eta} - \frac{\partial x}{\partial \eta}\frac{\partial y}{\partial \xi}} \times \frac{\partial y}{\partial \xi}\right), \tag{6.130}$$

$$\frac{\partial^2 \eta}{\partial y^2} = \frac{\partial}{\partial y}\left(+\frac{1}{\frac{\partial x}{\partial \xi}\frac{\partial y}{\partial \eta} - \frac{\partial x}{\partial \eta}\frac{\partial y}{\partial \xi}} \times \frac{\partial x}{\partial \xi}\right), \tag{6.131}$$

$$\frac{\partial^2 \eta}{\partial x \partial y} = \frac{\partial}{\partial y}\left(-\frac{1}{\frac{\partial x}{\partial \xi}\frac{\partial y}{\partial \eta} - \frac{\partial x}{\partial \eta}\frac{\partial y}{\partial \xi}} \times \frac{\partial y}{\partial \xi}\right), \tag{6.132}$$

where, for example, the following rule must be considered: $\frac{\partial}{\partial x}(f(\xi, \eta)) = \frac{\partial f}{\partial \xi}\frac{\partial \xi}{\partial x} + \frac{\partial f}{\partial \eta}\frac{\partial \eta}{\partial x}$.

Based on the derived equations, the triple matrix product $\boldsymbol{B}\boldsymbol{D}\boldsymbol{B}^{\mathrm{T}}$ (see Eq. (6.61)) can be numerically calculated to obtain the stiffness matrix.

Numerical Integration

The integration is performed as for a four-node plane elasticity element (see Sect. 5.3.2) based on GAUSS–LEGENDRE quadrature. For the domain integral of the stiffness matrix, one can write a 2 × 2 integration rule as indicated in Table 5.5:

$$\boldsymbol{K}^{\mathrm{e}} = \int\limits_{A} (\boldsymbol{B}\boldsymbol{D}\boldsymbol{B}^{\mathrm{T}})\mathrm{d}A = \boldsymbol{B}\boldsymbol{D}\boldsymbol{B}^{\mathrm{T}} J \times 1\Big|_{\left(-\frac{1}{\sqrt{3}},-\frac{1}{\sqrt{3}}\right)} + \boldsymbol{B}\boldsymbol{D}\boldsymbol{B}^{\mathrm{T}} J \times 1\Big|_{\left(\frac{1}{\sqrt{3}},-\frac{1}{\sqrt{3}}\right)} + \boldsymbol{B}\boldsymbol{D}\boldsymbol{B}^{\mathrm{T}} J \times 1\Big|_{\left(\frac{1}{\sqrt{3}},\frac{1}{\sqrt{3}}\right)} + \boldsymbol{B}\boldsymbol{D}\boldsymbol{B}^{\mathrm{T}} J \times 1\Big|_{\left(-\frac{1}{\sqrt{3}},\frac{1}{\sqrt{3}}\right)}, \tag{6.133}$$

where the Jacobian determinant is given in Eq. (A.49) as: $J = \frac{\partial x}{\partial \xi}\frac{\partial y}{\partial \eta} - \frac{\partial x}{\partial \eta}\frac{\partial y}{\partial \xi}$.

Evaluation of the Boundary Force Matrix

The right-hand side of the weak statement contains the boundary force matrix[8] according to Eq. (6.60). Let us look on the first boundary integral and its evaluation, for example, at the node number 1:

$$\boldsymbol{f}_t^{\mathrm{e}} = \int\limits_{s} \begin{bmatrix} N_1 \\ N_2 \\ \vdots \\ N_{12} \end{bmatrix} \begin{bmatrix} Q_x^{\mathrm{n}} & Q_y^{\mathrm{n}} \end{bmatrix} \begin{bmatrix} n_x \\ n_y \end{bmatrix} \mathrm{d}s - \cdots . \tag{6.134}$$

The expression for the boundary force matrix given in Eq. (6.134) needs to be evaluated for each node along the element boundary. For node 1, the interpolation function N_1 is equal to one and identically zero for all other nodes. In addition, all other interpolation functions are identically zero at node 1. Since each node has two different normal vectors, cf. Fig. 6.8, one may calculate the expression for node 1 in x-direction by evaluating the first row of the following system:

$$\begin{bmatrix} 1 \\ 0 \\ \vdots \end{bmatrix} \begin{bmatrix} Q_x^{\mathrm{n}} & Q_y^{\mathrm{n}} \end{bmatrix} \begin{bmatrix} -1 \\ 0 \end{bmatrix} 2b + \begin{bmatrix} 1 \\ 0 \\ \vdots \end{bmatrix} \begin{bmatrix} Q_x^{\mathrm{n}} & Q_y^{\mathrm{n}} \end{bmatrix} \begin{bmatrix} 0 \\ -1 \end{bmatrix} 2a , \tag{6.135}$$

or as

$$-1Q_x^{\mathrm{n}}(2b) - 1Q_y^{\mathrm{n}}(2a) , \tag{6.136}$$

[8] In the sense of generalized forces which includes moments.

Fig. 6.8 Normal vectors for the evaluation of the boundary force matrix at node 1

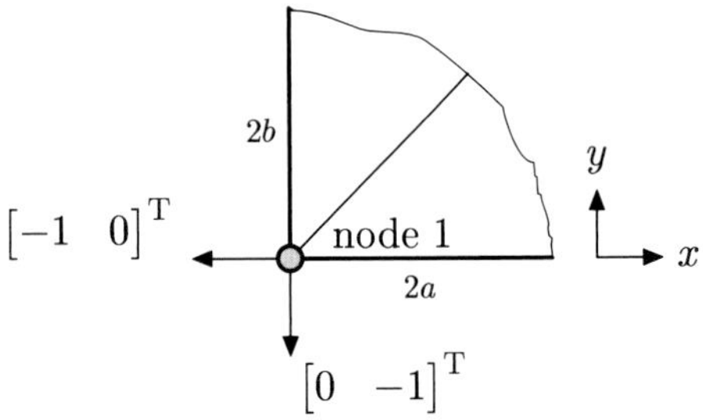

which is balanced by the external force F_{1z} at node 1. Similar results can be obtained for all other nodes and directions.

Evaluating the second boundary integral in Eq. (6.60), the boundary load matrix can be written in the case of node 1 as[9]:

$$\int_s (\boldsymbol{\mathcal{L}}_{1*}\boldsymbol{N}^{\mathrm{T}})^{\mathrm{T}}\,(\boldsymbol{M}^{\mathrm{n}})^{\mathrm{T}}\,\boldsymbol{n}\mathrm{d}s$$

$$= \left(\begin{bmatrix}\frac{\partial}{\partial x}\\ \frac{\partial}{\partial y}\end{bmatrix}\begin{bmatrix}N_{1u} & N_{1\varphi_x} & N_{1\varphi_y} & \dots & N_{4u} & N_{4\varphi_x} & N_{4\varphi_y}\end{bmatrix}\right)^{\mathrm{T}}\begin{bmatrix}M^{\mathrm{n}}_x & M^{\mathrm{n}}_{xy}\\ M^{\mathrm{n}}_{yx} & M^{\mathrm{n}}_y\end{bmatrix}^{\mathrm{T}}\begin{bmatrix}-1\\0\end{bmatrix}2b +$$
$$+ \left(\begin{bmatrix}\frac{\partial}{\partial x}\\ \frac{\partial}{\partial y}\end{bmatrix}\begin{bmatrix}N_{1u} & N_{1\varphi_x} & N_{1\varphi_y} & \dots & N_{4u} & N_{4\varphi_x} & N_{4\varphi_y}\end{bmatrix}\right)^{\mathrm{T}}\begin{bmatrix}M^{\mathrm{n}}_x & M^{\mathrm{n}}_{xy}\\ M^{\mathrm{n}}_{yx} & M^{\mathrm{n}}_y\end{bmatrix}^{\mathrm{T}}\begin{bmatrix}0\\-1\end{bmatrix}2a\,. \tag{6.137}$$

Performing two matrix multiplications gives:

$$\begin{bmatrix}\frac{\partial N_{1u}}{\partial x} & \frac{\partial N_{1\varphi_x}}{\partial x} & \frac{\partial N_{1\varphi_y}}{\partial x} & \dots & \frac{\partial N_{4u}}{\partial x} & \frac{\partial N_{4\varphi_x}}{\partial x} & \frac{\partial N_{4\varphi_y}}{\partial x}\\ \frac{\partial N_{1u}}{\partial y} & \frac{\partial N_{1\varphi_x}}{\partial y} & \frac{\partial N_{1\varphi_y}}{\partial y} & \dots & \frac{\partial N_{4u}}{\partial y} & \frac{\partial N_{4\varphi_x}}{\partial y} & \frac{\partial N_{4\varphi_y}}{\partial y}\end{bmatrix}^{\mathrm{T}}\begin{bmatrix}-M^{\mathrm{n}}_x\\ -M^{\mathrm{n}}_{xy}\end{bmatrix}2b +$$
$$+ \begin{bmatrix}\frac{\partial N_{1u}}{\partial x} & \frac{\partial N_{1\varphi_x}}{\partial x} & \frac{\partial N_{1\varphi_y}}{\partial x} & \dots & \frac{\partial N_{4u}}{\partial x} & \frac{\partial N_{4\varphi_x}}{\partial x} & \frac{\partial N_{4\varphi_y}}{\partial x}\\ \frac{\partial N_{1u}}{\partial y} & \frac{\partial N_{1\varphi_x}}{\partial y} & \frac{\partial N_{1\varphi_y}}{\partial y} & \dots & \frac{\partial N_{4u}}{\partial y} & \frac{\partial N_{4\varphi_x}}{\partial y} & \frac{\partial N_{4\varphi_y}}{\partial y}\end{bmatrix}^{\mathrm{T}}\begin{bmatrix}-M^{\mathrm{n}}_{yx}\\ -M^{\mathrm{n}}_y\end{bmatrix}2a\,. \tag{6.138}$$

Let us now consider the specific vlaues[10] of the derivatives of the shape functions at node 1, i.e.

$$\frac{\partial N_{1u}}{\partial x} = 0\,,\ \frac{\partial N_{1\varphi_x}}{\partial x} = 0\,,\ \frac{\partial N_{1\varphi_x}}{\partial y} = 1\,,\ \frac{\partial N_{1u}}{\partial y} = 0\,,\ \frac{\partial N_{1\varphi_y}}{\partial x} = -1\,,\ \frac{\partial N_{1\varphi_y}}{\partial y} = 0\,, \tag{6.139}$$

[9]It is advantageous for the further derivation to express $\boldsymbol{M}^{\mathrm{n}}$ as a (2×2)-matrix.

[10]Refer to supplementary Problem 6.9 for details.

which allows to state for node 1:

$$\begin{bmatrix} 0 & 0 \\ 0 & 1 \\ -1 & 0 \\ \vdots & \vdots \\ 0 & 0 \\ 0 & 0 \\ 0 & 0 \end{bmatrix} \begin{bmatrix} -M_x^{\mathrm{n}} \\ -M_{xy}^{\mathrm{n}} \end{bmatrix} 2b + \begin{bmatrix} 0 & 0 \\ 0 & 1 \\ -1 & 0 \\ \vdots & \vdots \\ 0 & 0 \\ 0 & 0 \\ 0 & 0 \end{bmatrix} \begin{bmatrix} -M_{yx}^{\mathrm{n}} \\ -M_y^{\mathrm{n}} \end{bmatrix} 2a \,, \tag{6.140}$$

or

$$\begin{bmatrix} 0 \\ -M_{xy}^{\mathrm{n}} 2b - M_y^{\mathrm{n}} 2a \\ M_x^{\mathrm{n}} 2b + M_{yx}^{\mathrm{n}} 2a \\ \vdots \\ 0 \\ 0 \\ 0 \end{bmatrix} \overset{\text{see Fig. 6.5a}}{=} \begin{bmatrix} 0 \\ M_{1x} \\ M_{1y} \\ \vdots \\ 0 \\ 0 \\ 0 \end{bmatrix} . \tag{6.141}$$

Summarizing the previous results for the boundary integrals in regards to forces and moments, the following boundary force matrix can be assembled

$$\boldsymbol{f}_t^{\mathrm{e}} = \begin{bmatrix} F_{1z} & M_{1x} & M_{1y} & \cdots & F_{4z} & M_{4x} & M_{4y} \end{bmatrix}^{\mathrm{T}} , \tag{6.142}$$

where an external forces or moments are positive if directed to the positive coordinate directions.

Evaluation of the Body Force Matrix

The right-hand side of the weak statement contains the distributed load $q_z(x, y)$ according to Eq. (6.60). Let us consider in the following the special case that the distributed load is constant, i.e. $q_z(x, y) \rightarrow q_z = \text{const}$. Introducing the column matrix of the interpolation functions in Eq. (6.63) gives:

$$\boldsymbol{f}_b^{\mathrm{e}} = \int\limits_A \begin{bmatrix} N_{1u} \\ N_{1\varphi x} \\ N_{1\varphi y} \\ N_{2u} \\ N_{2\varphi x} \\ N_{2\varphi y} \\ N_{3u} \\ N_{3\varphi x} \\ N_{3\varphi y} \\ N_{4u} \\ N_{4\varphi x} \\ N_{4\varphi y} \end{bmatrix} q_z \mathrm{d}A = q_z \int\limits_{-1}^{1}\int\limits_{-1}^{1} \begin{bmatrix} N_{1u} \\ N_{1\varphi x} \\ N_{1\varphi y} \\ N_{2u} \\ N_{2\varphi x} \\ N_{2\varphi y} \\ N_{3u} \\ N_{3\varphi x} \\ N_{3\varphi y} \\ N_{4u} \\ N_{4\varphi x} \\ N_{4\varphi y} \end{bmatrix} J \mathrm{d}\xi \mathrm{d}\eta = 4q_z ab \begin{bmatrix} \frac{1}{4} \\ \frac{b}{12} \\ -\frac{a}{12} \\ \frac{1}{4} \\ \frac{b}{12} \\ \frac{a}{12} \\ \frac{1}{4} \\ -\frac{b}{12} \\ \frac{a}{12} \\ \frac{1}{4} \\ -\frac{b}{12} \\ -\frac{a}{12} \end{bmatrix} , \tag{6.143}$$

where the Jacobian $J = ab$ can be taken from Example 5.2.

Let us summarize here the major steps which are required to calculate the elemental stiffness matrix.

❶ Introduce an elemental coordinate system (x, y).

❷ Express the coordinates (x_i, y_i) of the corner nodes i $(i = 1, \cdots, 4)$ in this elemental coordinate system.

❸ Calculate the partial derivatives of the Cartesian (x, y) coordinates with respect to the natural (ξ, η) coordinates, see Eqs. (6.119)–(6.122):

$$\frac{\partial x}{\partial \xi} = x_\xi = \frac{1}{4}\Big((-1+\eta)x_1 + (1-\eta)x_2 + (1+\eta)x_3 + (-1-\eta)x_4\Big),$$

$$\vdots$$

$$\frac{\partial y}{\partial \eta} = y_\eta = \frac{1}{4}\Big((-1+\xi)y_1 + (-1-\xi)y_2 + (1+\xi)y_3 + (1-\xi)y_4\Big).$$

❹ Calculate the partial derivatives of the natural (ξ, η) coordinates with respect to the Cartesian (x, y) coordinates. First-order derivatives according to Eqs. (6.123)–(6.126):

$$\frac{\partial \xi}{\partial x} = +\frac{1}{\frac{\partial x}{\partial \xi}\frac{\partial y}{\partial \eta} - \frac{\partial x}{\partial \eta}\frac{\partial y}{\partial \xi}} \times \frac{\partial y}{\partial \eta}, \quad \frac{\partial \xi}{\partial y} = -\frac{1}{\frac{\partial x}{\partial \xi}\frac{\partial y}{\partial \eta} - \frac{\partial x}{\partial \eta}\frac{\partial y}{\partial \xi}} \times \frac{\partial x}{\partial \eta},$$

$$\frac{\partial \eta}{\partial x} = -\frac{1}{\frac{\partial x}{\partial \xi}\frac{\partial y}{\partial \eta} - \frac{\partial x}{\partial \eta}\frac{\partial y}{\partial \xi}} \times \frac{\partial y}{\partial \xi}, \quad \frac{\partial \eta}{\partial y} = +\frac{1}{\frac{\partial x}{\partial \xi}\frac{\partial y}{\partial \eta} - \frac{\partial x}{\partial \eta}\frac{\partial y}{\partial \xi}} \times \frac{\partial x}{\partial \xi}.$$

Second-order derivatives according to Eqs. (6.127)–(6.132):

$$\frac{\partial^2 \xi}{\partial x^2} = \frac{\partial}{\partial x}\left(+\frac{1}{\frac{\partial x}{\partial \xi}\frac{\partial y}{\partial \eta} - \frac{\partial x}{\partial \eta}\frac{\partial y}{\partial \xi}} \times \frac{\partial y}{\partial \eta}\right),$$

$$\vdots$$

$$\frac{\partial^2 \eta}{\partial x \partial y} = \frac{\partial}{\partial y}\left(-\frac{1}{\frac{\partial x}{\partial \xi}\frac{\partial y}{\partial \eta} - \frac{\partial x}{\partial \eta}\frac{\partial y}{\partial \xi}} \times \frac{\partial y}{\partial \xi}\right).$$

❺ Calculate the $\boldsymbol{B}$-matrix and its transposed, see Eq. (6.66):

$$\boldsymbol{B}^{\mathrm{T}} = \begin{bmatrix} \frac{\partial^2 N_1}{\partial x^2} & \frac{\partial^2 N_2}{\partial x^2} & \cdots & \frac{\partial^2 N_n}{\partial x^2} \\ \frac{\partial^2 N_1}{\partial y^2} & \frac{\partial^2 N_2}{\partial y^2} & \cdots & \frac{\partial^2 N_n}{\partial y^2} \\ 2\frac{\partial^2 N_1}{\partial x \partial y} & 2\frac{\partial^2 N_2}{\partial x \partial y} & \cdots & 2\frac{\partial^2 N_n}{\partial x \partial y} \end{bmatrix},$$

where the second-order partial derivatives of the interpolation functions are given in Eq. (6.101).

❻ Calculate the triple matrix product $\boldsymbol{B}\boldsymbol{D}^{\mathrm{T}}\boldsymbol{B}$, where the plate elasticity matrix $\boldsymbol{D}$ is given by Eq. (6.46).

❼ Perform the numerical integration based on a 2×2 integration rule:

Table 6.4 Summary: derivation of principal finite element equation for plate elements

Strong formulation
$\boldsymbol{\mathcal{L}}_2^{\mathrm{T}}\left(\boldsymbol{D}\boldsymbol{\mathcal{L}}_2 u_z^0(x,y)\right) - q_z = 0$
Inner product
$\int_A \boldsymbol{W}^{\mathrm{T}}(x,y)\left(\boldsymbol{\mathcal{L}}_2^{\mathrm{T}}\left(\boldsymbol{D}\boldsymbol{\mathcal{L}}_2 u_z(x,y)\right) - q_z\right)\mathrm{d}A = 0$
Weak formulation
$\int_A (\boldsymbol{\mathcal{L}}_2\boldsymbol{W})^{\mathrm{T}}\boldsymbol{D}(\boldsymbol{\mathcal{L}}_2 u_z)\,\mathrm{d}A = \int_s \boldsymbol{W}^{\mathrm{T}}\left(\boldsymbol{Q}^{\mathrm{n}}\right)^{\mathrm{T}}\boldsymbol{n}\mathrm{d}s - \int_s (\boldsymbol{\mathcal{L}}_1\boldsymbol{W})^{\mathrm{T}}(\boldsymbol{M}^{\mathrm{n}})^{\mathrm{T}}\boldsymbol{n}\mathrm{d}s + \int_A \boldsymbol{W}^{\mathrm{T}} q_z \mathrm{d}A$
Principal finite element equation (quad 4 with 12 DOF)
$\underbrace{\int_A \underbrace{\left(\boldsymbol{\mathcal{L}}_2\boldsymbol{N}^{\mathrm{T}}\right)^{\mathrm{T}}}_{\boldsymbol{B}}\boldsymbol{D}\underbrace{\left(\boldsymbol{\mathcal{L}}_2\boldsymbol{N}^{\mathrm{T}}\right)}_{\boldsymbol{B}^{\mathrm{T}}}\mathrm{d}A}_{\boldsymbol{K}^{\mathrm{e}}}\begin{bmatrix} u_{1z} \\ \varphi_{1x} \\ \varphi_{1y} \\ \vdots \\ u_{4z} \\ \varphi_{4x} \\ \varphi_{4y} \end{bmatrix} = \begin{bmatrix} F_{1z} \\ M_{1x} \\ M_{1y} \\ \vdots \\ F_{4z} \\ M_{4x} \\ M_{4y} \end{bmatrix} + \int_A \boldsymbol{N} q_z \mathrm{d}A$

$$
\begin{aligned}
\int_A (\boldsymbol{B}\boldsymbol{D}\boldsymbol{B}^{\mathrm{T}})\mathrm{d}A &= \boldsymbol{B}\boldsymbol{D}\boldsymbol{B}^{\mathrm{T}} J \times 1\Big|_{\left(-\frac{1}{\sqrt{3}},-\frac{1}{\sqrt{3}}\right)} \\
&+ \boldsymbol{B}\boldsymbol{D}\boldsymbol{B}^{\mathrm{T}} J \times 1\Big|_{\left(\frac{1}{\sqrt{3}},-\frac{1}{\sqrt{3}}\right)} + \boldsymbol{B}\boldsymbol{D}\boldsymbol{B}^{\mathrm{T}} J \times 1\Big|_{\left(\frac{1}{\sqrt{3}},\frac{1}{\sqrt{3}}\right)} \\
&+ \boldsymbol{B}\boldsymbol{D}\boldsymbol{B}^{\mathrm{T}} J \times 1\Big|_{\left(-\frac{1}{\sqrt{3}},\frac{1}{\sqrt{3}}\right)} .
\end{aligned}
$$

❽ $\boldsymbol{K}$ obtained.

Let us summarize at the end of this section the major steps that were undertaken to transform the partial differential equation into the principal finite element equation, see Table 6.4.

6.3.3 Solved Classical Plate Element Problems

6.1 Example: One-element example of a cantilever plate

Given is a cantilever classical plate element as indicated in Fig. 6.9. The side lengths are equal to $2a$. The plate is loaded by two single forces $\frac{1}{2}F_0$ acting at the right-hand nodes of the element. The material is described based on the engineering constants YOUNG's modulus E and POISSON's ratio ν.

Fig. 6.9 Cantilever square plate element

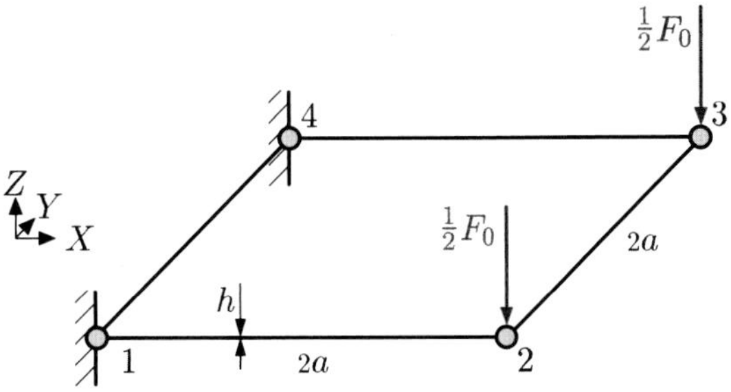

Use a single plate element in the following to model the problem and to calculate the nodal unknowns. Compare your results with the analytical solutions for a cantilever EULER–BERNOULLI and a TIMOSHENKO beam.

6.1 Solution

The solution procedure for the elemental stiffness matrix will follow the 8 steps introduced on p. 348.

❶ Introduce an elemental coordinate system (x, y, z).
Let us assume that the elemental coordinate system is located in the center of the plate.

❷ Express the coordinates (x_i, y_i) of the corner nodes i $(i = 1, \ldots, 4)$ in this elemental coordinate system.

$$(x_1, y_1, z_1) = (-a, -a)\,, \qquad (x_3, y_3, z_3) = (a, a)\,,$$
$$(x_2, y_2, z_2) = (a, -a)\,, \qquad (x_4, y_4, z_4) = (-a, a)\,.$$

❸ Calculate the partial derivatives of the Cartesian (x, y) coordinates with respect to the natural (ξ, η) coordinates, see Eqs. (6.119)–(6.122):

$$\frac{\partial x}{\partial \xi} = a\,, \qquad \frac{\partial y}{\partial \xi} = 0\,,$$
$$\frac{\partial x}{\partial \eta} = 0\,, \qquad \frac{\partial y}{\partial \eta} = a\,.$$

❹ Calculate the partial derivatives of the natural (ξ, η) coordinates with respect to the Cartesian (x, y) coordinates. First-order derivatives according to Eqs. (6.123)–(6.126):

$$\frac{\partial \xi}{\partial x} = \frac{1}{a}\,, \qquad \frac{\partial \xi}{\partial y} = 0\,,$$
$$\frac{\partial \eta}{\partial x} = 0\,, \qquad \frac{\partial \eta}{\partial y} = \frac{1}{a}\,.$$

Second-order derivatives according to Eqs. (6.127)–(6.132):

$$\frac{\partial^2 \xi}{\partial x^2} = 0\,, \qquad \frac{\partial^2 \xi}{\partial y^2} = 0\,, \qquad \frac{\partial^2 \xi}{\partial x \partial y} = 0\,,$$

$$\frac{\partial^2 \eta}{\partial x^2} = 0\,, \qquad \frac{\partial^2 \eta}{\partial y^2} = 0\,, \qquad \frac{\partial^2 \eta}{\partial x \partial y} = 0\,.$$

❺ Calculate the $\boldsymbol{B}$-matrix and its transposed, see Eq. (6.66):

$$\boldsymbol{B}^{\mathrm{T}} = \begin{bmatrix} \frac{3}{4}\frac{(1-\eta)\xi}{a^2} & 0 & \cdots & \frac{1}{4}\frac{(1+\eta)(1-3\xi)}{a} \\ \frac{3}{4}\frac{(1-\xi)\eta}{a^2} & -\frac{1}{4}\frac{b(1-\xi)(1-3\eta)}{a^2} & \cdots & 0 \\ -\frac{1}{4}\frac{3\eta^2+3\xi^2-4}{a^2} & -\frac{1}{4}\frac{b(1-\eta)(3\eta-1)}{a^2} & \cdots & -\frac{1}{4}\frac{(1-\xi)(-1-3\xi)}{a} \end{bmatrix},$$

where the second-order partial derivatives of the interpolation functions are given in Eq. (6.101).

❻ Calculate the triple matrix product $\mathrm{d}\boldsymbol{K} = \boldsymbol{B}\,\boldsymbol{D}^{\mathrm{T}}\boldsymbol{B}$, where the plate elasticity matrix $\boldsymbol{D}$ is given by Eq. (6.46).
The triple matrix product results in a 12×12 matrix with the following selected components:

$$\begin{aligned} \mathrm{d}K_{1\_1} = &-\frac{D}{32a^4}\Big(9\eta^4\nu - 18\eta^2\nu\xi^2 + 9\nu\xi^4 - 9\eta^4 + 36\eta^2\nu\xi - 54\eta^2\xi^2 + 36\eta\nu\xi^2 \\ &-9\xi^4 - 24\eta^2\nu + 36\eta^2\xi - 36\eta\nu\xi + 36\eta\xi^2 - 24\nu\xi^2 + 6\eta^2 + 6\xi^2 + 16\nu \\ &-16\Big)\,, \\ \mathrm{d}K_{1\_2} = &-\frac{bD}{32a^4}\Big(9\eta^4\nu - 9\eta^2\nu\xi^2 - 9\eta^4 - 6\eta^3\nu + 18\eta^2\nu\xi - 27\eta^2\xi^2 + 18\eta\nu\xi^2 \\ &+6\eta^3 - 15\eta^2\nu + 36\eta^2\xi - 24\eta\nu\,\xi + 12\eta\xi^2 - 9\nu\xi^2 - 3\eta^2 + 8\eta\nu - 12\xi\eta \\ &+6\xi\nu + 3\xi^2 - 2\eta + 4\nu - 4\Big)\,, \\ \mathrm{d}K_{2\_2} = &-\frac{b^2D}{32a^4}\Big(9\eta^4\nu - 9\eta^4 - 12\eta^3\nu - 18\eta^2\xi^2 + 12\eta^3 - 2\eta^2\nu + 36\eta^2\xi + 12\eta\,\xi^2 \\ &-16\eta^2 + 4\eta\nu - 24\xi\eta - 2\xi^2 + 8\eta + \nu + 4\xi - 3\Big)\,. \end{aligned}$$

❼ Perform the numerical integration based on a 2×2 integration rule:
The numerical integration results in a 12×12 matrix with the following selected components:

$$\begin{aligned} K_{1\_1} = &-D\frac{2\nu - 27}{10a^2}\,, \\ K_{1\_2} = &D\frac{b(4\nu + 11)}{10a^2}\,, \end{aligned}$$

$$K_{2\_2} = -D\frac{4b^2(\nu-6)}{15a^2}.$$

❽ $\boldsymbol{K}$ obtained.

The global system of equations, which includes the column matrix of unknowns and nodal forces, results in 12 equations for 12 unknowns. Introducing the boundary conditions, i.e. $u_{1Z} = u_{4Z} = 0$ and $\varphi_{1X} = \varphi_{1Y} = \varphi_{4X} = \varphi_{4Y} = 0$, results in a reduced system with 6 equations for 6 unknowns.

$$\begin{bmatrix} \frac{27-2\nu}{10a^2} & \frac{b(11+4\nu)}{10a^2} & \frac{11+4\nu}{10a} & \frac{-6+\nu}{5a^2} & \frac{b(11-\nu)}{10a^2} & \frac{2(1-\nu)}{5a} \\ \frac{b(11+4\nu)}{10a^2} & \frac{4b^2(6-\nu)}{15a^2} & \frac{b\nu}{a} & \frac{b(-11+\nu)}{10a^2} & \frac{b^2(9+\nu)}{15a^2} & 0 \\ \frac{11+4\nu}{10a} & \frac{b\nu}{a} & -\frac{24-4\nu}{15} & \frac{2(1-\nu)}{5a} & 0 & \frac{6+4\nu}{15} \\ \frac{-6+\nu}{5a^2} & \frac{b(-11+\nu)}{10a^2} & \frac{2(1-\nu)}{5a} & \frac{27-2\nu}{10a^2} & -\frac{b(11+4\nu)}{10a^2} & \frac{11+4\nu}{10a} \\ \frac{b(11-\nu)}{10a^2} & \frac{b^2(9+\nu)}{15a^2} & 0 & -\frac{b(11+4\nu)}{10a^2} & \frac{4b^2(6-\nu)}{15a^2} & -\frac{b\nu}{a} \\ \frac{2(1-\nu)}{5a} & 0 & \frac{6+4\nu}{15} & \frac{11+4\nu}{10a} & -\frac{b\nu}{a} & \frac{24-4\nu}{15} \end{bmatrix} \begin{bmatrix} u_{2Z} \\ \varphi_{2X} \\ \varphi_{2Y} \\ u_{3Z} \\ \varphi_{3X} \\ \varphi_{3Y} \end{bmatrix} = \begin{bmatrix} -\frac{F_0}{2} \\ 0 \\ 0 \\ -\frac{F_0}{2} \\ 0 \\ 0 \end{bmatrix}.$$

The solution of this system of equations gives:

$$u_{2Z} = u_{3Z} = -\frac{2(3\nu^2+2\nu-6)a^2F_0}{3D(3+2\nu)(-1+\nu)},$$

$$\varphi_{2X} = -\varphi_{3X} = \frac{\nu a^2 F_0}{D(2\nu^2+\nu-3)b},$$

$$\varphi_{2Y} = \varphi_{3Y} = \frac{(\nu^2+\nu-3)aF_0}{D(3+2\nu)(-1+\nu)}.$$

The analytical solution is obtained as $u_{z,\max} = -\frac{16Fa^2}{Eh^3}$ for the EULER–BERNOULLI beam and as $u_{z,\max} = -\frac{16Fa^2}{Eh^3} - \frac{6F}{5hG}$ for the TIMOSHENKO beam. It should be noted here that the EULER–BERNOULLI solution is obtained as a special case from u_{2Z} for $\nu \to 0$.

6.2 Example: Four-element example of a plate fixed at all four edges

Given is a classical plate which is fixed at all four sides, see Fig. 6.10. The side lengths are equal to $4a$. The plate is loaded by a single forces F_0 acting in the middle of the plate. The material is described based on the engineering constants YOUNG's modulus E and POISSON's ratio ν.

Use four classical plate elements (each $2a \times 2a \times h$) in the following to model the problem and to calculate the nodal unknowns in the middle of the plate.

6.2 Solution

The elemental stiffness matrix of a classical plate element with dimensions $2a \times 2a \times h$ can be taken from Example 6.1. To simplify the assemble of the global

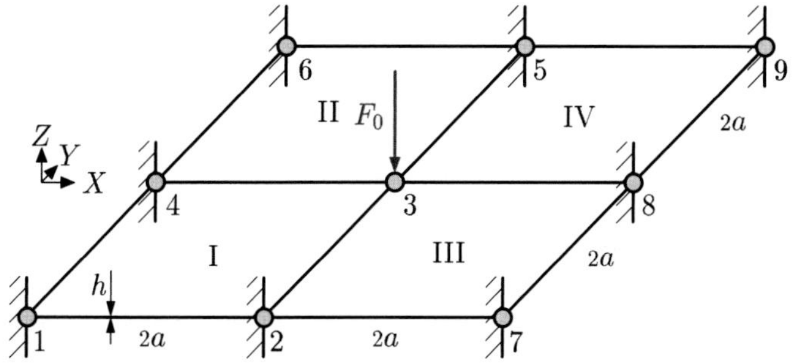

Fig. 6.10 Plate problem with four edges fixed

system of equations, it is advantageous to introduce the support conditions already on the level of the elemental stiffness matrices. Considering that only node 3 remains with three degrees of freedom (i.e., u_{3Z}, φ_{3X}, and φ_{3Y}), the following reduced 3×3 elemental stiffness matrices are obtained:

$$\boldsymbol{K}_{\mathrm{I}}^{\mathrm{red}} = \begin{bmatrix} \text{7-7} & \text{7-8} & \text{7-9} \\ \text{8-7} & \text{8-8} & \text{8-9} \\ \text{9-7} & \text{9-8} & \text{9-9} \end{bmatrix}, \quad \boldsymbol{K}_{\mathrm{II}}^{\mathrm{red}} = \begin{bmatrix} \text{4-4} & \text{4-5} & \text{4-6} \\ \text{5-4} & \text{5-5} & \text{5-6} \\ \text{6-4} & \text{6-5} & \text{6-6} \end{bmatrix},$$

$$\boldsymbol{K}_{\mathrm{III}}^{\mathrm{red}} = \begin{bmatrix} \text{10-10} & \text{10-11} & \text{10-12} \\ \text{11-10} & \text{11-11} & \text{11-12} \\ \text{12-10} & \text{12-11} & \text{12-12} \end{bmatrix}, \quad \boldsymbol{K}_{\mathrm{IV}}^{\mathrm{red}} = \begin{bmatrix} \text{1-1} & \text{1-2} & \text{1-3} \\ \text{2-1} & \text{2-2} & \text{2-3} \\ \text{3-1} & \text{3-2} & \text{3-3} \end{bmatrix}.$$

From these single elemental stiffness matrices, the reduced global system of equation is obtained as:

$$\begin{bmatrix} \frac{2D(27-2\nu)}{5a^2} & 0 & 0 \\ 0 & \frac{16D(6-\nu)}{15} & 0 \\ 0 & 0 & \frac{16D(6-\nu)}{15} \end{bmatrix} \begin{bmatrix} u_{3Z} \\ \varphi_{3X} \\ \varphi_{3Y} \end{bmatrix} = \begin{bmatrix} -F_0 \\ 0 \\ 0 \end{bmatrix}.$$

The solution of this system of equations can be obtained, for example, by inversion of the stiffness matrix. The solution matrix is finally obtained as:

$$u_{3Z} = \frac{-5a^2 F_0}{2D(27-2\nu)},$$

$$\varphi_{3X} = \varphi_{3Y} = 0.$$

The EULER–BERNOULLI solution for this problem is obtained as $u_{\mathrm{max}} = \frac{-F_0 L^3}{192EI} = -\frac{F_0 a^2}{Eh^3}$. The finite element solution reduces for $\nu \to 0$ to: $u_{3Z} = -\frac{10}{9}\frac{F_0 a^2}{Eh^3}$.

6.4 Supplementary Problems

6.3 Knowledge questions on plate elements

- State the required (a) geometrical parameters and (b) material parameters to define a rectangular plate element.
- State the DOF per node for a rectangular plate element.
- Which loads can be applied to a classical plate element?
- State possible advantages to model a beam bending problem with plate elements and not with 1D beam elements
- Which stress state is assumed for a classical plate?

6.4 Alternative definition of rotational angle

The rotational angle φ_x can be introduced in the xz-plane (see Fig. 6.11a) and φ_y in the yz-plane (see Fig. 6.11b). The angle φ_x is then considered positive if it leads to a positive displacement u_x at positive z-side of the neutral fibre. In the same sense, φ_y is considered positive if it leads to a positive displacement u_y at positive z-side of the neutral fibre.

Derive the relationship between the displacement u_x and the gradient $\frac{du_z}{dx}$ as well as the corresponding rotational angle. Repeat the derivations for the displacement u_y.

6.5 Derivation of stiffness matrix: application of Green-Gauss theorem

Starting from the inner product as given in Eq. (6.53), i.e.

$$\int_A W^{\mathrm{T}}(x, y) \left(\boldsymbol{\mathcal{L}}_2^{\mathrm{T}} \left(\boldsymbol{D}\boldsymbol{\mathcal{L}}_2 u_z(x, y)\right) - q_z\right) \mathrm{d}A \stackrel{!}{=} 0\,, \tag{6.144}$$

apply twice the GREEN–GAUSS theorem to derive the general expression for the stiffness matrix.

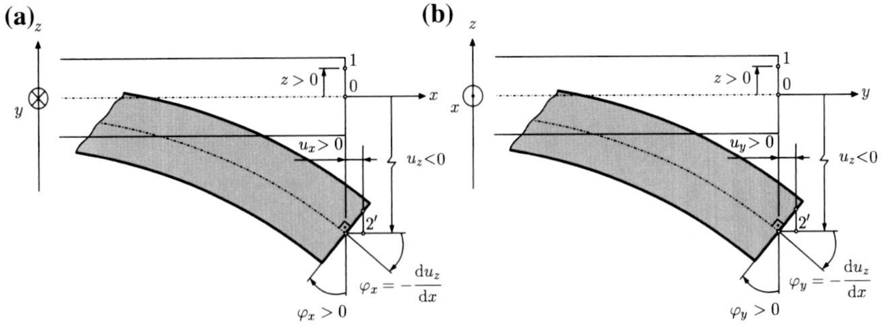

Fig. 6.11 Alternative definition of rotational angle: **a** xz-plane and **b** yz-plane

6.6 Derivation of boundary load matrix: application of Green-Gauss theorem

Starting from the inner product in the specific form of

$$\int_A W^{\mathrm{T}} \left(\boldsymbol{\mathcal{L}}_{1*}^{\mathrm{T}} \left[\boldsymbol{\mathcal{L}}_1^{\mathrm{T}} \left(\boldsymbol{D} \boldsymbol{\mathcal{L}}_2 u_z \right) \right] \right) = 0 \,, \tag{6.145}$$

apply twice the GREEN–GAUSS theorem to derive the general expression for the boundary load matrix.

6.7 Derivation of the weak form based on an alternative formulation of the partial differential equation

Consider the following form of the partial differential equation for a classical plate

$$\boldsymbol{\mathcal{L}}_2^{\mathrm{T}} \left(\frac{\boldsymbol{D}}{h} \boldsymbol{\mathcal{L}}_2 u_z(x, y) \right) - \frac{q_z}{h} = 0 \,, \tag{6.146}$$

where $\frac{q_z}{h}$ is the volume-specific distributed force. Derive the general formulation of the weak form.

6.8 Interpolation functions: angle between plate normal vector and different directions

Consider the interpolation function $N_{1\varphi_y}$ as given in Eq. (6.81). Derive the general expression of the normal vector in a surface point $(\xi, \eta, N_{1\varphi_y}(\xi, \eta))$. Calculate the angle between the normal vector in the points $(-1, -1) \vee (1, -1)$ and the following directional vectors: $(1, 0, 0)$, $(0, 1, 0)$, and $(1, 1, 0)$.

6.9 Interpolation functions: rate of change in direction of the Cartesian and natural axes

Consider the interpolation functions $N_{1\varphi_x}(\xi, \eta)$ and $N_{1\varphi_y}(\xi, \eta)$ as given in Eqs. (6.80) and (6.81). Calculate the functional values at all four nodes and the rate of change in direction of the Cartesian (x, y) and natural axes (ξ, η) expressed by the corresponding partial derivatives.

6.10 Interpolation functions in Cartesian coordinates

Consider a rectangular finite element with 4 nodes. The dimensions in x- and y-direction are $2a$ and $2b$ and the origin of the coordinate system is located in the center of the element. Since there are three conditions per node, i.e. one for the displacement and two for the rotation, the following 12-term polynomial can be used to describe the displacement $u_z^{\mathrm{e}}(x, y)$:

$$\begin{aligned} u_z^{\mathrm{e}}(x) = {} & a_1 + a_2 x + a_3 y + a_4 x^2 + a_5 xy + a_6 y^2 + a_7 x^3 \\ & + a_8 x^2 y + a_9 x y^2 + a_{10} y^3 + a_{11} x^3 y + a_{12} x y^3 \,. \end{aligned} \tag{6.147}$$

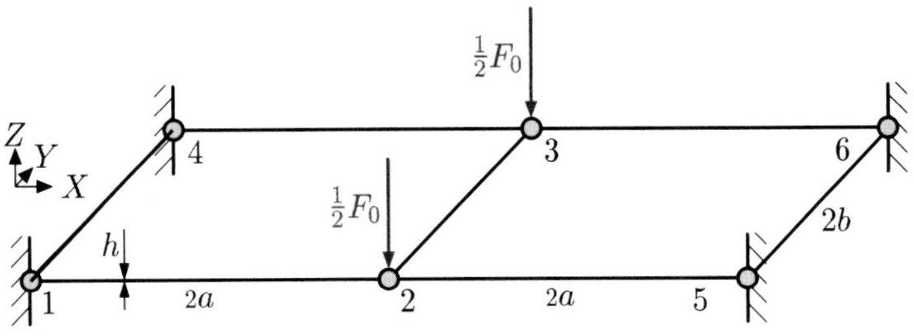

Fig. 6.12 Plate problem with two edges fixed

Derive the 12 interpolation functions $N_i(x, y)$ for this element in Cartesian coordinates.

6.11 Second-order derivatives of interpolation functions in Cartesian coordinates

Consider the interpolation functions $N_i(x, y)$ for a classical plate element in Cartesian coordinates, see supplementary Problem 6.10. Calculate the second-order derivatives in Cartesian coordinates, i.e. the $\boldsymbol{B}$-matrix.

6.12 Two-element example of a plate fixed at two edges

Given is a classical plate structure which is fixed at two sides, see Fig. 6.12. The side lengths of the entire structure are equal to $4a \times 2b$. The plate is loaded by two single forces $\frac{1}{2}F_0$ acting in the middle of the plate structure. The material is described based on the engineering constants YOUNG's modulus E and POISSON's ratio ν.

Use two classical plate elements (each $2a \times 2b \times h$) in the following to model the problem and to calculate the nodal unknowns in the middle of the plate structure. Simplify the results for the special case $\nu \to 0$ and compare these results with the EULER–BERNOULLI solution.

6.13 Symmetry solution for a plate fixed at all four edges

Reconsider Example 6.2 and derive the solution under consideration of the symmetry of the problem, i.e., based on single element ($2a \times 2a \times h$) and corresponding boundary conditions.

6.14 Investigation of displacement and slope consistency along boundaries

Investigate the interelement continuity of the displacement and slopes for a quad 4 plate element with 12 DOF.

References

1. Blaauwendraad J (2010) Plates and FEM: surprises and pitfalls. Springer, Dordrecht
2. Dawe DJ (1965) A finite element approach to plate vibration problems. J Mech Eng Sci 7:28–32

3. Hrabok MM, Hrudey TM (1984) A review and catalogue of plate bending finite elements. Comput Struct 19:479–495
4. Melosh RJ (1961) A stiffness matrix for the analysis of thin plates in bending. J Aerosp Sci 1:34–64
5. Melosh RJ (1963) Basis for derivation of matrices for the direct stiffness method. AIAA J 1:1631–1637
6. Reddy JN (2006) An introduction to the finite element method. McGraw Hill, Singapore
7. Timoshenko S, Woinowsky-Krieger S (1959) Theory of plates and shells. McGraw-Hill Book Company, New York
8. Ventsel E, Krauthammer T (2001) Thin plates and shells: theory, analysis, and applications. Marcel Dekker, New York
9. Wang CM, Reddy JN, Lee KH (2000) Shear deformable beams and plates: relationships with classical solution. Elsevier, Oxford

Chapter 7
Shear Deformable Plate Elements

Abstract This chapter starts with the analytical description of thick plate members. Thick plates are plates where the contribution of the shear force on the deformations is considered. Based on the three basic equations of continuum mechanics, i.e., the kinematics relationship, the constitutive law, and the equilibrium equation, the partial differential equations, which describes the physical problem, is derived. The weighted residual method is then used to derive the principal finite element equation for thick plate elements. The chapter exemplarily treats a four-node bilinear quadrilateral (quad 4) bending element.

7.1 Introduction

A thick plate is similarly defined as a thin plate (see Fig. 6.1). However, the condition that the thickness h is *much* smaller than the planar dimensions is weakened. The thickness is still smaller than a and b and not in the same range. The case of $h \approx a \approx b$ would rather refer to a three-dimensional element (see Chap. 8). The thick plate can be seen as a two-dimensional extension or generalization of the TIMOSHENKO beam and is also called the REISSNER–MINDLIN plate[1] in the finite element context.

7.2 Derivation of the Governing Differential Equation

7.2.1 Kinematics

Following the procedure outlined in Sect. 6.2.1, the relationships between the in-plane displacements and rotational angles are, see Fig. 7.1:

$$u_x = +z\phi_y \quad ; \quad u_y = -z\phi_x \,. \tag{7.1}$$

[1] Strictly speaking, there is a small difference between the plate theory according to REISSNER [6] and MINDLIN [4] and only for zero POISSON's ratio both derivations are the same.

A. Öchsner, *Computational Statics and Dynamics*,
https://doi.org/10.1007/978-981-15-1278-0_7

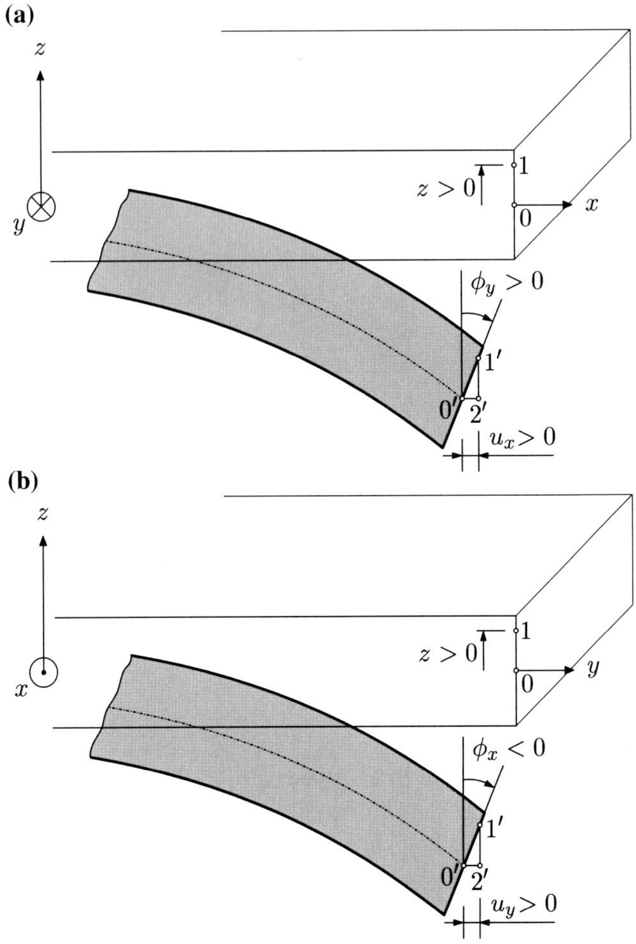

Fig. 7.1 Configuration for the derivation of kinematics relations: **a** xz-plane and **b** yz-plane. Note that the deformation is exaggerated for better illustration

Expanding the classical relationships for a plane stress state as given in Eqs. (6.8)–(6.10) by two through-thickness shear strains, the following five relations can be given:

$$\varepsilon_x = \frac{\partial u_x}{\partial x}\,;\ \varepsilon_y = \frac{\partial u_y}{\partial y}\,;\ \gamma_{xy} = \frac{\partial u_x}{\partial y} + \frac{\partial u_y}{\partial x}\,; \tag{7.2}$$

$$\gamma_{xz} = \frac{\partial u_x}{\partial z} + \frac{\partial u_z}{\partial x}\,;\ \gamma_{yz} = \frac{\partial u_y}{\partial z} + \frac{\partial u_z}{\partial y}\,. \tag{7.3}$$

Considering the results from Eq. (7.1), the five kinematics relationships can be specialized to:

$$\varepsilon_x = z\frac{\partial \phi_y}{\partial x} = z\kappa_x \; ; \; \varepsilon_y = -z\frac{\partial \phi_x}{\partial y} = z\kappa_y \; ; \; \gamma_{xy} = z\left(\frac{\partial \phi_y}{\partial y} - \frac{\partial \phi_x}{\partial x}\right) = z\kappa_{xy} \; ;$$

$$\gamma_{xz} = \phi_y + \frac{\partial u_z}{\partial x} \; ; \; \gamma_{yz} = -\phi_x + \frac{\partial u_z}{\partial y} . \tag{7.4}$$

In matrix notation, these three relationships can be written as

$$\begin{bmatrix} \frac{\partial \phi_y}{\partial x} \\ -\frac{\partial \phi_x}{\partial y} \\ \frac{\partial \phi_y}{\partial y} - \frac{\partial \phi_x}{\partial x} \\ \phi_y + \frac{\partial u_z}{\partial x} \\ -\phi_x + \frac{\partial u_z}{\partial y} \end{bmatrix} = \begin{bmatrix} 0 & 0 & \frac{\partial}{\partial x} \\ 0 & -\frac{\partial}{\partial y} & 0 \\ 0 & -\frac{\partial}{\partial x} & \frac{\partial}{\partial y} \\ \frac{\partial}{\partial x} & 0 & 1 \\ \frac{\partial}{\partial y} & -1 & 0 \end{bmatrix} \begin{bmatrix} u_z \\ \phi_x \\ \phi_y \end{bmatrix} , \tag{7.5}$$

or symbolically as

$$\boldsymbol{e} = \boldsymbol{\mathcal{L}}_1 \boldsymbol{u} . \tag{7.6}$$

One may find in the scholarly literature other definitions of the rotational angles [1, 5, 7, 8]. The angle ϕ_y is introduced in the xz-plane (see Fig. 7.1a) whereas ϕ_x is introduced in the yz-plane (see Fig. 7.1b). These definitions are closer to the classical definitions of the angles in the scope of finite elements but not conform with the definitions of the stress resultants (see M_x^{n} and M_y^{n} in Fig. 7.2). Other definitions assume, for example, that the rotational angle φ_x (now defined in the x–z plane) is positive if it leads to a positive displacement u_x at the positive z-side of the neutral axis. The same definition holds for the angle φ_y (now defined in the y–z plane).

7.2.2 Constitutive Equation

Let us start to assemble the constitutive equation based on the plane stress formulation for a *thin* plate as given in Table 6.2:

$$\begin{bmatrix} M_x^{\text{n}} \\ M_y^{\text{n}} \\ M_{xy}^{\text{n}} \end{bmatrix} = \underbrace{\underbrace{\frac{Eh^3}{12(1-\nu^2)}}_{D_{\text{b}}} \begin{bmatrix} 1 & \nu & 0 \\ \nu & 1 & 0 \\ 0 & 0 & \frac{1-\nu}{2} \end{bmatrix}}_{\boldsymbol{D}_{\text{b}}} \begin{bmatrix} \kappa_x \\ \kappa_y \\ \kappa_{xy} \end{bmatrix} , \tag{7.7}$$

or under consideration of the generalized strains $\boldsymbol{e}$ as (see Eq. (7.5)):

$$\begin{bmatrix} M^{\mathrm{n}}_x \\ M^{\mathrm{n}}_y \\ M^{\mathrm{n}}_{xy} \end{bmatrix} = \frac{Eh^3}{12(1-\nu^2)} \begin{bmatrix} 1 & \nu & 0 \\ \nu & 1 & 0 \\ 0 & 0 & \frac{1-\nu}{2} \end{bmatrix} \begin{bmatrix} \dfrac{\partial \phi_y}{\partial x} \\ -\dfrac{\partial \phi_x}{\partial y} \\ \dfrac{\partial \phi_y}{\partial y} - \dfrac{\partial \phi_x}{\partial x} \end{bmatrix}. \tag{7.8}$$

In extension to the equations for the TIMOSHENKO beam (see Eqs. (4.22) and (4.13)), the two through-thickness shear strains can be related to the normalized shear forces by:

$$\begin{bmatrix} -Q^{\mathrm{n}}_x \\ -Q^{\mathrm{n}}_y \end{bmatrix} = -\underbrace{k_{\mathrm{s}}Gh}_{D_{\mathrm{s}}} \begin{bmatrix} \gamma_{xz} \\ \gamma_{yz} \end{bmatrix} = \underbrace{-k_{\mathrm{s}}Gh \begin{bmatrix} 1 & 0 \\ 0 & 1 \end{bmatrix}}_{\boldsymbol{D}_{\mathrm{s}}} \begin{bmatrix} \gamma_{xz} \\ \gamma_{yz} \end{bmatrix}, \tag{7.9}$$

where the minus sign was only introduced for formal reasons to have a certain consistency in the further derivations (see also the constitutive equation for the TIMOSHENKO beam in Table 4.5).

Both equations for the constitutive contributions (see Eqs. (7.8) and (7.9)) can be combined to a single matrix form:

$$\begin{bmatrix} M^{\mathrm{n}}_x \\ M^{\mathrm{n}}_y \\ M^{\mathrm{n}}_{xy} \\ -Q^{\mathrm{n}}_x \\ -Q^{\mathrm{n}}_y \end{bmatrix} = \left[\begin{array}{c|c} \dfrac{Eh^3}{12(1-\nu^2)} \begin{bmatrix} 1 & \nu & 0 \\ \nu & 1 & 0 \\ 0 & 0 & \frac{1-\nu}{2} \end{bmatrix} & \begin{bmatrix} 0 & 0 \\ 0 & 0 \\ 0 & 0 \end{bmatrix} \\ \hline \begin{bmatrix} 0 & 0 & 0 \\ 0 & 0 & 0 \end{bmatrix} & -k_{\mathrm{s}}Gh \begin{bmatrix} 1 & 0 \\ 0 & 1 \end{bmatrix} \end{array}\right] \begin{bmatrix} \dfrac{\partial \phi_y}{\partial x} \\ -\dfrac{\partial \phi_x}{\partial y} \\ \dfrac{\partial \phi_y}{\partial y} - \dfrac{\partial \phi_x}{\partial x} \\ \gamma_{xz} \\ \gamma_{yz} \end{bmatrix}, \tag{7.10}$$

or symbolically as

$$\boldsymbol{s} = \boldsymbol{D}\boldsymbol{e}\,, \tag{7.11}$$

where $\boldsymbol{D}$ is the plate elasticity matrix.[2]

[2]This plate elasticity matrix should not be confused with the compliance matrix which is represented by the same symbol.

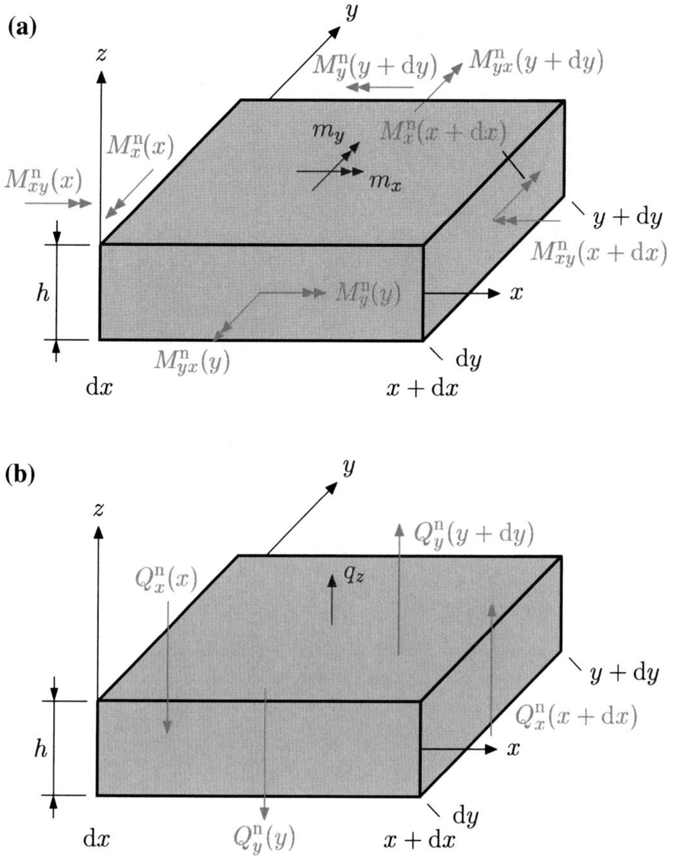

Fig. 7.2 Stress resultants acting on a thick plate element: **a** bending and twisting moments (the distributed moments m_i have the same positive direction as the rotational angles ϕ_i, see Fig. 7.1) and **b** shear forces. Positive directions are drawn

7.2.3 Equilibrium

The derivation of the equilibrium equations follows the line of reasoning which was introduced in Sect. 6.2.3 for thin plates. In addition, we consider in the following the area distributed moments $m_x(x, y)$ and $m_y(x, y)$, see Fig. 7.2.

The equilibrium condition will be determined in the following for the vertical forces. Assuming that the distributed load is constant ($q_z(x, y) \rightarrow q_z$) and that forces in the direction of the positive z-axis are considered positive, the following results:

$$-Q^{\rm n}_x(x){\rm d}y - Q^{\rm n}_y(y){\rm d}x + Q^{\rm n}_x(x+{\rm d}x){\rm d}y + Q^{\rm n}_y(y+{\rm d}y){\rm d}x + q_z{\rm d}x{\rm d}y = 0\,. \tag{7.12}$$

Evaluating the shear forces at $x + \mathrm{d}x$ and $y + \mathrm{d}y$ in a TAYLOR's series of first order as outlined in Eqs. (6.25) and (6.26), the following expression for the vertical force equilibrium can be obtained:

$$\frac{\partial Q_x^{\mathrm{n}}}{\partial x} + \frac{\partial Q_y^{\mathrm{n}}}{\partial y} + q_z = 0\,. \tag{7.13}$$

The equilibrium of moments around the reference axis at $x + \mathrm{d}x$ (positive if parallel to the y-axis) gives:

$$\begin{aligned} &M_x^{\mathrm{n}}(x + \mathrm{d}x)\mathrm{d}y - M_x^{\mathrm{n}}(x)\mathrm{d}y + M_{yx}^{\mathrm{n}}(y + \mathrm{d}y)\mathrm{d}x - M_{yx}^{\mathrm{n}}\mathrm{d}x \\ &- Q_y^{\mathrm{n}}(y)\mathrm{d}x\tfrac{\mathrm{d}x}{2} + Q_y^{\mathrm{n}}(y + \mathrm{d}y)\mathrm{d}x\tfrac{\mathrm{d}x}{2} - Q_x^{\mathrm{n}}(x)\mathrm{d}y\mathrm{d}x + q_z\mathrm{d}x\mathrm{d}y\tfrac{\mathrm{d}x}{2} + m_y\mathrm{d}x\mathrm{d}y = 0\,. \end{aligned} \tag{7.14}$$

Expanding the stress resultants at $x + \mathrm{d}x$ and $y + \mathrm{d}y$ into a TAYLOR's series of first order and neglecting the terms of third order gives finally:

$$\frac{\partial M_x^{\mathrm{n}}}{\partial x} + \frac{\partial M_{xy}^{\mathrm{n}}}{\partial y} - Q_x^{\mathrm{n}} + m_y = 0\,. \tag{7.15}$$

In a similar way, we can write the moment equilibrium around the x-axis (with the reference axis at $y + \mathrm{d}y$):

$$\begin{aligned} &- M_y^{\mathrm{n}}(y + \mathrm{d}y)\mathrm{d}x + M_y^{\mathrm{n}}(y)\mathrm{d}x - M_{xy}^{\mathrm{n}}(x + \mathrm{d}x)\mathrm{d}y + M_{xy}^{\mathrm{n}}(x)\mathrm{d}y \\ &- Q_x^{\mathrm{n}}(x + \mathrm{d}x)\mathrm{d}y\tfrac{\mathrm{d}y}{2} + Q_x^{\mathrm{n}}(x)\mathrm{d}y\tfrac{\mathrm{d}y}{2} + Q_y^{\mathrm{n}}(y)\mathrm{d}x\mathrm{d}y + m_x\mathrm{d}x\mathrm{d}y = 0\,. \end{aligned} \tag{7.16}$$

Expanding the stress resultants at $x + \mathrm{d}x$ and $y + \mathrm{d}y$ into a TAYLOR's series of first order and neglecting the terms of third order gives finally:

$$\frac{\partial M_y^{\mathrm{n}}}{\partial y} + \frac{\partial M_{xy}^{\mathrm{n}}}{\partial x} - Q_y^{\mathrm{n}} - m_x = 0\,. \tag{7.17}$$

The three equilibrium equations (see Eqs. (7.13), (7.15) and (7.17)) can be written in matrix notation as

$$\begin{bmatrix} 0 & 0 & 0 & \frac{\partial}{\partial x} & \frac{\partial}{\partial y} \\ 0 & -\frac{\partial}{\partial y} & -\frac{\partial}{\partial x} & 0 & -1 \\ \frac{\partial}{\partial x} & 0 & \frac{\partial}{\partial y} & 1 & 0 \end{bmatrix} \begin{bmatrix} M_x^{\mathrm{n}} \\ M_y^{\mathrm{n}} \\ M_{xy}^{\mathrm{n}} \\ -Q_x^{\mathrm{n}} \\ -Q_y^{\mathrm{n}} \end{bmatrix} + \begin{bmatrix} -q_z \\ m_x \\ m_y \end{bmatrix} = \begin{bmatrix} 0 \\ 0 \\ 0 \end{bmatrix}, \tag{7.18}$$

or symbolically as

$$\mathcal{L}_1^{\mathrm{T}} \boldsymbol{s} + \boldsymbol{b} = \boldsymbol{0} \,. \tag{7.19}$$

7.2.4 Differential Equation

Introducing the constitutive equation (7.11) and the kinematics equation (7.6) in the equilibrium equation (7.19) gives the general rule for the derivation of the differential equation as:

$$\mathcal{L}_1^{\mathrm{T}} \boldsymbol{D} \mathcal{L}_1 \boldsymbol{u} + \boldsymbol{b} = \boldsymbol{0} \,. \tag{7.20}$$

The first matrix multiplication, i.e. $\mathcal{L}_1^{\mathrm{T}} \boldsymbol{D}$, reads as:

$$\begin{bmatrix} 0 & 0 & 0 & \frac{\partial}{\partial x} & \frac{\partial}{\partial y} \\ 0 & -\frac{\partial}{\partial y} & -\frac{\partial}{\partial x} & 0 & -1 \\ \frac{\partial}{\partial x} & 0 & \frac{\partial}{\partial y} & 1 & 0 \end{bmatrix} \left[\begin{array}{c|c} D_{\mathrm{b}} \begin{bmatrix} 1 & \nu & 0 \\ \nu & 1 & 0 \\ 0 & 0 & \frac{1-\nu}{2} \end{bmatrix} & \begin{bmatrix} 0 & 0 \\ 0 & 0 \\ 0 & 0 \end{bmatrix} \\ \hline \begin{bmatrix} 0 & 0 & 0 \\ 0 & 0 & 0 \end{bmatrix} & -D_{\mathrm{s}} \begin{bmatrix} 1 & 0 \\ 0 & 1 \end{bmatrix} \end{array}\right] =$$

$$\begin{bmatrix} 0 & 0 & 0 & -D_{\mathrm{s}}\frac{\partial}{\partial x} & -D_{\mathrm{s}}\frac{\partial}{\partial y} \\ -D_{\mathrm{b}}\frac{\partial}{\partial y}\nu & -D_{\mathrm{b}}\frac{\partial}{\partial y} & -D_{\mathrm{b}}\frac{\partial}{\partial x}\frac{1-\nu}{2} & 0 & +D_{\mathrm{s}} \\ D_{\mathrm{b}}\frac{\partial}{\partial x} & D_{\mathrm{b}}\frac{\partial}{\partial x}\nu & D_{\mathrm{b}}\frac{\partial}{\partial y}\frac{1-\nu}{2} & -D_{\mathrm{s}} & 0 \end{bmatrix} . \tag{7.21}$$

The second matrix multiplication, i.e. $(\mathcal{L}_1^{\mathrm{T}} \boldsymbol{D})\mathcal{L}_1$, reads

$$\begin{bmatrix} 0 & 0 & 0 & -D_{\mathrm{s}}\frac{\partial}{\partial x} & -D_{\mathrm{s}}\frac{\partial}{\partial y} \\ -D_{\mathrm{b}}\frac{\partial}{\partial y}\nu & -D_{\mathrm{b}}\frac{\partial}{\partial y} & -D_{\mathrm{b}}\frac{\partial}{\partial x}\frac{1-\nu}{2} & 0 & +D_{\mathrm{s}} \\ D_{\mathrm{b}}\frac{\partial}{\partial x} & D_{\mathrm{b}}\frac{\partial}{\partial x}\nu & D_{\mathrm{b}}\frac{\partial}{\partial y}\frac{1-\nu}{2} & -D_{\mathrm{s}} & 0 \end{bmatrix} \begin{bmatrix} 0 & 0 & \frac{\partial}{\partial x} \\ 0 & -\frac{\partial}{\partial y} & 0 \\ 0 & -\frac{\partial}{\partial x} & \frac{\partial}{\partial y} \\ \frac{\partial}{\partial x} & 0 & 1 \\ \frac{\partial}{\partial y} & -1 & 0 \end{bmatrix} , \tag{7.22}$$

which finally results in the following matrix form of the differential equation:

$$\begin{bmatrix} -D_\mathrm{s}\left(\frac{\partial^2}{\partial x^2}+\frac{\partial^2}{\partial y^2}\right) & D_\mathrm{s}\frac{\partial}{\partial y} & -D_\mathrm{s}\frac{\partial}{\partial x} \\ D_\mathrm{s}\frac{\partial}{\partial y} & D_\mathrm{b}\left(\frac{1-\nu}{2}\frac{\partial^2}{\partial x^2}+\frac{\partial^2}{\partial y^2}\right)-D_\mathrm{s} & -\frac{1+\nu}{2}D_\mathrm{b}\frac{\partial^2}{\partial x\partial y} \\ -D_\mathrm{s}\frac{\partial}{\partial x} & -\frac{1+\nu}{2}D_\mathrm{b}\frac{\partial^2}{\partial x\partial y} & D_\mathrm{b}\left(\frac{\partial^2}{\partial x^2}+\frac{1-\nu}{2}\frac{\partial^2}{\partial y^2}\right)-D_\mathrm{s} \end{bmatrix}\times$$

$$\times\begin{bmatrix} u_z \\ \phi_x \\ \phi_y \end{bmatrix}+\begin{bmatrix} -q_z \\ m_x \\ m_y \end{bmatrix}=\begin{bmatrix} 0 \\ 0 \\ 0 \end{bmatrix}, \tag{7.23}$$

or symbolically as

$$\boldsymbol{\mathcal{L}}_1^\mathrm{T}\boldsymbol{D}\boldsymbol{\mathcal{L}}_1\boldsymbol{u}+\boldsymbol{b}=\boldsymbol{0}. \tag{7.24}$$

Table 7.1 summarizes the different formulations of the basic equations for a thick plate.

The general formulations of the basic equations for a thick plate as given in Table 7.1 can be slightly modified to avoid some esthetic appeals and to obtain more consistent representations of the finite element matrices (cf. the approach in Sect. 4.3.1). The kinematics equation remains unchanged while the minus signs in the constitutive equation (see the matrix of generalized strains) can be eliminated:

$$\begin{bmatrix} 1 & 0 & 0 & 0 & 0 \\ 0 & 1 & 0 & 0 & 0 \\ 0 & 0 & 1 & 0 & 0 \\ 0 & 0 & 0 & -1 & 0 \\ 0 & 0 & 0 & 0 & -1 \end{bmatrix}\begin{bmatrix} M_x^\mathrm{n} \\ M_y^\mathrm{n} \\ M_{xy}^\mathrm{n} \\ Q_x^\mathrm{n} \\ Q_y^\mathrm{n} \end{bmatrix}=\begin{bmatrix} 1 & 0 & 0 & 0 & 0 \\ 0 & 1 & 0 & 0 & 0 \\ 0 & 0 & 1 & 0 & 0 \\ 0 & 0 & 0 & -1 & 0 \\ 0 & 0 & 0 & 0 & -1 \end{bmatrix}\left[\begin{array}{c|c} D_\mathrm{b}\left[\cdots\right] & \left[\cdots\right] \\ \hline \left[\cdots\right] & D_\mathrm{s}\left[\cdots\right] \end{array}\right]\begin{bmatrix} \frac{\partial\phi_y}{\partial x} \\ -\frac{\partial\phi_x}{\partial y} \\ \frac{\partial\phi_y}{\partial y}-\frac{\partial\phi_x}{\partial x} \\ \gamma_{xz} \\ \gamma_{yz} \end{bmatrix}, \tag{7.25}$$

The diagonal matrix $\lceil 1\ 1\ 1\ -1\ -1 \rfloor$ can be eliminated from the last equation to obtain the modified constitutive law in matrix notation:

$$\begin{bmatrix} M_x^\mathrm{n} \\ M_y^\mathrm{n} \\ M_{xy}^\mathrm{n} \\ Q_x^\mathrm{n} \\ Q_y^\mathrm{n} \end{bmatrix}=\left[\begin{array}{c|c} D_\mathrm{b}\begin{bmatrix} 1 & \nu & 0 \\ \nu & 1 & 0 \\ 0 & 0 & \frac{1-\nu}{2} \end{bmatrix} & \begin{bmatrix} 0 & 0 \\ 0 & 0 \\ 0 & 0 \end{bmatrix} \\ \hline \begin{bmatrix} 0 & 0 & 0 \\ 0 & 0 & 0 \end{bmatrix} & D_\mathrm{s}\begin{bmatrix} 1 & 0 \\ 0 & 1 \end{bmatrix} \end{array}\right]\begin{bmatrix} \frac{\partial\phi_y}{\partial x} \\ -\frac{\partial\phi_x}{\partial y} \\ \frac{\partial\phi_y}{\partial y}-\frac{\partial\phi_x}{\partial x} \\ \gamma_{xz} \\ \gamma_{yz} \end{bmatrix}. \tag{7.26}$$

The next step is to have a closer look on the equilibrium equation, i.e.,

Table 7.1 Different formulations of the basic equations for a thick plate (bending perpendicular to the x–y plane). E: YOUNG's modulus; ν: POISSON's ratio; G: shear modulus; q_z: area-specific distributed force; m: area-specific distributed moment; h plate thickness; k_s: shear correction factor; M^n: length-specific moment; Q^n: length-specific shear force; $\boldsymbol{e}$: generalized strains; $\boldsymbol{s}$: generalized stresses

Specific formulation	General formulation
Kinematics	
$\begin{bmatrix} \frac{\partial\phi_y}{\partial x} \\ -\frac{\partial\phi_x}{\partial y} \\ \frac{\partial\phi_y}{\partial y} - \frac{\partial\phi_x}{\partial x} \\ \phi_y + \frac{\partial u_z}{\partial x} \\ -\phi_x + \frac{\partial u_z}{\partial y} \end{bmatrix} = \begin{bmatrix} 0 & 0 & \frac{\partial}{\partial x} \\ 0 & -\frac{\partial}{\partial y} & 0 \\ 0 & -\frac{\partial}{\partial x} & \frac{\partial}{\partial y} \\ \frac{\partial}{\partial x} & 0 & 1 \\ \frac{\partial}{\partial y} & -1 & 0 \end{bmatrix} \begin{bmatrix} u_z \\ \phi_x \\ \phi_y \end{bmatrix}$	$\boldsymbol{e} = \boldsymbol{\mathcal{L}}_1 \boldsymbol{u}$
Constitution	
$\begin{bmatrix} M_x^n \\ M_y^n \\ M_{xy}^n \\ -Q_x^n \\ -Q_y^n \end{bmatrix} = \left[\begin{array}{c\|c} \underbrace{\frac{Eh^3}{12(1-\nu^2)}}_{D_b} \begin{bmatrix} 1 & \nu & 0 \\ \nu & 1 & 0 \\ 0 & 0 & \frac{1-\nu}{2} \end{bmatrix} & \begin{bmatrix} 0 & 0 \\ 0 & 0 \\ 0 & 0 \end{bmatrix} \\ \hline \begin{bmatrix} 0 & 0 & 0 \\ 0 & 0 & 0 \end{bmatrix} & \underbrace{-k_s G h}_{D_s} \begin{bmatrix} 1 & 0 \\ 0 & 1 \end{bmatrix} \end{array}\right] \begin{bmatrix} \frac{\partial\phi_y}{\partial x} \\ -\frac{\partial\phi_x}{\partial y} \\ \frac{\partial\phi_y}{\partial y} - \frac{\partial\phi_x}{\partial x} \\ \phi_y + \frac{\partial u_z}{\partial x} \\ -\phi_x + \frac{\partial u_z}{\partial y} \end{bmatrix}$	$\boldsymbol{s} = \boldsymbol{D}\boldsymbol{e}$
Equilibrium	
$\begin{bmatrix} 0 & 0 & 0 & \frac{\partial}{\partial x} & \frac{\partial}{\partial y} \\ 0 & -\frac{\partial}{\partial y} & -\frac{\partial}{\partial x} & 0 & -1 \\ \frac{\partial}{\partial x} & 0 & \frac{\partial}{\partial y} & 1 & 0 \end{bmatrix} \begin{bmatrix} M_x^n \\ M_y^n \\ M_{xy}^n \\ -Q_x^n \\ -Q_y^n \end{bmatrix} + \begin{bmatrix} -q_z \\ m_x \\ m_y \end{bmatrix} = \begin{bmatrix} 0 \\ 0 \\ 0 \end{bmatrix}$	$\boldsymbol{\mathcal{L}}_1^T \boldsymbol{s} + \boldsymbol{b} = \boldsymbol{0}$
PDE	
$\begin{bmatrix} -D_s\left(\frac{\partial^2}{\partial x^2} + \frac{\partial^2}{\partial y^2}\right) & D_s \frac{\partial}{\partial y} & -D_s \frac{\partial}{\partial x} \\ D_s \frac{\partial}{\partial y} & D_b\left(\frac{1-\nu}{2}\frac{\partial^2}{\partial x^2} + \frac{\partial^2}{\partial y^2}\right) - D_s & -\frac{1+\nu}{2} D_b \frac{\partial^2}{\partial x \partial y} \\ -D_s \frac{\partial}{\partial x} & -\frac{1+\nu}{2} D_b \frac{\partial^2}{\partial x \partial y} & D_b\left(\frac{\partial^2}{\partial x^2} + \frac{1-\nu}{2}\frac{\partial^2}{\partial y^2}\right) - D_s \end{bmatrix} \begin{bmatrix} u_z \\ \phi_x \\ \phi_y \end{bmatrix} + \begin{bmatrix} -q_z \\ m_x \\ m_y \end{bmatrix} = \begin{bmatrix} 0 \\ 0 \\ 0 \end{bmatrix}$	$\boldsymbol{\mathcal{L}}_1^T \boldsymbol{D} \boldsymbol{\mathcal{L}}_1 \boldsymbol{u} + \boldsymbol{b} = \boldsymbol{0}$

$$\begin{bmatrix} 0 & 0 & 0 & \frac{\partial}{\partial x} & \frac{\partial}{\partial y} \\ 0 & -\frac{\partial}{\partial y} & -\frac{\partial}{\partial x} & 0 & -1 \\ \frac{\partial}{\partial x} & 0 & \frac{\partial}{\partial y} & 1 & 0 \end{bmatrix} \begin{bmatrix} M_x^n \\ M_y^n \\ M_{xy}^n \\ -Q_x^n \\ -Q_y^n \end{bmatrix} + \begin{bmatrix} -q_z \\ m_x \\ m_y \end{bmatrix} = \begin{bmatrix} 0 \\ 0 \\ 0 \end{bmatrix}, \tag{7.27}$$

or again re-written based on the diagonal matrices to extract the minus signs:

$$\begin{bmatrix} 0 & 0 & 0 & \frac{\partial}{\partial x} & \frac{\partial}{\partial y} \\ 0 & -\frac{\partial}{\partial y} & -\frac{\partial}{\partial x} & 0 & -1 \\ \frac{\partial}{\partial x} & 0 & \frac{\partial}{\partial y} & 1 & 0 \end{bmatrix} \begin{bmatrix} 1 & 0 & 0 & 0 & 0 \\ 0 & 1 & 0 & 0 & 0 \\ 0 & 0 & 1 & 0 & 0 \\ 0 & 0 & 0 & -1 & 0 \\ 0 & 0 & 0 & 0 & -1 \end{bmatrix} \begin{bmatrix} M_x^{\mathrm{n}} \\ M_y^{\mathrm{n}} \\ M_{xy}^{\mathrm{n}} \\ Q_x^{\mathrm{n}} \\ Q_y^{\mathrm{n}} \end{bmatrix} + \begin{bmatrix} -1 & 0 & 0 \\ 0 & 1 & 0 \\ 0 & 0 & 1 \end{bmatrix} \begin{bmatrix} q_z \\ m_x \\ m_y \end{bmatrix} = \begin{bmatrix} 0 \\ 0 \\ 0 \end{bmatrix}. \tag{7.28}$$

Let us now multiply the first two matrices and then multiply the resulting equation with the (3×3) diagonal matrix from the left-hand side:

$$\begin{bmatrix} -1 & 0 & 0 \\ 0 & 1 & 0 \\ 0 & 0 & 1 \end{bmatrix} \begin{bmatrix} 0 & 0 & 0 & -\frac{\partial}{\partial x} & -\frac{\partial}{\partial y} \\ 0 & -\frac{\partial}{\partial y} & -\frac{\partial}{\partial x} & 0 & 1 \\ \frac{\partial}{\partial x} & 0 & \frac{\partial}{\partial y} & -1 & 0 \end{bmatrix} \begin{bmatrix} 1 & 0 & 0 & 0 & 0 \\ 0 & 1 & 0 & 0 & 0 \\ 0 & 0 & 1 & 0 & 0 \\ 0 & 0 & 0 & -1 & 0 \\ 0 & 0 & 0 & 0 & -1 \end{bmatrix} \begin{bmatrix} M_x^{\mathrm{n}} \\ M_y^{\mathrm{n}} \\ M_{xy}^{\mathrm{n}} \\ Q_x^{\mathrm{n}} \\ Q_y^{\mathrm{n}} \end{bmatrix} +$$
$$\begin{bmatrix} -1 & 0 & 0 \\ 0 & 1 & 0 \\ 0 & 0 & 1 \end{bmatrix} \begin{bmatrix} -1 & 0 & 0 \\ 0 & 1 & 0 \\ 0 & 0 & 1 \end{bmatrix} \begin{bmatrix} q_z \\ m_x \\ m_y \end{bmatrix} = \begin{bmatrix} 0 \\ 0 \\ 0 \end{bmatrix}. \tag{7.29}$$

Or finally as the modified expression of the equilibrium equation:

$$\begin{bmatrix} 0 & 0 & 0 & \frac{\partial}{\partial x} & \frac{\partial}{\partial y} \\ 0 & -\frac{\partial}{\partial y} & -\frac{\partial}{\partial x} & 0 & 1 \\ \frac{\partial}{\partial x} & 0 & \frac{\partial}{\partial y} & -1 & 0 \end{bmatrix} \begin{bmatrix} M_x^{\mathrm{n}} \\ M_y^{\mathrm{n}} \\ M_{xy}^{\mathrm{n}} \\ Q_x^{\mathrm{n}} \\ Q_y^{\mathrm{n}} \end{bmatrix} + \begin{bmatrix} q_z \\ m_x \\ m_y \end{bmatrix} = \begin{bmatrix} 0 \\ 0 \\ 0 \end{bmatrix}. \tag{7.30}$$

Combining the three basic equations results again in the system of partial differential equations:

Table 7.2 Alternative formulations of the basic equations for a thick plate

Specific formulation	General formulation
Kinematics	
$\begin{bmatrix} \frac{\partial\phi_y}{\partial x} \\ -\frac{\partial\phi_x}{\partial y} \\ \frac{\partial\phi_y}{\partial y}-\frac{\partial\phi_x}{\partial x} \\ \phi_y+\frac{\partial u_z}{\partial x} \\ -\phi_x+\frac{\partial u_z}{\partial y}\end{bmatrix} = \begin{bmatrix} 0 & 0 & \frac{\partial}{\partial x} \\ 0 & -\frac{\partial}{\partial y} & 0 \\ 0 & -\frac{\partial}{\partial x} & \frac{\partial}{\partial y} \\ \frac{\partial}{\partial x} & 0 & 1 \\ \frac{\partial}{\partial y} & -1 & 0 \end{bmatrix}\begin{bmatrix}u_z\\ \phi_x \\ \phi_y\end{bmatrix}$	$\boldsymbol{e} = \boldsymbol{\mathcal{L}}_1\boldsymbol{u}$
Constitution	
$\begin{bmatrix} M_x^{\mathrm{n}} \\ M_y^{\mathrm{n}} \\ M_{xy}^{\mathrm{n}} \\ Q_x^{\mathrm{n}} \\ Q_y^{\mathrm{n}} \end{bmatrix} = \left[\begin{array}{c\|c} \underbrace{\frac{Eh^3}{12(1-\nu^2)}}_{D_{\mathrm{b}}} \begin{bmatrix} 1 & \nu & 0 \\ \nu & 1 & 0 \\ 0 & 0 & \frac{1-\nu}{2} \end{bmatrix} & \begin{bmatrix} 0 & 0 \\ 0 & 0 \\ 0 & 0 \end{bmatrix} \\ \hline \begin{bmatrix} 0 & 0 & 0 \\ 0 & 0 & 0 \end{bmatrix} & \underbrace{k_{\mathrm{s}}Gh}_{D_{\mathrm{s}}} \begin{bmatrix} 1 & 0 \\ 0 & 1 \end{bmatrix} \end{array}\right] \begin{bmatrix} \frac{\partial\phi_y}{\partial x} \\ -\frac{\partial\phi_x}{\partial y} \\ \frac{\partial\phi_y}{\partial y}-\frac{\partial\phi_x}{\partial x} \\ \phi_y+\frac{\partial u_z}{\partial x} \\ -\phi_x+\frac{\partial u_z}{\partial y}\end{bmatrix}$	$\boldsymbol{s}^* = \boldsymbol{D}^*\boldsymbol{e}$
Equilibrium	
$\begin{bmatrix} 0 & 0 & 0 & \frac{\partial}{\partial x} & \frac{\partial}{\partial y} \\ 0 & -\frac{\partial}{\partial y} & -\frac{\partial}{\partial x} & 0 & 1 \\ \frac{\partial}{\partial x} & 0 & \frac{\partial}{\partial y} & -1 & 0 \end{bmatrix} \begin{bmatrix} M_x^{\mathrm{n}} \\ M_y^{\mathrm{n}} \\ M_{xy}^{\mathrm{n}} \\ Q_x^{\mathrm{n}} \\ Q_y^{\mathrm{n}} \end{bmatrix} + \begin{bmatrix} q_z \\ m_x \\ m_y \end{bmatrix} = \begin{bmatrix} 0 \\ 0 \\ 0 \end{bmatrix}$	$\boldsymbol{\mathcal{L}}_{1*}^{\mathrm{T}}\boldsymbol{s}^* + \boldsymbol{b}^* = \boldsymbol{0}$
PDE	
$\begin{bmatrix} D_{\mathrm{s}}\left(\frac{\partial^2}{\partial x^2}+\frac{\partial^2}{\partial y^2}\right) & -D_{\mathrm{s}}\frac{\partial}{\partial y} & D_{\mathrm{s}}\frac{\partial}{\partial x} \\ D_{\mathrm{s}}\frac{\partial}{\partial y} & D_{\mathrm{b}}\left(\frac{1-\nu}{2}\frac{\partial^2}{\partial x^2}+\frac{\partial^2}{\partial y^2}\right)-D_{\mathrm{s}} & -\frac{1+\nu}{2}D_{\mathrm{b}}\frac{\partial^2}{\partial x\partial y} \\ -D_{\mathrm{s}}\frac{\partial}{\partial x} & -\frac{1+\nu}{2}D_{\mathrm{b}}\frac{\partial^2}{\partial x\partial y} & D_{\mathrm{b}}\left(\frac{\partial^2}{\partial x^2}+\frac{1-\nu}{2}\frac{\partial^2}{\partial y^2}\right)-D_{\mathrm{s}} \end{bmatrix} \begin{bmatrix} u_z \\ \phi_x \\ \phi_y \end{bmatrix} + \begin{bmatrix} q_z \\ m_x \\ m_y \end{bmatrix} = \begin{bmatrix} 0 \\ 0 \\ 0 \end{bmatrix}$	$\boldsymbol{\mathcal{L}}_{1*}^{\mathrm{T}}\boldsymbol{D}^*\boldsymbol{\mathcal{L}}_1\boldsymbol{u} + \boldsymbol{b}^* = \boldsymbol{0}$

$$\begin{bmatrix} D_{\mathrm{s}}\left(\frac{\partial^2}{\partial x^2}+\frac{\partial^2}{\partial y^2}\right) & -D_{\mathrm{s}}\frac{\partial}{\partial y} & D_{\mathrm{s}}\frac{\partial}{\partial x} \\ D_{\mathrm{s}}\frac{\partial}{\partial y} & D_{\mathrm{b}}\left(\frac{1-\nu}{2}\frac{\partial^2}{\partial x^2}+\frac{\partial^2}{\partial y^2}\right)-D_{\mathrm{s}} & -\frac{1+\nu}{2}D_{\mathrm{b}}\frac{\partial^2}{\partial x\partial y} \\ -D_{\mathrm{s}}\frac{\partial}{\partial x} & -\frac{1+\nu}{2}D_{\mathrm{b}}\frac{\partial^2}{\partial x\partial y} & D_{\mathrm{b}}\left(\frac{\partial^2}{\partial x^2}+\frac{1-\nu}{2}\frac{\partial^2}{\partial y^2}\right)-D_{\mathrm{s}} \end{bmatrix} \times$$

$$\times \begin{bmatrix} u_z \\ \phi_x \\ \phi_y \end{bmatrix} + \begin{bmatrix} q_z \\ m_x \\ m_y \end{bmatrix} = \begin{bmatrix} 0 \\ 0 \\ 0 \end{bmatrix}. \tag{7.31}$$

The modified basic equations, i.e., 'without the minus signs', are summarized in Table 7.2.

7.3 Finite Element Solution

7.3.1 Derivation of the Principal Finite Element Equation

Let us consider in the following the governing differential equation according to Eq. (7.24). This formulation assumes that the plate elasticity matrix $\boldsymbol{D}$ is constant and we obtain

$$\mathcal{L}_1^{\mathrm{T}} \boldsymbol{D} \mathcal{L}_1 \boldsymbol{u}^0 + \boldsymbol{b} = \boldsymbol{0}\,, \tag{7.32}$$

where $\boldsymbol{u}^0(x, y)$ represents the exact solution of the problem. The last equation which contains the exact solution of the problem is fulfilled at any location (x, y) of the plate and is called the *strong formulation* of the problem. Replacing the exact solution in Eq. (7.32) by an approximate solution $\boldsymbol{u}(x, y)$, a residual $\boldsymbol{r}$ is obtained:

$$\boldsymbol{r}(x, y) = \mathcal{L}_1^{\mathrm{T}} \boldsymbol{D} \mathcal{L}_1 \boldsymbol{u} + \boldsymbol{b} \neq \boldsymbol{0}\,. \tag{7.33}$$

As a consequence of the introduction of the approximate solution $\boldsymbol{u}(x, y)$, it is in general no longer possible to satisfy the differential equation at each location (x, y) of the plate. In the scope of the weighted residual method, it is alternatively requested that the differential equation is fulfilled over a certain area (and no longer at any location (x, y)) and the following integral statement is obtained

$$\int_A \boldsymbol{W}^{\mathrm{T}} \left(\mathcal{L}_1^{\mathrm{T}} \boldsymbol{D} \mathcal{L}_1 \boldsymbol{u} + \boldsymbol{b} \right) \mathrm{d}A \stackrel{!}{=} 0\,, \tag{7.34}$$

which is called the *inner product*. The function $\boldsymbol{W}(\boldsymbol{x}) = \left[W_{\phi_y}\ W_{\phi_x}\ W_{u_z}\right]^{\mathrm{T}}$ in Eq. (7.34) is called the weight function which distributes the error or the residual in the considered domain and $\boldsymbol{x} = \left[x\ y\right]^{\mathrm{T}}$ is the column matrix of Cartesian coordinates. Application of the GREEN–GAUSS theorem (cf. Sect. A.7) would shift the derivative to the weight functions $\boldsymbol{W}$. However, the matrix of differential operators, $\mathcal{L}_1^{\mathrm{T}}$, contains in addition to derivatives a constant value '1' and it is therefore appropriate to split the matrix into a part which contains all the derivatives, $\mathcal{L}_{1,a}^{\mathrm{T}}$, and a part with the constant value, $\mathcal{L}_{1,b}^{\mathrm{T}}$:

$$\underbrace{\begin{bmatrix} 0 & 0 & 0 & \frac{\partial}{\partial x} & \frac{\partial}{\partial y} \\ 0 & -\frac{\partial}{\partial y} & -\frac{\partial}{\partial x} & 0 & -1 \\ \frac{\partial}{\partial x} & 0 & \frac{\partial}{\partial y} & 1 & 0 \end{bmatrix}}_{\mathcal{L}_1^{\mathrm{T}}} = \underbrace{\begin{bmatrix} 0 & 0 & 0 & \frac{\partial}{\partial x} & \frac{\partial}{\partial y} \\ 0 & -\frac{\partial}{\partial y} & -\frac{\partial}{\partial x} & 0 & 0 \\ \frac{\partial}{\partial x} & 0 & \frac{\partial}{\partial y} & 0 & 0 \end{bmatrix}}_{\mathcal{L}_{1,a}^{\mathrm{T}}} + \underbrace{\begin{bmatrix} 0 & 0 & 0 & 0 & 0 \\ 0 & 0 & 0 & 0 & -1 \\ 0 & 0 & 0 & 1 & 0 \end{bmatrix}}_{\mathcal{L}_{1,b}^{\mathrm{T}}}\,. \tag{7.35}$$

Thus, we can write the inner product as:

$$\int_A \boldsymbol{W}^{\mathrm{T}} \left[\left(\boldsymbol{\mathcal{L}}_{1,a}^{\mathrm{T}} + \boldsymbol{\mathcal{L}}_{1,b}^{\mathrm{T}} \right) \boldsymbol{D} \boldsymbol{\mathcal{L}}_1 \boldsymbol{u} + \boldsymbol{b} \right] \mathrm{d}A \overset{!}{=} 0\,, \tag{7.36}$$

or

$$\int_A \boldsymbol{W}^{\mathrm{T}} \boldsymbol{\mathcal{L}}_{1,a}^{\mathrm{T}} \boldsymbol{D} \boldsymbol{\mathcal{L}}_1 \boldsymbol{u} \mathrm{d}A + \int_A \boldsymbol{W}^{\mathrm{T}} \boldsymbol{\mathcal{L}}_{1,b}^{\mathrm{T}} \boldsymbol{D} \boldsymbol{\mathcal{L}}_1 \boldsymbol{u} \mathrm{d}A + \int_A \boldsymbol{W}^{\mathrm{T}} \boldsymbol{b} \, \mathrm{d}A = 0\,. \tag{7.37}$$

Application of the GREEN–GAUSS theorem to the first integral gives:

$$\int_A \boldsymbol{W}^{\mathrm{T}} \boldsymbol{\mathcal{L}}_{1,a}^{\mathrm{T}} \boldsymbol{D} \boldsymbol{\mathcal{L}}_1 \boldsymbol{u} \, \mathrm{d}A = - \int_A \left(\boldsymbol{\mathcal{L}}_{1,a} \boldsymbol{W} \right)^{\mathrm{T}} \boldsymbol{D} \left(\boldsymbol{\mathcal{L}}_1 \boldsymbol{u} \right) \mathrm{d}A + \int_s \boldsymbol{W}^{\mathrm{T}} \underbrace{\left(\boldsymbol{D} \boldsymbol{\mathcal{L}}_1 \boldsymbol{u} \right)^{\mathrm{T}}}_{\boldsymbol{s}^{\mathrm{T}}} \boldsymbol{n} \mathrm{d}s\,. \tag{7.38}$$

and the weak formulation can be obtained as:

$$\int_A \left(\boldsymbol{\mathcal{L}}_{1,a} \boldsymbol{W} \right)^{\mathrm{T}} \boldsymbol{D} \left(\boldsymbol{\mathcal{L}}_1 \boldsymbol{u} \right) \mathrm{d}A - \int_A \boldsymbol{W}^{\mathrm{T}} \boldsymbol{\mathcal{L}}_{1,b}^{\mathrm{T}} \boldsymbol{D} \boldsymbol{\mathcal{L}}_1 \boldsymbol{u} \, \mathrm{d}A = \int_s \boldsymbol{W}^{\mathrm{T}} \boldsymbol{s}^{\mathrm{T}} \boldsymbol{n} \mathrm{d}s + \int_A \boldsymbol{W}^{\mathrm{T}} \boldsymbol{b} \, \mathrm{d}A\,. \tag{7.39}$$

The unknown deflection $u_z^{\mathrm{e}}(x, y)$ and rotational $\phi_x^{\mathrm{e}}(x, y)$, $\phi_y^{\mathrm{e}}(x, y)$ fields in an element are approximated based on the following approaches which assume four nodes with the deflection u_z and rotations φ_x, φ_y as the nodal independent unknowns:

$$\boldsymbol{u}^{\mathrm{e}}(\boldsymbol{x}) = \begin{bmatrix} u_z^{\mathrm{e}}(x, y) \\ \phi_x^{\mathrm{e}}(x, y) \\ \phi_y^{\mathrm{e}}(x, y) \end{bmatrix} = \boldsymbol{N}^{\mathrm{T}}(\boldsymbol{x}) \boldsymbol{u}_{\mathrm{p}}^{\mathrm{e}} =$$

$$= \left[\begin{array}{ccc|c|ccc} N_{1u} & 0 & 0 & \cdots & N_{4u} & 0 & 0 \\ 0 & N_{1\phi_x} & 0 & \cdots & 0 & N_{4\phi_x} & 0 \\ 0 & 0 & N_{1\phi_y} & \cdots & 0 & 0 & N_{4\phi_y} \end{array} \right] \begin{bmatrix} u_{1z} \\ \phi_{1x} \\ \phi_{1y} \\ \vdots \\ u_{4z} \\ \phi_{4x} \\ \phi_{4y} \end{bmatrix}. \tag{7.40}$$

The same approach is adopted for the weight functions:

$$\boldsymbol{W}(\boldsymbol{x}) = \boldsymbol{N}^{\mathrm{T}}(\boldsymbol{x}) \delta \boldsymbol{u}_{\mathrm{p}}\,, \tag{7.41}$$

Introducing the nodal approaches for the deformations and weight functions in the weak formulation (7.39) results in the weak formulation on element level after elimination of the virtual deformations $\delta \boldsymbol{u}_{\mathrm{p}}$:

$$\int_A (\boldsymbol{\mathcal{L}}_{1,a}\boldsymbol{N}^{\mathrm{T}})^{\mathrm{T}}\boldsymbol{D}(\boldsymbol{\mathcal{L}}_1\boldsymbol{N}^{\mathrm{T}})\mathrm{d}A\boldsymbol{u}_{\mathrm{p}}^{\mathrm{e}} - \int_A (\boldsymbol{\mathcal{L}}_{1,b}\boldsymbol{N}^{\mathrm{T}})^{\mathrm{T}}\boldsymbol{D}(\boldsymbol{\mathcal{L}}_1\boldsymbol{N}^{\mathrm{T}})\mathrm{d}A\boldsymbol{u}_{\mathrm{p}}^{\mathrm{e}} = \int_s \boldsymbol{N}\boldsymbol{s}^{\mathrm{T}}\boldsymbol{n}\mathrm{d}s + \int_A \boldsymbol{N}\boldsymbol{b}\,\mathrm{d}A\,, \tag{7.42}$$

where the evaluation of the integrals on the left-hand side results in the elemental stiffness matrix $\boldsymbol{K}^{\mathrm{e}}$ and the evaluation of the two integrals on the right-hand side results in the elemental load matrix $\boldsymbol{f}^{\mathrm{e}}$.

Thus, we can identify the following three element matrices from the principal finite element equation:

Stiffness matrix:

$$\boldsymbol{K}^{\mathrm{e}} = \int_A \left((\boldsymbol{\mathcal{L}}_{1,a}\boldsymbol{N}^{\mathrm{T}})^{\mathrm{T}}\boldsymbol{D}(\boldsymbol{\mathcal{L}}_1\boldsymbol{N}^{\mathrm{T}}) - (\boldsymbol{\mathcal{L}}_{1,b}\boldsymbol{N}^{\mathrm{T}})^{\mathrm{T}}\boldsymbol{D}(\boldsymbol{\mathcal{L}}_1\boldsymbol{N}^{\mathrm{T}})\right)\mathrm{d}A\,, \tag{7.43}$$

$$\text{Boundary force matrix:}\quad \boldsymbol{f}_t^{\mathrm{e}} = \int_s \boldsymbol{N}\boldsymbol{s}^{\mathrm{T}}\boldsymbol{n}\mathrm{d}s\,, \tag{7.44}$$

$$\text{Body force matrix:}\quad \boldsymbol{f}_b^{\mathrm{e}} = \int_A \boldsymbol{N}\boldsymbol{b}\mathrm{d}A\,. \tag{7.45}$$

Based on these abbreviations, the principal finite element equation for a single element can be written as:

$$\boldsymbol{K}^{\mathrm{e}}\boldsymbol{u}_{\mathrm{p}}^{\mathrm{e}} = \boldsymbol{f}_t^{\mathrm{e}} + \boldsymbol{f}_b^{\mathrm{e}}\,. \tag{7.46}$$

Let us now consider the alternative formulation of the governing differential equation according to Tale 7.2. This formulation assumes that the plate elasticity matrix $\boldsymbol{D}^*$ is constant and we obtain

$$\boldsymbol{\mathcal{L}}_{1*}^{\mathrm{T}}\boldsymbol{D}^*\boldsymbol{\mathcal{L}}_1\boldsymbol{u}^0 + \boldsymbol{b}^* = \boldsymbol{0}\,, \tag{7.47}$$

where $\boldsymbol{u}^0(x, y)$ represents the exact solution of the problem. The last equation which contains the exact solution of the problem is fulfilled at any location (x, y) of the plate and is called the *strong formulation* of the problem. Replacing the exact solution in Eq. (7.47) by an approximate solution $\boldsymbol{u}(x, y)$, a residual $\boldsymbol{r}$ is obtained:

$$\boldsymbol{r}(x, y) = \boldsymbol{\mathcal{L}}_{1*}^{\mathrm{T}}\boldsymbol{D}^*\boldsymbol{\mathcal{L}}_1\boldsymbol{u} + \boldsymbol{b}^* \neq \boldsymbol{0}\,. \tag{7.48}$$

It is again requested that the differential equation is fulfilled over a certain area (and no longer at any location (x, y)) and the following integral statement is obtained

$$\int_A \boldsymbol{W}^{\mathrm{T}} \left(\boldsymbol{\mathcal{L}}_{1^*}^{\mathrm{T}} \boldsymbol{D}^* \boldsymbol{\mathcal{L}}_1 \boldsymbol{u} + \boldsymbol{b}^* \right) \mathrm{d}A \stackrel{!}{=} 0 \,, \tag{7.49}$$

which is called the *inner product*. The column matrix $\boldsymbol{W}(\boldsymbol{x}) = \left[W_{\phi_y}\ W_{\phi_x}\ W_{u_z}\right]^{\mathrm{T}}$ in Eq. (7.34) contains again the weight functions and $\boldsymbol{x} = \left[x\ y\right]^{\mathrm{T}}$ is the column matrix of Cartesian coordinates. Application of the GREEN–GAUSS theorem (cf. Sect. A.7) would shift the derivative to the weight functions $\boldsymbol{W}$. However, the matrix of differential operators, $\boldsymbol{\mathcal{L}}_{1^*}^{\mathrm{T}}$, contains in addition to derivatives a constant value '1' and it is therefore appropriate to split the matrix into a part which contains all the derivatives, $\boldsymbol{\mathcal{L}}_{1^*,a}^{\mathrm{T}}$, and a part with the constant value, $\boldsymbol{\mathcal{L}}_{1^*,b}^{\mathrm{T}}$:

$$\underbrace{\begin{bmatrix} 0 & 0 & 0 & \frac{\partial}{\partial x} & \frac{\partial}{\partial y} \\ 0 & -\frac{\partial}{\partial y} & -\frac{\partial}{\partial x} & 0 & 1 \\ \frac{\partial}{\partial x} & 0 & \frac{\partial}{\partial y} & -1 & 0 \end{bmatrix}}_{\boldsymbol{\mathcal{L}}_{1^*}^{\mathrm{T}}} = \underbrace{\begin{bmatrix} 0 & 0 & 0 & \frac{\partial}{\partial x} & \frac{\partial}{\partial y} \\ 0 & -\frac{\partial}{\partial y} & -\frac{\partial}{\partial x} & 0 & 0 \\ \frac{\partial}{\partial x} & 0 & \frac{\partial}{\partial y} & 0 & 0 \end{bmatrix}}_{\boldsymbol{\mathcal{L}}_{1^*,a}^{\mathrm{T}}} + \underbrace{\begin{bmatrix} 0 & 0 & 0 & 0 & 0 \\ 0 & 0 & 0 & 0 & 1 \\ 0 & 0 & 0 & -1 & 0 \end{bmatrix}}_{\boldsymbol{\mathcal{L}}_{1^*,b}^{\mathrm{T}}} . \tag{7.50}$$

Thus, we can write the inner product as:

$$\int_A \boldsymbol{W}^{\mathrm{T}} \left[\left(\boldsymbol{\mathcal{L}}_{1^*,a}^{\mathrm{T}} + \boldsymbol{\mathcal{L}}_{1^*,b}^{\mathrm{T}} \right) \boldsymbol{D}^* \boldsymbol{\mathcal{L}}_1 \boldsymbol{u} + \boldsymbol{b}^* \right] \mathrm{d}A \stackrel{!}{=} 0 \,, \tag{7.51}$$

or

$$\int_A \boldsymbol{W}^{\mathrm{T}} \boldsymbol{\mathcal{L}}_{1^*,a}^{\mathrm{T}} \boldsymbol{D}^* \boldsymbol{\mathcal{L}}_1 \boldsymbol{u} \mathrm{d}A + \int_A \boldsymbol{W}^{\mathrm{T}} \boldsymbol{\mathcal{L}}_{1^*,b}^{\mathrm{T}} \boldsymbol{D}^* \boldsymbol{\mathcal{L}}_1 \boldsymbol{u} \mathrm{d}A + \int_A \boldsymbol{W}^{\mathrm{T}} \boldsymbol{b}^* \, \mathrm{d}A = 0 \,. \tag{7.52}$$

Application of the GREEN–GAUSS theorem to the first integral gives:

$$\int_A \boldsymbol{W}^{\mathrm{T}} \boldsymbol{\mathcal{L}}_{1^*,a}^{\mathrm{T}} \boldsymbol{D}^* \boldsymbol{\mathcal{L}}_1 \boldsymbol{u} \, \mathrm{d}A = - \int_A \left(\boldsymbol{\mathcal{L}}_{1^*,a} \boldsymbol{W} \right)^{\mathrm{T}} \boldsymbol{D}^* \left(\boldsymbol{\mathcal{L}}_1 \boldsymbol{u} \right) \mathrm{d}A + \\ + \int_s \boldsymbol{W}^{\mathrm{T}} \underbrace{\left(\boldsymbol{D}^* \boldsymbol{\mathcal{L}}_1 \boldsymbol{u} \right)^{\mathrm{T}} \boldsymbol{n}}_{(\boldsymbol{s}^*)^{\mathrm{T}}} \mathrm{d}s \,. \tag{7.53}$$

Thus, the weak formulation can be obtained as:

$$\int_A \left(\mathcal{L}_{1^*,a} \boldsymbol{W}\right)^{\mathrm{T}} \boldsymbol{D}^* \left(\mathcal{L}_1 \boldsymbol{u}\right) \mathrm{d}A - \int_A \boldsymbol{W}^{\mathrm{T}} \mathcal{L}_{1^*,b}^{\mathrm{T}} \boldsymbol{D}^* \mathcal{L}_1 \boldsymbol{u} \, \mathrm{d}A =$$
$$\int_s \boldsymbol{W}^{\mathrm{T}} (\boldsymbol{s}^*)^{\mathrm{T}} \boldsymbol{n} \mathrm{d}s + \int_A \boldsymbol{W}^{\mathrm{T}} \boldsymbol{b}^* \, \mathrm{d}A \,. \tag{7.54}$$

The last equation can be simplified (see the distributive law in the Appendix A.11.1) to

$$\int_A \left(\boldsymbol{W}^{\mathrm{T}} \mathcal{L}_{1^*,a}^{\mathrm{T}} - \boldsymbol{W}^{\mathrm{T}} \mathcal{L}_{1^*,b}^{\mathrm{T}}\right) \boldsymbol{D}^* \left(\mathcal{L}_1 \boldsymbol{u}\right) \mathrm{d}A = \int_s \boldsymbol{W}^{\mathrm{T}} (\boldsymbol{s}^*)^{\mathrm{T}} \boldsymbol{n} \mathrm{d}s + \int_A \boldsymbol{W}^{\mathrm{T}} \boldsymbol{b}^* \, \mathrm{d}A \,,$$
$$\int_A \boldsymbol{W}^{\mathrm{T}} \underbrace{\left(\mathcal{L}_{1^*,a}^{\mathrm{T}} - \mathcal{L}_{1^*,b}^{\mathrm{T}}\right)}_{\mathcal{L}_1^{\mathrm{T}}} \boldsymbol{D}^* \left(\mathcal{L}_1 \boldsymbol{u}\right) \mathrm{d}A = \int_s \boldsymbol{W}^{\mathrm{T}} (\boldsymbol{s}^*)^{\mathrm{T}} \boldsymbol{n} \mathrm{d}s + \int_A \boldsymbol{W}^{\mathrm{T}} \boldsymbol{b}^* \, \mathrm{d}A \,, \tag{7.55}$$

or finally

$$\int_A \left(\mathcal{L}_1 \boldsymbol{W}\right)^{\mathrm{T}} \boldsymbol{D}^* \left(\mathcal{L}_1 \boldsymbol{u}\right) \mathrm{d}A = \int_s \boldsymbol{W}^{\mathrm{T}} (\boldsymbol{s}^*)^{\mathrm{T}} \boldsymbol{n} \mathrm{d}s + \int_A \boldsymbol{W}^{\mathrm{T}} \boldsymbol{b}^* \, \mathrm{d}A \,. \tag{7.56}$$

The unknown deformation fields in an element are again approximated based on the approach

$$\boldsymbol{u}^{\mathrm{e}}(\boldsymbol{x}) = \boldsymbol{N}^{\mathrm{T}}(\boldsymbol{x}) \boldsymbol{u}_{\mathrm{p}}^{\mathrm{e}} \,,$$

and the same approach is adopted for the weight functions:

$$\boldsymbol{W}(\boldsymbol{x}) = \boldsymbol{N}^{\mathrm{T}}(\boldsymbol{x}) \delta \boldsymbol{u}_{\mathrm{p}} \,. \tag{7.57}$$

Introducing the nodal approaches for the deformations and weight functions in the weak formulation (7.54) results in the weak formulation on element level after elimination of the virtual deformations $\delta \boldsymbol{u}_{\mathrm{p}}$:

$$\int_A (\mathcal{L}_1 \boldsymbol{N}^{\mathrm{T}})^{\mathrm{T}} \boldsymbol{D}^* (\mathcal{L}_1 \boldsymbol{N}^{\mathrm{T}}) \mathrm{d}A \boldsymbol{u}_{\mathrm{p}}^{\mathrm{e}} = \int_s \boldsymbol{N} (\boldsymbol{s}^*)^{\mathrm{T}} \boldsymbol{n} \mathrm{d}s + \int_A \boldsymbol{N} \boldsymbol{b}^* \, \mathrm{d}A \,,$$

where the evaluation of the integrals on the left-hand side results in the elemental stiffness matrix $\boldsymbol{K}^{\mathrm{e}}$ and the evaluation of the two integrals[3] on the right-hand side results in the elemental load matrix $\boldsymbol{f}^{\mathrm{e}}$.

Thus, we can identify the following three element matrices from the principal finite element equation:

---

[3]The expression $(\boldsymbol{s}^*)^{\mathrm{T}} \boldsymbol{n}$ can be understood as the generalized tractions $\boldsymbol{t}^*$ as in the three-dimensional case [3].

$$\text{Stiffness matrix: } \boldsymbol{K}^{\mathrm{e}} = \int\limits_{A} \underbrace{(\boldsymbol{\mathcal{L}}_1 \boldsymbol{N}^{\mathrm{T}})^{\mathrm{T}}}_{\boldsymbol{B}} \boldsymbol{D}^* \underbrace{(\boldsymbol{\mathcal{L}}_1 \boldsymbol{N}^{\mathrm{T}})}_{\boldsymbol{B}^{\mathrm{T}}} \mathrm{d}A \,, \tag{7.58}$$

$$\text{Boundary force matrix: } \boldsymbol{f}_t^{\mathrm{e}} = \int\limits_{s} \boldsymbol{N}(\boldsymbol{s}^*)^{\mathrm{T}} \boldsymbol{n} \mathrm{d}s \,, \tag{7.59}$$

$$\text{Body force matrix: } \boldsymbol{f}_b^{\mathrm{e}} = \int\limits_{A} \boldsymbol{N}\boldsymbol{b}^* \mathrm{d}A \,. \tag{7.60}$$

Based on these abbreviations, the principal finite element equation for a single element can be written as:

$$\boldsymbol{K}^{\mathrm{e}} \boldsymbol{u}_{\mathrm{p}}^{\mathrm{e}} = \boldsymbol{f}_t^{\mathrm{e}} + \boldsymbol{f}_b^{\mathrm{e}} \,. \tag{7.61}$$

In the following, let us look at the $\boldsymbol{B}$-matrix according to Eq. (7.58), i.e. the matrix which contains the first-order derivatives of the interpolation functions. Application of the matrix of differential operators, i.e., $\boldsymbol{\mathcal{L}}_1$, according to Eq. (7.5) to the matrix of interpolation functions[4] gives:

$$\boldsymbol{B}^{\mathrm{T}} = \boldsymbol{\mathcal{L}}_1 \boldsymbol{N}^{\mathrm{T}} = \begin{bmatrix} 0 & 0 & \frac{\partial}{\partial x} \\ 0 & -\frac{\partial}{\partial y} & 0 \\ 0 & -\frac{\partial}{\partial x} & \frac{\partial}{\partial y} \\ \frac{\partial}{\partial x} & 0 & 1 \\ \frac{\partial}{\partial y} & -1 & 0 \end{bmatrix} \left[\begin{array}{ccc|c|ccc} N_1 & 0 & 0 & \cdots & N_n & 0 & 0 \\ 0 & N_1 & 0 & \cdots & 0 & N_n & 0 \\ 0 & 0 & N_1 & \cdots & 0 & 0 & N_n \end{array} \right] = \tag{7.62}$$

$$= \left[\begin{array}{ccc|c|ccc} 0 & 0 & \frac{\partial N_1}{\partial x} & & 0 & 0 & \frac{\partial N_n}{\partial x} \\ 0 & -\frac{\partial N_1}{\partial y} & 0 & & 0 & -\frac{\partial N_n}{\partial y} & 0 \\ 0 & -\frac{\partial N_1}{\partial x} & \frac{\partial N_1}{\partial y} & \cdots & 0 & -\frac{\partial N_n}{\partial x} & \frac{\partial N_n}{\partial y} \\ \frac{\partial N_1}{\partial x} & 0 & N_1 & & \frac{\partial N_n}{\partial x} & 0 & N_n \\ \frac{\partial N_1}{\partial y} & -N_1 & 0 & & \frac{\partial N_n}{\partial y} & -N_n & 0 \end{array} \right] = \boldsymbol{B}^{\mathrm{T}} \,, \tag{7.63}$$

[4]Let us assume here that $N_{\phi_y} = N_{\phi_x} = N_u$.

which is a $(5 \times 3n)$-matrix. The transposed, i.e. $(\boldsymbol{\mathcal{L}}_1 \boldsymbol{N}^{\mathrm{T}})^{\mathrm{T}}$, is thus a $(3n \times 5)$-matrix. Multiplication with the elasticity matrix, i.e. a (5×5)-matrix, results in $(\boldsymbol{\mathcal{L}}_1 \boldsymbol{N}^{\mathrm{T}})^{\mathrm{T}} \boldsymbol{D}$, which is a $(3n \times 5)$-matrix. The final multiplication, i.e. $(\boldsymbol{\mathcal{L}}_1 \boldsymbol{N}^{\mathrm{T}})^{\mathrm{T}} \boldsymbol{D} (\boldsymbol{\mathcal{L}}_1 \boldsymbol{N}^{\mathrm{T}})$ gives after integration the stiffness matrix with a dimension of $(3n \times 3n)$.

The integrations for the element matrices given in Eqs. (7.58)–(7.60) are approximated by numerical integration. To this end, the coordinates (x, y) are transformed to the natural coordinates (unit space) (ξ, η) where each coordinate ranges from -1 to 1. In the scope of the coordinate transformation, attention must be paid to the derivatives. For example, the derivative of the interpolation functions with respect to the x-coordinate is transformed in the following way:

$$\frac{\partial N_i}{\partial x} \rightarrow \frac{\partial N_i}{\partial \xi}\frac{\partial \xi}{\partial x} + \frac{\partial N_i}{\partial \eta}\frac{\partial \eta}{\partial x}. \tag{7.64}$$

Furthermore, the coordinate transformation requires that $\mathrm{d}A = \mathrm{d}x\mathrm{d}y \rightarrow \mathrm{d}A' = J\mathrm{d}\xi\mathrm{d}\eta$, where J is the Jacobian as given in the Appendix A.8.

7.3.2 *Rectangular Four-Node Plate Element*

This section will exemplarily focus on the simplest formulation of a four-node element, i.e., a formulation based on bilinear interpolation functions. More advanced topics such as shear locking, higher order formulations or integration techniques are not covered here and the interested reader should consult the literature [2, 5].

The schematic representation of a rectangular four-node thick plate element is given in Fig. 7.3 in the Cartesian and parametric space. As in the case of the other two-dimensional elements, the node numbering is counter-clockwise and each node has now three degrees of freedom.

Assuming bilinear interpolation functions, the relations from Chap. 5 can be taken, see Eqs. (5.57)–(5.60). Thus, the interpolations functions read as

$$N_1(\xi, \eta) = \frac{1}{4}(1 - \xi - \eta + \xi\eta) = \frac{1}{4}(1 - \xi)\,(1 - \eta)\,, \tag{7.65}$$

$$N_2(\xi, \eta) = \frac{1}{4}(1 + \xi - \eta - \xi\eta) = \frac{1}{4}(1 + \xi)\,(1 - \eta)\,, \tag{7.66}$$

$$N_3(\xi, \eta) = \frac{1}{4}(1 + \xi + \eta + \xi\eta) = \frac{1}{4}(1 + \xi)\,(1 + \eta)\,, \tag{7.67}$$

$$N_4(\xi, \eta) = \frac{1}{4}(1 - \xi + \eta - \xi\eta) = \frac{1}{4}(1 - \xi)\,(1 + \eta)\,. \tag{7.68}$$

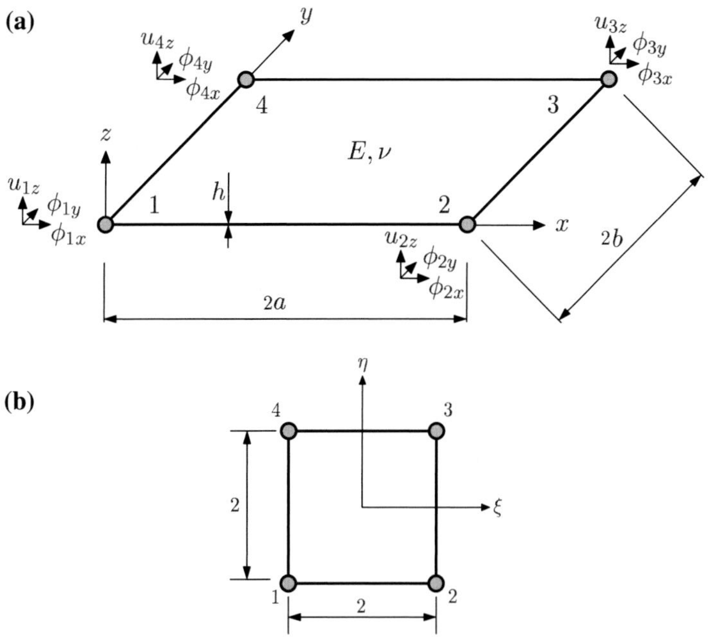

Fig. 7.3 Rectangular four-node thick plate element: **a** Cartesian and **b** parametric space

The derivatives with respect to the parametric coordinates can easily be obtained as:

$$\frac{\partial N_1(\xi,\eta)}{\partial \xi} = \frac{1}{4}(-1+\eta) \quad ; \quad \frac{\partial N_1(\xi,\eta)}{\partial \eta} = \frac{1}{4}(-1+\xi) \,, \tag{7.69}$$

$$\frac{\partial N_2(\xi,\eta)}{\partial \xi} = \frac{1}{4}(+1-\eta) \quad ; \quad \frac{\partial N_2(\xi,\eta)}{\partial \eta} = \frac{1}{4}(-1-\xi) \,, \tag{7.70}$$

$$\frac{\partial N_3(\xi,\eta)}{\partial \xi} = \frac{1}{4}(+1+\eta) \quad ; \quad \frac{\partial N_3(\xi,\eta)}{\partial \eta} = \frac{1}{4}(+1+\xi) \,, \tag{7.71}$$

$$\frac{\partial N_4(\xi,\eta)}{\partial \xi} = \frac{1}{4}(-1-\eta) \quad ; \quad \frac{\partial N_4(\xi,\eta)}{\partial \eta} = \frac{1}{4}(+1-\xi) \,. \tag{7.72}$$

Geometrical Derivatives

Let us assume the same interpolation for the global x- and y-coordinate as for the displacement (*isoparametric* element formulation), i.e. $\overline{N}_i = N_i$:

$$x(\xi,\eta) = \overline{N}_1(\xi,\eta) \times x_1 + \overline{N}_2(\xi,\eta) \times x_2 + \overline{N}_3(\xi,\eta) \times x_3 + \overline{N}_4(\xi,\eta) \times x_4 \,, \tag{7.73}$$

$$y(\xi,\eta) = \overline{N}_1(\xi,\eta) \times y_1 + \overline{N}_2(\xi,\eta) \times y_2 + \overline{N}_3(\xi,\eta) \times y_3 + \overline{N}_4(\xi,\eta) \times y_4 \,. \tag{7.74}$$

Remark: the *global* coordinates of the nodes $1, \ldots, 4$ can be used for $x_1, \ldots, x_4$ and $y_1, \ldots, y_4$.

Thus, the geometrical derivatives can easily be obtained as:

$$\frac{\partial x}{\partial \xi} = \frac{1}{4}\Big((-1+\eta)x_1 + (1-\eta)x_2 + (1+\eta)x_3 + (-1-\eta)x_4\Big)\,, \tag{7.75}$$

$$\frac{\partial y}{\partial \xi} = \frac{1}{4}\Big((-1+\eta)y_1 + (1-\eta)y_2 + (1+\eta)y_3 + (-1-\eta)y_4\Big)\,, \tag{7.76}$$

$$\frac{\partial x}{\partial \eta} = \frac{1}{4}\Big((-1+\xi)x_1 + (-1-\xi)x_2 + (1+\xi)x_3 + (1-\xi)x_4\Big)\,, \tag{7.77}$$

$$\frac{\partial y}{\partial \eta} = \frac{1}{4}\Big((-1+\xi)y_1 + (-1-\xi)y_2 + (1+\xi)y_3 + (1-\xi)y_4\Big)\,. \tag{7.78}$$

The calculation of the derivatives of the interpolation functions, i.e.

$$\frac{\partial N_i}{\partial x} \rightarrow \frac{\partial N_i}{\partial \xi}\frac{\partial \xi}{\partial x} + \frac{\partial N_i}{\partial \eta}\frac{\partial \eta}{\partial x}\,, \tag{7.79}$$

requires, however, the geometrical derivatives of the natural coordinates (ξ, η) with respect to the physical coordinates (x, y). These relations can be easily obtained from Eqs. (7.75)–(7.78) under consideration of the relationships provided in Sect. A.8:

$$\frac{\partial \xi}{\partial x} = +\frac{1}{\frac{\partial x}{\partial \xi}\frac{\partial y}{\partial \eta} - \frac{\partial x}{\partial \eta}\frac{\partial y}{\partial \xi}} \times \frac{\partial y}{\partial \eta}\,, \tag{7.80}$$

$$\frac{\partial \xi}{\partial y} = -\frac{1}{\frac{\partial x}{\partial \xi}\frac{\partial y}{\partial \eta} - \frac{\partial x}{\partial \eta}\frac{\partial y}{\partial \xi}} \times \frac{\partial x}{\partial \eta}\,, \tag{7.81}$$

$$\frac{\partial \eta}{\partial x} = -\frac{1}{\frac{\partial x}{\partial \xi}\frac{\partial y}{\partial \eta} - \frac{\partial x}{\partial \eta}\frac{\partial y}{\partial \xi}} \times \frac{\partial y}{\partial \xi}\,, \tag{7.82}$$

$$\frac{\partial \eta}{\partial y} = +\frac{1}{\frac{\partial x}{\partial \xi}\frac{\partial y}{\partial \eta} - \frac{\partial x}{\partial \eta}\frac{\partial y}{\partial \xi}} \times \frac{\partial x}{\partial \xi}\,. \tag{7.83}$$

Based on the derived equations, the triple matrix product $\boldsymbol{B}\boldsymbol{C}^{*}\boldsymbol{B}^{\mathrm{T}}$ (see Eq. (7.58)) can be numerically calculated to obtain the stiffness matrix.

Let us summarize here the major steps which are required to calculate the elemental stiffness matrix.

❶ Introduce an elemental coordinate system (x, y).
❷ Express the coordinates (x_i, y_i) of the corner nodes i $(i = 1, \ldots, 4)$ in this elemental coordinate system.
❸ Calculate the partial derivatives of the Cartesian (x, y) coordinates with respect to the natural (ξ, η) coordinates, see Eqs. (7.75)–(7.78):

$$\frac{\partial x}{\partial \xi} = x_{\xi} = \frac{1}{4}\Big((-1+\eta)x_1 + (1-\eta)x_2 + (1+\eta)x_3 + (-1-\eta)x_4\Big),$$

$$\vdots$$

$$\frac{\partial y}{\partial \eta} = y_{\eta} = \frac{1}{4}\Big((-1+\xi)y_1 + (-1-\xi)y_2 + (1+\xi)y_3 + (1-\xi)y_4\Big).$$

❹ Calculate the first-order partial derivatives of the natural (ξ, η) coordinates with respect to the Cartesian (x, y) coordinates according to Eqs. (7.80)–(7.83):

$$\frac{\partial \xi}{\partial x} = +\frac{1}{\frac{\partial x}{\partial \xi}\frac{\partial y}{\partial \eta} - \frac{\partial x}{\partial \eta}\frac{\partial y}{\partial \xi}} \times \frac{\partial y}{\partial \eta}, \quad \frac{\partial \xi}{\partial y} = -\frac{1}{\frac{\partial x}{\partial \xi}\frac{\partial y}{\partial \eta} - \frac{\partial x}{\partial \eta}\frac{\partial y}{\partial \xi}} \times \frac{\partial x}{\partial \eta},$$

$$\frac{\partial \eta}{\partial x} = -\frac{1}{\frac{\partial x}{\partial \xi}\frac{\partial y}{\partial \eta} - \frac{\partial x}{\partial \eta}\frac{\partial y}{\partial \xi}} \times \frac{\partial y}{\partial \xi}, \quad \frac{\partial \eta}{\partial y} = +\frac{1}{\frac{\partial x}{\partial \xi}\frac{\partial y}{\partial \eta} - \frac{\partial x}{\partial \eta}\frac{\partial y}{\partial \xi}} \times \frac{\partial x}{\partial \xi}.$$

❺ Calculate the $\boldsymbol{B}$-matrix and its transposed, see Eq. (7.63):

$$\boldsymbol{B}^{\mathrm{T}} = \left[\begin{array}{ccc|c|ccc} 0 & 0 & \frac{\partial N_1}{\partial x} & & 0 & 0 & \frac{\partial N_4}{\partial x} \\ 0 & -\frac{\partial N_1}{\partial y} & 0 & & 0 & -\frac{\partial N_4}{\partial y} & 0 \\ 0 & -\frac{\partial N_1}{\partial x} & \frac{\partial N_1}{\partial y} & \cdots & 0 & -\frac{\partial N_4}{\partial x} & \frac{\partial N_4}{\partial y} \\ \frac{\partial N_1}{\partial x} & 0 & N_1 & & \frac{\partial N_4}{\partial x} & 0 & N_4 \\ \frac{\partial N_1}{\partial y} & -N_1 & 0 & & \frac{\partial N_4}{\partial y} & -N_4 & 0 \end{array}\right],$$

where the first-order partial derivatives are $\frac{\partial N_1(\xi,\eta)}{\partial x} = \frac{\partial N_1}{\partial \xi}\frac{\partial \xi}{\partial x} + \frac{\partial N_1}{\partial \eta}\frac{\partial \eta}{\partial x}, \ldots$ and the derivatives of the interpolation functions are given in Eqs. (7.65)–(7.68), i.e., $\frac{\partial N_1}{\partial \xi} = \frac{1}{4}(-1+\eta), \ldots$
❻ Calculate the triple matrix product $\boldsymbol{B}\boldsymbol{D}^{*}\boldsymbol{B}^{\mathrm{T}}$, where the plate elasticity matrix $\boldsymbol{D}^{*}$ is given in Table 7.2.
❼ Perform the numerical integration based on n integration points:

$$\int_A (\boldsymbol{B}\boldsymbol{D}^{*}\boldsymbol{B}^{\mathrm{T}})\mathrm{d}A = \sum_{i=1}^{n} \boldsymbol{B}\boldsymbol{D}^{*}\boldsymbol{B}^{\mathrm{T}} J \times w_i\Big|_{\xi_i,\eta_i}.$$

❽ $\boldsymbol{K}$ obtained.

Table 7.3 Summary: derivation of principal finite element equation for thick plate elements (basic equations according to Table 7.2)

Strong formulation
$\mathcal{L}_{1*}^{\mathrm{T}} \boldsymbol{D}^* \mathcal{L}_1 \boldsymbol{u}^0 + \boldsymbol{b}^* = \boldsymbol{0}$
Inner product
$\int_A \boldsymbol{W}^{\mathrm{T}} \left(\mathcal{L}_{1*}^{\mathrm{T}} \boldsymbol{D}^* \mathcal{L}_1 \boldsymbol{u} + \boldsymbol{b}^* \right) \mathrm{d}A = 0$
Weak formulation
$\int_A (\mathcal{L}_1 \boldsymbol{W})^{\mathrm{T}} \boldsymbol{D}^* (\mathcal{L}_1 \boldsymbol{u}) \, \mathrm{d}A = \int_s \boldsymbol{W}^{\mathrm{T}} (\boldsymbol{s}^*)^{\mathrm{T}} \boldsymbol{n} \mathrm{d}s + \int_A \boldsymbol{W}^{\mathrm{T}} \boldsymbol{b}^* \mathrm{d}A$
Principal finite element equation (quad 4 with 12 DOF)
$\underbrace{\int_A \underbrace{\left(\mathcal{L}_1 \boldsymbol{N}^{\mathrm{T}}\right)^{\mathrm{T}}}_{\boldsymbol{B}} \boldsymbol{D}^* \underbrace{\left(\mathcal{L}_1 \boldsymbol{N}^{\mathrm{T}}\right)}_{\boldsymbol{B}^{\mathrm{T}}} \mathrm{d}A}_{\boldsymbol{K}^{\mathrm{e}}} \begin{bmatrix} u_{1z} \\ \phi_{1x} \\ \phi_{1y} \\ \vdots \\ u_{4z} \\ \phi_{4x} \\ \phi_{4y} \end{bmatrix} = \begin{bmatrix} F_{1z} \\ M_{1x} \\ M_{1y} \\ \vdots \\ F_{4z} \\ M_{4x} \\ M_{4y} \end{bmatrix} + \int_A \boldsymbol{N} \boldsymbol{b}^* \mathrm{d}A$

Let us summarize at the end of this section the major steps that were undertaken to transform the partial differential equation into the principal finite element equation, see Table 7.3.

7.3.3 *Solved Thick Plate Element Problems*

7.1 Example: Stiffness matrix for a rectangular four-node thick plate element

Given is a rectangular four-node thick plate element with dimensions $a \times b \times h$ as shown in Fig. 7.4.

Derive the expression for the elemental stiffness matrix based on (a) reduced one-point numerical integration and (b) analytical integration.

7.1 Solution

Following the eight steps outlined on p. 378, the following intermediate results can be mentioned:

Derivatives of Cartesian coordinates with respect to parametric coordinates:

$$\frac{\partial x}{\partial \xi} = \frac{a}{2}, \qquad \frac{\partial y}{\partial \xi} = 0, \tag{7.84}$$

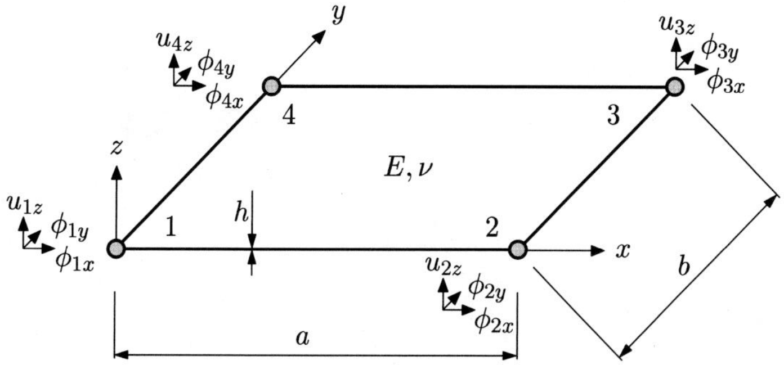

Fig. 7.4 Rectangular four-node thick plate element with dimensions $a \times b \times h$

$$\frac{\partial x}{\partial \eta} = 0\,, \qquad \frac{\partial y}{\partial \eta} = \frac{b}{2}. \tag{7.85}$$

Derivatives of parametric coordinates with respect to Cartesian coordinates:

$$\frac{\partial \xi}{\partial x} = \frac{2}{a}, \qquad \frac{\partial \xi}{\partial y} = 0\,, \tag{7.86}$$

$$\frac{\partial \eta}{\partial x} = 0\,, \qquad \frac{\partial \eta}{\partial y} = \frac{2}{b}. \tag{7.87}$$

Thus, the derivatives of the shape functions with respect to the Cartesian coordinates are given as follow:

$$\frac{\partial N_1}{\partial x} = \frac{\eta - 1}{2a}, \qquad \frac{\partial N_1}{\partial y} = \frac{\xi - 1}{2b}, \tag{7.88}$$

$$\frac{\partial N_2}{\partial x} = \frac{1 - \eta}{2a}, \qquad \frac{\partial N_2}{\partial y} = \frac{-\xi - 1}{2b}, \tag{7.89}$$

$$\frac{\partial N_3}{\partial x} = \frac{\eta + 1}{2a}, \qquad \frac{\partial N_3}{\partial y} = \frac{\xi + 1}{2b}, \tag{7.90}$$

$$\frac{\partial N_4}{\partial x} = \frac{-\eta - 1}{2a}, \qquad \frac{\partial N_4}{\partial y} = \frac{1 - \xi}{2b}. \tag{7.91}$$

Based on these results, the $\boldsymbol{B}$-matrix is obtained as shown in Eq. (F.364):

Table 7.4 Selected elements of the elemental stiffness matrix

K_{ij}	One-point integration	Analytical integration
K_{11}	$\frac{D_s b^2 + D_s a^2}{4ab}$	$\frac{D_s b^2 + D_s a^2}{3ab}$
K_{12}	$\frac{D_s a}{8}$	$\frac{D_s a}{6}$
K_{22}	$ab\left(\frac{D_b(1-\nu)}{8a^2}+\frac{1}{4b^2}+\frac{D_s}{16}\right)$	$ab\left(\frac{D_b(1-\nu)}{6a^2}+\frac{1}{3b^2}+\frac{D_s}{9}\right)$

$$
\boldsymbol{B}^{\mathrm{T}} =
\left[\begin{array}{cccccccccccc}
0 & 0 & \frac{\eta-1}{2a} & 0 & 0 & \frac{1-\eta}{2a} & 0 & 0 & \frac{\eta+1}{2a} & 0 & 0 & \frac{-\eta-1}{2a} \\
0 & -\frac{\xi-1}{2b} & 0 & 0 & -\frac{-\xi-1}{2b} & 0 & 0 & -\frac{\xi+1}{2b} & 0 & 0 & -\frac{1-\xi}{2b} & 0 \\
0 & -\frac{\eta-1}{2a} & \frac{\xi-1}{2b} & 0 & -\frac{1-\eta}{2a} & \frac{-\xi-1}{2b} & 0 & -\frac{\eta+1}{2a} & \frac{\xi+1}{2b} & 0 & -\frac{-\eta-1}{2a} & \frac{1-\xi}{2b} \\
\frac{\eta-1}{2a} & 0 & \frac{(1-\eta)(1-\xi)}{4} & \frac{1-\eta}{2a} & 0 & \frac{(1-\eta)(\xi+1)}{4} & \frac{\eta+1}{2a} & 0 & \frac{(\eta+1)(\xi+1)}{4} & \frac{-\eta-1}{2a} & 0 & \frac{(\eta+1)(1-\xi)}{4} \\
\frac{\xi-1}{2b} & -\frac{(1-\eta)(1-\xi)}{4} & 0 & \frac{-\xi-1}{2b} & -\frac{(1-\eta)(\xi+1)}{4} & 0 & \frac{\xi+1}{2b} & -\frac{(\eta+1)(\xi+1)}{4} & 0 & \frac{1-\xi}{2b} & -\frac{(\eta+1)(1-\xi)}{4} & 0
\end{array}\right]. \tag{7.92}
$$

Some of the elements of the elemental stiffness matrix are collected in Table 7.4.

7.4 Supplementary Problems

7.2 Knowledge questions on plate elements

- State the required (a) geometrical parameters and (b) material parameters to define a rectangular thick plate element.
- State the DOF per node for a rectangular thick plate element.
- Which loads can be applied to a thick plate element?
- State possible advantages to model a beam bending problem with plate elements and not with 1D beam elements.
- Which stress state is assumed for a thick plate?

7.3 Alternative definition of rotational angle

The rotational angle ϕ_x can be introduced in the xz-plane (see Fig. 7.5a) and ϕ_y in the yz-plane (see Fig. 7.5b). The angle ϕ_x is then considered positive if it leads to a

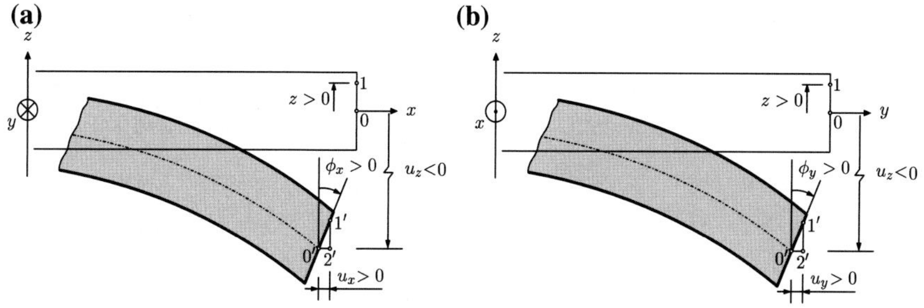

Fig. 7.5 Alternative definition of rotational angle: **a** xz-plane and **b** yz-plane

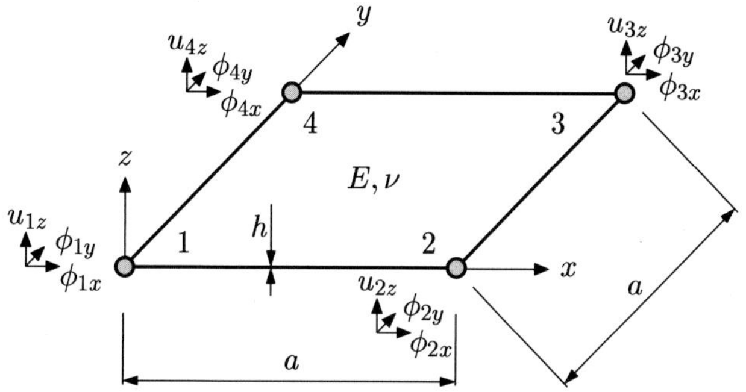

Fig. 7.6 Square four-node thick plate element with dimensions $a \times a \times h$

positive displacement u_x at positive z-side of the neutral fibre. In the same sense, ϕ_y is considered positive if it leads to a positive displacement u_y at positive z-side of the neutral fibre.

Derive the relationship between the displacement u_x and the corresponding rotational angle ϕ_x. Repeat the derivations for the displacement u_y.

7.4 Basic equations for alternative definition of rotational angle

Consider the definition of the rotational angles ϕ_x and ϕ_y as indicated in Fig. 7.5. Derive the basic equations of continuum mechanics, i.e., the kinematics, the constitutive and the equilibrium equations. Combine these equations to the corresponding differential equations.

7.5 Stiffness matrix for a square four-node thick plate element

Given is a square four-node thick plate element with dimensions $a \times a \times h$ as shown in Fig. 7.6.

Derive the expression for the elemental stiffness matrix based on (a) reduced one-point numerical integration and (b) analytical integration.

References

1. Blaauwendraad J (2010) Plates and FEM: surprises and pitfalls. Springer, Dordrecht
2. Cook RD, Malkus DS, Plesha ME, Witt RJ (2002) Concepts and applications of finite element analysis. Wiley, New York
3. Hjelmstad DK (2005) Fundamentals of structural mechanics. Springer, New York
4. Mindlin RD (1951) Influence of rotary inertia and shear on flexural motions isotropic, elastic plates. J Appl Mech-T ASME 18:1031–1036
5. Reddy JN (2006) An introduction to the finite element method. McGraw Hill, Singapore
6. Reissner E (1945) The effect of transverse shear deformation on the bending of elastic plates. J Appl Mech-T ASME 12:A68–A77
7. Ventsel E, Krauthammer T (2001) Thin plates and shells: theory, analysis, and applications. Marcel Dekker, New York
8. Wang CM, Reddy JN, Lee KH (2000) Shear deformable beams and plates: relationships with classical solution. Elsevier, Oxford

Chapter 8
Three-Dimensional Elements

Abstract This chapter starts with the analytical description of solid or three-dimensional members. Based on the three basic equations of continuum mechanics, i.e., the kinematics relationship, the constitutive law, and the equilibrium equation, the partial differential equation, which describes the physical problem, is derived. The weighted residual method is then used to derive the principal finite element equation for solid elements. The chapter exemplarily treats an eight-node trilinear hexahedron (hex 8)—solid element.

8.1 Derivation of the Governing Differential Equation

A solid element is defined as a three-dimensional member, as schematically shown in Fig. 8.1, where all dimensions have a similar magnitude. It can be seen as a three-dimensional extension or generalization of the plane elasticity element. The following derivations are restricted to some simplifications:

- the material is isotropic, homogenous and linear-elastic according to HOOKE's law for a three-dimensional stress and strain state,
- only members with 6 faces, 12 edges, 8 vertices (hexahedra) considered

The analogies between the rod, plane elasticity theories and three-dimensional elements are summarized in Table 8.1.

8.1.1 Kinematics

The kinematics or strain-displacement relations extract the strain field contained in a displacement field. Using engineering definitions of strain, the following relations can be obtained [1, 3]:

A. Öchsner, *Computational Statics and Dynamics*,
https://doi.org/10.1007/978-981-15-1278-0_8

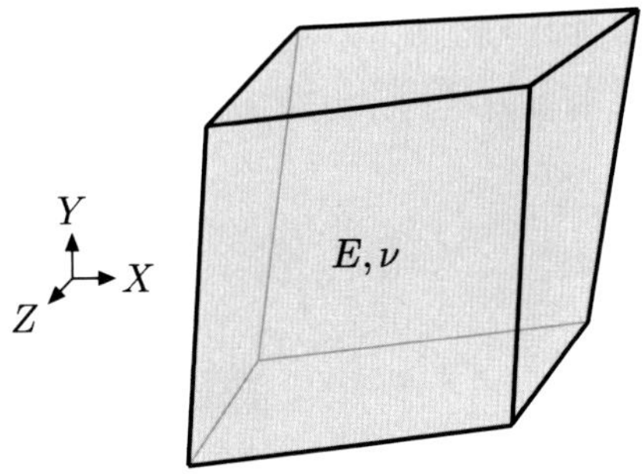

Fig. 8.1 General configuration for a three-dimensional problem

Table 8.1 Difference between rod, plane and three-dimensional element

Rod	Plane element	3D Element
1D	2D	3D
Deformation along principal axis	In-plane deformation	Spatial deformation
u_x	u_x, u_y	u_x, u_y, u_z

$$\varepsilon_x = \frac{\partial u_x}{\partial x}\,;\ \varepsilon_y = \frac{\partial u_y}{\partial y}\,;\ \varepsilon_z = \frac{\partial u_z}{\partial z}\,;\ \varepsilon_{xy} = \frac{1}{2}\left(\frac{\partial u_x}{\partial y} + \frac{\partial u_y}{\partial x}\right)\,; \tag{8.1}$$

$$\varepsilon_{xz} = \frac{1}{2}\left(\frac{\partial u_x}{\partial z} + \frac{\partial u_z}{\partial x}\right)\,;\ \varepsilon_{yz} = \frac{1}{2}\left(\frac{\partial u_y}{\partial z} + \frac{\partial u_z}{\partial y}\right)\,. \tag{8.2}$$

In matrix notation, these six relationships can be written as

$$\begin{bmatrix} \varepsilon_x \\ \varepsilon_y \\ \varepsilon_z \\ 2\varepsilon_{xy} \\ 2\varepsilon_{yz} \\ 2\varepsilon_{xz} \end{bmatrix} = \begin{bmatrix} \frac{\partial}{\partial x} & 0 & 0 \\ 0 & \frac{\partial}{\partial y} & 0 \\ 0 & 0 & \frac{\partial}{\partial z} \\ \frac{\partial}{\partial y} & \frac{\partial}{\partial x} & 0 \\ 0 & \frac{\partial}{\partial z} & \frac{\partial}{\partial y} \\ \frac{\partial}{\partial z} & 0 & \frac{\partial}{\partial x} \end{bmatrix} \begin{bmatrix} u_x \\ u_y \\ u_z \end{bmatrix}, \tag{8.3}$$

or symbolically as

$$\boldsymbol{\varepsilon} = \boldsymbol{\mathcal{L}}_1 \boldsymbol{u}\,. \tag{8.4}$$

8.1.2 Constitutive Equation

The generalized HOOKE's law for a linear-elastic isotropic material based on the YOUNG's modulus E and POISSON's ratio ν can be written for a constant temperature with all components as

$$\begin{bmatrix} \sigma_x \\ \sigma_y \\ \sigma_z \\ \sigma_{xy} \\ \sigma_{yz} \\ \sigma_{xz} \end{bmatrix} = \frac{E}{(1+\nu)(1-2\nu)} \begin{bmatrix} 1-\nu & \nu & \nu & 0 & 0 & 0 \\ \nu & 1-\nu & \nu & 0 & 0 & 0 \\ \nu & \nu & 1-\nu & 0 & 0 & 0 \\ 0 & 0 & 0 & \frac{1-2\nu}{2} & 0 & 0 \\ 0 & 0 & 0 & 0 & \frac{1-2\nu}{2} & 0 \\ 0 & 0 & 0 & 0 & 0 & \frac{1-2\nu}{2} \end{bmatrix} \begin{bmatrix} \varepsilon_x \\ \varepsilon_y \\ \varepsilon_z \\ 2\,\varepsilon_{xy} \\ 2\,\varepsilon_{yz} \\ 2\,\varepsilon_{xz} \end{bmatrix}, \tag{8.5}$$

or in matrix notation as

$$\boldsymbol{\sigma} = \boldsymbol{C}\boldsymbol{\epsilon}, \tag{8.6}$$

where $\boldsymbol{C}$ is the so-called elasticity matrix. It should be noted here that the engineering shear strain $\gamma_{ij} = 2\varepsilon_{ij}$ (for $i \neq j$) is used in the formulation of Eq. (8.5), see Sect. 4.1 for further details. Rearranging the elastic stiffness form given in Eq. (8.5) for the strains gives the elastic compliance form

$$\begin{bmatrix} \varepsilon_x \\ \varepsilon_y \\ \varepsilon_z \\ 2\,\varepsilon_{xy} \\ 2\,\varepsilon_{yz} \\ 2\,\varepsilon_{xz} \end{bmatrix} = \frac{1}{E} \begin{bmatrix} 1 & -\nu & -\nu & 0 & 0 & 0 \\ -\nu & 1 & -\nu & 0 & 0 & 0 \\ -\nu & -\nu & 1 & 0 & 0 & 0 \\ 0 & 0 & 0 & 2(1+\nu) & 0 & 0 \\ 0 & 0 & 0 & 0 & 2(1+\nu) & 0 \\ 0 & 0 & 0 & 0 & 0 & 2(1+\nu) \end{bmatrix} \begin{bmatrix} \sigma_x \\ \sigma_y \\ \sigma_z \\ \sigma_{xy} \\ \sigma_{yz} \\ \sigma_{xz} \end{bmatrix}, \tag{8.7}$$

or in matrix notation as

$$\boldsymbol{\epsilon} = \boldsymbol{D}\boldsymbol{\sigma}, \tag{8.8}$$

where $\boldsymbol{D} = \boldsymbol{C}^{-1}$ is the so-called elastic compliance matrix. The general characteristic of HOOKE's law in the form of Eqs. (8.6) and (8.8) is that two independent material parameters are used. In addition to YOUNG's modulus E and POISSON's ratio ν, other elastic parameters can be used to form the set of two independent material parameters and the following Table 8.2 summarizes the conversion between the common material parameters.

8.1.3 Equilibrium

Figure 8.2 shows the normal and shear stresses which are acting on a differential volume element in the x-direction. All forces are drawn in their positive direction at each cut face. A positive cut face is obtained if the outward surface normal is directed in the positive direction of the corresponding coordinate axis. This means that the right-hand face in Fig. 8.2 is positive and the force $(\sigma_x + \frac{\partial \sigma_x}{\partial x}\mathrm{d}x)\mathrm{d}y\mathrm{d}z$ is oriented in the positive x-direction. In a similar way, the top face is positive, i.e. the outward

Table 8.2 Conversion of elastic constants: λ, μ: LAMÉ's constants; K: bulk modulus; G: shear modulus; E: YOUNG's modulus; ν: POISSON's ratio [2]

	λ, μ	E, ν	μ, ν	E, μ	K, ν	G, ν	K, G
λ	λ	$\frac{\nu E}{(1+\nu)(1-2\nu)}$	$\frac{2\mu\nu}{1-2\nu}$	$\frac{\mu(E-2\mu)}{3\mu-E}$	$\frac{3K\nu}{1+\nu}$	$\frac{2G\nu}{1-2\nu}$	$K - \frac{2G}{3}$
μ	μ	$\frac{E}{2(1+\nu)}$	μ	μ	$\frac{3K(1-2\nu)}{2(1+\nu)}$	μ	μ
K	$\lambda + \frac{2}{3}\mu$	$\frac{E}{3(1-2\nu)}$	$\frac{2\mu(1+\nu)}{3(1-2\nu)}$	$\frac{\mu E}{3(3\mu-E)}$	K	$\frac{2G(1+\nu)}{3(1-2\nu)}$	K
E	$\frac{\mu(3\lambda+2\mu)}{\lambda+\mu}$	E	$2\mu(1+\nu)$	E	$3K(1-2\nu)$	$2G(1+\nu)$	$\frac{9KG}{3K+G}$
ν	$\frac{\lambda}{2(\lambda+\mu)}$	ν	ν	$\frac{E}{2\mu} - 1$	ν	ν	$\frac{3K-2G}{2(3K+G)}$
G	μ	$\frac{E}{2(1+\nu)}$	μ	G	$\frac{3K(1-2\nu)}{2(1+\nu)}$	G	G

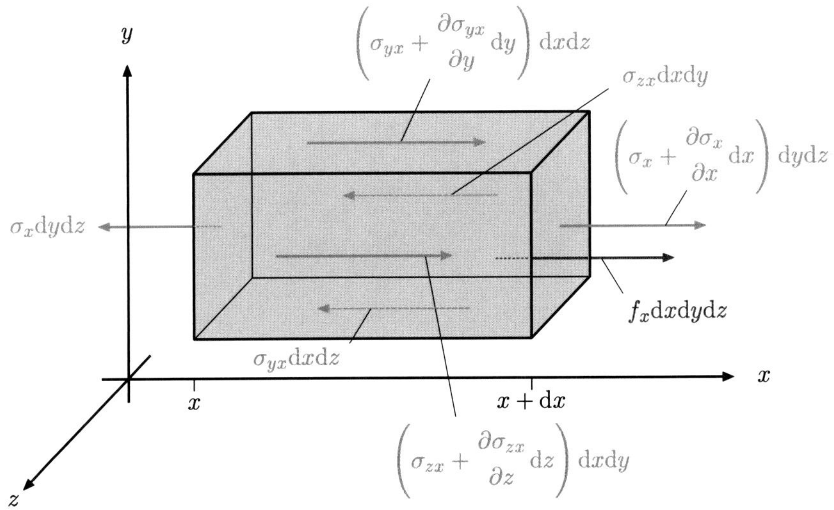

Fig. 8.2 Stress and body forces which act on a differential volume element in x-direction

surface normal is directed in the positive y-direction, and the shear force[1] is oriented in the positive x-direction. Since the volume element is assumed to be in equilibrium, forces resulting from stresses on the sides of the cuboid and from the body forces f_i $(i = x, y, z)$ must be balanced. These body forces are defined as forces per unit volume which can be produced by gravity,[2] acceleration, magnetic fields, and so on.

[1]In the case of a shear force σ_{ij}, the first index i indicates that the stress acts on a plane normal to the i-axis and the second index j denotes the direction in which the stress acts.

[2]If gravity is acting, the body force f results as the product of density times standard gravity: $f = \frac{F}{V} = \frac{mg}{V} = \frac{m}{V} g = \varrho g$. The units can be checked by consideration of $1\,\mathrm{N} = 1\,\frac{\mathrm{mkg}}{\mathrm{s}^2}$.

The static equilibrium of forces in the x-direction based on the seven force components—two normal forces, four shear forces and one body force—indicated in Fig. 8.2 gives after canceling with $\mathrm{d}V = \mathrm{d}x\mathrm{d}y\mathrm{d}z$:

$$\frac{\partial \sigma_x}{\partial x} + \frac{\partial \sigma_{yx}}{\partial y} + \frac{\partial \sigma_{zx}}{\partial z} + f_x = 0\,. \tag{8.9}$$

Based on the same approach, similar equations can be specified in the y- and z-direction:

$$\frac{\partial \sigma_y}{\partial y} + \frac{\partial \sigma_{yx}}{\partial x} + \frac{\partial \sigma_{yz}}{\partial z} + f_y = 0\,, \tag{8.10}$$

$$\frac{\partial \sigma_z}{\partial z} + \frac{\partial \sigma_{xz}}{\partial x} + \frac{\partial \sigma_{yz}}{\partial y} + f_z = 0\,. \tag{8.11}$$

These three balance equations can be written in matrix notation as

$$\begin{bmatrix} \frac{\partial}{\partial x} & 0 & 0 & \frac{\partial}{\partial y} & 0 & \frac{\partial}{\partial z} \\ 0 & \frac{\partial}{\partial y} & 0 & \frac{\partial}{\partial x} & \frac{\partial}{\partial z} & 0 \\ 0 & 0 & \frac{\partial}{\partial z} & 0 & \frac{\partial}{\partial y} & \frac{\partial}{\partial x} \end{bmatrix} \begin{bmatrix} \sigma_x \\ \sigma_y \\ \sigma_z \\ \sigma_{xy} \\ \sigma_{yz} \\ \sigma_{xz} \end{bmatrix} + \begin{bmatrix} f_x \\ f_y \\ f_z \end{bmatrix} = \begin{bmatrix} 0 \\ 0 \\ 0 \end{bmatrix}, \tag{8.12}$$

or in symbolic notation:

$$\boldsymbol{\mathcal{L}}_1^{\mathrm{T}} \boldsymbol{\sigma} + \boldsymbol{b} = \boldsymbol{0}\,, \tag{8.13}$$

where $\boldsymbol{\mathcal{L}}_1$ is the differential operator matrix and $\boldsymbol{b}$ the column matrix of body forces.

8.1.4 Differential Equation

The basic equations introduced in the previous three sections, i.e., the equilibrium, the constitutive and the kinematics equation, are summarized in the following Table 8.3 where in addition the tensor notation[3] is given.

For the solution of the 15 unknown spatial functions (3 components of the displacement vector, 6 components of the symmetric strain tensor and 6 components of the symmetric stress tensor), a set of 15 scalar field equations is available:

---

[3] A differentiation is there indicated by the use of a comma: The first index refers to the component and the comma indicates the partial derivative with respect to the second subscript corresponding to the relevant coordinate axis [1].

Table 8.3 Fundamental governing equations of a continuum in the three-dimensional case

Name	Matrix notation	Tensor notation
Equilibrium	$\boldsymbol{\mathcal{L}}_1^{\mathrm{T}}\boldsymbol{\sigma} + \boldsymbol{b} = \boldsymbol{0}$	$\sigma_{ij,i} + b_j = 0$
Constitution	$\boldsymbol{\sigma} = \boldsymbol{C}\boldsymbol{\epsilon}$	$\sigma_{ij} = C_{ijkl}\varepsilon_{kl}$
Kinematics	$\boldsymbol{\varepsilon} = \boldsymbol{\mathcal{L}}_1\boldsymbol{u}$	$\varepsilon_{ij} = \frac{1}{2}\left(u_{i,j} + u_{j,i}\right)$

- Equilibrium: 3,
- Constitution: 6,
- Kinematics: 6.

Furthermore, the boundary conditions are given:

$$\boldsymbol{u} \text{ on } \Gamma_u\,, \tag{8.14}$$

$$\boldsymbol{t} \text{ on } \Gamma_t\,, \tag{8.15}$$

where Γ_u is the part of the boundary where a displacement boundary condition is prescribed and Γ_t is the part of the boundary where a traction boundary condition, i.e. external force per unit area, is prescribed with $t_j = \sigma_{ij}n_j$, where n_j are the components of the normal vector.

The 15 scalar field equations can be combined to eliminate the stress and strain fields. As a result, three scalar field equations for the three scalar displacement fields are obtained. These equations are called the LAMÉ–NAVIER[4] equations and can be derived as follows:

Introducing the constitutive equation according to (8.6) in the equilibrium equation (8.13) gives:

$$\boldsymbol{\mathcal{L}}_1^{\mathrm{T}}\boldsymbol{C}\boldsymbol{\varepsilon} + \boldsymbol{b} = \boldsymbol{0}\,. \tag{8.16}$$

Introducing the kinematics relations in the last equation according to (8.4) finally gives the LAMÉ–NAVIER equations:

$$\boldsymbol{\mathcal{L}}_1^{\mathrm{T}}\boldsymbol{C}\boldsymbol{\mathcal{L}}_1\boldsymbol{u} + \boldsymbol{b} = \boldsymbol{0}\,. \tag{8.17}$$

Alternatively, the displacements may be substituted and the differential equations are obtained in terms of stresses. This formulation is known as the BELTRAMI–MICHELL[5] equations. If the body forces vanish ($\boldsymbol{b} = \boldsymbol{0}$), the partial differential equations in terms of stresses are called the BELTRAMI equations.

[4]Gabriel Léon Jean Baptiste LAMÉ (1795–1870), French mathematician.
Claude–Louis NAVIER (1785–1836), French engineer and physicist.

[5]Eugenio BELTRAMI (1835–1900), Italian mathematician.
John Henry MICHELL (1863–1940), Australian mathematician.

Table 8.4 summarizes different formulations of the basic equations for three-dimensional elasticity, once in their specific form and once in symbolic notation.

The following Table 8.5 shows a comparison between the basic equations for a rod, plane and 3D elasticity. It can be seen that the use of the differential operator $\mathcal{L}_1\{\dots\}$ allows to depict a simple analogy between these sets of equations.

8.2 Finite Element Solution

8.2.1 *Derivation of the Principal Finite Element Equation*

Let us assume in the following that the elasticity matrix in Eq. (8.17) is constant and that the exact solution is given by $\boldsymbol{u}^0$. Thus, the differential equation in terms of displacements can be written as:

$$\boldsymbol{\mathcal{L}}_1^{\mathrm{T}} \boldsymbol{C} \boldsymbol{\mathcal{L}}_1 \boldsymbol{u}^0 + \boldsymbol{b} = \boldsymbol{0}\,. \tag{8.18}$$

Replacing the exact solution by an approximate solution $\boldsymbol{u}$, a residual $\boldsymbol{r}$ is obtained:

$$\boldsymbol{r} = \boldsymbol{\mathcal{L}}_1^{\mathrm{T}} \boldsymbol{C} \boldsymbol{\mathcal{L}}_1 \boldsymbol{u} + \boldsymbol{b} \neq \boldsymbol{0}\,. \tag{8.19}$$

The inner product is obtained by weighting the residual and integration as

$$\int\limits_V \boldsymbol{W}^{\mathrm{T}} \left(\boldsymbol{\mathcal{L}}_1^{\mathrm{T}} \boldsymbol{C} \boldsymbol{\mathcal{L}}_1 \boldsymbol{u} + \boldsymbol{b} \right) \mathrm{d}V = 0\,, \tag{8.20}$$

where $\boldsymbol{W}(\boldsymbol{x}) = \begin{bmatrix} W_x & W_y & W_z \end{bmatrix}^{\mathrm{T}}$ is the column matrix of weight functions and $\boldsymbol{x} = \begin{bmatrix} x & y & z \end{bmatrix}^{\mathrm{T}}$ is the column matrix of Cartesian coordinates. Application of the GREEN–GAUSS theorem[6] (cf. Sect. A.7) gives the weak formulation as:

$$\int\limits_V (\boldsymbol{\mathcal{L}}_1 \boldsymbol{W})^{\mathrm{T}} \boldsymbol{C} \, (\boldsymbol{\mathcal{L}}_1 \boldsymbol{u}) \, \mathrm{d}V = \int\limits_A \boldsymbol{W}^{\mathrm{T}} \boldsymbol{t} \, \mathrm{d}A + \int\limits_V \boldsymbol{W}^{\mathrm{T}} \boldsymbol{b} \, \mathrm{d}V\,, \tag{8.21}$$

where the column matrix of traction forces $\boldsymbol{t} = \begin{bmatrix} t_x & t_y & t_z \end{bmatrix}^{\mathrm{T}}$ can be understood as the expression[7] $(\boldsymbol{C}\boldsymbol{\mathcal{L}}_1 \boldsymbol{u})^{\mathrm{T}} \boldsymbol{n} = \boldsymbol{\sigma}^{\mathrm{T}} \boldsymbol{n}$.

Any further development of Eq. (8.21) requires that the general expressions for the displacement and weight function, i.e. $\boldsymbol{u}$ and $\boldsymbol{W}$, are now approximated by some

[6] The supplementary problem 8.10 shows a possible way on how to use the theorem.

[7] Strictly speaking, the traction forces must be calculated based on the stress *tensor* as $t_i = \sigma_{ji} n_j$ and not based on the column matrix of stress components.

Table 8.4 Different formulations of the basic equations for three-dimensional elasticity. E: YOUNG's modulus; ν: POISSON's ratio; f: volume-specific force [5]

Specific formulation	General formulation
Kinematics	
$\begin{bmatrix} \varepsilon_x \\ \varepsilon_y \\ \varepsilon_z \\ 2\varepsilon_{xy} \\ 2\varepsilon_{yz} \\ 2\varepsilon_{xz} \end{bmatrix} = \begin{bmatrix} \frac{\partial}{\partial x} & 0 & 0 \\ 0 & \frac{\partial}{\partial y} & 0 \\ 0 & 0 & \frac{\partial}{\partial z} \\ \frac{\partial}{\partial y} & \frac{\partial}{\partial x} & 0 \\ 0 & \frac{\partial}{\partial z} & \frac{\partial}{\partial y} \\ \frac{\partial}{\partial z} & 0 & \frac{\partial}{\partial x} \end{bmatrix} \begin{bmatrix} u_x \\ u_y \\ u_z \end{bmatrix}$	$\boldsymbol{\varepsilon} = \boldsymbol{\mathcal{L}}_1 \boldsymbol{u}$
Constitution	
$\begin{bmatrix} \sigma_x \\ \sigma_y \\ \sigma_z \\ \sigma_{xy} \\ \sigma_{yz} \\ \sigma_{xz} \end{bmatrix} = \frac{E}{(1+\nu)(1-2\nu)} \begin{bmatrix} 1-\nu & \nu & \nu & 0 & 0 & 0 \\ \nu & 1-\nu & \nu & 0 & 0 & 0 \\ \nu & \nu & 1-\nu & 0 & 0 & 0 \\ 0 & 0 & 0 & \frac{1-2\nu}{2} & 0 & 0 \\ 0 & 0 & 0 & 0 & \frac{1-2\nu}{2} & 0 \\ 0 & 0 & 0 & 0 & 0 & \frac{1-2\nu}{2} \end{bmatrix} \begin{bmatrix} \varepsilon_x \\ \varepsilon_y \\ \varepsilon_z \\ 2\,\varepsilon_{xy} \\ 2\,\varepsilon_{yz} \\ 2\,\varepsilon_{xz} \end{bmatrix}$	$\boldsymbol{\sigma} = \boldsymbol{C}\boldsymbol{\varepsilon}$
Equilibrium	
$\begin{bmatrix} \frac{\partial}{\partial x} & 0 & 0 & \frac{\partial}{\partial y} & 0 & \frac{\partial}{\partial z} \\ 0 & \frac{\partial}{\partial y} & 0 & \frac{\partial}{\partial x} & \frac{\partial}{\partial z} & 0 \\ 0 & 0 & \frac{\partial}{\partial z} & 0 & \frac{\partial}{\partial y} & \frac{\partial}{\partial x} \end{bmatrix} \begin{bmatrix} \sigma_x \\ \sigma_y \\ \sigma_z \\ \sigma_{xy} \\ \sigma_{yz} \\ \sigma_{xz} \end{bmatrix} + \begin{bmatrix} f_x \\ f_y \\ f_z \end{bmatrix} = \begin{bmatrix} 0 \\ 0 \\ 0 \end{bmatrix}$	$\boldsymbol{\mathcal{L}}_1^{\mathrm{T}} \boldsymbol{\sigma} + \boldsymbol{b} = \boldsymbol{0}$
PDE	
$\frac{E}{(1+\nu)(1-2\nu)} \begin{bmatrix} \cdots\cdots\cdots \\ \cdots\cdots\cdots \\ \cdots\cdots\cdots \end{bmatrix} \begin{bmatrix} u_x \\ u_y \\ u_z \end{bmatrix} + \begin{bmatrix} f_x \\ f_y \\ f_z \end{bmatrix} = \begin{bmatrix} 0 \\ 0 \\ 0 \end{bmatrix}$	$\boldsymbol{\mathcal{L}}_1^{\mathrm{T}} \boldsymbol{C} \boldsymbol{\mathcal{L}}_1 \boldsymbol{u} + \boldsymbol{b} = \boldsymbol{0}$

with

$$\begin{bmatrix} \cdots\cdots\cdots \\ \cdots\cdots\cdots \\ \cdots\cdots\cdots \end{bmatrix} =$$

$$\begin{bmatrix} (1-\nu)\frac{\mathrm{d}^2}{\mathrm{d}x^2} + \left(\frac{1}{2}-\nu\right)\left(\frac{\mathrm{d}^2}{\mathrm{d}y^2} + \frac{\mathrm{d}^2}{\mathrm{d}z^2}\right) & \nu\frac{\mathrm{d}^2}{\mathrm{d}x\mathrm{d}y} + \left(\frac{1}{2}-\nu\right)\frac{\mathrm{d}^2}{\mathrm{d}x\mathrm{d}y} & \nu\frac{\mathrm{d}^2}{\mathrm{d}x\mathrm{d}z} + \left(\frac{1}{2}-\nu\right)\frac{\mathrm{d}^2}{\mathrm{d}x\mathrm{d}z} \\ \nu\frac{\mathrm{d}^2}{\mathrm{d}x\mathrm{d}y} + \left(\frac{1}{2}-\nu\right)\frac{\mathrm{d}^2}{\mathrm{d}x\mathrm{d}y} & (1-\nu)\frac{\mathrm{d}^2}{\mathrm{d}y^2} + \left(\frac{1}{2}-\nu\right)\left(\frac{\mathrm{d}^2}{\mathrm{d}x^2} + \frac{\mathrm{d}^2}{\mathrm{d}z^2}\right) & \nu\frac{\mathrm{d}^2}{\mathrm{d}y\mathrm{d}z} + \left(\frac{1}{2}-\nu\right)\frac{\mathrm{d}^2}{\mathrm{d}y\mathrm{d}z} \\ \nu\frac{\mathrm{d}^2}{\mathrm{d}x\mathrm{d}z} + \left(\frac{1}{2}-\nu\right)\frac{\mathrm{d}^2}{\mathrm{d}x\mathrm{d}z} & \nu\frac{\mathrm{d}^2}{\mathrm{d}y\mathrm{d}z} + \left(\frac{1}{2}-\nu\right)\frac{\mathrm{d}^2}{\mathrm{d}y\mathrm{d}z} & (1-\nu)\frac{\mathrm{d}^2}{\mathrm{d}z^2} + \left(\frac{1}{2}-\nu\right)\left(\frac{\mathrm{d}^2}{\mathrm{d}y^2} + \frac{\mathrm{d}^2}{\mathrm{d}x^2}\right) \end{bmatrix}$$

Table 8.5 Comparison of basic equations for rod, plane elasticity and three-dimensional elements

Rod	Plane elasticity	Three-dimensional
	Kinematics	
$\varepsilon_x(x) = \mathcal{L}_1\left(u_x(x)\right)$	$\boldsymbol{\varepsilon} = \boldsymbol{\mathcal{L}}_1 \boldsymbol{u}$	$\boldsymbol{\varepsilon} = \boldsymbol{\mathcal{L}}_1 \boldsymbol{u}$
	Constitution	
$\sigma_x(x) = C\varepsilon_x(x)$	$\boldsymbol{\sigma} = \boldsymbol{C}\boldsymbol{\varepsilon}$	$\boldsymbol{\sigma} = \boldsymbol{C}\boldsymbol{\varepsilon}$
	Equilibrium	
$\mathcal{L}_1\left(\sigma_x(x)\right) + b = 0$	$\boldsymbol{\mathcal{L}}_1^{\mathrm{T}}\boldsymbol{\sigma} + \boldsymbol{b} = \boldsymbol{0}$	$\boldsymbol{\mathcal{L}}_1^{\mathrm{T}}\boldsymbol{\sigma} + \boldsymbol{b} = \boldsymbol{0}$
	PDE	
$\mathcal{L}_1\left(C\mathcal{L}_1\left(u_x(x)\right)\right) + b = 0$	$\boldsymbol{\mathcal{L}}_1^{\mathrm{T}}\boldsymbol{C}\boldsymbol{\mathcal{L}}_1\boldsymbol{u} + \boldsymbol{b} = \boldsymbol{0}$	$\boldsymbol{\mathcal{L}}_1^{\mathrm{T}}\boldsymbol{C}\boldsymbol{\mathcal{L}}_1\boldsymbol{u} + \boldsymbol{b} = \boldsymbol{0}$

functional representations. The nodal approach for the displacements[8] can be generally written for a three-dimensional element with n nodes as:

$$u_x^{\mathrm{e}}(\boldsymbol{x}) = N_1 u_{1x} + N_2 u_{2x} + N_3 u_{3x} + \cdots + N_n u_{nx}\,, \tag{8.22}$$

$$u_y^{\mathrm{e}}(\boldsymbol{x}) = N_1 u_{1y} + N_2 u_{2y} + N_3 u_{3y} + \cdots + N_n u_{ny}\,, \tag{8.23}$$

$$u_z^{\mathrm{e}}(\boldsymbol{x}) = N_1 u_{1z} + N_2 u_{2z} + N_3 u_{3z} + \cdots + N_n u_{nz}\,, \tag{8.24}$$

or in matrix notation as:

$$\begin{bmatrix} u_x^{\mathrm{e}} \\ u_y^{\mathrm{e}} \\ u_z^{\mathrm{e}} \end{bmatrix} = \left[\begin{array}{ccc|ccc|ccc|c|ccc} N_1 & 0 & 0 & N_2 & 0 & 0 & N_3 & 0 & 0 & \cdots & N_n & 0 & 0 \\ 0 & N_1 & 0 & 0 & N_2 & 0 & 0 & N_3 & 0 & \cdots & 0 & N_n & 0 \\ 0 & 0 & N_1 & 0 & 0 & N_2 & 0 & 0 & N_3 & \cdots & 0 & 0 & N_n \end{array}\right] \left[\begin{array}{c} u_{1x} \\ u_{1y} \\ u_{1z} \\ \hline u_{2x} \\ u_{2y} \\ u_{2z} \\ \hline u_{3x} \\ u_{3y} \\ u_{3z} \\ \hline \vdots \\ \hline u_{nx} \\ u_{ny} \\ u_{nz} \end{array}\right]. \tag{8.25}$$

Introducing the notations

[8] The following derivations are written under the simplification that each node reveals only displacement DOF and no rotations.

$$N_i\mathbf{I} = \begin{bmatrix} N_i & 0 & 0 \\ 0 & N_i & 0 \\ 0 & 0 & N_i \end{bmatrix} \quad \text{and} \quad \boldsymbol{u}_{\text{p}i} = \begin{bmatrix} u_{ix} \\ u_{iy} \\ u_{iz} \end{bmatrix}, \tag{8.26}$$

Equation (8.25) can be written as

$$\begin{bmatrix} u_x^{\text{e}} \\ u_y^{\text{e}} \\ u_z^{\text{e}} \end{bmatrix} = \begin{bmatrix} N_1\mathbf{I} \; N_2\mathbf{I} \; N_3\mathbf{I} \cdots N_n\mathbf{I} \end{bmatrix} \begin{bmatrix} \boldsymbol{u}_{\text{p}1} \\ \boldsymbol{u}_{\text{p}2} \\ \boldsymbol{u}_{\text{p}3} \\ \vdots \\ \boldsymbol{u}_{\text{p}n} \end{bmatrix}, \tag{8.27}$$

or with $\boldsymbol{N}_i = N_i\mathbf{I}$ as

$$\boldsymbol{u}^{\text{e}}(\boldsymbol{x}) = \begin{bmatrix} u_x^{\text{e}} \\ u_y^{\text{e}} \\ u_z^{\text{e}} \end{bmatrix} = \begin{bmatrix} \boldsymbol{N}_1 \; \boldsymbol{N}_2 \; \boldsymbol{N}_3 \cdots \boldsymbol{N}_n \end{bmatrix} \begin{bmatrix} \boldsymbol{u}_{\text{p}1} \\ \boldsymbol{u}_{\text{p}2} \\ \boldsymbol{u}_{\text{p}3} \\ \vdots \\ \boldsymbol{u}_{\text{p}n} \end{bmatrix}. \tag{8.28}$$

The last equation can be written in abbreviated form as:

$$\boldsymbol{u}^{\text{e}}(\boldsymbol{x}) = \boldsymbol{N}^{\text{T}}(\boldsymbol{x})\boldsymbol{u}_{\text{p}}^{\text{p}}, \tag{8.29}$$

which is the same structure as in the case of the one-dimensional elements, cf. Eq. (2.21). The column matrix of the weight functions in Eq. (8.21) is approximated in a similar way as the unknown displacements:

$$\boldsymbol{W}(\boldsymbol{x}) = \boldsymbol{N}^{\text{T}}(\boldsymbol{x})\delta\boldsymbol{u}_{\text{p}}. \tag{8.30}$$

Introducing the approximations for $\boldsymbol{u}^{\text{e}}$ and $\boldsymbol{W}$ according to Eqs. (8.29) and (8.30) in the weak formulation gives:

$$\int\limits_V \left(\boldsymbol{\mathcal{L}}_1\boldsymbol{N}^{\text{T}}\delta\boldsymbol{u}_{\text{p}}\right)^{\text{T}} \boldsymbol{C}\left(\boldsymbol{\mathcal{L}}_1\boldsymbol{N}^{\text{T}}\boldsymbol{u}_{\text{p}}\right)\text{d}V = \int\limits_A (\boldsymbol{N}^{\text{T}}\delta\boldsymbol{u}_{\text{p}})^{\text{T}}\boldsymbol{t}\,\text{d}A + \int\limits_V (\boldsymbol{N}^{\text{T}}\delta\boldsymbol{u}_{\text{p}})^{\text{T}}\boldsymbol{b}\,\text{d}V, \tag{8.31}$$

which we can write under the consideration that the matrix of displacements and virtual displacements are not affected by the integration as:

$$\delta\boldsymbol{u}_{\text{p}}^{\text{T}} \int\limits_V \left(\boldsymbol{\mathcal{L}}_1\boldsymbol{N}^{\text{T}}\right)^{\text{T}} \boldsymbol{C}\left(\boldsymbol{\mathcal{L}}_1\boldsymbol{N}^{\text{T}}\right)\text{d}V\boldsymbol{u}_{\text{p}}^{\text{e}} = \delta\boldsymbol{u}_{\text{p}}^{\text{T}} \int\limits_A \boldsymbol{N}\boldsymbol{t}\,\text{d}A + \delta\boldsymbol{u}_{\text{p}}^{\text{T}} \int\limits_V \boldsymbol{N}\boldsymbol{b}\,\text{d}V, \tag{8.32}$$

which gives after elimination of $\delta \boldsymbol{u}_{\mathrm{p}}^{\mathrm{T}}$ the following statement for the principal finite element equation on element level as:

$$\int_V (\boldsymbol{\mathcal{L}}_1 \boldsymbol{N}^{\mathrm{T}})^{\mathrm{T}} \boldsymbol{C} (\boldsymbol{\mathcal{L}}_1 \boldsymbol{N}^{\mathrm{T}}) \, \mathrm{d}V \boldsymbol{u}_{\mathrm{p}}^{\mathrm{p}} = \int_A \boldsymbol{N} \boldsymbol{t} \, \mathrm{d}A + \int_V \boldsymbol{N} \boldsymbol{b} \, \mathrm{d}V \,. \tag{8.33}$$

Thus, we can identify the following three element matrices from the principal finite element equation:

$$\text{Stiffness matrix } (3n \times 3n)\text{:} \quad \boldsymbol{K}^{\mathrm{e}} = \int_V \underbrace{(\boldsymbol{\mathcal{L}}_1 \boldsymbol{N}^{\mathrm{T}})^{\mathrm{T}}}_{\boldsymbol{B}} \boldsymbol{C} \underbrace{(\boldsymbol{\mathcal{L}}_1 \boldsymbol{N}^{\mathrm{T}})}_{\boldsymbol{B}^{\mathrm{T}}} \, \mathrm{d}V \,, \tag{8.34}$$

$$\text{Boundary force matrix } (3n \times 1)\text{:} \quad \boldsymbol{f}_t^{\mathrm{e}} = \int_A \boldsymbol{N} \boldsymbol{t} \, \mathrm{d}A \,, \tag{8.35}$$

$$\text{Body force matrix } (3n \times 1)\text{:} \quad \boldsymbol{f}_b^{\mathrm{e}} = \int_V \boldsymbol{N} \boldsymbol{b} \, \mathrm{d}V \,. \tag{8.36}$$

Based on these abbreviations, the principal finite element equation for a single element can be written as:

$$\boldsymbol{K}^{\mathrm{e}} \boldsymbol{u}_{\mathrm{p}}^{\mathrm{e}} = \boldsymbol{f}_t^{\mathrm{e}} + \boldsymbol{f}_b^{\mathrm{e}} \,. \tag{8.37}$$

In the following, let us look at the $\boldsymbol{B}$-matrix, i.e. the matrix which contains the derivatives of the interpolation functions. Application of the matrix of differential operators according to Eq. (8.3) to the matrix of interpolation functions gives:

$$\boldsymbol{\mathcal{L}}_1 \boldsymbol{N}^{\mathrm{T}} = \begin{bmatrix} \frac{\partial}{\partial x} & 0 & 0 \\ 0 & \frac{\partial}{\partial y} & 0 \\ 0 & 0 & \frac{\partial}{\partial z} \\ \frac{\partial}{\partial y} & \frac{\partial}{\partial x} & 0 \\ 0 & \frac{\partial}{\partial z} & \frac{\partial}{\partial y} \\ \frac{\partial}{\partial z} & 0 & \frac{\partial}{\partial x} \end{bmatrix} \left[\begin{array}{ccc|ccc|c|ccc} N_1 & 0 & 0 & N_2 & 0 & 0 & \cdots & N_n & 0 & 0 \\ 0 & N_1 & 0 & 0 & N_2 & 0 & \cdots & 0 & N_n & 0 \\ 0 & 0 & N_1 & 0 & 0 & N_2 & \cdots & 0 & 0 & N_n \end{array} \right] \tag{8.38}$$

$$= \left[\begin{array}{ccc|ccc|c|ccc} \frac{\partial N_1}{\partial x} & 0 & 0 & \frac{\partial N_2}{\partial x} & 0 & 0 & & \frac{\partial N_n}{\partial x} & 0 & 0 \\ 0 & \frac{\partial N_1}{\partial y} & 0 & 0 & \frac{\partial N_2}{\partial y} & 0 & & 0 & \frac{\partial N_n}{\partial y} & 0 \\ 0 & 0 & \frac{\partial N_1}{\partial z} & 0 & 0 & \frac{\partial N_2}{\partial z} & & 0 & 0 & \frac{\partial N_n}{\partial z} \\ \frac{\partial N_1}{\partial y} & \frac{\partial N_1}{\partial x} & 0 & \frac{\partial N_2}{\partial y} & \frac{\partial N_2}{\partial x} & 0 & \cdots & \frac{\partial N_n}{\partial y} & \frac{\partial N_n}{\partial x} & 0 \\ 0 & \frac{\partial N_1}{\partial z} & \frac{\partial N_1}{\partial y} & 0 & \frac{\partial N_2}{\partial z} & \frac{\partial N_2}{\partial y} & & 0 & \frac{\partial N_n}{\partial z} & \frac{\partial N_n}{\partial y} \\ \frac{\partial N_1}{\partial z} & 0 & \frac{\partial N_1}{\partial x} & \frac{\partial N_2}{\partial z} & 0 & \frac{\partial N_2}{\partial x} & & \frac{\partial N_n}{\partial z} & 0 & \frac{\partial N_n}{\partial x} \end{array} \right] = \boldsymbol{B}^{\mathrm{T}} \,, \tag{8.39}$$

which is a $(6 \times 3n)$-matrix. The transposed, i.e. $(\boldsymbol{\mathcal{L}}_1 \boldsymbol{N}^{\mathrm{T}})^{\mathrm{T}}$, is thus a $(3n \times 6)$-matrix. Multiplication with the elasticity matrix, i.e. a (6×6)-matrix, results in $(\boldsymbol{\mathcal{L}}_1 \boldsymbol{N}^{\mathrm{T}})^{\mathrm{T}} \boldsymbol{C}$,

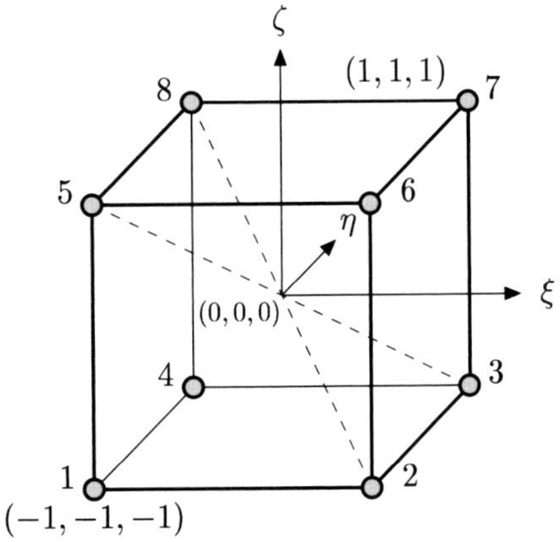

Fig. 8.3 Three-dimensional eight-node hexahedron in parametric space

which is a $(3n \times 6)$-matrix. The final multiplication, i.e. $(\boldsymbol{\mathcal{L}}_1 \boldsymbol{N}^{\mathrm{T}})^{\mathrm{T}} \boldsymbol{C} (\boldsymbol{\mathcal{L}}_1 \boldsymbol{N}^{\mathrm{T}})$ gives after integration the stiffness matrix with a dimension of $(3n \times 3n)$.

The integrations for the element matrices given in Eqs. (8.34)–(8.36) are approximated by numerical integration. To this end, the coordinates (x, y, z) are transformed to the natural coordinates (unit space) (ξ, η, ζ) where each coordinate ranges from -1 to 1. In the scope of the coordinate transformation, attention must be paid to the derivatives. For example, the derivative of the interpolation functions with respect to the x-coordinate is transformed in the following way:

$$\frac{\partial N_i}{\partial x} \rightarrow \frac{\partial N_i}{\partial \xi}\frac{\partial \xi}{\partial x} + \frac{\partial N_i}{\partial \eta}\frac{\partial \eta}{\partial x} + \frac{\partial N_i}{\partial \zeta}\frac{\partial \zeta}{\partial x}. \tag{8.40}$$

Furthermore, the coordinate transformation requires that $\mathrm{d}V = \mathrm{d}x\mathrm{d}y\mathrm{d}z \rightarrow \mathrm{d}V' = J\mathrm{d}\xi\mathrm{d}\eta\mathrm{d}\zeta$, where J is the Jacobian as given in the Appendix A.8.

8.2.2 *Hexahedron Solid Elements*

A simple representative of a three-dimensional finite element is an eight-node hexahedron (also called 'hex 8' or 'brick') as shown in Fig. 8.3. This element uses trilinear interpolation functions and the strains tend to be constant throughout the element. The stiffness matrix of this element is normally—in the case of full integration—calculated based on an eight-point GAUSS–LEGENDRE quadrature formula.

Let us derive the element formulation from the assumption that a linear displacement field is given in parametric space. For the x-component, we can write

$$u_x^{\mathrm{e}}(\xi, \eta, \zeta) = a_0 + a_1\xi + a_2\eta + a_3\zeta + a_4\xi\eta + a_5\eta\zeta + a_6\xi\zeta + a_7\xi\eta\zeta\,, \tag{8.41}$$

or in matrix notation

$$u_x^{\mathrm{e}}(\xi,\eta,\zeta) = \boldsymbol{\chi}^{\mathrm{T}}\boldsymbol{a} = \begin{bmatrix} 1 & \xi & \eta & \zeta & \xi\eta & \eta\zeta & \xi\zeta & \xi\eta\zeta \end{bmatrix} \begin{bmatrix} a_0 \\ a_1 \\ a_2 \\ a_3 \\ a_4 \\ a_5 \\ a_6 \\ a_7 \end{bmatrix} . \tag{8.42}$$

Evaluating Eq. (8.42) for all eight nodes of the hexahedron (cf. Fig. 8.3) gives:

$$\begin{aligned}
&\text{Node 1: } u_{1x} = u_x^{\mathrm{e}}(-1,-1,-1) = a_0 - a_1 - a_2 - a_3 + a_4 + a_5 + a_6 - a_7\,,\\
&\text{Node 2: } u_{2x} = u_x^{\mathrm{e}}(1,-1,-1) = a_0 + a_1 - a_2 - a_3 - a_4 + a_5 - a_6 + a_7\,,\\
&\qquad\qquad \vdots \qquad\qquad\qquad\qquad\qquad\qquad \vdots\\
&\text{Node 8: } u_{2x} = u_x^{\mathrm{e}}(-1,1,1) = a_0 - a_1 + a_2 + a_3 - a_4 + a_5 - a_6 - a_7\,,
\end{aligned}$$

or in matrix notation:

$$\begin{bmatrix} u_{1x} \\ u_{2x} \\ u_{3x} \\ u_{4x} \\ u_{5x} \\ u_{6x} \\ u_{7x} \\ u_{8x} \end{bmatrix} = \begin{bmatrix} 1 & -1 & -1 & -1 & 1 & 1 & 1 & -1 \\ 1 & 1 & -1 & -1 & -1 & 1 & -1 & 1 \\ 1 & 1 & 1 & -1 & 1 & -1 & -1 & -1 \\ 1 & -1 & 1 & -1 & -1 & -1 & 1 & 1 \\ 1 & -1 & -1 & 1 & 1 & -1 & -1 & 1 \\ 1 & 1 & -1 & 1 & -1 & -1 & 1 & -1 \\ 1 & 1 & 1 & 1 & 1 & 1 & 1 & 1 \\ 1 & -1 & 1 & 1 & -1 & 1 & -1 & -1 \end{bmatrix} \begin{bmatrix} a_0 \\ a_1 \\ a_2 \\ a_3 \\ a_4 \\ a_5 \\ a_6 \\ a_7 \end{bmatrix} . \tag{8.43}$$

Solving for $\boldsymbol{a}$ gives

$$\begin{bmatrix} a_0 \\ a_1 \\ a_2 \\ a_3 \\ a_4 \\ a_5 \\ a_6 \\ a_7 \end{bmatrix} = \frac{1}{8} \begin{bmatrix} 1 & 1 & 1 & 1 & 1 & 1 & 1 & 1 \\ -1 & 1 & 1 & -1 & -1 & 1 & 1 & -1 \\ -1 & -1 & 1 & 1 & -1 & -1 & 1 & 1 \\ -1 & -1 & -1 & -1 & 1 & 1 & 1 & 1 \\ 1 & -1 & 1 & -1 & 1 & -1 & 1 & -1 \\ 1 & 1 & -1 & -1 & -1 & -1 & 1 & 1 \\ 1 & -1 & -1 & 1 & -1 & 1 & 1 & -1 \\ -1 & 1 & -1 & 1 & 1 & -1 & 1 & -1 \end{bmatrix} \begin{bmatrix} u_{1x} \\ u_{2x} \\ u_{3x} \\ u_{4x} \\ u_{5x} \\ u_{6x} \\ u_{7x} \\ u_{8x} \end{bmatrix} , \tag{8.44}$$

or

$$\boldsymbol{a} = \boldsymbol{A}\boldsymbol{u}_{\mathrm{p}}\,. \tag{8.45}$$

The matrix of interpolation functions results as

$$\boldsymbol{N}^{\mathrm{T}} = \begin{bmatrix} N_1 & N_2 & N_3 & N_4 & N_5 & N_6 & N_7 & N_8 \end{bmatrix} = \boldsymbol{\chi}^{\mathrm{T}}\boldsymbol{A} \tag{8.46}$$

or

$$N_1 = \frac{1}{8}(1-\xi)(1-\eta)(1-\zeta)\,, \tag{8.47}$$

$$N_2 = \frac{1}{8}(1+\xi)(1-\eta)(1-\zeta)\,, \tag{8.48}$$

$$N_3 = \frac{1}{8}(1+\xi)(1+\eta)(1-\zeta)\,, \tag{8.49}$$

$$N_4 = \frac{1}{8}(1-\xi)(1+\eta)(1-\zeta)\,, \tag{8.50}$$

$$N_5 = \frac{1}{8}(-1+\xi)(-1+\eta)(1+\zeta)\,, \tag{8.51}$$

$$N_6 = \frac{1}{8}(1+\xi)(1-\eta)(1+\zeta)\,, \tag{8.52}$$

$$N_7 = \frac{1}{8}(1+\xi)(1+\eta)(1+\zeta)\,, \tag{8.53}$$

$$N_8 = \frac{1}{8}(1-\xi)(1+\eta)(1+\zeta)\,, \tag{8.54}$$

or in a more compact form as

$$N_i = \frac{1}{8}(1+\xi\xi_i)(1+\eta\eta_i)(1+\zeta\zeta_i)\,, \tag{8.55}$$

where ξ_i, η_i and ζ_i are the coordinates of the nodes in parametric space ($i = 1, \ldots, 8$), cf. Fig. 8.3. The derivatives with respect to the parametric coordinates can easily be obtained as:

$$\frac{\partial N_i}{\partial \xi} = \frac{1}{8}(\xi_i)(1+\eta\eta_i)(1+\zeta\zeta_i)\,, \tag{8.56}$$

$$\frac{\partial N_i}{\partial \eta} = \frac{1}{8}(1+\xi\xi_i)(\eta_i)(1+\zeta\zeta_i)\,, \tag{8.57}$$

$$\frac{\partial N_i}{\partial \zeta} = \frac{1}{8}(1+\xi\xi_i)(1+\eta\eta_i)(\zeta_i)\,. \tag{8.58}$$

The geometrical derivatives in Eq. (8.40), e.g. $\frac{\partial \xi}{\partial x}$, $\frac{\partial \eta}{\partial x}$, $\frac{\partial \zeta}{\partial x}$, can be calculated on the basis of

$$\begin{bmatrix} \frac{\partial \xi}{\partial x} & \frac{\partial \xi}{\partial y} & \frac{\partial \xi}{\partial z} \\ \frac{\partial \eta}{\partial x} & \frac{\partial \eta}{\partial y} & \frac{\partial \eta}{\partial z} \\ \frac{\partial \zeta}{\partial x} & \frac{\partial \zeta}{\partial y} & \frac{\partial \zeta}{\partial z} \end{bmatrix} = \frac{1}{J} \begin{bmatrix} \frac{\partial y}{\partial \eta}\frac{\partial z}{\partial \zeta} - \frac{\partial y}{\partial \zeta}\frac{\partial z}{\partial \eta} & -\frac{\partial x}{\partial \eta}\frac{\partial z}{\partial \zeta} + \frac{\partial x}{\partial \zeta}\frac{\partial z}{\partial \eta} & \frac{\partial x}{\partial \eta}\frac{\partial y}{\partial \zeta} - \frac{\partial x}{\partial \zeta}\frac{\partial y}{\partial \eta} \\ -\frac{\partial y}{\partial \xi}\frac{\partial z}{\partial \zeta} + \frac{\partial y}{\partial \zeta}\frac{\partial z}{\partial \xi} & \frac{\partial x}{\partial \xi}\frac{\partial z}{\partial \zeta} - \frac{\partial x}{\partial \zeta}\frac{\partial z}{\partial \xi} & -\frac{\partial x}{\partial \xi}\frac{\partial y}{\partial \zeta} + \frac{\partial x}{\partial \zeta}\frac{\partial y}{\partial \xi} \\ \frac{\partial y}{\partial \xi}\frac{\partial z}{\partial \eta} - \frac{\partial y}{\partial \eta}\frac{\partial z}{\partial \xi} & -\frac{\partial x}{\partial \xi}\frac{\partial z}{\partial \eta} + \frac{\partial x}{\partial \eta}\frac{\partial z}{\partial \xi} & \frac{\partial x}{\partial \xi}\frac{\partial y}{\partial \eta} - \frac{\partial x}{\partial \eta}\frac{\partial y}{\partial \xi} \end{bmatrix}, \tag{8.59}$$

where the Jacobian J is the determinant as given by

$$J = \left| \frac{\partial(x, y, z)}{\partial(\xi, \eta, \zeta)} \right| = x_\xi y_\eta z_\zeta + x_\eta y_\zeta z_\xi + x_\zeta y_\xi z_\eta - x_\eta y_\xi z_\zeta - x_\zeta y_\eta z_\xi - x_\xi y_\zeta z_\eta \,. \tag{8.60}$$

In Eq. (8.60), the abbreviation, e.g. x_ξ, stands for the partial derivative $\frac{\partial x}{\partial \xi}$ and so on. Let us assume the same interpolation for the global x, y and z-coordinate as for the displacement field (isoparametric element formulation), i.e. $\overline{N}_i = N_i$:

$$x(\xi, \eta, \zeta) = \sum_{i=1}^{8} \overline{N}_i(\xi, \eta, \zeta) x_i \,, \tag{8.61}$$

$$y(\xi, \eta, \zeta) = \sum_{i=1}^{8} \overline{N}_i(\xi, \eta, \zeta) y_i \,, \tag{8.62}$$

$$z(\xi, \eta, \zeta) = \sum_{i=1}^{8} \overline{N}_i(\xi, \eta, \zeta) z_i \,, \tag{8.63}$$

where the global coordinates of the nodes $1, \ldots, 8$ can be used for $x_1, \ldots, x_8$, and so on. Thus, the derivatives can easily be obtained as:

$$\frac{\partial x}{\partial \xi} = \sum_{i=1}^{8} \frac{\partial \overline{N}_i}{\partial \xi} x_i = \sum_{i=1}^{8} \frac{1}{8} (\xi_i)(1 + \eta\eta_i)(1 + \zeta\zeta_i) x_i \,, \tag{8.64}$$

$$\frac{\partial x}{\partial \eta} = \sum_{i=1}^{8} \frac{\partial \overline{N}_i}{\partial \eta} x_i = \sum_{i=1}^{8} \frac{1}{8} (1 + \xi\xi_i)(\eta_i)(1 + \zeta\zeta_i) x_i \,, \tag{8.65}$$

$$\frac{\partial x}{\partial \zeta} = \sum_{i=1}^{8} \frac{\partial \overline{N}_i}{\partial \zeta} x_i = \sum_{i=1}^{8} \frac{1}{8} (1 + \xi\xi_i)(1 + \eta\eta_i)(\zeta_i) x_i \,, \tag{8.66}$$

$$\frac{\partial y}{\partial \xi} = \sum_{i=1}^{8} \frac{\partial \overline{N}_i}{\partial \xi} y_i = \sum_{i=1}^{8} \frac{1}{8} (\xi_i)(1 + \eta\eta_i)(1 + \zeta\zeta_i) y_i \,, \tag{8.67}$$

$$\frac{\partial y}{\partial \eta} = \sum_{i=1}^{8} \frac{\partial \overline{N}_i}{\partial \eta} y_i = \sum_{i=1}^{8} \frac{1}{8} (1 + \xi\xi_i)(\eta_i)(1 + \zeta\zeta_i) y_i \,, \tag{8.68}$$

$$\frac{\partial y}{\partial \zeta} = \sum_{i=1}^{8} \frac{\partial \overline{N}_i}{\partial \zeta} y_i = \sum_{i=1}^{8} \frac{1}{8} (1 + \xi\xi_i)(1 + \eta\eta_i)(\zeta_i) y_i \,, \tag{8.69}$$

$$\frac{\partial z}{\partial \xi} = \sum_{i=1}^{8} \frac{\partial \overline{N}_i}{\partial \xi} z_i = \sum_{i=1}^{8} \frac{1}{8} (\xi_i)(1 + \eta\eta_i)(1 + \zeta\zeta_i) z_i \,, \tag{8.70}$$

Table 8.6 Derivatives of the interpolation functions in parametric space

Node	$\frac{\partial N_i}{\partial \xi}$	$\frac{\partial N_i}{\partial \eta}$	$\frac{\partial N_i}{\partial \zeta}$
1	$\frac{1}{8}(-1)(1-\eta)(1-\zeta)$	$\frac{1}{8}(1-\xi)(-1)(1-\zeta)$	$\frac{1}{8}(1-\xi)(1-\eta)(-1)$
2	$\frac{1}{8}(1)(1-\eta)(1-\zeta)$	$\frac{1}{8}(1+\xi)(-1)(1-\zeta)$	$\frac{1}{8}(1+\xi)(1-\eta)(-1)$
3	$\frac{1}{8}(1)(1+\eta)(1-\zeta)$	$\frac{1}{8}(1+\xi)(1)(1-\zeta)$	$\frac{1}{8}(1+\xi)(1+\eta)(-1)$
4	$\frac{1}{8}(-1)(1+\eta)(1-\zeta)$	$\frac{1}{8}(1-\xi)(1)(1-\zeta)$	$\frac{1}{8}(1-\xi)(1+\eta)(-1)$
5	$\frac{1}{8}(-1)(1-\eta)(1+\zeta)$	$\frac{1}{8}(1-\xi)(-1)(1+\zeta)$	$\frac{1}{8}(1-\xi)(1-\eta)(1)$
6	$\frac{1}{8}(1)(1-\eta)(1+\zeta)$	$\frac{1}{8}(1+\xi)(-1)(1+\zeta)$	$\frac{1}{8}(1+\xi)(1-\eta)(1)$
7	$\frac{1}{8}(1)(1+\eta)(1+\zeta)$	$\frac{1}{8}(1+\xi)(1)(1+\zeta)$	$\frac{1}{8}(1+\xi)(1+\eta)(1)$
8	$\frac{1}{8}(-1)(1+\eta)(1+\zeta)$	$\frac{1}{8}(1-\xi)(1)(1+\zeta)$	$\frac{1}{8}(1-\xi)(1+\eta)(1)$

$$\frac{\partial z}{\partial \eta} = \sum_{i=1}^{8} \frac{\partial \overline{N}_i}{\partial \eta} z_i = \sum_{i=1}^{8} \frac{1}{8}(1+\xi\xi_i)(\eta_i)(1+\zeta\zeta_i)z_i \,, \tag{8.71}$$

$$\frac{\partial z}{\partial \zeta} = \sum_{i=1}^{8} \frac{\partial \overline{N}_i}{\partial \zeta} z_i = \sum_{i=1}^{8} \frac{1}{8}(1+\xi\xi_i)(1+\eta\eta_i)(\zeta_i)z_i \,. \tag{8.72}$$

The derivatives of the interpolation functions with respect to the coordinates in parametric space are summarized in Table 8.6. Note that the derivatives (8.64)–(8.72) are simple constants and independent of ξ, η and ζ for a cuboid or a parallelepiped (cf. Fig. 8.4).

Looking at the example of a cube with edge length $2a$ (cf. Fig. 8.4a), one can derive that $\frac{\partial x}{\partial \xi} = \frac{\partial y}{\partial \eta} = \frac{\partial z}{\partial \zeta} = a$, whereas all other geometrical derivatives are zero. For a cuboid with edge length $2a$, $2b$, and $2c$ (cf. Fig. 8.4b), similar derivation gives $\frac{\partial x}{\partial \xi} = a$, $\frac{\partial y}{\partial \eta} = b$, and $\frac{\partial z}{\partial \zeta} = c$. Considering the parallelepiped shown in Fig. 8.4c where the lower nodes are moved by $-d$ in the negative x-direction and the upper nodes by $+d$ in the positive x-direction, one obtains $\frac{\partial x}{\partial \xi} = a$, $\frac{\partial y}{\partial \eta} = b$, $\frac{\partial z}{\partial \zeta} = c$, and $\frac{\partial x}{\partial \zeta} = d$ (other derivatives are zero). For other cases, the geometrical derivatives become dependent on the parametric coordinates. For example, Fig. 8.4d shows a distorted cuboid where node 1 is translated by d along the x-direction. For this case, one obtains:

$$\frac{\partial x}{\partial \xi} = a + \frac{1}{8}d - \frac{1}{8}\zeta d - \frac{1}{8}\eta d + \frac{1}{8}\eta\zeta d \,, \tag{8.73}$$

$$\frac{\partial x}{\partial \eta} = \frac{1}{8}d - \frac{1}{8}\zeta d - \frac{1}{8}\xi d + \frac{1}{8}\eta\zeta d \,, \tag{8.74}$$

$$\frac{\partial x}{\partial \zeta} = \frac{1}{8}d - \frac{1}{8}\eta d - \frac{1}{8}\xi d + \frac{1}{8}\eta\zeta d \,, \tag{8.75}$$

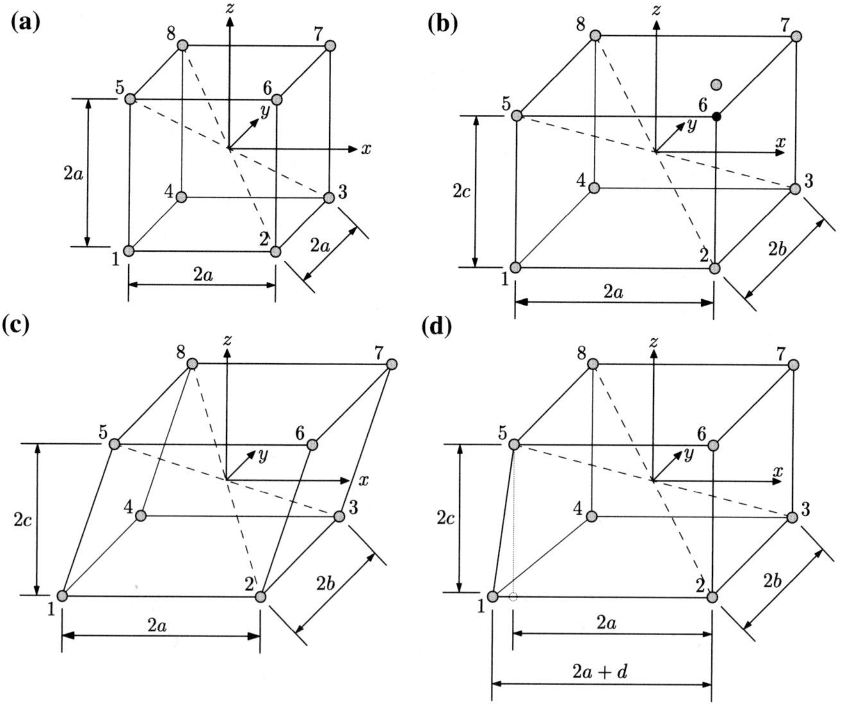

Fig. 8.4 Different shapes of an eight-node hexahedron in the (x, y, z)-space: **a** cube; **b** cuboid; **c** parallelepiped; **d** distorted cuboid

$$\frac{\partial y}{\partial \eta} = b\,, \tag{8.76}$$

$$\frac{\partial z}{\partial \zeta} = b\,. \tag{8.77}$$

On the basis of the derived equations, the element matrices given in Eqs. (8.34)–(8.36) can now be numerically evaluated. The integration is performed as in the case of the one-dimensional integrals based on GAUSS–LEGENDRE quadrature. For the domain integrals, one can write that

$$\int_V f(x, y, z)\mathrm{d}V = \int_{V'} f'(\xi, \eta, \zeta)\mathrm{d}V' = \int_{-1}^{1}\int_{-1}^{1}\int_{-1}^{1} f'(\xi, \eta, \zeta) J\,\mathrm{d}\xi\mathrm{d}\eta\mathrm{d}\zeta \tag{8.78}$$

$$= \sum_{i=1}^{n} f'(\xi, \eta, \zeta)_i J_i w_i\,, \tag{8.79}$$

Table 8.7 Integration rules for hexahedral elements [4]

Points	ξ_i	η_i	ζ_i	Weight w_i	Error
1	0	0	0	8	$\mathcal{O}(\xi^2)$
8	$\pm 1/\sqrt{3}$	$\pm 1/\sqrt{3}$	$\pm 1/\sqrt{3}$	1	$\mathcal{O}(\xi^4)$
27	0	0	0	$\left(\frac{8}{9}\right)^3$	
	$\pm\sqrt{0.6}$	0	0	$\left(\frac{5}{9}\right)\left(\frac{8}{9}\right)^2$	
	0	$\pm\sqrt{0.6}$	0	$\left(\frac{5}{9}\right)\left(\frac{8}{9}\right)^2$	$\mathcal{O}(\xi^6)$
	0	0	$\pm\sqrt{0.6}$	$\left(\frac{5}{9}\right)\left(\frac{8}{9}\right)^2$	
	$\pm\sqrt{0.6}$	$\pm\sqrt{0.6}$	0	$\left(\frac{8}{9}\right)\left(\frac{5}{9}\right)^2$	
	$\pm\sqrt{0.6}$	0	$\pm\sqrt{0.6}$	$\left(\frac{8}{9}\right)\left(\frac{5}{9}\right)^2$	
	0	$\pm\sqrt{0.6}$	$\pm\sqrt{0.6}$	$\left(\frac{8}{9}\right)\left(\frac{5}{9}\right)^2$	
	$\pm\sqrt{0.6}$	$\pm\sqrt{0.6}$	$\pm\sqrt{0.6}$	$\left(\frac{5}{9}\right)^3$	

where the Jacobian is given in Eq. (8.60), $(\xi, \eta, \zeta)_i$ are the coordinates of the integration of GAUSS points and w_i are the corresponding weight factors. The location of the integration points and values of associated weights are given in Table 8.7.

Evaluation of the Boundary Force Matrix

The right-hand side of the weak statement contains the boundary force matrix according to Eq. (8.35). To evaluate this expression, it might be more advantageous to replace again the column matrix of traction forces by the stress tensor[9] and normal vector. Thus, we can write Eq. (8.35) in the following form:

$$\boldsymbol{f}_t^{\mathrm{e}} = \int_A \underbrace{\begin{bmatrix} N_1 & 0 & 0 \\ 0 & N_1 & 0 \\ 0 & 0 & N_1 \\ \hline N_2 & 0 & 0 \\ 0 & N_2 & 0 \\ 0 & 0 & N_2 \\ \hline & \cdots & \\ & \cdots & \\ & \cdots & \\ \hline N_n & 0 & 0 \\ 0 & N_n & 0 \\ 0 & 0 & N_n \end{bmatrix}}_{\boldsymbol{N}} \underbrace{\begin{bmatrix} \sigma_{xx} & \sigma_{xy} & \sigma_{xz} \\ \sigma_{yx} & \sigma_{yy} & \sigma_{yz} \\ \sigma_{zx} & \sigma_{zy} & \sigma_{zz} \end{bmatrix} \begin{bmatrix} n_x \\ n_y \\ n_z \end{bmatrix}}_{\boldsymbol{t}} \mathrm{d}A\,, \tag{8.80}$$

[9]The stress *tensor* is used in the following since it makes the derivation easier, cf. Eq. (8.21) and the following comment and footnote.

Fig. 8.5 Normal vectors for the evaluation of the boundary force matrix at node 1

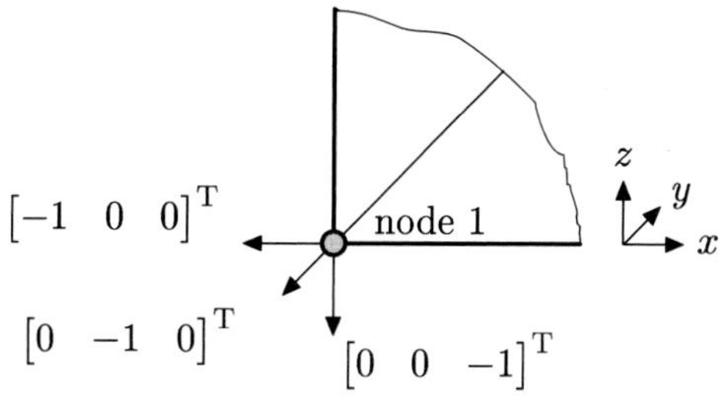

or

$$
\boldsymbol{f}_t^{\mathrm{e}} = \int_A \left[\begin{array}{ccc}
N_1\sigma_{xx} & N_1\sigma_{xy} & N_1\sigma_{xz} \\
N_1\sigma_{yx} & N_1\sigma_{yy} & N_1\sigma_{yz} \\
N_1\sigma_{zx} & N_1\sigma_{zy} & N_1\sigma_{zz} \\ \hline
N_2\sigma_{xx} & N_2\sigma_{xy} & N_2\sigma_{xz} \\
N_2\sigma_{yx} & N_2\sigma_{yy} & N_2\sigma_{yz} \\
N_2\sigma_{zx} & N_2\sigma_{zy} & N_2\sigma_{zz} \\ \hline
 & \cdots & \\
 & \cdots & \\
 & \cdots & \\ \hline
N_n\sigma_{xx} & N_n\sigma_{xy} & N_n\sigma_{xz} \\
N_n\sigma_{yx} & N_n\sigma_{yy} & N_n\sigma_{yz} \\
N_n\sigma_{zx} & N_n\sigma_{zy} & N_n\sigma_{zz}
\end{array}\right] \begin{bmatrix} n_x \\ n_y \\ n_z \end{bmatrix} \mathrm{d}A\,. \tag{8.81}
$$

The expression for the boundary force matrix given in Eq. (8.81) needs to be evaluated for each node along the element boundary. For node 1, the interpolation function N_1 is equal to one and identically zero for all other nodes. In addition, all other interpolation functions are identically zero at node 1. Since each node has three different normal vectors, cf. Fig. 8.5, one may calculate the expression for node 1 in x-direction by evaluating the first row of the following system:

$$
\left[\begin{array}{ccc}
\sigma_{xx} & \sigma_{xy} & \sigma_{xz} \\
\sigma_{yx} & \sigma_{yy} & \sigma_{yz} \\
\sigma_{zx} & \sigma_{zy} & \sigma_{zz} \\ \hline
0 & 0 & 0 \\ 0 & 0 & 0 \\ 0 & 0 & 0 \\ \hline
 & \cdots & \\ & \cdots & \\ & \cdots & \\ \hline
0 & 0 & 0 \\ 0 & 0 & 0 \\ 0 & 0 & 0
\end{array}\right] \begin{bmatrix} -1 \\ 0 \\ 0 \end{bmatrix} A +
\left[\begin{array}{ccc}
\sigma_{xx} & \sigma_{xy} & \sigma_{xz} \\
\sigma_{yx} & \sigma_{yy} & \sigma_{yz} \\
\sigma_{zx} & \sigma_{zy} & \sigma_{zz} \\ \hline
0 & 0 & 0 \\ 0 & 0 & 0 \\ 0 & 0 & 0 \\ \hline
 & \cdots & \\ & \cdots & \\ & \cdots & \\ \hline
0 & 0 & 0 \\ 0 & 0 & 0 \\ 0 & 0 & 0
\end{array}\right] \begin{bmatrix} 0 \\ -1 \\ 0 \end{bmatrix} A +
\left[\begin{array}{ccc}
\sigma_{xx} & \sigma_{xy} & \sigma_{xz} \\
\sigma_{yx} & \sigma_{yy} & \sigma_{yz} \\
\sigma_{zx} & \sigma_{zy} & \sigma_{zz} \\ \hline
0 & 0 & 0 \\ 0 & 0 & 0 \\ 0 & 0 & 0 \\ \hline
 & \cdots & \\ & \cdots & \\ & \cdots & \\ \hline
0 & 0 & 0 \\ 0 & 0 & 0 \\ 0 & 0 & 0
\end{array}\right] \begin{bmatrix} 0 \\ 0 \\ -1 \end{bmatrix} A\,, \tag{8.82}
$$

or as

$$-\sigma_{xx}A-\sigma_{xy}A-\sigma_{xz}A\,, \tag{8.83}$$

which is balanced by the external force F_{1x} at node 1. Similar results can be obtained for all other nodes and directions and the boundary force matrix can be written as

$$\boldsymbol{f}_t^{\mathrm{e}}=\begin{bmatrix}F_{1x} & F_{1y} & F_{1z} & F_{2x} & F_{2y} & F_{2z} & \ldots & F_{8x} & F_{8y} & F_{8z}\end{bmatrix}^{\mathrm{T}}, \tag{8.84}$$

where an external force is positive if directed to the positive coordinate direction.

Evaluation of the Body Force Matrix

The right-hand side of the weak statement contains the body force matrix according to Eq. (8.36). Let us consider in the following the special case that all the elements of the column matrix of body forces are constant, i.e. $\boldsymbol{b}=\begin{bmatrix}f & f & f\end{bmatrix}^{\mathrm{T}}$. Introducing the column matrix of the interpolation functions in Eq. (8.36) gives:

$$\boldsymbol{f}_b^{\mathrm{e}}=\int\limits_V \underbrace{\begin{bmatrix}N_1 & 0 & 0\\ 0 & N_1 & 0\\ 0 & 0 & N_1\\ \hline N_2 & 0 & 0\\ 0 & N_2 & 0\\ 0 & 0 & N_2\\ \hline & \ldots & \\ & \ldots & \\ & \ldots & \\ \hline N_8 & 0 & 0\\ 0 & N_8 & 0\\ 0 & 0 & N_8\end{bmatrix}}_{\boldsymbol{N}} \underbrace{\begin{bmatrix}f\\ f\\ f\end{bmatrix}}_{\boldsymbol{b}} \mathrm{d}V=\int\limits_V \begin{bmatrix}N_1 f\\ N_1 f\\ N_1 f\\ \hline N_2 f\\ N_2 f\\ N_2 f\\ \hline \ldots\\ \ldots\\ \ldots\\ \hline N_8 f\\ N_8 f\\ N_8 f\end{bmatrix} \mathrm{d}V = f\int\limits_V \begin{bmatrix}N_1\\ N_1\\ N_1\\ \hline N_2\\ N_2\\ N_2\\ \hline \ldots\\ \ldots\\ \ldots\\ \hline N_8\\ N_8\\ N_8\end{bmatrix} \mathrm{d}V\,. \tag{8.85}$$

Thus, the integration over the single interpolation functions must be performed. It should be noted here that the interpolation functions in Eq. (8.85) are functions of the Cartesian coordinates, i.e. $N_i=N_i(x,y,z)$. In order to continue the derivation, let us assume that the hexahedron element is a cube with edge length $2a$ as shown in Fig. 8.4a. For this regular shape, the Jacobian simplifies to $J=a^3$. To numerically integrate the integral, a coordinate transformation must be performed which reads:

$$\boldsymbol{f}_b^{\mathrm{e}}=f\int\limits_{-1}^{1}\int\limits_{-1}^{1}\int\limits_{-1}^{1} \begin{bmatrix}N_1(\xi,\eta,\zeta)\\ N_1(\xi,\eta,\zeta)\\ N_1(\xi,\eta,\zeta)\\ \hline \ldots\\ \ldots\\ \ldots\\ \hline N_8(\xi,\eta,\zeta)\\ N_8(\xi,\eta,\zeta)\\ N_8(\xi,\eta,\zeta)\end{bmatrix} a^3\mathrm{d}\xi\mathrm{d}\eta\mathrm{d}\zeta\,. \tag{8.86}$$

Approximation of the integral based on an eight-point GAUSS–LEGENDRE quadrature formula (cf. Table 8.7) gives:

$$f_b^e \approx f \sum_{i=1}^{8} \begin{bmatrix} N_1(\xi,\eta,\zeta) \\ N_1(\xi,\eta,\zeta) \\ N_1(\xi,\eta,\zeta) \\ \hline \cdots \\ \cdots \\ \cdots \\ \hline N_8(\xi,\eta,\zeta) \\ N_8(\xi,\eta,\zeta) \\ N_8(\xi,\eta,\zeta) \end{bmatrix}_i a^3 w_i = f 8a^3 \begin{bmatrix} \frac{1}{8} \\ \frac{1}{8} \\ \frac{1}{8} \\ \hline \cdots \\ \cdots \\ \cdots \\ \hline \frac{1}{8} \\ \frac{1}{8} \\ \frac{1}{8} \end{bmatrix}. \tag{8.87}$$

It can be concluded from the last equation that the equivalent nodal loads in the case of a constant body force matrix are obtained by first calculating the resultant force, i.e. body force times volume ($f \times V$), and then equally distributing this force to all the eight nodes ($\frac{f \times V}{8}$). It should be noted that the same result would be obtained based on analytical integration. The analytical integration for a cube with edge length $2a$ can be based on Eq. (8.85) and transforming the interpolation functions given in Eqs. (8.47)–(8.54) to the Cartesian space, for example, as $N_1(x,y,z) = \frac{1}{8}\left(1-\frac{x}{a}\right)\left(1-\frac{y}{a}\right)\left(1-\frac{z}{a}\right)$. Thus, analytical integration would require to evaluate the following expression:

$$f_b^e = f \int_{-a}^{a}\int_{-a}^{a}\int_{-a}^{a} \begin{bmatrix} N_1(x,y,z) \\ N_1(x,y,z) \\ N_1(x,y,z) \\ \hline \cdots \\ \cdots \\ \cdots \\ \hline N_8(x,y,z) \\ N_8(x,y,z) \\ N_8(x,y,z) \end{bmatrix} \mathrm{d}x\mathrm{d}y\mathrm{d}z\,. \tag{8.88}$$

Let us summarize here the major steps which are required to calculate the elemental stiffness matrix.

❶ Introduce an elemental coordinate system (x, y, z).
❷ Express the coordinates (x_i, y_i, z_i) of the corner nodes i $(i = 1, \ldots, 8)$ in this elemental coordinate system.
❸ Calculate the partial derivatives of the old Cartesian (x, y, z) coordinates with respect to the new natural (ξ, η, ζ) coordinates, see Eqs. (8.64)–(8.72):

$$\frac{\partial x}{\partial \xi} = \sum_{i=1}^{8} \frac{\partial N_i}{\partial \xi} x_i = \sum_{i=1}^{8} \frac{1}{8}(\xi_i)(1+\eta\eta_i)(1+\zeta\zeta_i)x_i \,,$$

$$\vdots$$

$$\frac{\partial z}{\partial \zeta} = \sum_{i=1}^{8} \frac{\partial N_i}{\partial \zeta} z_i = \sum_{i=1}^{8} \frac{1}{8}(1+\xi\xi_i)(1+\eta\eta_i)(\zeta_i)z_i \,.$$

❹ Calculate the partial derivatives of the new natural (ξ, η, ζ) coordinates with respect to the old Cartesian (x, y, z) coordinates, see Eq. (8.59):

$$\begin{bmatrix} \frac{\partial \xi}{\partial x} & \frac{\partial \xi}{\partial y} & \frac{\partial \xi}{\partial z} \\ \frac{\partial \eta}{\partial x} & \frac{\partial \eta}{\partial y} & \frac{\partial \eta}{\partial z} \\ \frac{\partial \zeta}{\partial x} & \frac{\partial \zeta}{\partial y} & \frac{\partial \zeta}{\partial z} \end{bmatrix} = \frac{1}{J} \begin{bmatrix} \frac{\partial y}{\partial \eta}\frac{\partial z}{\partial \zeta} - \frac{\partial y}{\partial \zeta}\frac{\partial z}{\partial \eta} & -\frac{\partial x}{\partial \eta}\frac{\partial z}{\partial \zeta} + \frac{\partial x}{\partial \zeta}\frac{\partial z}{\partial \eta} & \frac{\partial x}{\partial \eta}\frac{\partial y}{\partial \zeta} - \frac{\partial x}{\partial \zeta}\frac{\partial y}{\partial \eta} \\ -\frac{\partial y}{\partial \xi}\frac{\partial z}{\partial \zeta} + \frac{\partial y}{\partial \zeta}\frac{\partial z}{\partial \xi} & \frac{\partial x}{\partial \xi}\frac{\partial z}{\partial \zeta} - \frac{\partial x}{\partial \zeta}\frac{\partial z}{\partial \xi} & -\frac{\partial x}{\partial \xi}\frac{\partial y}{\partial \zeta} + \frac{\partial x}{\partial \zeta}\frac{\partial y}{\partial \xi} \\ \frac{\partial y}{\partial \xi}\frac{\partial z}{\partial \eta} - \frac{\partial y}{\partial \eta}\frac{\partial z}{\partial \xi} & -\frac{\partial x}{\partial \xi}\frac{\partial z}{\partial \eta} + \frac{\partial x}{\partial \eta}\frac{\partial z}{\partial \xi} & \frac{\partial x}{\partial \xi}\frac{\partial y}{\partial \eta} - \frac{\partial x}{\partial \eta}\frac{\partial y}{\partial \xi} \end{bmatrix} .$$

❺ Calculate the $\boldsymbol{B}$-matrix and its transposed, see Eq. (8.39):

$$\boldsymbol{B}^{\mathrm{T}} = \left[\begin{array}{ccc|ccc|c|ccc} \frac{\partial N_1}{\partial x} & 0 & 0 & \frac{\partial N_2}{\partial x} & 0 & 0 & & \frac{\partial N_8}{\partial x} & 0 & 0 \\ 0 & \frac{\partial N_1}{\partial y} & 0 & 0 & \frac{\partial N_2}{\partial y} & 0 & & 0 & \frac{\partial N_8}{\partial y} & 0 \\ 0 & 0 & \frac{\partial N_1}{\partial z} & 0 & 0 & \frac{\partial N_2}{\partial z} & \cdots & 0 & 0 & \frac{\partial N_8}{\partial z} \\ \frac{\partial N_1}{\partial y} & \frac{\partial N_1}{\partial x} & 0 & \frac{\partial N_2}{\partial y} & \frac{\partial N_2}{\partial x} & 0 & & \frac{\partial N_8}{\partial y} & \frac{\partial N_8}{\partial x} & 0 \\ 0 & \frac{\partial N_1}{\partial z} & \frac{\partial N_1}{\partial y} & 0 & \frac{\partial N_2}{\partial z} & \frac{\partial N_2}{\partial y} & & 0 & \frac{\partial N_8}{\partial z} & \frac{\partial N_8}{\partial y} \\ \frac{\partial N_1}{\partial z} & 0 & \frac{\partial N_1}{\partial x} & \frac{\partial N_2}{\partial z} & 0 & \frac{\partial N_2}{\partial x} & & \frac{\partial N_8}{\partial z} & 0 & \frac{\partial N_8}{\partial x} \end{array}\right] ,$$

where the partial derivatives are $\frac{\partial N_1(\xi,\eta,\zeta)}{\partial x} = \frac{\partial N_1}{\partial \xi}\frac{\partial \xi}{\partial x} + \frac{\partial N_1}{\partial \eta}\frac{\partial \eta}{\partial x} + \frac{\partial N_1}{\partial \zeta}\frac{\partial \zeta}{\partial x}, \ldots$ and the derivatives of the interpolation functions are given in Eqs. (8.56)–(8.58), i.e., $\frac{\partial N_i}{\partial \xi} = \frac{1}{8}(\xi_i)(1+\eta\eta_i)(1+\zeta\zeta_i), \ldots$

❻ Calculate the triple matrix product $\boldsymbol{B}\boldsymbol{C}^{\mathrm{T}}\boldsymbol{B}$, where the elasticity matrix $\boldsymbol{C}$ is given by Eq. (8.5).

❼ Perform the numerical integration based on a 8-point integration rule:

$$\int_V (\boldsymbol{B}\boldsymbol{C}\boldsymbol{B}^{\mathrm{T}})\mathrm{d}V = \sum_{1}^{8} \boldsymbol{B}\boldsymbol{C}\boldsymbol{B}^{\mathrm{T}} J \times 1 \Big|_{\left(\pm\frac{1}{\sqrt{3}}, \pm\frac{1}{\sqrt{3}}, \pm\frac{1}{\sqrt{3}}\right)} .$$

❽ $\boldsymbol{K}$ obtained.

Let us summarize at the end of this section the major steps that were undertaken to transform the partial differential equation into the principal finite element equation, see Table 8.8.

8.2.3 Solved Three-Dimensional Element Problems

8.1 Example: One-element example of a cantilever solid

Table 8.8 Summary: derivation of principal finite element equation for solid elements

Strong formulation
$\mathcal{L}_1^{\mathrm{T}} C \mathcal{L}_1 u^0 + b = 0$
Inner product
$\int_V W^{\mathrm{T}}(x)\left(\mathcal{L}_1^{\mathrm{T}} C \mathcal{L}_1 u + b\right) \mathrm{d}V = 0$
Weak formulation
$\int_V (\mathcal{L}_1 W)^{\mathrm{T}} C (\mathcal{L}_1 u) \mathrm{d}V = \int_A W^{\mathrm{T}} t \, \mathrm{d}A + \int_V W^{\mathrm{T}} b \, \mathrm{d}V$
Principal finite element equation (hex 8)
$\underbrace{\int_V \underbrace{\left(\mathcal{L}_1 N^{\mathrm{T}}\right)^{\mathrm{T}}}_{B} C \underbrace{\left(\mathcal{L}_1 N^{\mathrm{T}}\right)}_{B^{\mathrm{T}}} \mathrm{d}V}_{K^{\mathrm{e}}} \begin{bmatrix} u_{1x} \\ u_{1y} \\ u_{1z} \\ \vdots \\ u_{8x} \\ u_{8y} \\ u_{8z} \end{bmatrix} = \begin{bmatrix} F_{1x} \\ F_{1y} \\ F_{1z} \\ \vdots \\ F_{8x} \\ F_{8y} \\ F_{8z} \end{bmatrix} + \int_V N b \, \mathrm{d}V$

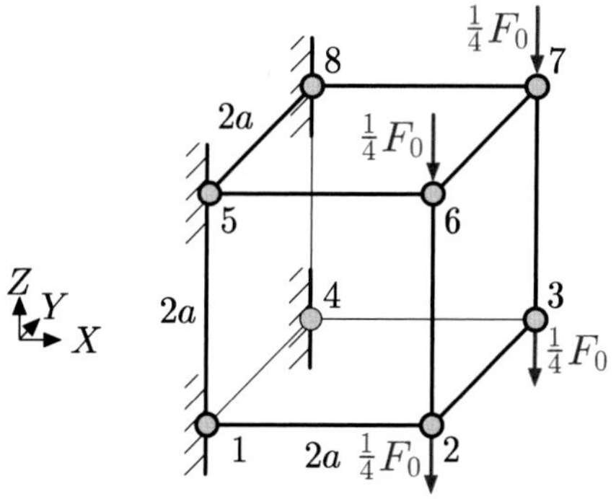

Fig. 8.6 Cantilever cubic solid element

Given is a cantilever cubic solid element as indicated in Fig. 8.6. The side lengths are equal to $2a$. The solid is loaded by four single forces $\frac{1}{4}F_0$ acting at the right-hand nodes of the element. The material is described based on the engineering constants YOUNG's modulus E and POISSON's ratio ν.

Use a single solid element in the following to model the problem and to calculate the nodal unknowns. Compare your results with the analytical solutions for a cantilever EULER–BERNOULLI and a TIMOSHENKO beam.

8.1 Solution

The solution procedure for the elemental stiffness matrix will follow the 8 steps introduced on p. 405.

❶ Introduce an elemental coordinate system (x, y, z).

Let us assume that the elemental coordinate system is located in the center of the cube.

❷ Express the coordinates (x_i, y_i, z_i) of the corner nodes i $(i = 1, \ldots, 8)$ in this elemental coordinate system.

$$(x_1, y_1, z_1) = (-a, -a, -a)\,, \qquad (x_5, y_5, z_5) = (-a, -a, a)\,,$$
$$(x_2, y_2, z_2) = (a, -a, -a)\,, \qquad (x_6, y_6, z_6) = (a, -a, a)\,,$$
$$(x_3, y_3, z_3) = (a, a, -a)\,, \qquad (x_7, y_7, z_7) = (a, a, a)\,,$$
$$(x_4, y_4, z_4) = (-a, a, -a)\,, \qquad (x_8, y_8, z_8) = (-a, a, a)\,.$$

❸ Calculate the partial derivatives of the old Cartesian (x, y, z) coordinates with respect to the new natural (ξ, η, ζ) coordinates, see Eqs. (8.64)–(8.72):

$$\frac{\mathrm{d}x}{\mathrm{d}\xi} = a\,, \qquad \frac{\mathrm{d}x}{\mathrm{d}\eta} = 0\,, \qquad \frac{\mathrm{d}x}{\mathrm{d}\zeta} = 0\,,$$
$$\frac{\mathrm{d}y}{\mathrm{d}\xi} = 0\,, \qquad \frac{\mathrm{d}y}{\mathrm{d}\eta} = a\,, \qquad \frac{\mathrm{d}y}{\mathrm{d}\zeta} = 0\,,$$
$$\frac{\mathrm{d}z}{\mathrm{d}\xi} = 0\,, \qquad \frac{\mathrm{d}z}{\mathrm{d}\eta} = 0\,, \qquad \frac{\mathrm{d}z}{\mathrm{d}\zeta} = a\,.$$

❹ Calculate the partial derivatives of the new natural (ξ, η, ζ) coordinates with respect to the old Cartesian (x, y, z) coordinates, see Eq. (8.59):

$$J = a^3\,,$$

$$\frac{\mathrm{d}\xi}{\mathrm{d}x} = \frac{1}{a}\,, \qquad \frac{\mathrm{d}\xi}{\mathrm{d}y} = 0\,, \qquad \frac{\mathrm{d}\xi}{\mathrm{d}z} = 0\,,$$
$$\frac{\mathrm{d}\eta}{\mathrm{d}x} = 0\,, \qquad \frac{\mathrm{d}\eta}{\mathrm{d}y} = \frac{1}{a}\,, \qquad \frac{\mathrm{d}\eta}{\mathrm{d}z} = 0\,,$$
$$\frac{\mathrm{d}\zeta}{\mathrm{d}x} = 0\,, \qquad \frac{\mathrm{d}\zeta}{\mathrm{d}y} = 0\,, \qquad \frac{\mathrm{d}\zeta}{\mathrm{d}z} = \frac{1}{a}\,.$$

❺ Calculate the $\boldsymbol{B}$-matrix and its transposed, see Eq. (8.39):

$$
\boldsymbol{B}^{\mathrm{T}} = \frac{1}{8a}\begin{bmatrix}
-(1-\eta)(1-\zeta) & 0 & 0 \\
0 & -(1-\xi)(1-\zeta) & 0 \\
0 & 0 & -(1-\xi)(1-\eta) \\
-(1-\xi)(1-\zeta) & -(1-\eta)(1-\zeta) & 0 \\
0 & -(1-\xi)(1-\eta) & -(1-\xi)(1-\zeta) \\
-(1-\xi)(1-\eta) & 0 & -(1-\eta)(1-\zeta)
\end{matrix}\right| \ldots
$$

$$
\ldots \left|\begin{matrix}
-(1+\eta)(1+\zeta) & 0 & 0 \\
0 & (1-\xi)(1+\zeta) & 0 \\
0 & 0 & (1-\xi)(1+\eta) \\
(1-\xi)(1+\zeta) & -(1+\eta)(1+\zeta) & 0 \\
0 & (1-\xi)(1+\eta) & (1-\xi)(1+\zeta) \\
(1-\xi)(1+\eta) & 0 & -(1+\eta)(1+\zeta)
\end{bmatrix}.
$$

❻ Calculate the triple matrix product $\boldsymbol{B}^{\mathrm{T}}\boldsymbol{C}\,\boldsymbol{B}$, where the elasticity matrix $\boldsymbol{C}$ is given by Eq. (8.5).

The triple matrix product results in a 24×24 matrix with the following selected components:

$$
\mathrm{d}K_{1\_1} = \frac{E}{64a^2\,(1+\nu)\,(1-2\,\nu)}\Big((1-\eta)^2\,(1-\zeta)^2\,(1-\nu) +
(1-\xi)^2\,(1-\zeta)^2\left(\tfrac{1}{2}-\nu\right) + (1-\xi)^2\,(1-\eta)^2\left(\tfrac{1}{2}-\nu\right)\Big)\,,
$$

$$
\mathrm{d}K_{1\_2} = \frac{E}{64a^2\,(1+\nu)\,(1-2\,\nu)}\Big((1-\xi)\,(1-\eta)\,(1-\zeta)^2\,\nu +
(1-\xi)\,(1-\eta)\,(1-\zeta)^2\left(\tfrac{1}{2}-\nu\right)\Big)\,,
$$

$$
\mathrm{d}K_{2\_2} = \frac{E}{64a^2\,(1+\nu)\,(1-2\,\nu)}\Big((1-\xi)^2\,(1-\zeta)^2\,(1-\nu) +
(1-\eta)^2\,(1-\zeta)^2\left(\tfrac{1}{2}-\nu\right) + (1-\xi)^2\,(1-\eta)^2\left(\tfrac{1}{2}-\nu\right)\Big)\,.
$$

❼ Perform the numerical integration based on a 8-point integration rule:

The numerical integration results in a 24×24 matrix with the following selected components:

$$
K_{1\_1} = \frac{2Ea(-2+3\nu)}{9(1+\nu)(-1+2\nu)},
$$

$$
K_{1\_2} = -\frac{Ea}{12(1+\nu)(-1+2\nu)},
$$

$$
K_{2\_2} = \frac{2Ea(-2+3\nu)}{9(1+\nu)(-1+2\nu)}.
$$

❽ $\boldsymbol{K}$ obtained.

The global system of equations, which includes the column matrix of unknowns and nodal forces, results in 24 equations for 24 unknowns. Introducing the boundary conditions, i.e. $u_x = u_y = u_z = 0$ at nodes 1, 4, 5 and 8, results in a reduced system with 12 equations for 12 unknowns. The solution of this system of equations gives:

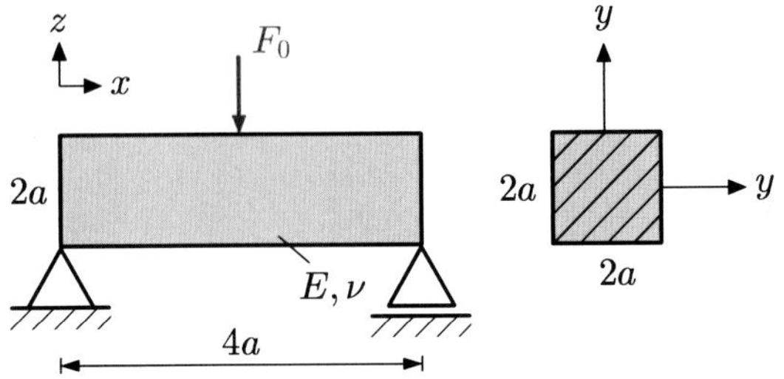

Fig. 8.7 Simply supported beam with square cross section

$$u_{2x} = u_{3x} = -u_{6x} = -u_{7x} = -\frac{3\,(18\nu - 13)\,(-1 + 2\nu)\,(1 + \nu)\,F_0}{Ea\left(60\nu^2 - 106\nu + 39\right)},$$

$$u_{2y} = -u_{3y} = -u_{6y} = u_{7y} = \frac{18\,(-1 + 2\nu)\,\nu\,(1 + \nu)\,F_0}{Ea\left(60\nu^2 - 106\nu + 39\right)},$$

$$u_{2z} = u_{3z} = u_{6z} = u_{7z} = -\frac{2\left(84\nu^2 - 119\nu + 39\right)(1 + \nu)\,F_0}{Ea\left(60\nu^2 - 106\nu + 39\right)}.$$

The analytical solution is obtained as $u_{z,\text{max}} = -\frac{32F_0}{Ea}$ for the EULER–BERNOULLI beam and as $u_{z,\text{max}} = -\frac{32F_0}{Ea} - \frac{24F_0(1+\nu)}{5Ea}$ for the TIMOSHENKO beam.

8.2 Advanced Example: Different 3D modeling approaches of a simply supported beam

Given is a simply supported beam as indicated in Fig. 8.7. The length of the beam is $4a$ and the square cross section has the dimensions $2a \times 2a$. The beam is loaded by a single force F_0 acting in the middle of the beam. Note that the problem is *not* symmetric.

Use two equally-sized solid elements in the following to model the problem and to calculate the nodal unknowns in the middle of the beam, i.e. for $x = 2a$. The modelling approach is based on two elements with nodes $1, \ldots, 12$, see Fig. 8.8, and different ways of introducing the acting force F_0. The solution should be given as a function of $F_0, a, E, \nu = 0.3$.

8.2 Solution

The elemental stiffness matrix of an eight-node hexahedron with dimensions $a \times a \times a$ is known from problem 8.1. The dimension of this matrix is 12×12. Considering two solid elements, a global system of equations with the dimension 36×36 must be assembled. Introducing the support conditions, i.e. $u_{1X} = u_{1Y} = u_{1Z} = u_{4X} = u_{4Y} = u_{4Z} = 0$ and $u_{9Z} = u_{10Z} = 0$ reduces this system to a dimension of 28×28. The results presented in Table 8.9 were obtained with a commercial computer algebra system.

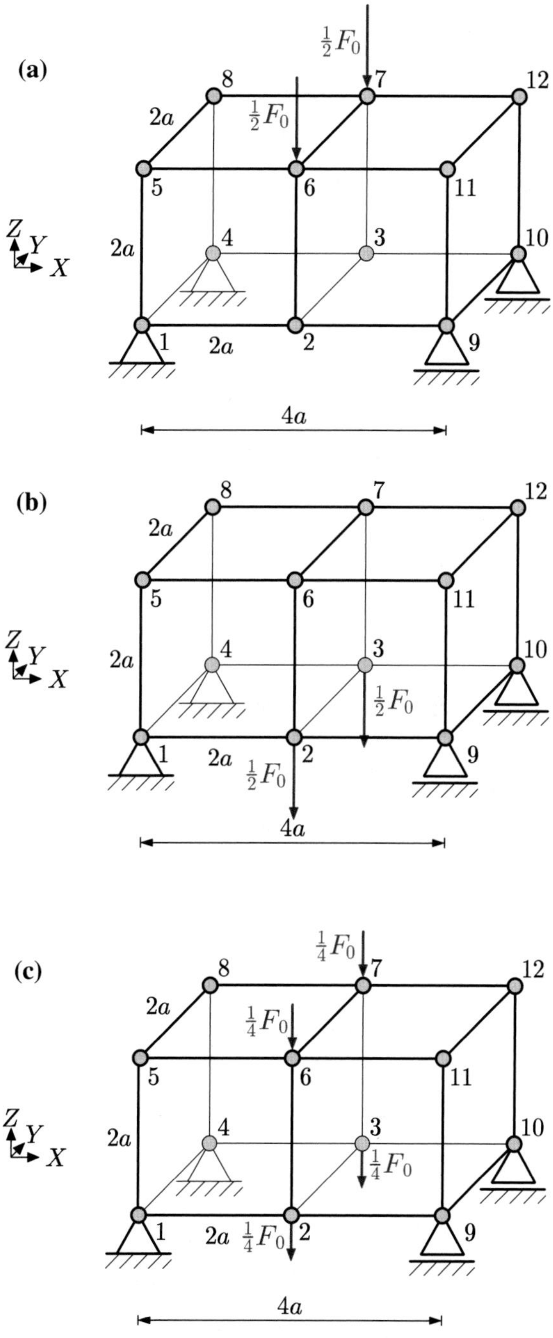

Fig. 8.8 Different modeling approaches for the beam shown in Fig. 8.7: **a** top load, **b** bottom load, and **c** equally distributed load

Table 8.9 Comparison of the results for the beam bending problem, see Fig. 8.8

Displacement	Top load	Equal load	Bottom load
u_{2X}	$0.551113\frac{F_0}{Ea}$	$0.519306\frac{F_0}{Ea}$	$0.487500\frac{F_0}{Ea}$
u_{2Y}	$0.052054\frac{F_0}{Ea}$	$0.107277\frac{F_0}{Ea}$	$0.162500\frac{F_0}{Ea}$
u_{2Z}	$-1.137500\frac{F_0}{Ea}$	$-1.381250\frac{F_0}{Ea}$	$-1.625000\frac{F_0}{Ea}$
u_{3X}	$0.551113\frac{F_0}{Ea}$	$0.519306\frac{F_0}{Ea}$	$0.487500\frac{F_0}{Ea}$
u_{3Y}	$-0.052054\frac{F_0}{Ea}$	$-0.107277\frac{F_0}{Ea}$	$-0.162500\frac{F_0}{Ea}$
u_{3Z}	$-1.137500\frac{F_0}{Ea}$	$-1.381250\frac{F_0}{Ea}$	$-1.625000\frac{F_0}{Ea}$
u_{6X}	$0.553561\frac{F_0}{Ea}$	$0.520531\frac{F_0}{Ea}$	$0.487500\frac{F_0}{Ea}$
u_{6Y}	$-0.122206\frac{F_0}{Ea}$	$-0.061103\frac{F_0}{Ea}$	$0.000000\frac{F_0}{Ea}$
u_{6Z}	$-1.385714\frac{F_0}{Ea}$	$-1.261607\frac{F_0}{Ea}$	$-1.137500\frac{F_0}{Ea}$
u_{7X}	$0.553561\frac{F_0}{Ea}$	$0.520531\frac{F_0}{Ea}$	$0.487500\frac{F_0}{Ea}$
u_{7Y}	$0.122206\frac{F_0}{Ea}$	$0.061103\frac{F_0}{Ea}$	$0.000000\frac{F_0}{Ea}$
u_{7Z}	$-1.385714\frac{F_0}{Ea}$	$-1.261607\frac{F_0}{Ea}$	$-1.137500\frac{F_0}{Ea}$

8.3 Supplementary Problems

8.3 Knowledge questions on three-dimensional elements

- How many material parameters are required for the three-dimensional HOOKE's law under the assumption of an isotropic and homogeneous material? Name possible material parameters.
- State the required (a) geometrical parameters and (b) material parameters to define a three-dimensional elasticity element (hex 8).
- State the DOF per node for a three-dimensional element (hex 8).
- State possible advantages to model a beam bending problem with three-dimensional elasticity elements and not with 1D beam elements

8.4 Hooke's law in terms of shear and bulk modulus

Derive HOOKE's law for a linear isotropic material in terms of shear modulus G and bulk modulus K in the elastic stiffness form ($\boldsymbol{\sigma} = \boldsymbol{\sigma}(\boldsymbol{\varepsilon})$) and the elastic compliance form $\boldsymbol{\varepsilon} = \boldsymbol{\varepsilon}(\boldsymbol{\sigma})$.

8.5 Hooke's law in terms of Lamé's constants

Derive HOOKE's law for a linear isotropic material in terms of Lamé's constants μ and λ in the elastic stiffness form ($\boldsymbol{\sigma} = \boldsymbol{\sigma}(\boldsymbol{\varepsilon})$) and the elastic compliance form $\boldsymbol{\varepsilon} = \boldsymbol{\varepsilon}(\boldsymbol{\sigma})$.

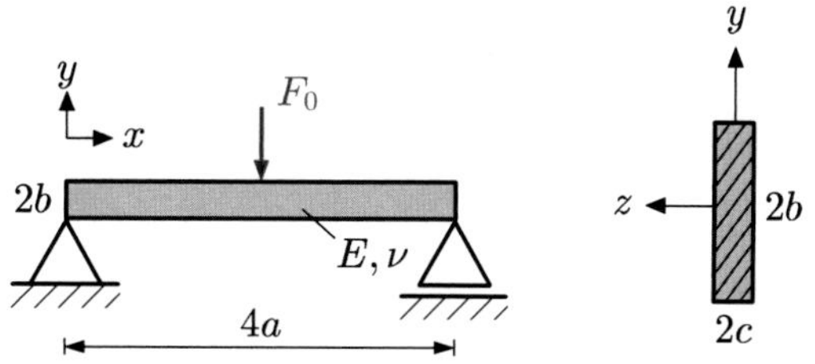

Fig. 8.9 Simply supported beam with rectangular cross section

8.6 Hooke's law for the plane stress state

Derive HOOKE's law for a two-dimensional plane stress state ($\sigma_z = \sigma_{yz} = \sigma_{xz} = 0$) in its elastic stiffness ($\boldsymbol{\sigma} = \boldsymbol{\sigma}(\boldsymbol{\epsilon})$) and elastic compliance ($\boldsymbol{\varepsilon} = \boldsymbol{\varepsilon}(\boldsymbol{\sigma})$) form in terms of YOUNG's modulus E and POISSON's ratio ν for a linear isotropic material.

8.7 Hooke's law for the plane strain state

Derive HOOKE's law for a two-dimensional plane strain state ($\varepsilon_z = \varepsilon_{yz} = \varepsilon_{xz} = 0$) in its elastic stiffness ($\boldsymbol{\sigma} = \boldsymbol{\sigma}(\boldsymbol{\epsilon})$) and elastic compliance ($\boldsymbol{\varepsilon} = \boldsymbol{\varepsilon}(\boldsymbol{\sigma})$) form in terms of YOUNG's modulus E and POISSON's ratio ν for a linear isotropic material.

8.8 Beltrami–Michell equations

Derive the mathematical expressions for the BELTRAMI–MICHELL equations.

8.9 Lamé–Navier equations in matrix notation

The LAMÉ–NAVIER equations are summarized in Table 8.4 in terms of YOUNG's modulus E and POISSON's ratio ν. Derive a similar expression based on the general components of the elasticity matrix $\boldsymbol{C}$. To simplify the expressions, isotropic material behavior should be assumed, i.e. the symmetric elasticity matrix contains only three different terms (named, for example, C_{11}, C_{12}, and C_{44}).

8.10 Green–Gauss theorem applied to equilibrium equation in x-direction

Write the equilibrium equation as given in Eq. (8.9) in matrix notation, i.e. $\boldsymbol{\mathcal{L}}_x^{\mathrm{T}}\boldsymbol{\sigma}_x + f_x = 0$, where $\boldsymbol{\mathcal{L}}_x^{\mathrm{T}} = \left[\frac{\partial}{\partial x}\ 0\ 0\ \frac{\partial}{\partial y}\ 0\ \frac{\partial}{\partial z}\right]$ and $\boldsymbol{\sigma}_x = \left[\sigma_x\ 0\ 0\ \sigma_{xy}\ 0\ \sigma_{xz}\right]^{\mathrm{T}}$. Under consideration of the constitutive equation and the strain-displacement relationship, replace the stress matrix $\boldsymbol{\sigma}_x$ by the displacement matrix $\boldsymbol{u} = \left[u_x\ u_y\ u_z\right]^{\mathrm{T}}$ and write the weighted residual statement for the x-axis in the form $\int_V W_x(\cdots)\mathrm{d}V = 0$. Use the GREEN–GAUSS theorem as given in Sect. A.7 to derive the weak form for the x-direction.

8.11 Green–Gauss theorem applied to derive general 3D weak form

How can the result from supplementary problem 8.10 be used to derive the general expression for the weak form as given in Eq. (8.21)?

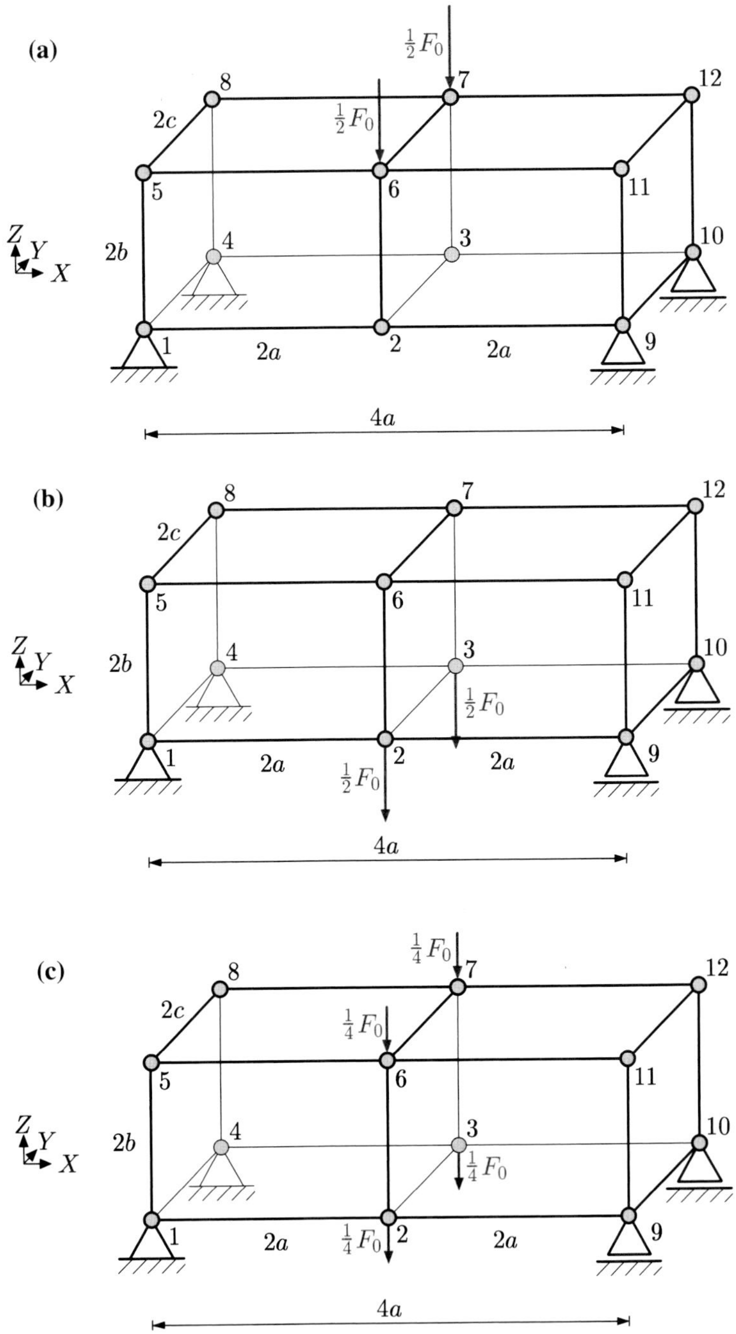

Fig. 8.10 Different modeling approaches for the beam shown in Fig. 8.9: **a** top load, **b** bottom load, and **c** equally distributed load

8.12 Body force matrix for gravity

Simplify the body force matrix as given in Eq. (8.87) for the special case of a body force due to standard gravity g acting in the negative y-direction.

8.13 Advanced Example: Different 3D modeling approaches of a simply supported beam

Given is a simply supported beam as indicated in Fig. 8.9. The length of the beam is $4a$ and the rectangular cross section has the dimensions $2b \times 2c$. The beam is loaded by a single force F_0 acting in the middle of the beam. Note that the problem is *not* symmetric.

Use two equally-sized solid elements in the following to model the problem and to calculate the nodal unknowns in the middle of the beam, i.e. for $x = 2a$. The modelling approach is based on two elements with nodes $1, \ldots, 12$, see Fig. 8.10, and different ways of introducing the acting force F_0. The solution should be given as a function of $F_0, a, b = 0.2a, c = 0.04a, E, \nu = 0.3$.

References

1. Chen WF, Han DJ (1988) Plasticity for structural engineers. Springer, New York
2. Chen WF, Saleeb AF (1982) Constitutive equations for engineering materials. Volume 1: elasticity and modelling. Wiley, New York
3. Eschenauer H, Olhoff N, Schnell W (1997) Applied structural mechanics: fundamentals of elasticity, load-bearing structures, structural optimization. Springer, Berlin
4. MacNeal RH (1994) Finite elements: their design and performance. Marcel Dekker, New York
5. Öchsner A (2014) Elasto-plasticity of frame structure elements: modeling and simulation of rods and beams. Springer, Berlin

Chapter 9
Principles of Linear Dynamics

Abstract This chapter gives a short introduction to NEWTON's laws of motion and the relationships between displacement, velocity, and acceleration for the special case of one-dimensional linear motion. The description is in the scope of classical analytical mechanics of point or spherical masses. This chapter must be seen as a preparation for the next chapter on transient finite element problems.

9.1 Newton's Laws of Motion

NEWTON's laws of motion describe the relationship between a body and the forces acting upon it. They can be expressed in different ways, for example, as [1, 3]:

• **Newton's first law**
The first law states that if the sum of all forces F_i acting on an object is zero, then the velocity v of the object is constant or zero:

$$\sum_i F_i = 0 \quad \Leftrightarrow \quad \frac{\mathrm{d}v}{\mathrm{d}t} = 0 \,. \tag{9.1}$$

NEWTON's first law of motion predicts the behavior of objects for which all existing forces are balanced.

• **Newton's second law**
The second law states that the sum of all forces F_i acting on a body is equal to the product of the mass m and acceleration a:

$$\sum_i F_i = m \times a \,. \tag{9.2}$$

NEWTON's second law of motion pertains to the behavior of objects for which all existing forces are *not* balanced.

A. Öchsner, *Computational Statics and Dynamics*,
https://doi.org/10.1007/978-981-15-1278-0_9

• Newton's third law

NEWTON's third law states that all forces exist in pairs: if one object A exerts a force F_A (*action*) on a second object B, then B simultaneously exerts a force F_B (*reaction*) on A, and the two forces are equal and opposite:

$$F_A = -F_B. \tag{9.3}$$

These forces act in pairs, the magnitudes are equal and they act in opposite directions.

9.2 Relationship Between Displacement, Velocity and Acceleration

The following equations are special cases for one-dimensional problems of linear motion. The relationship between displacement $u(t)$, velocity $v(t)$, and acceleration $a(t)$ can be expressed in differential form as:

$$v(t) = \dot{u}(t) = u(t)_{,t} = \frac{\mathrm{d}u(t)}{\mathrm{d}t}. \tag{9.4}$$

$$a(t) = \dot{v}(t) = v(t)_{,t} = \frac{\mathrm{d}v(t)}{\mathrm{d}t}, \tag{9.5}$$

$$= \ddot{u}(t) = u(t)_{,,t} = \frac{\mathrm{d}^2u(t)}{\mathrm{d}t^2}. \tag{9.6}$$

The time derivative in the above equations is also expressed by the short-hand notation with an 'over-dot'. In a similar way as in Eqs. (9.4)–(9.6), an integral formulation can be stated:

$$u(t) = \int_0^t \dot{u}(t)\mathrm{d}t = \int_0^t v(t)\mathrm{d}t\,, \tag{9.7}$$

$$v(t) = \dot{u}(t) = \int_0^t \ddot{u}(t)\mathrm{d}t = \int_0^t a(t)\mathrm{d}t\,. \tag{9.8}$$

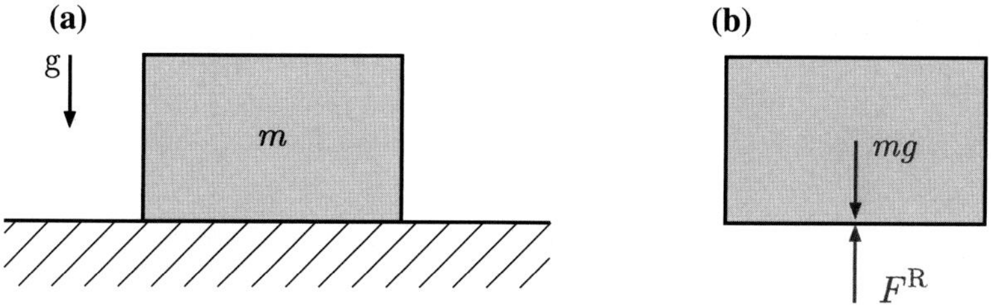

Fig. 9.1 Mass at rest on ground: **a** schematic sketch of the problem; **b** free body diagram

Fig. 9.2 Mass moving at constant velocity

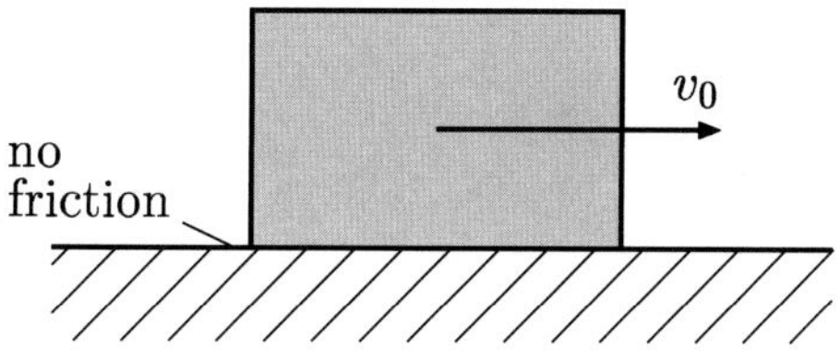

9.3 Solved Problems

9.1 Example: Mass at rest on ground
Given is a mass m at rest on the ground as shown in Fig. 9.1a. Apply NEWTON's third and first law to the problem under the assumption that the position of the mass does not vary over time.

9.1 Solution

Application of NEWTON's third law gives (see Fig. 9.1b):

$$mg = -F^{R} . \tag{9.9}$$

Application of NEWTON's first law gives (force equilibrium):

$$+ F^{R} - mg = 0 . \tag{9.10}$$

9.2 Example: Uniform linear motion of a mass with constant velocity
Given is a mass m which moves at constant velocity v_0 over a frictionless surface as shown in Fig. 9.2. This situation implies that no acceleration is acting on the mass. Determine the displacement as a function of time under the given initial condition $u(t = 0) = 0$.

9.2 Solution

Application of NEWTON's second law gives:

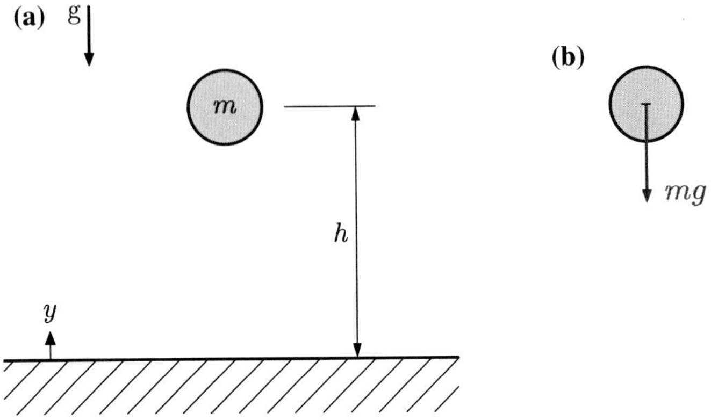

Fig. 9.3 Free fall with zero air resistance: **a** schematic sketch of the problem; **b** free body diagram

$$0 = ma = m\frac{\mathrm{d}v}{\mathrm{d}t} = m\frac{\mathrm{d}\ddot{u}}{\mathrm{d}t} \quad \text{or} \quad 0 = \frac{\mathrm{d}\ddot{u}}{\mathrm{d}t}. \tag{9.11}$$

Integration twice with respect to the time gives:

$$\dot{u}(t) = c_1 , \tag{9.12}$$

$$u(t) = c_1 t + c_2 . \tag{9.13}$$

Consideration of the boundary conditions ($t = 0$: $u = 0$ with $v_0 = \text{const.}$) gives $c_1 = v_0$ and $c_2 = 0$. The displacement results then as:

$$u(t) = v_0 \times t . \tag{9.14}$$

9.3 Example: Accelerated motion—free fall with zero air resistance
Given is a mass m which is at height h over ground at rest, see Fig. 9.3. Calculate the time to reach the ground level under the influence of gravity and the velocity at ground level. Neglect the influence of the air resistance.

9.3 Solution

Application of NEWTON's second law gives:

$$F = ma \quad \Leftrightarrow \quad -mg = m\frac{\mathrm{d}^2 u(t)}{\mathrm{d}t^2}. \tag{9.15}$$

Integration twice with respect to the time gives:

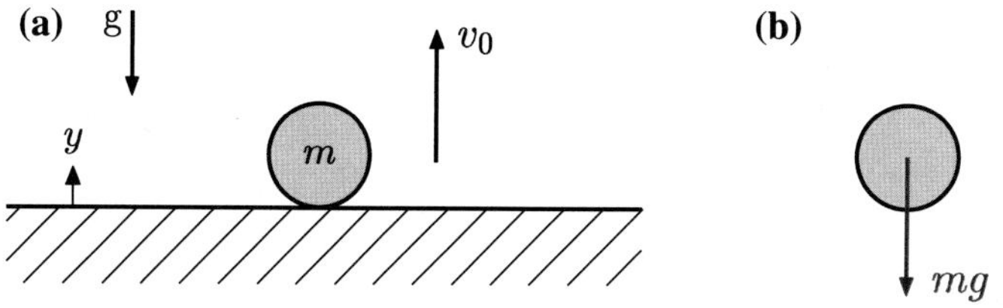

Fig. 9.4 Vertical throw upwards with zero air resistance: **a** schematic sketch of the problem; **b** free body diagram

$$-gt + c_1 = \frac{\mathrm{d}u(t)}{\mathrm{d}t}, \tag{9.16}$$

$$-\frac{1}{2}gt^2 + c_1 t + c_2 = u(t). \tag{9.17}$$

Consideration of the initial conditions ($t = 0$: $u(0) = h$ and $v(0) = 0$) gives $c_1 = 0$ and $c_2 = h$. The displacement results then as:

$$u(t) = h - \frac{gt^2}{2}, \tag{9.18}$$

and the velocity:

$$v(t) = -gt. \tag{9.19}$$

The time to reach the ground level results from the condition $0 = h - \frac{gt_h^2}{2}$:

$$t_h = \sqrt{\frac{2h}{g}}. \tag{9.20}$$

The velocity at ground level reads:

$$v(t_h) = -g\sqrt{\frac{2h}{g}} = -\sqrt{2gh} = v_h. \tag{9.21}$$

9.4 Example: Accelerated motion—vertical throw upwards with zero air resistance

Given is a mass m which is vertically thrown upwards from ground level with the initial velocity v_0, see Fig. 9.4. Calculate the maximum height, the time to reach the ground level from the maximum height, and the total flight time. Neglect the influence of the air resistance.

9.4 Solution

Application of NEWTON's second law gives:

$$F = ma \quad \Leftrightarrow \quad -mg = m\frac{\mathrm{d}^2u(t)}{\mathrm{d}t^2}. \tag{9.22}$$

Integration twice with respect to the time gives:

$$-gt + c_1 = \frac{\mathrm{d}u(t)}{\mathrm{d}t}, \tag{9.23}$$

$$-\frac{1}{2}gt^2 + c_1t + c_2 = u(t). \tag{9.24}$$

Consideration of the initial conditions ($t = 0$: $u(0) = 0$ and $v(0) = +v_0$) gives $c_1 = v_0$ and $c_2 = 0$. The displacement results then as:

$$u(t) = v_0t - \frac{gt^2}{2}, \tag{9.25}$$

and the velocity:

$$v(t) = v_0 - gt. \tag{9.26}$$

The maximum height of body results from the condition $v(t_\mathrm{m}) = 0$:

$$0 = v_0 - gt_\mathrm{m} \quad \Leftrightarrow \quad t_\mathrm{m} = \frac{v_0}{g}. \tag{9.27}$$

$$u(t_\mathrm{m}) = v_0\frac{v_0}{g} - \frac{1}{2}g\frac{v_0^2}{g^2} = \frac{v_0^2}{2g} = u_\mathrm{m}. \tag{9.28}$$

The time to reach ground level from u_m (see free fall):

$$t = \sqrt{\frac{2h}{g}} = \sqrt{\frac{2}{g}\frac{v_0^2}{2g}} = \frac{v_0}{g}. \tag{9.29}$$

Total flight time: $t_\mathrm{tot} = \dfrac{v_0}{g} + \dfrac{v_0}{g} = \dfrac{2v_0}{g}$.

9.5 Example: Accelerated motion—free fall with air resistance
Given is a mass m which is at height h over ground at rest, see Fig. 9.5. Calculate the velocity and the height as a function of time. Consider the influence of the air resistance for a slow and fast moving sphere. However, the influence of the buoyant force can be neglected.

9.5 Solution

To a reasonable approximation, air resistance tends to depend on either the first power of the speed (a linear resistance) or the second power (a quadratic resistance) [4]:

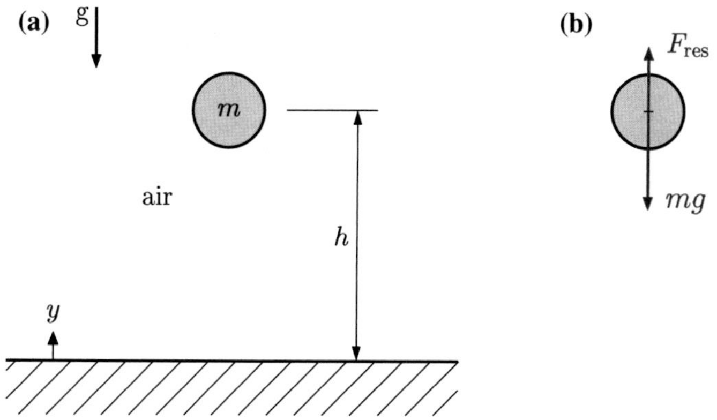

Fig. 9.5 Free fall under consideration of air resistance: **a** schematic sketch of the problem; **b** free body diagram

$$F_{res} = \begin{cases} bv & \text{for lower speed} \\ cv^2 & \text{for higher speed} \end{cases} . \tag{9.30}$$

In the previous equation, the coefficient of resistance b has the unit $\frac{\text{N}}{\text{m/s}} = \frac{\text{kg}}{\text{s}}$ while the coefficient of resistance c has the unit $\frac{\text{N}}{(\text{m/s})^2} = \frac{\text{kg}}{\text{m}}$.

• Considering lower velocities, application of NEWTON's second law gives:

$$-mg + b(-v) = m\frac{\mathrm{d}^2 u}{\mathrm{d}t^2} = m\frac{\mathrm{d}v}{\mathrm{d}t} . \tag{9.31}$$

The solution of this differential equation gives under consideration of the initial conditions ($t = 0$: $u(0) = h$ and $v(0) = 0$):

$$v(t) = \underbrace{-\frac{mg}{b}}_{v_\infty}\left(1 - \mathrm{e}^{-\frac{bt}{m}}\right) , \tag{9.32}$$

$$y(t) = h - \frac{gmt}{b} + \frac{gm^2}{b^2}\left(1 - \mathrm{e}^{-\frac{bt}{m}}\right) , \tag{9.33}$$

where v_∞ is the terminal speed.

• Considering higher velocities, we have

$$c = \frac{1}{2} c_\mathrm{D} \varrho A , \tag{9.34}$$

where c_{D} is the drag coefficient, ϱ is the density of the fluid and A is the cross-sectional area of the body. Application of NEWTON's second law gives:

$$-mg + \frac{1}{2} c_{\mathrm{D}} \varrho A v^2 = m \frac{\mathrm{d}^2 u}{\mathrm{d}t^2} = m \frac{\mathrm{d}v}{\mathrm{d}t}. \tag{9.35}$$

The solution of this differential equation gives under consideration of the initial conditions ($t = 0$: $u(0) = h$ and $v(0) = 0$):

$$v(t) = -\underbrace{\sqrt{\frac{2mg}{c_{\mathrm{D}} \varrho A}}}_{v_\infty} \tanh\left(\frac{gt}{\sqrt{\frac{2mg}{c_{\mathrm{D}} \varrho A}}}\right), \tag{9.36}$$

$$y(t) = h - \frac{v_\infty^2}{g} \ln \cosh\left(\frac{gt}{v_\infty}\right), \tag{9.37}$$

where v_∞ is the terminal speed.

9.4 Supplementary Problems

9.6 Knowledge questions on linear dynamics

- Explain the difference between statics and dynamics.
- State and explain NEWTON's first, second, and third law of motion.
- State two common approaches for the air resistance force F_{res} in a dynamic problem, i.e. for lower and higher velocities.
- Explain in a differential and integral formulation the relationship between displacement and velocity (one-dimensional linear motion).
- Explain in a differential and integral formulation the relationship between velocity and acceleration (one-dimensional linear motion).
- Explain the buoyant force which is experienced by a sphere in air.

9.7 Inclined throw

Consider a mass m which is thrown under an angle α as shown in Fig. 9.6. The initial velocity is equal to $\mathbf{v}_0 = \begin{bmatrix} v_x & v_y \end{bmatrix}^{\mathrm{T}} = v_0 \begin{bmatrix} \cos\alpha & \sin\alpha \end{bmatrix}^{\mathrm{T}}$ and the initial displacement vector is equal to $\mathbf{u} = \begin{bmatrix} 0 \; 0 \end{bmatrix}^{\mathrm{T}}$. Determine the velocity and displacement vector as a function of time. The air resistance and the buoyant force can be neglected.
Comment: The consideration of air resistance can be found in [2].

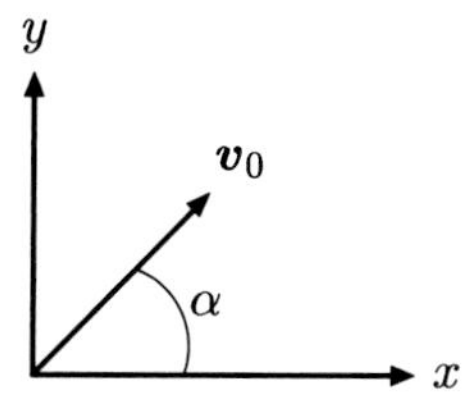

Fig. 9.6 Schematic sketch of an inclined throw problem

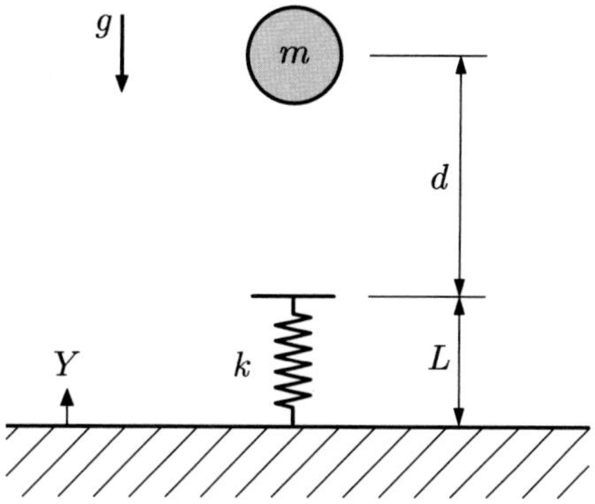

Fig. 9.7 Schematic sketch of an idealized impact test (drop tower)

9.8 Free fall under consideration of air resistance: simplification to frictionless case

Example 9.5 revealed the velocity and displacement relationships for lower speed as:

$$v(t) = -\frac{mg}{b}\left(1 - \mathrm{e}^{-\frac{bt}{m}}\right), \tag{9.38}$$

$$y(t) = h - \frac{gmt}{b} + \frac{gm^2}{b^2}\left(1 - \mathrm{e}^{-\frac{bt}{m}}\right). \tag{9.39}$$

Simplify these relationships for the frictionless case.

9.9 Idealized drop tower

Given is an idealized and simple model of an impact test which can be realized in a drop tower, see Fig. 9.7. The mass m undergoes a free fall to hit and deform an elastic spring (spring constant k) and then bounces back. Derive the functional equations for the position (y), the velocity (v), and the acceleration (a) of the mass m as a function of time (t). Disregard any losses due to air drag, friction, heat generation etc. The ideal process is fully reversible.

The following numerical values can be assumed to sketch the functions for the position, velocity, and acceleration: $g = -9.81\,\frac{\mathrm{m}}{\mathrm{s}^2}$, $k = 1000\,\frac{\mathrm{N}}{\mathrm{m}}$, $m = 0.3\,\mathrm{kg}$, $d = 1\,\mathrm{m}$, $L = 0.1\,\mathrm{m}$.

9.10 Refined drop tower model

Given is a refined model of an impact test which can be realized in a drop tower, see Fig. 9.8. The mass m undergoes a free fall to hit and deform an elastic spring (spring

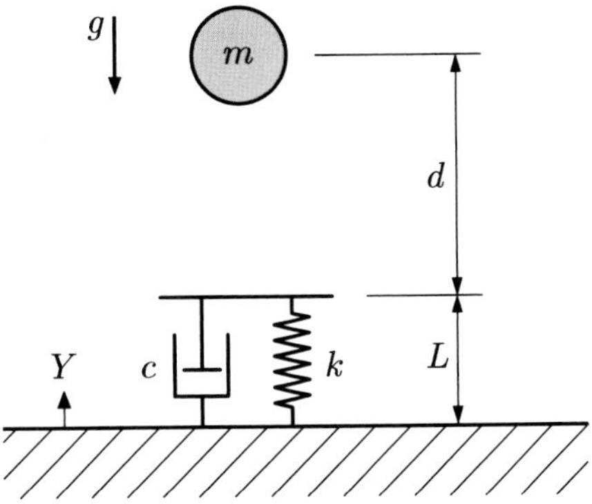

Fig. 9.8 Schematic sketch of a refined impact test model (drop tower)

constant k) in parallel with a viscous damper (viscous damping c). The mass may bounce back to a certain extend depending on the assigned parameters. Derive the functional equations for the position (y), the velocity (v), and the acceleration (a) of the mass m as a function of time (t).

The following numerical values can be assumed to sketch the functions for the position, velocity, and acceleration: $g = -9.81\,\frac{\text{m}}{\text{s}^2}$, $k = 1000\,\frac{\text{N}}{\text{m}}$, $c = 40\,\frac{\text{Ns}^2}{\text{m}}$ $m = 0.3\,\text{kg}$, $d = 1\,\text{m}$, $L = 0.1\,\text{m}$.

References

1. Hibbeler RC (2010) Engineering mechanics: dynamics. Prentice Hall, Upper Saddle River
2. Parker GW (1977) Projectile motion with air resistance quadratic in the speed. Am J Phys 45:606–610
3. Pytel A, Kiusalaas J (2010) Engineering mechanics: dynamics. Cengage Learning, Stamford
4. Timmerman P, van der Weele JP (1999) On the rise and fall of a ball with linear or quadratic drag. Am J Phys 67:538–546

Chapter 10
Integration Methods for Transient Problems

Abstract This chapter introduces to transient problems, i.e. problems where the state variables are time-dependent. The general treatment of transient problems is illustrated at the example of the rod element. Compared to the static case, the mass matrix and the solution procedure are one of the major differences. Furthermore, three different approaches to consider damping effects are briefly discussed.

10.1 Introduction

In extension to the explanations in Chap. 2, the *mass* of the rod is now considered and represented by its mass density ϱ. In addition, the distributed load $p_x(x, t)$ and the point load $F_x(t)$ are now functions of the *time* t, see Fig. 10.1.

10.2 Derivation of the Governing Differential Equation

10.2.1 Kinematics

The derivation of the kinematics relation is analogous to the approach presented in Sect. 2.2.1. Consideration in addition the time t gives:

$$\varepsilon_x(x, t) = \frac{\mathrm{d}u_x(x, t)}{\mathrm{d}x}. \tag{10.1}$$

10.2.2 Constitutive Equation

The constitutive description is based on HOOKE's law as presented in Sect. 2.2.2. Inclusion of the time gives for the relation between stress and strain:

A. Öchsner, *Computational Statics and Dynamics*,
https://doi.org/10.1007/978-981-15-1278-0_10

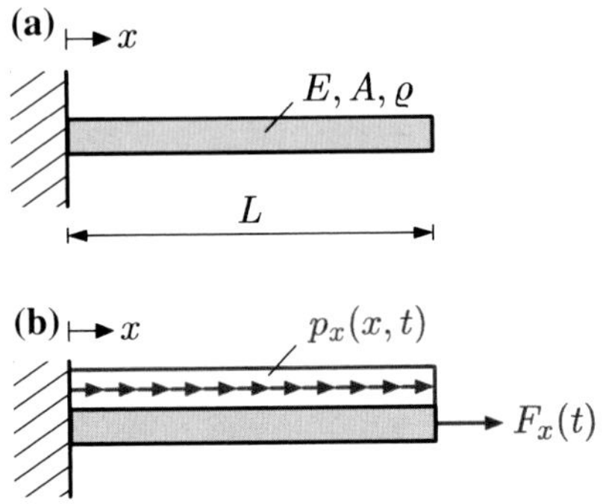

Fig. 10.1 General configuration of an axially loaded rod under consideration of time effects: **a** geometry and material property; **b** prescribed loads

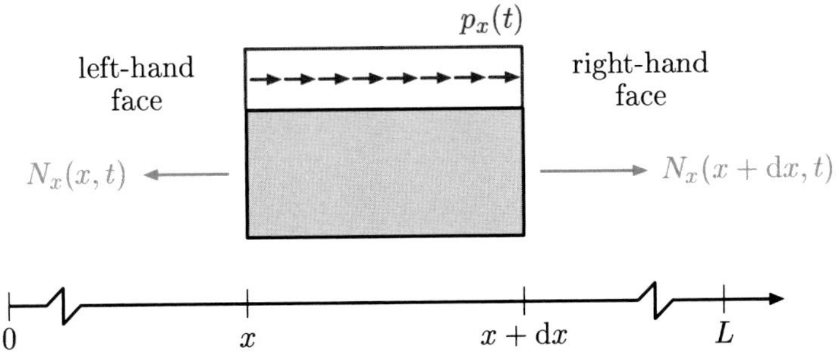

Fig. 10.2 Differential element of a rod under consideration of time effects with internal reactions and constant external distributed load

$$\sigma_x(x,t) = E\varepsilon_x(x,t). \tag{10.2}$$

10.2.3 Equilibrium

The relationship between the external forces and internal reactions, as presented in Sect. 2.2.3, must be extended under consideration of the acceleration. Consider a differential element of length dx where the distributed load $p_x(t)$ and the cross-sectional area A are constant along the x-axis, see Fig. 10.2.

Application of NEWTON's second law in the x-direction gives:

$$-N_x(x,t) + p_x(t)\mathrm{d}x + N_x(x+\mathrm{d}x,t) = m \times a_x(x,t)\,, \tag{10.3}$$

where the acceleration can be expressed as $a_x(x,t) = \frac{\mathrm{d}^2 u_x(x,t)}{\mathrm{d}t^2}$. A first-order TAYLOR's series expansion of $N_x(x+\mathrm{d}x)$ around point x, i.e.

Table 10.1 Fundamental governing equations of a rod for transient deformation along the x-axis

Expression	Equation
Kinematics	$\varepsilon_x(x,t) = \dfrac{\mathrm{d}u_x(x,t)}{\mathrm{d}x}$
Constitution	$\sigma_x(x,t) = E\varepsilon_x(x,t)$
Equilibrium	$\dfrac{\mathrm{d}N_x(x,t)}{\mathrm{d}x} = -p_x(x,t) + \dfrac{\gamma A}{\mathrm{g}} \times \dfrac{\mathrm{d}^2 u_x(x,t)}{\mathrm{d}t^2}$

$$N_x(x+\mathrm{d}x) \approx N_x(x) + \left.\frac{\mathrm{d}N_x}{\mathrm{d}x}\right|_x \mathrm{d}x\,, \tag{10.4}$$

gives finally:

$$\frac{\mathrm{d}N_x(x,t)}{\mathrm{d}x} = -p_x(x,t) + \frac{m}{\mathrm{d}x} \times \frac{\mathrm{d}^2 u_x(x,t)}{\mathrm{d}t^2}, \tag{10.5}$$

with $m = \varrho \times V = \varrho \times A \times \mathrm{d}x = \frac{\gamma}{\mathrm{g}} \times A \times \mathrm{d}x$. The mass and density related quantities are as follows[1]:

- ϱ: mass density (mass per unit volume) in $\frac{\mathrm{kg}}{\mathrm{m}^3}$,
- γ: weight density (weight per unit volume) in $\frac{\mathrm{N}}{\mathrm{m}^3}$,
- g: standard gravity or standard acceleration in $\frac{\mathrm{m}}{\mathrm{s}^2}$.

The three fundamental equations to describe the behavior of a rod element are summarized in Table 10.1.

10.2.4 Differential Equation

To derive the governing partial differential equation, the three fundamental equations given in Table 10.1 must be combined. Introducing the kinematics relation (10.1) into HOOKE's law (10.2) gives:

$$\sigma_x(x,t) = E\frac{\mathrm{d}u_x(x,t)}{\mathrm{d}x}. \tag{10.6}$$

Considering in the last equation that a normal stress is defined as an acting force N_x over a cross-sectional area A:

$$\frac{N_x(x,t)}{A} = E\frac{\mathrm{d}u_x(x,t)}{\mathrm{d}x}. \tag{10.7}$$

[1]Consider: $1\,\mathrm{N} = 1\,\frac{\mathrm{kg\,m}}{\mathrm{s}^2}$.

The last equation can be differentiated with respect to the x-coordinate to give:

$$\frac{\mathrm{d}N_x(x,t)}{\mathrm{d}x} = \frac{\mathrm{d}}{\mathrm{d}x}\left(EA\frac{\mathrm{d}u_x(x,t)}{\mathrm{d}x}\right), \tag{10.8}$$

where the derivative of the normal force can be replaced by the equilibrium equation (10.5) to obtain in the general case:

$$\frac{\mathrm{d}}{\mathrm{d}x}\left(E(x)A(x)\frac{\mathrm{d}u_x(x,t)}{\mathrm{d}x}\right) = -p_x(x,t) + \frac{\gamma A}{\mathrm{g}} \times \frac{\mathrm{d}^2 u_x(x,t)}{\mathrm{d}t^2}. \tag{10.9}$$

A common special case it obtained for $EA = \text{const.}$ and $p_x = 0$:

$$\underbrace{\frac{E\mathrm{g}}{\gamma}}_{a^2}\frac{\mathrm{d}^2 u(x,t)}{\mathrm{d}x^2} = \frac{\mathrm{d}^2 u(x,t)}{\mathrm{d}t^2}. \tag{10.10}$$

The analytical solution of this equation can be found, for example, in [3].

10.3 Finite Element Solution

10.3.1 Derivation of the Principal Finite Element Equation

Let us consider in the following the governing differential in the following form ($EA = \text{const.}$):

$$EA\frac{\mathrm{d}^2 u^0(x,t)}{\mathrm{d}x^2} + p(x,t) - \frac{\gamma A}{\mathrm{g}} \times \frac{\mathrm{d}^2 u^0(x,t)}{\mathrm{d}t^2} = 0, \tag{10.11}$$

where $u^0(x,t)$ represents here the *exact* solution of the problem. The last equation which contains the exact solution of the problem is fulfilled at each location x of the rod and is called the *strong formulation* of the problem. Replacing the exact solution in Eq. (10.11) by an approximate solution $u(x,t)$, a residual r is obtained:

$$r(x,t) = EA\frac{\mathrm{d}^2 u(x,t)}{\mathrm{d}x^2} + p(x,t) - \frac{\gamma A}{\mathrm{g}} \times \frac{\mathrm{d}^2 u(x,t)}{\mathrm{d}t^2} \neq 0. \tag{10.12}$$

As a consequence of the introduction of the approximate solution $u(x,t)$, it is in general no longer possible to satisfy the differential equation at each location x of the rod.

It is alternatively requested in the following that the differential equation is fulfilled over a certain length (an no longer at each location x) and the following integral statement is obtained

$$\int_0^L W(x,t)\left(EA\frac{\mathrm{d}^2u(x,t)}{\mathrm{d}x^2}+p(x,t)-\frac{\gamma A}{\mathrm{g}}\times\frac{\mathrm{d}^2u(x,t)}{\mathrm{d}t^2}\right)\mathrm{d}x \stackrel{!}{=} 0\,, \tag{10.13}$$

which is called the *inner product*. The function $W(x,t)$ in Eq. (10.13) is called the weight function which distributes the error or the residual in the considered domain. Integrating by parts of the first expression in the brackets of Eq. (10.13) gives

$$\int_0^L \underbrace{W}_{f}\ \underbrace{EA\frac{\mathrm{d}^2u(x,t)}{\mathrm{d}x^2}}_{g'}\mathrm{d}x = EA\left[W\frac{\mathrm{d}u}{\mathrm{d}x}\right]_0^L - EA\int_0^L \frac{\mathrm{d}W}{\mathrm{d}x}\frac{\mathrm{d}u}{\mathrm{d}x}\mathrm{d}x\,. \tag{10.14}$$

Under consideration of Eq. (10.13), the so-called *weak formulation* of the problem is obtained as:

$$\begin{aligned}EA\int_0^L \frac{\mathrm{d}W(x,t)}{\mathrm{d}x}\frac{\mathrm{d}u(x,t)}{\mathrm{d}x}\mathrm{d}x + \int_0^L W(x,t)\frac{\gamma A}{\mathrm{g}}\frac{\mathrm{d}^2u(x,t)}{\mathrm{d}t^2}\mathrm{d}x \\ = EA\left[W(x,t)\frac{\mathrm{d}u(x,t)}{\mathrm{d}x}\right]_0^L + \int_0^L W(x,t)p(x,t)\,\mathrm{d}x\,.\end{aligned} \tag{10.15}$$

Looking at the weak formulation, it can be seen that the integration by parts shifted one derivative from the approximate solution to the weight function and a symmetrical formulation with respect to the derivatives is obtained.

In order to continue the derivation of the principal finite element equation, the displacement $u(x,t)$ and the weight function $W(x,t)$ must be expressed by some functions. Again, the so-called nodal approach is applied where we now assume a decoupled formulation, i.e. time and spatial variations are assumed to be separated:

$$u^{\mathrm{e}}(x,t) = \boldsymbol{N}^{\mathrm{T}}(x)\,\boldsymbol{u}_{\mathrm{p}}(t)\,, \tag{10.16}$$

$$W(x,t) = \delta\boldsymbol{u}_{\mathrm{p}}^{\mathrm{T}}(t)\boldsymbol{N}(x)\,. \tag{10.17}$$

The required derivatives read as:

$$\frac{\mathrm{d}u^{\mathrm{e}}(x,t)}{\mathrm{d}x} = \frac{\mathrm{d}\boldsymbol{N}^{\mathrm{T}}(x)}{\mathrm{d}x}\boldsymbol{u}_{\mathrm{p}}(t)\ , \quad \frac{\mathrm{d}^2u^{\mathrm{e}}(x,t)}{\mathrm{d}t^2} = \boldsymbol{N}^{\mathrm{T}}(x)\frac{\mathrm{d}^2\boldsymbol{u}_{\mathrm{p}}(t)}{\mathrm{d}t^2}, \tag{10.18}$$

$$\frac{\mathrm{d}W(x,t)}{\mathrm{d}x} = \delta\boldsymbol{u}_{\mathrm{p}}(t)\frac{\mathrm{d}\boldsymbol{N}(x)}{\mathrm{d}x}. \tag{10.19}$$

Thus, the weak formulation can be written as:

$$\begin{aligned} EA\int_0^L \delta\boldsymbol{u}_{\mathrm{p}}(t)\frac{\mathrm{d}\boldsymbol{N}(x)}{\mathrm{d}x}\frac{\mathrm{d}\boldsymbol{N}^{\mathrm{T}}(x)}{\mathrm{d}x}\boldsymbol{u}_{\mathrm{p}}(t)\,\mathrm{d}x + \int_0^L \delta\boldsymbol{u}_{\mathrm{p}}(t)\boldsymbol{N}(x)\frac{\gamma A}{\mathrm{g}}\boldsymbol{N}^{\mathrm{T}}(x)\frac{\mathrm{d}^2\boldsymbol{u}_{\mathrm{p}}(t)}{\mathrm{d}t^2}\mathrm{d}x \\ = EA\left[\delta\boldsymbol{u}_{\mathrm{p}}(t)\boldsymbol{N}(x)\frac{\mathrm{d}u(x,t)}{\mathrm{d}x}\right]_0^L + \int_0^L \delta\boldsymbol{u}_{\mathrm{p}}(t)\boldsymbol{N}(x)p(x,t)\,\mathrm{d}x\,. \end{aligned} \tag{10.20}$$

The virtual displacements $\delta\boldsymbol{u}_{\mathrm{p}}(t)$ can be eliminated from the above equations to result in:

$$\begin{aligned} \underbrace{EA\int_0^L \frac{\mathrm{d}\boldsymbol{N}(x)}{\mathrm{d}x}\frac{\mathrm{d}\boldsymbol{N}^{\mathrm{T}}(x)}{\mathrm{d}x}\mathrm{d}x}_{\boldsymbol{K}}\ \boldsymbol{u}_{\mathrm{p}}(t) + \underbrace{\frac{\gamma A}{\mathrm{g}}\int_0^L \boldsymbol{N}(x)\boldsymbol{N}^{\mathrm{T}}(x)\,\mathrm{d}x}_{\boldsymbol{M}}\ \frac{\mathrm{d}^2\boldsymbol{u}_{\mathrm{p}}(t)}{\mathrm{d}t^2} \\ = \underbrace{EA\left[\boldsymbol{N}(x)\frac{\mathrm{d}u(x,t)}{\mathrm{d}x}\right]_0^L + \int_0^L \boldsymbol{N}(x)p(x,t)\,\mathrm{d}x}_{\boldsymbol{f}}\,, \end{aligned} \tag{10.21}$$

or in short:

$$\boldsymbol{K}\boldsymbol{u}_{\mathrm{p}}(t) + \boldsymbol{M}\frac{\mathrm{d}^2\boldsymbol{u}_{\mathrm{p}}(t)}{\mathrm{d}t^2} = \boldsymbol{f}(t)\,. \tag{10.22}$$

The matrices $\boldsymbol{K}$ and $\boldsymbol{f}$ are the same as in the static case. Thus, consideration requires only the mass matrix $\boldsymbol{M}$.

$$\begin{aligned} \boldsymbol{M} &= \frac{\gamma A}{\mathrm{g}}\int_0^L \boldsymbol{N}(x)\boldsymbol{N}^{\mathrm{T}}(x)\,\mathrm{d}x = \frac{\gamma A}{\mathrm{g}}\int_0^L \begin{bmatrix}(1-\frac{x}{L})\\ \frac{x}{L}\end{bmatrix}\begin{bmatrix}(1-\frac{x}{L}) & \frac{x}{L}\end{bmatrix}\mathrm{d}x = \\ &= \frac{\gamma A}{\mathrm{g}}\int_0^L \begin{bmatrix}(1-\frac{x}{L})^2 & \frac{x}{L}(1-\frac{x}{L})\\ \frac{x}{L}(1-\frac{x}{L}) & (\frac{x}{L})^2\end{bmatrix}\mathrm{d}x = \frac{\gamma A}{\mathrm{g}}\int_0^L \begin{bmatrix}\frac{L}{3} & \frac{L}{6}\\ \frac{L}{6} & \frac{L}{3}\end{bmatrix}\mathrm{d}x = \frac{\gamma A L}{6\mathrm{g}}\begin{bmatrix}2 & 1\\ 1 & 2\end{bmatrix}. \end{aligned} \tag{10.23}$$

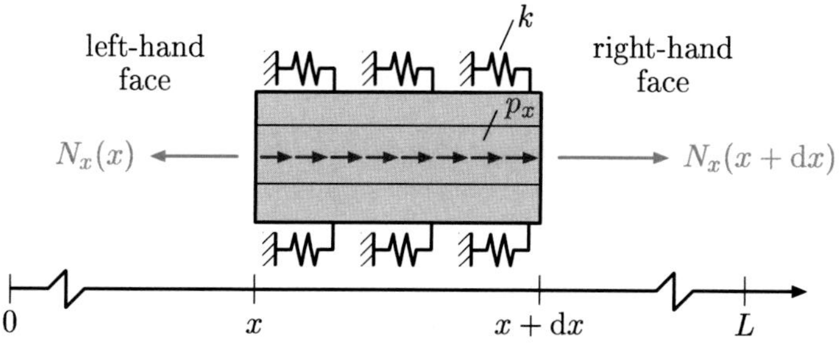

Fig. 10.3 Consideration of damping I: continuously distributed springs

Thus, the principal finite element equation for a dynamic problem reads in components as:

$$\frac{\gamma AL}{6g}\begin{bmatrix}2 & 1\\ 1 & 2\end{bmatrix}\begin{bmatrix}\ddot{u}_{1x}\\ \ddot{u}_{2x}\end{bmatrix}+\frac{EA}{L}\begin{bmatrix}1 & -1\\ -1 & 1\end{bmatrix}\begin{bmatrix}u_{1x}\\ u_{2x}\end{bmatrix}=\begin{bmatrix}F_{1x}\\ F_{2x}\end{bmatrix}+\int\limits_0^L\begin{bmatrix}N_1\\ N_2\end{bmatrix}p_x(x,t)\,\mathrm{d}x\,. \tag{10.24}$$

10.3.2 Consideration of Damping

10.3.2.1 Continuously Distributed Springs

Consider a differential element of length $\mathrm{d}x$ where the distributed load p_x and the cross-sectional area A are constant. A resistance force $F_{\mathrm{res}} = -ku_x$ is acting against any translation of the rod element, see Fig. 10.3. It should be noted here that the spring constant k has in this case the unit of force per unit area: $\frac{\mathrm{N}}{\mathrm{m}^2}$.

Application of NEWTON's second law in the x-direction gives:

$$-N_x(x) + p_x\mathrm{d}x + N_x(x+\mathrm{d}x) - F_{\mathrm{res}} = m \times a_x\,. \tag{10.25}$$

Consideration of a first-order TAYLOR's series expansion of $N_x(x + \mathrm{d}x)$ and the definition of the acceleration as well as the resistance force gives:

$$\frac{\mathrm{d}N_x(x)}{\mathrm{d}x} = -p_x(x) + \frac{\gamma A}{g} \times \frac{\mathrm{d}^2u_x(x,t)}{\mathrm{d}t^2} + ku_x(x,t)\,. \tag{10.26}$$

Combining the last equation with the kinematics and constitutive equation gives the governing partial differential equations as:

$$\frac{\mathrm{d}}{\mathrm{d}x}\left(E(x)A(x)\frac{\mathrm{d}u_x(x,t)}{\mathrm{d}x}\right) = -p_x(x) + \frac{\gamma A}{\mathrm{g}} \times \frac{\mathrm{d}^2u_x(x,t)}{\mathrm{d}t^2} + ku_x(x,t)\,. \quad (10.27)$$

The derivation of the principal finite element equation is again based on the weighted residual method. The inner product reads for constant tensile stiffness EA:

$$\int_0^L W(x,t)\left(EA\frac{\mathrm{d}^2u_x(x,t)}{\mathrm{d}x^2} + p_x(x) - \frac{\gamma A}{\mathrm{g}} \times \frac{\mathrm{d}^2u_x(x,t)}{\mathrm{d}t^2} - ku_x(x,t)\right)\mathrm{d}x = 0\,. \quad (10.28)$$

Integration by parts gives the following weak formulation as:

$$\begin{aligned} EA\int_0^L \frac{\mathrm{d}W(x,t)}{\mathrm{d}x}\frac{\mathrm{d}u_x(x,t)}{\mathrm{d}x}\mathrm{d}x + \int_0^L kW(x,t)u_x(x,t)\,\mathrm{d}x \\ + \int_0^L W(x,t)\frac{\gamma A}{\mathrm{g}}\frac{\mathrm{d}u_x^2(x,t)}{\mathrm{d}t^2}\mathrm{d}x = EA\left[W(x,t)\frac{\mathrm{d}u_x(x,t)}{\mathrm{d}x}\right]_0^L \\ + \int_0^L W(x,t)p_x(x,t)\,\mathrm{d}x\,. \end{aligned} \quad (10.29)$$

Compared to the previous case, only the following expressions requires some additional consideration:

$$\int_0^L kW(x,t)u_x(x,t)\,(d)x = k\int_0^L \delta\boldsymbol{u}_{\mathrm{p}}^{\mathrm{T}}(t)\boldsymbol{N}(x)\boldsymbol{N}^{\mathrm{T}}(x)\boldsymbol{u}_{\mathrm{p}}(t)\,\mathrm{d}x$$

$$\Rightarrow k\int_0^L \boldsymbol{N}(x)\boldsymbol{N}^{\mathrm{T}}(x)\,\mathrm{d}x\boldsymbol{u}_{\mathrm{p}}(t) \Rightarrow k\int_0^L \begin{bmatrix}(1-\frac{x}{L})\\ \frac{x}{L}\end{bmatrix}\begin{bmatrix}(1-\frac{x}{L}) & \frac{x}{L}\end{bmatrix}\mathrm{d}x =$$

$$= k\int_0^L \begin{bmatrix}(1-\frac{x}{L})^2 & \frac{x}{L}(1-\frac{x}{L})\\ \frac{x}{L}(1-\frac{x}{L}) & (\frac{x}{L})^2\end{bmatrix}\mathrm{d}x = k\int_0^L \begin{bmatrix}\frac{L}{3} & \frac{L}{6}\\ \frac{L}{6} & \frac{L}{3}\end{bmatrix}\mathrm{d}x = \frac{kL}{6}\begin{bmatrix}2 & 1\\ 1 & 2\end{bmatrix}.$$

Thus, the principal finite element equation for a dynamic problem with spring-type resistance reads in components as:

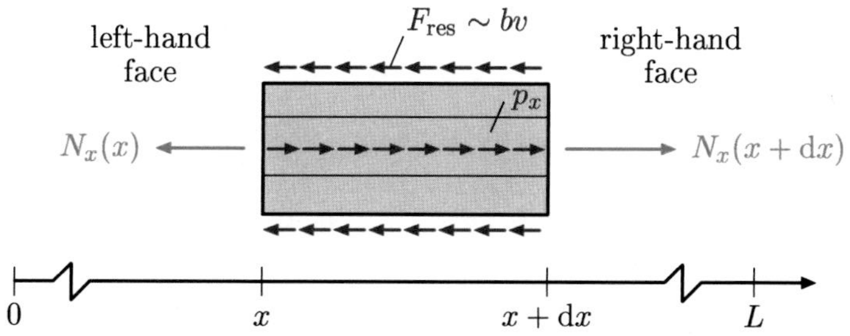

Fig. 10.4 Consideration of damping II: fluid resistance at low velocity

$$\frac{\gamma AL}{6g}\begin{bmatrix}2 & 1\\ 1 & 2\end{bmatrix}\begin{bmatrix}\ddot{u}_{1x}\\ \ddot{u}_{2x}\end{bmatrix} + \left(\frac{EA}{L}\begin{bmatrix}1 & -1\\ -1 & 1\end{bmatrix} + \frac{kL}{6}\begin{bmatrix}2 & 1\\ 1 & 2\end{bmatrix}\right)\begin{bmatrix}u_{1x}\\ u_{2x}\end{bmatrix} = \begin{bmatrix}F_{1x}\\ F_{2x}\end{bmatrix} + \int_0^L \begin{bmatrix}N_1\\ N_2\end{bmatrix} p_x(x)\,dx\,. \tag{10.30}$$

10.3.2.2 Fluid Resistance at Low Velocity

Consider a differential element of length dx where the distributed load p_x and the cross-sectional area A are constant. A resistance force $F_{res} = -bv_x$ is acting against any translation of the rod element, see Fig. 10.4. It should be noted here that the coefficient b has in this case the unit $\frac{Ns}{m^2}$.

Application of NEWTON's second law in the x-direction gives:

$$-N_x(x) + p_x dx + N_x(x + dx) - bv_x dx = m \times a_x\,. \tag{10.31}$$

Consideration of a first-order TAYLOR's series expansion of $N_x(x + dx)$ and the definition of the acceleration as well as the resistance force gives:

$$\frac{dN_x(x)}{dx} = -p_x(x) + \frac{\gamma A}{g} \times \frac{d^2u_x(x,t)}{dt^2} + b\frac{du_x(x,t)}{dt}\,. \tag{10.32}$$

Combining the last equation with the kinematics and constitutive equation gives the governing partial differential equations as:

$$\frac{d}{dx}\left(E(x)A(x)\frac{du_x(x,t)}{dx}\right) = -p_x(x) + \frac{\gamma A}{g} \times \frac{d^2u_x(x,t)}{dt^2} + b\frac{du_x(x,t)}{dt}\,. \tag{10.33}$$

The derivation of the principal finite element equation is again based on the weighted residual method. The inner product reads for constant tensile stiffness EA:

$$\int_0^L W(x,t)\left(EA\frac{\mathrm{d}^2u_x(x,t)}{\mathrm{d}x^2}+p_x(x)-\frac{\gamma A}{\mathrm{g}}\times\frac{\mathrm{d}^2u_x(x,t)}{\mathrm{d}t^2}-b\frac{\mathrm{d}u_x(x,t)}{\mathrm{d}t}\right)\mathrm{d}x=0\,. \tag{10.34}$$

Integration by parts gives the following weak formulation as:

$$EA\int_0^L \frac{\mathrm{d}W(x,t)}{\mathrm{d}x}\frac{\mathrm{d}u_x(x,t)}{\mathrm{d}x}\mathrm{d}x+\int_0^L bW(x,t)\frac{\mathrm{d}u_x(x,t)}{\mathrm{d}t}\mathrm{d}x$$
$$+\int_0^L W(x,t)\frac{\gamma A}{\mathrm{g}}\frac{\mathrm{d}u_x^2(x,t)}{\mathrm{d}t^2}\mathrm{d}x=EA\left[W(x,t)\frac{\mathrm{d}u_x(x,t)}{\mathrm{d}x}\right]_0^L$$
$$+\int_0^L W(x,t)p_x(x,t)\,\mathrm{d}x\,. \tag{10.35}$$

Compared to the previous case, only the following expressions requires some additional consideration:

$$\int_0^L bW(x,t)\frac{\mathrm{d}u_x(x,t)}{\mathrm{d}t}(d)x=b\int_0^L \delta\boldsymbol{u}_\mathrm{p}^\mathrm{T}(t)\boldsymbol{N}(x)\boldsymbol{N}^\mathrm{T}(x)\frac{\mathrm{d}\boldsymbol{u}_\mathrm{p}(t)}{\mathrm{d}t}\mathrm{d}x$$
$$\Rightarrow b\int_0^L \boldsymbol{N}(x)\boldsymbol{N}^\mathrm{T}(x)\frac{\mathrm{d}\boldsymbol{u}_\mathrm{p}(t)}{\mathrm{d}t}\Rightarrow b\int_0^L\begin{bmatrix}(1-\frac{x}{L})\\ \frac{x}{L}\end{bmatrix}\begin{bmatrix}(1-\frac{x}{L}) & \frac{x}{L}\end{bmatrix}\mathrm{d}x= \tag{10.36}$$
$$=b\int_0^L\begin{bmatrix}(1-\frac{x}{L})^2 & \frac{x}{L}(1-\frac{x}{L})\\ \frac{x}{L}(1-\frac{x}{L}) & (\frac{x}{L})^2\end{bmatrix}\mathrm{d}x=b\int_0^L\begin{bmatrix}\frac{L}{3} & \frac{L}{6}\\ \frac{L}{6} & \frac{L}{3}\end{bmatrix}\mathrm{d}x=\frac{bL}{6}\begin{bmatrix}2 & 1\\ 1 & 2\end{bmatrix}.$$

Thus, the principal finite element equation for a dynamic problem with spring-type resistance reads in components as:

$$\frac{\gamma AL}{6\mathrm{g}}\begin{bmatrix}2 & 1\\ 1 & 2\end{bmatrix}\begin{bmatrix}\ddot{u}_{1x}\\ \ddot{u}_{2x}\end{bmatrix}+\frac{bAL}{6}\begin{bmatrix}2 & 1\\ 1 & 2\end{bmatrix}\begin{bmatrix}\dot{u}_{1x}\\ \dot{u}_{2x}\end{bmatrix}+$$
$$\frac{EA}{L}\begin{bmatrix}1 & -1\\ -1 & 1\end{bmatrix}\begin{bmatrix}u_{1x}\\ u_{2x}\end{bmatrix}=\begin{bmatrix}F_{1x}\\ F_{2x}\end{bmatrix}+\int_0^L\begin{bmatrix}N_1\\ N_2\end{bmatrix}p_x(x)\,\mathrm{d}x\,. \tag{10.37}$$

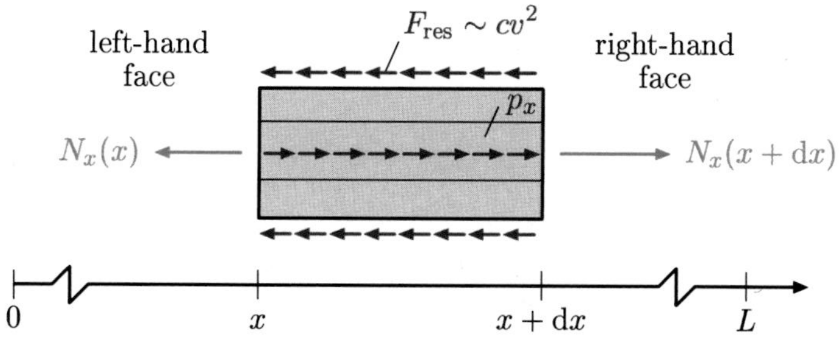

Fig. 10.5 Consideration of damping III: fluid resistance at high velocity

10.3.2.3 Fluid Resistance at High Velocity

Consider a differential element of length $\mathrm{d}x$ where the distributed load p_x and the cross-sectional area A are constant. A resistance force $F_{\rm res} = -cv_x^2$ is acting against any translation of the rod element, see Fig. 10.5. It should be noted here that the drag coefficient c has in this case the unit $\frac{\mathrm{Ns}^2}{\mathrm{m}^3}$.

Application of NEWTON's second law in the x-direction gives:

$$-N_x(x) + p_x\mathrm{d}x + N_x(x+\mathrm{d}x) - c(v_x)^2\mathrm{d}x = m \times a_x \,. \tag{10.38}$$

10.3.3 Transient Solution Schemes

The principal finite element equation for a dynamic rod element (without damping effects) reads according to Eqs. (10.24) and (10.22) in components as:

$$\frac{\gamma AL}{6\mathrm{g}}\begin{bmatrix} 2 & 1 \\ 1 & 2 \end{bmatrix}\begin{bmatrix} \ddot{u}_{1x} \\ \ddot{u}_{2x} \end{bmatrix} + \frac{EA}{L}\begin{bmatrix} 1 & -1 \\ -1 & 1 \end{bmatrix}\begin{bmatrix} u_{1x} \\ u_{2x} \end{bmatrix} = \begin{bmatrix} F_{1x} \\ F_{2x} \end{bmatrix} + \int_0^L \begin{bmatrix} N_1 \\ N_2 \end{bmatrix} p_x(x)\,\mathrm{d}x \,, \tag{10.39}$$

or in short as:

$$\boldsymbol{M}\frac{\mathrm{d}^2\boldsymbol{u}_{\rm p}(t)}{\mathrm{d}t^2} + \boldsymbol{K}\boldsymbol{u}_{\rm p}(t) = \boldsymbol{f} \,. \tag{10.40}$$

Let us recall first some fundamental mathematics which is required for the derivation of the solution scheme. A TAYLOR's series expansion of a function $f(x)$ with respect to x_0 is given by:

$$f(x) = f(x_0) + \left(\frac{\mathrm{d}f}{\mathrm{d}x}\right)_{x_0}(x - x_0) + \frac{1}{2!}\left(\frac{\mathrm{d}^2 f}{\mathrm{d}x^2}\right)_{x_0}(x - x_0)^2 + \cdots$$
$$+ \frac{1}{k!}\left(\frac{\mathrm{d}^k f}{\mathrm{d}x^k}\right)_{x_0}(x - x_0)^k , \tag{10.41}$$

where $k! = \prod_{n=1}^{k} n$ denotes the factorial of k (for example, $4! = 1 \times 2 \times 3 \times 4 = 24$).

Based on TAYLOR's series expansions of the function $u(t)$, approximate expressions for the derivatives $\frac{\mathrm{d}u(t)}{\mathrm{d}t} = \dot{u}(t)$ and $\frac{\mathrm{d}^2u(t)}{\mathrm{d}t^2} = \ddot{u}(t)$ can be derived. For sufficient smooth functions $u(t)$, a TAYLOR's series expansion around time t_i gives [4]:

$$u_{i+1} = u_i + \left(\frac{\mathrm{d}u}{\mathrm{d}t}\right)_i \Delta t + \left(\frac{\mathrm{d}^2u}{\mathrm{d}t^2}\right)_i \frac{\Delta t^2}{2} + \cdots + \left(\frac{\mathrm{d}^ku}{\mathrm{d}t^k}\right)_i \frac{\Delta t^k}{k!}, \tag{10.42}$$

$$u_{i-1} = u_i - \left(\frac{\mathrm{d}u}{\mathrm{d}t}\right)_i \Delta t + \left(\frac{\mathrm{d}^2u}{\mathrm{d}t^2}\right)_i \frac{\Delta t^2}{2} - \cdots + \left(\frac{\mathrm{d}^ku}{\mathrm{d}t^k}\right)_i \frac{\Delta t^k}{k!}, \tag{10.43}$$

where $u_{i+1} = u(t_{i+1})$ and $u_{i-1} = u(t_{i-1})$. The infinite series of Eqs. (10.42) and (10.43) are truncated for practical use after a certain number of terms. As a result of this approximation, the so-called truncation errors occurs.

Subtracting of Eq. (10.43) from (10.42) gives

$$u_{i+1} - u_{i-1} = 2\left(\frac{\mathrm{d}u}{\mathrm{d}t}\right)_i \Delta t + \frac{1}{3}\left(\frac{\mathrm{d}^3u}{\mathrm{d}t^3}\right)_i \Delta t^3 + \cdots \tag{10.44}$$

or rearranged for the first order derivative

$$\left(\frac{\mathrm{d}u}{\mathrm{d}t}\right)_i = \frac{u_{i+1} - u_{i-1}}{2\Delta t} \underbrace{- \frac{1}{6}\left(\frac{\mathrm{d}^3u}{\mathrm{d}t^3}\right)_i \Delta t^2 - \cdots}_{O(\Delta t^2)}, \tag{10.45}$$

which gives the centered difference or centered EULER approximation of the first order derivative with a truncation error is of order Δt^2 (second order accurate approximation).

The symbol 'O' in Eq. (10.45) reads 'order of' and states that if the first order derivative of $u(t)$ is approximated by the first expression on the right-hand side of Eq. (10.45), then the truncation error is of order of Δt^2.

Summing up the expression of Eqs. (10.42) and (10.43) gives

$$u_{i+1} + u_{i-1} = 2u_i + \left(\frac{\mathrm{d}^2u}{\mathrm{d}t^2}\right)_i \Delta t^2 + \frac{1}{12}\left(\frac{\mathrm{d}^4u}{\mathrm{d}t^4}\right)_i \Delta t^4 + \cdots , \tag{10.46}$$

Table 10.2 Finite difference approximations for various time differentiations, partly adapted from [1, 2]. FD = forward difference, BD = backward difference, CD = centered difference

Derivative	Finite difference approximation	Type	Error
$\left(\frac{du}{dt}\right)_i$	$\frac{u_{i+1}-u_i}{\Delta t}$	FD	$O(\Delta t)$
	$\frac{-3u_i+4u_{i+1}-u_{i+2}}{2\Delta t}$	"	$O(\Delta t^2)$
	$\frac{u_i-u_{i-1}}{\Delta t}$	BD	$O(\Delta t)$
	$\frac{3u_i-4u_{i-1}+u_{i-2}}{2\Delta t}$	"	$O(\Delta t^2)$
	$\frac{u_{i+1}-u_{i-1}}{2\Delta t}$	CD	$O(\Delta t^2)$
$\left(\frac{d^2u}{dt^2}\right)_i$	$\frac{u_{i+2}-2u_{i+1}+u_i}{\Delta t^2}$	FD	$O(\Delta t)$
	$\frac{-u_{i+3}+4u_{i+2}-5u_{i+1}+2u_i}{\Delta t^2}$	"	$O(\Delta t^2)$
	$\frac{u_i-2u_{i-1}+u_{i-2}}{\Delta t^2}$	BD	$O(\Delta t)$
	$\frac{2u_i-5u_{i-1}+4u_{i-2}-u_{i-3}}{\Delta t^2}$	"	$O(\Delta t^2)$
	$\frac{u_{i+1}-2u_i+u_{i-1}}{\Delta t^2}$	CD	$O(\Delta t^2)$

or rearranged for the second order derivative:

$$\left(\frac{d^2u}{dt^2}\right)_i = \frac{u_{i+1}-2u_i+u_{i-1}}{\Delta t^2} \underbrace{-\frac{1}{12}\left(\frac{d^4u}{dt^4}\right)_i \Delta t^2 - \cdots}_{O(\Delta t^2)}, \tag{10.47}$$

which gives the centered difference or centered EULER approximation of the second order derivative with a truncation error of order Δt^2 (second order accurate approximation). Some common expressions for derivatives of different order and the respective truncation errors are summarized in Table 10.2.

If the centered difference approximation (second order accuracy) for the time derivative ($u \rightarrow \boldsymbol{u}$) is inserted into the equation of motion according to Eq. (10.40) at the point of time t_i, one obtains:

$$\boldsymbol{M}\frac{\boldsymbol{u}_{i+1}-2\boldsymbol{u}_i+\boldsymbol{u}_{i-1}}{\Delta t^2} + \boldsymbol{K}\boldsymbol{u}_i = \boldsymbol{f}_i\,, \tag{10.48}$$

from which the displacements $\boldsymbol{u}_{i+1} = \boldsymbol{u}(t_{i+1})$ can be calculated, if the displacement at the previous points of time t_i and t_{i-1} are known:

$$\boldsymbol{M}\ (\boldsymbol{u}_{i+1} - 2\boldsymbol{u}_i + \boldsymbol{u}_{i-1}) = \Delta t^2 \boldsymbol{f}_i - \Delta t^2 \boldsymbol{K}\boldsymbol{u}_i\,, \tag{10.49}$$

$$\boldsymbol{M}\ (\boldsymbol{u}_{i+1}) = \Delta t^2 \boldsymbol{f}_i - \Delta t^2 \boldsymbol{K}\boldsymbol{u}_i + 2\boldsymbol{M}\boldsymbol{u}_i - \boldsymbol{M}\boldsymbol{u}_{i-1}\,, \tag{10.50}$$

or finally:

$$\boldsymbol{u}_{i+1} = \boldsymbol{M}^{-1}\left[\Delta t^2 \boldsymbol{f}_i - \Delta t^2\left(\boldsymbol{K} - \frac{2\boldsymbol{M}}{\Delta t^2}\right)\boldsymbol{u}_i - \boldsymbol{M}\boldsymbol{u}_{i-1}\right]. \tag{10.51}$$

To start a computation, it is required to have an approximation for $\boldsymbol{u}_{i-1} = \boldsymbol{u}(t_{i-1})$, especially for $i = 0$, i.e. $\boldsymbol{u}_{-1}$ which is known as a fictitious time step.

To overcome this problem, let us consider the approximations of the first and second order derivatives (see Eqs. (10.45) and (10.47)):

$$\left(\frac{\mathrm{d}\boldsymbol{u}}{\mathrm{d}t}\right)_i = \frac{\boldsymbol{u}_{i+1} - \boldsymbol{u}_{i-1}}{2\Delta t}, \tag{10.52}$$

$$\left(\frac{\mathrm{d}^2\boldsymbol{u}}{\mathrm{d}t^2}\right)_i = \frac{\boldsymbol{u}_{i+1} - 2\boldsymbol{u}_i + \boldsymbol{u}_{i-1}}{\Delta t^2}. \tag{10.53}$$

These two equations can be written as:

$$2\Delta t\left(\frac{\mathrm{d}\boldsymbol{u}}{\mathrm{d}t}\right)_i + \boldsymbol{u}_{i-1} = \boldsymbol{u}_{i+1}\,, \tag{10.54}$$

$$\Delta t^2\left(\frac{\mathrm{d}^2\boldsymbol{u}}{\mathrm{d}t^2}\right)_i + 2\boldsymbol{u}_i - \boldsymbol{u}_{i-1} = \boldsymbol{u}_{i+1}\,. \tag{10.55}$$

The last two equations can be equated to obtain:

$$2\Delta t\left(\frac{\mathrm{d}\boldsymbol{u}}{\mathrm{d}t}\right)_i + \boldsymbol{u}_{i-1} = \Delta t^2\left(\frac{\mathrm{d}^2\boldsymbol{u}}{\mathrm{d}t^2}\right)_i + 2\boldsymbol{u}_i - \boldsymbol{u}_{i-1}\,, \tag{10.56}$$

or finally for the fictitious time step:

$$\boldsymbol{u}_{i-1} = \boldsymbol{u}_i - \Delta t\left(\frac{\mathrm{d}\boldsymbol{u}}{\mathrm{d}t}\right)_i + \frac{\Delta t^2}{2}\left(\frac{\mathrm{d}^2\boldsymbol{u}}{\mathrm{d}t^2}\right)_i. \tag{10.57}$$

Let us summarize at the end of this section the recommended solution steps for a transient finite element problem ('hand calculation'):

① Given: $\boldsymbol{u}_0$, $\dot{\boldsymbol{u}}_0$ and $\boldsymbol{f}(t)$.
② If $\ddot{\boldsymbol{u}}_0$ is not given, solve at $t=0$: $\ddot{\boldsymbol{u}}_0 = \boldsymbol{M}^{-1}\left(\boldsymbol{f}_0 - \boldsymbol{K}\boldsymbol{u}_0\right)$.
③ Solve Eq. (10.57) for $i=0$: $\boldsymbol{u}_{-1} = \boldsymbol{u}_0 - \Delta t\left(\frac{\mathrm{d}\boldsymbol{u}}{\mathrm{d}t}\right)_0 + \frac{\Delta t^2}{2}\left(\frac{\mathrm{d}^2\boldsymbol{u}}{\mathrm{d}t^2}\right)_0$.
④ Use Eq. (10.51) to solve for $\boldsymbol{u}_1$:
$\boldsymbol{u}_1 = \boldsymbol{M}^{-1}\left[\Delta t^2\boldsymbol{f}_0 - \Delta t^2\left(\boldsymbol{K} - \frac{2\boldsymbol{M}}{\Delta t^2}\right)\boldsymbol{u}_0 - \boldsymbol{M}\boldsymbol{u}_{-1}\right]$.
⑤ With $\boldsymbol{u}_0$ given as initial condition and $\boldsymbol{u}_1$ from step 4, use Eq. (10.51) to obtain:
$\boldsymbol{u}_2 = \boldsymbol{M}^{-1}\left[\Delta t^2\boldsymbol{f}_1 - \Delta t^2\left(\boldsymbol{K} - \frac{2\boldsymbol{M}}{\Delta t^2}\right)\boldsymbol{u}_1 - \boldsymbol{M}\boldsymbol{u}_0\right]$.
⑥ Use Eq. (10.40) to solve for $\ddot{\boldsymbol{u}}_1$: $\ddot{\boldsymbol{u}}_1 = \boldsymbol{M}^{-1}\left(\boldsymbol{f}_1 - \boldsymbol{K}\boldsymbol{u}_1\right)$.
⑦ Use Eq. (10.45) to solve for $\dot{\boldsymbol{u}}_1$: $\dot{\boldsymbol{u}}_1 = \frac{\boldsymbol{u}_2 - \boldsymbol{u}_0}{2\Delta t}$.
⑧ Repeat steps 5–7 to obtain displacement, acceleration, and velocity for all other time steps.

A more general approach is the integration scheme according to NEWMARK for transient problems [1]. This scheme has been adopted in numerous finite element codes. The velocity reads

$$\dot{\boldsymbol{u}}_{n+1} = \dot{\boldsymbol{u}}_n + \Delta t\ddot{\boldsymbol{u}}_\gamma\,, \tag{10.58}$$

where the weighted acceleration is

$$\ddot{\boldsymbol{u}}_\gamma = (1-\gamma)\ddot{\boldsymbol{u}}_n + \gamma\ddot{\boldsymbol{u}}_{n+1} \qquad (0 \le \gamma \le 1)\,. \tag{10.59}$$

The parameter γ is often chosen to be $\frac{1}{2}$:

$$\ddot{\boldsymbol{u}}_{\frac{1}{2}} = \ddot{\boldsymbol{u}}_\mathrm{m} = \frac{1}{2}\left(\ddot{\boldsymbol{u}}_n + \ddot{\boldsymbol{u}}_{n+1}\right)\,, \tag{10.60}$$

which represents the constant average acceleration in the time interval $[n, n+1]$. The displacement reads as:

$$\boldsymbol{u}_{n+1} = \boldsymbol{u}_n + \Delta t\dot{\boldsymbol{u}}_n + \frac{1}{2}(\Delta t)^2\,\ddot{\boldsymbol{u}}_\beta\,, \tag{10.61}$$

where

$$\ddot{\boldsymbol{u}}_\beta = (1-2\beta)\ddot{\boldsymbol{u}}_n + 2\beta\ddot{\boldsymbol{u}}_{n+1} \qquad (0 \le 2\beta \le 1)\,. \tag{10.62}$$

To find $\boldsymbol{u}_{n+1}$, we multiply Eq. (10.61) with $\boldsymbol{M}$ and insert the following expression for the acceleration $\ddot{\boldsymbol{u}}_{n+1} = \boldsymbol{M}^{-1}\left(\boldsymbol{F}_{n+1} - \boldsymbol{K}\boldsymbol{u}_{n+1}\right)$:

$$\begin{aligned}\boldsymbol{M}\boldsymbol{u}_{n+1} = \boldsymbol{M}\boldsymbol{u}_n + \Delta t\boldsymbol{M}\dot{\boldsymbol{u}}_n + \frac{1}{2}(\Delta t)^2\,\boldsymbol{M}(1-2\beta)\ddot{\boldsymbol{u}}_n + \\ (\Delta t)^2\,\boldsymbol{M}\beta\boldsymbol{M}^{-1}\left(\boldsymbol{F}_{n+1} - \boldsymbol{K}\boldsymbol{u}_{n+1}\right)\,,\end{aligned} \tag{10.63}$$

which can be rearranged for:

$$\boldsymbol{u}_{n+1} = \left(\boldsymbol{K} + \frac{\boldsymbol{M}}{\beta(\Delta t)^2}\right)^{-1} \left\{\boldsymbol{F}_{n+1} + \frac{\boldsymbol{M}}{\beta(\Delta t)^2}\left(\boldsymbol{u}_n + \Delta t\dot{\boldsymbol{u}}_n + \left(\frac{1}{2} - \beta\right)(\Delta t)^2\ddot{\boldsymbol{u}}_n\right)\right\}. \tag{10.64}$$

Equations (10.61) and (10.62) can be combined to result in:

$$\ddot{\boldsymbol{u}}_{n+1} = \frac{1}{\beta(\Delta)^2}\left(\boldsymbol{u}_{n+1} - \boldsymbol{u}_n - \Delta t\dot{\boldsymbol{u}}_n - (\Delta t)^2\left(\frac{1}{2} - \beta\right)\ddot{\boldsymbol{u}}_n\right). \tag{10.65}$$

The complete NEWMARK scheme can be summarized in the following manner:

① Given: $\boldsymbol{u}_0$, $\dot{\boldsymbol{u}}_0$ and $\boldsymbol{f}(t)$.
② If $\ddot{\boldsymbol{u}}_0$ is not given, solve at $t = 0$: $\ddot{\boldsymbol{u}}_0 = \boldsymbol{M}^{-1}\left(\boldsymbol{f}_0 - \boldsymbol{K}\boldsymbol{u}_0\right)$.
③ Use Eq. (10.64) to solve for $\ddot{\boldsymbol{u}}_1$:
$\boldsymbol{u}_1 = \left(\boldsymbol{K} + \frac{\boldsymbol{M}}{\beta(\Delta t)^2}\right)^{-1}\left\{\boldsymbol{F}_1 + \frac{\boldsymbol{M}}{\beta(\Delta t)^2}\left(\boldsymbol{u}_0 + \Delta t\dot{\boldsymbol{u}}_0 + \left(\frac{1}{2} - \beta\right)(\Delta t)^2\ddot{\boldsymbol{u}}_0\right)\right\}$.
④ Use Eq. (10.65) to solve for $\ddot{\boldsymbol{u}}_1$:
$\ddot{\boldsymbol{u}}_1 = \frac{1}{\beta(\Delta)^2}\left(\boldsymbol{u}_1 - \boldsymbol{u}_0 - \Delta t\dot{\boldsymbol{u}}_0 - (\Delta t)^2\left(\frac{1}{2} - \beta\right)\ddot{\boldsymbol{u}}_0\right)$.
⑤ Use Eqs. (10.58) and (10.59) to solve for $\dot{\boldsymbol{u}}_1$:
$\dot{\boldsymbol{u}}_1 = \dot{\boldsymbol{u}}_0 + \Delta t\left((1-\gamma)\ddot{\boldsymbol{u}}_0 + \gamma\ddot{\boldsymbol{u}}_1\right)$.
⑥ Repeat steps 3–5 to obtain displacement, acceleration, and velocity for all other time steps.

10.3.4 Solved Problems

10.1 Example: Transient analysis of a rod element

Given is a rod which is loaded by a time-dependent force $F_x(t)$, see Fig. 10.6. Use a single rod element and the following initial conditions $u(0) = \dot{u}(0) = 0$ to apply the solution scheme from p. 441. Further values are $F_x(0) = 2000$, $t^* = 0.2$, $\frac{EA}{L} = 100$, $m = \frac{\gamma AL}{g} = 90$, $\Delta t = 0.05$. Assume consistent units.

10.1 Solution

Let us start in the common manner, i.e. state the non-reduced system of equations. Since we consider only one single element:

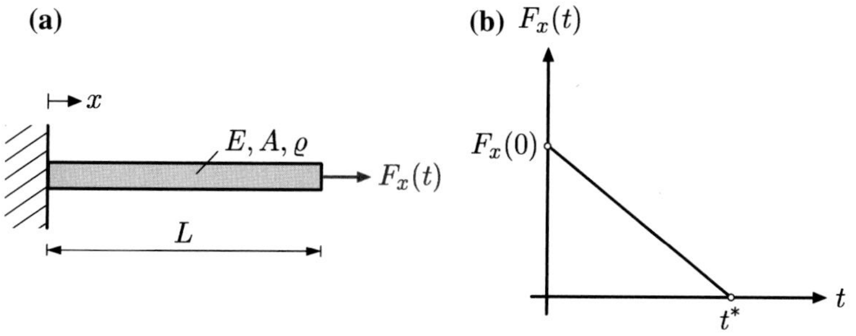

Fig. 10.6 Transient analysis of a rod element: **a** schematic sketch of the problem; **b** force-time relationship

$$\frac{\gamma AL}{6g}\begin{bmatrix}2 & 1\\ 1 & 2\end{bmatrix}\begin{bmatrix}\ddot{u}_{1x}\\ \ddot{u}_{2x}\end{bmatrix}+\frac{EA}{L}\begin{bmatrix}1 & -1\\ -1 & 1\end{bmatrix}\begin{bmatrix}u_{1x}\\ u_{2x}\end{bmatrix}=\begin{bmatrix}-R_1\\ F_x(t)\end{bmatrix}. \tag{10.66}$$

Consideration of the boundary condition at the left-hand end at $x = 0$, i.e. $u_{1x} = \ddot{u}_{1x} = 0$, gives:

$$\underbrace{\frac{\gamma AL}{3g}}_{M}\times\ddot{u}_{2x}+\underbrace{\frac{EA}{L}}_{K}\times u_{2x}=\underbrace{F_x(t)}_{f}. \tag{10.67}$$

① Given: $u_0 = u_{2x}(t_0) = 0$, $\dot{u}_0 = \dot{u}_{2x}(t_0) = 0$ and $f(t) = F_x(t) = 2000 - \frac{2000}{0.2} \times t$.
② $\ddot{u}_0$ is not given: $\ddot{u}_0 = \frac{1}{M}(F_x(0) - Ku_0) = 66.6\bar{6}$.
③ Solve Eq. (10.57) for $i = 0$: $u_{-1} = u_0 - \Delta t\dot{u}_0 + \frac{\Delta t^2}{2}\ddot{u}_0 = 0.083\bar{3}$.
④ Use Eq. (10.51) to solve for u_1: $u_1 = \frac{1}{M}\left[\Delta t^2 F_x(0) - \Delta t^2\left(K - \frac{2M}{\Delta t^2}\right)u_0 - Mu_{-1}\right] = 0.083\bar{3}$.
⑤ Use Eq. (10.40) to solve for $\ddot{u}_1$: $\ddot{u}_1 = \frac{1}{M}(F_x(t_1) - Ku_1) = 49.72\bar{2}$.
⑥ Use Eq. (10.45) to solve for $\dot{u}_1$: $\dot{u}_1 = \frac{u_2 - u_0}{2\Delta t} = 2.90972\bar{2}$.
⑦ Repeat steps 5–7 to obtain displacement, acceleration, and velocity for all other time steps.

10.2 Example: Transient analysis of a rod structure—discretization via two finite elements, consistent and lumped mass approach

Consider a cantilever rod which is discretized with two finite elements, see Fig. 10.7. The geometry and the material behavior are described by $A = 650$, $L = 2540$, $E = 7.8 \times 10^{-9}$, and $E = 210000$. A constant load of $F_x = 4450$ is applied on the right-hand end of the rod. Assume consistent units.

Calculate the nodal displacements, velocities and accelerations for $0 \leq t \leq 0.001$ and $\Delta t = 0.00025$ based on the solution scheme from p. 441.

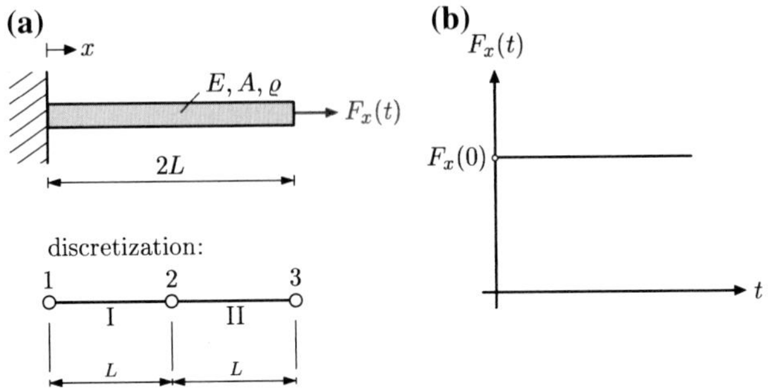

Fig. 10.7 Transient analysis of a rod structure: **a** schematic sketch of the problem and discretization; **b** force-time relationship

10.2 Solution

(a) Consistent mass approach for $\boldsymbol{M}$

Element I:

$$\underbrace{\frac{\varrho AL}{6}}_{\frac{m}{6}}\begin{bmatrix}2 & 1\\ 1 & 2\end{bmatrix}\begin{bmatrix}\ddot{u}_{1x}\\ \ddot{u}_{2x}\end{bmatrix}+\frac{EA}{L}\begin{bmatrix}1 & -1\\ -1 & 1\end{bmatrix}\begin{bmatrix}u_{1x}\\ u_{2x}\end{bmatrix}=\begin{bmatrix}-R_1\\ 0\end{bmatrix}. \tag{10.68}$$

Element II:

$$\underbrace{\frac{\varrho AL}{6}}_{\frac{m}{6}}\begin{bmatrix}2 & 1\\ 1 & 2\end{bmatrix}\begin{bmatrix}\ddot{u}_{2x}\\ \ddot{u}_{3x}\end{bmatrix}+\frac{EA}{L}\begin{bmatrix}1 & -1\\ -1 & 1\end{bmatrix}\begin{bmatrix}u_{2x}\\ u_{3x}\end{bmatrix}=\begin{bmatrix}0\\ F_x(t)\end{bmatrix}. \tag{10.69}$$

Or combined:

$$\underbrace{\frac{\varrho AL}{6}}_{\frac{m}{6}}\begin{bmatrix}2 & 1 & 0\\ 1 & 4 & 1\\ 0 & 1 & 2\end{bmatrix}\begin{bmatrix}\ddot{u}_{1x}\\ \ddot{u}_{2x}\\ \ddot{u}_{3x}\end{bmatrix}+\frac{EA}{L}\begin{bmatrix}1 & -1 & 0\\ -1 & 2 & -1\\ 0 & -1 & 1\end{bmatrix}\begin{bmatrix}u_{1x}\\ u_{2x}\\ u_{3x}\end{bmatrix}=\begin{bmatrix}-R_1\\ 0\\ F_x(t)\end{bmatrix}. \tag{10.70}$$

Consideration of the boundary condition at the left-hand end, i.e. $u_{1x}=\ddot{u}_{1x}=0$, gives:

$$\underbrace{\frac{\varrho AL}{6}}_{\frac{m}{6}}\begin{bmatrix}4 & 1\\ 1 & 2\end{bmatrix}\begin{bmatrix}\ddot{u}_{2x}\\ \ddot{u}_{3x}\end{bmatrix}+\frac{EA}{L}\begin{bmatrix}2 & -1\\ -1 & 1\end{bmatrix}\begin{bmatrix}u_{2x}\\ u_{3x}\end{bmatrix}=\begin{bmatrix}0\\ F_x(t)\end{bmatrix}. \tag{10.71}$$

① Given: $\boldsymbol{u}_0 = \boldsymbol{u}(t_0) = \begin{bmatrix}0 & 0\end{bmatrix}^{\mathrm{T}}$, $\dot{\boldsymbol{u}}_0 = \dot{\boldsymbol{u}}(t_0) = \begin{bmatrix}0 & 0\end{bmatrix}^{\mathrm{T}}$ and $F_x(t) = 4450$.

② $\ddot{\boldsymbol{u}}_0$ is not given: $\ddot{\boldsymbol{u}}_0 = \boldsymbol{M}^{-1}\left(\boldsymbol{f}_0 - \boldsymbol{K}\boldsymbol{u}_0\right)$.

$$\begin{bmatrix}\ddot{u}_{2x}\\ \ddot{u}_{3x}\end{bmatrix}=\left(\frac{\varrho AL}{6}\begin{bmatrix}4 & 1\\ 1 & 2\end{bmatrix}\right)^{-1}\left(\begin{bmatrix}0\\ 4450\end{bmatrix}-\frac{EA}{L}\begin{bmatrix}2 & -1\\ -1 & 1\end{bmatrix}\begin{bmatrix}0\\ 0\end{bmatrix}\right)$$
$$=\begin{bmatrix}-296190.786803\\ 1184763.147210\end{bmatrix}$$

③ Solve Eq. (10.57) for $i = 0$: $\boldsymbol{u}_{-1} = \boldsymbol{u}_0 - \Delta t\dot{\boldsymbol{u}}_0 + \frac{\Delta t^2}{2}\ddot{\boldsymbol{u}}_0 = \begin{bmatrix}-0.009256\\ 0.037024\end{bmatrix}$.

④ Use Eq. (10.51) to solve for $\boldsymbol{u}_1$:
$$\boldsymbol{u}_1 = \boldsymbol{M}^{-1}\left[\Delta t^2\boldsymbol{f}_0 - \Delta t^2\left(\boldsymbol{K} - \tfrac{2\boldsymbol{M}}{\Delta t^2}\right)\boldsymbol{u}_0 - \boldsymbol{M}\boldsymbol{u}_{-1}\right] = \begin{bmatrix}-0.009256\\ 0.037024\end{bmatrix}.$$

⑤ With $\boldsymbol{u}_0$ given as initial condition and $\boldsymbol{u}_1$ from step 4, use Eq. (10.51) to obtain:
$$\boldsymbol{u}_2 = \boldsymbol{M}^{-1}\left[\Delta t^2\boldsymbol{f}_1 - \Delta t^2\left(\boldsymbol{K} - \tfrac{2\boldsymbol{M}}{\Delta t^2}\right)\boldsymbol{u}_1 - \boldsymbol{M}\boldsymbol{u}_0\right] = \begin{bmatrix}-0.001847\\ 0.094295\end{bmatrix}.$$

⑥ Use Eq. (10.40) to solve for $\ddot{\boldsymbol{u}}_1$: $\ddot{\boldsymbol{u}}_1 = \boldsymbol{M}^{-1}\left(\boldsymbol{f}_1 - \boldsymbol{K}\boldsymbol{u}_1\right)\begin{bmatrix}266644.033500\\ 323956.951453\end{bmatrix}$.

⑦ Use Eq. (10.45) to solve for $\dot{\boldsymbol{u}}_1$: $\dot{\boldsymbol{u}}_1 = \frac{\boldsymbol{u}_2-\boldsymbol{u}_0}{2\Delta t} = \begin{bmatrix}-3.693344\\ 188.590012\end{bmatrix}$.

⑧ Repeat steps 5–7 to obtain displacement, acceleration, and velocity for all other time steps.

(b) Lumped mass approach for $\boldsymbol{M}$

In the case of a lumped mass approach, the mass is concentrated on the nodes and not continuously distributed over the element.

Element I:

$$\underbrace{\varrho AL}_{m}\begin{bmatrix}\frac{1}{2} & 0\\ 0 & \frac{1}{2}\end{bmatrix}\begin{bmatrix}\ddot{u}_{1x}\\ \ddot{u}_{2x}\end{bmatrix}+\frac{EA}{L}\begin{bmatrix}1 & -1\\ -1 & 1\end{bmatrix}\begin{bmatrix}u_{1x}\\ u_{2x}\end{bmatrix}=\begin{bmatrix}-R_1\\ 0\end{bmatrix}. \tag{10.72}$$

Element II:

$$\underbrace{\varrho AL}_{m}\begin{bmatrix}\frac{1}{2} & 0\\ 0 & \frac{1}{2}\end{bmatrix}\begin{bmatrix}\ddot{u}_{2x}\\ \ddot{u}_{3x}\end{bmatrix}+\frac{EA}{L}\begin{bmatrix}1 & -1\\ -1 & 1\end{bmatrix}\begin{bmatrix}u_{2x}\\ u_{3x}\end{bmatrix}=\begin{bmatrix}0\\ F_x(t)\end{bmatrix}. \tag{10.73}$$

Or combined:

$$\underbrace{\varrho AL}_{m}\begin{bmatrix}\frac{1}{2} & 0 & 0\\ 0 & 1 & 0\\ 0 & 0 & \frac{1}{2}\end{bmatrix}\begin{bmatrix}\ddot{u}_{1x}\\ \ddot{u}_{2x}\\ \ddot{u}_{3x}\end{bmatrix}+\frac{EA}{L}\begin{bmatrix}1 & -1 & 0\\ -1 & 2 & -1\\ 0 & -1 & 1\end{bmatrix}\begin{bmatrix}u_{1x}\\ u_{2x}\\ u_{3x}\end{bmatrix}=\begin{bmatrix}-R_1\\ 0\\ F_x(t)\end{bmatrix}. \tag{10.74}$$

Consideration of the boundary condition at the left-hand end, i.e. $u_{1x} = \ddot{u}_{1x} = 0$, gives:

$$\underbrace{\varrho AL}_{m}\begin{bmatrix}1 & 0\\ 0 & \frac{1}{2}\end{bmatrix}\begin{bmatrix}\ddot{u}_{2x}\\ \ddot{u}_{3x}\end{bmatrix}+\frac{EA}{L}\begin{bmatrix}2 & -1\\ -1 & 1\end{bmatrix}\begin{bmatrix}u_{2x}\\ u_{3x}\end{bmatrix}=\begin{bmatrix}0\\ F_x(t)\end{bmatrix}. \tag{10.75}$$

① Given: $\boldsymbol{u}_0 = \boldsymbol{u}(t_0) = \begin{bmatrix}0 & 0\end{bmatrix}^{\mathrm{T}}$, $\dot{\boldsymbol{u}}_0 = \dot{\boldsymbol{u}}(t_0) = \begin{bmatrix}0 & 0\end{bmatrix}^{\mathrm{T}}$ and $F_x(t) = 4450$.

② $\ddot{\boldsymbol{u}}_0$ is not given: $\ddot{\boldsymbol{u}}_0 = \boldsymbol{M}^{-1}\left(\boldsymbol{f}_0 - \boldsymbol{K}\boldsymbol{u}_0\right)$.

$$\begin{bmatrix}\ddot{u}_{2x}\\ \ddot{u}_{3x}\end{bmatrix} = \left(\varrho AL\begin{bmatrix}1 & 0\\ 0 & \frac{1}{2}\end{bmatrix}\right)^{-1}\left(\begin{bmatrix}0\\ 4450\end{bmatrix} - \frac{EA}{L}\begin{bmatrix}2 & -1\\ -1 & 1\end{bmatrix}\begin{bmatrix}0\\ 0\end{bmatrix}\right)$$

$$= \begin{bmatrix}0.0\\ 691111.835873\end{bmatrix}$$

③ Solve Eq. (10.57) for $i = 0$: $\boldsymbol{u}_{-1} = \boldsymbol{u}_0 - \Delta t\dot{\boldsymbol{u}}_0 + \frac{\Delta t^2}{2}\ddot{\boldsymbol{u}}_0 = \begin{bmatrix}0\\ 0.021597\end{bmatrix}$.

④ Use Eq. (10.51) to solve for $\boldsymbol{u}_1$:

$$\boldsymbol{u}_1 = \boldsymbol{M}^{-1}\left[\Delta t^2\boldsymbol{f}_0 - \Delta t^2\left(\boldsymbol{K} - \tfrac{2\boldsymbol{M}}{\Delta t^2}\right)\boldsymbol{u}_0 - \boldsymbol{M}\boldsymbol{u}_{-1}\right] = \begin{bmatrix}0\\ 0.021597\end{bmatrix}.$$

⑤ With $\boldsymbol{u}_0$ given as initial condition and $\boldsymbol{u}_1$ from step 4, use Eq. (10.51) to obtain:

$$\boldsymbol{u}_2 = \boldsymbol{M}^{-1}\left[\Delta t^2\boldsymbol{f}_1 - \Delta t^2\left(\boldsymbol{K} - \tfrac{2\boldsymbol{M}}{\Delta t^2}\right)\boldsymbol{u}_1 - \boldsymbol{M}\boldsymbol{u}_0\right] = \begin{bmatrix}0.005633\\ 0.075123\end{bmatrix}.$$

⑥ Use Eq. (10.51) to solve for $\ddot{\boldsymbol{u}}_1$: $\ddot{\boldsymbol{u}}_1 = \boldsymbol{M}^{-1}\left(\boldsymbol{f}_1 - \boldsymbol{K}\boldsymbol{u}_1\right)\begin{bmatrix}90127.144428\\ 510857.547017\end{bmatrix}$.

⑦ Use Eq. (10.45) to solve for $\dot{\boldsymbol{u}}_1$: $\dot{\boldsymbol{u}}_1 = \frac{\boldsymbol{u}_2 - \boldsymbol{u}_0}{2\Delta t} = \begin{bmatrix}11.265893\\ 150.246173\end{bmatrix}$.

⑧ Repeat steps 5–7 to obtain displacement, acceleration, and velocity for all other time steps.

The advantage of the lumped (i.e., a diagonal matrix, see Eq. (10.76)) mass lies in the fact that its inversion is simply obtained by inverting the diagonal elements (see Eq. (10.77)). This fact speeds up the computational analysis scheme which requires many times the computation of the inverse mass matrix.

$$\boldsymbol{M} = \begin{bmatrix}a_{11} & 0 & \cdots & 0\\ 0 & a_{22} & \cdots & 0\\ \vdots & \vdots & \ddots & \vdots\\ 0 & 0 & \cdots & a_{nn}\end{bmatrix}, \tag{10.76}$$

$$M^{-1} = \begin{bmatrix} \frac{1}{a_{11}} & 0 & \cdots & 0 \\ 0 & \frac{1}{a_{22}} & \cdots & 0 \\ \vdots & \vdots & \ddots & \vdots \\ 0 & 0 & \cdots & \frac{1}{a_{nn}} \end{bmatrix}. \tag{10.77}$$

10.4 Supplementary Problems

10.3 Knowledge questions on transient finite element problems

- State the general form of the principal finite element equation for (a) a linear-static and (b) a dynamic problem. Name two major differences.
- Sate the difference between the consistent and lumped mass approach.
- Explain three different approaches to consider damping and their effect and the principal finite element equation.
- Explain the difference in the spatial and time discretization.
- Explain the major characteristics of the NEWMARK scheme.

10.4 Consistent and lumped mass approach

Given is a rod element of length L and cross-sectional area A. The mass is represented by the element density ϱ. Derive the formulation for (a) the consistent finite element mass matrix and (b) the lumped finite element mass matrix. Calculate in addition the inverse mass matrices.

References

1. Bathe K-J (1996) Finite element procedures. Prentice-Hall, Upper Saddle River
2. Collatz L (1966) The numerical treatment of differential equations. Springer, Berlin
3. Inman DJ (2008) Engineering vibration. Pearson Education, Upper Saddle River
4. Öchsner A (2014) Elasto-plasticity of frame structure elements: modelling and simulation of rods and beams. Springer, Berlin

Appendix A
Mathematics

A.1 Greek Alphabet

See Table A.1.

A.2 Frequently Used Constants

$$\pi = 3.14159\,,$$
$$e = 2.71828\,,$$
$$\sqrt{2} = 1.41421\,,$$
$$\sqrt{3} = 1.73205\,,$$
$$\sqrt{5} = 2.23606\,,$$
$$\sqrt{e} = 1.64872\,,$$
$$\sqrt{\pi} = 1.77245\,.$$

A.3 Special Products

$$(x+y)^2 = x^2 + 2xy + y^2\,, \tag{A.1}$$
$$(x-y)^2 = x^2 - 2xy + y^2\,, \tag{A.2}$$
$$(x+y)^3 = x^3 + 3x^2y + 3xy^2 + y^3\,, \tag{A.3}$$
$$(x-y)^3 = x^3 - 3x^2y + 3xy^2 - y^3\,, \tag{A.4}$$
$$(x+y)^4 = x^4 + 4x^3y + 6x^2y^2 + 4xy^3 + y^4\,, \tag{A.5}$$

A. Öchsner, *Computational Statics and Dynamics*,
https://doi.org/10.1007/978-981-15-1278-0

Table A.1 The Greek alphabet

Name	Small letters	Capital letters
Alpha	α	A
Beta	β	B
Gamma	γ	Γ
Delta	δ	Δ
Epsilon	ϵ	E
Zeta	ζ	Z
Eta	η	H
Theta	θ, ϑ	Θ
Iota	ι	I
Kappa	κ	K
Lambda	λ	Λ
My	μ	M
Ny	ν	N
Xi	ξ	Ξ
Omikron	o	O
Pi	π	Π
Rho	ρ, ϱ	P
Sigma	σ	Σ
Tau	τ	T
Ypsilon	υ	Υ
Phi	ϕ, φ	Φ
Chi	χ	X
Psi	ψ	Ψ
Omega	ω	Ω

$$(x-y)^4 = x^4 - 4x^3y + 6x^2y^2 - 4xy^3 + y^4 \,. \tag{A.6}$$

A.4 Trigonometric Functions

Definition on a Right-Angled Triangle

The triangle ABC is in C right-angled and has edges of length a, b, c. The trigonometric functions of the angle α are defined in the following manner (see Fig. A.1):

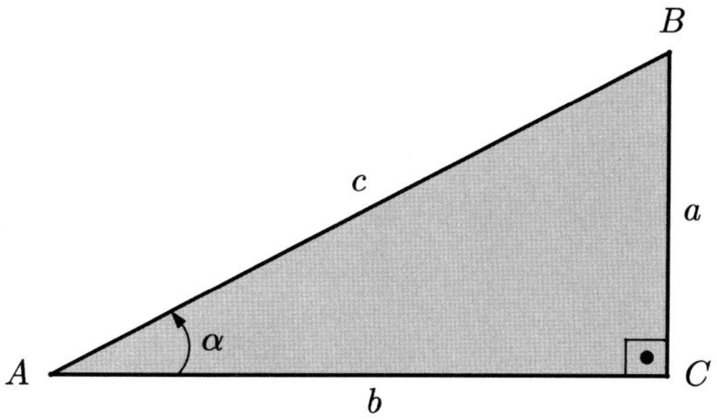

Fig. A.1 Right-angled triangle

$$\text{sine of } \alpha = \sin\alpha = \frac{a}{c} = \frac{\text{opposite}}{\text{hypotenuse}}, \tag{A.7}$$

$$\text{cosine of } \alpha = \cos\alpha = \frac{b}{c} = \frac{\text{adjacent}}{\text{hypotenuse}}, \tag{A.8}$$

$$\text{tangent of } \alpha = \tan\alpha = \frac{a}{b} = \frac{\text{opposite}}{\text{adjacent}}, \tag{A.9}$$

$$\text{cotangent of } \alpha = \cot\alpha = \frac{b}{a} = \frac{\text{adjacent}}{\text{opposite}}, \tag{A.10}$$

$$\text{secant of } \alpha = \sec\alpha = \frac{c}{b} = \frac{\text{hypotenuse}}{\text{adjacent}}, \tag{A.11}$$

$$\text{cosecant of } \alpha = \csc\alpha = \frac{c}{a} = \frac{\text{hypotenuse}}{\text{opposite}}. \tag{A.12}$$

Addition Formulae

$$\sin(\alpha \pm \beta) = \sin\alpha\cos\beta \pm \cos\alpha\sin\beta\,, \tag{A.13}$$

$$\cos(\alpha \pm \beta) = \cos\alpha\cos\beta \mp \sin\alpha\sin\beta\,, \tag{A.14}$$

$$\tan(\alpha \pm \beta) = \frac{\tan\alpha \pm \tan\beta}{1 \mp \tan\alpha\tan\beta}, \tag{A.15}$$

$$\cot(\alpha \pm \beta) = \frac{\cot\alpha\cot\beta \mp 1}{\cot\beta \pm \cot\beta}. \tag{A.16}$$

Identity Formula

$$\sin^2\alpha + \cos^2\alpha = 1\,. \tag{A.17}$$

Analytic Values for Different Angles (see Table A.2)

Table A.2 Analytical values of sine, cosine, tangent and cotangent for different angles

α in degree	α in radian	$\sin\alpha$	$\cos\alpha$	$\tan\alpha$	$\cot\alpha$
$0°$	0	0	1	0	$\pm\infty$
$30°$	$\frac{1}{6}\pi$	$\frac{1}{2}$	$\frac{\sqrt{3}}{2}$	$\frac{\sqrt{3}}{3}$	$\sqrt{3}$
$45°$	$\frac{1}{4}\pi$	$\frac{\sqrt{2}}{2}$	$\frac{\sqrt{2}}{2}$	1	1
$60°$	$\frac{1}{3}\pi$	$\frac{\sqrt{3}}{2}$	$\frac{1}{2}$	$\sqrt{3}$	$\frac{\sqrt{3}}{3}$
$90°$	$\frac{1}{2}\pi$	1	0	$\pm\infty$	0
$120°$	$\frac{2}{3}\pi$	$\frac{\sqrt{3}}{2}$	$-\frac{1}{2}$	$-\sqrt{3}$	$-\frac{\sqrt{3}}{3}$
$135°$	$\frac{3}{4}\pi$	$\frac{\sqrt{2}}{2}$	$-\frac{\sqrt{2}}{2}$	1	1
$150°$	$\frac{5}{6}\pi$	$\frac{1}{2}$	$-\frac{\sqrt{3}}{2}$	$-\frac{\sqrt{3}}{3}$	$-\sqrt{3}$
$180°$	π	0	-1	0	$\pm\infty$
$210°$	$\frac{7}{6}\pi$	$-\frac{1}{2}$	$-\frac{\sqrt{3}}{2}$	$\frac{\sqrt{3}}{3}$	$\sqrt{3}$
$225°$	$\frac{5}{4}\pi$	$-\frac{\sqrt{2}}{2}$	$-\frac{\sqrt{2}}{2}$	1	1
$240°$	$\frac{4}{3}\pi$	$-\frac{\sqrt{3}}{2}$	$-\frac{1}{2}$	$\sqrt{3}$	$\frac{\sqrt{3}}{3}$
$270°$	$\frac{3}{2}\pi$	-1	0	$\pm\infty$	0
$300°$	$\frac{5}{3}\pi$	$-\frac{\sqrt{3}}{2}$	$\frac{1}{2}$	$-\sqrt{3}$	$-\frac{\sqrt{3}}{3}$
$315°$	$\frac{7}{4}\pi$	$-\frac{\sqrt{2}}{2}$	$\frac{\sqrt{2}}{2}$	-1	-1
$330°$	$\frac{11}{6}\pi$	$-\frac{1}{2}$	$\frac{\sqrt{3}}{2}$	$-\frac{\sqrt{3}}{3}$	$-\sqrt{3}$
$360°$	2π	0	1	0	$\pm\infty$

Table A.3 Recursion formulae for trigonometric functions

	$-\alpha$	$90° \pm \alpha$	$180° \pm \alpha$	$270° \pm \alpha$	$k(360°) \pm \alpha$
		$\frac{\pi}{2} \pm \alpha$	$\pi \pm \alpha$	$\frac{3\pi}{2} \pm \alpha$	$2k\pi \pm \alpha$
sin	$-\sin\alpha$	$\cos\alpha$	$\mp\sin\alpha$	$-\cos\alpha$	$\pm\sin\alpha$
cos	$\cos\alpha$	$\mp\sin\alpha$	$-\cos\alpha$	$\pm\sin\alpha$	$\cos\alpha$
tan	$-\tan\alpha$	$\mp\cot\alpha$	$\pm\tan\alpha$	$\mp\cot\alpha$	$\pm\tan\alpha$
csc	$-\csc\alpha$	$\sec\alpha$	$\mp\csc\alpha$	$-\sec\alpha$	$\pm\csc\alpha$
sec	$\sec\alpha$	$\mp\csc\alpha$	$-\sec\alpha$	$\pm\csc\alpha$	$\sec\alpha$
cot	$-\cot\alpha$	$\mp\tan\alpha$	$\pm\cot\alpha$	$\mp\tan\alpha$	$\pm\cot\alpha$

Recursion Formulae

Typical recursion formulae are summarized in Table A.3.

A.5 Derivatives

- $\dfrac{\mathrm{d}}{\mathrm{d}x}\left(\dfrac{1}{x}\right) = -\dfrac{1}{x^2}$

- $\frac{\mathrm{d}}{\mathrm{d}x}x^n = n \times x^{n-1}$
- $\frac{\mathrm{d}}{\mathrm{d}x}\sqrt[n]{x} = \frac{1}{n \times \sqrt[n]{x^{n-1}}}$
- $\frac{\mathrm{d}}{\mathrm{d}x}\sin(x) = \cos(x)$
- $\frac{\mathrm{d}}{\mathrm{d}x}\cos(x) = -\sin(x)$
- $\frac{\mathrm{d}}{\mathrm{d}x}\ln(x) = \frac{1}{x}$
- $\frac{\mathrm{d}}{\mathrm{d}x}|x| = \begin{cases} -1 \text{ for } x < 0 \\ \ \ 1 \text{ for } x > 0 \end{cases}$
- $\frac{\mathrm{d}}{\mathrm{d}x}(f(x) \times g(x)) = \frac{\mathrm{d}f(x)}{\mathrm{d}x}g(x) + f(x)\frac{\mathrm{d}g(x)}{\mathrm{d}x}$ (product rule)
- $\frac{\mathrm{d}}{\mathrm{d}x}\left(\frac{f(x)}{g(x)}\right) = \frac{\mathrm{d}f(x)/\mathrm{d}x \times g(x) - f(x) \times \mathrm{d}g(x)/\mathrm{d}x}{[g(x)]^2}$ (quotient rule)

A.6 Integrals

The indefinite integral or antiderivative $F(x) = \int f(x)\mathrm{d}x + c$ of a function $f(x)$ is a differentiable function $F(x)$ whose derivative is equal to $f(x)$, i.e., $\frac{\mathrm{d}F(x)}{\mathrm{d}x} = f(x)$. The definite integral of a continuous real-valued function $f(x)$ on a closed interval $[a, b]$, i.e., $\int_a^b f(d)\mathrm{d}x = F(b) - F(a)$, is represented by the area under the curve $f(x)$ from $x = a$ to $x = b$.

Some selected antiderivatives (c: arbitrary constant of integration):

- $\int e^x \mathrm{d}x = e^x + c$
- $\int \sqrt{x}\mathrm{d}x = \frac{2}{3}x^{\frac{3}{2}} + c$
- $\int \sin(x)\mathrm{d}x = -\cos(x) + c$
- $\int \cos(x)\mathrm{d}x = \sin(x) + c$
- $\int \sin(\alpha x) \cdot \cos(\alpha x)\mathrm{d}x = \frac{1}{2\alpha}\sin^2(\alpha x) + c$
- $\int \sin^2(\alpha x)\mathrm{d}x = \frac{1}{2}(x - \sin(\alpha x)\cos(\alpha x)) + c = \frac{1}{2}(x - \frac{1}{2\alpha}\sin(2\alpha x)) + c$
- $\int \cos^2(\alpha x)\mathrm{d}x = \frac{1}{2}(x + \sin(\alpha x)\cos(\alpha x)) + c = \frac{1}{2}(x + \frac{1}{2\alpha}\sin(2\alpha x)) + c$

A.1 Example: Indefinite and definite integral

Calculate the indefinite and definite integral of $f(x) = x^2 + 1$. The definite integral is to be calculated in the interval $[1, 2]$. Furthermore, give a graphical interpretation of the definite integral.

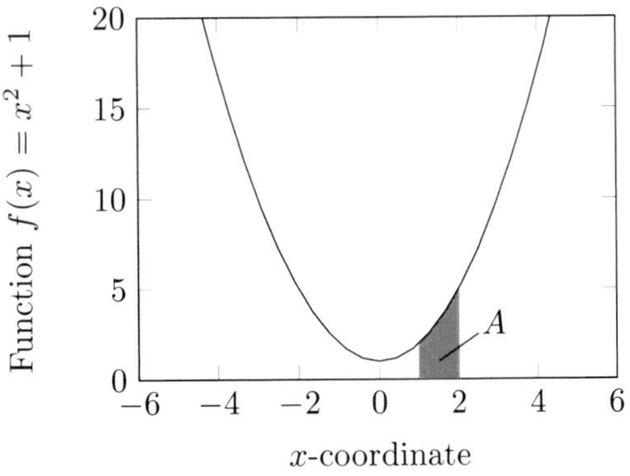

Fig. A.2 Graphical representation of the definite integral as the area A under the graph of $f(x)$

A.1 Solution

Indefinite Integral:

$$F(x) = \int (x^2 + 1)\mathrm{d}x = \tfrac{x^3}{3} + x + c\,. \tag{A.18}$$

Definite Integral:

$$\int_1^2 (x^2 + 1)\mathrm{d}x = \left[\tfrac{x^3}{3} + x\right]_1^2 = \tfrac{10}{3}\,. \tag{A.19}$$

The graphical interpretation of the definite integral is shown in Fig. A.2.

A.7 Integration by Parts

- One-dimensional case:

$$\int_a^b f(x)\tfrac{\mathrm{d}g}{\mathrm{d}x}\mathrm{d}x = f(x)g(x)|_a^b - \int_a^b \tfrac{\mathrm{d}f(x)}{\mathrm{d}x} g(x)\mathrm{d}x$$

$$= f(x)g(x)|_b - f(x)g(x)|_a - \int_a^b \tfrac{\mathrm{d}f(x)}{\mathrm{d}x} g(x)\mathrm{d}x\,. \tag{A.20}$$

- Two-dimensional case (plane): $\mathrm{d}\Omega = \mathrm{d}A = \mathrm{d}x\mathrm{d}y$

$$\int_\Omega f\tfrac{\mathrm{d}g}{\mathrm{d}x}\mathrm{d}\Omega = \int_\Gamma f g n_x\, \mathrm{d}\Gamma - \int_\Omega \tfrac{\mathrm{d}f}{\mathrm{d}x} g\, \mathrm{d}\Omega\,. \tag{A.21}$$

- Three-dimensional case (space): $\mathrm{d}\Omega = \mathrm{d}x\mathrm{d}y\mathrm{d}z$

$$\int_{\Omega} f \tfrac{\mathrm{d}g}{\mathrm{d}x} \mathrm{d}\Omega = \int_{\Gamma} f g n_x \, \mathrm{d}\Gamma - \int_{\Omega} \tfrac{\mathrm{d}f}{\mathrm{d}x} g \, \mathrm{d}\Omega \,. \tag{A.22}$$

Remark: n_x is the cosine between the outward normal and the x-direction.

Or more general (Ω is the domain and Γ refers to the boundary):

$$\int_{\Omega} f g_{,i} \mathrm{d}\Omega = \int_{\Gamma} f g n_i \, \mathrm{d}\Gamma - \int_{\Omega} f_{,i} g \, \mathrm{d}\Omega \,, \tag{A.23}$$

$$\int_{\Omega} f_{,i} g \, \mathrm{d}\Omega = \int_{\Gamma} f g n_i \, \mathrm{d}\Gamma - \int_{\Omega} f g_{,i} \, \mathrm{d}\Omega \,. \tag{A.24}$$

• GREEN–GAUSS theorem: α: scalar function; $\boldsymbol{b}$: vector function; $\nabla = \left[\frac{\partial}{\partial x} \; \frac{\partial}{\partial y} \; \frac{\partial}{\partial z} \right]^{\mathrm{T}}$: Nabla operator.

$$\int_{\Omega} \alpha \nabla^{\mathrm{T}} \boldsymbol{b} \, \mathrm{d}\Omega = \int_{\Gamma} \alpha \boldsymbol{b}^{\mathrm{T}} \boldsymbol{n} \mathrm{d}\Gamma - \int_{\Omega} (\nabla^{\mathrm{T}} \alpha) \boldsymbol{b} \, \mathrm{d}\Omega \,, \tag{A.25}$$

where $\boldsymbol{n}$ is the unit outward vector acting on the boundary surface.

– Special case: $\boldsymbol{b} \rightarrow \beta \boldsymbol{b}$ (β: scalar)

$$\int_{\Omega} \alpha \nabla^{\mathrm{T}} (\beta \boldsymbol{b}) \, \mathrm{d}\Omega = \int_{\Gamma} \alpha (\beta \boldsymbol{b})^{\mathrm{T}} \boldsymbol{n} \mathrm{d}\Gamma - \int_{\Omega} (\nabla^{\mathrm{T}} \alpha)(\beta \boldsymbol{b}) \, \mathrm{d}\Omega \,. \tag{A.26}$$

– Special case: $\boldsymbol{b} \rightarrow \boldsymbol{K}\boldsymbol{b}$ ($\boldsymbol{K}$: matrix)

$$\int_{\Omega} \alpha \nabla^{\mathrm{T}} (\boldsymbol{K}\boldsymbol{b}) \, \mathrm{d}\Omega = \int_{\Gamma} \alpha (\boldsymbol{K}\boldsymbol{b})^{\mathrm{T}} \boldsymbol{n} \mathrm{d}\Gamma - \int_{\Omega} (\nabla^{\mathrm{T}} \alpha)(\boldsymbol{K}\boldsymbol{b}) \, \mathrm{d}\Omega \,. \tag{A.27}$$

In the notation of the finite element method, especially the weighted residual method, the following formulation is convenient:

$$\int_{V} \boldsymbol{W}^{\mathrm{T}} \boldsymbol{\mathcal{L}}_1^{\mathrm{T}} (\boldsymbol{C} \boldsymbol{\mathcal{L}}_1 \boldsymbol{u}) \, \mathrm{d}V = \int_{A} \boldsymbol{W}^{\mathrm{T}} (\boldsymbol{C} \boldsymbol{\mathcal{L}}_1 \boldsymbol{u})^{\mathrm{T}} \boldsymbol{n} \mathrm{d}A - \int_{V} (\boldsymbol{\mathcal{L}}_1 \boldsymbol{W})^{\mathrm{T}} (\boldsymbol{C} \boldsymbol{\mathcal{L}}_1 \boldsymbol{u}) \, \mathrm{d}V \,. \tag{A.28}$$

A.2 Example: One-dimensional integration by parts

Calculate the definite integral of

$$\int_0^1 x^2 \mathrm{e}^x \, \mathrm{d}x \,. \tag{A.29}$$

A.2 Solution

$$\int_0^1 \underbrace{x^2}_{f} \underbrace{e^x}_{g'} dx = \left[x^2 e^x\right]_0^1 - \int_0^1 2x e^x dx \,. \tag{A.30}$$

With

$$\int_0^1 \underbrace{x}_{f} \underbrace{e^x}_{g'} dx = \left[x e^x\right]_0^1 - \int_0^1 1 e^x dx \,. \tag{A.31}$$

$$\begin{aligned} \int_0^1 x^2 e^x dx &= \left[x^2 e^x - 2x e^x\right]_0^1 + 2 \int_0^1 e^x dx \\ &= e - 2e + 2\left[e^x\right]_0^1 = e - 2 \approx 0.71828 \,. \end{aligned} \tag{A.32}$$

A.3 Example: One-dimensional integration by parts

Calculate the definite integral of

$$\int_0^1 x^3 \frac{d^2(x^3)}{dx^2} dx \,. \tag{A.33}$$

A.3 Solution

$$\int_0^1 \underbrace{x^3}_{f} \underbrace{\frac{d^2(x^3)}{dx^2}}_{g'} dx = \left[x^3 \frac{d(x^3)}{dx}\right]_0^1 - \int_0^1 3x^2 \frac{d(x^3)}{dx} dx \,. \tag{A.34}$$

With

$$\int_0^1 \underbrace{3x^2}_{f} \underbrace{\frac{d(x^3)}{dx}}_{g'} dx = \left[3x^2 x^3\right]_0^1 - \int_0^1 6x\, x^3 dx \,. \tag{A.35}$$

$$\int_0^1 x^3 \frac{d^2(x^3)}{dx^2} dx = \left[x^3 \frac{d(x^3)}{dx} - 3x^5\right]_0^1 + \int_0^1 6x^4 dx \tag{A.36}$$

$$= \left[3x^5 - 3x^5 + \frac{6}{5}x^5\right]_0^1 = \frac{6}{5} \,. \tag{A.37}$$

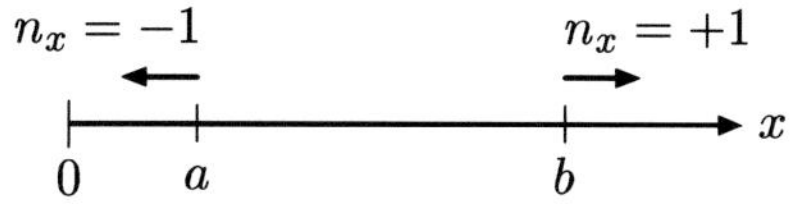

Fig. A.3 Domain and boundary representation for a one-dimensional problem

Without integration by parts (just to check the result):

$$\int_0^1 x^3 \frac{\mathrm{d}^2(x^3)}{\mathrm{d}x^2}\mathrm{d}x = \int_0^1 x^3 6x\,\mathrm{d}x = \int_0^1 6x^4\mathrm{d}x = \left[\frac{6}{5}x^5\right]_0^1 = \frac{6}{5}. \tag{A.38}$$

A.4 Example: Simplification of the Green–Gauss theorem

Simplify the GREEN–GAUSS theorem given in Eq. (A.25) to the one-dimensional case given in Eq. (A.20).

A.4 Solution

The consideration of the simplifications $\nabla^{\mathrm{T}} \rightarrow \frac{\mathrm{d}\ldots}{\mathrm{d}x}$, $\alpha \rightarrow \alpha(x)$ and $\boldsymbol{b} \rightarrow b(x)$ in Eq. (A.25) gives:

$$\int_{\Omega} \alpha(x)\frac{\mathrm{d}b(x)}{\mathrm{d}x}\mathrm{d}\Omega = \int_{\Gamma} \alpha(x)b(x)n_x\,\mathrm{d}\Gamma - \int_{\Omega} \frac{\mathrm{d}\alpha(x)}{\mathrm{d}x}b(x)\,\mathrm{d}x\,. \tag{A.39}$$

Assuming a domain $\Omega = [a, b]$ with boundaries at $x = a$ and $x = b$, the normal vectors at the boundaries result as $n_x|_a = -1$ and $n_x|_b = +1$, see Fig. A.3.

Thus, Eq. (A.39) can be transformed into:

$$\int_a^b \alpha(x)\frac{\mathrm{d}b(x)}{\mathrm{d}x}\mathrm{d}\Omega = \alpha(x)b(x)(+1)|_b + \alpha(x)b(x)(-1)|_a - \int_a^b \frac{\mathrm{d}\alpha(x)}{\mathrm{d}x}b(x)\,\mathrm{d}x\,. \tag{A.40}$$

It should be noted here that the evaluation of the boundary integral in Eq. (A.39) requires the evaluation of this integral at all boundaries and summing up the obtained values.

A.8 Integration and Coordinate Transformation

(a) One-Dimensional Case:

Let $T : \mathbb{R} \rightarrow \mathbb{R}$ given by $x = g(u)$ be a one-dimensional transformation from S to R. If g has a continuous partial derivative such that the Jacobian determinant is never zero, then

$$\int_R f(x)\mathrm{d}x = \int_S f(g(u)) \left|\frac{\mathrm{d}x}{\mathrm{d}u}\right| \mathrm{d}u\,, \tag{A.41}$$

where the Jacobian determinant[1] is $J = \left|\frac{\mathrm{d}x}{\mathrm{d}u}\right| = \frac{\mathrm{d}x}{\mathrm{d}u} = x_u$. Pay attention to the fact that the symbol $|\ldots|$ represents the determinant and should not be confused with the absolute value.

(b) Two-Dimensional Case:

Let $T : \mathbb{R}^2 \to \mathbb{R}^2$ given by $x = g(u, v)$ and $y = h(u, v)$ be a transformation on the plane that is one from a region S to a region R. If g and h have continuous partial derivatives such that the Jacobian determinant is never zero, then

$$\iint_R f(x, y)\mathrm{d}y\mathrm{d}x = \iint_S f(g(u, v), h(u, v)) \left|\frac{\partial(x, y)}{\partial(u, v)}\right| \mathrm{d}u\,\mathrm{d}v\,, \tag{A.42}$$

where the Jacobian determinant $J = \left|\frac{\partial(x, y)}{\partial(u, v)}\right| = \begin{vmatrix} x_u & x_v \\ y_u & y_v \end{vmatrix} = x_u \cdot y_v - x_v \cdot y_u = \frac{\partial x}{\partial u} \cdot \frac{\partial y}{\partial v} - \frac{\partial x}{\partial v} \cdot \frac{\partial y}{\partial u}$.

(c) Three-Dimensional Case:

Let $T : \mathbb{R}^3 \to \mathbb{R}^3$ given by $x = g(u, v, w)$, $y = h(u, v, w)$ and $z = k(u, v, w)$ be a transformation on the space that is one from a space S to a space R. If g, h and k have continuous partial derivatives such that the Jacobian determinant is never zero, then

$$\iiint_R f(x, y, z)\mathrm{d}z\mathrm{d}y\mathrm{d}x = \iiint_S f(g(u, v, w), h(u, v, w), k(u, v, w)) \left|\frac{\partial(x, y, z)}{\partial(u, v, w)}\right| \mathrm{d}w\,\mathrm{d}u\,\mathrm{d}v\,, \tag{A.43}$$

where the Jacobian determinant is $J = \left|\frac{\partial(x, y, z)}{\partial(u, v, w)}\right| = \begin{vmatrix} x_u & x_v & x_w \\ y_u & y_v & y_w \\ z_u & z_v & z_w \end{vmatrix}$.

Remark: A useful fact is that the Jacobian determinant of the inverse transformation is the reciprocal of the Jacobian determinant of the original transformation, e.g.

[1] A scalar a can be viewed as a 1×1 matrix and the corresponding determinant is just the number a itself.

$$\left|\frac{\partial(x,y)}{\partial(u,v)}\right| = \left|\frac{\partial(u,v)}{\partial(x,y)}\right|^{-1}, \tag{A.44}$$

or

$$\left|\frac{\partial(u,v)}{\partial(x,y)}\right| = \left|\frac{\partial(x,y)}{\partial(u,v)}\right|^{-1}. \tag{A.45}$$

A useful relation in the scope of coordinate transformations holds for the inverse Jacobian matrix together with its determinant in the following way:

$$\underbrace{\begin{bmatrix} u_x & u_y & u_z \\ v_x & v_y & v_z \\ w_x & w_y & w_z \end{bmatrix}}_{J} = \underbrace{\begin{bmatrix} x_u & x_v & x_w \\ y_u & y_v & y_w \\ z_u & z_v & z_w \end{bmatrix}^{-1}}_{J'^{-1}} = \frac{1}{J}\cdot\begin{bmatrix} y_v z_w - y_w z_v & -x_v z_w + x_w z_v & x_v y_w - x_w y_v \\ -y_u z_w + y_w z_u & x_u z_w - x_w z_u & -x_u y_w + x_w y_u \\ y_u z_v - y_v z_u & -x_u z_v + x_v z_u & x_u y_v - x_v y_u \end{bmatrix}, \tag{A.46}$$

where the Jacobian determinant J is given by

$$J = \left|\frac{\partial(x,y,z)}{\partial(u,v,w)}\right| = x_u y_v z_w + x_v y_w z_u + x_w y_u z_v - x_v y_u z_w - x_w y_v z_u - x_u y_w z_v\,. \tag{A.47}$$

For the 2D case, Eq. (A.46) can be simplified to

$$\underbrace{\begin{bmatrix} u_x & u_y \\ v_x & v_y \end{bmatrix}}_{J} = \underbrace{\begin{bmatrix} x_u & x_v \\ y_u & y_v \end{bmatrix}^{-1}}_{J'^{-1}} = \frac{1}{J}\cdot\begin{bmatrix} y_v & -x_v \\ -y_u & x_u \end{bmatrix}, \tag{A.48}$$

where the Jacobian determinant J is given by

$$J = \left|\frac{\partial(x,y)}{\partial(u,v)}\right| = \begin{vmatrix} x_u & x_v \\ y_u & y_v \end{vmatrix} = x_u y_v - x_v y_u\,. \tag{A.49}$$

For the 1D case, Eq. (A.46) can be simplified to

$$\underbrace{\begin{bmatrix} u_x \end{bmatrix}}_{J} = \underbrace{\begin{bmatrix} x_u \end{bmatrix}^{-1}}_{J'^{-1}} = \frac{1}{J}, \tag{A.50}$$

where the Jacobian determinant J is given by

$$J = \left|\frac{\mathrm{d}x}{\mathrm{d}u}\right| = x_u\,. \tag{A.51}$$

A.5 Example: One-dimensional integration and coordinate transformation for a constant

Given is the integral $\int_0^L \frac{EA}{L^2} \times 1 \mathrm{d}x$. Change the variable from $0 \leq x \leq L$ to $-1 \leq \xi \leq 1$ and calculate the definite integral.

A.5 Solution

The transformation between the variables can be written as $\xi = \frac{2x}{L} - 1$ or as $\mathrm{d}\xi = \frac{2}{L}\mathrm{d}x$. Thus,

$$\int_0^L \frac{EA}{L^2} \times 1 \, \mathrm{d}x = \frac{EA}{L^2} \int_0^L 1 \, \mathrm{d}x = \frac{EA}{L^2} \int_{-1}^{1} 1 \times \frac{L}{2} \, \mathrm{d}\xi$$
$$= \frac{EA}{L^2} \times \frac{L}{2} [\xi]_{-1}^{1} = \frac{EA}{2L}(1+1) = \frac{EA}{L}. \tag{A.52}$$

A.6 Example: One-dimensional integration and coordinate transformation for a linear function

Given is the integral $\int_0^L \frac{EA}{L^2} \times x \, \mathrm{d}x$. Change the variable from $0 \leq x \leq L$ to $-1 \leq \xi \leq 1$ and calculate the definite integral.

A.6 Solution

The transformation between the variables can be written as $\xi = \frac{2x}{L} - 1$ or as $\mathrm{d}\xi = \frac{2}{L}\mathrm{d}x$.

Solution without transformation:

$$\int_0^L \frac{EA}{L^2} \times x \, \mathrm{d}x = \frac{EA}{L^2} \left[\frac{x^2}{2} \right]_0^L = \frac{EA}{L^2} \times \frac{L^2}{2} = \frac{EA}{2}. \tag{A.53}$$

Solution with transformation:

$$\int_0^L \frac{EA}{L^2} \times x \, \mathrm{d}x = \frac{EA}{L^2} \int_0^L x \, \mathrm{d}x = \frac{EA}{L^2} \int_{-1}^{1} \underbrace{\frac{L}{2}(\xi + 1)}_{x} \times \frac{L}{2} \, \mathrm{d}\xi$$
$$= \frac{EA}{L^2} \frac{L^2}{4} \left[\frac{\xi^2}{2} + \xi \right]_{-1}^{1} = \frac{EA}{4} \left\{ \left(\tfrac{1}{2} + 1\right) - \left(\tfrac{1}{2} - 1\right) \right\} = \frac{EA}{2}. \tag{A.54}$$

A.7 Example: One-dimensional integration and coordinate transformation for a polynomial of order two

Given is the integral $\int_1^2 (x^2 + 1) \, \mathrm{d}x$. Change the variable from $1 \leq x \leq 2$ to $-1 \leq \xi \leq 1$ and calculate the definite integral.

A.7 Solution

The transformation between the variables can be written as $\xi = \frac{1}{2}(\xi + 3)$ or as $\mathrm{d}\xi = 2\mathrm{d}x$.

Solution without transformation: see Example A.1.

Solution with transformation:

$$\int_{1}^{2} (x^2 + 1)\,\mathrm{d}x = \int_{-1}^{1} \left(\tfrac{1}{4}(\xi + 3)^2 + 1\right) \times \tfrac{1}{2}\mathrm{d}\xi$$

$$= \int_{-1}^{1} \left(\tfrac{1}{8}\xi^2 + \tfrac{3}{4}\xi + \tfrac{13}{8}\right)\mathrm{d}\xi = \frac{10}{4}. \tag{A.55}$$

A.8 Example: Two-dimensional integration and coordinate transformation

Given is the integral $\int_{-a}^{a}\int_{-b}^{b} x^2y^2\mathrm{d}x\mathrm{d}y$. Change the variables from $-a \le x \le a$ and $-b \le y \le b$ to $-1 \le \xi \le 1$ and $-1 \le \eta \le 1$ for the transformations $\xi = \frac{x}{a}$ and $\eta = \frac{y}{b}$. Calculate the definite integral.

A.8 Solution

Solution without transformation:

$$\int_{-a}^{a}\int_{-b}^{b} x^2y^2\mathrm{d}x\mathrm{d}y = \int_{-a}^{a} x^2\left[\frac{y^3}{3}\right]_{-b}^{b}\mathrm{d}x = \int_{-a}^{a} x^2\left(\frac{b^3}{3} + \frac{b^3}{3}\right)\mathrm{d}x$$

$$= \frac{2b^3}{3}\int_{-a}^{a} x^2\mathrm{d}x = \frac{2b^3}{3}\left[\frac{x^3}{3}\right]_{-a}^{a} = \frac{4}{9}a^3b^3. \tag{A.56}$$

Solution with transformation:

$$J = \frac{\partial x}{\partial \xi}\frac{\partial y}{\partial \eta} - \frac{\partial x}{\partial \eta}\frac{\partial y}{\partial \xi} = ab - 0. \tag{A.57}$$

$$\int_{-1}^{1}\int_{-1}^{1} (a\xi)^2(b\eta)^2 J\,\mathrm{d}\xi\mathrm{d}\eta = a^3b^3\int_{-1}^{1} \xi^2\left[\frac{\eta^3}{3}\right]_{-1}^{1}\mathrm{d}\xi =$$

$$= a^3b^3\int_{-1}^{1} \frac{2\xi^2}{3}\mathrm{d}\xi = \frac{2}{3}a^3b^3\left[\frac{\xi^3}{3}\right]_{-1}^{1} =$$

$$= \frac{4}{9}a^3b^3. \tag{A.58}$$

A.9 Example: Two-dimensional integration and coordinate transformation: paraboloid

Given is the integral $\int_1^2 \int_1^2 (x^2 + y^2 + 1)\mathrm{d}x\mathrm{d}y$. Change the variables from $1 \le x \le 2$ and $1 \le y \le 2$ to $-1 \le \xi \le 1$ and $-1 \le \eta \le 1$ and calculate the definite integral. Compare the result with the integration based on the Cartesian coordinates (x, y).

A.9 Solution

The relationship between the Cartesian (x, y) and the natural coordinates (ξ, η) can be derived as:

$$\xi = 2\left(x - \tfrac{3}{2}\right) \quad \text{and} \quad \eta = 2\left(y - \tfrac{3}{2}\right) \quad \text{or} \tag{A.59}$$

$$x = \tfrac{1}{2}(\xi + 3) \quad \text{and} \quad y = \tfrac{1}{2}(\eta + 3)\,. \tag{A.60}$$

Thus, the transformed integral results finally in:

$$= \int\limits_{\xi=-1}^{+1} \int\limits_{\eta=-1}^{+1} \underbrace{\left(\tfrac{1}{16}\xi^2 + \tfrac{3}{8}\xi + \tfrac{1}{16}\eta^2 + \tfrac{3}{8}\eta + \tfrac{11}{8}\right)}_{g(\xi,\eta)} \mathrm{d}\xi\mathrm{d}\eta\,. \tag{A.61}$$

The analytical integration of Eq. (A.61) gives $\frac{17}{3}$.

A.9 Numerical Integration

A.9.1 Simpson's Rule

To derive the SIMPSON's[2] formula, three points (ξ_1, f_1), (ξ_2, f_2) and (ξ_3, f_3) are selected (cf. Fig. A.4) and the general equation for a parabola, i.e.

$$f(\xi) = a + b \times \xi + c \times \xi^2\,, \tag{A.62}$$

is separately evaluated for each of these three points:

$$f_1 = f(\xi_1) = a + b \times \xi_1 + c \times \xi_1^2\,, \tag{A.63}$$

$$f_2 = f(\xi_2) = a + b \times \xi_2 + c \times \xi_2^2\,, \tag{A.64}$$

$$f_3 = f(\xi_3) = a + b \times \xi_3 + c \times \xi_3^2\,. \tag{A.65}$$

Solving the last system of equations for the unknown coefficients a, b and c yields:

[2] Thomas SIMPSON (1710–1761), English mathematician.

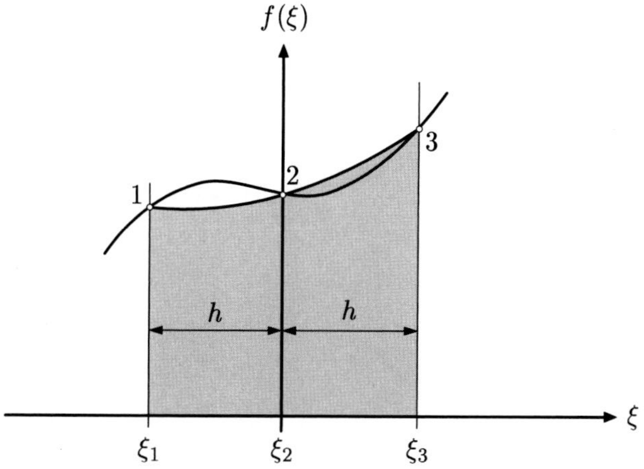

Fig. A.4 Schematic sketch for the derivation of the SIMPSON's integration rule

$$a = \frac{\xi_1^2\xi_2 f_3 - \xi_2\xi_3^2 f_1 + \xi_2^2\xi_3 f_1 - \xi_1\xi_2^2 f_3 + \xi_1\xi_3^2 f_2 - \xi_1^2\xi_3 f_2}{-\xi_2\xi_3^2 - \xi_1\xi_2^2 + \xi_1\xi_3^2 + \xi_1^2\xi_2 - \xi_1^2\xi_3 + \xi_2^2\xi_3}, \tag{A.66}$$

$$b = \frac{\xi_1^2 f_2 - \xi_1^2 f_3 + \xi_2^2 f_3 + \xi_3 f_1^2 - \xi_3^2 f_2 - \xi_2^2 f_1}{-\xi_2\xi_3^2 - \xi_1\xi_2^2 + \xi_1\xi_3^2 + \xi_1^2\xi_2 - \xi_1^2\xi_3 + \xi_2^2\xi_3}, \tag{A.67}$$

$$c = -\frac{\xi_2 f_3 + \xi_3 f_1 + \xi_1 f_2 - \xi_1 f_3 - \xi_2 f_1 - \xi_3 f_2}{-\xi_2\xi_3^2 - \xi_1\xi_2^2 + \xi_1\xi_3^2 + \xi_1^2\xi_2 - \xi_1^2\xi_3 + \xi_2^2\xi_3}. \tag{A.68}$$

Introducing these values for a, b and c in the general form of the parabola according to Eq. (A.62), substituting $\xi_2 = \xi_1 + h,\ \xi_3 = \xi_1 + 2h$ and rearranging yields:

$$f(\xi) = f_1 + \frac{(\xi - \xi_1)(f_2 - f_1)}{h} + \frac{(\xi - \xi_1)(\xi - \xi_1 - h)(f_3 - 2f_2 + f_1)}{2h^2}. \tag{A.69}$$

After a brief calculation we get the area under the parabola as

$$\int_{\xi_1}^{\xi_3} f(\xi)\,\mathrm{d}\xi = \frac{h}{3} \times (f_1 + 4f_2 + f_3), \tag{A.70}$$

which is also called SIMPSON's one-third rule. For the natural coordinate range, i.e. $\xi_1 = -1 < \xi < \xi_3 = 1$, Eq. (A.70) can be written as:

$$\int_{-1}^{1} f(\xi)\,\mathrm{d}\xi = \frac{1}{3} \times \Big(f(-1) + 4f(0) + f(1)\Big). \tag{A.71}$$

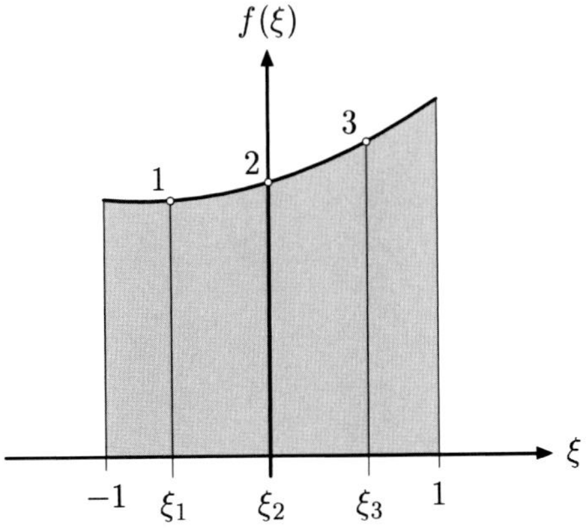

Fig. A.5 Schematic sketch for the derivation of the GAUSS–LEGENDRE integration rule

When we add up a sequence of n of these segments, we find

$$\int_{\xi_1}^{\xi_n} f(\xi)\mathrm{d}\xi = \frac{h}{3} \times \big(\underbrace{f_1 + 4f_2 + f_3}_{1^{\text{st}}\text{ segment}} + \underbrace{f_3 + 4f_4 + f_5}_{2^{\text{nd}}\text{ segment}} + \cdots + \underbrace{f_{n-2} + 4f_{n-1} + f_n}_{n^{\text{th}}\text{ segment}} \big)\,, \tag{A.72}$$

$$= \frac{h}{3}\big(f_1 + 4f_2 + 2f_3 + 4f_4 + 2f_5 + \cdots + 2f_{n-2} + 4f_{n-1} + f_n\big)\,, \tag{A.73}$$

where h is the constant distance between two consecutive grid points, i.e. $h = \xi_{i+1} - \xi_i$. If the entire domain extends over the length L, one gets $h = \frac{L}{n-1}$, or for the special case $-1 \leq \xi \leq +1$: $h = \frac{2}{n-1}$.

The last equation can be written in a general form as a function of the natural coordinates ξ as

$$\int_{-1}^{1} f(\xi)\,\mathrm{d}\xi \approx \sum_{i=1}^{n} f(\xi_i) \times w(\xi_i)\,. \tag{A.74}$$

Values for the abscissae ξ and weights w are given in Table A.4.

A.9.2 Gauss–Legendre Quadrature

The principal idea of GAUSS–LEGENDRE quadrature can be introduced by taking not only the weights w_i, but also the location of the abscissae ξ_i as unknowns (cf. Fig. A.5) and to require that an arbitrary polynomial of a certain order is exactly integrated.

Table A.4 Values for the abscissae ξ_i and weights w_i for SIMPSON integration rule, $\int_{-1}^{1} f(\xi)\mathrm{d}\xi \approx \sum_i f(\xi_i)w_i$, for n grid points ($i = 1, 2, \ldots, n-1, n$)

No. points n	Abscissae ξ_i	Weights w_i
3	0	$\frac{4}{3}$
	± 1	$\frac{1}{3}$
5	0	$\frac{1}{3}$
	± 0.5	$\frac{2}{3}$
	± 1	$\frac{1}{6}$
9	0	$\frac{1}{6}$
	± 0.25	$\frac{1}{3}$
	± 0.5	$\frac{1}{6}$
	± 0.75	$\frac{1}{3}$
	± 1	$\frac{1}{12}$
17	0	$\frac{1}{12}$
	± 0.125	$\frac{1}{6}$
	± 0.25	$\frac{1}{12}$
	± 0.375	$\frac{1}{6}$
	± 0.5	$\frac{1}{12}$
	± 0.625	$\frac{1}{6}$
	± 0.75	$\frac{1}{12}$
	± 0.875	$\frac{1}{6}$
	± 1	$\frac{1}{24}$
$n = 2^m + 1 (m = 1, 2, 3, \ldots)(\wedge\, n > 3)$	0	$\ldots$
	$\vdots$	$\vdots$
	$\ldots$	$2 \times \frac{2}{3(n-1)}$
	$\ldots$	$4 \times \frac{2}{3(n-1)}$
	± 1	$\frac{2}{3(n-1)}$

Additionally, let us assume that the weights and abscissae are symmetric around the midpoint ($\xi = 0$) of the range of integration ($-1 \leq \xi \leq 1$). Similar to the previous section (cf. Fig. A.4), let us assume three functional evaluations of $f(\xi)$ in the scope of this derivation. Thus, the integral can be approximated by

$$\int_{-1}^{1} f(\xi)\,\mathrm{d}\xi \approx f(\xi_1) \times w_1 + f(\xi_2) \times w_2 + f(\xi_3) \times w_3\,, \tag{A.75}$$

and additionally requiring that the formula gives exact expressions for integrating a fifth order polynomial, i.e.

$$\int_{-1}^{1} (a_0 + a_1\xi + a_2\xi^2 + a_3\xi^3 + a_4\xi^4 + a_5\xi^5)\mathrm{d}\xi\,, \tag{A.76}$$

where $a_0, \ldots, a_5$ are arbitrary coefficients. Since integration is additive, it will sufficient to require that Eq. (A.76) is exact for the single functions $f(\xi) = 1, \xi, \ldots, \xi^5$. Thus, to determine the six unknowns ξ_1, ξ_2, ξ_3 and w_1, w_2, w_3, the following six integral equations need to be evaluated:

$$\int_{-1}^{1} 1\mathrm{d}\xi = [\xi]_{-1}^{1} = 2 = w_1 + w_2 + w_3\,, \tag{A.77}$$

$$\int_{-1}^{1} \xi\mathrm{d}\xi = \left[\tfrac{1}{2}\,\xi^2\right]_{-1}^{1} = 0 = w_1\xi_1 + w_2\xi_2 + w_3\xi_3\,, \tag{A.78}$$

$$\int_{-1}^{1} \xi^2\mathrm{d}\xi = \left[\tfrac{1}{3}\,\xi^3\right]_{-1}^{1} = \tfrac{2}{3} = w_1\xi_1^2 + w_2\xi_2^2 + w_3\xi_3^2\,, \tag{A.79}$$

$$\int_{-1}^{1} \xi^3\mathrm{d}\xi = \left[\tfrac{1}{4}\,\xi^4\right]_{-1}^{1} = 0 = w_1\xi_1^3 + w_2\xi_2^3 + w_3\xi_3^3\,, \tag{A.80}$$

$$\int_{-1}^{1} \xi^4\mathrm{d}\xi = \left[\tfrac{1}{5}\,\xi^5\right]_{-1}^{1} = \tfrac{2}{5} = w_1\xi_1^4 + w_2\xi_2^4 + w_3\xi_3^4\,, \tag{A.81}$$

$$\int_{-1}^{1} \xi^5\mathrm{d}\xi = \left[\tfrac{1}{6}\,\xi^6\right]_{-1}^{1} = 0 = w_1\xi_1^5 + w_2\xi_2^5 + w_3\xi_3^5\,. \tag{A.82}$$

These six simultaneous nonlinear equations can be solved to obtain:

$$\xi_1 = -0.7745966692\,, \qquad w_1 = 0.5555555556\,, \tag{A.83}$$

$$\xi_2 = 0.0\,, \qquad w_2 = 0.8888888889\,, \tag{A.84}$$

$$\xi_3 = 0.7745966692\,, \qquad w_3 = 0.5555555556\,. \tag{A.85}$$

In Table A.5, coefficients and arguments for n-point GAUSS–LEGENDRE quadrature rules are given for the integral of the form

$$\int_{-1}^{1} f(\xi)\,\mathrm{d}\xi \approx \sum_{i=1}^{n} f(\xi_i) \times w(\xi_i)\,. \tag{A.86}$$

Table A.5 Values for the abscissae ξ_i and weights w_i for GAUSS–LEGENDRE integration rule, $\int_{-1}^{1} f(\xi)\mathrm{d}\xi \approx \sum_i f(\xi_i)w_i$

No. points n	Abscissae ξ_i	Weights w_i
1	0	2
2	$\pm 1/\sqrt{3}$	1
3	$\pm\sqrt{0.6}$	$\frac{5}{9}$
	0	$\frac{8}{9}$
4	$\pm\frac{1}{35}\sqrt{525+70\sqrt{30}}$	$\frac{1}{36}\left(18-\sqrt{30}\right)$
	$\pm\frac{1}{35}\sqrt{525-70\sqrt{30}}$	$\frac{1}{36}\left(18+\sqrt{30}\right)$
5	0	$\frac{128}{225}$
	$\pm\frac{1}{21}\sqrt{245-14\sqrt{70}}$	$\frac{1}{900}\left(322+13\sqrt{70}\right)$
	$\pm\frac{1}{21}\sqrt{245+14\sqrt{70}}$	$\frac{1}{900}\left(322-13\sqrt{70}\right)$

The general theory of GAUSS–LEGENDRE integration is based on so-called LEGENDRE polynomials. Without going into detail of this theory, useful formulae for the numerical determination of abscissas and weights can be specified. The best way to generate an explicit formulae for a LEGENDRE polynomial $P_n(\xi)$ is to use recurrence relations [1, 3]. These recurrence relationships $P_i(\xi)$ with $1 \leq i \leq n-1$ are particularly useful for computer evaluation of the roots of the LEGENDRE polynomials, their derivatives and thus the corresponding weights. Two such relations that are widely used are:

$$(i+1)P_{i+1}(\xi) = (2i+1)\xi P_i(\xi) - iP_{i-1}(\xi) \quad \text{for} \quad i \geq 1\,, \tag{A.87}$$

and for the derivative of the LEGENDRE polynomials

$$(\xi^2-1)\underbrace{\frac{\mathrm{d}P_i(\xi)}{\mathrm{d}\xi}}_{P_i'} = i\,(\xi P_i(\xi) - P_{i-1}(\xi)) \quad \text{for} \quad i \geq 1\,, \tag{A.88}$$

where the first two polynomials are given as $P_0 = 1$ and $P_1 = \xi$. The weights for a n-point integration rule are given by

$$w_i(\xi_i) = \frac{2}{n\cdot P_{n-1}(\xi_i)P_n'(\xi_i)} = \frac{2}{(1-\xi_i^2)(P_n'(\xi_i))^2} = \frac{2(1-\xi_i^2)}{(n+1)^2(P_{n+1}(\xi_i))^2}\,, \tag{A.89}$$

where the roots of P_n are consecutively taken for the ξ_i in Eq. (A.89). Constructing a LEGENDRE polynomial of degree n based on Eq. (A.87) and numerically determining the roots (e.g. based on NEWTON's method) which are equal to the abscissas, gives together with Eqs. (A.88)–(A.89) the complete set for a n-point GAUSS–LEGENDRE integration rule. The general n-point GAUSS–LEGENDRE rule is exact for polynomial functions of degree $\leq 2n-1$. Let us return to the example given at the beginning of this section where the idea of GAUSS–LEGENDRE integration has been introduced based on a 3-point rule. The evaluation of the recurrence relation (A.87) gives:

$$\begin{aligned} & P_0 = 1 \quad \text{(known)}, \\ & P_1 = \xi \quad \text{(known)}, \\ i = 1: \quad & 2P_2 = 3\xi P_1 - 1P_0 \\ \rightarrow \quad & P_2 = \tfrac{1}{2}\left(3\xi^2 - 1\right), \\ i = 2: \quad & 3P_3 = 5\xi P_2 - 2P_1 \\ \rightarrow \quad & P_3 = \tfrac{1}{2}\left(5\xi^3 - 3\xi\right). \end{aligned}$$

The roots of P_3 can numerically be calculated as $\xi_1 = -\sqrt{15}/5$, $\xi_2 = 0$ and $\xi_3 = \sqrt{15}/5$. The respective derivatives follow from Eq. (A.88) as:

$$\begin{aligned} & P_0' = 0 \quad \text{(known)}, \\ i = 1: \quad & (\xi^2 - 1)P_1' = 1 \cdot (\xi \cdot P_1 - P_0) \\ \rightarrow \quad & P_1' = 1, \\ i = 2: \quad & (\xi^2 - 1)P_2' = 2 \cdot (\xi \cdot P_2 - P_1) \\ \rightarrow \quad & P_2' = 3\xi, \\ i = 3: \quad & (\xi^2 - 1)P_3' = 3 \cdot (\xi \cdot P_3 - P_2) \\ \rightarrow \quad & P_3' = \tfrac{3}{2(\xi^2-1)}\left(5\xi^4 - 6\xi^2 + 1\right) = \tfrac{1}{2}\left(15\xi^2 - 3\right). \end{aligned}$$

Finally, the weights can be expressed according to Eq. (A.89) as:

$$w_3 = \frac{2}{3 \cdot P_2 \cdot P_3'} = \frac{2}{\frac{9}{2}\frac{3\xi^2-1}{\xi^2-1}\left(\frac{5}{2}\xi^3 - \frac{3}{2}\xi^2 - \frac{3}{2}\xi + \frac{1}{2}\right)}, \tag{A.90}$$

which gives for the roots of the LEGENDRE polynomial:

$$w_3\left(-\frac{\sqrt{15}}{5}\right) = \frac{5}{9}, \; w_3(0) = \frac{8}{9} \text{ and } w_3\left(\frac{\sqrt{15}}{5}\right) = \frac{5}{9}.$$

Details are given in Tables A.6, A.7 and A.8

Table A.6 LEGENDRE polynomials $P_{i+1}(\xi) = \frac{1}{i+1} \cdot ((2i+1)\xi P_i - i P_{i-1})$ with $i \geq 1$

$P_0 = 1\,,$	(A.91)
$P_1 = \xi\,,$	(A.92)
$P_2 = \frac{1}{2}(3\xi^2 - 1)\,,$	(A.93)
$P_3 = \frac{1}{2}(5\xi^3 - 3\xi)\,,$	(A.94)
$P_4 = \frac{1}{8}(35\xi^4 - 30\xi^2 + 3)\,,$	(A.95)
$P_5 = \frac{1}{8}(63\xi^5 - 70\xi^3 + 15\xi)\,,$	(A.96)
$P_6 = \frac{1}{16}(231\xi^6 - 315\xi^4 + 105\xi^2 - 5)\,,$	(A.97)
$P_7 = \frac{1}{16}(429\xi^7 - 693\xi^5 + 315\,\xi^3 - 35\xi)\,,$	(A.98)
$P_8 = \frac{1}{128}(6435\xi^8 - 12012\xi^6 + 6930\xi^4 - 1260\xi^2 + 35)\,,$	(A.99)
$P_9 = \frac{1}{128}(12155\xi^9 - 25740\xi^7 + 18018\xi^5 - 4620\xi^3 + 315\xi)\,,$	(A.100)
$P_{10} = \frac{1}{256}(46189\xi^{10} - 109395\xi^8 + 90090\xi^6 - 30030\xi^4 + 3465\xi^2 - 63)\,.$	(A.101)

Table A.7 Derivatives of LEGENDRE polynomials $\frac{\mathrm{d}P_i(\xi)}{\mathrm{d}\xi} = P_i(\xi)' = \frac{i}{\xi^2-1} \cdot (\xi P_i - P_{i-1})$ with $i \geq 1$

$P_0' = 0\,,$	(A.102)
$P_1' = 1\,,$	(A.103)
$P_2' = 3\xi\,,$	(A.104)
$P_3' = \frac{1}{2}(15\xi^2 - 3)\,,$	(A.105)
$P_4' = \frac{5}{2}\,(7\,\xi^2 - 3)\,\xi\,,$	(A.106)
$P_5' = \frac{1}{8}(315\xi^4 - 210\xi^2 + 15)\,,$	(A.107)
$P_6' = \frac{21}{8}\,(33\,\xi^4 - 30\,\xi^2 + 5)\,\xi\,,$	(A.108)
$P_7' = \frac{1}{16}(3003\xi^6 - 3465\xi^4 + 945\xi^2 - 35)\,,$	(A.109)
$P_8' = \frac{9}{16}\,(715\,\xi^6 - 1001\,\xi^4 + 385\,\xi^2 - 35)\,\xi\,,$	(A.110)
$P_9' = \frac{1}{128}(109395\xi^8 - 180180\xi^6 + 90090\xi^4 - 13860\xi^2 + 315)\,,$	(A.111)
$P_{10}' = \frac{55}{128}\,(4199\,\xi^8 - 7956\,\xi^6 + 4914\,\xi^4 - 1092\,\xi^2 + 63)\,\xi\,.$	(A.112)

A.10 Example: One-dimensional numerical integration

Calculate the integral $\int_{-1}^{1} x^2\,\mathrm{d}x$ analytically and numerically based on a 1-point and 2-point *Gauss–Legendre* integration rule.

A.10 Solution

Analytical solution:

$$\int\limits_{-1}^{1} x^2\,\mathrm{d}x = \left[\frac{x^3}{3}\right]_{-1}^{1} = \frac{1}{3} + \frac{1}{3} = \frac{2}{3}. \tag{A.113}$$

Table A.8 Abscissae and weight factors for GAUSS–LEGENDRE integration: $\int_{-1}^{1} f(\xi)\mathrm{d}\xi \approx \sum_{i=1}^{n} w_i f(\xi_i)$

$\pm\xi_i$	w_i
$n = 1$	
0.000000000000000	1.000000000000000
$n = 2$	
0.577350269189626	1.000000000000000
$n = 3$	
0.000000000000000	0.888888888888889
0.774596669241483	0.555555555555556
$n = 4$	
0.339981043584856	0.652145154862546
0.861136311594053	0.347854845137454
$n = 5$	
0.000000000000000	0.568888888888889
0.538469310105683	0.478628670499367
0.906179845938664	0.236926885056189
$n = 6$	
0.238619186083197	0.467913934572691
0.661209386466265	0.360761573048139
0.932469514203152	0.171324492379171
$n = 7$	
0.000000000000000	0.417959183673469
0.405845151377397	0.381830050505119
0.741531185599395	0.279705391489277
0.949107912342759	0.129484966168869
$n = 8$	
0.183434642495650	0.362683783378362
0.525532409916329	0.313706645877887
0.796666477413627	0.222381034453375
0.960289856497536	0.101228536290377
$n = 9$	
0.000000000000000	0.330239355001260
0.324253423403809	0.312347077040003
0.613371432700591	0.260610696402936
0.836031107326636	0.180648160694858
0.968160239507626	0.081274388361575
$n = 10$	
0.148874338981631	0.295524224714753
0.433395394129247	0.269266719309996
0.679409568299024	0.219086362515982
0.865063366688985	0.149451349150581
0.973906528517172	0.066671344308688

(continued)

Table A.8 (continued)

$\pm\xi_i$	w_i
$n = 12$	
0.125233408511469	0.249147045813403
0.367831498998180	0.233492536538355
0.587317954286618	0.203167426723066
0.769902674194305	0.160078328543346
0.904117256370475	0.106939325995318
0.981560634246719	0.047175336386512
$n = 16$	
0.095012509837637	0.189450610455068
0.281603550779259	0.182603415044924
0.458016777657227	0.169156519395002
0.617876244402644	0.149595988816577
0.755404408355003	0.124628971255534
0.865631202387832	0.095158511682493
0.944575023073233	0.062253523938648
0.989400934991650	0.027152459411754
$n = 20$	
0.076526521133497	0.152753387130726
0.227785851141645	0.149172986472604
0.373706088715420	0.142096109318382
0.510867001950827	0.131688638449177
0.636053680726515	0.118194531961518
0.746331906460151	0.101930119817241
0.839116971822219	0.083276741576705
0.912234428251326	0.062672048334109
0.963971927277914	0.040601429800387
0.993128599185095	0.017614007139152
$n = 24$	
0.064056892862606	0.127938195346752
0.191118867473616	0.125837456346828
0.315042679696163	0.121670472927803
0.433793507626045	0.115505668053726
0.545421471388840	0.107444270115966
0.648093651936976	0.097618652104114
0.740124191578554	0.086190161531953
0.820001985973903	0.073346481411080
0.886415527004401	0.059298584915437
0.938274552002733	0.044277438817416
0.974728555971310	0.028531388628934
0.995187219997021	0.012341229799987

1-point integration rule:

$$\int_{-1}^{1} x^2\,\mathrm{d}x \approx (0)^2 \times 2 = 0 \quad (\text{rel. error } 100\%)\,. \tag{A.114}$$

2-point integration rule:

$$\int_{-1}^{1} x^2\,\mathrm{d}x \approx \left(-\frac{1}{\sqrt{3}}\right)^2 \times 1 + \left(\frac{1}{\sqrt{3}}\right)^2 \times 1 = \frac{2}{3} \quad (\text{rel. error } 0\%)\,. \tag{A.115}$$

A.11 Example: One-dimensional numerical integration of a polynomial of order two

Calculate the integral $\int_{-1}^{1}\left(\frac{1}{8}\xi^2 + \frac{3}{4}\xi + \frac{13}{8}\right)\mathrm{d}\xi$ analytically and numerically based on a 1-point and 2-point *Gauss–Legendre* integration rule.

A.11 Solution

Analytical solution: see Example A.7
1-point integration rule:

$$\int_{-1}^{1} f(\xi)\mathrm{d}\xi \approx \left(\tfrac{1}{8}\xi_0^2 + \tfrac{3}{4}\xi_0 + \tfrac{13}{8}\right)_{\xi_0=0} \times 2 = \tfrac{13}{4} \quad (\text{rel. error } 2.5\%)\,. \tag{A.116}$$

2-point integration rule:

$$\int_{-1}^{1} f(\xi)\mathrm{d}\xi \approx \left(\tfrac{1}{8}\xi_1^2 + \tfrac{3}{4}\xi_1 + \tfrac{13}{8}\right)_{\xi_1=-\frac{1}{\sqrt{3}}} \times 1 + \left(\tfrac{1}{8}\xi_2^2 + \tfrac{3}{4}\xi_2 + \tfrac{13}{8}\right)_{\xi_2=+\frac{1}{\sqrt{3}}} \times 1$$
$$= \tfrac{10}{3} \quad (\text{rel. error } 0\%)\,. \tag{A.117}$$

A.12 Example: Two-dimensional numerical integration of a parabolid

Calculate the integral $\int_{-1}^{1}\int_{-1}^{1}\left(\frac{1}{16}\xi^2 + \frac{3}{8}\xi + \frac{1}{16}\eta^2 + \frac{3}{8}\eta + \frac{11}{8}\right)\mathrm{d}\xi\mathrm{d}\eta$ numerically based on a 1-point and 2-point *Gauss–Legendre* integration rule.

A.12 Solution

1-point integration rule:

$$\int_{-1}^{+1}\int_{-1}^{+1} g(\xi,\eta)\,\mathrm{d}\xi\mathrm{d}\eta \approx \left(\tfrac{1}{16}\xi_0^2 + \tfrac{3}{8}\xi_0 + \tfrac{1}{16}\eta_0^2 + \tfrac{3}{8}\eta_0 + \tfrac{11}{8}\right)_{\xi_0=0,\eta_0=0} \times 4$$
$$= \tfrac{11}{2}\,. \tag{A.118}$$

2-point integration rule:

$$\int_{-1}^{+1}\int_{-1}^{+1} g(\xi,\eta)\,\mathrm{d}\xi\mathrm{d}\eta \approx$$
$$\left(\tfrac{1}{16}\xi_1^2 + \tfrac{3}{8}\xi_1 + \tfrac{1}{16}\eta_1^2 + \tfrac{3}{8}\eta_1 + \tfrac{11}{8}\right)_{\xi_1=\frac{1}{\sqrt{3}},\eta_1=\frac{1}{\sqrt{3}}} \times 1$$
$$+\left(\tfrac{1}{16}\xi_2^2 + \tfrac{3}{8}\xi_2 + \tfrac{1}{16}\eta_2^2 + \tfrac{3}{8}\eta_2 + \tfrac{11}{8}\right)_{\xi_2=-\frac{1}{\sqrt{3}},\eta_2=\frac{1}{\sqrt{3}}} \times 1$$
$$+\left(\tfrac{1}{16}\xi_3^2 + \tfrac{3}{8}\xi_3 + \tfrac{1}{16}\eta_3^2 + \tfrac{3}{8}\eta_3 + \tfrac{11}{8}\right)_{\xi_3=-\frac{1}{\sqrt{3}},\eta_3=-\frac{1}{\sqrt{3}}} \times 1$$
$$\left(\tfrac{1}{16}\xi_4^2 + \tfrac{3}{8}\xi_4 + \tfrac{1}{16}\eta_4^2 + \tfrac{3}{8}\eta_4 + \tfrac{11}{8}\right)_{\xi_4=\frac{1}{\sqrt{3}},\eta_4=-\frac{1}{\sqrt{3}}} \times 1$$
$$= \tfrac{17}{3}\,. \tag{A.119}$$

A.10 Taylor's Series Expansion

A TAYLOR's series expansion of $f(x)$ with respect to x_0 is given by:

$$f(x) = f(x_0) + \left(\frac{\mathrm{d}f}{\mathrm{d}x}\right)_{x_0}(x-x_0) + \frac{1}{2!}\left(\frac{\mathrm{d}^2 f}{\mathrm{d}x^2}\right)_{x_0}(x-x_0)^2 + \cdots + \frac{1}{k!}\left(\frac{\mathrm{d}^k f}{\mathrm{d}x^k}\right)_{x_0}(x-x_0)^k\,. \tag{A.120}$$

The first order approximation takes the first two terms in the series and approximates the function as:

$$f(x) = f(x_0 + \mathrm{d}x) \approx f(x_0) + \left(\frac{\mathrm{d}f}{\mathrm{d}x}\right)_{x_0}(x-x_0)\,. \tag{A.121}$$

Recall from calculus that the derivative gives the slope of the tangent line at a given point and that the point-slope form is given by $f(x) - f(x_0) = m \times (x - x_0)$. Thus,

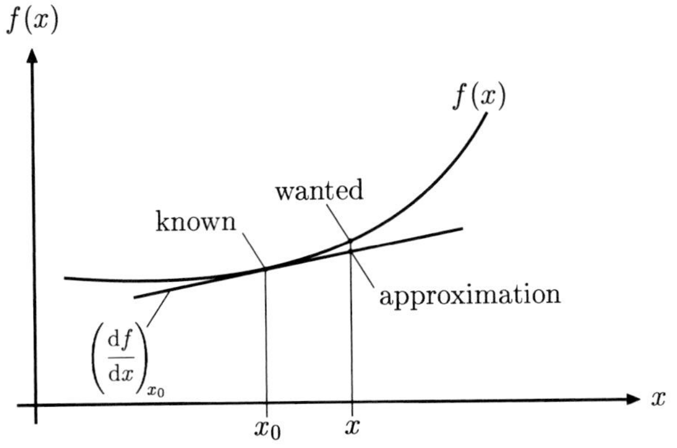

Fig. A.6 Approximation of a function $f(x)$ by a first order TAYLOR's series

we can conclude that the first order approximation gives us the equation of a straight line passing through the point $(x_0, f(x_0))$ with a slope of $m = f'(x_0) = (\mathrm{d}f/\mathrm{d}x)_{x_0}$, cf. Fig. A.6.

A.13 Example: Taylor's series expansion of a function

Calculate the first and second order Taylor's series approximation of the function $f(x) = x^2$ at the functional value $x = 2$ based on the value at $x_0 = 1$. The exact solution is $f(x = 2) = 2^2 = 4$.

A.13 Solution

First order approximation:

$$f(x=2) \approx f(x_0=1) + 2x|_{x_0} \times (x - x_0) = 1 + 2 \times 1(2-1) = 3\,. \quad \text{(A.122)}$$

$$\text{rel. error} \left|\frac{4-3}{3}\right| = \left|\frac{1}{4}\right| = 0.25 = 25\%\,. \quad \text{(A.123)}$$

Second order approximation:

$$f(x=2) \approx f(x_0=1) + 2x|_{x_0} \times (x - x_0) + \quad \text{(A.124)}$$

$$+ \frac{1}{2} \times 2\bigg|_{x_0} (x - x_0)^2 = 1 + 2 + 1 = 4 \quad \checkmark \quad \text{(A.125)}$$

A.11 Matrix Operations

A matrix $\boldsymbol{A}$ is defined as a set of quantities a_{ij} ordered as follows

$$\boldsymbol{A} = \begin{array}{c} \rightarrow j \\ \begin{array}{c}\downarrow \\ i\end{array}\begin{bmatrix} a_{11} & a_{12} & \dots & a_{1n} \\ a_{21} & a_{22} & \dots & a_{2n} \\ \vdots & \vdots & \ddots & \vdots \\ a_{m1} & a_{m2} & \dots & a_{mn} \end{bmatrix}\end{array} \tag{A.126}$$

and denoted by a bold upper case letter. As indicated, the position of a term a_{ij} is defined by indices i and j: the first index determines the row and the second index determines the column in the matrix. For example, a_{23} is the term in the second row and in the third column. The matrix (A.158) is of order $m \times n$. A matrix is said to be square if $m = n$, and rectangular if $m \neq n$. An $m \times 1$ matrix is called a column matrix and denoted by a bold lower case variable. For example, the column matrix $\boldsymbol{b}$

$$\boldsymbol{b} = \begin{bmatrix} b_1 \\ b_2 \\ \vdots \\ b_m \end{bmatrix} \tag{A.127}$$

is defined as the matrix of order $m \times 1$.

The transposed matrix of a matrix $\boldsymbol{A}$ is denoted $\boldsymbol{A}^{\mathrm{T}}$ and represents the matrix with the columns and rows interchanged, i.e.

$$(\boldsymbol{A}^{\mathrm{T}})_{ij} = (\boldsymbol{A})_{ji} \,. \tag{A.128}$$

If we use symmetric matrices for which the elements are symmetric with respect to the main diagonal, then

$$\boldsymbol{A}^{\mathrm{T}} = \boldsymbol{A} \,. \tag{A.129}$$

The transposed of the matrix (A.127) results in a row matrix:

$$\boldsymbol{b}^{\mathrm{T}} = \begin{bmatrix} b_1 \; b_2 \; \dots \; b_m \end{bmatrix} . \tag{A.130}$$

The product of two transposes satisfies

$$(\boldsymbol{A}\boldsymbol{B})^{\mathrm{T}} = \boldsymbol{B}^{\mathrm{T}}\boldsymbol{A}^{\mathrm{T}} . \tag{A.131}$$

Proof:

$$(\boldsymbol{B}^{\mathrm{T}}\boldsymbol{A}^{\mathrm{T}})_{ij} = (b^{\mathrm{T}})_{ik}(a^{\mathrm{T}})_{kj} = b_{ki}a_{jk} = a_{jk}b_{ki} = (\boldsymbol{AB})_{ji} = (\boldsymbol{AB})^{\mathrm{T}}_{ij} .$$

The inverse of a square matrix $\boldsymbol{A}$ is a matrix $\boldsymbol{A}^{-1}$ such that

$$\boldsymbol{A}\boldsymbol{A}^{-1} = \mathbf{I} , \tag{A.132}$$

where $\mathbf{I}$ is the identity matrix. A square matrix $\boldsymbol{A}$ has an inverse if the determinant $|\boldsymbol{A}| \neq 0$. A matrix possessing an inverse is called nonsingular, or invertible. The inverse of a product $\boldsymbol{AB}$ of matrices $\boldsymbol{A}$ and $\boldsymbol{B}$ can be expressed in terms of $\boldsymbol{A}^{-1}$ and $\boldsymbol{B}^{-1}$ as

$$(\boldsymbol{AB})^{-1} = \boldsymbol{B}^{-1}\boldsymbol{A}^{-1} . \tag{A.133}$$

Proof:

$$\boldsymbol{C} \equiv \boldsymbol{AB} .$$

Then

$$\boldsymbol{B} = \boldsymbol{A}^{-1}\boldsymbol{AB} = \boldsymbol{A}^{-1}\boldsymbol{C}$$

and

$$\boldsymbol{A} = \boldsymbol{ABB}^{-1} = \boldsymbol{CB}^{-1} .$$

Therefore,

$$\boldsymbol{C} = \boldsymbol{AB} = (\boldsymbol{CB}^{-1})(\boldsymbol{A}^{-1}\boldsymbol{C}) = \boldsymbol{CB}^{-1}\boldsymbol{A}^{-1}\boldsymbol{C}$$

so

$$\boldsymbol{CB}^{-1}\boldsymbol{A}^{-1} = \mathbf{I} \quad \text{and} \quad \boldsymbol{B}^{-1}\boldsymbol{A}^{-1} = \boldsymbol{C}^{-1} = (\boldsymbol{AB})^{-1} . \tag{A.134}$$

A.11.1 Matrix Multiplication

The product $\boldsymbol{C}$ of two matrices $\boldsymbol{A}$ and $\boldsymbol{B}$ is defined by

$$c_{ij} = a_{ik} \times b_{jk} , \tag{A.135}$$

where k is summed over for all possible values of i and j. Writing out the product explicitly,

$$\begin{bmatrix} c_{11} & c_{12} & \dots & c_{1p} \\ c_{21} & c_{22} & \dots & c_{2p} \\ \vdots & \vdots & \ddots & \vdots \\ c_{n1} & c_{n2} & \dots & c_{np} \end{bmatrix} = \begin{bmatrix} a_{11} & a_{12} & \dots & a_{1m} \\ a_{21} & a_{22} & \dots & a_{2m} \\ \vdots & \vdots & \ddots & \vdots \\ a_{n1} & a_{n2} & \dots & a_{nm} \end{bmatrix} \times \begin{bmatrix} b_{11} & b_{12} & \dots & b_{1p} \\ b_{21} & b_{22} & \dots & b_{2p} \\ \vdots & \vdots & \ddots & \vdots \\ b_{m1} & b_{m2} & \dots & b_{mp} \end{bmatrix} , \tag{A.136}$$

where

$$c_{11} = a_{11}b_{11} + a_{12}b_{21} + \cdots + a_{1m}b_{m1}$$
$$c_{12} = a_{11}b_{12} + a_{12}b_{22} + \cdots + a_{1m}b_{m2}$$
$$c_{1p} = a_{11}b_{1p} + a_{12}b_{2p} + \cdots + a_{1m}b_{mp}$$
$$c_{21} = a_{21}b_{11} + a_{22}b_{21} + \cdots + a_{2m}b_{m1} \tag{A.137}$$
$$\vdots$$
$$c_{np} = a_{n1}b_{1p} + a_{n2}b_{2p} + \cdots + a_{nm}b_{mp} \, .$$

It can be seen from the rule given in (A.137) that the element c_{ij} of the matrix C is obtained by multiplying row i of matrix A with column j of matrix B:

$$\begin{bmatrix} & & \\ & c_{ij} & \\ & & \end{bmatrix} = \begin{matrix} i \end{matrix} \begin{bmatrix} - & - & - & - & - \\ & & & & \end{bmatrix} \times \begin{bmatrix} & & j & \\ & & | & \\ & & | & \\ & & | & \\ & & | & \end{bmatrix} . \tag{A.138}$$

The matrix multiplication is not commutative

$$A \times B \neq B \times A \tag{A.139}$$

but obeys the following laws:

$$A \times (B \times C) = (A \times B) \times C \qquad \text{associative law}, \tag{A.140}$$
$$A \times (B + C) = A \times B + A \times C \qquad \text{distributive law (left distributivity)}, \tag{A.141}$$
$$(A + B) \times C = A \times C + B \times C \qquad \text{distributive law (right distributivity)}. \tag{A.142}$$

A.14 Example: Triple matrix product

Calculate the triple matrix product $A \times B \times C$ with

$$A = \begin{bmatrix} a_{11} & a_{11} \\ a_{21} & a_{22} \\ a_{31} & a_{32} \end{bmatrix}, B = \begin{bmatrix} b_{11} & b_{11} \\ b_{21} & b_{22} \end{bmatrix}, C = \begin{bmatrix} c_{11} & c_{12} & c_{13} \\ c_{21} & c_{22} & c_{23} \end{bmatrix}. \tag{A.143}$$

A.14 Solution

The problem can be solved based on $(A \times B) \times C$:

$$A \times B = \underbrace{\begin{bmatrix} a_{11} & a_{12} \\ a_{21} & a_{22} \\ a_{31} & a_{32} \end{bmatrix}}_{(3 \times 2)} \times \underbrace{\begin{bmatrix} b_{11} & b_{11} \\ b_{21} & b_{22} \end{bmatrix}}_{2 \times 2} = \underbrace{\begin{bmatrix} a_{11}b_{11} + a_{12}b_{21} & a_{11}b_{12} + a_{12}b_{22} \\ a_{21}b_{11} + a_{22}b_{21} & a_{21}b_{12} + a_{22}b_{22} \\ a_{31}b_{11} + a_{32}b_{21} & a_{31}b_{12} + a_{32}b_{22} \end{bmatrix}}_{3 \times 2}$$
$$\tag{A.144}$$

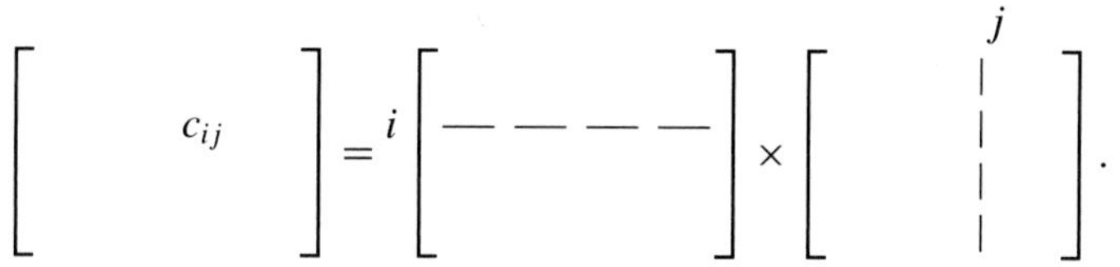

$$(A \times B) \times C = \underbrace{\begin{bmatrix} a_{11}b_{11} + a_{12}b_{21} & a_{11}b_{12} + a_{12}b_{22} \\ a_{21}b_{11} + a_{22}b_{21} & a_{21}b_{12} + a_{22}b_{22} \\ a_{31}b_{11} + a_{32}b_{21} & a_{31}b_{12} + a_{32}b_{22} \end{bmatrix}}_{3\times 2} \times \underbrace{\begin{bmatrix} c_{11} & c_{12} & c_{13} \\ c_{21} & c_{22} & c_{23} \end{bmatrix}}_{2\times 3} =$$

$$= \begin{bmatrix} (a_{11}b_{11}+a_{12}b_{21})c_{11} + (a_{11}b_{12}+a_{12}b_{22})c_{21} & (a_{11}b_{11}+a_{12}b_{21})c_{12} + (a_{11}b_{12}+a_{12}b_{22})c_{22} & (a_{11}b_{11}+a_{12}b_{21})c_{13} + (a_{11}b_{12}+a_{12}b_{22})c_{23} \\ (a_{21}b_{11}+a_{22}b_{21})c_{11} + (a_{21}b_{12}+a_{22}b_{22})c_{21} & (a_{21}b_{11}+a_{22}b_{21})c_{12} + (a_{21}b_{12}+a_{22}b_{22})c_{22} & (a_{21}b_{11}+a_{22}b_{21})c_{13} + (a_{21}b_{12}+a_{22}b_{22})c_{23} \\ (a_{31}b_{11}+a_{32}b_{21})c_{11} + (a_{31}b_{12}+a_{32}b_{22})c_{21} & (a_{31}b_{11}+a_{32}b_{21})c_{12} + (a_{31}b_{12}+a_{32}b_{22})c_{22} & (a_{31}b_{11}+a_{32}b_{21})c_{13} + (a_{31}b_{12}+a_{32}b_{22})c_{23} \end{bmatrix} \quad \text{(A.145)}$$

$$= (3 \times 3)\,.$$

A.11.2 Scalar Product

The scalar product[3] of two vectors $\boldsymbol{a}$ and $\boldsymbol{b}$ is given by:

$$\boldsymbol{a} : \boldsymbol{b} = \begin{bmatrix} a_1 \\ a_2 \\ \vdots \\ a_n \end{bmatrix} : \begin{bmatrix} b_1 \\ b_2 \\ \vdots \\ b_n \end{bmatrix} = a_1 b_1 + a_2 b_2 + \cdots + a_n b_n\,. \quad \text{(A.146)}$$

The same result is obtained by the following matrix multiplication:

$$\boldsymbol{a}^{\mathrm{T}} \times \boldsymbol{b} = \begin{bmatrix} a_1 & a_2 & \ldots & a_n \end{bmatrix} \times \begin{bmatrix} b_1 \\ b_2 \\ \vdots \\ b_n \end{bmatrix} = a_1 b_1 + a_2 b_2 + \cdots + a_n b_n\,. \quad \text{(A.147)}$$

If vectors $\boldsymbol{a}$ and $\boldsymbol{b}$ are defined in a Cartesian coordinate system (x, y, z), then

$$\boldsymbol{a} : \boldsymbol{b} = |\boldsymbol{a}| \times |\boldsymbol{b}| \cdot \cos\alpha \quad \text{(A.148)}$$

and it holds the geometric interpretation that the scalar product of the vectors $\boldsymbol{a}$ and $\boldsymbol{b}$ is equal to the product of the length of $\boldsymbol{a}$ projected onto $\boldsymbol{b}$, or visa versa. For example, the length of vector $\boldsymbol{a}$ is obtained as $|\boldsymbol{a}| = \sqrt{a_x^2 + a_y^2 + a_z^2}$.

[3] In the literature, the alternative designation *dot product* and *inner product* can be found.

A.11.3 Dyadic Product

The dyadic product of two vectors $\boldsymbol{a}$ and $\boldsymbol{b}$ is given by

$$\boldsymbol{a} \otimes \boldsymbol{b} = \begin{bmatrix} a_1 \\ a_2 \\ \vdots \\ a_n \end{bmatrix} \otimes \begin{bmatrix} b_1 \\ b_2 \\ \vdots \\ b_n \end{bmatrix} = \begin{bmatrix} a_1b_1 & a_1b_2 & \ldots & a_1b_n \\ a_2b_1 & a_2b_2 & \ldots & a_2b_n \\ \vdots & \vdots & \ddots & \vdots \\ a_nb_1 & \vdots & \ldots & a_nb_n \end{bmatrix}. \tag{A.149}$$

The same result is obtained by the following matrix multiplication:

$$\boldsymbol{a} \times \boldsymbol{b}^{\mathrm{T}} = \begin{bmatrix} a_1 \\ a_2 \\ \vdots \\ a_n \end{bmatrix} \times \begin{bmatrix} b_1 & b_2 & \ldots & b_n \end{bmatrix} = \begin{bmatrix} a_1b_1 & a_1b_2 & \ldots & a_1b_n \\ a_2b_1 & a_2b_2 & \ldots & a_2b_n \\ \vdots & \vdots & \ddots & \vdots \\ a_nb_1 & \vdots & \ldots & a_nb_n \end{bmatrix}. \tag{A.150}$$

A.11.4 Inverse of Matrices

Equation for a 2 × 2 matrix:

$$\boldsymbol{A}^{-1} = \begin{bmatrix} a & b \\ c & d \end{bmatrix}^{-1} = \frac{1}{ad - bc} \times \begin{bmatrix} d & -b \\ -c & a \end{bmatrix}. \tag{A.151}$$

Equation for a 3 × 3 matrix:

$$\boldsymbol{A}^{-1} = \begin{bmatrix} a & b & c \\ d & e & f \\ g & h & i \end{bmatrix}^{-1} = \frac{1}{\det(\boldsymbol{A})} \begin{bmatrix} ei - fh & ch - bi & bf - ce \\ fg - di & ai - cg & cd - af \\ dh - eg & bg - ah & ae - bd \end{bmatrix}, \tag{A.152}$$

where

$$\det(\boldsymbol{A}) = aei + bfg + cdh - ceg - afh - bdi\,. \tag{A.153}$$

Equation for a 4 × 4 matrix:

$$\boldsymbol{A}^{-1} = \begin{bmatrix} a & b & c & d \\ e & f & g & h \\ i & j & k & l \\ m & n & o & p \end{bmatrix}^{-1} = \frac{1}{\det(\boldsymbol{A})} \begin{bmatrix} b_{11} & b_{12} & b_{13} & b_{14} \\ b_{21} & b_{22} & b_{23} & b_{24} \\ b_{31} & b_{32} & b_{33} & b_{34} \\ b_{41} & b_{42} & b_{43} & b_{44} \end{bmatrix}, \tag{A.154}$$

where

$$\begin{aligned}\det(\boldsymbol{A}) = & afkp + agln + ahjo + belo + bgip + bhkm \\ & + cejp + cflm + chin + dekn + dfio + dgjm \\ & - aflo - agjp - ahkn - bekp - bglm - bhio \\ & - celn - cfip - chjm - dejo - dfkm - dgin\,,\end{aligned} \tag{A.155}$$

and

$$\begin{aligned} b_{11} &= fkp + gln + hjo - flo - gjp - hkn\,, \\ b_{12} &= blo + cjp + dkn - bkp - cln - djo\,, \\ b_{13} &= bgp + chn + dfo - bho - cfp - dgn\,, \\ b_{14} &= bhk + cfl + dgj - bgl - chj - dfk\,, \\ b_{21} &= elo + gip + hkm - ekp - glm - hio\,, \\ b_{22} &= akp + clm + dio - alo - cip - dkm\,, \\ b_{23} &= aho + cep + dgm - agp - chm - deo\,, \\ b_{24} &= agl + chi + dek - ahk - cel - dgi\,, \\ b_{31} &= ejp + flm + hin - eln - fip - hjm\,, \\ b_{32} &= aln + bip + djm - ajp - blm - din\,, \\ b_{33} &= afp + bhm + den - ahn - bep - dfm\,, \\ b_{34} &= ahj + bel + dfi - afl - bhi - dej\,, \\ b_{41} &= ekn + fio + gjm - ejo - fkm - gin\,, \\ b_{42} &= ajo + bkm + cin - akn - bio - cjm\,, \\ b_{43} &= agn + beo + cfm - afo - bgm - cen\,, \\ b_{44} &= afk + bgi + cej - agj - bek - cfi\,. \end{aligned} \tag{A.156}$$

Inverse of a Diagonal Matrix

A square $n \times n$ *diagonal* matrix $\boldsymbol{A}$ is defined as follows where only the elements on the main diagonal are unequal to zero:

$$\boldsymbol{A} = \begin{bmatrix} a_{11} & 0 & \dots & 0 \\ 0 & a_{22} & \dots & 0 \\ \vdots & \vdots & \ddots & \vdots \\ 0 & 0 & \dots & a_{nn} \end{bmatrix}. \tag{A.157}$$

Then the inverse $\boldsymbol{A}^{-1}$ is given by:

$$A^{-1} = \begin{bmatrix} \frac{1}{a_{11}} & 0 & \dots & 0 \\ 0 & \frac{1}{a_{22}} & \dots & 0 \\ \vdots & \vdots & \ddots & \vdots \\ 0 & 0 & \dots & \frac{1}{a_{nn}} \end{bmatrix}, \tag{A.158}$$

provided that none of the diagonal elements are zero.

Blockwise Inversion

$$\begin{bmatrix} \boldsymbol{A} & \boldsymbol{B} \\ \boldsymbol{C} & \boldsymbol{D} \end{bmatrix}^{-1} = \begin{bmatrix} \boldsymbol{A}^{-1} + \boldsymbol{A}^{-1}\boldsymbol{B}(\boldsymbol{D} - \boldsymbol{C}\boldsymbol{A}^{-1}\boldsymbol{B})^{-1}\boldsymbol{C}\boldsymbol{A}^{-1} & -\boldsymbol{A}^{-1}\boldsymbol{B}(\boldsymbol{D} - \boldsymbol{C}\boldsymbol{A}^{-1}\boldsymbol{B})^{-1} \\ -(\boldsymbol{D} - \boldsymbol{C}\boldsymbol{A}^{-1}\boldsymbol{B})^{-1}\boldsymbol{C}\boldsymbol{A}^{-1} & (\boldsymbol{D} - \boldsymbol{C}\boldsymbol{A}^{-1}\boldsymbol{B})^{-1} \end{bmatrix}, \tag{A.159}$$

where $\boldsymbol{A}$ and $\boldsymbol{D}$ must be square, so that they can be inverted. This is the case if and only if $\boldsymbol{A}$ and $\boldsymbol{D} - \boldsymbol{C}\boldsymbol{A}^{-1}\boldsymbol{B}$ are nonsingular. Alternatively, the inversion procedure can be expressed if $\boldsymbol{D}$ and $\boldsymbol{A} - \boldsymbol{B}\boldsymbol{D}^{-1}\boldsymbol{C}$ are nonsingular as:

$$\begin{bmatrix} \boldsymbol{A} & \boldsymbol{B} \\ \boldsymbol{C} & \boldsymbol{D} \end{bmatrix}^{-1} = \begin{bmatrix} (\boldsymbol{A} - \boldsymbol{B}\boldsymbol{D}^{-1}\boldsymbol{C})^{-1} & -(\boldsymbol{A} - \boldsymbol{B}\boldsymbol{D}^{-1}\boldsymbol{C})^{-1}\boldsymbol{B}\boldsymbol{D}^{-1} \\ -\boldsymbol{D}^{-1}\boldsymbol{C}(\boldsymbol{A} - \boldsymbol{B}\boldsymbol{D}^{-1}\boldsymbol{C})^{-1} & \boldsymbol{D}^{-1} + \boldsymbol{D}^{-1}\boldsymbol{C}(\boldsymbol{A} - \boldsymbol{B}\boldsymbol{D}^{-1}\boldsymbol{C})^{-1}\boldsymbol{B}\boldsymbol{D}^{-1} \end{bmatrix}. \tag{A.160}$$

A.12 Solution of Linear Systems of Equations

A.12.1 Elimination of Variables

This method is also known as GAUSSian elimination or row reduction and eliminates gradually the unknowns in an equation by operations such as summation, substraction and multiplying of equations. The following examples illustrates the procedure:

$$\begin{array}{rl} x + 3y - 2z = 6 & (1) \\ 2x + 5y + 5z = 5 & (2) \\ 3x + 4y + 2z = 9 & (3) \end{array}$$

$$\begin{array}{rrl} & x + 3y - 2z = 6 & (1) \\ (2) - 2(1): & -\ y + 9z = -7 & (2) \\ (3) - 3(1): & -\ 5y + 8z = -9 & (3) \end{array}$$

$$\begin{array}{rrl} & x + 3y - 2z = 6 & (1) \\ -1(2): & +\ y - 9z = 7 & (2) \\ -1(3): & +\ 5y - 8z = 9 & (3) \end{array}$$

$$\begin{array}{rrrrrrl} & x & + 3y & - & 2z & = & 6 \quad (1) \\ & & + \; y & - & 9z & = & 7 \quad (2) \\ (3) - 5(2): & & & & + 37z & = & -26 \quad (3) \end{array}$$

From (3): $z = -\frac{26}{37}$; in (2): $y = \frac{25}{37}$; in (1): $x = \frac{95}{37}$.

A.12.2 Matrix Solution

The system of equations

$$\begin{aligned} x + 3y - 2z &= 6 \quad (1) \\ 2x + 5y + 5z &= 5 \quad (2) \\ 3x + 4y + 2z &= 9 \quad (3) \end{aligned}$$

can be written in matrix notation as

$$\begin{bmatrix} 1 & 3 & -2 \\ 2 & 5 & 5 \\ 3 & 4 & 2 \end{bmatrix} \begin{bmatrix} x \\ y \\ z \end{bmatrix} = \begin{bmatrix} 6 \\ 5 \\ 9 \end{bmatrix}, \tag{A.161}$$

or in abbreviated form as:

$$\boldsymbol{A}\boldsymbol{x} = \boldsymbol{b}. \tag{A.162}$$

Multiplication from the left-hand side with the inverse of the coefficient matrix, i.e., $\boldsymbol{A}^{-1}$, allows to solve the system of equations:

$$\boldsymbol{A}^{-1}\boldsymbol{A}\boldsymbol{x} = \boldsymbol{A}^{-1}\boldsymbol{b}, \tag{A.163}$$

$$\mathbf{I}\boldsymbol{x} = \boldsymbol{A}^{-1}\boldsymbol{b}, \tag{A.164}$$

$$\boldsymbol{x} = \boldsymbol{A}^{-1}\boldsymbol{b}. \tag{A.165}$$

Considering Eq. (A.161), the inverse of the coefficient matrix is obtained as:

$$\boldsymbol{A}^{-1} = \frac{1}{37} \begin{bmatrix} -10 & -14 & 25 \\ 11 & 8 & -9 \\ -7 & 5 & -1 \end{bmatrix}, \tag{A.166}$$

and multiplication with the right-hand column matrix $\boldsymbol{b}$ gives the unknowns as:

$$\boldsymbol{x} = \frac{1}{37} \begin{bmatrix} 95 \\ 25 \\ -26 \end{bmatrix}. \tag{A.167}$$

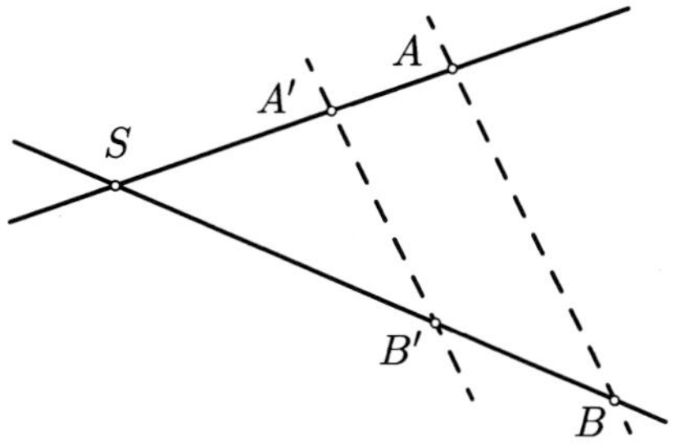

Fig. A.7 Intercept theorem

In the scope of the finite element method, the coefficient matrix $\boldsymbol{A}$ is equal to the global stiffness matrix $\boldsymbol{K}$. Since this matrix is symmetric and may have a special triangular structure, more sophisticated solution procedures can be introduced.

A.13 Elementary Geometry

• Intercept theorem (see Fig. A.7)

$$\overline{SA} : \overline{SA'} = \overline{SB} : \overline{SB'}, \tag{A.168}$$

$$\overline{AB} : \overline{A'B'} = \overline{SA} : \overline{SA'}. \tag{A.169}$$

A.14 Analytical Geometry

A.14.1 Straight-Line Equations

• Point-slope form:
Given is a point (x_1, y_1) and a slope m:

$$y - y_1 = m(x - x_1), \tag{A.170}$$

$$y = (y_1 - mx_1) + m \times x. \tag{A.171}$$

• Slope-intercept form:
Given is a slope m and the y-intercept b:

$$y = b + m \times x. \tag{A.172}$$

• Two-point form:

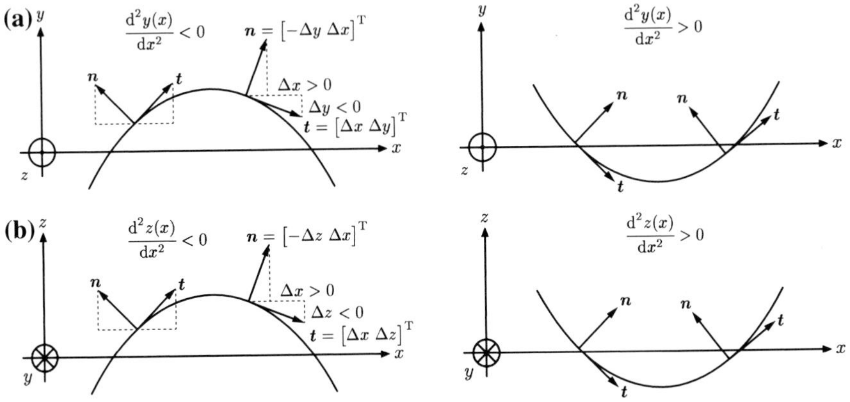

Fig. A.8 Sign of the second derivative of a curve: **a** x–y plane: a curve which is bending away from normal vector $\boldsymbol{n}$ gives a negative sign (left) and a curve which is bending towards the normal vector $\boldsymbol{n}$ gives a positive sign (right); **b** x–z plane: a curve which is bending away from normal vector $\boldsymbol{n}$ gives a negative sign (left) and a curve which is bending towards the normal vector $\boldsymbol{n}$ gives a positive sign (right)

Given are the points (x_1, y_1) and (x_2, y_2):

$$\frac{y - y_1}{x - x_1} = \frac{y_1 - y_2}{x_1 - x_2}, \tag{A.173}$$

$$y = \left(y_1 - \frac{y_1 - y_2}{x_1 - x_2} x_1 \right) + \frac{y_1 - y_2}{x_1 - x_2} \times x \,. \tag{A.174}$$

A.14.2 *Sign of Second Derivative of a Curve*

The sign of the second derivative of a curve is illustrated in Fig. A.8.

A.14.3 *Area of a Polygon*

The area of a polygon, i.e. a closed plane shape with straight lines and vertices (listed counterclockwise[4] around the perimeter), can be obtained based on the surveyor's formula [2]:

[4]The counterclockwise numbering ensures a positive expression.

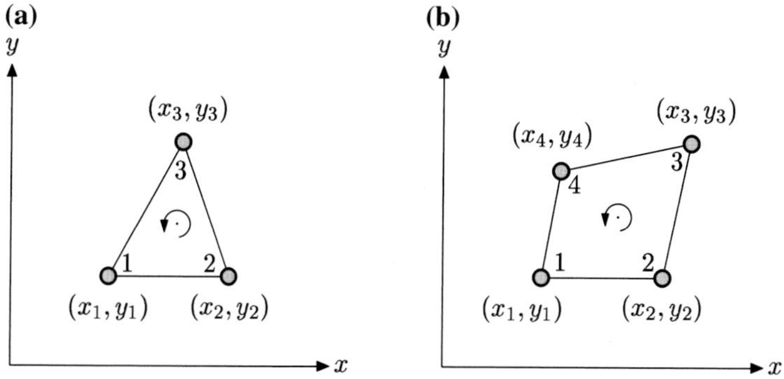

Fig. A.9 Simple polygon shapes: **a** triangle and **b** rectangle

$$A = \frac{1}{2}\left\{ \begin{vmatrix} x_0 & x_1 \\ y_0 & y_1 \end{vmatrix} + \begin{vmatrix} x_1 & x_2 \\ y_1 & y_2 \end{vmatrix} + \cdots + \begin{vmatrix} x_{n-2} & x_{n-1} \\ y_{n-2} & y_{n-1} \end{vmatrix} + \begin{vmatrix} x_{n-1} & x_0 \\ y_{n-1} & y_0 \end{vmatrix} \right\}, \tag{A.175}$$

where the (x_i, y_i) are the plane coordinates of the vertices. Equation (A.175) can be simplified to the case of triangles and rectangles as (see Fig. A.9):

$$A^{\triangle} = \frac{1}{2}\left\{ \begin{vmatrix} x_1 & x_2 \\ y_1 & y_2 \end{vmatrix} + \begin{vmatrix} x_2 & x_3 \\ y_2 & y_3 \end{vmatrix} + \begin{vmatrix} x_3 & x_1 \\ y_3 & y_1 \end{vmatrix} \right\}, \tag{A.176}$$

$$A^{\square} = \frac{1}{2}\left\{ \begin{vmatrix} x_1 & x_2 \\ y_1 & y_2 \end{vmatrix} + \begin{vmatrix} x_2 & x_3 \\ y_2 & y_3 \end{vmatrix} + \begin{vmatrix} x_3 & x_4 \\ y_3 & y_4 \end{vmatrix} + \begin{vmatrix} x_4 & x_1 \\ y_4 & y_1 \end{vmatrix} \right\}, \tag{A.177}$$

or after evaluation of the 2×2 determinants:

$$A^{\triangle} = \frac{1}{2}(x_1 y_2 - x_2 y_1 + x_2 y_3 - x_3 y_2 + x_3 y_1 - x_1 y_3), \tag{A.178}$$

$$A^{\square} = \frac{1}{2}(x_1 y_2 - x_2 y_1 + x_2 y_3 - x_3 y_2 + x_3 y_4 - x_4 y_3 + x_4 y_1 - x_1 y_4). \tag{A.179}$$

Appendix B
Mechanics

B.1 Centroids

The coordinates (z_S, y_S) of the centroid S of the plane surface shown in Fig. B.1 can be expressed as

$$z_S = \frac{\int z\,\mathrm{d}A}{\int \mathrm{d}A}, \tag{B.1}$$

$$y_S = \frac{\int y\,\mathrm{d}A}{\int \mathrm{d}A}, \tag{B.2}$$

where the integrals $\int z\,\mathrm{d}A$ and $\int y\,\mathrm{d}A$ are known as the first moments of area.[5] In the case of surfaces composed of n simple shapes, the integrals can be replaced by summations to obtain:

$$z_S = \frac{\sum\limits_{i=1}^{n} z_i A_i}{\sum\limits_{i=1}^{n} A_i}, \tag{B.3}$$

$$y_S = \frac{\sum\limits_{i=1}^{n} y_i A_i}{\sum\limits_{i=1}^{n} A_i}. \tag{B.4}$$

[5] A better expression would be moment of surface since area means strictly speaking the measure of the size of the surface which is different to the surface itself.

A. Öchsner, *Computational Statics and Dynamics*,
https://doi.org/10.1007/978-981-15-1278-0

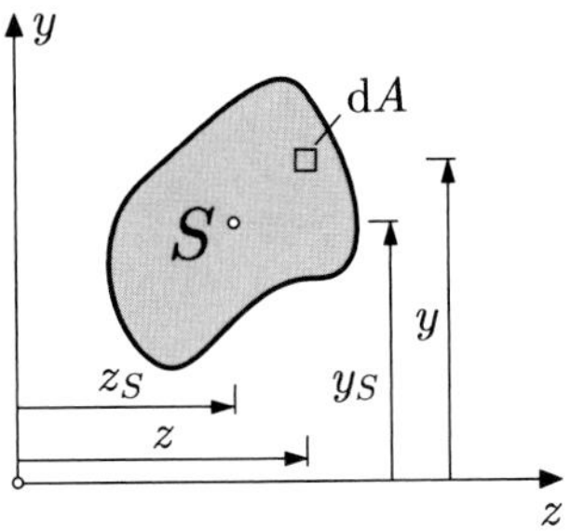

Fig. B.1 Plane surface with centroid S

B.2 Second Moment of Area

The second moment of area[6] or the second area moment is a geometrical property of a surface which reflects how its area elements are distributed with regard to an arbitrary axis. The second moments of area for an arbitrary surface with respect to an arbitrary Cartesian coordinate system (see Fig. B.1) are generally defined as:

$$I_y = \int_A z^2 \mathrm{d}A\,, \tag{B.5}$$

$$I_z = \int_A y^2 \mathrm{d}A\,. \tag{B.6}$$

These quantities are normally used in the context of plane bending of symmetrical cross sections. For unsymmetrical bending, the product moment of area is additionally required:

$$I_{yz} = -\int_A yz \mathrm{d}A\,. \tag{B.7}$$

B.3 Parallel-Axis Theorem

The parallel-axis theorem gives the relationship between the second moment of area with respect to a centroidal axis (z_1, y_1) and the second moment of area with respect to any parallel axis[7] (z, y). For the rectangular shown in Fig. B.2, the relations can be expressed as:

[6]The second moment of area is also called in the literature the second moment of inertia. However, the expression moment of inertia is in context of properties of surfaces misleading since no mass or movement is involved.

[7]This arbitrary axis can be for example the axis trough the common centroid S of a composed surface.

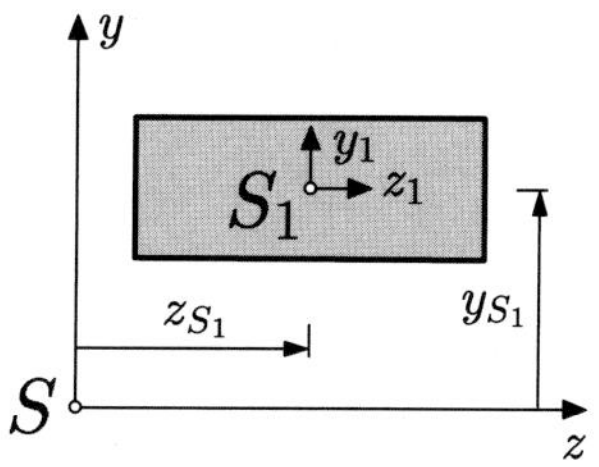

Fig. B.2 Configuration for the parallel-axis theorem

$$I_y = I_{y_1} + z_{S_1}^2 \times A_1 \, , \tag{B.8}$$

$$I_z = I_{z_1} + y_{S_1}^2 \times A_1 \, , \tag{B.9}$$

$$I_{zy} = I_{z_1 y_1} - z_{S_1} y_{S_1} \times A_1 \, . \tag{B.10}$$

Appendix C
Units and Conversion

C.1 SI Base Units

The International System of Units (SI)[8] must be used in scientific publications to express physical units. This system consists of the seven base quantities—length, mass, time, electric current, thermodynamic temperature, amount of substance, and luminous intensity—and their respective base units are the meter, kilogram, second, ampere, kelvin, mole, and candela.[9]

C.2 Coherent SI derived Units

A coherent SI derived unit is defined uniquely as a product of powers of base units that include no numerical factor other than 1. Table C.1 gives some examples of derived units and their expression in terms of base units.

C.3 Consistent Units

The application of a finite element code does normally not require that a specific system of units is selected. A finite element code keeps *consistent* units throughout an analysis and requires only that a user assigns the absolute measure without specifying a specific unit. Thus, the units considered by the user during the pre-processing phase are maintained for the post-processing phase. The user must assure that the considered units are consistent, i.e. they fit each other. The following Table C.2 shows an example of consistent units

[8]The original name is known in French as: Système International d'Unités.

[9]More information on units can be found in the brochures of the Bureau International des Poids et Mesures (BIPM): www.bipm.org/en/si.

A. Öchsner, *Computational Statics and Dynamics*,
https://doi.org/10.1007/978-981-15-1278-0

Table C.1 Example of coherent SI derived units

Quantity	Name	Coherent derived unit		
		Symbol	In terms of other SI units	In terms of SI base units
Celsius temperature	Degree Celsius	℃		K
Energy, work	Joule	J	N m	$m^2\,kg\,s^{-2}$
Force	Newton	N		$m\,kg\,s^{-2}$
Plane angle	Radian	rad	1	m/m
Power	Watt	W	J/s	$m^2\,kg\,s^{-3}$
Pressure, stress	Pascal	Pa	N/m^2	$m^{-1}\,kg\,s^{-2}$

Table C.2 Example of consistent units

Property	Unit
Length	mm
Area	mm^2
Force	N
Pressure	$MPa = \frac{N}{mm^2}$
Moment	Nmm
Moment of inertia	mm^4
E-modulus	$MPa = \frac{N}{mm^2}$
Density	$\frac{Ns^2}{mm^4}$
Time	s
Mass	10^3kg

Pay attention to the unit of density. The following example shows the conversion of the density of steel:

$$\varrho_{St} = 7.8\,\frac{kg}{dm^3} = 7.8 \times 10^3\,\frac{kg}{m^3} = 7.8 \times 10^{-6}\,\frac{kg}{mm^3}. \tag{C.1}$$

With

$$1\,N = 1\,\frac{m\,kg}{s^2} = 1 \times 10^3\,\frac{mm\,kg}{s^2} \quad \text{und} \quad 1\,kg = 1 \times 10^{-3}\,\frac{Ns^2}{mm} \tag{C.2}$$

follows the consistent density to:

$$\varrho_{St} = 7.8 \times 10^{-9}\,\frac{Ns^2}{mm^4}. \tag{C.3}$$

Table C.3 Example of consistent English units

Property	Unit
Length	in
Area	in^2
Force	lbf
Pressure	$psi = \frac{lbf}{in^2}$
Moment	lbf in
Moment of inertia	in^4
E-Modulus	$psi = \frac{lbf}{in^2}$
Density	$\frac{lbf\,s^2}{in^4}$
Time	s

C.4 Conversion of Important English Units to the Metric System

Since literature reports time after time also other units, Table C.3 shows an example of consistent English units. The conversion of important U.S. customary units and British Imperial units is given in Table C.4.

Pay attention to the conversion of the density:

Table C.4 Conversion of important U.S. customary units and British Imperial units ('English units') to metric units (m: Meter; cm: Centimeter; g: Gram; N: Newton; J: Joule; W: Watt)

Type	English unit	Conversion
Length	Inch	1 in = 0.025400 m
	Foot	1 ft = 0.304800 m
	Yard	1 yd = 0.914400 m
	Mile (statute)	1 mi = 1609.344 m
	Mile (nautical)	1 nm = 1852.216 m
Area	Square inch	1 sq in = 1 in^2 = 6.45160 cm^2
	Square foot	1 sq ft = 1 ft^2 = 0.092903040 m^2
	Square yard	1 sq yd = 1 yd^2 = 0.836127360 m^2
	Square mile	1 sq mi = 1 mi^2 = 2589988.110336 m^2
	Acre	1 ac = 4046.856422400 m^2

(continued)

Table C.4 (continued)

Type	English unit	Conversion
Volume	Cubic inch	1 cu in = 1 in^3 = 0.000016387064 m^3
	Cubic foot	1 cu ft = 1 ft^3 = 0.028316846592 m^3
	Cubic yard	1 cu yd = 1 yd^3 = 0.764554857984 m^3
Mass	Ounce	1 oz = 28.349523125 g
	Pound (mass)	1 lb_m = 453.592370 g
	Short ton	1 sh to = 907184.74 g
	Long ton	1 lg to = 1016046.9088 g
Force	Pound-force	1 lbf = 1 lb_F = 4.448221615260500 N
	Poundal	1 pdl = 0.138254954376 N
Stress	Pound-force per square inch	1 psi = 1 $\frac{\text{lbf}}{\text{in}^2}$ = 6894.75729316837 $\frac{\text{N}}{\text{m}^2}$
	Pound-force per square foot	1 $\frac{\text{lbf}}{\text{ft}^2}$ = 47.880258980336 $\frac{\text{N}}{\text{m}^2}$
Energy	British thermal unit	1 Btu = 1055.056 J
	Calorie	1 cal = 4185.5 J
Power	Horsepower	1 hp = 745.699871582270 W

$$\varrho_{\text{St}} = 0.282\,\frac{\text{lb}}{\text{in}^3} = 0.282\,\frac{1}{\text{in}^3} \times 0.00259\,\frac{\text{lbf s}^2}{\text{in}} = 0.73038 \times 10^{-3}\,\frac{\text{lbf s}^2}{\text{in}^4}. \tag{C.4}$$

Appendix D
Triangular Elements

Let us consider in the following the three-node planar element as shown in Fig. D.1. For simplicity, the thickness (t) is assumed constant. Important for the following derivations is that the node numbering is counterclockwise. Furthermore, all the following derivations are restricted to a linear three-node element under the plane stress assumption. This element is known in literature as the constant strain triangle or CST element.

In order to derive the interpolation functions, let us assume in the following a linear displacement field in the Cartesian x–y space

$$u_x^{\mathrm{e}}(x, y) = a_1 + a_2 x + a_3 y \,, \tag{D.1}$$

$$u_y^{\mathrm{e}}(x, y) = a_4 + a_5 x + a_6 y \,, \tag{D.2}$$

or combined in a single matrix equation:

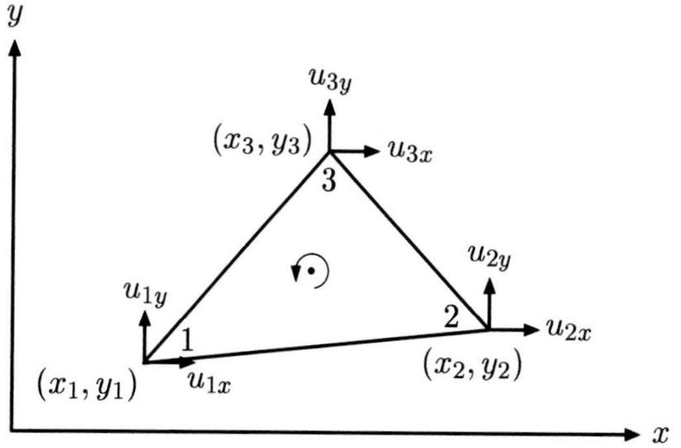

Fig. D.1 Three-node planar element in the Cartesian space (x, y)

A. Öchsner, *Computational Statics and Dynamics*,
https://doi.org/10.1007/978-981-15-1278-0

$$\begin{bmatrix} u_x^{\mathrm{e}} \\ u_y^{\mathrm{e}} \end{bmatrix} = \begin{bmatrix} 1 & x & y & 0 & 0 & 0 \\ 0 & 0 & 0 & 1 & x & y \end{bmatrix} \begin{bmatrix} a_1 \\ a_2 \\ a_3 \\ a_4 \\ a_5 \\ a_6 \end{bmatrix} . \tag{D.3}$$

Evaluating Eqs. (D.1) and (D.2) for the three nodes shown in Fig. D.1 gives for the x- and y-direction:

$$\text{Node 1}: u_{1x} = u_x^{\mathrm{e}}(x = x_1, y = y_1) = a_1 + a_2 x_1 + a_3 y_1 , \tag{D.4}$$
$$\text{Node 2}: u_{2x} = u_x^{\mathrm{e}}(x = x_2, y = y_2) = a_1 + a_2 x_2 + a_3 y_2 , \tag{D.5}$$
$$\text{Node 3}: u_{3x} = u_x^{\mathrm{e}}(x = x_3, y = y_3) = a_1 + a_2 x_3 + a_3 y_3 , \tag{D.6}$$

$$-----$$

$$\text{Node 1}: u_{1y} = u_x^{\mathrm{e}}(x = x_1, y = y_1) = a_4 + a_5 x_1 + a_6 y_1 , \tag{D.7}$$
$$\text{Node 2}: u_{2y} = u_x^{\mathrm{e}}(x = x_2, y = y_2) = a_4 + a_5 x_2 + a_6 y_2 , \tag{D.8}$$
$$\text{Node 3}: u_{3y} = u_x^{\mathrm{e}}(x = x_3, y = y_3) = a_4 + a_5 x_3 + a_6 y_3 . \tag{D.9}$$

The first three equations of (D.4)–(D.9), i.e., for the x-direction, can be expressed in matrix notation as:

$$\underbrace{\begin{bmatrix} u_{1x} \\ u_{2x} \\ u_{3x} \end{bmatrix}}_{\boldsymbol{u}_{\mathrm{p},x}} = \underbrace{\begin{bmatrix} 1 & x_1 & y_1 \\ 1 & x_2 & y_2 \\ 1 & x_3 & y_3 \end{bmatrix}}_{\boldsymbol{X}} \underbrace{\begin{bmatrix} a_1 \\ a_2 \\ a_3 \end{bmatrix}}_{\boldsymbol{a}} . \tag{D.10}$$

Solving for the coefficients $\boldsymbol{a}$ gives:

$$\begin{bmatrix} a_1 \\ a_2 \\ a_3 \end{bmatrix} = \frac{1}{x_1(y_2 - y_3) + x_2(y_3 - y_1) + x_3(y_1 - y_2)} \times$$

$$\times \begin{bmatrix} x_2 y_3 - x_3 y_2 & x_3 y_1 - x_1 y_3 & x_1 y_2 - x_2 y_2 \\ y_2 - y_3 & y_3 - y_1 & y_1 - y_2 \\ x_3 - x_2 & x_1 - x_3 & x_2 - x_1 \end{bmatrix} \begin{bmatrix} u_{1x} \\ u_{2x} \\ u_{3x} \end{bmatrix} \tag{D.11}$$

$$= \underbrace{\frac{1}{2A^{\triangle}} \times \begin{bmatrix} x_2 y_3 - x_3 y_2 & x_3 y_1 - x_1 y_3 & x_1 y_2 - x_2 y_2 \\ y_2 - y_3 & y_3 - y_1 & y_1 - y_2 \\ x_3 - x_2 & x_1 - x_3 & x_2 - x_1 \end{bmatrix}}_{\boldsymbol{A} = \boldsymbol{X}^{-1}} \begin{bmatrix} u_{1x} \\ u_{2x} \\ u_{3x} \end{bmatrix} , \tag{D.12}$$

$$\tag{D.13}$$

where $A^{\triangle}$ is the area of the triangle (see Appendix A.14.3). In abbreviated form, we can write the last equation as:

$$\boldsymbol{a} = \boldsymbol{A}\boldsymbol{u}_{\mathrm{p}x} = \boldsymbol{X}^{-1}\boldsymbol{u}_{\mathrm{p},x}\,. \tag{D.14}$$

The column matrix of interpolation functions results as (see Eq. (5.56)):

$$\boldsymbol{N}_{\mathrm{e}}^{\mathrm{T}} = \boldsymbol{\chi}^{\mathrm{T}}\boldsymbol{A} = \begin{bmatrix}1 & x & y\end{bmatrix}\frac{1}{2A^{\triangle}} \times \begin{bmatrix} x_2y_3 - x_3y_2 & x_3y_1 - x_1y_3 & x_1y_2 - x_2y_2 \\ y_2 - y_3 & y_3 - y_1 & y_1 - y_2 \\ x_3 - x_2 & x_1 - x_3 & x_2 - x_1 \end{bmatrix}, \tag{D.15}$$

or written as three single equations

$$N_{1x}(x, y) = \frac{1}{2A^{\triangle}}[(x_2y_3 - x_3y_2)1 + (y_2 - y_3)x + (x_3 - x_2)y]\,, \tag{D.16}$$

$$N_{2x}(x, y) = \frac{1}{2A^{\triangle}}[(x_3y_1 - x_1y_3)1 + (y_3 - y_1)x + (x_1 - x_3)y]\,, \tag{D.17}$$

$$N_{3x}(x, y) = \frac{1}{2A^{\triangle}}[(x_1y_2 - x_2y_1)1 + (y_1 - y_2)x + (x_2 - x_1)y]\,. \tag{D.18}$$

The evaluation of the y-direction results in the same interpolation functions, i.e., $N_{iy} = N_{ix}$ $(i = 1, 2, 3)$. Thus, we can write for the displacement field in Cartesian coordinates:

$$u_x(x, y) = N_{1x}u_{1x} + N_{2x}u_{2x} + N_{3x}u_{3x}\,, \tag{D.19}$$

$$u_y(x, y) = N_{1y}u_{1y} + N_{2y}u_{2y} + N_{3y}u_{3y}\,, \tag{D.20}$$

or with $N_{ix} = N_{iy} = N_i$:

$$u_x(x, y) = N_1u_{1x} + N_2u_{2x} + N_3u_{3x}\,, \tag{D.21}$$

$$u_y(x, y) = N_1u_{1y} + N_2u_{2y} + N_3u_{3y}\,. \tag{D.22}$$

One may combine the last two equations in matrix form to obtain the following representation:

$$\begin{bmatrix} u_x(x, y) \\ u_y(x, y) \end{bmatrix} = \begin{bmatrix} N_1 & 0 & N_2 & 0 & N_3 & 0 \\ 0 & N_1 & 0 & N_2 & 0 & N_3 \end{bmatrix} \begin{bmatrix} u_{1x} \\ u_{1y} \\ u_{2x} \\ u_{2y} \\ u_{3x} \\ u_{3y} \end{bmatrix}. \tag{D.23}$$

The last equation can be written in abbreviated form as:

$$\boldsymbol{u}^{\mathrm{e}}(\boldsymbol{x}) = \boldsymbol{N}^{\mathrm{T}}(\boldsymbol{x})\boldsymbol{u}_{\mathrm{p}}^{\mathrm{p}}\,. \tag{D.24}$$

The graphical representation of the three interpolation functions in Cartesian coordinates is given in Fig. D.2. It can be seen from this figure that the classical condition for interpolation functions is fulfilled, i.e., that N_i is equal to 1 at node i and 0 at the other nodes.

Let us assume the same interpolation for the global x- and y-coordinate as for the displacement (*isoparametric* element formulation), i.e. $\overline{N}_i = N_i$:

$$x(x, y) = \overline{N}_1(x, y) \times x_1 + \overline{N}_2(x, y) \times x_2 + \overline{N}_3(x, y) \times x_3 \,, \tag{D.25}$$

$$y(x, y) = \overline{N}_1(x, y) \times y_1 + \overline{N}_2(x, y) \times y_2 + \overline{N}_3(x, y) \times y_3 \,, \tag{D.26}$$

or as:

$$x(x, y) = N_1(x, y) \times x_1 + N_2(x, y) \times x_2 + N_3(x, y) \times x_3 \,, \tag{D.27}$$

$$y(x, y) = N_1(x, y) \times y_1 + N_2(x, y) \times y_2 + N_3(x, y) \times y_3 \,. \tag{D.28}$$

Alternatively, the above derivation can be based on natural coordinates. In the case of triangular elements, the so-called triangular coordinates[10] can be stated for any point P (see Fig. D.3) as:

$$\xi = \frac{A_1}{A}, \quad \eta = \frac{A_2}{A}, \quad \zeta = \frac{A_3}{A}. \tag{D.29}$$

The area A_1, for example, is obtained by connecting the two nodes other than 1 with point P, see Fig. D.3a. Furthermore, it should be noted that each triangular coordinate ranges between 0 and 1, see Fig. D.3b, c.

Now, the three nodes of the triangle shown in Fig. D.1 can be expressed based on these triangular coordinates as (see the graphical representation of the three-node element in the natural space ξ–η in Fig. D.4):

$$\text{Node 1}: \ \xi = 1, \eta = 0, \zeta = 0 \,, \tag{D.30}$$

$$\text{Node 2}: \ \xi = 0, \eta = 1, \zeta = 0 \,, \tag{D.31}$$

$$\text{Node 3}: \ \xi = 0, \eta = 0, \zeta = 1 \,. \tag{D.32}$$

This can be generalized to express the coordinate of the reference point $P(x_P, y_P)$ in triangular coordinates as:

$$x_P = x_1\xi + x_2\eta + x_3\zeta \,, \tag{D.33}$$

$$y_P = y_1\xi + y_2\eta + y_3\zeta \,, \tag{D.34}$$

where (x_i, y_i) are the Cartesian coordinates of node i. Furthermore, we can deduct from the definition of the triangular coordinates (see Eq. (D.29)) that:

---

[10] Alternatively named as area coordinates or barycentric coordinates.

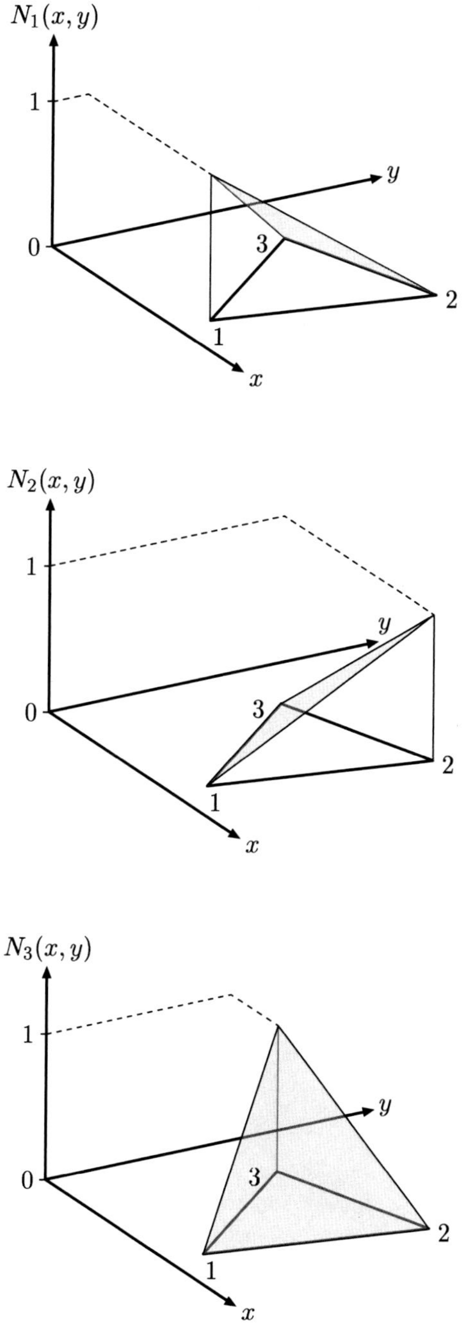

Fig. D.2 Graphical representation of the three interpolation functions for the three-node planar element in Cartesian space. Note that the interpolation functions are only plotted over the area of the triangle 123

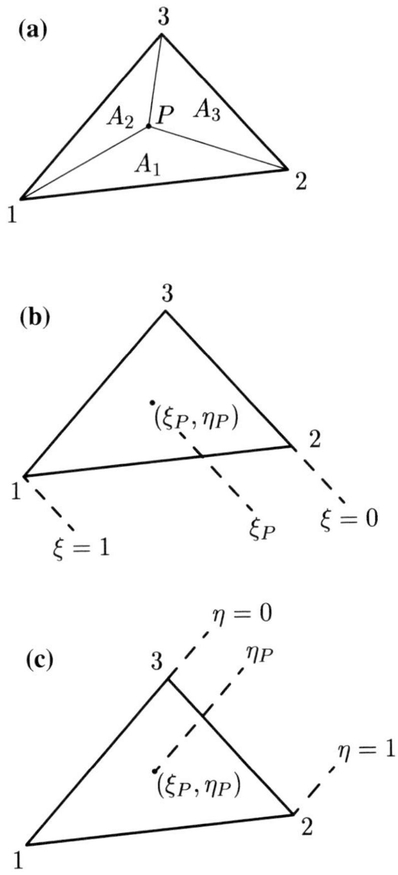

Fig. D.3 Graphical representation of triangular coordinates: **a** definition of subareas A_i, **b** range of coordinate ξ, **c** range of coordinate η

$$\xi + \eta + \zeta = \frac{A_1}{A} + \frac{A_2}{A} + \frac{A_3}{A} = \frac{1}{A}(A_1 + A_2 + A_3) = 1\,. \tag{D.35}$$

Introducing this relationship in Eqs. (D.33)–(D.34), we get

$$x_P = x_1\xi + x_2\eta + x_3\,(1 - \xi - \eta)\ , \tag{D.36}$$

$$y_P = y_1\xi + y_2\eta + y_3\,(1 - \xi - \eta)\ , \tag{D.37}$$

and a comparison with Eqs. (D.27)–(D.28) allows under the assumption of an *isoparametric* element formulation the deduction of the three interpolation functions in triangular coordinates as:

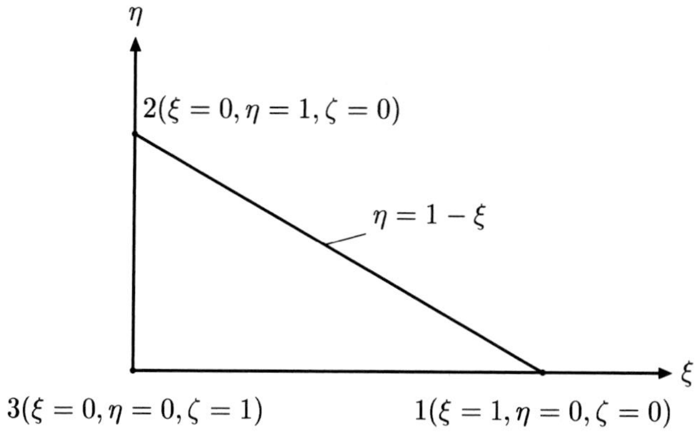

Fig. D.4 Three-node planar element in the natural space (ξ, η)

$$N_1(\xi,\eta) = \xi\,, \tag{D.38}$$
$$N_2(\xi,\eta) = \eta\,, \tag{D.39}$$
$$N_3(\xi,\eta) = 1 - \xi - \eta\,. \tag{D.40}$$

Equations (D.35) and (D.33)–(D.34) can be combined in matrix notation to obtain:

$$\begin{bmatrix} 1 \\ x_P \\ y_P \end{bmatrix} = \begin{bmatrix} 1 & 1 & 1 \\ x_1 & x_2 & x_3 \\ y_1 & y_2 & y_3 \end{bmatrix} \begin{bmatrix} \xi \\ \eta \\ \zeta \end{bmatrix}. \tag{D.41}$$

The last equation can be solved for the triangular coordinates as a function the Cartesian coordinates of the reference point P[11]:

$$\begin{bmatrix} \xi \\ \eta \\ \zeta \end{bmatrix} = \frac{1}{x_1(y_2-y_3)+x_2(y_3-y_1)+x_3(y_1-y_2)} \times$$
$$\times \begin{bmatrix} x_2y_3-x_3y_2 & y_2-y_3 & x_3-x_2 \\ x_3y_1-x_1y_3 & y_3-y_1 & x_1-x_3 \\ x_1y_2-x_2y_2 & y_1-y_2 & x_2-x_1 \end{bmatrix} \begin{bmatrix} 1 \\ x_P \\ y_P \end{bmatrix} \tag{D.42}$$
$$= \frac{1}{2A} \begin{bmatrix} x_2y_3-x_3y_2 & y_2-y_3 & x_3-x_2 \\ x_3y_1-x_1y_3 & y_3-y_1 & x_1-x_3 \\ x_1y_2-x_2y_2 & y_1-y_2 & x_2-x_1 \end{bmatrix} \begin{bmatrix} 1 \\ x_P \\ y_P \end{bmatrix}. \tag{D.43}$$

Let us now focus on the derivation of the elemental stiffness matrix. From Eq. (5.38), we can state the general form as

[11]If we skip den index 'P', then we simply have the relationship between the triangular and the Cartesian coordinates.

$$\boldsymbol{K}^{\mathrm{e}} = \int_V \underbrace{\left(\boldsymbol{\mathcal{L}}_1 \boldsymbol{N}^{\mathrm{T}}\right)^{\mathrm{T}}}_{\boldsymbol{B}} \boldsymbol{C} \underbrace{\left(\boldsymbol{\mathcal{L}}_1 \boldsymbol{N}^{\mathrm{T}}\right)}_{\boldsymbol{B}^{\mathrm{T}}} \mathrm{d}V\,, \tag{D.44}$$

where the differential operator matrix is given according to Eq. (5.2) as

$$\boldsymbol{\mathcal{L}}_1 = \begin{bmatrix} \frac{\partial}{\partial x} & 0 \\ 0 & \frac{\partial}{\partial y} \\ \frac{\partial}{\partial y} & \frac{\partial}{\partial x} \end{bmatrix}, \tag{D.45}$$

and the elasticity matrix follows for the plane stress assumption according to Eq. (5.4) as:

$$\boldsymbol{C} = \frac{E}{1-\nu^2} \begin{bmatrix} 1 & \nu & 0 \\ \nu & 1 & 0 \\ 0 & 0 & \frac{1-\nu}{2} \end{bmatrix} = \begin{bmatrix} C_{11} & C_{12} & C_{13} \\ C_{21} & C_{22} & C_{23} \\ C_{31} & C_{32} & C_{33} \end{bmatrix}. \tag{D.46}$$

If the derivation is based on the Cartesian coordinates, the set of interpolation functions given in Eqs. (D.16)–(D.18) should be considered. Thus, the $\boldsymbol{B}$-matrix can be stated as:

$$\boldsymbol{B} = \boldsymbol{\mathcal{L}}_1 \boldsymbol{N}^{\mathrm{T}}(\boldsymbol{x}) \tag{D.47}$$

$$= \begin{bmatrix} \frac{\partial}{\partial x} & 0 \\ 0 & \frac{\partial}{\partial y} \\ \frac{\partial}{\partial y} & \frac{\partial}{\partial x} \end{bmatrix} \begin{bmatrix} N_1 & 0 & N_2 & 0 & N_3 & 0 \\ 0 & N_1 & 0 & N_2 & 0 & N_3 \end{bmatrix} \tag{D.48}$$

$$= \begin{bmatrix} \frac{\partial N_1(x,y)}{\partial x} & 0 & \frac{\partial N_2 x,y)}{\partial x} & 0 & \frac{\partial N_3(x,y)}{\partial x} & 0 \\ 0 & \frac{\partial N_1(x,y)}{\partial y} & 0 & \frac{\partial N_2(x,y)}{\partial y} & 0 & \frac{\partial N_3(x,y)}{\partial y} \\ \frac{\partial N_1(x,y)}{\partial y} & \frac{\partial N_1(x,y)}{\partial x} & \frac{\partial N_2(x,y)}{\partial y} & \frac{\partial N_2(x,y)}{\partial x} & \frac{\partial N_3(x,y)}{\partial y} & \frac{\partial N_3(x,y)}{\partial x} \end{bmatrix}. \tag{D.49}$$

The partial derivatives of the interpolation functions with respect to the Cartesian coordinates can easily obtained as:

$$\frac{\partial N_1(x,y)}{\partial x} = \frac{1}{2A}(y_2 - y_3) = \frac{y_{23}}{2A}, \quad \frac{\partial N_1(x,y)}{\partial y} = \frac{1}{2A}(x_3 - x_2) = \frac{x_{32}}{2A}, \tag{D.50}$$

$$\frac{\partial N_2(x,y)}{\partial x} = \frac{1}{2A}(y_3 - y_1) = \frac{y_{31}}{2A}, \quad \frac{\partial N_2(x,y)}{\partial x} = \frac{1}{2A}(x_1 - x_2) = \frac{x_{13}}{2A}, \tag{D.51}$$

$$\frac{\partial N_3(x,y)}{\partial x} = \frac{1}{2A}(y_1 - y_2) = \frac{y_{12}}{2A}, \quad \frac{\partial N_3(x,y)}{\partial y} = \frac{1}{2A}(x_2 - x_1) = \frac{y_{21}}{2A}, \tag{D.52}$$

where the abbreviations $x_i - x_j = x_{ij}$ and $y_i - y_j = y_{ij}$ were used. Thus, we can write the $\boldsymbol{B}$-matrix as:

$$\boldsymbol{B} = \boldsymbol{\mathcal{L}}_1 \boldsymbol{N}^{\mathrm{T}}(\boldsymbol{x}) \tag{D.53}$$

$$= \frac{1}{2A}\begin{bmatrix} (y_2-y_3) & 0 & (y_3-y_1) & 0 & (y_1-y_2) & 0 \\ 0 & (x_3-x_2) & 0 & (x_1-x_3) & 0 & (x_2-x_1) \\ (x_3-x_2) & (y_2-y_3) & (x_1-x_3) & (y_3-y_1) & (x_2-x_1) & (y_1-y_2) \end{bmatrix} \tag{D.54}$$

$$= \frac{1}{2A}\begin{bmatrix} y_{23} & 0 & y_{31} & 0 & y_{12} & 0 \\ 0 & x_{32} & 0 & x_{13} & 0 & x_{21} \\ x_{32} & y_{23} & x_{13} & y_{31} & x_{21} & y_{12} \end{bmatrix}. \tag{D.55}$$

Now we can state the stiffness matrix —under the assumption that the triangle has a constant thickness t and the result that the all derivatives of interpolation functions result in absolute terms[12]—as:

$$\boldsymbol{K}^{\mathrm{e}} = \int_V \boldsymbol{B}\boldsymbol{C}\boldsymbol{B}^{\mathrm{T}}\mathrm{d}V = t\int_A \boldsymbol{B}\boldsymbol{C}\boldsymbol{B}^{\mathrm{T}}\mathrm{d}A = tA\boldsymbol{B}\boldsymbol{C}\boldsymbol{B}^{\mathrm{T}} \tag{D.56}$$

$$= \frac{t}{4A}\begin{bmatrix} y_{23} & 0 & x_{32} \\ 0 & x_{32} & y_{23} \\ y_{31} & 0 & x_{13} \\ 0 & x_{13} & y_{31} \\ y_{12} & 0 & x_{21} \\ 0 & x_{21} & y_{12} \end{bmatrix}\begin{bmatrix} C_{11} & C_{12} & C_{13} \\ C_{21} & C_{22} & C_{23} \\ C_{31} & C_{32} & C_{33} \end{bmatrix}\begin{bmatrix} y_{23} & 0 & y_{31} & 0 & y_{12} & 0 \\ 0 & x_{32} & 0 & x_{13} & 0 & x_{21} \\ x_{32} & y_{23} & x_{13} & y_{31} & x_{21} & y_{12} \end{bmatrix}. \tag{D.57}$$

Let us now focus on the derivation based on the triangular coordinates. The partial derivatives of the interpolation functions (expressed in triangular coordinates) with respect to the Cartesian coordinates can be written under consideration of the chain rule as follows:

$$\frac{\partial N_1(\xi,\eta)}{\partial x} = \frac{\partial N_1}{\partial \xi}\frac{\partial \xi}{\partial x} + \frac{\partial N_1}{\partial \eta}\frac{\partial \eta}{\partial x}, \tag{D.58}$$

$$\frac{\partial N_1(\xi,\eta)}{\partial y} = \frac{\partial N_1}{\partial \xi}\frac{\partial \xi}{\partial y} + \frac{\partial N_1}{\partial \eta}\frac{\partial \eta}{\partial y}, \tag{D.59}$$

$$\vdots \tag{D.60}$$

$$\frac{\partial N_3(\xi,\eta)}{\partial x} = \frac{\partial N_3}{\partial \xi}\frac{\partial \xi}{\partial x} + \frac{\partial N_3}{\partial \eta}\frac{\partial \eta}{\partial x}, \tag{D.61}$$

$$\frac{\partial N_3(\xi,\eta)}{\partial y} = \frac{\partial N_3}{\partial \xi}\frac{\partial \xi}{\partial y} + \frac{\partial N_3}{\partial \eta}\frac{\partial \eta}{\partial y}. \tag{D.62}$$

Partial derivatives of the interpolation functions with respect to the triangular coordinates can be easily obtained from Eqs. (D.38)–(D.40) as:

---

[12]These absolute terms result in a constant strain distribution within the element (see Eqs. (5.3) and (D.24)) and this justifies the name constant strain triangle.

$$\frac{\partial N_1(\xi,\eta)}{\partial \xi} = 1\,, \qquad \frac{\partial N_1(\xi,\eta)}{\partial \eta} = 0\,, \tag{D.63}$$

$$\frac{\partial N_2(\xi,\eta)}{\partial \xi} = 0\,, \qquad \frac{\partial N_2(\xi,\eta)}{\partial \eta} = 1\,, \tag{D.64}$$

$$\frac{\partial N_3(\xi,\eta)}{\partial \xi} = -1\,, \qquad \frac{\partial N_3(\xi,\eta)}{\partial \eta} = -1\,. \tag{D.65}$$

The geometrical derivatives result from Eq. (D.43) as:

$$\frac{\partial \xi(x,y)}{\partial x} = \frac{y_2 - y_3}{2A}\,, \qquad \frac{\partial \xi(x,y)}{\partial y} = \frac{x_3 - x_3}{2A}\,, \tag{D.66}$$

$$\frac{\partial \eta(x,y)}{\partial x} = \frac{y_3 - y_1}{2A}\,, \qquad \frac{\partial \eta(x,y)}{\partial y} = \frac{x_1 - x_3}{2A}\,. \tag{D.67}$$

Thus, we can finally write the partial derivatives as follows:

$$\frac{\partial N_1(\xi,\eta)}{\partial x} = \frac{y_2 - y_3}{2A}\,, \qquad \frac{\partial N_1(\xi,\eta)}{\partial y} = \frac{x_3 - x_2}{2A}\,, \tag{D.68}$$

$$\frac{\partial N_2(\xi,\eta)}{\partial x} = \frac{y_3 - y_1}{2A}\,, \qquad \frac{\partial N_2(\xi,\eta)}{\partial y} = \frac{x_1 - x_3}{2A}\,, \tag{D.69}$$

$$\frac{\partial N_3(\xi,\eta)}{\partial x} = \frac{y_1 - y_2}{2A}\,, \qquad \frac{\partial N_3(\xi,\eta)}{\partial y} = \frac{x_2 - x_1}{2A}\,. \tag{D.70}$$

These derivatives result in the same $\boldsymbol{B}$-matrix as in Eq. (D.55) and the same elemental stiffness matrix is obtained.

Appendix E
Summary of Stiffness Matrices

E.1 One-Dimensional Elements

- Linear rod element ($E, A:$ constat):

$$\boldsymbol{K}^{\mathrm{e}} = \frac{EA}{L}\begin{bmatrix} 1 & -1 \\ -1 & 1 \end{bmatrix}. \tag{E.1}$$

- Quadratic rod element ($E, A:$ constant):

$$\boldsymbol{K}^{\mathrm{e}} = \frac{EA}{3L}\begin{bmatrix} 7 & -8 & 1 \\ -8 & 16 & -8 \\ 1 & -8 & 7 \end{bmatrix}. \tag{E.2}$$

- Euler–Bernoulli beam element ($E, I_y:$ constant):

$$\boldsymbol{K}^{\mathrm{e}} = \frac{EI_y}{L^3}\begin{bmatrix} 12 & -6L & -12 & -6L \\ -6L & 4L^2 & 6L & 2L^2 \\ -12 & 6L & 12 & 6L \\ -6L & 2L^2 & 6L & 4L^2 \end{bmatrix}. \tag{E.3}$$

- Generalized beam element ($E, A, I_y:$ constant):

A. Öchsner, *Computational Statics and Dynamics*,
https://doi.org/10.1007/978-981-15-1278-0

$$\begin{bmatrix} \frac{EA}{L} & 0 & 0 & -\frac{EA}{L} & 0 & 0 \\ 0 & \frac{12EI}{L^3} & -\frac{6EI}{L^2} & 0 & -\frac{12EI}{L^3} & -\frac{6EI}{L^2} \\ 0 & -\frac{6EI}{L^2} & \frac{4EI}{L} & 0 & \frac{6EI}{L^2} & \frac{2EI}{L} \\ -\frac{EA}{L} & 0 & 0 & \frac{EA}{L} & 0 & 0 \\ 0 & -\frac{12EI}{L^3} & \frac{6EI}{L^2} & 0 & \frac{12EI}{L^3} & \frac{6EI}{L^2} \\ 0 & -\frac{6EI}{L^2} & \frac{2EI}{L} & 0 & \frac{6EI}{L^2} & \frac{4EI}{L} \end{bmatrix}. \tag{E.4}$$

- Timoshenko beam element (E, G, A, I_y, k_s: constant), $\alpha = \frac{4EI_y}{k_sAG}$, linear interpolation for displacement and rotational field, analytical integration:

$$\boldsymbol{K}^{\mathrm{e}} = \frac{k_sAG}{4L} \begin{bmatrix} 4 & -2L & -4 & -2L \\ -2L & \frac{4}{3}L^2 + \alpha & 2L & \frac{4}{6}L^2 - \alpha \\ -4 & 2L & 4 & 2L \\ -2L & \frac{4}{6}L^2 - \alpha & 2L & \frac{4}{3}L^2 + \alpha \end{bmatrix}. \tag{E.5}$$

- Timoshenko beam element (E, G, A, I_y, k_s: constant), $\alpha = \frac{4EI_y}{k_sAG}$, linear interpolation for displacement and rotational field, numerical one-point integration:

$$\boldsymbol{K}^{\mathrm{e}} = \frac{k_sAG}{4L} \begin{bmatrix} 4 & -2L & -4 & -2L \\ -2L & L^2 + \alpha & 2L & L^2 - \alpha \\ -4 & 2L & 4 & 2L \\ -2L & L^2 - \alpha & 2L & L^2 + \alpha \end{bmatrix}. \tag{E.6}$$

E.2 Two-Dimensional Elements

• Plane elasticity element (E, ν: constant), rectangular $2a \times 2b \times t$, plane *stress* formulation:

$$
\boldsymbol{K}^{\mathrm{e}} = \frac{Et}{1-\nu^2}
\begin{bmatrix}
-\frac{a^2\nu-a^2-2b^2}{6ab} & \frac{\nu+1}{8} & -\frac{a^2\nu-a^2+4b^2}{12ab} & \frac{3\nu-1}{8} & \frac{a^2\nu-a^2-2b^2}{12ab} & -\frac{\nu+1}{8} & \frac{a^2\nu-a^2+b^2}{6ab} & \frac{1-3\nu}{8} \\
\frac{\nu+1}{8} & \frac{-b^2\nu+2a^2+b^2}{6ab} & \frac{1-3\nu}{8} & \frac{b^2\nu+a^2-b^2}{6ab} & -\frac{1+\nu}{8} & -\frac{-b^2\nu+2a^2+b^2}{12ab} & \frac{3\nu-1}{8} & -\frac{b^2\nu+4a^2-b^2}{12ab} \\
-\frac{a^2\nu-a^2+4b^2}{12ab} & \frac{1-3\nu}{8} & -\frac{a^2\nu-a^2-2b^2}{6ab} & -\frac{1+\nu}{8} & \frac{a^2\nu-a^2+b^2}{6ab} & \frac{3\nu-1}{8} & \frac{a^2\nu-a^2-2b^2}{12ab} & \frac{1+\nu}{8} \\
\frac{3\nu-1}{8} & \frac{b^2\nu+a^2-b^2}{6ab} & -\frac{1+\nu}{8} & \frac{-b^2\nu+2a^2+b^2}{6ab} & \frac{1-3\nu}{8} & -\frac{b^2\nu+4a^2-b^2}{12ab} & \frac{1+\nu}{8} & -\frac{-b^2\nu+2a^2+b^2}{12ab} \\
\frac{a^2\nu-a^2-2b^2}{12ab} & -\frac{1+\nu}{8} & \frac{a^2\nu-a^2+b^2}{6ab} & \frac{1-3\nu}{8} & -\frac{a^2\nu-a^2-2b^2}{6ab} & \frac{1+\nu}{8} & -\frac{a^2\nu-a^2+4b^2}{12ab} & \frac{3\nu-1}{8} \\
-\frac{1+\nu}{8} & -\frac{-b^2\nu+2a^2+b^2}{12ab} & \frac{3\nu-1}{8} & -\frac{b^2\nu+4a^2-b^2}{12ab} & \frac{\nu+1}{8} & \frac{-b^2\nu+2a^2+b^2}{6ab} & \frac{1-3\nu}{8} & \frac{b^2\nu+a^2-b^2}{6ab} \\
\frac{a^2\nu-a^2+b^2}{6ab} & \frac{3\nu-1}{8} & \frac{a^2\nu-a^2-2b^2}{12ab} & \frac{1+\nu}{8} & -\frac{a^2\nu-a^2+4b^2}{12ab} & \frac{1-3\nu}{8} & -\frac{a^2\nu-a^2-2b^2}{6ab} & -\frac{\nu+1}{8} \\
\frac{1-3\nu}{8} & -\frac{b^2\nu+4a^2-b^2}{12ab} & \frac{1+\nu}{8} & -\frac{-b^2\nu+2a^2+b^2}{12ab} & \frac{3\nu-1}{8} & \frac{b^2\nu+a^2-b^2}{6ab} & -\frac{\nu+1}{8} & \frac{-b^2\nu+2a^2+b^2}{6ab}
\end{bmatrix}
\tag{E.7}
$$

• Plane elasticity element (E, ν: constant), rectangular $2a \times 2b \times t$, plane *strain* formulation:

$$
\boldsymbol{K}^{\mathrm{e}} = \frac{Et}{(1+\nu)(1-2\nu)} \times
\begin{bmatrix}
-\frac{2a^2\nu+2b^2\nu-a^2-2b^2}{6ab} & \frac{1}{8} & -\frac{2a^2\nu-4b^2\nu-a^2+4b^2}{12ab} & \frac{4\nu-1}{8} & \frac{2a^2\nu+2b^2\nu-a^2-2b^2}{12ab} & -\frac{1}{8} & \frac{2a^2\nu-b^2\nu-a^2+b^2}{6ab} & \frac{1-4\nu}{8} \\
\frac{1}{8} & -\frac{2a^2\nu+2b^2\nu-2a^2-b^2}{6ab} & \frac{1-4\nu}{8} & -\frac{a^2\nu-2b^2\nu-a^2+b^2}{6ab} & -\frac{1}{8} & \frac{2a^2\nu+2b^2\nu-2a^2-b^2}{12ab} & \frac{4\nu-1}{8} & \frac{4a^2\nu-2b^2\nu-4a^2+b^2}{12ab} \\
-\frac{2a^2\nu-4b^2\nu-a^2+4b^2}{12ab} & \frac{1-4\nu}{8} & -\frac{2a^2\nu+2b^2\nu-a^2-2b^2}{6ab} & -\frac{1}{8} & \frac{2a^2\nu-b^2\nu-a^2+b^2}{6ab} & \frac{4\nu-1}{8} & \frac{2a^2\nu+2b^2\nu-a^2-2b^2}{12ab} & \frac{1}{8} \\
\frac{4\nu-1}{8} & -\frac{a^2\nu-2b^2\nu-a^2+b^2}{6ab} & -\frac{1}{8} & -\frac{2a^2\nu+2b^2\nu-2a^2-b^2}{6ab} & \frac{1-4\nu}{8} & \frac{4a^2\nu-2b^2\nu-4a^2+b^2}{12ab} & \frac{1}{8} & \frac{2a^2\nu+2b^2\nu-2a^2-b^2}{12ab} \\
\frac{2a^2\nu+2b^2\nu-a^2-2b^2}{12ab} & -\frac{1}{8} & \frac{2a^2\nu-b^2\nu-a^2+b^2}{6ab} & \frac{1-4\nu}{8} & -\frac{2a^2\nu+2b^2\nu-a^2-2b^2}{6ab} & \frac{1}{8} & -\frac{2a^2\nu-4b^2\nu-a^2+4b^2}{12ab} & \frac{4\nu-1}{8} \\
-\frac{1}{8} & \frac{2a^2\nu+2b^2\nu-2a^2-b^2}{12ab} & \frac{4\nu-1}{8} & \frac{4a^2\nu-2b^2\nu-4a^2+b^2}{12ab} & \frac{1}{8} & \frac{2a^2\nu+2b^2\nu-2a^2-b^2}{6ab} & \frac{1-4\nu}{8} & -\frac{a^2\nu-2b^2\nu-a^2+b^2}{6ab} \\
\frac{2a^2\nu-b^2\nu-a^2+b^2}{6ab} & \frac{4\nu-1}{8} & \frac{2a^2\nu+2b^2\nu-a^2-2b^2}{12ab} & \frac{1}{8} & -\frac{2a^2\nu-4b^2\nu-a^2+4b^2}{12ab} & \frac{1-4\nu}{8} & -\frac{2a^2\nu+2b^2\nu-a^2-2b^2}{6ab} & -\frac{1}{8} \\
\frac{1-4\nu}{8} & \frac{4a^2\nu-2b^2\nu-4a^2+b^2}{12ab} & \frac{1}{8} & \frac{2a^2\nu+2b^2\nu-2a^2-b^2}{12ab} & \frac{4\nu-1}{8} & -\frac{a^2\nu-2b^2\nu-a^2+b^2}{6ab} & -\frac{1}{8} & -\frac{2a^2\nu+2b^2\nu-2a^2-b^2}{6ab}
\end{bmatrix}
\tag{E.8}
$$

• Classical plate element (E, ν: constant), rectangular $2a \times 2b \times h$:

$$\boldsymbol{K}^{\mathrm{e}} = \begin{bmatrix} \boldsymbol{K}_{11} & \boldsymbol{K}_{12} \\ \boldsymbol{K}_{21} & \boldsymbol{K}_{22} \end{bmatrix}. \tag{E.9}$$

$$\boldsymbol{K}_{11} = D \begin{bmatrix} \frac{7a^2b^2-2a^2b^2\nu+10a^4+10b^4}{10a^3b^3} & \frac{4b^2\nu+10a^2+b^2}{10ab^2} & -\frac{4a^2\nu+a^2+10b^2}{10a^2b} & \frac{2a^2b^2\nu-7a^2b^2+5a^4-10b^4}{10a^3b^3} & \frac{-4b^2\nu+5a^2-b^2}{10ab^2} & \frac{a^2\nu-a^2-10b^2}{10a^2b} \\ \frac{4b^2\nu+10a^2+b^2}{10ab^2} & \frac{4(-b^2\nu+5a^2+b^2)}{15ab} & -\nu & \frac{-4b^2\nu+5a^2-b^2}{10ab^2} & \frac{2(2b^2\nu+5a^2-2b^2)}{15ab} & 0 \\ -\frac{4a^2\nu+a^2+10b^2}{10a^2b} & -\nu & -\frac{4(a^2\nu-a^2-5b^2)}{15ab} & -\frac{a^2\nu-a^2-10b^2}{10a^2b} & 0 & \frac{a^2\nu-a^2+10b^2}{15ab} \\ \frac{2a^2b^2\nu+5a^4-7a^2b^2-10b^4}{10a^3b^3} & \frac{-4b^2\nu+5a^2-b^2}{10ab^2} & -\frac{a^2\nu-a^2-10b^2}{10a^2b} & \frac{-2a^2b^2\nu+10a^4+7a^2b^2+10b^2}{10a^3b^3} & \frac{4b^2\nu+10a^2+b^2}{10ab^2} & \frac{4a^2\nu+a^2+10b^2}{10a^2b} \\ \frac{-4b^2\nu+5a^2-b^2}{10ab^2} & \frac{2(2b^2\nu+5a^2-2b^2)}{15ab} & 0 & \frac{4b^2\nu+10a^2+b^2}{10ab^2} & \frac{4(-b^2\nu+5a^2+b^2)}{15ab} & \nu \\ \frac{a^2\nu-a^2-10b^2}{10a^2b} & 0 & \frac{a^2\nu-a^2+10b^2}{15ab} & \frac{4a^2\nu+a^2+10b^2}{10a^2b} & \nu & -\frac{4(a^2\nu-a^2-5b^2)}{15ab} \end{bmatrix} \tag{E.10}$$

$$\boldsymbol{K}_{12} = D \begin{bmatrix} -\frac{2a^2b^2\nu+5a^2-7a^2b^2+5b^4}{10a^3b^3} & \frac{b^2\nu+5a^2-b^2}{10ab^2} & -\frac{a^2\nu-a^2+5b^2}{10a^2b} & -\frac{-2a^2b^2\nu+10a^4+7a^2b^2-5b^4}{10a^3b^3} & \frac{-b^2\nu+10a^2+b^2}{10ab^2} & \frac{4a^2\nu+a^2-5b^2}{10a^2b} \\ -\frac{b^2\nu+5a^2-b^2}{10ab^2} & \frac{-b^2\nu+5a^2+b^2}{15ab} & 0 & -\frac{-b^2\nu+10a^2+b^2}{10ab^2} & \frac{b^2\nu+10a^2-b^2}{15ab} & 0 \\ \frac{a^2\nu-a^2+5b^2}{10a^2b} & 0 & -\frac{a^2\nu-a^2-5b^2}{15ab} & \frac{4a^2\nu+a^2-5b^2}{10a^2b} & 0 & \frac{2(2a^2\nu-2a^2+5b^2)}{15ab} \\ -\frac{-2a^2b^2\nu+10a^4+7a^2b^2-5b^4}{10a^3b^3} & \frac{-b^2\nu+10a^2+b^2}{10ab^2} & -\frac{4a^2\nu+a^2-5b^2}{10a^2b} & -\frac{2a^2b^2\nu+5a^4-7a^2b^2+5b^4}{10a^3b^3} & \frac{b^2\nu+5a^2-b^2}{10ab^2} & \frac{a^2\nu-a^2+5b^2}{10a^2b} \\ -\frac{-b^2\nu+10a^2+b^2}{10ab^2} & \frac{b^2\nu+10a^2-b^2}{15ab} & 0 & -\frac{b^2\nu+5a^2-b^2}{10ab^2} & \frac{-b^2\nu+5a^2+b^2}{15ab} & 0 \\ -\frac{4a^2\nu+a^2-5b^2}{10a^2b} & 0 & \frac{2(2a^2\nu-2a^2+5b^2)}{15ab} & -\frac{a^2\nu-a^2+5b^2}{10a^2b} & 0 & -\frac{a^2\nu-a^2-5b^2}{15ab} \end{bmatrix} \tag{E.11}$$

$$
\boldsymbol{K}_{22} = D \begin{bmatrix}
\frac{-2a^2b^2\nu+10a^4+7a^2b^2+10b^4}{10a^3b^3} & -\frac{4b^2\nu+10a^2+b^2}{10ab^2} & \frac{4a^2\nu+a^2+10b^2}{10a^2b} & \frac{2a^2b^2\nu+5a^4-7a^2b^2-10b^4}{10a^3b^3} & -\frac{-4b^2\nu+5a^2-b^2}{10ab^2} & -\frac{a^2\nu-a^2-10b^2}{10a^2b} \\
-\frac{4b^2\nu+10a^2+b^2}{10ab^2} & \frac{4(-b^2\nu+5a^2+b^2)}{15ab} & -\nu & -\frac{-4b^2\nu+5a^2-b^2}{10ab^2} & \frac{2(2b^2\nu+5a^2-2b^2)}{15ab} & 0 \\
\frac{4a^2\nu+a^2+10b^2}{10a^2b} & -\nu & -\frac{4(a^2\nu-a^2-5b^2)}{15ab} & \frac{a^2\nu-a^2-10b^2}{10a^2b} & 0 & \frac{a^2\nu-a^2+10b^2}{15ab} \\
\frac{2a^2b^2\nu+5a^4-7a^2b^2-10b^4}{10a^3b^3} & -\frac{-4b^2\nu+5a^2-b^2}{10ab^2} & \frac{a^2\nu-a^2-10b^2}{10a^2b} & \frac{-2a^2b^2\nu+10a^4+7a^2b^2+10b^4}{10a^3b^3} & -\frac{4b^2\nu+10a^2+b^2}{10ab^2} & -\frac{4a^2\nu+a^2+10b^2}{10a^2b} \\
-\frac{-4b^2\nu+5a^2-b^2}{10ab^2} & \frac{2(2b^2\nu+5a^2-2b^2)}{15ab} & 0 & -\frac{4b^2\nu+10a^2+b^2}{10ab^2} & \frac{4(-b^2\nu+5a^2+b^2)}{15ab} & \nu \\
-\frac{a^2\nu-a^2-10b^2}{10a^2b} & 0 & \frac{a^2\nu-a^2+10b^2}{15ab} & -\frac{4a^2\nu+a^2+10b^2}{10a^2b} & \nu & -\frac{4(a^2\nu-a^2-5b^2)}{15ab}
\end{bmatrix} \tag{E.12}
$$

• Thick plate element (E, ν, G: constant), rectangular $2a\times 2b\times h$, full integration (2×2):

$$\boldsymbol{K}^{\mathrm{e}} = \begin{bmatrix} \boldsymbol{K}_{11} & \boldsymbol{K}_{12} \\ \boldsymbol{K}_{21} & \boldsymbol{K}_{22} \end{bmatrix}. \tag{E.13}$$

$$\boldsymbol{K}_{11} = \begin{bmatrix}
-\frac{8D_{\mathrm{s}}a^2b^2+3D_{\mathrm{b}}a^2\nu-3D_{\mathrm{b}}a^2-6D_{\mathrm{b}}b^2}{18ab} & -\frac{D_{\mathrm{b}}(\nu+1)}{8} & \frac{D_{\mathrm{s}}b}{3} & -\frac{8D_{\mathrm{s}}a^2b^2+3D_{\mathrm{b}}a^2\nu-3D_{\mathrm{b}}a^2+12D_{\mathrm{b}}b^2}{36ab} & \frac{-3D_{\mathrm{b}}\nu+D_{\mathrm{b}}}{8} & -\frac{D_{\mathrm{s}}b}{3} \\
-\frac{D_{\mathrm{b}}(\nu+1)}{8} & \frac{-8D_{\mathrm{s}}a^2b^2-3D_{\mathrm{b}}b^2\nu+6D_{\mathrm{b}}a^2+3D_{\mathrm{b}}b^2}{18ab} & -\frac{D_{\mathrm{s}}a}{3} & \frac{3D_{\mathrm{b}}\nu-D_{\mathrm{b}}}{8} & \frac{-4D_{\mathrm{s}}a^2b^2+3D_{\mathrm{b}}b^2\nu+3D_{\mathrm{b}}a^2-3D_{\mathrm{b}}b^2}{18ab} & -\frac{D_{\mathrm{s}}a}{6} \\
\frac{D_{\mathrm{s}}b}{3} & -\frac{D_{\mathrm{s}}a}{3} & -\frac{D_{\mathrm{s}}(a^2+b^2)}{3ab} & \frac{D_{\mathrm{s}}b}{3} & -\frac{D_{\mathrm{s}}a}{6} & -\frac{D_{\mathrm{s}}(a^2-2b^2)}{6ab} \\
-\frac{8D_{\mathrm{s}}a^2b^2+3D_{\mathrm{b}}a^2\nu-3D_{\mathrm{b}}a^2+12D_{\mathrm{b}}b^2}{36ab} & \frac{3D_{\mathrm{b}}\nu-D_{\mathrm{b}}}{8} & \frac{D_{\mathrm{s}}b}{3} & -\frac{8D_{\mathrm{s}}a^2b^2+3D_{\mathrm{b}}a^2\nu-3D_{\mathrm{b}}a^2-6D_{\mathrm{b}}b^2}{18ab} & \frac{D_{\mathrm{b}}(\nu+1)}{8} & -\frac{D_{\mathrm{s}}b}{3} \\
\frac{-3D_{\mathrm{b}}\nu+D_{\mathrm{b}}}{8} & \frac{-4D_{\mathrm{s}}a^2b^2+3D_{\mathrm{b}}b^2\nu+3D_{\mathrm{b}}a^2-3D_{\mathrm{b}}b^2}{18ab} & -\frac{D_{\mathrm{s}}a}{6} & \frac{D_{\mathrm{b}}(\nu+1)}{8} & \frac{-8D_{\mathrm{s}}a^2b^2-3D_{\mathrm{b}}b^2\nu+6D_{\mathrm{b}}a^2+3D_{\mathrm{b}}b^2}{18ab} & -\frac{D_{\mathrm{s}}a}{3} \\
-\frac{D_{\mathrm{s}}b}{3} & -\frac{D_{\mathrm{s}}a}{6} & -\frac{D_{\mathrm{s}}(a^2-2b^2)}{6ab} & -\frac{D_{\mathrm{s}}b}{3} & -\frac{D_{\mathrm{s}}a}{3} & -\frac{D_{\mathrm{s}}(a^2+b^2)}{3ab}
\end{bmatrix} \tag{E.14}$$

$$
\boldsymbol{K}_{12} = \left[\begin{matrix}
-\frac{4D_s a^2b^2+3D_b a^2\nu-3D_b a^2+6D_b b^2}{36ab} & \frac{3D_b\nu-D_b}{8} & -\frac{D_s b}{6} & \frac{-4D_s a^2b^2+3D_b a^2\nu-3D_b a^2+3D_b b^2}{18ab} \\
\frac{3D_b\nu-D_b}{8} & -\frac{4D_s a^2b^2+3D_b b^2\nu+6D_b a^2-3D_b b^2}{36ab} & \frac{D_s a}{6} & -\frac{D_b(3\nu-1)}{8} \\
\frac{D_s b}{6} & -\frac{D_s a}{6} & \frac{D_s(a^2+b^2)}{6ab} & \frac{D_s b}{6} \\
-\frac{4D_s a^2b^2+3D_b a^2\nu-3D_b a^2-3D_b b^2}{18ab} & -\frac{D_b(\nu+1)}{8} & -\frac{D_s b}{6} & \frac{-4D_s a^2b^2+3D_b a^2\nu-3D_b a^2-6D_b b^2}{36ab} \\
\frac{D_b(\nu+1)}{8} & -\frac{8D_s a^2b^2-3D_b b^2\nu+12D_b a^2+3D_b b^2}{36ab} & \frac{D_s a}{3} & \frac{-D_b\nu-D_b}{8} \\
-\frac{D_s b}{6} & -\frac{D_s a}{3} & \frac{D_s(2a^2-b^2)}{6ab} & -\frac{D_s b}{6}
\end{matrix}\right.
$$

$$
\left.\begin{matrix}
\frac{D_b(3\nu-1)}{8} & \frac{D_s b}{6} \\
-\frac{8D_s a^2b^2+3D_b b^2\nu+12D_b a^2-3D_b b^2}{36ab} & \frac{D_s a}{3} \\
-\frac{D_s a}{3} & \frac{D_s(2a^2-b^2)}{6ab} \\
\frac{-D_b\nu-D_b}{8} & \frac{D_s b}{6} \\
-\frac{4D_s a^2b^2-3D_b b^2\nu+6D_b a^2+3D_b b^2}{36ab} & \frac{D_s a}{6} \\
-\frac{D_s a}{6} & \frac{D_s(a^2+b^2)}{6ab}
\end{matrix}\right] \tag{E.15}
$$

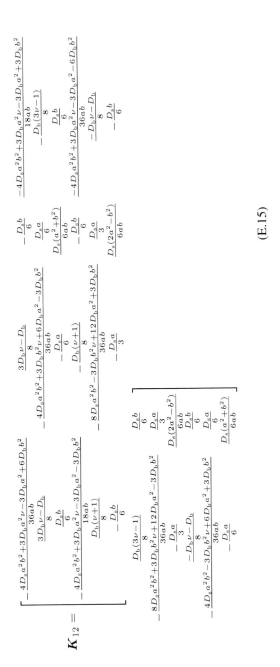

$$
\boldsymbol{K}_{22} = \begin{bmatrix}
-\frac{8D_sa^2b^2+3D_ba^2\nu-3D_ba^2-6D_bb^2}{18ab} & -\frac{D_b(\nu+1)}{8} & -\frac{D_sb}{3} & \frac{-8D_sa^2b^2+3D_ba^2\nu-3D_ba^2-12D_bb^2}{36ab} & \frac{-D_b\nu-D_b}{8} & \frac{D_sb}{3} \\
-\frac{D_b(\nu+1)}{8} & \frac{-8D_sa^2b^2-3D_bb^2\nu+6D_ba^2+3D_bb^2}{18ab} & \frac{D_sa}{3} & \frac{D_b\nu+D_b}{8} & \frac{-4D_sa^2b^2-3D_bb^2\nu+3D_ba^2+3D_bb^2}{18ab} & \frac{D_sa}{6} \\
-\frac{D_sb}{3} & \frac{D_sa}{3} & -\frac{D_s(a^2+b^2)}{3ab} & -\frac{D_sb}{3} & \frac{D_sa}{6} & -\frac{D_s(a^2-2b^2)}{6ab} \\
\frac{-8D_sa^2b^2+3D_ba^2\nu-3D_ba^2-12D_bb^2}{36ab} & \frac{D_b\nu+D_b}{8} & -\frac{D_sb}{3} & -\frac{8D_sa^2b^2+3D_ba^2\nu-3D_ba^2-6D_bb^2}{18ab} & \frac{D_b(\nu+1)}{8} & \frac{D_sb}{3} \\
\frac{-D_b\nu-D_b}{8} & \frac{-4D_sa^2b^2-3D_bb^2\nu+3D_ba^2+3D_bb^2}{18ab} & \frac{D_sa}{6} & \frac{D_b(\nu+1)}{8} & \frac{-8D_sa^2b^2-3D_bb^2\nu+6D_ba^2+3D_bb^2}{18ab} & \frac{D_sa}{3} \\
\frac{D_sb}{3} & \frac{D_sa}{6} & -\frac{D_s(a^2-2b^2)}{6ab} & \frac{D_sb}{3} & \frac{D_sa}{3} & -\frac{D_s(a^2+b^2)}{3ab}
\end{bmatrix}
\tag{E.16}
$$

E.3 Three-Dimensional Elements

- Hexaeder element (E, ν: constant), cuboid with $2a \times 2b \times 2c$:

$$\boldsymbol{K}^{\mathrm{e}} = \begin{bmatrix} \boldsymbol{K}_{11} & \boldsymbol{K}_{12} & \cdots & \boldsymbol{K}_{18} \\ \boldsymbol{K}_{21} & \boldsymbol{K}_{22} & \cdots & \boldsymbol{K}_{28} \\ \vdots & \vdots & & \vdots \\ \boldsymbol{K}_{81} & \boldsymbol{K}_{82} & \cdots & \boldsymbol{K}_{88} \end{bmatrix}. \tag{E.17}$$

$$\boldsymbol{K}_{11} = \frac{E}{(1+\nu)(1-2\nu)} \times \begin{bmatrix} -\frac{2a^2b^2\nu+2c^2a^2\nu+2c^2b^2\nu-a^2b^2-c^2a^2-2c^2b^2}{9abc} & \frac{b}{12} & \frac{c}{12} \\ \frac{b}{12} & -\frac{2a^2b^2\nu+2c^2a^2\nu+2c^2b^2\nu-2a^2b^2-c^2a^2-c^2b^2}{9abc} & \frac{a}{12} \\ \frac{c}{12} & \frac{a}{12} & -\frac{2a^2b^2\nu+2c^2a^2\nu+2c^2b^2\nu-a^2b^2-2c^2a^2-c^2b^2}{9abc} \end{bmatrix} \tag{E.18}$$

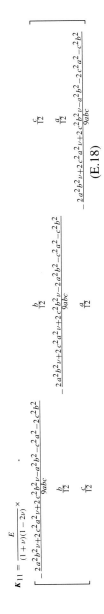

Appendix F
Extrapolation from Integration Points to Nodes

It was already highlighted in Table 2.13 that some quantities are evaluated at the nodes (i.e., deformations and reactions) and other quantities are elemental functions (i.e., stresses and strains). These elemental functions are only calculated at the integration points. If these quantities should be displayed at the nodes,[13] an extrapolation from the integration points to the nodes must be performed (so-called stress recovery). This procedure will be explained in the following based on a four-node plane stress element for the case of the stress extrapolation, see Fig. F.1. It is obvious that the same procedures can be applied for the strain extrapolation.

In general, one may distinguish the following three approaches.

- **Average Method**: The average value of all the integration points is computed and assigned to the nodes. Thus, all nodes have an equal value assigned to them.
- **Translate Method**: Copies the value from the integration point to the closest corresponding node. Where there are fewer integration points than nodes, averaging of neighboring integration points occurs.
- **Linear Method**: Extrapolation by averaging the integration point values to the centroid of the element and then performing a linear extrapolation through the integration point to the node.

In the case of a single integration point (reduced integration element), all the four nodes are assigned with the same value of the integration point and there is no difference between these three methods.

Let us now return to the four-node plane stress element which is shown in Fig. F.1 and explore some linear extrapolation methods. The corresponding coordinates of the nodes and integration points can be expressed as indicated in Table F.1.

The average stress value of all the integration points, $\bar{\sigma}_{\mathrm{g}}$, can be calculated based on the single values $\sigma_{\mathrm{g}1} \ldots \sigma_{\mathrm{g}4}$ as:

$$\bar{\sigma}_{\mathrm{g}} = \frac{\sigma_{\mathrm{g}1} + \sigma_{\mathrm{g}2} + \sigma_{\mathrm{g}3} + \sigma_{\mathrm{g}4}}{4}. \tag{F.1}$$

[13]This is also required in a post-processor where values are only displayed at nodes.

A. Öchsner, *Computational Statics and Dynamics*,
https://doi.org/10.1007/978-981-15-1278-0

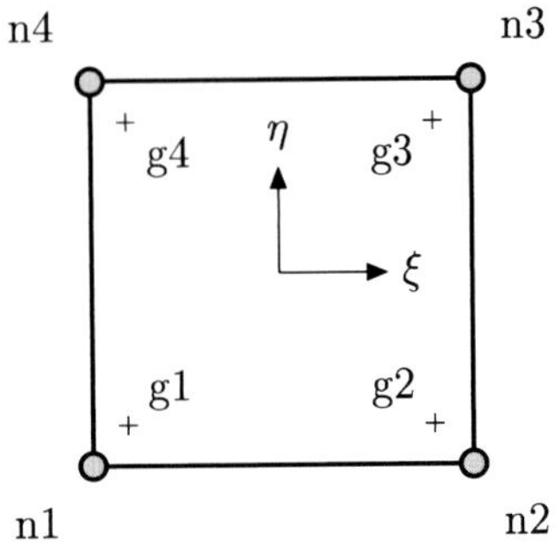

Fig. F.1 Representation of a four-node planar bilinear quadrilateral (quad 4) with full integration in the parametric ξ–η space: nodes are symbolized by circles (∘) and integration points by crosses (+)

Table F.1 Coordinates of nodes and integration points for quad4 element, cf. Fig. F.1

i	Node ni		Integration point gi	
	ξ-coordinate	η-coordinate	ξ-coordinate	η-coordinate
1	-1	-1	$-\frac{1}{\sqrt{3}}$	$-\frac{1}{\sqrt{3}}$
2	1	-1	$\frac{1}{\sqrt{3}}$	$-\frac{1}{\sqrt{3}}$
3	1	1	$\frac{1}{\sqrt{3}}$	$\frac{1}{\sqrt{3}}$
4	-1	1	$-\frac{1}{\sqrt{3}}$	$\frac{1}{\sqrt{3}}$

Then, a linear geometrical extrapolation from the center through the nearest integration point to the node gives, for example, in the case of the nodal stress value at node 1:

$$\sigma_{\text{n1}} = \bar{\sigma}_{\text{g}} + \frac{\sqrt{(\xi_{\text{c}} - \xi_{\text{n1}})^2 + (\eta_{\text{c}} - \eta_{\text{n1}})^2}}{\sqrt{(\xi_{\text{c}} - \xi_{\text{g1}})^2 + (\eta_{\text{c}} - \eta_{\text{g1}})^2}} \times \left(\sigma_{\text{g1}} - \bar{\sigma}_{\text{g}}\right) . \tag{F.2}$$

Similar equations can be written for the remaining nodal stress components $\sigma_{\text{n2}} \dots \sigma_{\text{n4}}$ and all four equations can be summarized in matrix form as:

$$\begin{bmatrix} \sigma_{\text{n1}} \\ \sigma_{\text{n2}} \\ \sigma_{\text{n3}} \\ \sigma_{\text{n4}} \end{bmatrix} = \frac{1}{4} \begin{bmatrix} 1+3\sqrt{3} & 1-\sqrt{3} & 1-\sqrt{3} & 1-\sqrt{3} \\ 1-\sqrt{3} & 1+3\sqrt{3} & 1-\sqrt{3} & 1-\sqrt{3} \\ 1-\sqrt{3} & 1-\sqrt{3} & 1+3\sqrt{3} & 1-\sqrt{3} \\ 1-\sqrt{3} & 1-\sqrt{3} & 1-\sqrt{3} & 1+3\sqrt{3} \end{bmatrix} \begin{bmatrix} \sigma_{\text{g1}} \\ \sigma_{\text{g2}} \\ \sigma_{\text{g3}} \\ \sigma_{\text{g4}} \end{bmatrix} . \tag{F.3}$$

The transformation of Eq. (F.2) into the matrix form of Eq. (F.3) is not obvious. With $\xi_{\text{c}} = \eta_{\text{c}} = 0$, $\xi_{\text{n1}} = \eta_{\text{n1}} = -1$ and $\xi_{\text{g1}} = \eta_{\text{g1}} = -\frac{1}{\sqrt{3}}$ we can rearrange Eq. (F.2) into the following form:

$$
\begin{aligned}
\sigma_{\text{n1}} &= \bar{\sigma}_{\text{g}} + \frac{\sqrt{(0+1)^2+(0+1)^2}}{\sqrt{(0+\frac{1}{\sqrt{3}})^2+(0+\frac{1}{\sqrt{3}})^2}} \times \left(\sigma_{\text{g1}} - \bar{\sigma}_{\text{g}}\right) \\
&= \frac{1}{4}(\sigma_{\text{g1}}+\sigma_{\text{g2}}+\sigma_{\text{g3}}+\sigma_{\text{g4}}) + \frac{\sqrt{3}}{4} \times \left(3\sigma_{\text{g1}}-\sigma_{\text{g2}}-\sigma_{\text{g3}}-\sigma_{\text{g4}}\right) \\
&= \frac{1}{4}\left((1+3\sqrt{3})\sigma_{\text{g1}} + (1-\sqrt{3})\sigma_{\text{g2}} + (1-\sqrt{3})\sigma_{\text{g3}} + (1-\sqrt{3})\sigma_{\text{g4}}\right). \quad \text{(F.4)}
\end{aligned}
$$

Another extrapolation approach was proposed by HINTON and CAMPBELL in [4] where a local interpolation/extrapolation scheme based on the following bilinear surface was applied:

$$
\sigma^{\text{e}}(\xi,\eta) = a_1 + a_2\xi + a_3\eta + a_4\xi\eta = \begin{bmatrix} 1 \; \xi \; \eta \; \xi\eta \end{bmatrix} \begin{bmatrix} a_1 \\ a_2 \\ a_3 \\ a_4 \end{bmatrix} = \boldsymbol{\chi}^{\text{T}}\boldsymbol{a}\,. \quad \text{(F.5)}
$$

Evaluating Eq. (F.5) for all four integration points gi of the quadrilateral element gives

$$\text{Int. point 1:}\ \sigma_{\text{g1}} = \sigma^{\text{e}}(\xi = -\tfrac{1}{\sqrt{3}}, \eta = -\tfrac{1}{\sqrt{3}}) = a_1 - \tfrac{a_2}{\sqrt{3}} - \tfrac{a_3}{\sqrt{3}} + \tfrac{a_4}{3}\,, \quad \text{(F.6)}$$

$$\text{Int. point 2:}\ \sigma_{\text{g2}} = \sigma^{\text{e}}(\xi = +\tfrac{1}{\sqrt{3}}, \eta = -\tfrac{1}{\sqrt{3}}) = a_1 + \tfrac{a_2}{\sqrt{3}} - \tfrac{a_3}{\sqrt{3}} - \tfrac{a_4}{3}\,, \quad \text{(F.7)}$$

$$\text{Int. point 3:}\ \sigma_{\text{g3}} = \sigma^{\text{e}}(\xi = +\tfrac{1}{\sqrt{3}}, \eta = +\tfrac{1}{\sqrt{3}}) = a_1 + \tfrac{a_2}{\sqrt{3}} + \tfrac{a_3}{\sqrt{3}} + \tfrac{a_4}{3}\,, \quad \text{(F.8)}$$

$$\text{Int. point 4:}\ \sigma_{\text{g4}} = \sigma^{\text{e}}(\xi = -\tfrac{1}{\sqrt{3}}, \eta = +\tfrac{1}{\sqrt{3}}) = a_1 - \tfrac{a_2}{\sqrt{3}} + \tfrac{a_3}{\sqrt{3}} - \tfrac{a_4}{3}\,, \quad \text{(F.9)}$$

or in matrix notation:

$$
\begin{bmatrix} \sigma_{\text{g1}} \\ \sigma_{\text{g2}} \\ \sigma_{\text{g3}} \\ \sigma_{\text{g4}} \end{bmatrix} = \underbrace{\begin{bmatrix} 1 & -\frac{\sqrt{3}}{3} & -\frac{\sqrt{3}}{3} & \frac{1}{3} \\ 1 & \frac{\sqrt{3}}{3} & -\frac{\sqrt{3}}{3} & -\frac{1}{3} \\ 1 & \frac{\sqrt{3}}{3} & \frac{\sqrt{3}}{3} & \frac{1}{3} \\ 1 & -\frac{\sqrt{3}}{3} & \frac{\sqrt{3}}{3} & -\frac{1}{3} \end{bmatrix}}_{\boldsymbol{X}} \begin{bmatrix} a_1 \\ a_2 \\ a_3 \\ a_4 \end{bmatrix}. \quad \text{(F.10)}
$$

Solving for $\boldsymbol{a}$ gives:

$$
\begin{bmatrix} a_1 \\ a_2 \\ a_3 \\ a_4 \end{bmatrix} = \frac{1}{4} \begin{bmatrix} 1 & 1 & 1 & 1 \\ -\sqrt{3} & \sqrt{3} & \sqrt{3} & -\sqrt{3} \\ -\sqrt{3} & -\sqrt{3} & \sqrt{3} & \sqrt{3} \\ 3 & -3 & 3 & -3 \end{bmatrix} \begin{bmatrix} \sigma_{\text{g1}} \\ \sigma_{\text{g2}} \\ \sigma_{\text{g3}} \\ \sigma_{\text{g4}} \end{bmatrix}, \quad \text{(F.11)}
$$

or

$$\boldsymbol{a} = \boldsymbol{A}\boldsymbol{\sigma}_{\mathrm{g}} = \boldsymbol{X}^{-1}\boldsymbol{\sigma}_{\mathrm{g}}\,. \tag{F.12}$$

The system of Eq. (F.11) allows to calculate each coefficient a_i which can be used in Eq. (F.5). The evaluation of Eq. (F.5) for each of the nodes ni gives finally the following system of equations to determine the nodal stress values:

$$\begin{bmatrix} \sigma_{\mathrm{n}1} \\ \sigma_{\mathrm{n}2} \\ \sigma_{\mathrm{n}3} \\ \sigma_{\mathrm{n}4} \end{bmatrix} = \frac{1}{2} \begin{bmatrix} 2+\sqrt{3} & -1 & 2-\sqrt{3} & -1 \\ -1 & 2+\sqrt{3} & -1 & 2-\sqrt{3} \\ 2-\sqrt{3} & -1 & 2+\sqrt{3} & -1 \\ -1 & 2-\sqrt{3} & -1 & 2+\sqrt{3} \end{bmatrix} \begin{bmatrix} \sigma_{\mathrm{g}1} \\ \sigma_{\mathrm{g}2} \\ \sigma_{\mathrm{g}3} \\ \sigma_{\mathrm{g}4} \end{bmatrix}. \tag{F.13}$$

This approach can be understood as well as a local least squares fit over single elements. A slightly different way of deriving the extrapolation scheme by HINTON and CAMPBELL is to assume a nodal approach of the stress field as in the case of the displacement field (see Eq. (5.27)):

$$\sigma^{\mathrm{e}}(\xi,\eta) = N_1\sigma_{\mathrm{n}1} + N_2\sigma_{\mathrm{n}2} + N_3\sigma_{\mathrm{n}3} + N_4\sigma_{\mathrm{n}4}\,, \tag{F.14}$$

where the four interpolations functions N_i are given by Eqs. (5.57)–(5.60) and the supporting points are the values at the nodes. Evaluating this expression for the first integration point gives:

$$\begin{aligned} \sigma_{\mathrm{g}1}x = {} & \frac{1}{4}\left(1+\frac{1}{\sqrt{3}}\right)\left(1+\frac{1}{\sqrt{3}}\right)\sigma_{\mathrm{n}1}x + \frac{1}{4}\left(1-\frac{1}{\sqrt{3}}\right)\left(1+\frac{1}{\sqrt{3}}\right)\sigma_{\mathrm{n}2}x \\ & + \frac{1}{4}\left(1-\frac{1}{\sqrt{3}}\right)\left(1-\frac{1}{\sqrt{3}}\right)\sigma_{\mathrm{n}3}x + \frac{1}{4}\left(1+\frac{1}{\sqrt{3}}\right)\left(1-\frac{1}{\sqrt{3}}\right)\sigma_{\mathrm{n}4}x\,. \end{aligned} \tag{F.15}$$

Similar expressions can be written for the other three integration points and a system of four equations for the unknowns four nodal values can be written. Finally, Eq. (F.13) is again obtained.

In the case of a mesh where elements are connected via nodes, a nodal averaging is performed based on the extrapolated elemental values, see Fig. F.2. In other words, the integration point values are first extrapolated to the node and then an averaging is performed

$$\sigma_{\mathrm{n}} = \frac{\sum\limits_{i=1}^{\mathrm{IV}} \sigma_{\mathrm{g}}^{i}}{4}\,, \tag{F.16}$$

where σ_{n} is the averaged nodal value and the σ_{g}^{i} are the—from the integration points to the node—extrapolated values.

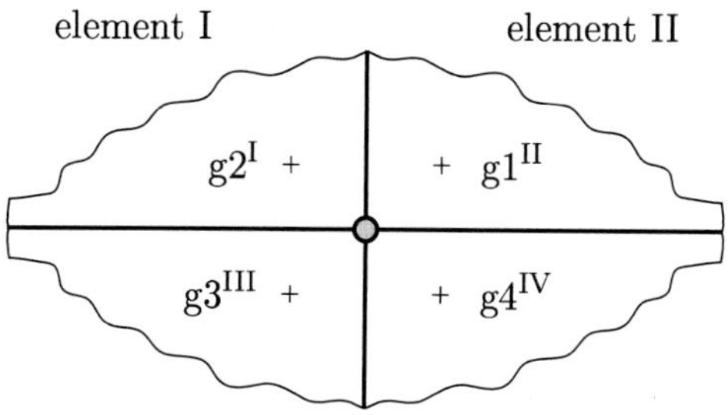

Fig. F.2 Averaging of values at a common node of an element

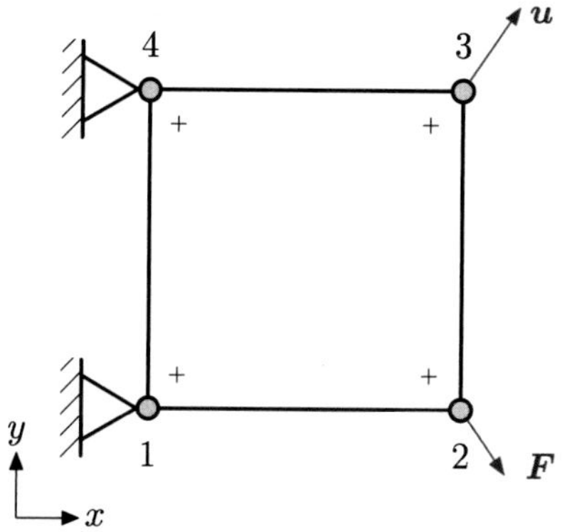

Fig. F.3 Square plane stress element

F.1 Example: Stress extrapolation for a square plane stress element

Given is a square plane stress element with full integration (2 × 2) as shown in Fig. F.3. The nodal coordinates are 1(0, 0), 2(4, 0), 3(4, 4) and 4(0, 4). The constant thickness equals $t = 1.1$. The right-hand nodes are loaded by an external force $\boldsymbol{F} = \left[10, -12\right]^{\mathrm{T}}$ at node 2 and a prescribed displacement $\boldsymbol{u} = \left[0.012, 0.014\right]^{\mathrm{T}}$ at node 3. A linear-elastic finite element calculation resulted for the elastic properties $E = 200{,}000$ and $\nu = 0.3$ in the following stress values in the x-direction at the integration points: $\sigma_{\mathrm{g}1x} = 260.331$, $\sigma_{\mathrm{g}2x} = 308.159$, $\sigma_{\mathrm{g}3x} = 613.070$ and $\sigma_{\mathrm{g}4x} = 565.242$.

Calculate the nodal stress values for the x-component based on:

- a linear extrapolation based on the ξ–η space (see Eq. (F.3));
- a linear extrapolation based on the x–y space (for comparison reasons);
- a local least squares fit extrapolation in the ξ–η space (see Eq. (F.13)).

F.1 Solution

• The application of Eq. (F.3) gives immediately: $\sigma_{\mathrm{n}1x} = 131.220$, $\sigma_{\mathrm{n}2x} = 214.061$, $\sigma_{\mathrm{n}3x} = 742.181$ and $\sigma_{\mathrm{n}4x} = 659.341$.

• The linear extrapolation based on the x–y space requires to rewrite Eq. (F.3) in the following form based on Cartesian coordinates:

$$\sigma_{n1} = \bar{\sigma}_g + \frac{\sqrt{(x_c - x_{n1})^2 + (y_c - y_{n1})^2}}{\sqrt{(x_c - x_{g1})^2 + (y_c - y_{g1})^2}} \times \left(\sigma_{g1} - \bar{\sigma}_g\right) , \tag{F.17}$$

where the center of the element is obtained from:

$$x_c = \frac{1}{4}(x_{n1} + x_{n2} + x_{n3} + x_{n4}) , \tag{F.18}$$

$$y_c = \frac{1}{4}(y_{n1} + y_{n2} + y_{n3} + y_{n4}) . \tag{F.19}$$

The coordinates of the integration points in Cartesian coordinates can be obtained from Eqs. (5.65) and (5.66). As an example, the x-coordinate of the first integration point is obtained as:

$$\begin{aligned} x_{g1}\left(\xi = \tfrac{-1}{\sqrt{3}}, \eta = \tfrac{-1}{\sqrt{3}}\right) &= N_1\left(\tfrac{-1}{\sqrt{3}}, \tfrac{-1}{\sqrt{3}}\right) \times \underbrace{0}_{x_1} + N_2\left(\tfrac{-1}{\sqrt{3}}, \tfrac{-1}{\sqrt{3}}\right) \times \underbrace{4}_{x_2} + \\ &+ N_3\left(\tfrac{-1}{\sqrt{3}}, \tfrac{-1}{\sqrt{3}}\right) \times \underbrace{4}_{x_3} + N_4\left(\tfrac{-1}{\sqrt{3}}, \tfrac{-1}{\sqrt{3}}\right) \times \underbrace{0}_{x_4} \\ &= 2\left(1 - \tfrac{\sqrt{3}}{3}\right) = 0.845299 . \end{aligned} \tag{F.20}$$

Based on the same procedure, the complete set of coordinates is: g1(0.845299, 0.845299), g2(3.154701, 0.845299), g3(3.154701, 3.154701) and g4(0.845299, 3.154701). Application of Eq. (F.17) gives the same results as in the ξ–η space.
• The application of Eq. (F.13) gives immediately: $\sigma_{n1x} = 131.220$, $\sigma_{n2x} = 214.061$, $\sigma_{n3x} = 742.181$ and $\sigma_{n4x} = 659.341$.

F.2 Example: Stress extrapolation for a distorted plane stress element

Given is a distorted plane stress element with full integration (2×2) as shown in Fig. F.4. The nodal coordinates are 1(0, 0), 2(4, −0.5), 3(4.5, 5.5) and 4(0, 4). The constant thickness equals $t = 1.1$. The right-hand nodes are loaded by an external force $\boldsymbol{F} = \left[10, -12\right]^T$ at node 2 and a prescribed displacement $\boldsymbol{u} = \left[0.012, 0.014\right]^T$ at node 3. A linear-elastic finite element calculation resulted for the elastic properties $E = 200{,}000$ and $\nu = 0.3$ in the following stress values in the x-direction at the integration points: $\sigma_{g1x} = 106.720$, $\sigma_{g2x} = 166.995$, $\sigma_{g3x} = 441.752$ and $\sigma_{g4x} = 452.815$.

Calculate the nodal stress values for the x-component based on:

- a linear extrapolation based on the ξ–η space (see Eq. (F.3));
- a linear extrapolation based on the x–y space;
- a local least squares fit extrapolation in the ξ–η space (see Eq. (F.13)).

F.2 Solution

• The application of Eq. (F.3) gives immediately: $\sigma_{n1x} = -28.966$, $\sigma_{n2x} = 75.433$, $\sigma_{n3x} = 551.326$ and $\sigma_{n4x} = 570.488$.

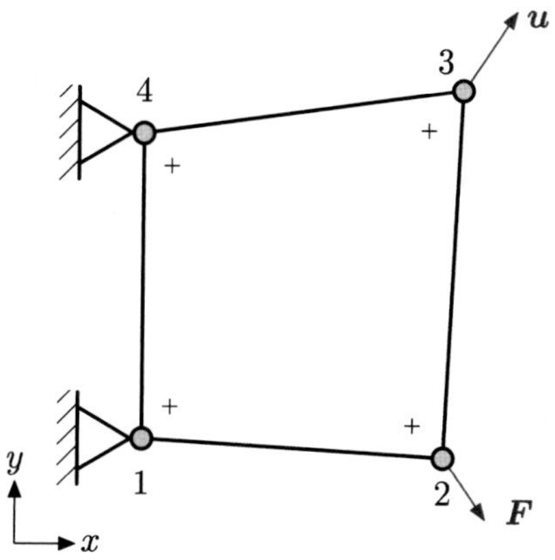

Fig. F.4 Distorted plane stress element

• The linear extrapolation based on the x–y space can be performed as described in Example F.1 and gives: $\sigma_{n1x} = 609.608$, $\sigma_{n2x} = 619.450$, $\sigma_{n3x} = 564.946$ and $\sigma_{n4x} = 657.832$.
• The application of Eq. (F.13) gives immediately: $\sigma_{n1x} = -51.579$, $\sigma_{n2x} = 98.047$, $\sigma_{n3x} = 528.713$ and $\sigma_{n4x} = 593.101$.

References

1. Abramowitz M, Stegun IA (1965) Handbook of mathematical functions. Dover Publications, New York
2. Braden B (1986) The surveyor's area formula. Coll Math J 17:326–337
3. Hildebrand FB (1956) Introduction to numerical analysis. McGraw-Hill, New York
4. Hinton E, Campbell JS (1974) Local and global smoothing of discontinuous finite element functions using a least squares method. Int J Numer Methods Eng 8:461–480

Appendix G
Answers to Supplementary Problems

G.1 Problems from Chap. 2

2.9 Simplified model of a tower under dead weight (analytical approach)
The load can be expressed based on the vertical force equilibrium as

$$N_x(X) = -\varrho g A(L - x)\,, \tag{G.1}$$

or as distributed load as:

$$p_x(x) = -\frac{\mathrm{d}N_x(x)}{\mathrm{d}x} = -\varrho g A = p_0\,. \tag{G.2}$$

Alternatively, Eq. (G.2) could be obtained by normalizing the weight of the tower by its length:

$$p_x(x) = -\frac{F_\mathrm{G}}{L} = -\frac{mg}{L} = -\varrho g A = p_0\,. \tag{G.3}$$

Further results:

$$\sigma_x(x) = -\varrho g(L - x)\,, \tag{G.4}$$

$$u_x(x) = \frac{1}{EA}\left(+\tfrac{1}{2}\varrho A g x^2 - \varrho A g L x\right)\,, \tag{G.5}$$

$$L' = L + u_x(L) = L\left(1 - \tfrac{\varrho g L}{2E}\right)\,, \tag{G.6}$$

$$L_\mathrm{max} = \tfrac{\sigma_\mathrm{max}}{\varrho g}\,. \tag{G.7}$$

2.10 Analytical solution for a rod problem
Case (a)

$$u_x(x) = \frac{u_0}{L}x\,,\ \varepsilon_x(x) = \frac{u_0}{L}\,,\ \sigma_x(x) = \frac{u_0}{L}E\,. \tag{G.8}$$

A. Öchsner, *Computational Statics and Dynamics*,
https://doi.org/10.1007/978-981-15-1278-0

Case (b)

$$u_x(x) = \frac{F_0}{EA} \times x \, , \varepsilon_x(x) = \frac{F_0}{EA} \, , \sigma_x(x) = \frac{F_0}{A} \, . \tag{G.9}$$

2.11 Weighted residual method based on general formulation of partial differential equation

$$\int_{\Omega} (\mathcal{L}_1 W)^{\mathrm{T}} C \, (\mathcal{L}_1 u_x) \, \mathrm{d}\Omega = \int_{\Gamma} W^{\mathrm{T}} \, (C \mathcal{L}_1 u_x)^{\mathrm{T}} n_x \, \mathrm{d}\Gamma + \int_{\Omega} W^{\mathrm{T}} b \, \mathrm{d}\Omega \, . \tag{G.10}$$

Comment: $C = E$, $\mathrm{d}\Omega = A\mathrm{d}x$, $\mathrm{d}\Gamma = \mathrm{d}A$, $b = \frac{p_x(x)}{A}$.

2.12 Weighted residual method with arbitrary distributed load for a rod

$$\int_0^L W^{\mathrm{T}}(x) \left(EA \frac{\mathrm{d}^2 u_x(x)}{\mathrm{d}x^2} + p_x(x) \right) \mathrm{d}x = 0 \tag{G.11}$$

$$\int_0^L \frac{\mathrm{d}W^{\mathrm{T}}}{\mathrm{d}x} EA \frac{\mathrm{d}u_x}{\mathrm{d}x} \mathrm{d}x = \int_0^L W^{\mathrm{T}} p_x(x) \mathrm{d}x + EA \left[W^{\mathrm{T}} \frac{\mathrm{d}u_x}{\mathrm{d}x} \right]_0^L \tag{G.12}$$

$$\cdots = \delta \boldsymbol{u}_{\mathrm{p}}^{\mathrm{T}} \int_0^L \boldsymbol{N} \, p_x(x) \mathrm{d}x + \cdots \tag{G.13}$$

$$\cdots = \int_0^L \begin{bmatrix} N_1 \\ N_2 \end{bmatrix} q_y(x) \mathrm{d}x + \cdots \tag{G.14}$$

The additional expression on the right-hand side gives the equivalent nodal loads for a distributed load according to Eq. (2.53).

2.13 Numerical integration and coordinate transformation

The arbitrary coordinate range $x_1 \leq x \leq x_2$ (see Fig. G.1a) is first translated to the origin of the coordinate system (see Fig. G.1b) based on the following transformation:

$$\xi = x - \left(x_1 + \frac{x_2 - x_1}{2} \right) . \tag{G.15}$$

This transformation gives the new interval $-\frac{x_2 - x_1}{2} \leq x \leq +\frac{x_2 - x_1}{2}$. However, the width of the interval remains unchanged, i.e. $x_2 - x_1$. In order to create the required

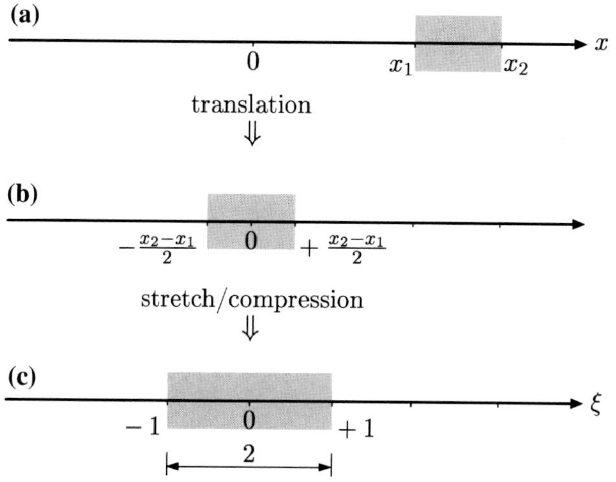

Fig. G.1 Transformation of a coordinate range for numerical integration: **a** original interval; **b** after translation; **c** after stretching

new coordinate range, i.e. $-1 \leq \xi \leq +1$ (see Fig. G.1c), the existing interval (see Fig. G.1b) must be stretched or compressed depending on its actual width. This can be done by multiplying with the stretch factor $\frac{2}{x_2-x_1}$. A width smaller than 2 results in a stretching while a width larger than 2 results in a compression:

$$\xi = \frac{2}{x_2 - x_1} \times \left[x - \left(x_1 + \frac{x_2 - x_1}{2} \right) \right] = \frac{2}{x_2 - x_1} \times (x - x_1) - 1 \,. \tag{G.16}$$

2.14 Finite element solution for a rod problem

Case (a)

$$u_x(x) = \frac{F_0}{EA} \times x \,, \varepsilon_x(x) = \frac{F_0}{EA} \,, \sigma_x(x) = \frac{F_0}{A} \,. \tag{G.17}$$

Case (b)

$$u_x(x) = \frac{u_0}{L} \times x \,, \varepsilon_x(x) = \frac{u_0}{L} \,, \sigma_x(x) = \frac{u_0 E}{L} \,. \tag{G.18}$$

2.15 Finite element approximation with a single linear rod element

The analytical solutions are given in [2]. The finite element solutions are obtained as:

$$\text{(a)} \quad u_x(x) = \frac{F_0 x}{EA} \,, \qquad u_x(L) = \frac{F_0 L}{EA} \,, \tag{G.19}$$

$$\text{(b)}\ \ u_x(x) = \frac{p_0}{EA}\left(\frac{xL}{2}\right), \qquad u_x(L) = \frac{p_0 L^2}{2EA}, \tag{G.20}$$

$$\text{(c)}\ \ u_x(x) = \frac{p_0}{EA}\left(\frac{xL}{3}\right), \qquad u_x(L) = \frac{p_0 L^2}{3EA}. \tag{G.21}$$

The finite element approximation is equal to the exact solution only in the case of the single force. In the case of distributed loads, the same solution is only obtained at the nodes but the distribution between the nodes is different.

2.16 Different formulations for the displacement field of a linear rod element

$$u^{\mathrm{e}}(x) = N_1(x)u_1 + N_2(x)u_2 \tag{G.22}$$

$$= \left(1 - \frac{x}{L}\right)u_1 + \frac{x}{L}u_2 \tag{G.23}$$

$$= \underbrace{u_1}_{a_0} + \underbrace{\frac{1}{L}(u_2 - u_1)\,x}_{a_1} \tag{G.24}$$

$$= a_0 + a_1 x \quad \text{for} \quad 0 \le x \le L\,, \tag{G.25}$$

or

$$u^{\mathrm{e}}(\xi) = a_0 + a_1\xi \quad \text{for} \quad -1 \le \xi \le +1\,. \tag{G.26}$$

2.17 Finite element approximation with a single quadratic rod element

The analytical solutions are given in [2]. The finite element solutions are obtained as:

$$\text{(a)}\ \ u_{2x} = \frac{F_0 L}{2EA}, \qquad u_{3x} = \frac{F_0 L}{EA}, \qquad u_x(x) = \frac{F_0 x}{EA}, \tag{G.27}$$

$$\text{(b)}\ \ u_{2x} = \frac{3p_0 L^2}{8EA}, \qquad u_{3x} = \frac{p_0 L^2}{2EA}, \qquad u_x(x) = \frac{p_0}{EA}\left(xL - \frac{x^2}{2}\right), \tag{G.28}$$

$$\text{(c)}\ \ u_{2x} = \frac{11 p_0 L^2}{48EA}, \qquad u_{3x} = \frac{p_0 L^2}{3EA}, \qquad u_x(x) = \frac{p_0}{EA}\left(\frac{7}{12}xL - \frac{x^2}{4}\right). \tag{G.29}$$

The finite element solution at the nodes is in all cases equal to the analytical solution. However, only in case (a) and (b) both are the same between the nodes.

2.18 Equivalent nodal loads for a quadratic distribution (linear rod element)

$$\text{(a)}\ F_{1x} = \frac{p_0^* L^3}{12}, \quad F_{2x} = \frac{p_0^* L^3}{4}, \tag{G.30}$$

$$\text{(b)}\ F_{1x} = \frac{p_0 L}{12}, \quad F_{2x} = \frac{p_0 L}{4}. \tag{G.31}$$

2.19 Derivation of interpolation functions for a quadratic rod element

$$N^{\mathrm{T}} = \chi^{\mathrm{T}} A = \begin{bmatrix} 1 & \xi & \xi^2 \end{bmatrix} \begin{bmatrix} 0 & 1 & 0 \\ -\frac{1}{2} & 0 & \frac{1}{2} \\ \frac{1}{2} & -1 & \frac{1}{2} \end{bmatrix}. \tag{G.32}$$

2.20 Derivation of the Jacobian determinant for a quadratic rod element

$$J = \frac{\mathrm{d}x}{\mathrm{d}\xi} = \left(-\tfrac{1}{2} + \xi\right) 0 + (-2\xi)\, \tfrac{L}{2} + \left(\tfrac{1}{2} + \xi\right) L = \tfrac{L}{2}\,. \tag{G.33}$$

2.21 Comparison of the stress distribution for a linear and quadratic rod element with linear increasing load

(a) analytical solution

$$\sigma(x) = \frac{p_0 L}{A} \left(\frac{1}{2} - \frac{1}{2} \left(\frac{x}{L} \right)^2 \right). \tag{G.34}$$

(b) single *linear* rod element

$$\sigma^{\mathrm{e}}(x) = \frac{p_0 L}{3A}. \tag{G.35}$$

(c) single *quadratic* rod element

$$\sigma^{\mathrm{e}}(x) = \frac{p_0 L}{A} \left[\left(4 - \frac{8x}{L} \right) \frac{11}{48} + \left(-1 + \frac{4x}{L} \right) \frac{16}{48} \right]. \tag{G.36}$$

The different distributions are compared in Fig. G.2.

2.22 Derivation of interpolation functions and stiffness matrix for a quadratic rod element with unevenly distributed nodes

$$N_1(\xi) = \frac{b}{2(b-1)} - \frac{\xi}{2} - \frac{\xi^2}{2(b-1)}, \tag{G.37}$$

$$N_2(\xi) = -\frac{1}{b^2 - 1} + \frac{\xi^2}{b^2 - 1}, \tag{G.38}$$

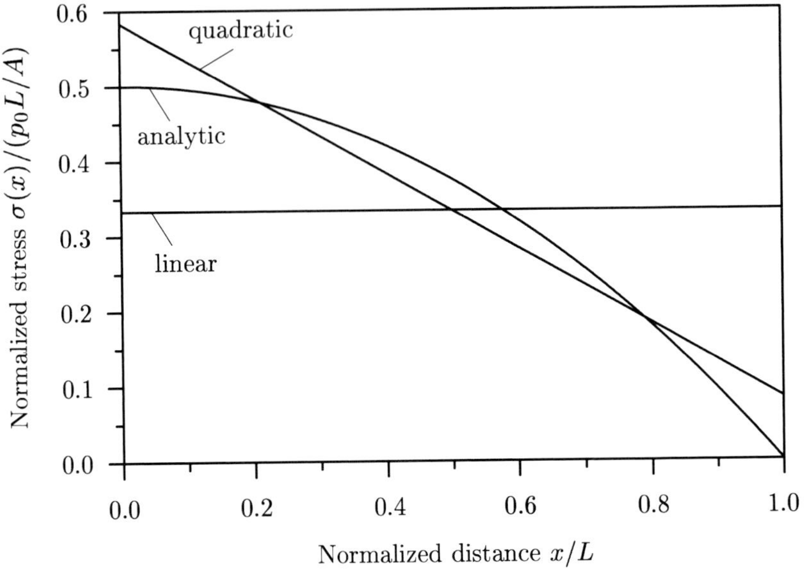

Fig. G.2 Stress distribution for rod element with linear increasing load

$$N_3(\xi) = \frac{b}{2(1+b)} + \frac{\xi}{2} + \frac{\xi^2}{2(1+b)}\,. \tag{G.39}$$

$$\boldsymbol{K} = \frac{EA}{L}\begin{bmatrix} \frac{7-6b+3b^2}{3(b-1)^2} & -\frac{8}{3(b^2-1)(b-1)} & -\frac{3b^2+1}{3(b^2-1)} \\ -\frac{8}{3(b^2-1)(b-1)} & \frac{16}{3(b^2-1)^2} & \frac{8}{3(1+b)(b^2-1)} \\ -\frac{3b^2+1}{3(b^2-1)} & \frac{8}{3(1+b)(b^2-1)} & \frac{7+6b+3b^2}{3(1+b)^2} \end{bmatrix}. \tag{G.40}$$

For a value of $b = 0.9$ ($\xi = -0.9$), the following matrix is obtained:

$$\boldsymbol{K}_{b=0.9} = \frac{EA}{L}\begin{bmatrix} 134.3333333 & -140.3508772 & 6.017543860 \\ -140.3508772 & 147.7377654 & -7.386888275 \\ 6.017543860 & -7.386888275 & 1.369344414 \end{bmatrix}. \tag{G.41}$$

If the inner node is close to the left or right-hand boundary, the stiffness matrix will contain very small and very large values. This can result in numerical problems ($\rightarrow$ inversion of the matrix). As a rule of thumb, the inner node should be at least $\frac{1}{4}$ from the outer nodes.

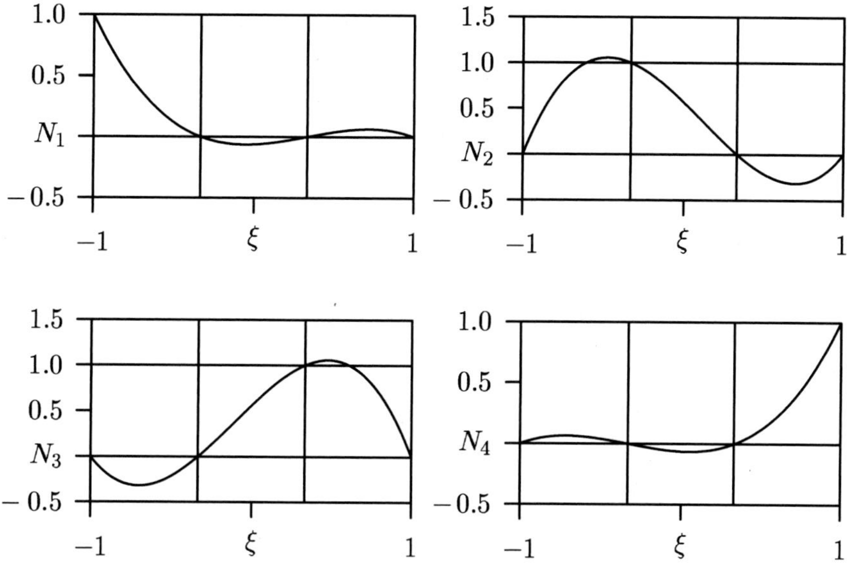

Fig. G.3 Interpolation functions for a cubic rod element with equidistant nodes plotted over the natural coordinate (ξ)

2.23 Derivation of interpolation functions for a cubic rod element

$$\boldsymbol{N}^{\mathrm{T}} = \boldsymbol{\chi}^{\mathrm{T}}\boldsymbol{A} = \begin{bmatrix} 1 & \xi & \xi^2 & \xi^3 \end{bmatrix} \begin{bmatrix} -\frac{1}{16} & \frac{9}{16} & \frac{9}{16} & -\frac{1}{16} \\ \frac{1}{16} & -\frac{27}{16} & \frac{27}{16} & -\frac{1}{16} \\ \frac{9}{16} & -\frac{9}{16} & -\frac{9}{16} & \frac{9}{16} \\ -\frac{9}{16} & \frac{27}{16} & -\frac{27}{16} & \frac{9}{16} \end{bmatrix}. \tag{G.42}$$

The graphical representation of the four interpolation functions is given in Fig. G.3.

2.24 Structure composed of three linear rod elements

Displacement matrix:

$$\text{(a)}\ \boldsymbol{u} = \frac{LF_0}{EA}\begin{bmatrix} 0 & \frac{1}{3} & \frac{5}{6} & \frac{11}{6} \end{bmatrix}^{\mathrm{T}}, \tag{G.43}$$

$$\text{(b)}\ \boldsymbol{u} = u_0\begin{bmatrix} 0 & \frac{2}{11} & \frac{5}{11} & 1 \end{bmatrix}^{\mathrm{T}}. \tag{G.44}$$

Reaction forces:

$$\text{(a)}\ F_1^{\mathrm{R}} = -F_0\,, \tag{G.45}$$

$$\text{(b)}\ F_1^{\mathrm{R}} = -\frac{6EAu_0}{11L},\quad F_4^{\mathrm{R}} = +\frac{6EAu_0}{11L}. \tag{G.46}$$

2.25 Finite element approximation of a rod with four elements: comparison of displacement, strain and stress distribution with analytical solution

Analytical solution:

$$EAu(X) = -\frac{p_0 X^2}{2} + c_1 X + c_2 \,. \tag{G.47}$$

Boundary conditions: $u(0) = 0$ and $N_X(L) = 0$.

$$u(X) = \frac{p_0 L^2}{EA}\left(\frac{X}{L} - \frac{1}{2}\left(\frac{X}{L}\right)^2\right), \tag{G.48}$$

$$\varepsilon(X) = \frac{p_0 L}{EA}\left(1 - \frac{X}{L}\right), \tag{G.49}$$

$$\sigma(X) = \frac{p_0 L}{A}\left(1 - \frac{X}{L}\right). \tag{G.50}$$

Finite element solution:

$$u_2 = \frac{7p_0 L^2}{32EA}, \quad u_3 = \frac{3p_0 L^2}{8EA}, \quad u_4 = \frac{15p_0 L^2}{32EA}, \quad u_5 = \frac{p_0 L^2}{2EA}, \tag{G.51}$$

$$\varepsilon_{\mathrm{I}} = \frac{7p_0 L}{8EA}, \quad \varepsilon_{\mathrm{II}} = \frac{5p_0 L}{8EA}, \quad \varepsilon_{\mathrm{III}} = \frac{3p_0 L}{8EA}, \quad \varepsilon_{\mathrm{IV}} = \frac{p_0 L}{8EA}, \tag{G.52}$$

$$\varepsilon_1 = \varepsilon_{\mathrm{I}}, \; \varepsilon_2 = \frac{\varepsilon_{\mathrm{I}} + \varepsilon_{\mathrm{II}}}{2}, \; \varepsilon_3 = \frac{\varepsilon_{\mathrm{II}} + \varepsilon_{\mathrm{III}}}{2}, \; \varepsilon_4 = \frac{\varepsilon_{\mathrm{III}} + \varepsilon_{\mathrm{IV}}}{2}, \; \varepsilon_5 = \varepsilon_{\mathrm{IV}}, \tag{G.53}$$

$$\sigma_{\mathrm{I}} = \frac{7p_0 L}{8A}, \quad \sigma_{\mathrm{II}} = \frac{5p_0 L}{8A}, \quad \sigma_{\mathrm{III}} = \frac{3p_0 L}{8A}, \quad \sigma_{\mathrm{IV}} = \frac{p_0 L}{8A}, \tag{G.54}$$

$$\sigma_1 = \sigma_{\mathrm{I}}, \; \sigma_2 = \frac{\sigma_{\mathrm{I}} + \sigma_{\mathrm{II}}}{2}, \; \sigma_3 = \frac{\sigma_{\mathrm{II}} + \sigma_{\mathrm{III}}}{2}, \; \sigma_4 = \frac{\sigma_{\mathrm{III}} + \sigma_{\mathrm{IV}}}{2}, \; \sigma_5 = \sigma_{\mathrm{IV}}. \tag{G.55}$$

Displacement, strain and stress distributions are shown in Fig. G.4.

2.26 Elongation of a bi-material rod: finite element solution and comparison with analytical solution

The analytical solution of this problem is discussed in [2].
Finite element solution:

$$u_1 = 0\,, \tag{G.56}$$

$$u_2 = \frac{3k_{\mathrm{I}} p_0 L^2 + k_{\mathrm{II}} p_0 L^2 + 4k_{\mathrm{I}} k_{\mathrm{II}} u_0}{8k_{\mathrm{I}}(k_{\mathrm{I}} + k_{\mathrm{II}})}, \tag{G.57}$$

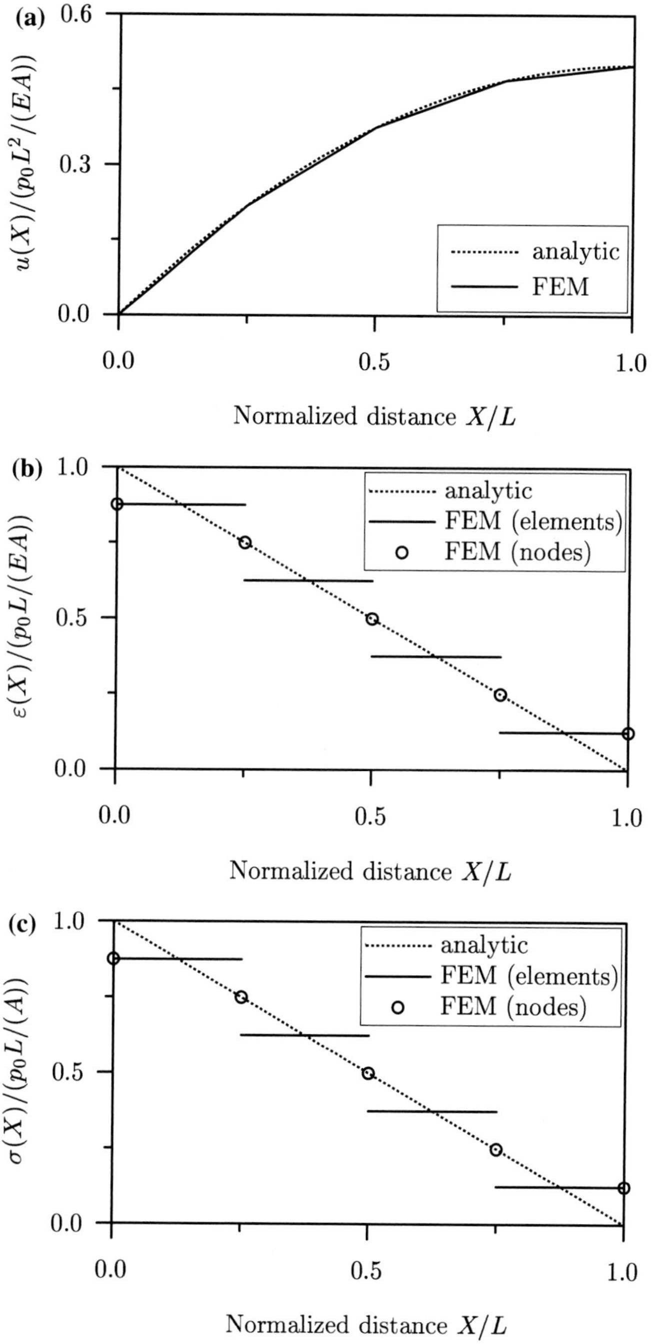

Fig. G.4 Rod element discretized by four elements: **a** displacement; **b** strain, and **c** stress

$$u_3 = \frac{p_0 L^2 + 2k_{\text{II}} u_0}{2(k_{\text{I}} + k_{\text{II}})}, \tag{G.58}$$

$$u_4 = \frac{p_0 L^2 + 2k_{\text{I}} u_0 + 4k_{\text{II}} u_0}{4(k_{\text{I}} + k_{\text{II}})}, \tag{G.59}$$

$$u_5 = u_0 \,. \tag{G.60}$$

$$\varepsilon_{\text{I}} = \frac{3k_{\text{I}} p_0 L^2 + k_{\text{II}} p_0 L^2 + 4k_{\text{I}} k_{\text{II}} u_0}{4Lk_{\text{I}}(k_{\text{I}} + k_{\text{II}})}, \tag{G.61}$$

$$\varepsilon_{\text{II}} = \frac{k_{\text{I}} p_0 L^2 - k_{\text{II}} p_0 L^2 + 4k_{\text{I}} k_{\text{II}} u_0}{4Lk_{\text{I}}(k_{\text{I}} + k_{\text{II}})}, \tag{G.62}$$

$$\varepsilon_{\text{III}} = \frac{-\, p_0 L^2 + 2k_{\text{I}} u_0}{2L(k_{\text{I}} + k_{\text{II}})}, \tag{G.63}$$

$$\varepsilon_{\text{IV}} = \frac{-\, p_0 L^2 + 2k_{\text{I}} u_0}{2L(k_{\text{I}} + k_{\text{II}})}. \tag{G.64}$$

$$\sigma_{\text{I}} = \frac{E_{\text{I}}(3k_{\text{I}} p_0 L^2 + k_{\text{II}} p_0 L^2 + 4k_{\text{I}} k_{\text{II}} u_0)}{4Lk_{\text{I}}(k_{\text{I}} + k_{\text{II}})}, \tag{G.65}$$

$$\sigma_{\text{II}} = \frac{E_{\text{I}}(k_{\text{I}} p_0 L^2 - k_{\text{II}} p_0 L^2 + 4k_{\text{I}} k_{\text{II}} u_0)}{4Lk_{\text{I}}(k_{\text{I}} + k_{\text{II}})}, \tag{G.66}$$

$$\sigma_{\text{III}} = \frac{E_{\text{II}}(-p_0 L^2 + 2k_{\text{I}} u_0)}{2L(k_{\text{I}} + k_{\text{II}})}, \tag{G.67}$$

$$\sigma_{\text{IV}} = \frac{E_{\text{II}}(-p_0 L^2 + 2k_{\text{I}} u_0)}{2L(k_{\text{I}} + k_{\text{II}})}. \tag{G.68}$$

The values of strain and stress at inner nodes are obtained by averaging the elemental values. The distribution of stress, strain and displacement is shown in Fig. G.5.

2.27 Stress distribution for a fixed-fixed rod structure

Solution matrix:

$$\boldsymbol{u} = \frac{LF_0}{EA} \begin{bmatrix} \frac{1}{6} & \frac{1}{3} & \frac{1}{2} & \frac{1}{3} & \frac{1}{6} \end{bmatrix}^{\text{T}}. \tag{G.69}$$

The difference between the elemental stress values and the averaged nodal values is shown in Fig. G.6.

2.28 Linear rod element with variable cross section: derivation of stiffness matrix

(a) $d = d(x)$ linear

$$\boldsymbol{K}^{\text{e}} = \frac{E}{L} \times \frac{\pi(d_1^2 + d_1 d_2 + d_2^2)}{12} \begin{bmatrix} 1 & -1 \\ -1 & 1 \end{bmatrix}, \tag{G.70}$$

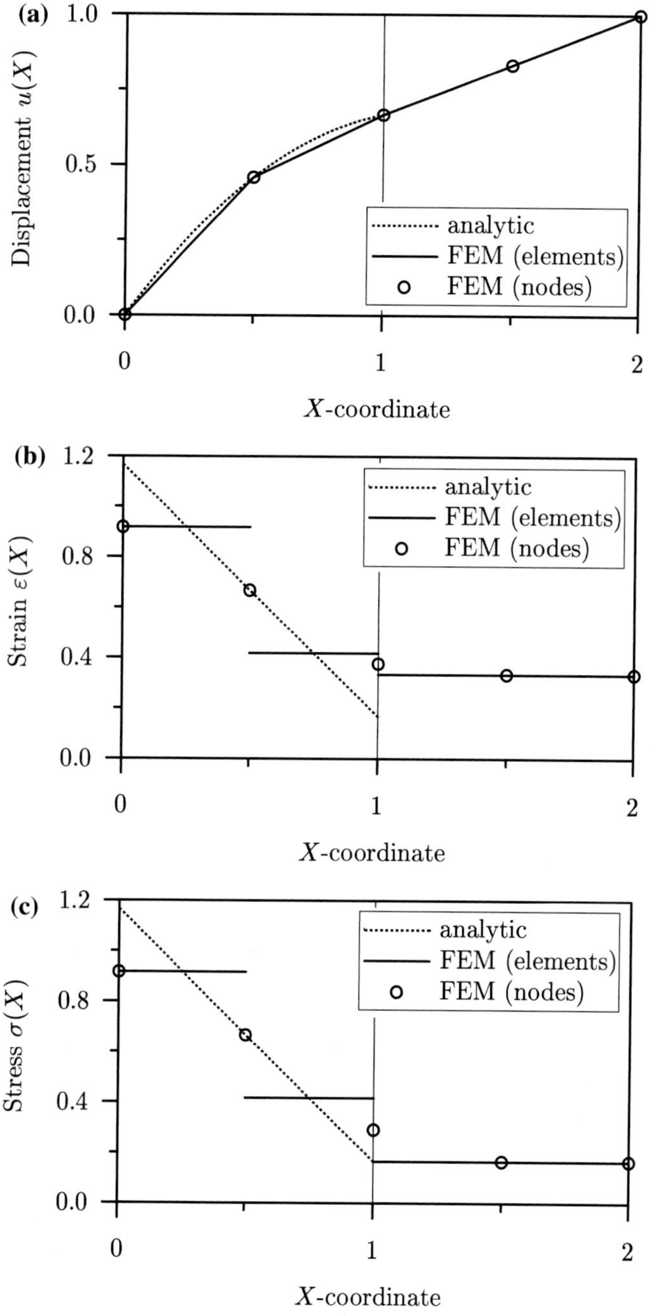

Fig. G.5 Bi-material rod discretized by four elements: **a** displacement; **b** strain, and **c** stress

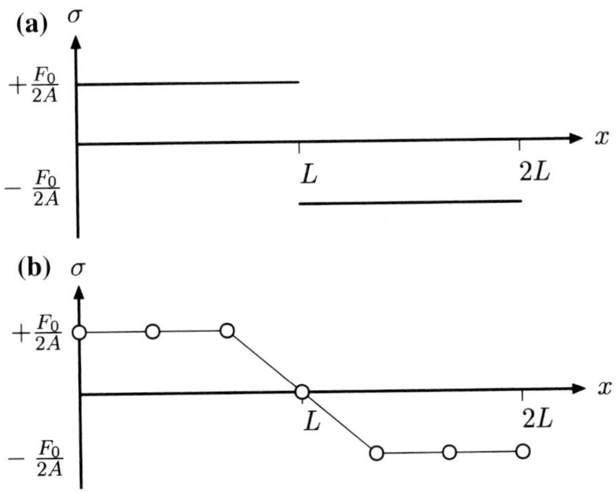

Fig. G.6 **a** Stress distribution based on elemental values and **b** based on nodal averaging

or by replacing the diameters with the cross-section areas, i.e. $d_i = 2\sqrt{\frac{A_i}{\pi}}$,

$$\boldsymbol{K}^{\mathrm{e}} = \frac{E}{L} \times \frac{\pi(A_1 + A_2 + \sqrt{A_1 A_2})}{3} \begin{bmatrix} 1 & -1 \\ -1 & 1 \end{bmatrix}. \tag{G.71}$$

(b) $A = A(x)$ linear

$$\boldsymbol{K}^{\mathrm{e}} = \frac{E}{L} \times \frac{A_1 + A_2}{2} \begin{bmatrix} 1 & -1 \\ -1 & 1 \end{bmatrix}. \tag{G.72}$$

Two-point GAUSS integration gives for both cases the same result as the analytical integration!

2.29 Quadratic rod element with variable cross section: derivation of stiffness matrix

(a) $d = d(x)$ linear

$$\boldsymbol{K}^{\mathrm{e}} = \frac{\pi E}{60L} \times \begin{bmatrix} 23d_1^2 + 9d_1 d_2 + 3d_2^2 & -2(13d_1^2 + 4d_1 d_2 + 3d_2^2) & 3d_1^2 - d_1 d_2 + 3d_2^2 \\ -2(13d_1^2 + 4d_1 d_2 + 3d_2^2) & 8(4d_1^2 + 2d_1 d_2 + 4d_2^2) & -2(3d_1^2 + 4d_1 d_2 + 13d_2^2) \\ 3d_1^2 - d_1 d_2 + 3d_2^2 & -2(3d_1^2 + 4d_1 d_2 + 13d_2^2) & 3d_1^2 + 9d_1 d_2 + 23d_2^2 \end{bmatrix}. \tag{G.73}$$

(b) $A = A(x)$ linear

$$\boldsymbol{K}^{\mathrm{e}} = \frac{E}{3L}\begin{bmatrix} \frac{11A_1+3A_2}{2} & -2(3A_1+A_2) & \frac{A_1+A_2}{2} \\ -2(3A_1+A_2) & 8(A_1+A_2) & -2(A_1+3A_2) \\ \frac{A_1+A_2}{2} & -2(A_1+3A_2) & \frac{3A_1+11A_2}{2} \end{bmatrix}. \tag{G.74}$$

2.30 Linear rod element with variable cross section: comparison of displacements between FE and analytical solution for a single element

(a) Force boundary condition

• Finite element solution:

$$u_2 = \frac{2LF_0}{E(A_1+A_2)}. \tag{G.75}$$

For the case $A_2 = 2A_1 = 2A$ we obtain:

$$u_2 = \frac{2LF_0}{E(A+2A)} = \frac{2}{3}\frac{LF_0}{EA} = 0.\bar{6}\,\frac{LF_0}{EA}. \tag{G.76}$$

For the case $A_2 = A_1 = A$ we obtain:

$$u_2 = \frac{LF_0}{EA}. \tag{G.77}$$

• Analytical solution:

$$u(x) = \frac{F_0 L}{E(A_2-A_1)} \times \ln\left(1 + \frac{x}{L}\left(\frac{A_2}{A_1} - 1\right)\right), \tag{G.78}$$

or at the right-hand end:

$$u(L) = \frac{F_0 L}{E(A_2-A_1)} \ln\left(\frac{A_2}{A_1}\right). \tag{G.79}$$

For the case $A_2 = 2A_1 = 2A$ we obtain:

$$u(L) = \frac{F_0 L}{EA} \ln(2) \approx 0.693\,\frac{LF_0}{EA}. \tag{G.80}$$

For the case $A_2 = A_1 = A$ we obtain under consideration of $\lim \frac{\ln x}{x-1}\Big|_{x\to 1} = 1$ where $x = \frac{A_2}{A_1}$:

$$u(L) = \frac{LF_0}{EA}. \tag{G.81}$$

The finite element and the analytical solution is different which comes from the fact that EA is not constant!
(b) Displacement boundary condition
• Finite element solution:

$$u(x) = \frac{x}{L} \times u_0 \,, \tag{G.82}$$

$$u(L) = u_0 \,. \tag{G.83}$$

The finite element solution is independent of the cross section ratio.
• Analytical solution:

$$u(x) = \frac{u_0}{\ln\left(\frac{A_2}{A_1}\right)} \times \ln\left(1 + \frac{x}{L}\left(\frac{A_2}{A_1} - 1\right)\right) \,, \tag{G.84}$$

or at the right-hand end:

$$u(L) = u_0 \,. \tag{G.85}$$

2.31 Quadratic rod element with variable cross section: comparison of end displacement between FE and analytical solution for single element

Finite element solution:

$$u_{3x} = \frac{A_1 + A_2}{A_1^2 + 4A_1A_2 + A_2^2} \times \frac{3LF_0}{E} \,. \tag{G.86}$$

For the case $A_2 = 2A_1 = 2A$ we obtain:

$$u_{3x} = \frac{9}{13} \times \frac{LF_0}{EA} \approx 0.6923 \times \frac{LF_0}{EA} \,. \tag{G.87}$$

2.32 Subdivided structure with variable cross section: comparison of displacements and stresses between FE and analytical solution for four elements

• Finite element solution:

$$\boldsymbol{u} = \frac{F_0 L}{EA_1} \times \begin{bmatrix} 0 & \frac{5}{18} & \frac{40}{63} & \frac{143}{126} & \frac{124}{63} \end{bmatrix}^{\mathrm{T}} \,. \tag{G.88}$$

The following matrix states the constant stresses in each element and should not be confused with the averaged nodal values:

$$\boldsymbol{\sigma} = \frac{F_0}{A_1} \times \begin{bmatrix} \frac{10}{9} & \frac{10}{7} & \frac{2}{1} & \frac{10}{3} \end{bmatrix}^{\mathrm{T}} \,. \tag{G.89}$$

• Analytical solutions:

$$\frac{u(X)}{\frac{F_0 L}{EA_1}} = \frac{1}{\left(\frac{A_5}{A_1} - 1\right)} \times \ln\left(1 + \frac{X}{L}\left(\frac{A_5}{A_1} - 1\right)\right). \tag{G.90}$$

$$\boldsymbol{u} = \frac{F_0 L}{EA_1} \times \left[0 \;\; \frac{5}{4} \times \ln\left(\frac{4}{5}\right) \;\; \frac{5}{4} \times \ln\left(\frac{3}{5}\right) \;\; \frac{5}{4} \times \ln\left(\frac{2}{5}\right) \;\; \frac{5}{4} \times \ln(5)\right]^{\mathrm{T}}. \tag{G.91}$$

$$\frac{\varepsilon(X)}{\frac{F_0}{EA_1}} = \frac{1}{1 + \frac{X}{L}\left(\frac{A_5}{A_1} - 1\right)}. \tag{G.92}$$

$$\frac{\sigma(X)}{\frac{F_0}{A_1}} = \frac{1}{1 + \frac{X}{L}\left(\frac{A_5}{A_1} - 1\right)}. \tag{G.93}$$

A comparison between the finite element and the analytical solution is given in Fig. G.7.

2.33 Submodel of a structure with variable cross section

The following matrix states the constant stresses in each element and should not be confused with the averaged nodal values:

$$\boldsymbol{\sigma} = \frac{F_0}{A_1} \times \left[1.016366 \; 1.059615 \; 1.106709 \; 1.158184 \; 1.214681\right]^{\mathrm{T}}. \tag{G.94}$$

The stress distribution based on the submodel and a comparison with the analytical solution is presented in Fig. G.8.

2.34 Rod with elastic embedding: stiffness matrix

$$\int_0^L W(x)\left(EA\frac{\mathrm{d}^2 u_x}{\mathrm{d}x^2} - k u_x\right)\mathrm{d}x = 0\,, \tag{G.95}$$

$$\boldsymbol{K}^{\mathrm{e}} = \cdots + k\int_0^L \boldsymbol{N}(x)\boldsymbol{N}(x)^{\mathrm{T}}\mathrm{d}x\,. \tag{G.96}$$

(a) Linear interpolation functions:

$$\boldsymbol{K}^{\mathrm{e}} = \frac{EA}{L}\begin{bmatrix} 1 & -1 \\ -1 & 1 \end{bmatrix} + \frac{kL}{6}\begin{bmatrix} 2 & 1 \\ 1 & 2 \end{bmatrix}. \tag{G.97}$$

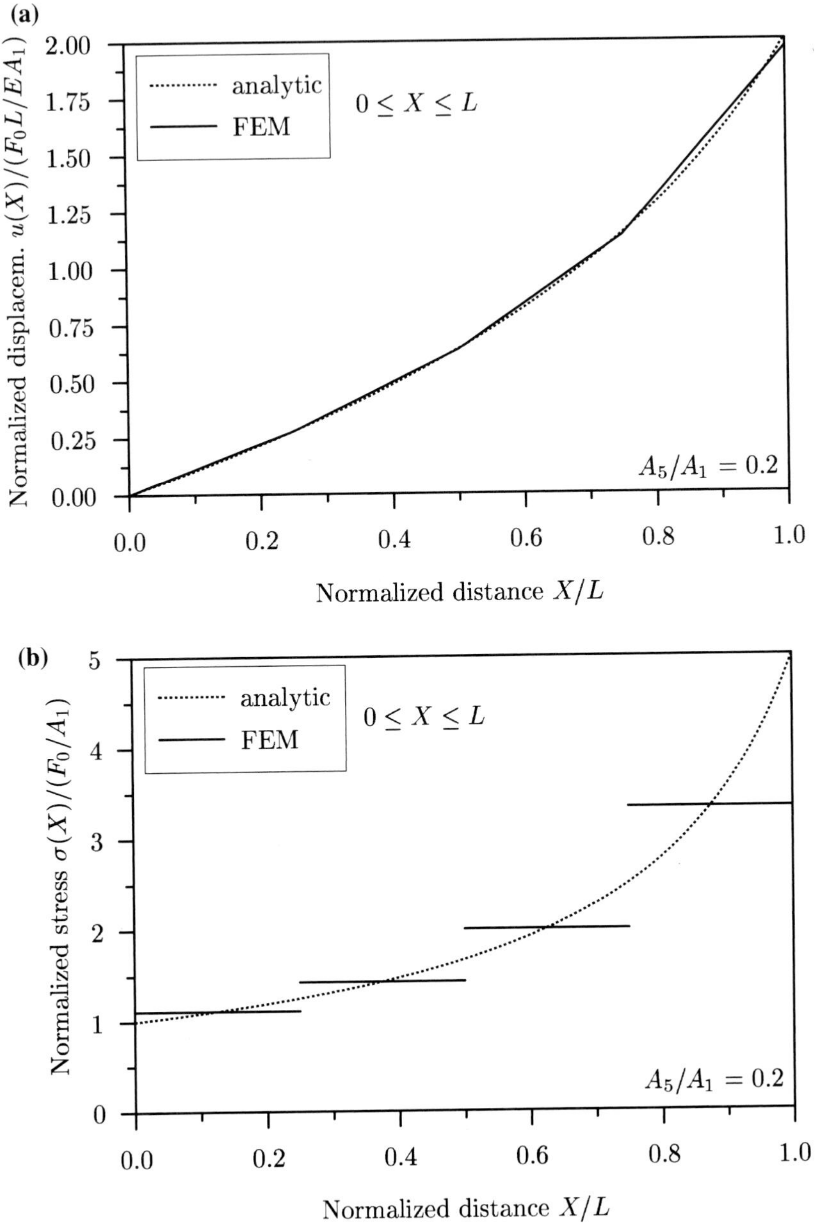

Fig. G.7 Comparison between FE and analytical solution: **a** displacements; **b** stresses

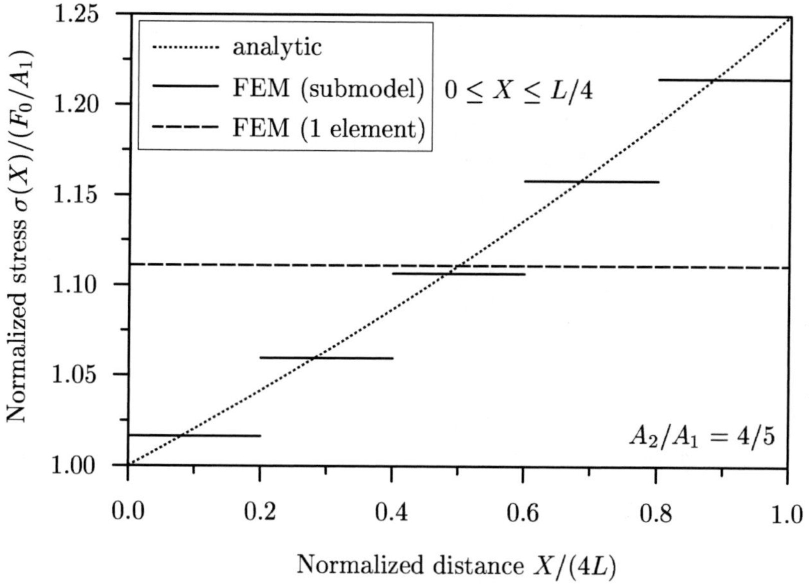

Fig. G.8 Stress distribution based on the submodel and comparison with the analytical solution and the result from the coarse mesh

(b) Quadratic interpolation functions:

$$\boldsymbol{K}^{\mathrm{e}} = \frac{EA}{3L}\begin{bmatrix} 7 & -8 & 1 \\ -8 & 16 & -8 \\ 1 & -8 & 7 \end{bmatrix} + \frac{kL}{30}\begin{bmatrix} 4 & 2 & -1 \\ 2 & 16 & 2 \\ -1 & 2 & 4 \end{bmatrix}. \tag{G.98}$$

2.35 Rod with elastic embedding: single force case

(a) Linear interpolation functions:

$$\left(\frac{EA}{L} + \frac{kL}{3}\right) u_2 = F_0\,, \tag{G.99}$$

$$u_2 = \frac{F_0}{\frac{EA}{L} + \frac{kL}{3}}\,, \tag{G.100}$$

$$u_2|_{k\to 0} = \frac{F_0 L}{EA}\,, \quad u_2|_{EA\to 0} = \frac{3F_0}{kL}\,. \tag{G.101}$$

Comparison between FE and analytical solution:

$$u_x(L)_{\mathrm{FE}} = 0.50 F_0\,, \tag{G.102}$$

$$u_x(L)_{\mathrm{analyt}} = 0.542304 F_0\,. \tag{G.103}$$

(b) Quadratic interpolation functions:

$$\begin{bmatrix} \frac{16EA}{3L}+\frac{16kL}{30} & -\frac{8EA}{3L}+\frac{2kL}{30} \\ -\frac{8EA}{3L}+\frac{2kL}{30} & \frac{7EA}{3L}+\frac{4kL}{30} \end{bmatrix} \begin{bmatrix} u_2 \\ u_3 \end{bmatrix} = \begin{bmatrix} 0 \\ F_0 \end{bmatrix}, \tag{G.104}$$

$$u_2 = \frac{24\left(10EA + kL^2\right)LF_0}{240E^2A^2 + 104EAkL^2 + 3k^2L^4}, \tag{G.105}$$

$$u_2|_{k\to 0} = \frac{F_0 L}{EA}, \quad u_2|_{EA\to 0} = \frac{8F_0}{kL}. \tag{G.106}$$

Comparison between FE and analytical solution:

$$u_x(L)_{\text{FE}} = 0.538860 F_0\,, \tag{G.107}$$

$$u_x(L)_{\text{analyt}} = 0.542304 F_0\,. \tag{G.108}$$

2.36 Plane truss structure arranged in a square

(a) Force boundary condition

$$u_{2X} = \frac{(4+\sqrt{2})LF_0}{2(2+\sqrt{2})EA}\,, \quad u_{2Y} = -\frac{\sqrt{2}\,LF_0}{2(2+\sqrt{2})EA}, \tag{G.109}$$

$$F^{\text{R}}_{1X} = -\frac{(4+\sqrt{2})F_0}{2(2+\sqrt{2})}\,, \quad F^{\text{R}}_{1Y} = 0\,, \quad F^{\text{R}}_{3X} = -\frac{\sqrt{2}\,F_0}{2(2+\sqrt{2})}, \tag{G.110}$$

$$F^{\text{R}}_{3Y} = -\frac{\sqrt{2}\,F_0}{2(2+\sqrt{2})}\,, \quad F^{\text{R}}_{4X} = 0\,, \quad F^{\text{R}}_{4Y} = \frac{\sqrt{2}\,F_0}{2(2+2\sqrt{2})}, \tag{G.111}$$

$$F_{\text{I}} = \frac{(4+\sqrt{2})F_0}{2(2+\sqrt{2})}\,, \quad F_{\text{II}} = \frac{F_0}{2+\sqrt{2}}\,, \quad F_{\text{III}} = -\frac{\sqrt{2}\,F_0}{2(2+\sqrt{2})}. \tag{G.112}$$

(b) Displacement boundary condition

$$u_{2X} = u\,, \quad u_{2Y} = -\frac{\sqrt{2}\,u_0}{4+\sqrt{2}}, \tag{G.113}$$

$$F^{\text{R}}_{1X} = -\frac{EAu_0}{L}\,, \quad F^{\text{R}}_{1Y} = 0\,, \quad F^{\text{R}}_{2X} = \frac{2(2+\sqrt{2})EAu_0}{(4+\sqrt{2})L}, \tag{G.114}$$

$$F_{2Y}^{\mathrm{R}} = 0 \ , \quad F_{3X}^{\mathrm{R}} = -\frac{\sqrt{2}\,EAu_0}{(4+\sqrt{2})L} \ , \quad F_{3Y}^{\mathrm{R}} = -\frac{\sqrt{2}\,EAu_0}{(4+\sqrt{2})L}, \tag{G.115}$$

$$F_{4X}^{\mathrm{R}} = 0 \ , \quad F_{4Y}^{\mathrm{R}} = \frac{\sqrt{2}\,EAu_0}{(4+\sqrt{2})L}, \tag{G.116}$$

$$F_{\mathrm{I}} = \frac{EAu_0}{L} \ , \quad F_{\mathrm{II}} = \frac{2EAu_0}{(4+\sqrt{2})L} \ , \quad F_{\mathrm{III}} = -\frac{\sqrt{2}EAu_0}{(4+\sqrt{2})L}. \tag{G.117}$$

2.37 Plane truss structure arranged in a triangle

(a) Force boundary condition

$$u_{4X} = \frac{\sqrt{2}}{1+\sqrt{2}} \times \frac{LF_X}{EA}, \ u_{4Y} = -\sqrt{2} \times \frac{LF_Y}{EA}, \tag{G.118}$$

$$F_{1X}^{\mathrm{R}} = -\frac{F_X}{2(1+\sqrt{2})} - \frac{F_Y}{2}, \ F_{1Y}^{\mathrm{R}} = \frac{F_X}{2(1+\sqrt{2})} + \frac{F_Y}{2}, \tag{G.119}$$

$$F_{2X}^{\mathrm{R}} = -\frac{\sqrt{2}F_X}{1+\sqrt{2}}, \ F_{2Y}^{\mathrm{R}} = 0, \tag{G.120}$$

$$F_{3X}^{\mathrm{R}} = -\frac{F_X}{2(1+\sqrt{2})} + \frac{F_Y}{2}, \ F_{3Y}^{\mathrm{R}} = -\frac{F_X}{2(1+\sqrt{2})} + \frac{F_Y}{2}, \tag{G.121}$$

$$F_{\mathrm{I}} = \frac{\sqrt{2}F_X}{2(1+\sqrt{2})} + \frac{\sqrt{2}F_y}{2}, \ F_{\mathrm{II}} = \frac{\sqrt{2}F_X}{1+\sqrt{2}}, \ F_{\mathrm{III}} = \frac{\sqrt{2}F_X}{2(1+\sqrt{2})} - \frac{\sqrt{2}F_y}{2}. \tag{G.122}$$

(b) Displacement boundary condition

$$F_{1X}^{\mathrm{R}} = -\frac{EA(u_X - u_Y)}{2\sqrt{2}L}, \ F_{1Y}^{\mathrm{R}} = \frac{EA(u_X - u_Y)}{2\sqrt{2}L}, \tag{G.123}$$

$$F_{2X}^{\mathrm{R}} = -\frac{EAu_X}{L}, \ F_{2Y}^{\mathrm{R}} = 0, \tag{G.124}$$

$$F_{3X}^{\mathrm{R}} = -\frac{EA(u_X + u_Y)}{2\sqrt{2}L}, \ F_{3Y}^{\mathrm{R}} = -\frac{EA(u_X + u_Y)}{2\sqrt{2}L}, \tag{G.125}$$

$$F_{4X}^{\mathrm{R}} = \frac{(1+\sqrt{2})EAu_X}{\sqrt{2}L}, \ F_{4Y}^{\mathrm{R}} = \frac{EAu_Y}{\sqrt{2}L}, \tag{G.126}$$

$$F_{\mathrm{I}} = \frac{EA}{2L}(u_X - u_Y)\,, \quad F_{\mathrm{II}} = \frac{EAu_X}{L}\,, \quad F_{\mathrm{III}} = \frac{EA}{2L}(u_X + u_Y)\,. \tag{G.127}$$

2.38 Plane truss structure with two rod elements

• Free body diagram
Assume all reaction forces acting in positive directions.
• Global system of equations

$$\frac{EA}{L}\begin{bmatrix} \frac{1}{2\sqrt{2}} & -\frac{1}{2\sqrt{2}} & -\frac{1}{2\sqrt{2}} & \frac{1}{2\sqrt{2}} & 0 & 0 \\ -\frac{1}{2\sqrt{2}} & \frac{1}{2\sqrt{2}} & \frac{1}{2\sqrt{2}} & -\frac{1}{2\sqrt{2}} & 0 & 0 \\ -\frac{1}{2\sqrt{2}} & \frac{1}{2\sqrt{2}} & \frac{1}{2\sqrt{2}}+1 & -\frac{1}{2\sqrt{2}} & -1 & 0 \\ \frac{1}{2\sqrt{2}} & -\frac{1}{2\sqrt{2}} & -\frac{1}{2\sqrt{2}} & \frac{1}{2\sqrt{2}} & 0 & 0 \\ 0 & 0 & -1 & 0 & 1 & 0 \\ 0 & 0 & 0 & 0 & 0 & 0 \end{bmatrix}\begin{bmatrix} u_{1X} \\ u_{1Y} \\ u_{2X} \\ u_{2Y} \\ u_{3X} \\ u_{3Y} \end{bmatrix} = \begin{bmatrix} F^{\mathrm{R}}_{1X} \\ F^{\mathrm{R}}_{1Y} \\ F_0 \\ F^{\mathrm{R}}_{2Y} \\ F^{\mathrm{R}}_{3X} \\ F^{\mathrm{R}}_{3Y} \end{bmatrix}. \tag{G.128}$$

• Reduced system of equations under consideration of the BCs

$$\frac{EA}{L}\begin{bmatrix} \frac{1}{2\sqrt{2}}+1 & -\frac{1}{2\sqrt{2}} \\ 0 & \frac{L}{EA} \end{bmatrix}\begin{bmatrix} u_{2X} \\ u_{2Y} \end{bmatrix} = \begin{bmatrix} F_0 \\ -u_0 \end{bmatrix}. \tag{G.129}$$

• Nodal displacements at node 2

$$\begin{bmatrix} u_{2X} \\ u_{2Y} \end{bmatrix} = \begin{bmatrix} \frac{2\sqrt{2}}{1+2\sqrt{2}}\left(\frac{LF_0}{EA} - \frac{u_0}{2\sqrt{2}}\right) \\ -u_0 \end{bmatrix}. \tag{G.130}$$

• All reaction forces

$$F^{\mathrm{R}}_{1X} = -\frac{\sqrt{2}\,(u_0\,EA + LF_0)}{L\left(\sqrt{2}+4\right)}\,, \quad F^{\mathrm{R}}_{1Y} = \frac{\sqrt{2}\,(u_0\,EA + LF_0)}{L\left(\sqrt{2}+4\right)}\,, \tag{G.131}$$

$$F^{\mathrm{R}}_{3X} = \frac{\sqrt{2}u_0\,EA - 4\,LF_0}{L\left(\sqrt{2}+4\right)}\,, \quad F^{\mathrm{R}}_{3Y} = 0\,, \tag{G.132}$$

$$F^{\mathrm{R}}_{2Y} = -\frac{\sqrt{2}\,(u_0\,EA + LF_0)}{L\left(\sqrt{2}+4\right)}\,. \tag{G.133}$$

• Check if the global force equilibrium is fulfilled

$$\sum F_X = 0 \checkmark \ \sum F_Y = 0 \checkmark\,. \tag{G.134}$$

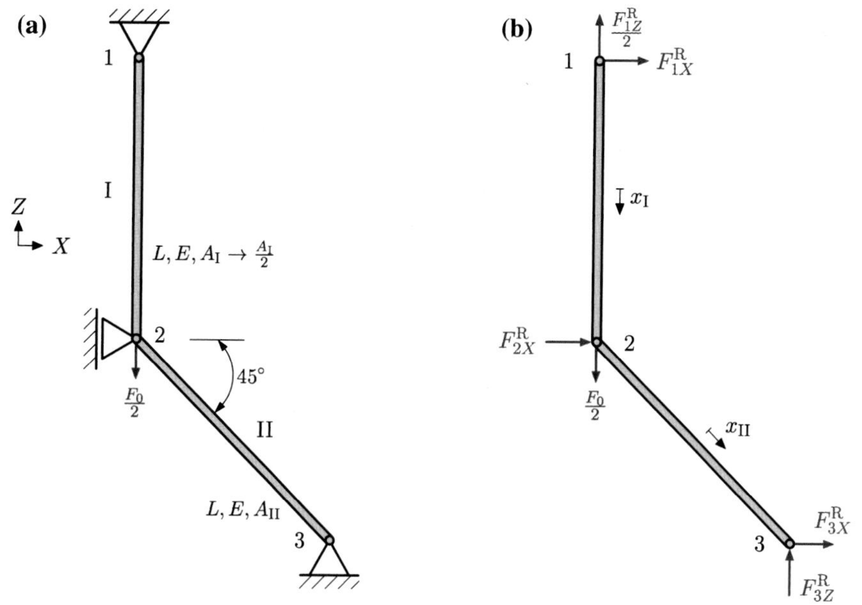

Fig. G.9 Star truss: **a** equivalent statical system under consideration of symmetry and **b** free-body diagram

2.39 Truss structure in star formation

The problem is symmetric in regards to geometry and loading conditions with respect to the Z-axis. The equivalent system can be obtained by, see Fig. G.9:

- replacing element III by a vertical roller support at node 2,
- assuming that the new cross sectional-area of rod I is equal to $\frac{A_{\mathrm{I}}}{2}$,
- applying at node 2 a force of $\frac{F_0}{2}$.

Element I is rotated by $\alpha_{\mathrm{I}} = +90°$ while element II is rotated by $\alpha_{\mathrm{II}} = +45°$. The non-reduced global system of equations is obtained as:

$$\frac{E}{2L}\begin{bmatrix} 0 & 0 & 0 & 0 & 0 & 0 \\ 0 & A_{\mathrm{I}} & 0 & -A_{\mathrm{I}} & 0 & 0 \\ 0 & 0 & A_{\mathrm{II}} & -A_{\mathrm{II}} & -A_{\mathrm{II}} & A_{\mathrm{II}} \\ 0 & -A_{\mathrm{I}} & -A_{\mathrm{II}} & A_{\mathrm{I}}+A_{\mathrm{II}} & A_{\mathrm{II}} & -A_{\mathrm{II}} \\ 0 & 0 & -A_{\mathrm{II}} & A_{\mathrm{II}} & A_{\mathrm{II}} & -A_{\mathrm{II}} \\ 0 & 0 & A_{\mathrm{II}} & -A_{\mathrm{II}} & -A_{\mathrm{II}} & A_{\mathrm{II}} \end{bmatrix} \begin{bmatrix} u_{1X} \\ u_{1Z} \\ u_{2X} \\ u_{2Z} \\ u_{3X} \\ u_{3Z} \end{bmatrix} = \begin{bmatrix} R_{1X} \\ \frac{R_{1Z}}{2} \\ R_{2X} \\ -\frac{F_0}{2} \\ R_{3X} \\ R_{3Z} \end{bmatrix}. \tag{G.135}$$

Consideration of the support conditions gives the reduced system of equations:

$$\frac{E}{2L}(A_{\mathrm{I}} + A_{\mathrm{II}})\,u_{2Z} = -\frac{F_0}{2}, \tag{G.136}$$

or solved for the unknown nodal displacement:

$$u_{2Z} = -\frac{F_0 L}{E\,(A_\text{I} + A_\text{II})}\,. \tag{G.137}$$

The special case $A_\text{I} = A_\text{II} = A$ yields: $u_{2Z} = -\frac{F_0 L}{2EA}$.

For this case, the reactions are obtained as: $\frac{F^\text{R}_{1Z}}{2} = \frac{F_0}{4}$ and $F^\text{R}_{3Z} = \frac{F_0}{4}$.

G.2 Problems from Chap. 3

3.11 Cantilever beam with a distributed load: analytical solution

$$u_z(x) = \frac{q_0 x^2}{24EI_y}\left(x^2 - 4Lx + 6L^2\right)\,. \tag{G.138}$$

$$\varphi_y(x) = -\frac{\mathrm{d}u_z(x)}{\mathrm{d}x} = -\frac{q_0 x}{6EI_y}\left(x^2 - 3Lx + 3L^2\right)\,. \tag{G.139}$$

3.12 Cantilever beam with a point load: analytical solution

$$u_z(x) = \frac{1}{EI_y}\left(-\frac{1}{6}F_0 x^3 + \frac{1}{2}F_0 L x^2\right). \tag{G.140}$$

3.13 Cantilever beam with different end loads and deformations: analytical solution

Case (a): Single force F_0 at $x = L$

$$u_z(x) = \frac{F_0 L^3}{EI}\left\{\frac{1}{6}\left(\frac{x}{L}\right)^3 - \frac{1}{2}\left(\frac{x}{L}\right)^2\right\}, \tag{G.141}$$

$$\varphi_y(x) = \frac{F_0 L^2}{EI}\left\{-\frac{1}{2}\left(\frac{x}{L}\right)^2 + \left(\frac{x}{L}\right)\right\}, \tag{G.142}$$

$$M_y(x) = F_0 L\left\{-\left(\frac{x}{L}\right) + 1\right\}, \tag{G.143}$$

$$Q_z(x) = -F_0\,. \tag{G.144}$$

Case (b): Single moment M_0 at $x = L$

$$u_z(x) = \frac{M_0 L^2}{EI}\left\{\frac{1}{2}\left(\frac{x}{L}\right)^2\right\}, \tag{G.145}$$

$$\varphi_y(x) = -\frac{M_0 L}{EI}\left(\frac{x}{L}\right), \tag{G.146}$$

$$M_y(x) = -M_0\,, \tag{G.147}$$

$$Q_z(x) = 0\,. \tag{G.148}$$

Case (c): Displacement u_0 at $x = L$

$$u_z(x) = \left\{\frac{1}{2}\left(\frac{x}{L}\right)^3 - \frac{3}{2}\left(\frac{x}{L}\right)^2\right\} u_0\,, \tag{G.149}$$

$$\varphi_y(x) = \left\{-\frac{3}{2}\left(\frac{x}{L}\right)^2 + 3\left(\frac{x}{L}\right)\right\}\frac{u_0}{L}, \tag{G.150}$$

$$M_y(x) = \frac{3EIu_0}{L^2}\left\{-\left(\frac{x}{L}\right) + 1\right\}, \tag{G.151}$$

$$Q_z(x) = -\frac{3EIu_0}{L^3}. \tag{G.152}$$

Case (d): Rotation φ_0 at $x = L$

$$u_z(x) = \frac{\varphi_0 L}{2}\left(\frac{x}{L}\right)^2, \tag{G.153}$$

$$\varphi_y(x) = -\varphi_0\left(\frac{x}{L}\right), \tag{G.154}$$

$$M_y(x) = -\frac{\varphi_0 EI}{L}, \tag{G.155}$$

$$Q_z(x) = 0\,. \tag{G.156}$$

The constants of integration for all cases are summarized in Table G.1.

3.14 Simply supported beam with centered single force: analytical solution

The general solution can be written in the range $0 \leq x \leq \frac{L}{2}$ as:

$$EI_y u_z(x) = \tfrac{1}{6}c_1 x^3 + \tfrac{1}{2}c_2 x^2 + c_3 x + c_4\,. \tag{G.157}$$

Consideration of the boundary conditions, i.e. $u_z(0) = 0$, $EI_y \frac{\mathrm{d}^2 u_z}{\mathrm{d}x^2}(0) = -M_y(0) = 0$, $\frac{\mathrm{d}u_z}{\mathrm{d}x}\left(\frac{L}{2}\right) = 0$ and $EI_y \frac{\mathrm{d}^3 u_z}{\mathrm{d}x^3}\left(\frac{L}{2}\right) = -Q_z\left(\frac{L}{2}\right) = \frac{F_0}{2}$, allows the determination of the

Table G.1 Constants of integration for the problems shown in Fig. 3.48

Case	c_1	c_2	c_3	c_4
(a)	F_0	$-F_0L$	0	0
(b)	0	M_0	0	0
(c)	$\frac{3EI_yu_0}{L^3}$	$-\frac{3EI_yu_0}{L^2}$	0	0
(d)	0	$\frac{EI_y\varphi_0}{L}$	0	0

constants of integration as:

$$c_1 = \frac{F_0}{2}\,, \quad c_2 = 0\,, \tag{G.158}$$

$$c_3 = -\frac{L^2F_0}{16}\,, \quad c_4 = 0\,. \tag{G.159}$$

Thus, the following function for the vertical deflection is obtained:

$$u_z(x) = -\frac{F_0x}{48EI_y}\left(3L^2 - 4x^2\right)\,. \tag{G.160}$$

3.15 Simply supported beam under pure bending load: analytical solution

The general solution can be written in the range $0 \leq x \leq L$ as:

$$EI_yu_z(x) = \frac{1}{2}M_0x^2 + c_1x + c_2\,. \tag{G.161}$$

Consideration of the boundary conditions, i.e. $u_z(0) = 0$ and $u_z(L) = 0$, gives $c_1 = -\frac{1}{2}M_0L$ and $c_2 = 0$. Thus, the following function for the vertical deflection is obtained:

$$u_z(x) = \frac{M_0}{2EI_y}\left(x^2 - Lx\right) \quad \text{and} \quad u_z\left(\frac{L}{2}\right) = -\frac{M_0L^2}{8EI_y}\,. \tag{G.162}$$

3.16 Bernoulli beam fixed at both ends: analytical solution

The general solution is given by:

$$u_z(x) = \frac{1}{EI_y}\left(\frac{q_0x^4}{24} + \frac{c_1x^3}{6} + \frac{c_2x^2}{2} + c_3x + c_4\right)\,. \tag{G.163}$$

(a) Single force case ($0 \leq x \leq \frac{L}{2}$)
Boundary conditions: $u_z(0) = 0$, $\varphi_y(0) = 0$, $Q_z(0) = \frac{F_0}{2}$, $\varphi_y(\frac{L}{2}) = 0$.

$$u_z(x) = \frac{1}{EI_y}\left(-\frac{F_0x^3}{12} + \frac{F_0Lx^2}{16}\right), \varphi_y(x) = \frac{1}{EI_y}\left(\frac{F_0x^2}{4} - \frac{F_0Lx}{8}\right). \quad \text{(G.164)}$$

$$u_{z,\max} = \frac{F_0L^3}{192EI_y}, \varphi_{y,\max} = \varphi_y(\tfrac{L}{4}) = -\frac{F_0L^2}{64EI_y}. \quad \text{(G.165)}$$

(b) Distributed load case
Boundary conditions: $u_z(0) = 0$, $\varphi_y(0) = 0$, $Q_z(0) = \frac{q_0L}{2}$, $\varphi_y(\frac{L}{2}) = 0$.

$$u_z(x) = \frac{1}{EI_y}\left(\frac{q_0x^4}{24} - \frac{q_0Lx^3}{12} + \frac{q_0L^2x^2}{24}\right), \quad \text{(G.166)}$$

$$\varphi_y(x) = -\frac{1}{EI_y}\left(\frac{q_0x^3}{6} - \frac{q_0Lx^2}{4} + \frac{q_0L^2x}{12}\right). \quad \text{(G.167)}$$

$$u_{z,\max} = \frac{q_0L^4}{384EI_y}, \varphi_{y,\max} = \varphi_y\left(\tfrac{3-\sqrt{3}}{6}L\right) = -\frac{\sqrt{3}q_0L^3}{216EI_y}. \quad \text{(G.168)}$$

3.17 Cantilever Bernoulli beam with triangular shaped distributed load: analytical solution

$$q_z(x) = q_0\left(1 - \frac{x}{L}\right), \quad \text{(G.169)}$$

$$Q_z(x) = -\frac{q_0L}{2}\left(1 - \frac{x}{L}\right)^2, \quad \text{(G.170)}$$

$$M_y(x) = \frac{q_0L^2}{6}\left(1 - \frac{x}{L}\right)^3. \quad \text{(G.171)}$$

Thus, the bending line is obtained as:

$$u_z(x) = -\frac{q_0L^4}{EI_y}\left(\frac{1}{120}\left[1 - \frac{x}{L}\right]^5 + \frac{1}{24}\left[\frac{x}{L}\right]^1 - \frac{1}{120}\right). \quad \text{(G.172)}$$

3.18 Weighted residual method based on general formulation of partial differential equation

$$\int_\Omega W^{\mathrm{T}}\mathcal{L}_2^{\mathrm{T}}(EI_y\mathcal{L}_2(u_z))\mathrm{d}\Omega = \int_\Gamma \left\{-(\mathcal{L}_1W)^{\mathrm{T}}\left[(EI_y\mathcal{L}_2(u_z))\right]n_x + \right.$$
$$\left. + (W^{\mathrm{T}}\mathcal{L}_1^{\mathrm{T}})\left[EI_y\mathcal{L}_2(u_x)\right]n_x\right\}\mathrm{d}\Gamma + \int_\Omega W^{\mathrm{T}}q_z\mathrm{d}\Omega, \quad \text{(G.173)}$$

or with $EI_y = \text{const.}$ and $\mathcal{L}_3 = \frac{\mathrm{d}^3}{\mathrm{d}x^3}$:

$$EI_y \int_{\Omega} (W^{\mathrm{T}}\mathcal{L}_2^{\mathrm{T}})(\mathcal{L}_2(u_z))\mathrm{d}\Omega = EI_y \int_{\Gamma} \Big\{-W^{\mathrm{T}}\mathcal{L}_3(u_z) + \\ + (\mathcal{L}_1 W)^{\mathrm{T}}(\mathcal{L}_2(u_x))\Big\}\, n_x \mathrm{d}\Gamma + \int_{\Omega} W^{\mathrm{T}} q_z \mathrm{d}\Omega\,. \tag{G.174}$$

3.19 Weighted residual method with arbitrary distributed load for a beam

$$\int_0^L W^{\mathrm{T}}(x)\left(EI_y \frac{\mathrm{d}^4 u_z(x)}{\mathrm{d}x^4} - q_z(x)\right)\mathrm{d}x = 0 \tag{G.175}$$

$$\int_0^L EI_y \frac{\mathrm{d}^2 W^{\mathrm{T}}}{\mathrm{d}x^2}\frac{\mathrm{d}^2 u_z}{\mathrm{d}x^2}\mathrm{d}x = \int_0^L W^{\mathrm{T}} q_z(x)\mathrm{d}x + \left[-W^{\mathrm{T}}\frac{\mathrm{d}^3 u_z}{\mathrm{d}x^3} + \frac{\mathrm{d}W^{\mathrm{T}}}{\mathrm{d}x}\frac{\mathrm{d}^2 u_z}{\mathrm{d}x^2}\right]_0^L \tag{G.176}$$

$$\cdots = \delta \boldsymbol{u}_{\mathrm{p}}^{\mathrm{T}} \int_0^L \boldsymbol{N}^{\mathrm{T}} q_z(x)\mathrm{d}x + \cdots \tag{G.177}$$

$$\cdots = \int_0^L \begin{bmatrix} N_{1u} \\ N_{1\varphi} \\ N_{2u} \\ N_{2\varphi} \end{bmatrix} q_z(x)\mathrm{d}x + \cdots \tag{G.178}$$

The additional expression on the right-hand side results in the equivalent nodal loads for a distributed load according to Eq. (3.93).

3.20 Stiffness matrix for bending in the x–y plane

For bending in the x–y plane it must be considered that the rotation is defined by $\varphi_z(x) = \frac{\mathrm{d}u_y(x)}{\mathrm{d}x}$. Thus, the following interpolation functions can be derived:

$$N_{1u}^{xy} = 1 - 3\left(\frac{x}{L}\right)^2 + 2\left(\frac{x}{L}\right)^3, \tag{G.179}$$

$$N_{1\varphi}^{xy} = +x - 2\frac{x^2}{L} + \frac{x^3}{L^2}, \tag{G.180}$$

$$N_{2u}^{xy} = 3\left(\frac{x}{L}\right)^2 - 2\left(\frac{x}{L}\right)^3, \tag{G.181}$$

$$N_{2\varphi}^{xy} = -\frac{x^2}{L} + \frac{x^3}{L^2}. \tag{G.182}$$

A comparison with the interpolation functions for bending in the x–y plane according to Eqs. (3.57)–(3.60) yields that the interpolation functions for the rotations are multiplied by (-1).

3.21 Investigation of displacement and slope consistency along boundaries

Evaluation of Eqs. (3.108) and (3.109) for $x = 0$ gives:

$$u_y(x=0) = a_0\,,\; \varphi_y(x=0) = -a_1\,. \tag{G.183}$$

The two DOF at node 1 allow to uniquely define the displacement and slope and both quantities are continuous along an interelement boundary. The same result is obtained at the right-hand boundary ($x = L$) under consideration of all four elemental DOF.

3.22 Bending moment distribution for a cantilever beam

The solution of the 2×2 reduced system of equations can be obtained as:

$$\begin{bmatrix} u_{2z} \\ \varphi_{2y} \end{bmatrix} = \frac{L}{12EI_y}\begin{bmatrix} -4L^2F_0 \\ +6LF_0 \end{bmatrix}\,. \tag{G.184}$$

$$M_y(\xi) = EI_y\left\{0 + 0 + \frac{6}{L^2}[\,\xi\,]\,u_{2z} + \frac{1}{L}[1+3\xi]\,\varphi_{2y}\right\}\,. \tag{G.185}$$

Or under consideration of the result for u_{2z} and φ_{2y}:

$$M_y(\xi) = \frac{F_0L}{2}(1-\xi)\,. \tag{G.186}$$

Numerical values at the integration points:

$$M_y(\xi_1) = 443.65\,, \tag{G.187}$$

$$M_y(\xi_2) = 250.00\,, \tag{G.188}$$

$$M_y(\xi_3) = 56.35\,. \tag{G.189}$$

3.23 Beam with variable cross-sectional area

The second moments of area are obtained as:

$$I_z(x) = \frac{\pi}{64}\left(d_1 + (d_2 - d_1)\frac{x}{L}\right)^4 \text{ (circle)}\,, \tag{G.190}$$

$$I_z(x) = \frac{b}{12}\left(d_1 + (d_2 - d_1)\frac{x}{L}\right)^3 \text{ (rectangle)}\,. \tag{G.191}$$

For the circular and rectangular cross section one obtains:

$$\boldsymbol{k}^{\mathrm{e}} = \frac{\pi E}{64L^3}\begin{bmatrix} \frac{12(11d_2^4+11d_1^4+5d_2^3d_1+3d_2^2d_1^2+5d_2\,d_1^3)}{35} & -\frac{2(19d_2^4+47\,d_1^4+8d_2^3d_1+9d_2^2d_1^2+22d_2\,d_1^3)L}{35} & -\frac{12(11d_2^4+11d_1^4+5d_2^3d_1+3d_2^2d_1^2+5d_2\,d_1^3)}{35} & -\frac{2(47d_2^4+19d_1^4+22d_2^3d_1+9d_2^2d_1^2+8d_2\,d_1^3)L}{35} \\ -\frac{2(19\,d_2^4+47\,d_1^4+8\,d_2^3d_1+9\,d_2^2d_1^2+22\,d_2\,d_1^3)L}{35} & \frac{4(3\,d_2^4+17\,d_1^4+2\,d_2^3d_1+4\,d_2^2d_1^2+9\,d_2\,d_1^3)L^2}{35} & +\frac{2(19\,d_2^4+47\,d_1^4+8\,d_2^3d_1+9\,d_2^2d_1^2+22\,d_2\,d_1^3)L}{35} & \frac{2(13\,d_2^4+13\,d_1^4+4\,d_2^3d_1+d_2^2d_1^2+4\,d_2\,d_1^3)L^2}{35} \\ -\frac{12(11\,d_2^4+11\,d_1^4+5\,d_2^3d_1+3\,d_2^2d_1^2+5\,d_2\,d_1^3)}{35} & +\frac{2(19\,d_2^4+47\,d_1^4+8\,d_2^3d_1+9\,d_2^2d_1^2+22\,d_2\,d_1^3)L}{35} & \frac{12(11\,d_2^4+11\,d_1^4+5\,d_2^3d_1+3\,d_2^2d_1^2+5\,d_2\,d_1^3)}{35} & +\frac{2(47\,d_2^4+19\,d_1^4+22\,d_2^3d_1+9\,d_2^2d_1^2+8\,d_2\,d_1^3)L}{L} \\ -\frac{2(47\,d_2^4+19\,d_1^4+22\,d_2^3d_1+9\,d_2^2d_1^2+8\,d_2\,d_1^3)L}{35} & \frac{2(13\,d_2^4+13\,d_1^4+4\,d_2^3d_1+d_2^2d_1^2+4\,d_2\,d_1^3)L^2}{35} & +\frac{2(47\,d_2^4+19\,d_1^4+22\,d_2^3d_1+9\,d_2^2d_1^2+8\,d_2\,d_1^3)L}{35} & \frac{4(17\,d_2^4+3\,d_1^4+9\,d_2^3d_1+4\,d_2^2d_1^2+2\,d_2\,d_1^3)L^2}{35} \end{bmatrix} \tag{G.192}$$

$$\boldsymbol{k}^{\mathrm{e}} = \frac{bE}{12L^3}\begin{bmatrix} \frac{3(7\,d_2^3+3\,d_2^2d_1+3\,d_2\,d_1^2+7\,d_1^3)}{5} & -\frac{3(2\,d_2^3+d_2^2d_1+2\,d_2\,d_1^2+5\,d_1^3)L}{5} & -\frac{3(7\,d_2^3+3\,d_2^2d_1+3\,d_2\,d_1^2+7\,d_1^3)}{5} & -\frac{3(5\,d_2^3+2\,d_2^2d_1+d_2\,d_1^2+2\,d_1^3)L}{5} \\ -\frac{3(2\,d_2^3+d_2^2d_1+2\,d_2\,d_1^2+5\,d_1^3)L}{5} & \frac{(2\,d_2^3+2\,d_2^2d_1+5\,d_2\,d_1^2+11\,d_1^3)L^2}{5} & +\frac{3(2\,d_2^3+d_2^2d_1+2\,d_2\,d_1^2+5\,d_1^3)L}{5} & \frac{(4\,d_2^3+d_2^2d_1+d_2\,d_1^2+4\,d_1^3)L^2}{5} \\ -\frac{3(7\,d_2^3+3\,d_2^2d_1+3\,d_2\,d_1^2+7\,d_1^3)}{5} & +\frac{3(2\,d_2^3+d_2^2d_1+2\,d_2\,d_1^2+5\,d_1^3)L}{5} & \frac{3(7\,d_2^3+3\,d_2^2d_1+3\,d_2\,d_1^2+7\,d_1^3)}{5} & +\frac{3(5\,d_2^3+2\,d_2^2d_1+d_2\,d_1^2+2\,d_1^3)L}{5} \\ -\frac{3(5\,d_2^3+2\,d_2^2d_1+d_2\,d_1^2+2\,d_1^3)L}{5} & \frac{(4\,d_2^3+d_2^2d_1+d_2\,d_1^2+4\,d_1^3)L^2}{5} & +\frac{3(5\,d_2^3+2\,d_2^2d_1+d_2\,d_1^2+2\,d_1^3)L}{5} & \frac{(11\,d_2^3+5\,d_2^2d_1+2\,d_2\,d_1^2+2\,d_1^3)L^2}{5} \end{bmatrix} \tag{G.193}$$

Table G.2 Equivalent nodal loads for quadratic distributed load

$q(x) = -q_0 x^2$	$q(x) = -q_0 \left(\frac{x}{L}\right)^2$
$F_{1y} = -\frac{q_0 L^3}{15}$	$F_{1y} = -\frac{q_0 L}{15}$
$M_{1z} = +\frac{q_0 L^4}{60}$	$M_{1z} = +\frac{q_0 L^2}{60}$
$F_{2y} = -\frac{4 q_0 L^3}{15}$	$F_{2y} = -\frac{4 q_0 L}{15}$
$M_{1z} = -\frac{q_0 L^4}{30}$	$M_{1z} = -\frac{q_0 L^2}{30}$

3.24 Equivalent nodal loads for quadratic distributed load

See Table G.2

3.25 Beam with variable cross section loaded by a single force

Analytical solution:

$$EI_y(x)\frac{\mathrm{d}^2 u_z(x)}{\mathrm{d}x^2} = -M_y(x)\,, \tag{G.194}$$

$$\frac{E\pi h^4}{64}\left(2-\frac{x}{L}\right)^4 \frac{\mathrm{d}^2 u_z(x)}{\mathrm{d}x^2} = -F_0(L-x)\,. \tag{G.195}$$

$$u_z(x) = \frac{F_0 L}{E\pi h^4}\left(\frac{64L^3}{2(-2L+x)} + \frac{64L^4}{6(-2L+x)}\right) + \frac{16L}{3}x + \frac{40L^2}{3}\,. \tag{G.196}$$

$$u_z(L) = -\frac{8}{3}\frac{F_0 L^3}{E\pi h^4} \approx -2.666667\frac{F_0 L^3}{E\pi h^4}\,. \tag{G.197}$$

Finite element solution:

$$u_z(L) = -\frac{7360}{2817}\frac{F_0 L^3}{E\pi h^4} \approx -2.612709\frac{F_0 L^3}{E\pi h^4}\,. \tag{G.198}$$

3.26 Beam on elastic foundation: stiffness matrix

$$\int_0^L W(x)\left(EI_y\frac{\mathrm{d}^4 u_z}{\mathrm{d}x^4} + k u_z\right)\mathrm{d}x = 0\,, \tag{G.199}$$

$$\boldsymbol{K}^{\mathrm{e}} = \cdots + k\int_0^L \boldsymbol{N}(x)\boldsymbol{N}(x)^{\mathrm{T}}\mathrm{d}x\,, \tag{G.200}$$

$$\boldsymbol{K}^{\mathrm{e}} = \frac{EI_y}{L^3}\begin{bmatrix} 12 & -6L & -12 & -6L \\ -6L & 4L^2 & 6L & 2L^2 \\ -12 & 6L & 12 & 6L \\ -6L & 2L^2 & 6L & 4L^2 \end{bmatrix} + \frac{kL}{420}\begin{bmatrix} 156 & -22L & 54 & 13L \\ -22L & 4L^2 & -13L & -3L^2 \\ 54 & -13L & 156 & 22L \\ 13L & -3L^2 & 22L & 4L^2 \end{bmatrix}. \tag{G.201}$$

3.27 Beam on elastic foundation: single force case

The reduced system of equations:

$$\begin{bmatrix} 12\frac{EI_y}{L^3} + \frac{156}{420}kL & +6\frac{EI_yL}{L^3} + \frac{22}{420}kL^2 \\ +6\frac{EI_yL}{L^3} + \frac{22}{420}kL^2 & 4\frac{EI_yL^2}{L^3} + \frac{4}{420}kL^3 \end{bmatrix}\begin{bmatrix} u_{2z} \\ \varphi_{2y} \end{bmatrix} = \begin{bmatrix} -F_0 \\ 0 \end{bmatrix}. \tag{G.202}$$

Deflection and rotation of the beam at $x = L$:

$$u_{2z} = -\frac{12\left(420EI_y + kL^4\right)L^3F_0}{15120E^2{I_y}^2 + 1224EI_ykL^4 + k^2L^8}, \tag{G.203}$$

$$\varphi_{2y} = +\frac{6\left(1260EI_y + 11\,kL^4\right)L^2F_0}{15120E^2{I_y}^2 + 1224EI_ykL^4 + k^2L^8}. \tag{G.204}$$

Special case 1, $k = 0$:

$$u_{2z}|_{k=0} = -\frac{F_0L^3}{3EI_y}, \quad \varphi_{2y}|_{k=0} = +\frac{F_0L^2}{2EI_y}. \tag{G.205}$$

Special case 2, $EI_y = 0$:

$$u_{2z}|_{EI_y=0} = -\frac{12F_0}{kL}, \quad \varphi_{2y}|_{EI_y=0} = +\frac{66F_0}{kL^2}. \tag{G.206}$$

Comparison between FE and analytical solution:

$$u_z(L)_{\mathrm{FE}} = -0.253994 \times F_0\,, \tag{G.207}$$

$$u_z(L)_{\mathrm{analyt}} = -0.254166 \times F_0\,. \tag{G.208}$$

3.28 Beam on nonlinear elastic foundation: stiffness matrix

$$k(u_y) = k_0 - \frac{u_y}{u_1}(k_0 - k_1) = k_0\Big(1 - u_y \times \underbrace{\frac{1 - k_1/k_0}{u_1}}_{\alpha_{01}}\Big), \tag{G.209}$$

$$\boldsymbol{K}^{\mathrm{e}} = \cdots + k_0\int_0^L \left(1 - u_y(x) \times \alpha_{01}\right)\boldsymbol{N}(x)\boldsymbol{N}(x)^{\mathrm{T}}\mathrm{d}x\,, \tag{G.210}$$

where $u_y(x)$ can be taken from Table 3.10.

$$
\boldsymbol{K}^{\mathrm{e}} = \frac{EI_z}{L^3}\begin{bmatrix} 12 & -6L & -12 & -6L \\ -6L & 4L^2 & 6L & 2L^2 \\ -12 & 6L & 12 & 6L \\ -6L & 2L^2 & 6L & 4L^2 \end{bmatrix} + \frac{k_0 L}{420}\begin{bmatrix} 156 & -22L & 54 & 13L \\ -22L & 4L^2 & -13L & -3L^2 \\ 54 & -13L & 156 & 22L \\ 13L & -3L^2 & 22L & 4L^2 \end{bmatrix} -
$$

$$
- \frac{k_0 L \times \alpha_{01}}{2520}\begin{bmatrix} -43\varphi_{2z}L + 162u_{2y} + 774u_{1y} + 97\varphi_{1z}L & -L\left(-9\varphi_{2z}L + 35u_{2y} + 97u_{1y} + 16\varphi_{1z}L\right) & -35\varphi_{2z}L + 162u_{2y} + 162u_{1y} + 35\varphi_{1z}L & +L\left(-8\varphi_{2z}L + 35u_{2y} + 43u_{1y} + 9\varphi_{1z}L\right) \\ -L\left(-9\varphi_{2}zL + 35u_{2y} + 97u_{1y} + 16\varphi_{1z}L\right) & L^2\left(-2\varphi_{2z}L + 8u_{2y} + 16u_{1y} + 3\varphi_{1z}L\right) & -L\left(8\varphi_{1z}L + 35u_{1y} + 43u_{2y} - 9\varphi_{2z}L\right) & -L^2\left(2\varphi_{1z}L + 9u_{1y} + 9u_{2y} - 2\varphi_{2z}L\right) \\ -35\varphi_{2z}L + 162u_{2y} + 162u_{1y} + 35\varphi_{1z}L & -L\left(8\varphi_{1z}L + 35u_{1y} + 43u_{2y} - 9\varphi_{2z}L\right) & 43\varphi_{1z}L + 162u_{1y} + 774u_{2y} - 97\varphi_{2z}L & +L\left(9\varphi_{1z}L + 35u_{1y} + 97u_{2y} - 16\varphi_{2z}L\right) \\ +L\left(-8\varphi_{2z}L + 35u_{2y} + 43u_{1y} + 9\varphi_{1z}L\right) & -L^2\left(2\varphi_{1z}L + 9u_{1y} + 9u_{2y} - 2\varphi_{2z}L\right) & +L\left(9\varphi_{1z}L + 35u_{1y} + 97u_{2y} - 16\varphi_{2z}L\right) & L^2\left(2\varphi_{1z}L + 8u_{1y} + 16u_{2y} - 3\varphi_{2z}L\right) \end{bmatrix}. \tag{G.211}
$$

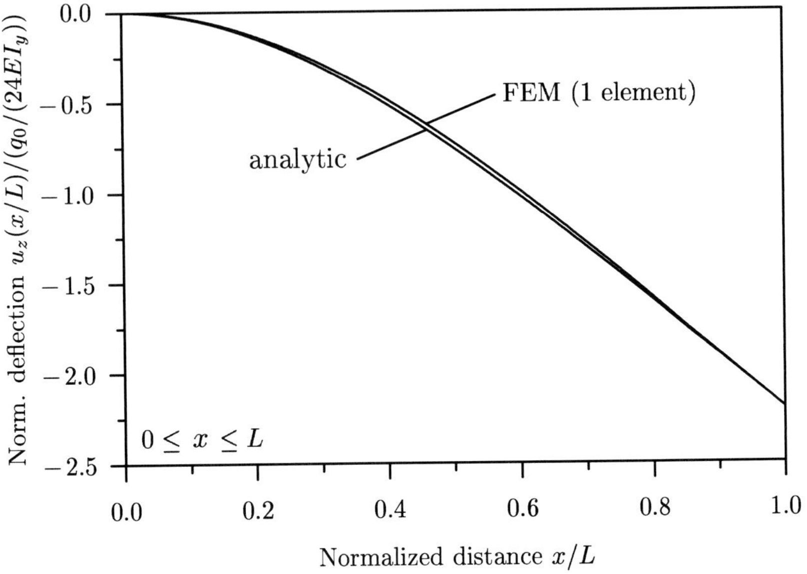

Fig. G.10 Comparison of finite element and analytical solution for a cantilever beam with triangular shaped distributed load

3.29 Cantilever beam with triangular shaped distributed load

Finite element solution:

$$\frac{u_z^{\mathrm{e}}\left(\frac{x}{L}\right)}{\left(\frac{q_0}{24EI_y}\right)} = -\left(\frac{18}{5}L^4\left(\frac{x}{L}\right)^2 - \frac{7}{5}L^4\left(\frac{x}{L}\right)^3\right). \tag{G.212}$$

Analytical solution:

$$\frac{u_z\left(\frac{x}{L}\right)}{\left(\frac{q_0}{24EI_y}\right)} = -\left(4L^4\left(\frac{x}{L}\right)^2 - 2L^4\left(\frac{x}{L}\right)^3 + \frac{L^4}{5}\left(\frac{x}{L}\right)^5\right). \tag{G.213}$$

A comparison between the finite element and the analytical solution is shown in Fig. G.10.

3.30 Cantilever beam with triangular shaped distributed load and roller support

Bending line:

$$u_z(x) = \frac{q_0 L^4}{EI}\left(-\frac{1}{120}\left[\frac{x}{L}\right]^5 + \frac{3}{80}\left[\frac{x}{L}\right]^3 - \frac{7}{240}\left[\frac{x}{L}\right]^2\right). \tag{G.214}$$

Bending moment:

$$M_y(x) = q_0 L^2 \left(\frac{1}{6}\left[\frac{x}{L}\right]^3 - \frac{9}{40}\left[\frac{x}{L}\right]^1 + \frac{7}{120} \right). \tag{G.215}$$

Rotation (FE and the analytical approach result in the same value at the right-hand end):

$$\varphi_y(x = L) = -\frac{q_0 L^3}{8EI}. \tag{G.216}$$

3.31 Finite element approximation with a single beam element

(a) FE solution for the simply supported beam

$$u_z(0) = 0\,, \quad u_z(L) = 0\,, \tag{G.217}$$

$$u_z(a) = -\frac{F_0}{6EI} \times \frac{2a^2b^2(a^2 + ab + b^2)}{(a + b)^3}. \tag{G.218}$$

The analytical solution gives at $x = 0$ and $x = L$ the same solution as the FE approach but at the location of the force: $u_z(a) = -\frac{F_0 a^2 b^2}{3EI(a+b)}$.
(b) FE solution for the cantilever beam

$$u_z(0) = 0\,, \quad u_z(L) = -\frac{F_0}{12EI} \times (4a^3 + 6a^2 b)\,, \tag{G.219}$$

$$u_z(a) = -\frac{F_0}{EI} \times \frac{a^4(a^2 + 3ab + 3b^2)}{3(a + b)^3}. \tag{G.220}$$

The analytical solution gives at $x = 0$ and $x = L$ the same solution as the FE approach but at the location of the force: $u_z(a) = -\frac{F_0 a^3}{3EI}$.

3.32 Cantilever beam: moment curvature relationship

$$u_z(0) = 0\,, \quad u_z(L) = -\frac{5F_0L^3}{6EI_y}, \quad u_z(2L) = -\frac{8F_0L^3}{3EI_y}, \tag{G.221}$$

$$\varphi_y(0) = 0\,, \quad \varphi_y(L) = +\frac{3F_0L^2}{2EI_y}, \quad \varphi_y(2L) = +\frac{2F_0L^2}{EI_y}, \tag{G.222}$$

$$\kappa_y(0) = +\frac{2F_0L}{EI_y}, \quad \kappa_y(L) = +\frac{F_0L}{EI_y}, \quad \kappa_y(2L) = 0\,, \tag{G.223}$$

$$M_y(0) = +2F_0L\,, \quad M_y(L) = +F_0L\,, \quad M_y(2L) = 0\,, \tag{G.224}$$

$$\frac{M_y}{\kappa_y}(0) = EI_y\,, \quad \frac{M_y}{\kappa_y}(L) = EI_y\,, \quad \frac{M_y}{\kappa_y}(2L) = \frac{0}{0}\,. \tag{G.225}$$

The finite element solution is identical with the analytical solution.

3.33 Fixed-end beam with distributed load and displacement boundary condition

Reduced system of equations (one possible formulation):

$$\frac{EI}{L^3}\begin{bmatrix} \frac{L^3}{EI} & 0 \\ 0 & 8L^2 \end{bmatrix}\begin{bmatrix} u_{2Z} \\ \varphi_{2Y} \end{bmatrix} = \begin{bmatrix} -u_0 \\ -\frac{q_0L^2}{20} + \frac{q_0L^2}{20} \end{bmatrix}. \tag{G.226}$$

Solution:

$$\begin{bmatrix} u_{2Z} \\ \varphi_{2Y} \end{bmatrix} = \begin{bmatrix} -u_0 \\ 0 \end{bmatrix}. \tag{G.227}$$

3.34 Fixed-end beam with distributed load and single force load

Reduced system of equations:

$$\frac{EI}{L^3}\begin{bmatrix} 24 & 0 \\ 0 & 8L^2 \end{bmatrix}\begin{bmatrix} u_{2Z} \\ \varphi_{2Y} \end{bmatrix} = \begin{bmatrix} -F_0 - \frac{14q_0L}{20} \\ -\frac{q_0L^2}{20} + \frac{q_0L^2}{20} \end{bmatrix}. \tag{G.228}$$

Solution:

$$\begin{bmatrix} u_{2Z} \\ \varphi_{2Y} \end{bmatrix} = -\frac{L^3}{24EI}\begin{bmatrix} F_0 + \frac{7q_0L}{10} \\ 0 \end{bmatrix}. \tag{G.229}$$

3.35 Cantilever stepped beam with two sections

(a) $I_{\text{I}} = 2I$ and $I_{\text{II}} = I$:

$$u_{2z} = -\frac{5F_0L^3}{12EI}\,, \qquad u_{3z} = -\frac{3F_0L^3}{2EI}\,, \tag{G.230}$$

$$\varphi_{2y} = +\frac{3F_0L^2}{4EI}\,, \qquad \varphi_{3y} = +\frac{5F_0L^2}{4EI}\,, \tag{G.231}$$

$$F^{\text{R}}_{1z} = F_0\,, \quad M^{\text{R}}_{1y} = -2LF_0\,, \tag{G.232}$$

$$M_1 = +2LF_0\,, \quad M_2 = +LF_0\,, \quad M_3 = 0\,, \tag{G.233}$$

$$M_{\mathrm{I}} = -0.5F_0L(\xi - 3)\,,\ M_{\mathrm{II}} = -0.5F_0L(\xi - 1)\,, \tag{G.234}$$

$$Q_{\mathrm{I}} = -F_0\,,\ Q_{\mathrm{II}} = -F_0\,, \tag{G.235}$$

$$\sigma_{\mathrm{I}}(0) = +\frac{LF_0}{J}z\,,\ \sigma_{\mathrm{I}}(0.5L) = +\frac{3LF_0}{4J}z\,,\ \sigma_{\mathrm{I}}(L) = +\frac{LF_0}{2J}z\,, \tag{G.236}$$

$$\sigma_{\mathrm{II}}(L) = +\frac{LF_0}{J}z\,,\ \sigma_{\mathrm{II}}(1.5L) = +\frac{LF_0}{2J}z\,,\ \sigma_{\mathrm{II}}(2L) = 0\,z\,. \tag{G.237}$$

(b) $I_{\mathrm{I}} = I_{\mathrm{II}} = I$:

$$u_{2z} = -\frac{5F_0L^3}{6EI}\,, \qquad u_{3z} = -\frac{8F_0L^3}{3EI}\,, \tag{G.238}$$

$$\varphi_{2y} = +\frac{3F_0L^2}{2EI}\,, \qquad \varphi_{3y} = +\frac{2F_0L^2}{EI}\,, \tag{G.239}$$

$$F_{1z}^{\mathrm{R}} = F_0\,,\ M_{1y}^{\mathrm{R}} = -2LF_0\,, \tag{G.240}$$

$$M_1 = +2LF_0\,,\ M_2 = +LF_0\,,\ M_3 = 0\,, \tag{G.241}$$

$$M_{\mathrm{I}} = -0.5F_0L(\xi - 3)\,,\ M_{\mathrm{II}} = -0.5F_0L(\xi - 1)\,, \tag{G.242}$$

$$Q_{\mathrm{I}} = -F_0\,,\ Q_{\mathrm{II}} = -F_0\,, \tag{G.243}$$

$$\sigma_{\mathrm{I}}(0) = +\frac{2LF_0}{J}z\,,\ \sigma_{\mathrm{I}}(0.5L) = +\frac{3LF_0}{2J}z\,,\ \sigma_{\mathrm{I}}(L) = +\frac{LF_0}{J}z\,, \tag{G.244}$$

$$\sigma_{\mathrm{II}}(L) = +\frac{LF_0}{J}z\,,\ \sigma_{\mathrm{II}}(1.5L) = +\frac{LF_0}{2J}z\,,\ \sigma_{\mathrm{II}}(2L) = 0\,z\,. \tag{G.245}$$

3.36 Cantilever stepped beam with three sections

A single stiffness matrix of an Euler–Bernoulli beam element can be stated as:

$$\boldsymbol{K}_i^{\mathrm{e}} = \frac{E_iI_i}{L_i^3}\begin{bmatrix} 12 & -6L_i & -12 & -6L_i \\ -6L_i & 4L_i^2 & 6L_i & 2L_i^2 \\ -12 & 6L_i & 12 & 6L_i \\ -6L_i & 2L_i^2 & 6L_i & 4L_i^2 \end{bmatrix}. \tag{G.246}$$

Assembling the three elemental stiffness matrices $\boldsymbol{K}_i^{\mathrm{e}}$ under the consideration of $E_{\mathrm{I}} = E_{\mathrm{II}} = E_{\mathrm{III}} = E$ and $L_{\mathrm{I}} = L_{\mathrm{II}} = L_{\mathrm{III}} = \frac{L}{3}$ results in the following global stiffness matrix:

$$\boldsymbol{K} =$$
$$E \begin{bmatrix} \left(\frac{324 I_{\text{II}}}{L^3} + \frac{324 I_{\text{I}}}{L^3}\right) & \left(\frac{54 I_{\text{I}}}{L^2} - \frac{54 I_{\text{II}}}{L^2}\right) & -\frac{324 I_{\text{II}}}{L^3} & -\frac{54 I_{\text{II}}}{L^2} & 0 & 0 \\ \left(\frac{54 I_{\text{I}}}{L^2} - \frac{54 I_{\text{II}}}{L^2}\right) & \left(\frac{12 I_{\text{II}}}{L} + \frac{12 I_{\text{I}}}{L}\right) & \frac{54 I_{\text{II}}}{L^2} & \frac{6 I_{\text{II}}}{L} & 0 & 0 \\ -\frac{324 I_{\text{II}}}{L^3} & \frac{54 I_{\text{II}}}{L^2} & \left(\frac{324 I_{\text{III}}}{L^3} + \frac{324 I_{\text{II}}}{L^3}\right) & \left(\frac{54 I_{\text{II}}}{L^2} - \frac{54 I_{\text{III}}}{L^2}\right) & -\frac{324 I_{\text{III}}}{L^3} & -\frac{54 I_{\text{III}}}{L^2} \\ -\frac{54 I_{\text{II}}}{L^2} & \frac{6 I_{\text{II}}}{L} & \left(\frac{54 I_{\text{II}}}{L^2} - \frac{54 I_{\text{III}}}{L^2}\right) & \left(\frac{12 I_{\text{III}}}{L} + \frac{12 I_{\text{II}}}{L}\right) & \frac{54 I_{\text{III}}}{L^2} & \frac{6 I_{\text{III}}}{L} \\ 0 & 0 & -\frac{324 I_{\text{III}}}{L^3} & \frac{54 I_{\text{III}}}{L^2} & \frac{324 I_{\text{III}}}{L^3} & \frac{54 I_{\text{III}}}{L^2} \\ 0 & 0 & -\frac{54 I_{\text{III}}}{L^2} & \frac{6 I_{\text{III}}}{L} & \frac{54 I_{\text{III}}}{L^2} & \frac{12 I_{\text{III}}}{L} \end{bmatrix} . \tag{G.247}$$

The last equation already considered that all degrees of freedom are zero at node 1. The solution of the linear system of equations can be obtained, for example, by inverting the global stiffness matrix and multiplying with the right-hand side, i.e. $\boldsymbol{u} = \boldsymbol{K}^{-1} \boldsymbol{f}$, to obtain the column matrix of nodal unknowns:

$$\begin{bmatrix} u_{2Z} \\ \varphi_{2Y} \\ u_{3Z} \\ \varphi_{3Y} \\ u_{4Z} \\ \varphi_{4Y} \end{bmatrix} = \begin{bmatrix} -\frac{4 F_0 L^3}{81 E I_{\text{I}}} \\ \frac{5 F_0 L^2}{18 E I_{\text{I}}} \\ -\frac{F_0 (23 I_{\text{II}} + 5 I_{\text{I}}) L^3}{162 E I_I I_{\text{II}}} \\ \frac{F_0 (5 I_{\text{II}} + 3 I_{\text{I}}) L^2}{18 E I_{\text{I}} I_{\text{II}}} \\ -\frac{F_0 (19 I_{\text{II}} I_{\text{III}} + 7 I_{\text{I}} I_{\text{III}} + I_I I_{\text{II}}) L^3}{81 E I_I I_{\text{II}} I_{\text{III}}} \\ \frac{F_0 (5 I_{\text{II}} I_{\text{III}} + 3 I_{\text{I}} I_{\text{III}} + I_I I_{\text{II}}) L^2}{18 E I_{\text{I}} I_{\text{II}} I_{\text{III}}} \end{bmatrix} . \tag{G.248}$$

3.37 Simply supported stepped beam with three sections

A single stiffness matrix of an EULER–BERNOULLI beam element can be stated as:

$$\boldsymbol{K}_i^{\text{e}} = \frac{E_i I_i}{L_i^3} \begin{bmatrix} 12 & -6L_i & -12 & -6L_i \\ -6L_i & 4L_i^2 & 6L_i & 2L_i^2 \\ -12 & 6L_i & 12 & 6L_i \\ -6L_i & 2L_i^2 & 6L_i & 4L_i^2 \end{bmatrix} . \tag{G.249}$$

Assembling the four elemental stiffness matrices $\boldsymbol{K}_i^{\text{e}}$ under the consideration of $E_{\text{I}} = \cdots = E_{\text{IV}} = E$, $L_{\text{I}} = \cdots = L_{\text{IV}} = \frac{L}{4}$, $I_{\text{IV}} = I_{\text{I}}$, and $I_{\text{III}} = I_{\text{II}}$ results in the following global stiffness matrix:

$\boldsymbol{K} =$

$$
E\begin{bmatrix}
\frac{16I_{\text{I}}}{L} & \frac{96I_{\text{I}}}{L^2} & \frac{8I_{\text{I}}}{L} & 0 & 0 & 0 & 0 & 0 \\
\frac{96I_{\text{I}}}{L^2} & \left(\frac{768I_{\text{II}}}{L^3}+\frac{768I_{\text{I}}}{L^3}\right) & \left(\frac{96I_{\text{I}}}{L^2}-\frac{96I_{\text{II}}}{L^2}\right) & -\frac{768I_{\text{II}}}{L^3} & -\frac{96I_{\text{II}}}{L^2} & 0 & 0 & 0 \\
\frac{8I_{\text{I}}}{L} & \left(\frac{96I_{\text{I}}}{L^2}-\frac{96I_{\text{II}}}{L^2}\right) & \left(\frac{16I_{\text{II}}}{L}+\frac{16I_{\text{I}}}{L}\right) & \frac{96I_{\text{II}}}{L^2} & \frac{8I_{\text{II}}}{L} & 0 & 0 & 0 \\
0 & -\frac{768I_{\text{II}}}{L^3} & \frac{96I_{\text{II}}}{L^2} & \frac{1536I_{\text{II}}}{L^3} & 0 & -\frac{768I_{\text{II}}}{L^3} & -\frac{96I_{\text{II}}}{L^2} & 0 \\
0 & -\frac{96I_{\text{II}}}{L^2} & \frac{8I_{\text{II}}}{L} & 0 & \frac{32I_{\text{II}}}{L} & \frac{96I_{\text{II}}}{L^2} & \frac{8I_{\text{II}}}{L} & 0 \\
0 & 0 & 0 & -\frac{768I_{\text{II}}}{L^3} & \frac{384I_{\text{II}}}{L^3} & \left(\frac{768I_{\text{II}}}{L^3}+\frac{768I_{\text{I}}}{L^3}\right) & \left(\frac{96I_{\text{II}}}{L^2}-\frac{96I_{\text{I}}}{L^2}\right) & -\frac{96I_{\text{I}}}{L^2} \\
0 & 0 & 0 & -\frac{96I_{\text{II}}}{L^2} & \frac{8I_{\text{II}}}{L} & \left(\frac{96I_{\text{II}}}{L^2}-\frac{96I_{\text{I}}}{L^2}\right) & \left(\frac{16I_{\text{II}}}{L}+\frac{16I_{\text{I}}}{L}\right) & \frac{8I_{\text{I}}}{L} \\
0 & 0 & 0 & 0 & 0 & -\frac{96I_{\text{I}}}{L^2} & \frac{8I_{\text{I}}}{L} & \frac{16I_{\text{I}}}{L}
\end{bmatrix}. \tag{G.250}
$$

The last equation already considered that all degrees of freedom are zero at node 1. The solution of the linear system of equations can be obtained, for example, by inverting the global stiffness matrix and multiplying with the right-hand side, i.e. $\boldsymbol{u} = \boldsymbol{K}^{-1}\boldsymbol{f}$, to obtain the column matrix of nodal unknowns:

$$
\begin{bmatrix} u_{2Z} \\ \varphi_{2Y} \\ u_{3Z} \\ \varphi_{3Y} \\ u_{4Z} \\ \varphi_{4Y} \end{bmatrix} =
\begin{bmatrix}
\frac{F_0\,(I_{\text{II}}+3I_{\text{I}})\,L^2}{64E\,I_{\text{I}}I_{\text{II}}} \\
-\frac{F_0\,(2I_{\text{II}}+9I_{\text{I}})\,L^3}{768E\,I_{\text{I}}I_{\text{II}}} \\
\frac{3F_0L^2}{64E\,I_{\text{II}}} \\
-\frac{F_0\,(I_{\text{II}}+7I_{\text{I}})\,L^3}{384E\,I_{\text{I}}I_{\text{II}}} \\
0 \\
-\frac{F_0\,(2I_{\text{II}}+9I_{\text{I}})\,L^3}{768E\,I_{\text{I}}I_{\text{II}}} \\
-\frac{3F_0L^2}{64E\,I_{\text{II}}} \\
-\frac{F_0\,(I_{\text{II}}+3I_{\text{I}})\,L^2}{64E\,I_{\text{I}}I_{\text{II}}}
\end{bmatrix}. \tag{G.251}
$$

3.38 Overhang beam with distributed load and single force

$$
\varphi_{1Y} = -\frac{L_{\text{I}}(-L_{\text{I}}^2 q_0 + 4L_{\text{II}}F_0)}{24EI}\;;\;\varphi_{2Y} = +\frac{L_{\text{I}}(-L_{\text{I}}^2 q_0 + 8L_{\text{II}}F_0)}{24EI}, \tag{G.252}
$$

$$
u_{3Z} = -\frac{L_{\text{II}}(-L_{\text{I}}^3 q_0 + 8L_{\text{I}}L_{\text{II}}F_0 + 8L_{\text{II}}^2F_0)}{24EI}, \tag{G.253}
$$

$$
\varphi_{3Y} = +\frac{-\,L_{\text{I}}^3 q_0 + 12L_{\text{II}}^2F_0 + 8L_{\text{I}}L_{\text{II}}F_0}{24EI}, \tag{G.254}
$$

$$F^{\mathrm{R}}_{1Z} = -\frac{L_{\mathrm{II}}F_0}{L_{\mathrm{I}}} + \frac{L_{\mathrm{I}}q_0}{2} \quad ; \quad F^{\mathrm{R}}_{2Z} = \frac{(L_{\mathrm{I}}+L_{\mathrm{II}})F_0}{L_{\mathrm{I}}} + \frac{L_{\mathrm{I}}q_0}{2}, \tag{G.255}$$

$$u_Z(X = \tfrac{1}{2}L_{\mathrm{I}}) = \frac{L_{\mathrm{I}}^2(-L_{\mathrm{I}}^2 q_0 + 6L_{\mathrm{II}}F_0)}{96EI}, \tag{G.256}$$

$$M_Y(X = \tfrac{1}{2}L_{\mathrm{I}}) = -\frac{L_{\mathrm{I}}^2 q_0}{12} + \frac{L_2 F_0}{2} \quad ; \quad Q_Z(X = \tfrac{1}{2}L_{\mathrm{I}}) = +\frac{L_{\mathrm{II}}F_0}{L_{\mathrm{I}}}, \tag{G.257}$$

$$u_Z(\tfrac{1}{2}L_{\mathrm{I}}) = \frac{L_{\mathrm{I}}L_{\mathrm{II}}(-L_{\mathrm{I}}^3 q_0 - 4L_{\mathrm{I}}^2 L_{\mathrm{II}} q_0 - 4L_{\mathrm{I}}L_{\mathrm{II}}^2 q_0 - L_{\mathrm{II}}^3 q_0 + 8L_{\mathrm{I}}L_{\mathrm{III}}F_0 + 4L_{\mathrm{II}}L_{\mathrm{III}}F_0)}{24(L_{\mathrm{I}}+L_{\mathrm{II}})EI}. \tag{G.258}$$

3.39 Beam structure with a gap

Force to close the gap:

$$F_0^* = \frac{6EI\delta}{5L^3}. \tag{G.259}$$

Deflection and rotation at the free end:

– $u_{2z}(x = L) < \delta$:

$$u_{3z} = -\frac{8L^3 F_0}{3EI}, \quad \varphi_{3z} = +\frac{2L^2 F_0}{EI}. \tag{G.260}$$

– $u_{2z}(x = L) \geq \delta$:

$$u_{3z} = -\frac{30\delta EI + 7L^3 F_0}{12EI}, \quad \varphi_{3z} = +\frac{3(2\delta EI + L^3 F_0)}{4LEI}. \tag{G.261}$$

Maximum normalized stress $\frac{\sigma_x(z_{\max})}{z_{\max}}$ at the nodes 1, 2 and 3:

– $u_{2z}(x = L) < \delta$:

$$\frac{2LF_0}{I}, \quad \frac{LF_0}{I}, \quad 0. \tag{G.262}$$

– $u_{2z}(x = L) \geq \delta$:

$$-\frac{-6\delta EI + L^3 F_0}{2L^2 I}, \quad \frac{LF_0}{I}, \quad 0. \tag{G.263}$$

3.40 Advanced example: beam element with nonlinear bending stiffness

$$a = EI_0\,, b = \frac{EI_0}{\kappa_1}(4\beta_{05} - \beta_1 - 3)\,, c = -\frac{4EI_0}{\kappa_1^2}\left(\beta_{05} - \frac{1}{2}\beta_1 - \frac{1}{2}\right). \tag{G.264}$$

$$EI(\varepsilon) = EI_0\Big(1 - \underbrace{\frac{(3+\beta_1-4\beta_{05})}{\kappa_1}}_{\alpha_1}\cdot\kappa - \underbrace{\frac{4(-\frac{1}{2}-\frac{1}{2}\beta_1+\beta_{05})}{\kappa_1^2}}_{\alpha_2}\cdot\kappa^2\Big). \tag{G.265}$$

$$\boldsymbol{K}^{\mathrm{e}} = (3.282) - \frac{4EI_0 \times \alpha_2}{5L^7}\begin{bmatrix} K_{11} & K_{12} & K_{13} & K_{14} \\ & K_{22} & K_{23} & K_{24} \\ & & K_{33} & K_{34} \\ \text{sym.} & & & K_{44} \end{bmatrix}. \tag{G.266}$$

$$K_{11} = 324\,u_1\,\varphi_1\,L + 324\,u_1\,\varphi_2\,L + 132\,\varphi_1\varphi_2 L^2 - 324\,u_2\,\varphi_2\,L - 324\,u_2\varphi_1 L + 324\,u_1^2 - 648\,u_1\,u_2 + 96\,\varphi_1^2 L^2 + 324\,{u_2}^2 + 96\,\varphi_2^2 L^2\,,$$

$$K_{12} = -3\,(64\,u_1\,\varphi_1\,L + 44\,u_1\,\varphi_2\,L + 22\,\varphi_1\varphi_2 L^2 - 44\,u_2\,\varphi_2\,L - 64\,u_2\varphi_1\,L + 54\,u_1^2 - 108\,u_1\,u_2 + 21\,\varphi_1^2 L^2 + 54\,{u_2}^2 + 11\,\varphi_2^2 L^2)L\,,$$

$$K_{13} = -324\,u_1\,\varphi_1\,L - 324\,u_1\,\varphi_2\,L - 132\,\varphi_1\varphi_2 L^2 + 324\,u_2\,\varphi_2\,L + 324\,u_2\varphi_1\,L - 324\,u_1^2 + 648\,u_1\,u_2 - 96\,\varphi_1^2 L^2 - 324\,{u_2}^2 - 96\,\varphi_2^2 L^2\,,$$

$$K_{14} = -3\,(44\,u_1\,\varphi_1\,L + 64\,u_1\,\varphi_2\,L + 22\,\varphi_1\varphi_2 L^2 - 64\,u_2\,\varphi_2\,L - 44\,u_2\varphi_1\,L + 54\,u_1^2 - 108\,u_1\,u_2 + 11\,\varphi_1^2 L^2 + 54\,{u_2}^2 + 21\,\varphi_2^2 L^2)L\,,$$

$$K_{22} = 2\,(63\,u_1\,\varphi_1\,L + 33\,u_1\,\varphi_2\,L + 19\,\varphi_1\varphi_2 L^2 - 33\,u_2\,\varphi_2\,L - 63\,u_2\varphi_1\,L + 48\,u_1^2 - 96\,u_1\,u_2 + 22\,\varphi_1^2 L^2 + 48\,{u_2}^2 + 7\,\varphi_2^2 L^2)L^2\,,$$

$$K_{23} = +3\,(64\,u_1\,\varphi_1\,L + 44\,u_1\,\varphi_2\,L + 22\,\varphi_1\varphi_2 L^2 - 44\,u_2\,\varphi_2\,L - 64\,u_2\varphi_1\,L + 54\,u_1^2 - 108\,u_1\,u_2 + 21\,\varphi_1^2 L^2 + 54\,{u_2}^2 + 11\,\varphi_2^2 L^2)L\,,$$

$$K_{24} = (66\,u_1\,\varphi_1\,L + 66\,u_1\,\varphi_2\,L + 28\,\varphi_1\varphi_2 L^2 - 66\,u_2\,\varphi_2\,L - 66\,u_2\varphi_1\,L + 66\,u_1^2 - 132\,u_1\,u_2 + 19\,\varphi_1^2 L^2 + 66\,{u_2}^2 + 19\,\varphi_2^2 L^2)L^2\,,$$

$$K_{33} = 324\,u_1\,\varphi_1\,L + 324\,u_1\,\varphi_2\,L + 132\,\varphi_1\varphi_2 L^2 - 324\,u_2\,\varphi_2\,L - 324\,u_2\varphi_1\,L + 324\,u_1^2 - 648\,u_1\,u_2 + 96\,\varphi_1^2 L^2 + 324\,{u_2}^2 + 96\,\varphi_2^2 L^2\,,$$

$$K_{34} = +3\,(44\,u_1\,\varphi_1\,L + 64\,u_1\,\varphi_2\,L + 22\,\varphi_1\varphi_2 L^2 - 64\,u_2\,\varphi_2\,L - 44\,u_2\varphi_1\,L + 54\,u_1^2 - 108\,u_1\,u_2 + 11\,\varphi_1^2 L^2 + 54\,{u_2}^2 + 21\,\varphi_2^2 L^2)L\,,$$

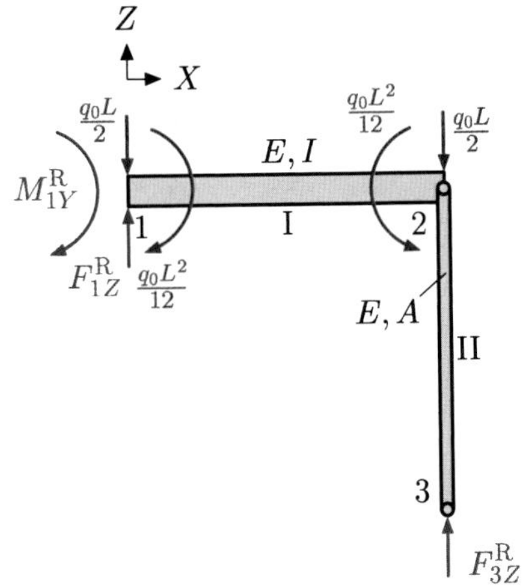

Fig. G.11 Free body diagram of the frame structure problem

$$
\begin{aligned}
K_{44} = {} & 2\,(33\,u_1\,\varphi_1\,L + 63\,u_1\,\varphi_2\,L + 19\,\varphi_1\varphi_2 L^2 - 63\,u_2\,\varphi_2\,L - 33\,u_2\varphi_1\,L \\
& + 48\,u_1^2 - 96\,u_1\,u_2 + 7\,\varphi_1^2 L^2 + 48\,u_2{}^2 + 22\,\varphi_2^2 L^2)L^2\,.
\end{aligned}
$$

3.41 Plane beam rod structure

$$
\begin{bmatrix} u_{2X} \\ u_{2Z} \\ \varphi_{2Y} \end{bmatrix} = \begin{bmatrix} -\dfrac{4L^3F}{E\left(2AL^2+3I\right)} \\ -\dfrac{8L^3F}{E\left(2AL^2+3I\right)} \\ +\dfrac{6L^2F}{E\left(2AL^2+3I\right)} \end{bmatrix}, \tag{G.267}
$$

$$
\begin{bmatrix} R_{1X} \\ R_{1Z} \\ M_{1Y} \end{bmatrix} = \begin{bmatrix} \dfrac{2AL^2F}{2AL^2+3I} \\ \dfrac{3IF}{2AL^2+3I} \\ \dfrac{-6ILF}{2AL^2+3I} \end{bmatrix} \quad ; \quad \begin{bmatrix} R_{3X} \\ R_{3Z} \end{bmatrix} = \begin{bmatrix} -\dfrac{2AL^2F}{2AL^2+3I} \\ \dfrac{2AL^2F}{2AL^2+3I} \end{bmatrix}. \tag{G.268}
$$

3.42 Plane beam-rod structure with distributed load

The free body diagram is shown in Fig. G.11.

The global system of equation without consideration of the boundary conditions is obtained as:

$$\begin{array}{c} \\ \\ \\ \\ \\ \end{array}\begin{bmatrix} 12\frac{EI_Y}{L^3} & -6L\frac{EI_Y}{L^3} & -12\frac{EI_Y}{L^3} & -6L\frac{EI_Y}{L^3} & 0 \\ -6L\frac{EI_Y}{L^3} & 4L^2\frac{EI_Y}{L^3} & 6L\frac{EI_Y}{L^3} & 2L^2\frac{EI_Y}{L^3} & 0 \\ -12\frac{EI_Y}{L^3} & 6L\frac{EI_Y}{L^3} & 12\frac{EI_Y}{L^3}+\frac{EA}{L} & 6L\frac{EI_Y}{L^3} & -\frac{EA}{L} \\ -6L\frac{EI_Y}{L^3} & 2L^2\frac{EI_Y}{L^3} & 6L\frac{EI_Y}{L^3} & 4L^2\frac{EI_Y}{L^3} & 0 \\ 0 & 0 & -\frac{EA}{L} & 0 & \frac{EA}{L} \end{bmatrix} \begin{bmatrix} u_{1Z} \\ \varphi_{1Y} \\ u_{2Z} \\ \varphi_{2Y} \\ u_{3Z} \end{bmatrix} = \begin{bmatrix} -\frac{q_0 L}{2} + F^{\mathrm{R}}_{1Z} \\ \frac{q_0 L^2}{12} + M^{\mathrm{R}}_{1Y} \\ -\frac{q_0 L}{2} \\ -\frac{q_0 L^2}{12} \\ F^{\mathrm{R}}_{3Z} \end{bmatrix}. \tag{G.269}$$

(column labels: u_{1Z}, φ_{1Y}, u_{2Z}, φ_{2Y}, φ_{3Z})

Introduction of the boundary conditions, i.e. $u_{1Z} = u_{3Z} = 0$ and $\varphi_{1Y} = 0$, gives the following reduced system of equations:

$$\begin{bmatrix} 12\frac{EI_Y}{L^3} + \frac{EA}{L} & 6L\frac{EI_Y}{L^3} \\ 6L\frac{EI_Y}{L^3} & 4L^2\frac{EI_Y}{L^3} \end{bmatrix} \begin{bmatrix} u_{2Z} \\ \varphi_{2Y} \end{bmatrix} = \begin{bmatrix} -\frac{q_0 L}{2} \\ -\frac{q_0 L^2}{12} \end{bmatrix}. \tag{G.270}$$

The solution of this system of equations can be obtained by calculating the inverse of the stiffness matrix to give:

$$\begin{bmatrix} u_{2Z} \\ \varphi_{2Y} \end{bmatrix} = \frac{1}{4L^2\frac{EI_Y}{L^3}\left(12\frac{EI_Y}{L^3} + \frac{EA}{L}\right) - 36L^2\frac{E^2 I_Y^2}{L^6}} \begin{bmatrix} 4L^2\frac{EI_Y}{L^3} & -6L\frac{EI_Y}{L^3} \\ -6L\frac{EI_Y}{L^3} & 12\frac{EI_Y}{L^3} + \frac{EA}{L} \end{bmatrix} \begin{bmatrix} -\frac{q_0 L}{2} \\ -\frac{q_0 L^2}{12} \end{bmatrix}, \tag{G.271}$$

or simplified as:

$$\begin{bmatrix} u_{2Z} \\ \varphi_{2Y} \end{bmatrix} = \frac{L^3 q_0}{12\frac{EI_Y}{L} + 4EAL} \begin{bmatrix} -\frac{3}{2} \\ \frac{2}{L} - \frac{EAL}{12EI_Y} \end{bmatrix}. \tag{G.272}$$

• Special case: No rod ⇔ $EA = 0$.
It follows from Eq. (G.270) or from the general solution (G.272) that:

$$\begin{bmatrix} u_{2Z} \\ \varphi_{2Y} \end{bmatrix} = \frac{L^4 q_0}{12EI_Y} \begin{bmatrix} -\frac{3}{2} \\ \frac{2}{L} \end{bmatrix}. \tag{G.273}$$

This result is equal to the analytical solution.
• Special case: No beam ⇔ $EI_Y = 0$.
Equation (G.270) must be reduced under consideration of the additional boundary condition $\varphi_{2Y} = 0$:

$$\left[12\frac{EI_Y}{L^3} + \frac{EA}{L}\right]\left[u_{2Z}\right] = \left[-\frac{q_0 L}{2}\right], \tag{G.274}$$

which gives under consideration of $EI_Y = 0$ the following displacement at node 2:

$$u_{2Z} = -\frac{q_0 L^2}{2EA}. \tag{G.275}$$

This result is equal to the analytical solution.

3.43 Plane beam-rod structure with a triangular shaped distributed load

Reduced system of equations:

$$\frac{EI}{L^3}\begin{bmatrix} 192 + \frac{2AL^2}{I} & 0 \\ 0 & 16L^2 \end{bmatrix}\begin{bmatrix} u_{2Z} \\ \varphi_{2Y} \end{bmatrix} = \begin{bmatrix} -\frac{7}{40}\, q_0 L \\ -\frac{q_0}{80}\, L^2 \end{bmatrix}. \tag{G.276}$$

Solution of the system:

$$u_{2Z} = -\frac{7 q_0 L^4}{40EI\left(192 + \frac{2AL^2}{I}\right)}, \tag{G.277}$$

$$\varphi_{2Y} = -\frac{q_0 L^3}{1280EI}. \tag{G.278}$$

Normalized bending line under consideration of $\frac{AL^2}{I} = 20$:

- Element I: $0 \le x_{\mathrm{I}} \le \frac{L}{2}$

$$\frac{u_z^{\mathrm{e}}(x_{\mathrm{I}})}{\frac{q_0 L^4}{EI}} = \left[3\left(\frac{x_{\mathrm{I}}}{\frac{L}{2}}\right)^2 - 2\left(\frac{x_{\mathrm{I}}}{\frac{L}{2}}\right)^3\right] \times \left(-\frac{7}{9280}\right) + \\ + \left[\left(\frac{x_{\mathrm{I}}}{\frac{L}{2}}\right)^2 - \left(\frac{x_{\mathrm{I}}}{\frac{L}{2}}\right)^3\right] \times \left(-\frac{1}{2560}\right). \tag{G.279}$$

- Element II: $0 \le x_{\mathrm{II}} \le \frac{L}{2}$

$$\frac{u_z^{\mathrm{e}}(x_{\mathrm{II}})}{\frac{q_0 L^4}{EI}} = \left[1 - 3\left(\frac{x_{\mathrm{II}}}{\frac{L}{2}}\right)^2 + 2\left(\frac{x_{\mathrm{II}}}{\frac{L}{2}}\right)^3\right] \times \left(-\frac{7}{9280}\right) + \\ + \left[-\left(\frac{x_{\mathrm{II}}}{\frac{L}{2}}\right)^1 + 2\left(\frac{x_{\mathrm{II}}}{\frac{L}{2}}\right)^2 - \left(\frac{x_{\mathrm{II}}}{\frac{L}{2}}\right)^3\right] \times \left(-\frac{1}{2560}\right). \tag{G.280}$$

The normalized deflection is shown in Fig. G.12.
Special case: No rod $\Leftrightarrow EA = 0$.

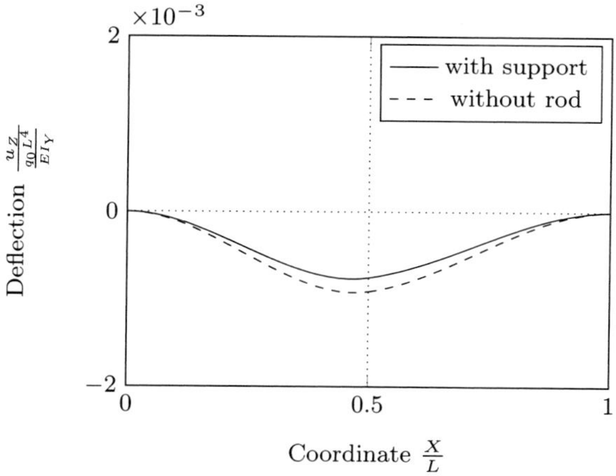

Fig. G.12 Beam deflection along the major axis

$$u_{2Z} = -\frac{7q_0L^4}{7680EI}, \tag{G.281}$$

$$\varphi_{2Y} = -\frac{q_0L^3}{1280EI}. \tag{G.282}$$

3.44 Plane generalized beam-rod structure with different distributed loads

• Reduced system of equations:

$$E \times \begin{bmatrix} \frac{2A}{L} & 0 & 0 \\ 0 & \frac{24I}{L^3} + \frac{2A}{L} & 0 \\ 0 & 0 & \frac{8I}{L} \end{bmatrix} \begin{bmatrix} u_{2X} \\ u_{2Z} \\ \varphi_{2Y} \end{bmatrix} = \begin{bmatrix} \frac{p_0L}{2} \\ -\frac{7q_0L}{20} \\ -\frac{q_0L^2}{20} \end{bmatrix}. \tag{G.283}$$

The reduced stiffness matrix is a diagonal matrix and it can be easily inverted by replacing each element in the main diagonal with its reciprocal.

• Unknown deformations at node 2:

$$\begin{bmatrix} u_{2X} \\ u_{2Z} \\ \varphi_{2Y} \end{bmatrix} = \frac{1}{E} \times \begin{bmatrix} \frac{p_0L^2}{4A} \\ -\frac{7q_0L^4}{20(24I + 2AL^2)} \\ -\frac{q_0L^3}{160I} \end{bmatrix}. \tag{G.284}$$

• All the reactions at the supports:

$$F_{1X}^{\mathrm{R}} = -\frac{p_0 L}{4}, \quad F_{1Z}^{\mathrm{R}} = \frac{21 q_0 L}{5\left(24 + \frac{2AL^2}{I}\right)} + \frac{15 q_0 L}{80}, \tag{G.285}$$

$$M_{1Y}^{\mathrm{R}} = -\frac{21 q_0 L^2}{10\left(24 + \frac{2AL^2}{I}\right)} - \frac{11 q_o L^2}{240}, \tag{G.286}$$

$$F_{3X}^{\mathrm{R}} = -\frac{3 p_0 L}{4}, \quad F_{3Z}^{\mathrm{R}} = \frac{21 q_0 L}{5\left(24 + \frac{2AL^2}{I}\right)} - \frac{3 q_0 L}{80}, \tag{G.287}$$

$$M_{3Y}^{\mathrm{R}} = \frac{21 q_0 L^2}{10\left(24 + \frac{2AL^2}{I}\right)} - \frac{q_o L^2}{80}, \tag{G.288}$$

$$F_{4X}^{\mathrm{R}} = 0, \quad F_{4Z}^{\mathrm{R}} = \frac{7 q_0 L}{10\left(\frac{24I}{AL^2} + 2\right)}. \tag{G.289}$$

• Normal stress distribution in each element:

$$\sigma_{x_{\mathrm{I}}}(x_{\mathrm{I}}) = \frac{E}{L} u_{2X} + E\left(0 + 0 + \left[-\frac{6}{L^2} + \frac{12 x_{\mathrm{I}}}{L^3}\right] u_{2Z} + \left[-\frac{2}{L} + \frac{6 x_{\mathrm{I}}}{L^2}\right] \varphi_{2Y}\right) z_{\mathrm{I}}, \tag{G.290}$$

$$\sigma_{x_{\mathrm{II}}}(x_{\mathrm{II}}) = -\frac{E}{L} u_{3X} + E\left(\left[\frac{6}{L^2} - \frac{12 x_{\mathrm{II}}}{L^3}\right] u_{2Z} + \left[-\frac{4}{L} + \frac{6 x_{\mathrm{II}}}{L^2}\right] \varphi_{2Y} + 0 + 0\right) z_{\mathrm{II}}, \tag{G.291}$$

$$\sigma_{x_{\mathrm{III}}}(x_{\mathrm{III}}) = -\frac{E}{\frac{L}{2}} u_{3Z}. \tag{G.292}$$

• Horizontal ($u_X(x_i)$) and vertical displacement ($u_Z(x_i)$) (see Fig. G.13) distributions of the generalized beam:
Simplified solution:

$$\begin{bmatrix} u_{2X} \\ u_{2Z} \\ \varphi_{2Y} \end{bmatrix} = \frac{1}{E} \times \begin{bmatrix} \frac{p_0 L^2}{4A} \\ -\frac{q_0 L^4}{80I} \\ -\frac{q_0 L^3}{160I} \end{bmatrix}. \tag{G.293}$$

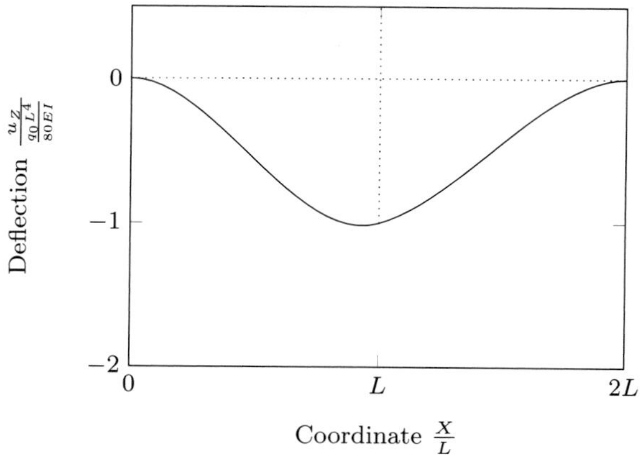

Fig. G.13 Deflection of the generalized beam along the major axis

Element I:

$$\frac{u_Z(x_{\mathrm{I}})}{\frac{q_0L^4}{80EI}} = -\left[3\left(\frac{x_{\mathrm{I}}}{L}\right)^2 - 2\left(\frac{x_{\mathrm{I}}}{L}\right)^3\right] \times 1 - \left[\left(\frac{x_{\mathrm{I}}}{L}\right)^2 - \left(\frac{x_{\mathrm{I}}}{L}\right)^3\right] \times \frac{1}{2}, \tag{G.294}$$

$$\frac{u_X(x_{\mathrm{I}})}{\frac{p_0L^2}{4EA}} = \frac{x_{\mathrm{I}}}{L}. \tag{G.295}$$

Element II:

$$\begin{aligned}\frac{u_Z(x_{\mathrm{II}})}{\frac{q_0L^4}{80EI}} = &-\left[1 - 3\left(\frac{x_{\mathrm{II}}}{L}\right)^2 + 2\left(\frac{x_{\mathrm{II}}}{L}\right)^3\right] \times 1 \\ &-\left[-\left(\frac{x_{\mathrm{II}}}{L}\right) + 2\left(\frac{x_{\mathrm{II}}}{L}\right)^2 - \left(\frac{x_{\mathrm{II}}}{L}\right)^3\right] \times \frac{1}{2},\end{aligned} \tag{G.296}$$

$$\frac{u_X(x_{\mathrm{II}})}{\frac{p_0L^2}{4EA}} = 1 - \frac{x_{\mathrm{II}}}{L}. \tag{G.297}$$

3.45 Plane generalized beam structure with different distributed loads

• Nodal deformations

$$\begin{bmatrix}\varphi_{1Y} \\ \varphi_{2Y}\end{bmatrix} = \frac{q_0L_{\mathrm{I}}^2}{20EI} \times \frac{1}{\frac{3}{L_{\mathrm{I}}^2} + \frac{4}{L_{\mathrm{I}}L_{\mathrm{II}}}} \times \begin{bmatrix}\frac{4}{3L_{\mathrm{I}}} + \frac{1}{L_{\mathrm{II}}} \\ -\frac{7}{6L_{\mathrm{I}}}\end{bmatrix}. \tag{G.298}$$

• Horizontal reaction force at node 1

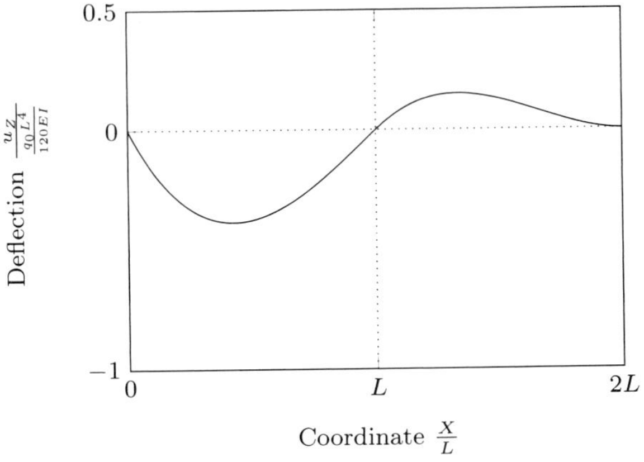

Fig. G.14 Deflection of the generalized beam along the major axis

$$F^{\mathrm{R}}_{1X} = 0\,. \tag{G.299}$$

• Nodal deformations for special case $L_{\mathrm{I}} = L_{\mathrm{II}} = L$

$$\begin{bmatrix} \varphi_{1Y} \\ \varphi_{2Y} \end{bmatrix} = \frac{q_0 L^3}{60EI} \times \begin{bmatrix} 1 \\ -\frac{1}{2} \end{bmatrix}. \tag{G.300}$$

• Nodal deformations for the special case $L_{\mathrm{I}} = L$ and $L_{\mathrm{II}} \to 0$
Further reduction of the reduced system of equations with $\varphi_{2Y} = 0$ yields:

$$\varphi_{1Y} = \frac{1}{80} \times \frac{q_0 L_{\mathrm{I}}^3}{EI}. \tag{G.301}$$

• Bending line $u_Z(X)$ in the range $0 \le X \le L_{\mathrm{I}} + L_{\mathrm{II}}$ for the special case $L_{\mathrm{I}} = L_{\mathrm{II}} = L$ (see Fig. G.14)

$$\frac{u_Z(x_{\mathrm{I}})}{\frac{q_0 L^4}{120EI}} = \left[-\left(\frac{x_{\mathrm{I}}}{L}\right) + 2\left(\frac{x_{\mathrm{I}}}{L}\right)^2 - \left(\frac{x_{\mathrm{I}}}{L}\right)^3\right] \times 2 + \left[\left(\frac{x_{\mathrm{I}}}{L}\right)^2 - \left(\frac{x_{\mathrm{I}}}{L}\right)^3\right] \times (-1)\,, \tag{G.302}$$

$$\frac{u_Z(x_{\mathrm{II}})}{\frac{q_0 L^4}{120EI}} = \left[-\left(\frac{x_{\mathrm{II}}}{L}\right) + 2\left(\frac{x_{\mathrm{II}}}{L}\right)^2 - \left(\frac{x_{\mathrm{II}}}{L}\right)^3\right] \times (-1)\,. \tag{G.303}$$

3.46 Stiffness matrix for a generalized beam element for different rotation angles at the nodes

$$\boldsymbol{K}^{\mathrm{e}}_{XZ}(\alpha_1,\alpha_2)=\begin{bmatrix}\boldsymbol{K}_{11} & \boldsymbol{K}_{12}\\ \boldsymbol{K}_{21} & \boldsymbol{K}_{22}\end{bmatrix}, \tag{G.304}$$

where the submatrices are given under consideration of the abbreviations $\mathrm{s}=\sin$ and $\mathrm{c}=\cos$ as:

$$\boldsymbol{K}_{11}=\begin{bmatrix}\frac{12I}{L^3}\mathrm{s}^2\alpha_1+\frac{A}{L}\mathrm{c}^2\alpha_1 & -\left(-\frac{12I}{L^3}+\frac{A}{L}\right)\mathrm{s}\alpha_1\mathrm{c}\alpha_1 & -\frac{6I}{L^2}\mathrm{s}\alpha_1\\ -\left(-\frac{12I}{L^3}+\frac{A}{L}\right)\mathrm{s}\alpha_1\mathrm{c}\alpha_1 & \frac{12I}{L^3}\mathrm{c}^2\alpha_1+\frac{A}{L}\mathrm{s}^2\alpha_1 & -\frac{6I}{L^2}\mathrm{c}\alpha_1\\ -\frac{6I}{L^2}\mathrm{s}\alpha_1 & -\frac{6I}{L^2}\mathrm{c}\alpha_1 & \frac{4I}{L}\end{bmatrix},$$

$$\boldsymbol{K}_{12}=\begin{bmatrix}-\frac{12I}{L^3}\mathrm{s}\alpha_1\mathrm{s}\alpha_2-\frac{A}{L}\mathrm{c}\alpha_1\mathrm{c}\alpha_2 & -\frac{12I}{L^3}\mathrm{s}\alpha_1\mathrm{c}\alpha_2+\frac{A}{L}\mathrm{c}\alpha_1\mathrm{s}\alpha_2 & -\frac{6I}{L^2}\mathrm{s}\alpha_1\\ -\frac{12I}{L^3}\mathrm{c}\alpha_1\mathrm{s}\alpha_2+\frac{A}{L}\mathrm{s}\alpha_1\mathrm{c}\alpha_2 & -\frac{12I}{L^3}\mathrm{c}\alpha_1\mathrm{c}\alpha_2-\frac{A}{L}\mathrm{s}\alpha_1\mathrm{s}\alpha_2 & -\frac{6I}{L^2}\mathrm{c}\alpha_1\\ \frac{6I}{L^2}\mathrm{s}\alpha_2 & +\frac{6I}{L^2}\mathrm{c}\alpha_2 & \frac{2I}{L}\end{bmatrix},$$

$$\boldsymbol{K}_{21}=\begin{bmatrix}-\frac{12I}{L^3}\mathrm{s}\alpha_1\mathrm{s}\alpha_2-\frac{A}{L}\mathrm{c}\alpha_1\mathrm{c}\alpha_2 & -\frac{12I}{L^3}\mathrm{c}\alpha_1\mathrm{s}\alpha_2+\frac{A}{L}\mathrm{s}\alpha_1\mathrm{c}\alpha_2 & \frac{6I}{L^2}\mathrm{s}\alpha_2\\ -\frac{12I}{L^3}\mathrm{s}\alpha_1\mathrm{c}\alpha_2+\frac{A}{L}\mathrm{c}\alpha_1\mathrm{s}\alpha_2 & -\frac{12I}{L^3}\mathrm{c}\alpha_1\mathrm{c}\alpha_2-\frac{A}{L}\mathrm{s}\alpha_1\mathrm{s}\alpha_2 & +\frac{6I}{L^2}\mathrm{c}\alpha_2\\ -\frac{6I}{L^2}\mathrm{s}\alpha_1 & -\frac{6I}{L^2}\mathrm{c}\alpha_1 & \frac{2I}{L}\end{bmatrix},$$

$$
\boldsymbol{K}_{22} = \begin{bmatrix}
\frac{12I}{L^3}\mathrm{s}^2\alpha_2 + \frac{A}{L}\mathrm{c}^2\alpha_2 & -\left(-\frac{12I}{L^3}+\frac{A}{L}\right)\mathrm{s}\alpha_2\mathrm{c}\alpha_2 & \frac{6I}{L^2}\mathrm{s}\alpha_2 \\
-\left(-\frac{12I}{L^3}+\frac{A}{L}\right)\mathrm{s}\alpha_2\mathrm{c}\alpha_2 & \frac{12I}{L^3}\mathrm{c}^2\alpha_2 + \frac{A}{L}\mathrm{s}^2\alpha_2 & +\frac{6I}{L^2}\mathrm{c}\alpha_2 \\
\frac{6I}{L^2}\mathrm{s}\alpha_2 & +\frac{I}{L^2}\mathrm{c}\alpha_2 & \frac{4I}{L}
\end{bmatrix}.
$$

3.47 Mechanical properties of a square frame structure

(a) [quarter model, BC: $\frac{F_0}{2}$]

$$u_{3Z} = \frac{1}{384}\frac{\left(5AL^2 + 96\,I\right)LF_0}{EIA}, \tag{G.305}$$

$$E_{\text{struct}} = \frac{\sigma}{\varepsilon} = \frac{192EIA}{\left(5AL^2 + 96I\right)Ld}. \tag{G.306}$$

For $A = \frac{\pi d^2}{4}$ and $I = \frac{\pi d^4}{64}$:

$$E_{\text{struct}} = \frac{3Ed^3\pi}{\left(6d^2 + 5L^2\right)L}. \tag{G.307}$$

(b) [quarter model, BC: $\frac{F_0}{2}$]

$$u_{2Z} = \frac{1}{48}\frac{\left(AL^2 + 12\,I\right)LF_0}{EIA}, \tag{G.308}$$

$$E_{\text{struct}} = \frac{\sigma}{\varepsilon} = \frac{24EIA}{\left(AL^2 + 12I\right)Ld}. \tag{G.309}$$

For $A = \frac{\pi d^2}{4}$ and $I = \frac{\pi d^4}{64}$:

$$E_{\text{struct}} = \frac{3Ed^3\pi}{2\left(3d^2 + 4L^2\right)L}. \tag{G.310}$$

3.48 Square frame structure: different ways of load application

(a) [quarter model]

$$u_{2Z} = \frac{1}{4}\frac{LF_0}{EA}, \tag{G.311}$$

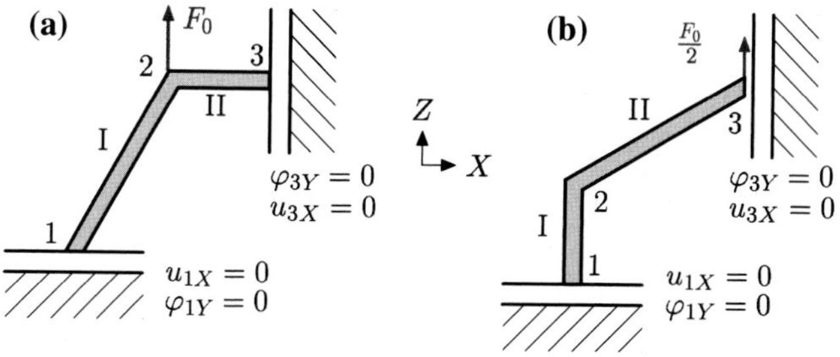

Fig. G.15 Honeycomb structure approximated under consideration of symmetry: **a** flat orientation and **b** pointy or angled orientation

$$E_{\text{struct}} = \frac{\sigma}{\varepsilon} = \frac{2EA}{Ld}. \tag{G.312}$$

For $A = \frac{\pi d^2}{4}$ and $I = \frac{\pi d^4}{64}$:

$$E_{\text{struct}} = \frac{dE\pi}{2L}. \tag{G.313}$$

3.49 Mechanical properties of idealized honeycomb structure

The problem under consideration of the symmetry is shown in Fig. G.15.
(a)

$$u_{2Z} = \frac{1}{24} \frac{\left(AL^2 + 18\,I\right) LF_0}{EIA}, \tag{G.314}$$

$$E_{\text{struct}} = \frac{\sigma}{\varepsilon} = \frac{\frac{F_0}{\frac{1}{2}(\sqrt{3}+1)Ld}}{\frac{\frac{1}{24}\frac{(AL^2+18\,I)LF_0}{EIA}}{\frac{1}{2}\sqrt{3}L}} = \frac{24EIA}{\left(1+\sqrt{3}\right)\left(AL^2+18\,I\right)Ld}. \tag{G.315}$$

For $A = \frac{\pi d^2}{4}$ and $I = \frac{\pi d^4}{64}$:

$$E_{\text{struct}} = \frac{3Ed^3\pi}{\left(1+\sqrt{3}\right)\left(8L^2+9d^2\right)L}. \tag{G.316}$$

(b)

$$u_{3Z} = \frac{1}{16} \frac{\left(AL^2 + 6\,I\right) LF_0}{EIA}, \tag{G.317}$$

$$E_{\text{struct}} = \frac{\sigma}{\varepsilon} = \frac{\frac{F_0/2}{\frac{1}{2}\sqrt{3}Ld}}{\frac{\frac{1}{16}\frac{(AL^2+6\,I)LF_0}{EIA}}{L}} = \frac{16EIA}{\sqrt{3}\left(AL^2+6\,I\right)Ld}. \tag{G.318}$$

For $A = \frac{\pi d^2}{4}$ and $I = \frac{\pi d^4}{64}$:

$$E_{\text{struct}} = \frac{2}{\sqrt{3}} \times \frac{E d^3 \pi}{\left(8L^2 + 3d^2\right) L}. \tag{G.319}$$

3.50 Bridge structure (computational problem)

Since the solution is obtained numerically,[14] the obtained numbers must be carefully interpreted. Looking at case (a) load (1), i.e. the symmetrical loading, the displacement at node 3 in the horizontal direction must be zero. However, there is a small value calculated, cf. Table G.3. This value must be related to the vertical displacement at node 3 and one can see that u_{3X} is several orders of magnitude smaller than u_{3Z}. Thus, from a practical point of view, u_{3X} can be considered as equal to zero.

Some rotations in Table G.4 could not be evaluated (indicated by '—') since the rod elements do not have a rotational degree of freedom.

General expressions for the nodal unknowns for case (c) load (1) (Table G.5):

$$\begin{aligned}
&u_{2X} = 0\,, \; u_{2Z} = -2.414214 \frac{F_0 L}{EA}\,, \; u_{3X} = 0\,, \; u_{3Z} = -4.828427 \frac{F_0 L}{EA}\,, \\
&u_{4X} = 0\,, \; u_{4Z} = -2.414214 \frac{F_0 L}{EA}\,, \; u_{6X} = -\frac{F_0 L}{EA}\,, \; u_{6Z} = u_{2Z}\,, \\
&u_{7X} = 0\,, \; u_{7Z} = -4.828427 \frac{F_0 L}{EA}\,, \; u_{8X} = -\frac{F_0 L}{EA}\,, \; u_{8Z} = u_{6Z}\,.
\end{aligned} \tag{G.320}$$

General expressions for the nodal unknowns for case (c) load (2):

$$\begin{aligned}
&u_{2X} = 0.25 \frac{F_0 L}{EA}\,, \; u_{2Z} = -3.871320 \frac{F_0 L}{EA}\,, \; u_{3X} = 0.5 \frac{F_0 L}{EA}\,, \; u_{3Z} = -2.414214 \frac{F_0 L}{EA}\,, \\
&u_{4X} = u_{2X}\,, \; u_{4Z} = -0.957107 \frac{F_0 L}{EA}\,, \; u_{6X} = -u_{4X}\,, \; u_{6Z} = u_{4Z}\,, \\
&u_{7X} = u_{2X}\,, \; u_{7Z} = u_{3Z}\,, \; u_{8X} = 0.75 \frac{F_0 L}{EA}\,, \; u_{8Z} = -2.871320 \frac{F_0 L}{EA}\,.
\end{aligned} \tag{G.321}$$

General expressions for the nodal unknowns for case (d) load (1) (see Table G.6 for details) :

$$\begin{aligned}
&\varphi_{1Y} = -\frac{F_0 L^2}{EI}\,, \; u_{2X} = 0\,, \; u_{2Z} = -0.916667 \frac{F_0 L^3}{EI}\,, \; \varphi_{2Y} = -0.75 \frac{F_0 L^2}{EI}\,, \\
&u_{3X} = 0\,, \; u_{3Z} = -1.333333 \frac{F_0 L^3}{EI}\,, \; \varphi_{3Y} = 0\,, \; u_{4X} = 0\,, \\
&u_{4Z} = -0.916667 \frac{F_0 L^3}{EI}\,, \; \varphi_{4Y} = 0.75 \frac{F_0 L^2}{EI}\,, \; \varphi_{5Y} = \frac{F_0 L^2}{EI}\,.
\end{aligned} \tag{G.322}$$

[14]The actual results were obtained with Maple 8.0 and the accuracy setting 'Digits := 25'.

Table G.3 Numerical results (case a) for the bridge structure problem with $L = 4000\,\text{mm}$, $E = 200{,}000\,\text{MPa}$, $I = 80\,\text{mm}^4$ and $A = 10\,\text{mm}^2$

Node Nr.	u_X	u_Z	φ_Y
Case (a) Load (1)			
1	0.0	0.0	$0.105623 \times 10^{-5} F_0$
2	$-0.110022 \times 10^{-7} F_0$	$-0.004828 F_0$	$0.111306 \times 10^{-5} F_0$
3	$-0.855248 \times 10^{-8} F_0$	$-0.009657 F_0$	$0.141264 \times 10^{-11} F_0$
4	$0.244971 \times 10^{-8} F_0$	$-0.004828 F_0$	$-0.111306 \times 10^{-5} F_0$
5	0.0	0.0	$-0.105623 \times 10^{-5} F_0$
6	$-0.002000 F_0$	$-0.004828 F_0$	$-0.100792 \times 10^{-5} F_0$
7	$-0.427622 \times 10^{-8} F_0$	$-0.009657 F_0$	$0.126408 \times 10^{-11} F_0$
8	$0.002000 F_0$	$-0.004828 F_0$	$0.100792 \times 10^{-5} F_0$
Case (a) Load (2)			
1	0.0	0.0	$0.210602 \times 10^{-5} F_0$
2	$0.000500 F_0$	$-0.007743 F_0$	$0.460909 \times 10^{-6} F_0$
3	$0.001000 F_0$	$-0.004828 F_0$	$-0.607748 \times 10^{-6} F_0$
4	$0.000500 F_0$	$-0.001914 F_0$	$-0.488052 \times 10^{-6} F_0$
5	0.0	0.0	$-0.368358 \times 10^{-6} F_0$
6	$-0.000500 F_0$	$-0.001914 F_0$	$-0.466897 \times 10^{-6} F_0$
7	$0.0005 F_0$	$-0.004828 F_0$	$-0.379865 \times 10^{-6} F_0$
8	$0.001500 F_0$	$-0.005743 F_0$	$+0.107569 \times 10^{-6} F_0$
Case (a) Load (3)			
1	0.0	0.0	$0.022606 F_0$
2	$0.000435 F_0$	$-0.004693 F_0$	$-0.014409 F_0$
3	$0.000693 F_0$	$-0.002386 F_0$	$0.001727 F_0$
4	$0.000349 F_0$	$-0.000898 F_0$	$-0.000287 F_0$
5	0.0	0.0	$0.000108 F_0$
6	$-0.000165 F_0$	$-0.000878 F_0$	$-0.000118 F_0$
7	$0.000315 F_0$	$-0.002383 F_0$	$-0.000205 F_0$
8	$0.000777 F_0$	$-0.003443 F_0$	$-0.000380 F_0$
Case (a) Load (4)			
1	0.0	0.0	$-0.003528 F_0$
2	$0.000265 F_0$	$-0.006822 F_0$	$0.012608 F_0$
3	$0.000671 F_0$	$-0.007799 F_0$	$-0.008826 F_0$
4	$0.000323 F_0$	$-0.003309 F_0$	$0.001471 F_0$
5	0.0	0.0	$-0.000542 F_0$
6	$-0.001319 F_0$	$-0.003409 F_0$	$0.000532 F_0$
7	$0.000280 F_0$	$-0.007783 F_0$	$0.001513 F_0$
8	$0.001967 F_0$	$-0.005759 F_0$	$-0.000788 F_0$

Table G.4 Numerical results (case b) for the bridge structure problem with $L = 4000\,\text{mm}$, $E = 200{,}000\,\text{MPa}$, $I = 80\,\text{mm}^4$ and $A = 10\,\text{mm}^2$

Node Nr.	u_X	u_Z	φ_Y
Case (b) Load (1)			
1	0.0	0.0	$0.103467 \times 10^{-5} F_0$
2	0.0	$-0.004828 F_0$	$0.155198 \times 10^{-5} F_0$
3	0.0	$-0.009657 F_0$	−0.0
4	0.0	$-0.004828 F_0$	$-0.155198 \times 10^{-5} F_0$
5	0.0	0.0	$-0.103467 \times 10^{-5} F_0$
6	$-0.002000 F_0$	$-0.004828 F_0$	—
7	0.0	$-0.009657 F_0$	—
8	$0.002000 F_0$	$-0.004828 F_0$	—
Case (b) Load (2)			
1	0.0	0.0	$0.264480 \times 10^{-5} F_0$
2	$0.000500 F_0$	$-0.007743 F_0$	$0.517336 \times 10^{-6} F_0$
3	$0.001000 F_0$	$-0.004828 F_0$	$-0.109282 \times 10^{-5} F_0$
4	$0.000500 F_0$	$-0.001914 F_0$	$-0.517336 \times 10^{-6} F_0$
5	0.0	0.0	$-0.459164 \times 10^{-6} F_0$
6	$-0.000500 F_0$	$-0.001914 F_0$	—
7	$0.000500 F_0$	$-0.004828 F_0$	—
8	$0.001500 F_0$	$-0.005743 F_0$	—
Case (b) Load (3)			
1	0.0	0.0	$0.045761 F_0$
2	$0.000344 F_0$	$-0.004935 F_0$	$-0.02901 F_0$
3	$0.000687 F_0$	$-0.002156 F_0$	$0.007812 F_0$
4	$0.000344 F_0$	$-0.000928 F_0$	$-0.002232 F_0$
5	0.0	0.0	$0.001116 F_0$
6	$-0.000103 F_0$	$-0.000848 F_0$	—
7	$0.000344 F_0$	$-0.002156 F_0$	—
8	$0.000790 F_0$	$-0.003480 F_0$	—
Case (b) Load (4)			
1	0.0	0.0	$-0.012275 F_0$
2	$0.000344 F_0$	$-0.007091 F_0$	$0.024555 F_0$
3	$0.000687 F_0$	$-0.008019 F_0$	$-0.023438 F_0$
4	$0.000344 F_0$	$-0.003084 F_0$	$0.006695 F_0$
5	0.0	0.0	$-0.003349 F_0$
6	$-0.001317 F_0$	$-0.003325 F_0$	—
7	$0.000344 F_0$	$-0.008019 F_0$	—
8	$0.002004 F_0$	$-0.005957 F_0$	—

Table G.5 Numerical results (case c) for the bridge structure problem with $L = 4000\,\text{mm}$, $E = 200{,}000\,\text{MPa}$, $I = 80\,\text{mm}^4$ and $A = 10\,\text{mm}^2$

Node Nr.	u_X	u_Z	φ_Y
Case (c) Load (1)			
1	0.0	0.0	—
2	0.0	$-0.004828 F_0$	—
3	0.0	$-0.009657 F_0$	—
4	0.0	$-0.004828 F_0$	—
5	0.0	0.0	—
6	$-0.002000 F_0$	$-0.004828 F_0$	—
7	0.0	$-0.009657 F_0$	—
8	$0.002000 F_0$	$-0.004828 F_0$	—
Case (c) Load (2)			
1	0.0	0.0	—
2	$0.000500 F_0$	$-0.007743 F_0$	—
3	$0.001000 F_0$	$-0.004828 F_0$	—
4	$0.000500 F_0$	$-0.001914 F_0$	—
5	0.0	0.0	—
6	$-0.000500 F_0$	$-0.001914_0 F$	—
7	$0.000500 F_0$	$-0.004828 F_0$	—
8	$0.001500 F_0$	$-0.005743 F_0$	—

It should be noted here that the case (d) corresponds to a simply supported beam under bending.

G.3 Problems from Chap. 4

4.4 Calculation of the shear stress distribution in a rectangular cross section

Starting point for the derivation of the shear stress distribution is, for example, the cantilever beam with shear force as shown in Fig. G.16. Furthermore, only rectangular cross sections of size $b \times h$ will be considered in the following.

An infinitesimal beam element (in horizontal direction) of this configuration is shown in Fig. G.17. The internal reactions are drawn according the sign convention in Fig. 3.12. Under the assumption that no distributed load is acting, it follows from the vertical force equilibrium in Eq. (3.30) that $Q_z(x) \approx Q_z(x + \mathrm{d}x)$.

The next step is to replace the internal reactions, i.e. the bending moment and the shear force, by the corresponding normal and shear stresses. To this end, a part of the infinitesimal beam element is cut off at $z = z'$, see Fig. G.18. The infinitesimal vertical dimension of this element is now $\mathrm{d}z'$.

Table G.6 Numerical results (case d) for the bridge structure problem with $L = 4000\,\text{mm}$, $E = 200{,}000\,\text{MPa}$, $I = 80\,\text{mm}^4$ and $A = 10\,\text{mm}^2$

Node Nr.	u_X	u_Z	φ_Y
Case (d) Load (1)			
1	0.0	0.0	$1.000000 F_0$
2	0.0	$-3666.666667 F_0$	$0.750000 F_0$
3	0.0	$-5333.333333 F_0$	0.0
4	0.0	$-3666.666667 F_0$	$-0.750000 F_0$
5	0.0	0.0	$-1.000000 F_0$
Case (d) Load (2)			
1	0.0	0.0	$0.875000 F_0$
2	0.0	$-3000 F_0$	$0.500000 F_0$
3	0.0	$-3666.666667 F_0$	$-0.125000 F_0$
4	0.0	$-2333.333333 F_0$	$-0.500000 F_0$
5	0.0	0.0	$-0.625000 F_0$
Case (d) Load (3)			
1	0.0	0.0	$0.546875 F_0$
2	0.0	$-1687.500000 F_0$	$0.234375 F_0$
3	0.0	$-1958.333333 F_0$	$-0.078125 F_0$
4	0.0	$-1229.166667 F_0$	$-0.265625 F_0$
5	0.0	0.0	$-0.328125 F_0$
Case (d) Load (4)			
1	0.0	0.0	$1.015625 F_0$
2	0.0	$-3645.833333 F_0$	$0.703125 F_0$
3	0.0	$-4875.000000 F_0$	$-0.109375 F_0$
4	0.0	$-3187.500000 F_0$	$-0.671875 F_0$
5	0.0	0.0	$-0.859375 F_0$

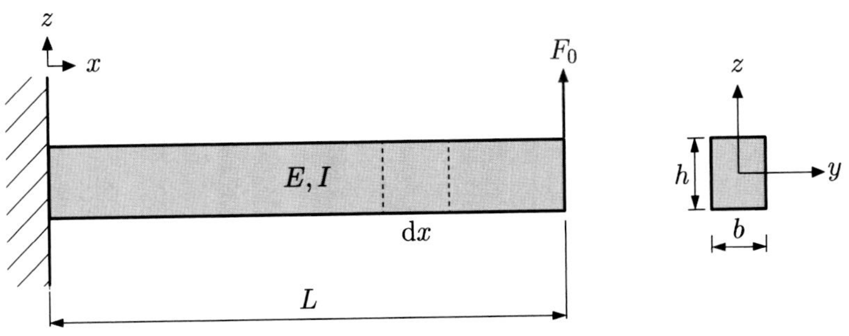

Fig. G.16 General configuration of a cantilever beam with shear force

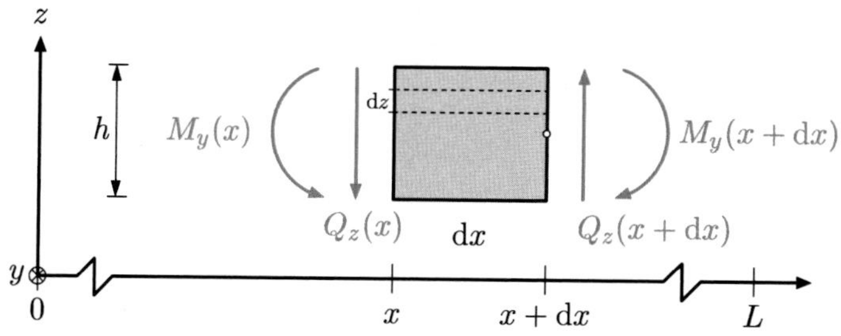

Fig. G.17 Infinitesimal beam element dx in the x–z plane with internal reactions

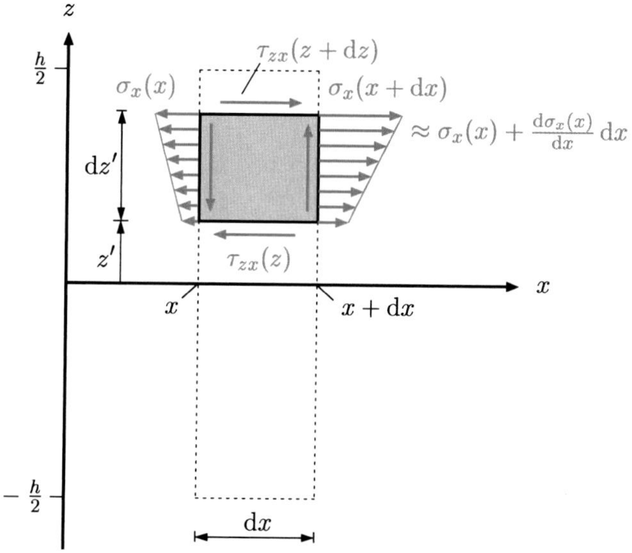

Fig. G.18 Infinitesimal beam element of size dx × dz'. The general configuration is shown in Fig. G.16

Force equilibrium for this infinitesimal element in the x-direction gives:

$$-\sigma_x(x)b\mathrm{d}z + \sigma_x(x+\mathrm{d}x)b\mathrm{d}z - \tau_{zx}(z)b\mathrm{d}x + \tau_{zx}(z+\mathrm{d}z)b\mathrm{d}x = 0\,, \qquad \text{(G.323)}$$

or simplified after a TAYLOR's series expansion of the stresses at $(x + \mathrm{d}x)$ and $(z + \mathrm{d}z)$:

$$\frac{\mathrm{d}\sigma_x(x)}{\mathrm{d}x}\mathrm{d}xb\mathrm{d}z + \frac{\mathrm{d}\tau_{zx}(z)}{\mathrm{d}z}\mathrm{d}zb\mathrm{d}x = 0\,, \qquad \text{(G.324)}$$

or

$$\frac{\mathrm{d}\sigma_x(x)}{\mathrm{d}x} + \frac{\mathrm{d}\tau_{zx}(z)}{\mathrm{d}z} = 0\,. \tag{G.325}$$

Rearranged for

$$\mathrm{d}\tau_{zx}(z) = -\frac{\mathrm{d}\sigma_x(x)}{\mathrm{d}x}\mathrm{d}z \tag{G.326}$$

an integration gives the general result:

$$\tau_{zx}(z) = -\int\limits_0^z \frac{\mathrm{d}\sigma_x(x)}{\mathrm{d}x}\mathrm{d}z' + c\,. \tag{G.327}$$

It follows with

$$\frac{\mathrm{d}\sigma_x(x)}{\mathrm{d}x} = \frac{\mathrm{d}}{\mathrm{d}x}\left(\frac{M_y(x)}{I_y}\times z\right) = \frac{z}{I_y}\times\frac{\mathrm{d}M_y(x)}{\mathrm{d}x} = \frac{Q_z(x)}{I_y}\times z \tag{G.328}$$

for the shear stress distribution:

$$\tau_{zx}(z) = -\int\limits_0^z \frac{Q_z(x)}{I_y}\times z'\mathrm{d}z' + c = -\frac{Q_z(x)}{I_y}\int\limits_0^z z'\mathrm{d}z' + c = -\frac{Q_z(x)}{2I_y}z^2 + c\,. \tag{G.329}$$

The constant of integration c follows from the condition that the shear stresses vanish at the free surface, i.e. $\tau_{zx}(z = \frac{h}{2}) = 0$, as:

$$c = \frac{Q_z(x)}{2I_y}\left(\frac{h}{2}\right)^2\,. \tag{G.330}$$

Thus, the shear stress distribution in beam with a rectangular cross section is finally obtained as follows:

$$\tau_{zx}(z) = \frac{Q_z(x)}{2I_y}\left[\left(\frac{h}{2}\right)^2 - z^2\right]\,. \tag{G.331}$$

The maximum shear stress is obtained for $z = 0$:

$$\tau_{zx,\max} = \frac{Q_z(x)h^2}{8I_y} = \frac{3Q_z(x)}{2bh} = \frac{3Q_z(x)}{2A}\,. \tag{G.332}$$

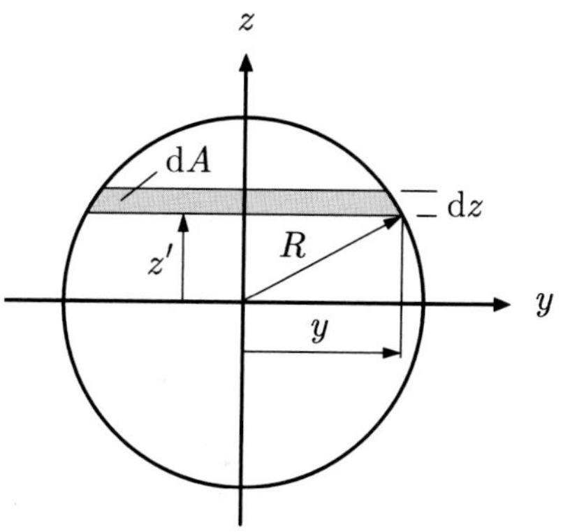

Fig. G.19 Circular cross section used for the derivation of the shear stress τ_{xz}

4.5 Calculation of the shear stress distribution in a circular cross section
Starting point is again the infinitesimal beam element given in Fig. 3.12. However, the cross section is now as indicated in Fig. G.19.

Force equilibrium in the x-direction gives:

$$\int \sigma_x(x)\,\mathrm{d}A - \int \left(\sigma_x(x) + \frac{\mathrm{d}\sigma_x(x)}{\mathrm{d}x}\mathrm{d}x \right) \mathrm{d}A + \tau_{xz} 2y\,\mathrm{d}x = 0\,. \tag{G.333}$$

It results from Eq. (3.26) after differentiation with respect to the x-coordinate:

$$\frac{\mathrm{d}\sigma_x(x)}{\mathrm{d}x} = +\frac{z}{I_y}\frac{\mathrm{d}M_y(x)}{\mathrm{d}x} \overset{(3.35)}{=} \frac{Q_z(x) \times z}{I_y}\,. \tag{G.334}$$

Thus,

$$\tau_{xy} = \frac{Q_z(x)}{2yI_y}\int z\,\mathrm{d}A\,. \tag{G.335}$$

Considering that $\mathrm{d}A = 2y\mathrm{d}z$ and $y = \sqrt{R^2 - z^2}$, the final distribution is obtained as:

$$\tau_{xy} = \frac{Q_z(x)}{3I_y}\left(R^2 - (z')^2\right)\,. \tag{G.336}$$

Maximum shear stress for $z' = 0$:

$$\tau_{xz,\max} = \frac{Q_z(x)R^2}{3I_y} = \frac{4Q_z(x)}{3\pi R^2} = \frac{4Q_z(x)}{3A}\,. \tag{G.337}$$

4.6 Calculation of the shear correction factor for rectangular cross section

$$\int_{\Omega} \frac{1}{2G} \tau_{xz}^2 \mathrm{d}\Omega \overset{!}{=} \int_{\Omega_s} \frac{1}{2G} \left(\frac{Q_z}{A_s} \right)^2 \mathrm{d}\Omega_s \,, \tag{G.338}$$

$$k_s = \frac{Q_z}{A \int\limits_A \tau_{xz}^2 \mathrm{d}A} = \frac{5}{6} \,. \tag{G.339}$$

4.7 Differential equation under consideration of distributed moment

Shear force: no difference, meaning $\frac{\mathrm{d}Q_z(x)}{\mathrm{d}x} = -q_z(x)$.

Bending moment:

$$M_y(x + \mathrm{d}x) - M_y(x) - Q_z(x)\mathrm{d}x + \frac{1}{2} q_z \mathrm{d}x^2 + m_y \mathrm{d}x = 0 \,. \tag{G.340}$$

$$\frac{\mathrm{d}M_y(x)}{\mathrm{d}x} = +Q_z(x) - m_y \,, \tag{G.341}$$

$$\frac{\mathrm{d}^2 M_y(x)}{\mathrm{d}x^2} + \frac{\mathrm{d}m_y(x)}{\mathrm{d}x} = -q_z(x) \,. \tag{G.342}$$

Differential equations:

$$\frac{\mathrm{d}}{\mathrm{d}x} \left(EI_y \frac{\mathrm{d}\phi_y}{\mathrm{d}x} \right) - k_s AG \left(\frac{\mathrm{d}u_z}{\mathrm{d}x} + \phi_y \right) = -m_y(x) \,, \tag{G.343}$$

$$\frac{\mathrm{d}}{\mathrm{d}x} \left[k_s AG \left(\frac{\mathrm{d}u_z}{\mathrm{d}x} + \phi_y \right) \right] = -q_z(x) \,. \tag{G.344}$$

4.8 Differential equations for Timoshenko beam

$$u_z(x) = \frac{1}{EI_y} \left(\frac{q_z x^4}{24} + c_1 \frac{x^3}{6} + c_2 \frac{x^2}{2} + c_3 x + c_4 \right) \,, \tag{G.345}$$

$$\phi_y(x) = -\frac{1}{EI_y} \left(\frac{q_z x^3}{6} + c_1 \frac{x^2}{2} + c_2 x + c_3 \right) - \frac{q_z x}{k_s AG} - \frac{c_1}{k_s AG} + \frac{m_y}{k_s AG} \,. \tag{G.346}$$

4.9 Analytical calculation of the distribution of the deflection and rotation for a cantilever beam under point load

Boundary conditions:

$$u_z(x = 0) = 0 \;\;,\;\; \phi_y(x = 0) = 0 \,, \tag{G.347}$$

$$M_y(x = 0) = -F_0 L \;\;,\;\; Q_z(x = 0) = F_0 \,. \tag{G.348}$$

Integration constants:

$$c_1 = -F_0 \; ; \; c_2 = F_0 L \; ; \; c_3 = \frac{EI_y}{k_s AG} F_0 \; ; \; c_4 = 0 \,. \tag{G.349}$$

Course of the displacement:

$$u_z(x) = \frac{1}{EI_y}\left(-F_0\frac{x^3}{6} + F_0 L\frac{x^2}{2} + \frac{EI_y F_0}{k_s AG}x\right) . \tag{G.350}$$

Course of the rotation:

$$\phi_y(x) = -\frac{1}{EI_y}\left(-F_0\frac{x^2}{2} + F_0 L x\right) . \tag{G.351}$$

Maximal bending:

$$u_z(x = L) = \frac{1}{EI_y}\left(\frac{F_0 L^3}{3} + \frac{EI_y F_0 L}{k_s AG}\right) . \tag{G.352}$$

Rotation at the loading point:

$$\phi_y(x = L) = -\frac{F_0 L^2}{2EI_y} . \tag{G.353}$$

Limit value:

$$u_z(x = L) = \frac{4F_0}{b}\left(\frac{L}{h}\right)^3 + \frac{F_0}{k_s bG}\left(\frac{L}{h}\right) . \tag{G.354}$$

$$u_z(L)|_{h \ll L} \to \frac{4F_0}{b}\left(\frac{L}{h}\right)^3 = \frac{F_0 L^3}{3EI_y} , \tag{G.355}$$

$$u_z(L)|_{h \gg L} \to \frac{F_0}{k_s bG}\left(\frac{L}{h}\right) = \frac{F_0 L}{k_s AG} . \tag{G.356}$$

4.10 Analytical calculation of various quantities for a cantilever beam under point load

- Deflection $u_z(x)$:

$$u_z(x) = \frac{1}{EI_y}\left(-F_0\frac{x^3}{6} + F_0 L\frac{x^2}{2} + \frac{EI_y F_0}{k_s AG}x\right) . \tag{G.357}$$

- Rotation $\phi_y(x)$:

$$\phi_y(x) = -\frac{1}{EI_y}\left(-F_0\frac{x^2}{2} + F_0 Lx\right). \tag{G.358}$$

- Bending moment distribution $M_y(x)$:

$$M_y(x) = +EI_y\frac{\mathrm{d}\phi_y(x)}{\mathrm{d}x} = -F_0(L-x). \tag{G.359}$$

- Shear force distribution $Q_z(x)$:

$$Q_z(x) = +\frac{\mathrm{d}M_y(x)}{\mathrm{d}x} = \frac{\mathrm{d}\phi_y^2(x)}{\mathrm{d}x^2} = F_0. \tag{G.360}$$

- Absolute maximum normal strain $|\varepsilon_{x,\max}(x)|$:

$$\varepsilon_x(x) = +z\frac{\mathrm{d}\phi_y(x)}{\mathrm{d}x} \quad \rightarrow \quad |\varepsilon_{x,\max}(x)| = -\frac{F_0}{EI_y}(L-x)|z_{\max}|. \tag{G.361}$$

- Curvature $\kappa_y(x)$:

$$\kappa_y(x) = \frac{\mathrm{d}\phi_y(x)}{\mathrm{d}x} = -\frac{F_0}{EI_y}(L-x). \tag{G.362}$$

- Absolute maximum normal stress $|\sigma_{x,\max}(x)|$:

$$\sigma_x(x) = E\varepsilon_x \quad \rightarrow \quad |\sigma_{x,\max}(x)| = -\frac{F_0}{I_y}(L-x)|z_{\max}|. \tag{G.363}$$

- Absolute maximum shear stress $|\tau_{xz,\max}(x)|$:

$$\tau_{xz} = \frac{Q_z(x)}{A_\mathrm{s}} = \frac{Q_z(x)}{k_\mathrm{s}A} = \frac{F_0}{k_\mathrm{s}A}. \tag{G.364}$$

- Absolute maximum shear strain $|\gamma_{xz,\max}(x)|$:

$$\gamma xz = \frac{\tau_{xz}}{G} = \frac{F_0}{k_\mathrm{s}AG}. \tag{G.365}$$

Numerical values for the special case are provided in Table G.7.

4.11 Analytical calculation of the normalized deflection for beams with shear contribution

$$I_y = \frac{bh^3}{12},\ A = hb,\ k_\mathrm{s} = \frac{5}{6},\ G = \frac{E}{2(1+\nu)}. \tag{G.366}$$

Table G.7 Numerical results for $h = 0.5$, $E = 200{,}000$, $\nu = 0.3$, $F_0 = 100$, and $L = 2h$ at both ends

Quantity	$x = 0$	$x = L$
u_z	0	0.03824
ϕ_y	0	–0.048
M_y	–100	0
Q_z	100	100
$\lvert\varepsilon_{x,\max}\rvert$	–0.024	0
κ_y	–0.096	0
$\lvert\sigma_{x,\max}\rvert$	-4800	0
γ_{xz}	0.00624	0.00624
τ_{xz}	480	480

$$u_{z,\,\text{norm}} = \frac{1}{3} + \frac{1+\nu}{5}\left(\frac{h}{L}\right)^2 , \tag{G.367}$$

$$u_{z,\,\text{norm}} = \frac{1}{8} + \frac{1+\nu}{10}\left(\frac{h}{L}\right)^2 , \tag{G.368}$$

$$u_{z,\,\text{norm}} = \frac{1}{48} + \frac{1+\nu}{20}\left(\frac{h}{L}\right)^2 . \tag{G.369}$$

4.12 Cantilever beam loaded by a single force

Boundary conditions:

$$u_z(x = L) = 0 \ , \quad \phi_y(x = L) = 0 , \tag{G.370}$$

$$M_y(x = 0) = 0 \ , \quad Q_z(x = 0) = F_0 . \tag{G.371}$$

Integration constants:

$$c_1 = -F_0 \ ; \ c_2 = 0 \ ; \ c_3 = \frac{EI_yF_0}{k_sAG} + \frac{F_0L^2}{2} \ ; \ c_4 = -\frac{EI_yF_0L}{k_sAG} - \frac{F_0L^3}{3} . \tag{G.372}$$

Displacement distribution:

$$u_z(x) = \frac{F_0L^3}{6EI_y}\left(-\frac{x^3}{L^3} + \frac{3x}{L} - 2\right) + \frac{F_0L^3(1+\nu)}{6k_sEI_y}\left(\frac{h^2}{L^2}\right)\left(\frac{x}{L} - 1\right) . \tag{G.373}$$

4.13 Simply supported beam in the elastic range loaded by a distributed load

Boundary conditions:

$$u_z(x=-L)=0\ ,\quad u_z(x=+L)=0\,, \tag{G.374}$$
$$M_y(x=-L)=0\ ,\quad M_y(x=+L)=0\,. \tag{G.375}$$

Integration constants:

$$c_1=0\ ;\ c_2=-\frac{EI_y q_0}{k_s AG}-\frac{q_0 L^2}{2}\ ;\ c_3=0\ ;\ c_4=\frac{EI_y q_0 L^2}{2k_s AG}+\frac{5q_0L^4}{24}\,. \tag{G.376}$$

Displacement distribution:

$$u_z(x)=\frac{q_0L^4}{24EI_y}\left(\frac{x^4}{L^4}-\frac{6x^2}{L^2}+5\right)+\frac{q_0L^4(1+\nu)}{12k_sEI_y}\left(\frac{h^2}{L^2}\right)\left(1-\frac{x^2}{L^2}\right)\,. \tag{G.377}$$

4.14 Timoshenko beam element with quadratic interpolation functions for the displacement and linear interpolation functions for the rotation

The nodal displacement at the middle node as a function of the other unknowns results in:

$$u_{2z}=\frac{u_{1z}+u_{3z}}{2}+\frac{-\phi_{1y}+\phi_{3y}}{8}L+\frac{1}{32}\frac{6L}{k_sAG}\int_0^L q_z(x)N_{2u}(x)\mathrm{d}x\,. \tag{G.378}$$

The additional load matrix on the right-hand side results in:

$$\dots=\dots+\begin{bmatrix}\int_0^L q_z(x)N_{1u}\mathrm{d}x+\frac{1}{2}\int_0^L q_z(x)N_{2u}\mathrm{d}x\\ -\frac{1}{8}L\int_0^L q_z(x)N_{2u}\mathrm{d}x\\ \int_0^L q_z(x)N_{3u}\mathrm{d}x+\frac{1}{2}\int_0^L q_z(x)N_{2u}\mathrm{d}x\\ +\frac{1}{8}L\int_0^L q_z(x)N_{2u}\mathrm{d}x\end{bmatrix}. \tag{G.379}$$

With $\int_0^L N_{1u}\mathrm{d}x=\frac{L}{6}$, $\int_0^L N_{2u}\mathrm{d}x=\frac{2L}{3}$ and $\int_0^L N_{3u}\mathrm{d}x=\frac{L}{6}$ the following results for a constant distributed load q_0:

$$\cdots = \cdots + \begin{bmatrix} \frac{1}{2}q_0 L \\ -\frac{1}{12}q_0 L^2 \\ \frac{1}{2}q_0 L \\ +\frac{1}{12}q_0 L^2 \end{bmatrix}. \tag{G.380}$$

This result is identical with the equivalent distributed load for an EULER–BERNOULLI beam. For this see Table 3.7.

4.15 Timoshenko beam element with cubic interpolation functions for the displacement and quadratic interpolation functions for the rotation

The element is exact!
Deformation in the x–z plane:

$$\frac{2EI_y}{L^3(1+12\Lambda)}\begin{bmatrix} 6 & -3L & -6 & -3L \\ -3L & 2L^2(1+3\Lambda) & 3L & L^2(1-6\Lambda) \\ -6 & 3L & 6 & 3L \\ -3L & L^2(1-6\Lambda) & 3L & 2L^2(1+3\Lambda) \end{bmatrix}\begin{bmatrix} u_{1z} \\ \phi_{1y} \\ u_{2z} \\ \phi_{2y} \end{bmatrix} = \begin{bmatrix} F_{1z} \\ M_{1y} \\ F_{2z} \\ M_{2y} \end{bmatrix}. \tag{G.381}$$

Deformation in the x–y plane:

$$\frac{2EI_z}{L^3(1+12\Lambda)}\begin{bmatrix} 6 & 3L & -6 & 3L \\ 3L & 2L^2(1+3\Lambda) & -3L & L^2(1-6\Lambda) \\ -6 & -3L & 6 & -3L \\ 3L & L^2(1-6\Lambda) & -3L & 2L^2(1+3\Lambda) \end{bmatrix}\begin{bmatrix} u_{1y} \\ \phi_{1z} \\ u_{2y} \\ \phi_{2z} \end{bmatrix} = \begin{bmatrix} F_{1y} \\ M_{1z} \\ F_{2y} \\ M_{2z} \end{bmatrix}. \tag{G.382}$$

4.16 Plane beam-rod structure with Timoshenko element

The reduced system of equations (i.e., under consideration of the boundary conditions $u_{1Z} = u_{3Z} = 0$ and $\varphi_{1Y} = 0$) reads:

$$\begin{bmatrix} 4\frac{k_s AG}{4L} + \frac{EA}{L} & 2L\frac{k_s AG}{4L} \\ 2L\frac{k_s AG}{4L} & (L^2+\alpha)\frac{k_s AG}{4L} \end{bmatrix}\begin{bmatrix} u_{2Z} \\ \varphi_{2Y} \end{bmatrix} = \begin{bmatrix} -F_0 \\ 0 \end{bmatrix}. \tag{G.383}$$

The solution of this system is obtained as:

$$\begin{bmatrix} u_{2Z} \\ \varphi_{2Y} \end{bmatrix} = \frac{1}{4\alpha\frac{k_s AG}{4L} + (L^2+\alpha)\frac{EA}{L}}\begin{bmatrix} -F_0(L^2+\alpha) \\ 2F_0 L \end{bmatrix}. \tag{G.384}$$

- Special case: No rod $\Leftrightarrow$ $EA = 0$:

$$\begin{bmatrix} u_{2Z} \\ \varphi_{2Y} \end{bmatrix} = \frac{L}{\alpha k_s AG}\begin{bmatrix} -F_0(L^2+\alpha) \\ 2F_0 L \end{bmatrix}. \tag{G.385}$$

- Special case: No beam $\Leftrightarrow k_s AG = 0$ (starting from the reduced system with $\varphi_{2Y} = 0$):

$$u_{2Z} = -\frac{F_0 L}{EA}. \tag{G.386}$$

G.4 Problems from Chap. 5

5.7 Uniform representation of plane stress and plane strain

Plane strain case:

$$\frac{E'}{1-\nu'^2}\begin{bmatrix} 1 & \nu' & 0 \\ \nu' & 1 & 0 \\ 0 & 0 & \frac{1-\nu'^2}{2} \end{bmatrix} = \frac{E}{\frac{(1-2\nu+\nu^2-\nu^2)}{(1-\nu)^2}(1-\nu^2)(1-\nu)}\begin{bmatrix} 1-\nu & \nu & 0 \\ \nu & 1-\nu & 0 \\ 0 & 0 & \frac{1}{2}(1-2\nu) \end{bmatrix}. \tag{G.387}$$

Using the identity $(1-\nu^2) = (1-\nu)(1+\nu)$ results in the formulation of Eq. (5.9).

5.8 Plane elasticity bending problem

The reduced 4×4 stiffness matrix is obtained as:

$$\boldsymbol{K} = \begin{bmatrix} 3273.809528 & -1339.285715 & -526.5567755 & -103.0219780 \\ -1339.285715 & 4761.904762 & 103.0219782 & -3800.366300 \\ -526.5567755 & 103.0219782 & 3273.809528 & 1339.285715 \\ -103.0219780 & -3800.366300 & 1339.285715 & 4761.904762 \end{bmatrix}. \tag{G.388}$$

The unknown displacements are obtained as $\boldsymbol{u} = \boldsymbol{K}^{\mathrm{T}}\boldsymbol{f}$:

$$u_{2x} = -0.045818\,,$$
$$u_{2y} = -0.120727\,,$$
$$u_{3x} = 0.045818\,,$$
$$u_{3y} = -0.120727\,.$$

G.5 Problems from Chap. 6

6.4 Alternative definition of rotational angle

$$u_x = +z\varphi_x = -z\frac{\mathrm{d}u_z}{\mathrm{d}x}\,,\quad u_y = +z\varphi_y = -z\frac{\mathrm{d}u_z}{\mathrm{d}y}. \tag{G.389}$$

6.5 Derivation of stiffness matrix: application of Green–Gauss theorem

Let us reformulate Eq. (6.53) based on $\mathcal{L}_2^{\mathrm{T}} = (\mathcal{L}_1\mathcal{L}_{1*})^{\mathrm{T}} = \mathcal{L}_{1*}^{\mathrm{T}}\mathcal{L}_1^{\mathrm{T}}$ where $\mathcal{L}_{1*}^{\mathrm{T}} = \left[\frac{\partial}{\partial x}\ \frac{\partial}{\partial y}\right]$ and $\mathcal{L}_1$ as given by Eqs. (5.17) and (5.18):

$$\int_A W^{\mathrm{T}}\left(\mathcal{L}_{1*}^{\mathrm{T}}\left[\mathcal{L}_1^{\mathrm{T}}\left(\boldsymbol{D}\mathcal{L}_2 u_z\right)\right]\right)\mathrm{d}A \cdots = \cdots . \tag{G.390}$$

First application of the GREEN–GAUSS theorem gives:

$$-\int_A (\mathcal{L}_{1*}W)^{\mathrm{T}}\left[\mathcal{L}_1^{\mathrm{T}}\left(\boldsymbol{D}\mathcal{L}_2 u_z\right)\right]\mathrm{d}A \cdots = \cdots . \tag{G.391}$$

Second application of the GREEN–GAUSS theorem gives:

$$\int_A (\mathcal{L}_1\mathcal{L}_{1*}W)^{\mathrm{T}}\boldsymbol{D}\mathcal{L}_2 u_z\,\mathrm{d}A \cdots = \cdots , \tag{G.392}$$

or with $\mathcal{L}_2 = \mathcal{L}_1\mathcal{L}_{1*}$:

$$\int_A (\mathcal{L}_2 W)^{\mathrm{T}}\boldsymbol{D}\mathcal{L}_2 u_z\,\mathrm{d}A \cdots = \cdots . \tag{G.393}$$

Under consideration of $(\mathcal{L}_2 W)^{\mathrm{T}} = ((\mathcal{L}_2\boldsymbol{N}^{\mathrm{T}})\delta\boldsymbol{u}_{\mathrm{p}})^{\mathrm{T}} = \delta\boldsymbol{u}_{\mathrm{p}}^{\mathrm{T}}(\mathcal{L}_2\boldsymbol{N}^{\mathrm{T}})^{\mathrm{T}}$, the formulation given in Eq. (6.55) is obtained.

6.6 Derivation of boundary load matrix: application of Green–Gauss theorem

First application of the GREEN–GAUSS theorem gives:

$$-\int_A (\mathcal{L}_{1*}W)^{\mathrm{T}}\left[\mathcal{L}_1^{\mathrm{T}}\left(\boldsymbol{D}\mathcal{L}_2 u_z\right)\right]\mathrm{d}A + \int_s W^{\mathrm{T}}\left[\mathcal{L}_1^{\mathrm{T}}\left(\boldsymbol{D}\mathcal{L}_2 u_z\right)\right]^{\mathrm{T}}\boldsymbol{n}\mathrm{d}s = 0\,. \tag{G.394}$$

Second application of the GREEN–GAUSS theorem gives:

$$\int_A (\mathcal{L}_1\mathcal{L}_{1*}W)^{\mathrm{T}}\boldsymbol{D}\mathcal{L}_2 u_z\mathrm{d}A - \int_s (\mathcal{L}_{1*}W)^{\mathrm{T}}\left[\boldsymbol{D}\mathcal{L}_2 u_z\right]^{\mathrm{T}}\boldsymbol{n}\mathrm{d}s$$
$$+\int_s W^{\mathrm{T}}\left[\mathcal{L}_1^{\mathrm{T}}\left(\boldsymbol{D}\mathcal{L}_2 u_z\right)\right]^{\mathrm{T}}\boldsymbol{n}\mathrm{d}s = 0\,. \tag{G.395}$$

Or rearranged:

$$\int_A (\underbrace{\boldsymbol{\mathcal{L}}_1\boldsymbol{\mathcal{L}}_{1^*}}_{\boldsymbol{\mathcal{L}}_2} W)^{\mathrm{T}} \boldsymbol{D}\boldsymbol{\mathcal{L}}_2 u_z \mathrm{d}A = + \int_s (\boldsymbol{\mathcal{L}}_{1^*} W)^{\mathrm{T}} \underbrace{\left[\boldsymbol{D}\boldsymbol{\mathcal{L}}_2 u_z\right]^{\mathrm{T}}}_{-(\boldsymbol{M}^{\mathrm{n}})^{\mathrm{T}}} \boldsymbol{n} \mathrm{d}s - \int_s W^{\mathrm{T}} \underbrace{\left[\boldsymbol{\mathcal{L}}_1^{\mathrm{T}} (\boldsymbol{D}\boldsymbol{\mathcal{L}}_2 u_z)\right]^{\mathrm{T}}}_{-(\boldsymbol{Q}^{\mathrm{n}})^{\mathrm{T}}} \boldsymbol{n} \mathrm{d}s = 0 . \tag{G.396}$$

It is advantageous for the further derivation to express $\boldsymbol{M}^{\mathrm{n}}$ as a (2×2)-matrix.

6.7 Derivation of the weak form based on an alternative formulation of the partial differential equation

The weak form is obtained as follows:

$$\int_V (\boldsymbol{\mathcal{L}}_2 W)^{\mathrm{T}} \frac{\boldsymbol{D}}{h} (\boldsymbol{\mathcal{L}}_2 u_z) \,\mathrm{d}V = \int_A W^{\mathrm{T}} \left(\boldsymbol{Q}^{\mathrm{n}}\right)^{\mathrm{T}} \boldsymbol{n} \mathrm{d}A - \int_A (\boldsymbol{\mathcal{L}}_{1^*} W)^{\mathrm{T}} (\boldsymbol{M}^{\mathrm{n}})^{\mathrm{T}} \boldsymbol{n} \mathrm{d}A + \int_V W^{\mathrm{T}} \frac{q_z}{h} \mathrm{d}V . \tag{G.397}$$

6.8 Interpolation functions: angle between plate normal vector and different directions

A normal vector $\boldsymbol{n}_N$ to the surface of $N_{1\varphi_y}$ at point $(\xi, \eta, N_{1\varphi_y}(\xi, \eta))$ can be expressed as

$$\boldsymbol{n}_N = \left[-\frac{\partial N_{1\varphi_y}(\xi, \eta)}{\partial \xi}, -\frac{\partial N_{1\varphi_y}(\xi, \eta)}{\partial \eta}, 1 \right]^{\mathrm{T}} , \tag{G.398}$$

whereas the angle (see Eq. (A.148)) between this normal vector and an arbitrary vector $\boldsymbol{v} = \left[v_x, v_y, v_z\right]^{\mathrm{T}}$ is given by:

$$\cos \measuredangle(\boldsymbol{n}, \boldsymbol{v}) = \frac{\boldsymbol{n}\boldsymbol{v}}{|\boldsymbol{n}||\boldsymbol{v}|} . \tag{G.399}$$

The calculate angles between the normal vector in the points $(-1, -1) \vee (1, -1)$ and the directional vectors $(1, 0, 0)$, $(0, 1, 0)$, and $(1, 1, 0)$ are summarized in Table G.8.

6.9 Interpolation functions: rate of change in direction of the Cartesian and natural axes

The functional values of $N_{1\varphi_x}$ and $N_{1\varphi_y}$ are summarized in Table G.9.

The partial derivatives are summarized in Table G.10.

6.10 Interpolation functions in Cartesian coordinates

$$N_{1u} = N_1 = \frac{(b-y)(a-x)\left(2b^2a^2 - a^2by - a^2y^2 - ab^2x - b^2x^2\right)}{8a^3b^3} , \tag{G.400}$$

Table G.8 Angles between surface normal vector and vector $\boldsymbol{v}$

$\boldsymbol{v}$	$\boldsymbol{n}(-1,-1)=\left[1,0,1\right]^{\mathrm{T}}$	$\boldsymbol{n}(+1,-1)=\left[0,0,1\right]^{\mathrm{T}}$
$\left[1,0,0\right]^{\mathrm{T}}$	45°	90°
$\left[0,1,0\right]^{\mathrm{T}}$	90°	90°
$\left[1,1,0\right]^{\mathrm{T}}$	60°	90°

Table G.9 Functional values of $N_{1\varphi_x}$ and $N_{1\varphi_y}$ at the four nodes

Node	$N_{1\varphi_x}(\xi,\eta)$	$N_{1\varphi_y}(\xi,\eta)$
$1(-1,-1)$	0	0
$2(+1,-1)$	0	0
$3(+1,+1)$	0	0
$4(-1,+1)$	0	0

Table G.10 Partial derivatives of $N_{1\varphi_x}$ and $N_{1\varphi_y}$ at the four nodes

Node	$\frac{\partial N_{1\varphi_x}(\xi,\eta)}{\partial\xi}$	$\frac{\partial N_{1\varphi_x}(\xi,\eta)}{\partial x}$	$\frac{\partial N_{1\varphi_x}(\xi,\eta)}{\partial\eta}$	$\frac{\partial N_{1\varphi_x}(\xi,\eta)}{\partial y}$
$1(-1,-1)$	0	0	b	1
$2(+1,-1)$	0	0	0	0
$3(+1,+1)$	0	0	0	0
$4(-1,+1)$	0	0	0	0
Node	$\frac{\partial N_{1\varphi_y}(\xi,\eta)}{\partial\xi}$	$\frac{\partial N_{1\varphi_y}(\xi,\eta)}{\partial x}$	$\frac{\partial N_{1\varphi_y}(\xi,\eta)}{\partial\eta}$	$\frac{\partial N_{1\varphi_y}(\xi,\eta)}{\partial y}$
$1(-1,-1)$	–a	–1	0	0
$2(+1,-1)$	0	0	0	0
$3(+1,+1)$	0	0	0	0
$4(-1,+1)$	0	0	0	0

$$N_{1\varphi_x}=N_2=\frac{(b+y)(b-y)^2(a-x)}{8ab}, \tag{G.401}$$

$$N_{1\varphi_y}=N_3=-\frac{(a+x)(a-x)^2(b-y)}{8ab}, \tag{G.402}$$

$$N_{2u}=N_4=\frac{(b-y)(a+x)\left(2\,b^2a^2-a^2by-a^2y^2+ab^2x-b^2x^2\right)}{8a^3b^3}, \tag{G.403}$$

$$N_{2\varphi_x}=N_5=\frac{(b+y)(b-y)^2(a+x)}{8ab}, \tag{G.404}$$

$$N_{2\varphi_y}=N_6=\frac{(a-x)(a+x)^2(b-y)}{8ab}, \tag{G.405}$$

$$N_{3u}=N_7=\frac{(b+y)(a+x)\left(2\,b^2a^2+a^2by-a^2y^2+ab^2x-b^2x^2\right)}{8a^3b^3}, \tag{G.406}$$

$$N_{3\varphi_x}=N_8=-\frac{(b-y)(b+y)^2(a+x)}{8ab}, \tag{G.407}$$

$$N_{3\varphi_y}=N_9=\frac{(a-x)(a+x)^2(b+y)}{8ab}, \tag{G.408}$$

$$N_{4u} = N_{10} = \frac{(b+y)(a-x)\left(2b^2a^2+a^2by-a^2y^2-ab^2x-b^2x^2\right)}{8a^3b^3}\,, \tag{G.409}$$

$$N_{4\varphi_x} = N_{11} = -\frac{(b-y)(b+y)^2(a-x)}{8ab}\,, \tag{G.410}$$

$$N_{4\varphi_y} = N_{12} = -\frac{(a+x)(a-x)^2(b+y)}{8ab}\,. \tag{G.411}$$

6.11 Second-order derivatives of interpolation functions in Cartesian coordinates

$$\boldsymbol{B} = \frac{1}{4}\begin{bmatrix} \frac{3(b-y)x}{ba^3} & \frac{3(a-x)y}{ab^3} & \frac{4a^2b^2-3a^2y^2-3b^2x^2}{a^3b^3} \\ 0 & -\frac{(a-x)(b-3y)}{ab^2} & \frac{(b-y)(b+3y)}{ab^2} \\ \frac{(b-y)(a-3x)}{a^2b} & 0 & -\frac{(a-x)(a+3x)}{a^2b} \\ \vdots & \vdots & \vdots \\ \frac{(b+y)(a-3x)}{a^2b} & 0 & \frac{(a-x)(a+3x)}{a^2b} \end{bmatrix}. \tag{G.412}$$

6.12 Two-element example of a plate fixed at two edges

$$u_{2Z} = u_{3Z} = -\frac{4(2a^2+b^2-b^2\nu)a^3F_0}{bEh^3(4a^2+2b^2-a^2\nu^2-2b^2\nu)}\,, \tag{G.413}$$

$$\varphi_{2X} = -\varphi_{3X} = \frac{\nu a^3 F_0}{2D(4a^2+2b^2-a^2\nu^2-2b^2\nu)}\,, \tag{G.414}$$

$$\varphi_{2Y} = \varphi_{3Y} = 0\,. \tag{G.415}$$

The special case $\nu \to 0$ gives $u_Z = -\frac{2a^3F_0}{Ebh^3}$ and $\varphi_X = \varphi_Y = 0$ which is equal to the EULER–BERNOULLI solution.

6.13 Symmetry solution for a plate fixed at all four edges

Reduced system of equations:

$$K_{7\text{-}7} \times u_{3Z} = -\frac{F_0}{4}\,. \tag{G.416}$$

Solution:

$$u_{3Z} = \frac{10a^2}{D(27-2\nu)} \times \frac{F_0}{4} = -\frac{5a^2F_0}{2D(27-2\nu)}\,. \tag{G.417}$$

6.14 Investigation of displacement and slope consistency along boundaries

Consider the boundary $(x, y = 0)$, i.e. between node 1 and 2 in Fig. 6.6a. Evaluation of Eqs. (6.69), (6.72) and (6.74) in Cartesian coordinates for $y = 0$ gives:

$$u_z^{\mathrm{e}}(y=0) = a_1 + a_2x + a_4x^2 + a_7x^7\,, \tag{G.418}$$

$$\varphi_x^{\mathrm{e}}(y=0) = a_3 + a_5x + a_8x^2 + a_{11}x^3\,, \tag{G.419}$$

$$\varphi_y^{\mathrm{e}}(y=0) = -(a_2 + 2a_4x + 3a_7x^2)\,. \tag{G.420}$$

Four DOF from node 1 and 2 can be used to determine a_1, a_2, a_4 and a_7 and thus u_z^{e} and φ_y^{e} (which are continuous along element boundaries). However, the remaining two DOF do not allow to uniquely define the four constants a_3, a_5, a_8 and a_{11} for φ_x^{e}. Thus, a slope discontinuity occurs for φ_x^{e}.

G.6 Problems from Chap. 7

7.3 Alternative definition of rotational angle

$$u_x = +z\phi_x \ , \ \ u_y = +z\phi_y \, . \tag{G.421}$$

7.4 Basic equations for alternative definition of rotational angle

The basic equations are summarized in Table G.11.

7.5 Stiffness matrix for a square four-node thick plate element

The $\boldsymbol{B}$-matrix is obtained as shown in the following equation:

$$\boldsymbol{B}^{\mathrm{T}} = \begin{bmatrix} 0 & 0 & \frac{\eta-1}{2a} & 0 & 0 & \frac{1-\eta}{2a} & 0 & 0 & \frac{\eta+1}{2a} & 0 & 0 & \frac{-\eta-1}{2a} \\ 0 & -\frac{\xi-1}{2a} & 0 & 0 & -\frac{-\xi-1}{2a} & 0 & 0 & -\frac{\xi+1}{2a} & 0 & 0 & -\frac{1-\xi}{2a} & 0 \\ 0 & -\frac{\eta-1}{2a} & \frac{\xi-1}{2a} & 0 & -\frac{1-\eta}{2a} & \frac{-\xi-1}{2a} & 0 & -\frac{\eta+1}{2a} & \frac{\xi+1}{2a} & 0 & -\frac{-\eta-1}{2a} & \frac{1-\xi}{2a} \\ \frac{\eta-1}{2a} & 0 & \frac{(1-\eta)(1-\xi)}{4} & \frac{1-\eta}{2a} & 0 & \frac{(1-\eta)(\xi+1)}{4} & \frac{\eta+1}{2a} & 0 & \frac{(\eta+1)(\xi+1)}{4} & \frac{-\eta-1}{2a} & 0 & \frac{(\eta+1)(1-\xi)}{4} \\ \frac{\xi-1}{2a} & -\frac{(1-\eta)(1-\xi)}{4} & 0 & \frac{-\xi-1}{2a} & -\frac{(1-\eta)(\xi+1)}{4} & 0 & \frac{\xi+1}{2a} & -\frac{(\eta+1)(\xi+1)}{4} & 0 & \frac{1-\xi}{2a} & -\frac{(\eta+1)(1-\xi)}{4} & 0 \end{bmatrix} . \tag{G.422}$$

G.7 Problems from Chap. 8

8.4 Hooke's law in terms of shear and bulk modulus

Elastic stiffness form:

Table G.11 Different formulations of the basic equations for a thick plate

Specific formulation	General formulation
Kinematics	
$\begin{bmatrix} \frac{\partial\phi_x}{\partial x} \\ \frac{\partial\phi_y}{\partial y} \\ \frac{\partial\phi_x}{\partial y}+\frac{\partial\phi_y}{\partial x} \\ \phi_x+\frac{\partial u_z}{\partial x} \\ \phi_y+\frac{\partial u_z}{\partial y}\end{bmatrix} = \begin{bmatrix} \frac{\partial}{\partial x} & 0 & 0 \\ 0 & \frac{\partial}{\partial y} & 0 \\ \frac{\partial}{\partial y} & \frac{\partial}{\partial x} & 0 \\ 1 & 0 & \frac{\partial}{\partial x} \\ 0 & 1 & \frac{\partial}{\partial y}\end{bmatrix}\begin{bmatrix}\phi_x \\ \phi_y \\ u_z\end{bmatrix}$	$\boldsymbol{e} = \boldsymbol{\mathcal{L}}_1\boldsymbol{u}$
Constitution	
$\begin{bmatrix} M_x^{\text{n}} \\ M_y^{\text{n}} \\ M_{xy}^{\text{n}} \\ -Q_x^{\text{n}} \\ -Q_y^{\text{n}}\end{bmatrix} = \left[\begin{array}{c\|c} \frac{Eh^3}{12(1-\nu^2)}\begin{bmatrix} 1 & \nu & 0 \\ \nu & 1 & 0 \\ 0 & 0 & \frac{1-\nu}{2}\end{bmatrix} & \begin{bmatrix} 0 & 0 \\ 0 & 0 \\ 0 & 0\end{bmatrix} \\ \hline \begin{bmatrix} 0 & 0 & 0 \\ 0 & 0 & 0\end{bmatrix} & -k_{\text{s}}Gh\begin{bmatrix} 1 & 0 \\ 0 & 1\end{bmatrix}\end{array}\right]\begin{bmatrix} \frac{\partial\phi_x}{\partial x} \\ \frac{\partial\phi_y}{\partial y} \\ \frac{\partial\phi_x}{\partial y}+\frac{\partial\phi_y}{\partial x} \\ \phi_x+\frac{\partial u_z}{\partial x} \\ \phi_y+\frac{\partial u_z}{\partial y}\end{bmatrix}$	$\boldsymbol{s} = \boldsymbol{D}\boldsymbol{e}$
Equilibrium	
$\begin{bmatrix} \frac{\partial}{\partial x} & 0 & \frac{\partial}{\partial y} & 1 & 0 \\ 0 & \frac{\partial}{\partial y} & \frac{\partial}{\partial x} & 0 & 1 \\ 0 & 0 & 0 & \frac{\partial}{\partial x} & \frac{\partial}{\partial y}\end{bmatrix}\begin{bmatrix} M_x^{\text{n}} \\ M_y^{\text{n}} \\ M_{xy}^{\text{n}} \\ -Q_x^{\text{n}} \\ -Q_y^{\text{n}}\end{bmatrix} + \begin{bmatrix} m_y \\ m_x \\ -q_z\end{bmatrix} = \begin{bmatrix} 0 \\ 0 \\ 0\end{bmatrix}$	$\boldsymbol{\mathcal{L}}_1^{\text{T}}\boldsymbol{s} + \boldsymbol{b} = \boldsymbol{0}$
PDE	
$\begin{bmatrix} D_{\text{b}}\left(\frac{\partial^2}{\partial x^2}+\frac{1-\nu}{2}\frac{\partial^2}{\partial y^2}\right)-D_{\text{s}} & \frac{1+\nu}{2}D_{\text{b}}\frac{\partial^2}{\partial x\partial y} & -D_{\text{s}}\frac{\partial}{\partial x} \\ \frac{1+\nu}{2}D_{\text{b}}\frac{\partial^2}{\partial x\partial y} & D_{\text{b}}\left(\frac{1-\nu}{2}\frac{\partial^2}{\partial x^2}+\frac{\partial^2}{\partial y^2}\right)-D_{\text{s}} & -D_{\text{s}}\frac{\partial}{\partial y} \\ -D_{\text{s}}\frac{\partial}{\partial x} & -D_{\text{s}}\frac{\partial}{\partial y} & -D_{\text{s}}\left(\frac{\partial^2}{\partial x^2}+\frac{\partial^2}{\partial y^2}\right)\end{bmatrix}$ $\begin{bmatrix}\phi_x \\ \phi_y \\ u_z\end{bmatrix} + \begin{bmatrix} m_x \\ m_y \\ -q_z\end{bmatrix} = \begin{bmatrix} 0 \\ 0 \\ 0\end{bmatrix}$	$\boldsymbol{\mathcal{L}}_1^{\text{T}}\boldsymbol{D}\boldsymbol{\mathcal{L}}_1\boldsymbol{u} + \boldsymbol{b} = \boldsymbol{0}$

$$\begin{bmatrix}\sigma_x \\ \sigma_y \\ \sigma_z \\ \sigma_{xy} \\ \sigma_{yz} \\ \sigma_{xz}\end{bmatrix} = \begin{bmatrix} K+\frac{4}{3}G & K-\frac{2}{3}G & K-\frac{2}{3}G & 0 & 0 & 0 \\ K-\frac{2}{3}G & K+\frac{4}{3}G & K-\frac{2}{3}G & 0 & 0 & 0 \\ K-\frac{2}{3}G & K-\frac{2}{3}G & K+\frac{4}{3}G & 0 & 0 & 0 \\ 0 & 0 & 0 & G & 0 & 0 \\ 0 & 0 & 0 & 0 & G & 0 \\ 0 & 0 & 0 & 0 & 0 & G\end{bmatrix}\begin{bmatrix}\varepsilon_x \\ \varepsilon_y \\ \varepsilon_z \\ 2\varepsilon_{xy} \\ 2\varepsilon_{yz} \\ 2\varepsilon_{xz}\end{bmatrix}. \tag{G.423}$$

Elastic compliance form:

$$\begin{bmatrix} \varepsilon_x \\ \varepsilon_y \\ \varepsilon_z \\ 2\,\varepsilon_{xy} \\ 2\,\varepsilon_{yz} \\ 2\,\varepsilon_{xz} \end{bmatrix} = \frac{1}{18KG} \begin{bmatrix} 6K+2G & -3K+2G & -3K+2G & 0 & 0 & 0 \\ -3K+2G & 6K+2G & -3K+2G & 0 & 0 & 0 \\ -3K+2G & -3K+2G & 6K+2G & 0 & 0 & 0 \\ 0 & 0 & 0 & 18K & 0 & 0 \\ 0 & 0 & 0 & 0 & 18K & 0 \\ 0 & 0 & 0 & 0 & 0 & 18K \end{bmatrix} \begin{bmatrix} \sigma_x \\ \sigma_y \\ \sigma_z \\ \sigma_{xy} \\ \sigma_{yz} \\ \sigma_{xz} \end{bmatrix}. \tag{G.424}$$

8.5 Hooke's law in terms of Lamé's constants

Elastic stiffness form:

$$\begin{bmatrix} \sigma_x \\ \sigma_y \\ \sigma_z \\ \sigma_{xy} \\ \sigma_{yz} \\ \sigma_{xz} \end{bmatrix} = \begin{bmatrix} \lambda+2\mu & \lambda & \lambda & 0 & 0 & 0 \\ \lambda & \lambda+2\mu & \lambda & 0 & 0 & 0 \\ \lambda & \lambda & \lambda+2\mu & 0 & 0 & 0 \\ 0 & 0 & 0 & \mu & 0 & 0 \\ 0 & 0 & 0 & 0 & \mu & 0 \\ 0 & 0 & 0 & 0 & 0 & \mu \end{bmatrix} \begin{bmatrix} \varepsilon_x \\ \varepsilon_y \\ \varepsilon_z \\ 2\,\varepsilon_{xy} \\ 2\,\varepsilon_{yz} \\ 2\,\varepsilon_{xz} \end{bmatrix}. \tag{G.425}$$

Elastic compliance form:

$$\begin{bmatrix} \varepsilon_x \\ \varepsilon_y \\ \varepsilon_z \\ 2\,\varepsilon_{xy} \\ 2\,\varepsilon_{yz} \\ 2\,\varepsilon_{xz} \end{bmatrix} = \begin{bmatrix} \frac{\lambda+\mu}{\mu(3\lambda+2\mu)} & -\frac{\lambda}{2\mu(3\lambda+2\mu)} & -\frac{\lambda}{2\mu(3\lambda+2\mu)} & 0 & 0 & 0 \\ -\frac{\lambda}{2\mu(3\lambda+2\mu)} & \frac{\lambda+\mu}{\mu(3\lambda+2\mu)} & -\frac{\lambda}{2\mu(3\lambda+2\mu)} & 0 & 0 & 0 \\ -\frac{\lambda}{2\mu(3\lambda+2\mu)} & -\frac{\lambda}{2\mu(3\lambda+2\mu)} & \frac{\lambda+\mu}{\mu(3\lambda+2\mu)} & 0 & 0 & 0 \\ 0 & 0 & 0 & \frac{1}{\mu} & 0 & 0 \\ 0 & 0 & 0 & 0 & \frac{1}{\mu} & 0 \\ 0 & 0 & 0 & 0 & 0 & \frac{1}{\mu} \end{bmatrix} \begin{bmatrix} \sigma_x \\ \sigma_y \\ \sigma_z \\ \sigma_{xy} \\ \sigma_{yz} \\ \sigma_{xz} \end{bmatrix}. \tag{G.426}$$

8.6 Hooke's law for the plane stress state

Condition:

$$\sigma_z = \frac{E}{(1+\nu)(1-2\nu)}\left[\nu\varepsilon_x + \nu\varepsilon_y + (1-\nu)\varepsilon_z\right] \stackrel{!}{=} 0\,, \tag{G.427}$$

$$\rightarrow\ \varepsilon_z = -\frac{\nu}{1-\nu}\left(\varepsilon_x + \varepsilon_y\right). \tag{G.428}$$

Elastic stiffness form:

$$\begin{bmatrix} \sigma_x \\ \sigma_y \\ \sigma_{xy} \end{bmatrix} = \frac{E}{1-\nu^2} \begin{bmatrix} 1 & \nu & 0 \\ \nu & 1 & 0 \\ 0 & 0 & \frac{1-\nu}{2} \end{bmatrix} \begin{bmatrix} \varepsilon_x \\ \varepsilon_y \\ 2\,\varepsilon_{xy} \end{bmatrix}. \tag{G.429}$$

Elastic compliance form:

$$\begin{bmatrix} \varepsilon_x \\ \varepsilon_y \\ 2\,\varepsilon_{xy} \end{bmatrix} = \frac{1}{E} \begin{bmatrix} 1 & -\nu & 0 \\ -\nu & 1 & 0 \\ 0 & 0 & 2(\nu+1) \end{bmatrix} \begin{bmatrix} \sigma_x \\ \sigma_y \\ \sigma_{xy} \end{bmatrix}. \tag{G.430}$$

Furthermore:

$$\varepsilon_z = -\frac{\nu}{1-\nu}(\varepsilon_x + \varepsilon_y) = -\frac{\nu}{E}(\sigma_x + \sigma_y). \tag{G.431}$$

8.7 Hooke's law for the plane strain state

Elastic stiffness form:

$$\begin{bmatrix} \sigma_x \\ \sigma_y \\ \sigma_{xy} \end{bmatrix} = \frac{E}{(1+\nu)(1-2\nu)} \begin{bmatrix} 1-\nu & \nu & 0 \\ \nu & 1-\nu & 0 \\ 0 & 0 & \frac{1-2\nu}{2} \end{bmatrix} \begin{bmatrix} \varepsilon_x \\ \varepsilon_y \\ 2\,\varepsilon_{xy} \end{bmatrix}. \tag{G.432}$$

Elastic compliance form:

$$\begin{bmatrix} \varepsilon_x \\ \varepsilon_y \\ 2\,\varepsilon_{xy} \end{bmatrix} = \frac{1-\nu^2}{E} \begin{bmatrix} 1 & -\frac{\nu}{1-\nu} & 0 \\ -\frac{\nu}{1-\nu} & 1 & 0 \\ 0 & 0 & \frac{2}{1-\nu} \end{bmatrix} \begin{bmatrix} \sigma_x \\ \sigma_y \\ \sigma_{xy} \end{bmatrix}. \tag{G.433}$$

Furthermore:

$$\sigma_z = \nu\,(\sigma_x + \sigma_y). \tag{G.434}$$

8.8 Beltrami–Michell equations

$$\boldsymbol{\mathcal{L}}^{\mathrm{T}}\boldsymbol{\sigma} + \boldsymbol{b} = \boldsymbol{0}. \tag{G.435}$$

8.9 Lamé–Navier equations in matrix notation

Thee symmetric elasticity matrix for isotropic material behavior can be written, for example, as:

$$\boldsymbol{C} = \begin{bmatrix} C_{11} & C_{12} & C_{12} & 0 & 0 & 0 \\ C_{12} & C_{11} & C_{12} & 0 & 0 & 0 \\ C_{12} & C_{12} & C_{11} & 0 & 0 & 0 \\ 0 & 0 & 0 & C_{44} & 0 & 0 \\ 0 & 0 & 0 & 0 & C_{44} & 0 \\ 0 & 0 & 0 & 0 & 0 & C_{44} \end{bmatrix}. \tag{G.436}$$

Thus, the LAMÉ–NAVIER equations can finally be stated as

$$\begin{bmatrix} \frac{C_{11}\partial^2}{\partial x^2}+\frac{C_{44}\partial^2}{\partial y^2}+\frac{C_{44}\partial^2}{\partial z^2} & \frac{C_{12}\partial^2}{\partial x\partial y}+\frac{C_{44}\partial^2}{\partial y\partial x} & \frac{C_{12}\partial^2}{\partial x\partial z}+\frac{C_{44}\partial^2}{\partial z\partial x} \\ \frac{C_{12}\partial^2}{\partial y\partial x}+\frac{C_{44}\partial^2}{\partial x\partial y} & \frac{C_{11}\partial^2}{\partial y^2}+\frac{C_{44}\partial^2}{\partial x^2}+\frac{C_{44}\partial^2}{\partial z^2} & \frac{C_{12}\partial^2}{\partial y\partial z}+\frac{C_{44}\partial^2}{\partial z\partial y} \\ \frac{C_{12}\partial^2}{\partial z\partial x}+\frac{C_{44}\partial^2}{\partial x\partial z} & \frac{C_{12}\partial^2}{\partial z\partial y}+\frac{C_{44}\partial^2}{\partial y\partial z} & \frac{C_{11}\partial^2}{\partial z^2}+\frac{C_{44}\partial^2}{\partial y^2}+\frac{C_{44}\partial^2}{\partial x^2} \end{bmatrix} \begin{bmatrix} u_x \\ u_y \\ u_z \end{bmatrix} + \begin{bmatrix} f_x \\ f_y \\ f_z \end{bmatrix} = \begin{bmatrix} 0 \\ 0 \\ 0 \end{bmatrix}. \tag{G.437}$$

8.10 Green–Gauss theorem applied to equilibrium equation in x-direction

$$\underbrace{\begin{bmatrix} \sigma_x \\ 0 \\ 0 \\ \sigma_{xy} \\ 0 \\ \sigma_{xy} \end{bmatrix}}_{\sigma_x} = \underbrace{\frac{E}{(1+\nu)(1-2\nu)} \begin{bmatrix} 1-\nu & \nu & \nu & 0 & 0 & 0 \\ 0 & 0 & 0 & 0 & 0 & 0 \\ 0 & 0 & 0 & 0 & 0 & 0 \\ 0 & 0 & 0 & \frac{1-2\nu}{2} & 0 & 0 \\ 0 & 0 & 0 & 0 & 0 & 0 \\ 0 & 0 & 0 & 0 & 0 & \frac{1-2\nu}{2} \end{bmatrix}}_{C_x} \underbrace{\begin{bmatrix} \varepsilon_x \\ \varepsilon_y \\ \varepsilon_z \\ 2\,\varepsilon_{xy} \\ 0 \\ 2\,\varepsilon_{xz} \end{bmatrix}}_{\varepsilon_x}, \tag{G.438}$$

$$\underbrace{\begin{bmatrix} \varepsilon_x \\ \varepsilon_y \\ \varepsilon_z \\ 2\varepsilon_{xy} \\ 0 \\ 2\varepsilon_{xz} \end{bmatrix}}_{\varepsilon_x} = \underbrace{\begin{bmatrix} \frac{\partial}{\partial x} & 0 & 0 \\ 0 & \frac{\partial}{\partial y} & 0 \\ 0 & 0 & \frac{\partial}{\partial z} \\ \frac{\partial}{\partial y} & \frac{\partial}{\partial x} & 0 \\ 0 & 0 & 0 \\ \frac{\partial}{\partial z} & 0 & \frac{\partial}{\partial x} \end{bmatrix}}_{\boldsymbol{\mathcal{L}}_{\mathrm{m},x}} \cdot \underbrace{\begin{bmatrix} u_x \\ u_y \\ u_z \end{bmatrix}}_{\boldsymbol{u}}, \tag{G.439}$$

Weighted residual statement in x-direction with $\boldsymbol{\mathcal{L}}_x^{\mathrm{T}} = \left[\frac{\partial}{\partial x}\;0\;0\;\frac{\partial}{\partial y}\;0\;\frac{\partial}{\partial z}\right]$:

$$\int_V \left(W_x \boldsymbol{\mathcal{L}}_x^{\mathrm{T}} \left(\boldsymbol{C}_x \boldsymbol{\mathcal{L}}_{\mathrm{m},x}\boldsymbol{u}\right) + W_x f_x\right) \mathrm{d}V = 0. \tag{G.440}$$

Application of the GREEN–GAUSS theorem gives:

$$\int_V \left(\boldsymbol{\mathcal{L}}_x^{\mathrm{T}} W_x\right) \boldsymbol{C}_x \left(\boldsymbol{\mathcal{L}}_{\mathrm{m},x}\boldsymbol{u}\right) \mathrm{d}V = \int_A W_x \left(\boldsymbol{C}_x \boldsymbol{\mathcal{L}}_{\mathrm{m},x}\boldsymbol{u}\right)^{\mathrm{T}} \boldsymbol{n} \mathrm{d}A + \int_V W_x f_x \mathrm{d}V. \tag{G.441}$$

8.11 Green–Gauss theorem applied to derive general 3D weak form

Write the differential equations for each direction separately. For example, this reads for the x-direction before multiplication:

Table G.12 Comparison of the results for the beam bending problem, see Fig. 8.10

Displacement	Top load	Equal load	Bottom load
u_{2X}	$44.856276\frac{F_0}{Ea}$	$44.039990\frac{F_0}{Ea}$	$43.223704\frac{F_0}{Ea}$
u_{2Y}	$0.424065\frac{F_0}{Ea}$	$0.500309\frac{F_0}{Ea}$	$0.576553\frac{F_0}{Ea}$
u_{2Z}	$-299.671372\frac{F_0}{Ea}$	$-301.397385\frac{F_0}{Ea}$	$-303.123398\frac{F_0}{Ea}$
u_{3X}	$44.856291\frac{F_0}{Ea}$	$44.040005\frac{F_0}{Ea}$	$43.223719\frac{F_0}{Ea}$
u_{3Y}	$-0.424455\frac{F_0}{Ea}$	$-0.500702\frac{F_0}{Ea}$	$-0.576947\frac{F_0}{Ea}$
u_{3Z}	$-299.671325\frac{F_0}{Ea}$	$-301.397337\frac{F_0}{Ea}$	$-303.123349\frac{F_0}{Ea}$
u_{6X}	$44.974206\frac{F_0}{Ea}$	$44.139045\frac{F_0}{Ea}$	$43.303884\frac{F_0}{Ea}$
u_{6Y}	$-0.448065\frac{F_0}{Ea}$	$-0.350925\frac{F_0}{Ea}$	$-0.253786\frac{F_0}{Ea}$
u_{6Z}	$-300.793437\frac{F_0}{Ea}$	$-300.232405\frac{F_0}{Ea}$	$-299.671372\frac{F_0}{Ea}$
u_{7X}	$44.974219\frac{F_0}{Ea}$	$44.139057\frac{F_0}{Ea}$	$43.303896\frac{F_0}{Ea}$
u_{7Y}	$0.447199\frac{F_0}{Ea}$	$0.350055\frac{F_0}{Ea}$	$0.252911\frac{F_0}{Ea}$
u_{7Z}	$-300.793390\frac{F_0}{Ea}$	$-300.232357\frac{F_0}{Ea}$	$-299.671324\frac{F_0}{Ea}$

$$\begin{bmatrix} \frac{\partial}{\partial x} & 0 & 0 & \frac{\partial}{\partial y} & 0 & \frac{\partial}{\partial z} \\ 0 & 0 & 0 & 0 & 0 & 0 \\ 0 & 0 & 0 & 0 & 0 & 0 \end{bmatrix} \frac{E}{(1+\nu)(1-2\nu)} \begin{bmatrix} 1-\nu & \nu & \nu & 0 & 0 & 0 \\ 0 & 0 & 0 & 0 & 0 & 0 \\ 0 & 0 & 0 & 0 & 0 & 0 \\ 0 & 0 & 0 & \frac{1-2\nu}{2} & 0 & 0 \\ 0 & 0 & 0 & 0 & 0 & 0 \\ 0 & 0 & 0 & 0 & 0 & \frac{1-2\nu}{2} \end{bmatrix} \begin{bmatrix} \frac{\partial}{\partial x} & 0 & 0 \\ 0 & \frac{\partial}{\partial y} & 0 \\ 0 & 0 & \frac{\partial}{\partial z} \\ \frac{\partial}{\partial y} & \frac{\partial}{\partial x} & 0 \\ 0 & 0 & 0 \\ \frac{\partial}{\partial z} & 0 & \frac{\partial}{\partial x} \end{bmatrix} \begin{bmatrix} u_x \\ u_y \\ u_z \end{bmatrix} + \begin{bmatrix} f_x \\ 0 \\ 0 \end{bmatrix} = \begin{bmatrix} 0 \\ 0 \\ 0 \end{bmatrix} . \tag{G.442}$$

Or finally as:

$$\begin{bmatrix} (1-\nu)\frac{\mathrm{d}^2}{\mathrm{d}x^2} + \left(\frac{1}{2}-\nu\right)\left(\frac{\mathrm{d}^2}{\mathrm{d}y^2} + \frac{\mathrm{d}^2}{\mathrm{d}z^2}\right) & \nu\frac{\mathrm{d}^2}{\mathrm{d}x\mathrm{d}y} + \left(\frac{1}{2}-\nu\right)\frac{\mathrm{d}^2}{\mathrm{d}x\mathrm{d}y} & \nu\frac{\mathrm{d}^2}{\mathrm{d}x\mathrm{d}z} + \left(\frac{1}{2}-\nu\right)\frac{\mathrm{d}^2}{\mathrm{d}x\mathrm{d}z} \\ & 0 & \\ & 0 & \end{bmatrix} \begin{bmatrix} u_x \\ u_y \\ u_z \end{bmatrix} + \begin{bmatrix} f_x \\ 0 \\ 0 \end{bmatrix} = \begin{bmatrix} 0 \\ 0 \\ 0 \end{bmatrix} . \tag{G.443}$$

Derive the weak statement for all three coordinate directions separately and superimpose these three equations.

8.12 Body force matrix for gravity

$$\boldsymbol{f}_b^{\mathrm{e}} = \varrho g \times 8a^3 \left[0 \; -\tfrac{1}{8} \; 0 \,|\, \cdots \,|\, 0 \; -\tfrac{1}{8} \; 0\right]^{\mathrm{T}} .$$

8.13 Advanced Example: Different 3D modeling approaches of a simply supported beam

The results are presented in Table G.12 were obtained with a commercial computer algebra system.

G.8 Problems from Chap. 9

9.7 Inclined throw

$$\boldsymbol{v}(t) = \begin{bmatrix} v_0 \cos\alpha \\ v_0 \sin\alpha - gt \end{bmatrix}, \tag{G.444}$$

$$\boldsymbol{u}(t) = \begin{bmatrix} v_0 t \cos\alpha \\ v_0 t \sin\alpha - \frac{g}{2}t^2 \end{bmatrix}. \tag{G.445}$$

9.8 Free fall under consideration of air resistance: simplification to frictionless case

- $v(t)$: Use the TAYLOR's series expansion $e^{-\frac{bt}{m}} \approx 1 - \frac{bt}{m}$ to obtain $v(t) = -gt$.
- $y(t)$: Use the TAYLOR's series expansion $e^{-\frac{bt}{m}} \approx 1 - \frac{bt}{m} + \frac{b^2t^2}{2m^2}$ to obtain $y(t) = h - \frac{gt^2}{2}$.

9.9 Idealized drop tower

The problem is split into four phases:

1. Free fall: The description is based on a local coordinate x_1 at $X = 0$, which is pointing in the positive direction of the global coordinate.
2. Mass in contact with the spring (compression): The description is based on a local coordinate x_2 at $X = L$, which is pointing in the positive direction of the global coordinate.
3. Mass in contact with the spring (extension): The description is based on a local coordinate x_3 at $X = L$, which is pointing in the positive direction of the global coordinate.
4. Mass moving upwards (no more contact with the spring): The description is based on a local coordinate x_4 at $X = L$, which is pointing in the positive direction of the global coordinate.

Phase 1:

DE
$$\frac{d^2 u_1(t_1)}{dt_1^2} = +g, \tag{G.446}$$

BC
$$u_1(0) = d + L, \, v_1(0) = 0. \tag{G.447}$$

The solution of the DE gives under consideration of the BC the following distributions of coordinate, velocity, and acceleration:

$$u_1(t_1) = +\frac{gt_1^2}{2} + d + L, \tag{G.448}$$

$$v_1(t_1) = +gt_1, \tag{G.449}$$

$$a_1(t_1) = +g. \tag{G.450}$$

Phase 2:

$$\text{DE} \qquad \frac{\mathrm{d}^2 u_2(t_2)}{\mathrm{d}t_2^2} = -\frac{k}{m} u_2(t_2) + g\,, \tag{G.451}$$

$$\text{BC} \qquad u_2(t_2 = 0) = 0\,,\, v_1(t_2 = 0) = -\sqrt{2gd}\,. \tag{G.452}$$

The solution of the DE gives under consideration of the BC the following distributions of coordinate, velocity, and acceleration:

$$u_2(t_2) = -\sqrt{\frac{-2mgd}{k}} \sin\left(\sqrt{\frac{k}{m}} t_2\right) - \frac{gm}{k} \cos\left(\sqrt{\frac{k}{m}} t_2\right) + \frac{gm}{k}, \tag{G.453}$$

$$v_2(t_2) = -\sqrt{2gd} \cos\left(\sqrt{\frac{k}{m}} t_2\right) + \sqrt{\frac{m}{k}} g \sin\left(\sqrt{\frac{k}{m}} t_2\right), \tag{G.454}$$

$$a_2(t_2) = \sqrt{\frac{-2gdk}{m}} \sin\left(\sqrt{\frac{k}{m}} t_2\right) + g \cos\left(\sqrt{\frac{k}{m}} t_2\right). \tag{G.455}$$

Time for maximal deformation of spring: $v_2(t_2^*) = 0$:

$$t_2^* = \sqrt{\frac{m}{k}} \arctan\left(\sqrt{\frac{-2gdk}{m}} \frac{1}{g}\right) + \sqrt{\frac{m}{k}} \pi\,. \tag{G.456}$$

Phase 3:

$$\text{DE} \qquad \frac{\mathrm{d}^2 u_3(t_3)}{\mathrm{d}t_2^3} = -\frac{k}{m} u_3(t_3) + g\,, \tag{G.457}$$

$$\text{BC} \qquad u_3(t_3 = 0) = u_2(t_2^*)\,,\, v_3(t_3 = 0) = 0\,. \tag{G.458}$$

The solution of the DE gives under consideration of the BC the following distributions of coordinate, velocity, and acceleration:

$$u_3(t_3) = -\sqrt{\frac{-2mgd}{k}} \cos\left(\sqrt{\frac{k}{m}} t_3\right) \sin\left(\pi + \arctan\left(\sqrt{\frac{-2gdk}{m}} \frac{1}{g}\right)\right)$$
$$- \frac{gm}{k} \cos\left(\sqrt{\frac{k}{m}} t_3\right) \cos\left(\pi + \arctan\left(\sqrt{\frac{-2gdk}{m}} \frac{1}{g}\right)\right) + \frac{gm}{k}, \tag{G.459}$$

$$v_3(t_3) = \sqrt{2gd}\sin\left(\sqrt{\frac{k}{m}}t_3\right)\sin\left(\pi + \arctan\left(\sqrt{-\frac{2gdk}{m}}\frac{1}{g}\right)\right)$$
$$+\sqrt{\frac{m}{k}}g\sin\left(\sqrt{\frac{k}{m}}t_3\right)\cos\left(\pi + \arctan\left(\sqrt{-\frac{2gdk}{m}}\frac{1}{g}\right)\right), \tag{G.460}$$

$$a_3(t_3) = \sqrt{-\frac{2gdk}{m}}\cos\left(\sqrt{\frac{k}{m}}t_3\right)\sin\left(\pi + \arctan\left(\sqrt{-\frac{2gdk}{m}}\frac{1}{g}\right)\right)$$
$$+g\cos\left(\sqrt{\frac{k}{m}}t_3\right)\cos\left(\pi + \arctan\left(\sqrt{-\frac{2gdk}{m}}\frac{1}{g}\right)\right). \tag{G.461}$$

Phase 4:

DE $$\frac{\mathrm{d}^2 u_4(t_4)}{\mathrm{d}t_4^2} = +g\,, \tag{G.462}$$

BC $$u_4(0) = L\,,\ v_4(0) = v_3(t_3^*)\,. \tag{G.463}$$

The solution of the DE gives under consideration of the BC the following distributions of coordinate, velocity, and acceleration:

$$u_4(t_4) = +\frac{gt_4^2}{2} + v_3(t_3^*)t_4 + L\,, \tag{G.464}$$
$$v_4(t_4) = +gt_4 + v_3(t_3^*)\,, \tag{G.465}$$
$$a_4(t_4) = +g\,. \tag{G.466}$$

The graphical representations of coordinate, velocity, and acceleration are shown in Fig. G.20 as a function of the global time t.

9.10 Refined drop tower model

The problem can be split again into four phases (see the solution to additional problem 9.9). Phases 1 and 4 can be treated in a very similar way. However, the DE for phase 3 and 4 reads

$$\frac{\mathrm{d}^2 u_{3,4}(t_{3,4})}{\mathrm{d}t_{3,4}^3} = -\frac{k}{m}u_{3,4}(t_{3,4}) - \frac{c}{m}\frac{\mathrm{d}u_{3,4}(t_{3,4})}{\mathrm{d}t_{3,4}} + g\,, \tag{G.467}$$

where the general solution of the DE is gives as:

$$u_{3,4}(t_{3,4}) = c_1 \mathrm{e}^{\frac{1}{2}\frac{(-c+\sqrt{c^2-4km})t_{3,4}}{m}} + c_2 \mathrm{e}^{-\frac{1}{2}\frac{(+c+\sqrt{c^2-4km})t_{3,4}}{m}} + \frac{gm}{k}. \tag{G.468}$$

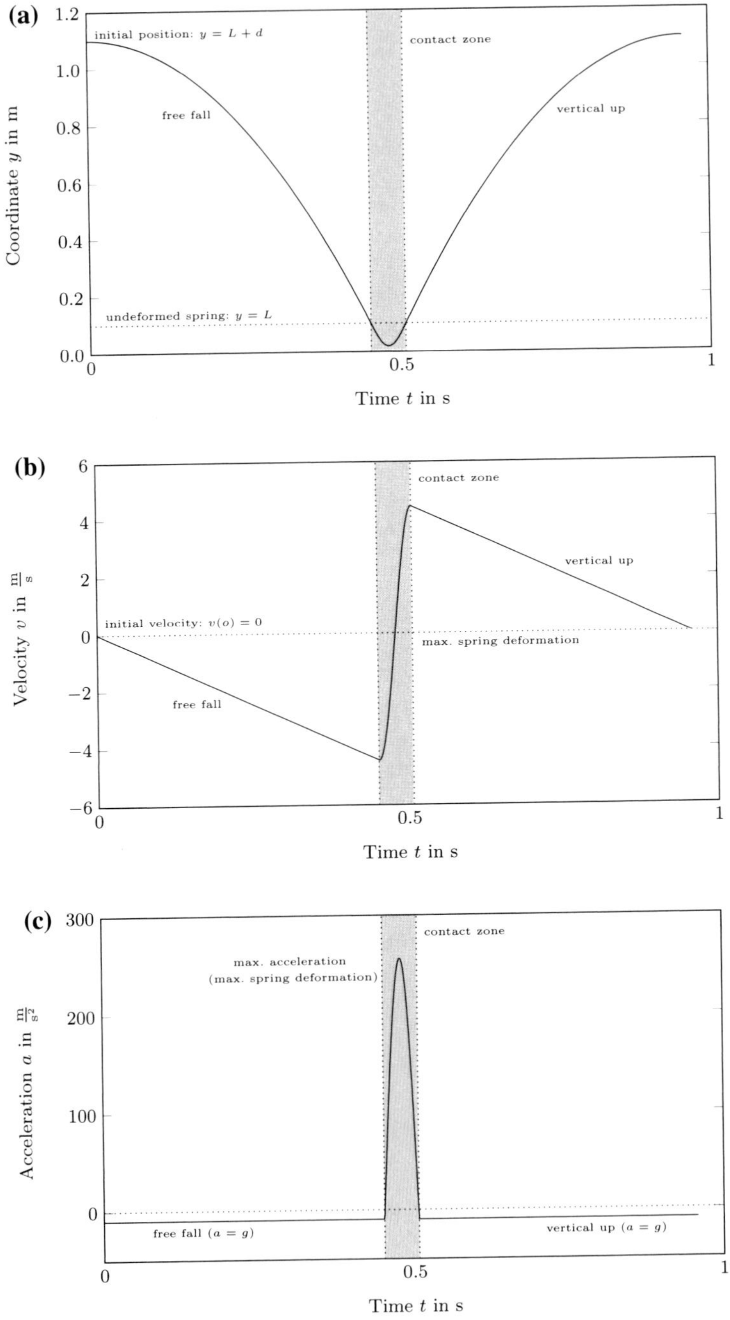

Fig. G.20 Ideal drop test results: **a** y-coordinate, **b** velocity, and **c** acceleration

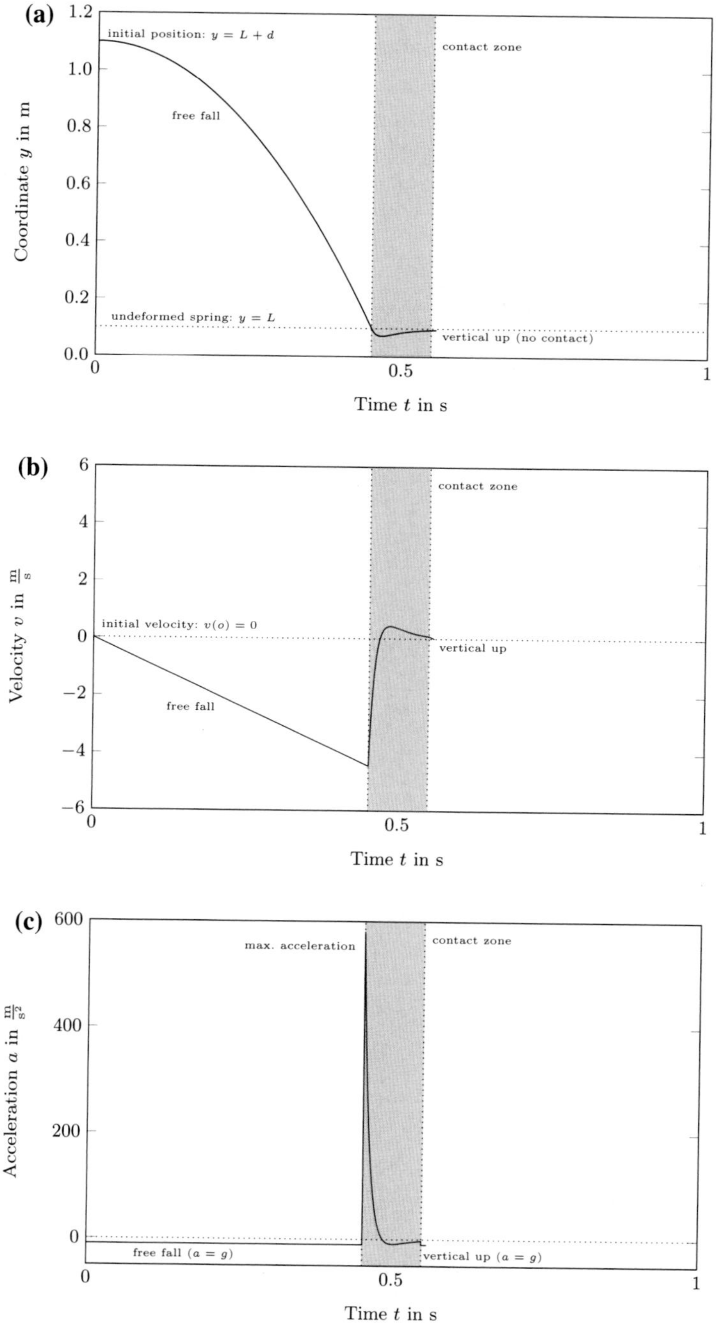

Fig. G.21 Refined drop test results: **a** y-coordinate, **b** velocity, and **c** acceleration

An interesting question is the transition from phase 3 to 4. A good overview on the possible conditions for the transition between contact (i.e., phase 3 in our case) and non-contact (i.e., phase 4 in our case) can be found in [1]. We implemented the conditions that the contact force between the mass m and the spring-damper system becomes zero, i.e. $F_c + F_k \stackrel{!}{=} 0$. The graphical representations of coordinate, velocity, and acceleration are shown in Fig. G.21 as a function of the global time t.

G.9 Problems from Chap. 10

10.4 Consistent and lumped mass approach

(a) Consistent mass:

$$\boldsymbol{M} = \frac{\varrho A L}{6}\begin{bmatrix} 2 & 1 \\ 1 & 2 \end{bmatrix}, \boldsymbol{M}^{-1} = \frac{2}{\varrho A L}\begin{bmatrix} 2 & -1 \\ -1 & 2 \end{bmatrix}. \tag{G.469}$$

(b) Lumped mass:

$$\boldsymbol{M} = \varrho A L\begin{bmatrix} \frac{1}{2} & 0 \\ 0 & \frac{1}{2} \end{bmatrix}, \boldsymbol{M}^{-1} = \frac{1}{\varrho A L}\begin{bmatrix} 2 & 0 \\ 0 & 2 \end{bmatrix}. \tag{G.470}$$

References

1. Jönsson A, Bathelt J, Broman G (2005) Implications of modelling one-dimensional impact by using a spring and damper element. P I Mech Eng K-J Mul 219:299–305
2. Öchsner A (2014) Elasto-plasticity of frame structure elements: modeling and simulation of rods and beams. Springer, Berlin

Index

A. Öchsner, *Computational Statics and Dynamics*,
https://doi.org/10.1007/978-981-15-1278-0

Printed in the United States
By Bookmasters